KU-107-184

Probability and
Statistical Inference

FOURTH EDITION

Probability and Statistical Inference

ROBERT V. HOGG
UNIVERSITY OF IOWA

ELLIOT A. TANIS
HOPE COLLEGE

MACMILLAN PUBLISHING COMPANY
NEW YORK

Maxwell Macmillan Canada
TORONTO

Maxwell Macmillan International
NEW YORK   OXFORD   SINGAPORE   SYDNEY

Editor: Robert W. Pirtle
Production Supervisor: Elaine W. Wetterau
Production Manager: Su Levine
Text and Cover Design: Robert Freese
Illustrations: York Graphic Services, Inc.

This book was set in Times Roman and Optima by York Graphic Services, Inc., printed and bound by Arcata Graphics Company. The cover was printed by Arcata Graphics Company.

Copyright © 1993 by Macmillan Publishing Company, a division of Macmillan, Inc.

PRINTED IN THE UNITED STATES OF AMERICA

All rights reserved. No part of this book may be reproduced or transmitted in any form or by any means, electronic or mechanical, including photocopying, recording, or any information storage and retrieval system, without permission in writing from the publisher.

Earlier editions copyright © 1988, 1983, and 1977 by Macmillan Publishing Company

Macmillan Publishing Company
866 Third Avenue, New York, New York 10022

Macmillan Publishing Company is part
of the Maxwell Communication Group of Companies.

Maxwell Macmillan Canada, Inc.
1200 Eglinton Avenue East
Suite 200
Don Mills, Ontario M3C 3N1

Library of Congress Cataloging in Publication Data

Hogg, Robert V.
Probability and statistical inference / Robert V. Hogg, Elliot A. Tanis.—4th ed.
p. cm.
Includes index.
ISBN 0-02-355821-0
1. Probabilities. 2. Mathematical statistics. I. Tanis, Elliot A. II. Title.
QA273.H694 1993
519.2—dc20
92-6115
CIP

Printing: 2 3 4 5 6 7 8 Year: 3 4 5 6 7 8 9 0 1

Preface

We are pleased with the reception that was given to the first three editions of *Probability and Statistical Inference*. The fourth edition is still designed for use in a course having from three to six semester hours of credit. No previous study of statistics is assumed, and a standard course in calculus provides an adequate mathematical background. Certain sections have been starred and they are not needed in subsequent sections. This, however, does not mean that these starred sections are unimportant, and we hope many of you will study them.

Although we still view this book as the basis of a junior or senior level course in the mathematics of probability and statistics that many departments of mathematics and/or statistics teach, we have tried to respond to statisticians' recent view of what is good statistical education. In particular, many believe in a data-oriented approach and, if time permits, will assign projects involving the collection of data by teams composed of from four to eight students. Accordingly, Chapter 1 includes good descriptive statistics, exploratory data analysis, the method of least squares, and correlation. Such things as stem-and-leaf displays, box-and-whisker diagrams, quantile-quantile plots, time-sequences (including digidots), and scatter plots are included. These graphical tools are then used throughout the rest of the text.

Chapter 2 provides some basic concepts in probability, much of which is illustrated by the traditional counting problems, although some of this material can be omitted if the instructor so desires. Chapters 3 and 4 contain, respectively, basic discrete and continuous distributions, including their moment-generating functions.

The difficult concept of sampling distribution theory is introduced in Chapter 5 in a very reasonable way. The students learn about the characteristics of the distribution of sums of independent random variables. This, in turn, leads to the Central Limit Theorem and approximations for the binomial distribution, among others. The t- and F-distributions are then introduced. We look upon Section 5.8 as a transition from probability to statistical inference. So often students do not recognize the importance of understanding variability. We point this out and illustrate it with some of the thoughts of W. Edwards Deming and Shewhart's control charts.

Chapter 6 considers confidence intervals and point estimation, including maximum likelihood estimators. In a starred section, the asymptotic distribution of the maximum likelihood estimator is considered. Our discussion of tests of statistical hypotheses in Chapter 7 has been simplified and is more consistent with present-day usage. The considerations of best critical regions and likelihood ratio tests are in starred sections.

Linear models, along with some analysis of variance and regression, are in Chapter 8. Chapter 9 contains multivariate distributions, including the correlation coefficient of two random variables and chi-square goodness of fit tests. Chapter 10 provides a good introduction to nonparametric methods.

In the past many students have asked us to review certain mathematical techniques. We have responded by including Appendix A, which contains helpful notions concerning algebra of sets and certain summations as well as suggestions about limits, series, integration, and multivariate calculus. We believe that we have included some basic mathematical notions here for those who feel that their mathematics is a little "rusty."

Throughout the book, many more figures and real applications have been added. These should help the student understand statistics and what statistical methods can accomplish. For some exercises, it is assumed that calculators or computers are available; thus the solutions will not always involve "nice" numbers. Solutions using a computer are given if a complicated data set is involved. Also, we include answers to almost all of the odd-numbered exercises.

At least 85% of the book could be covered in a two-semester sequence: Probability (much of Chapters 1–5) and Introduction to Mathematical Statistics (much of Chapters 6–10). In a four-semester hour course at the University of Iowa, omitting the starred sections and certain other topics, we cover the first seven chapters plus a few selected topics in the last three chapters. This material could also be covered in a course over two quarters, each with three credit hours.

We are indebted to the *Biometrika* Trustees for permission to include Tables IV and VII, which are abridgments and adaptations of tables published in *Biometrika Tables for Statisticians*. We are also grateful to the Literary Executor of the late Sir Ronald A. Fisher, F.R.S., to Dr. Frank Yates, F.R.S., and to Longman Group Ltd, London, for permission to use Table III from their book *Statistical Tables for Biological, Agricultural, and Medical Research* (6th ed., 1974), reproduced as our Table VI.

We wish to thank our colleagues, students, and friends for many suggestions and for their generosity in supplying data for exercises and examples. Over the years, Elias Ionas has found several corrections; we encourage others to do likewise so that we can correct future printings. Also, we thank Mrs. Julie DeYoung and Mrs. Lori McDowell for their help with the typing, Professor Todd Swanson for his preparation of answers in this book and solutions of even-numbered problems for *The Solutions Manual,* and the University of Iowa and Hope College for providing time, encouragement, and a Faculty Study in Van Wylen Library. Finally, our families, through four editions, have been most understanding during the preparation of all of this material; we truly appreciate their patience and needed their love.

R. V. H.

E. A. T.

Contents

4

Continuous Distributions   191

5

Sampling Distribution Theory   250

6

Estimation   334

7

Tests of Statistical Hypotheses   394

8
Linear Models   456

9
Multivariate Distributions   511

10
Nonparametric Methods   589

Appendix A: Review of Selected Mathematical Techniques   647

References   667

Appendix Tables   669

Answers to Odd-Numbered Exercises   699

Probability and
Statistical Inference

1

Summary and Display of Data

1.1 Basic Concepts

The discipline of statistics deals with the *collection* and *analysis of data*. When measurements are taken, even seemingly under the same conditions, the results usually vary. Despite this variability, a statistician tries to find a pattern; yet due to the "noise," not all of the points lie on the pattern. In the face of this variability, he or she must still do the best to describe the pattern. Accordingly, statisticians know that mistakes will be made in data analysis, and they try to minimize those errors as much as possible and then give bounds on the possible errors. By considering these bounds, decision makers can decide how much confidence they want to place on these data and the analysis of them. If the bounds are wide, perhaps more data should be collected. If they are small, however, the person involved in the study might want to make a decision and proceed accordingly.

Variability is a fact of life, and proper statistical methods can help us understand data collected under inherent variability. Because of this variability, many decisions have to be made that involve uncertainties. In medical research, interest may center on the effectiveness of a new vaccine for mumps; an agronomist must decide if an increase in yield can be attributed to a new strain of wheat; a meteorologist is interested in predicting the probability of rain; the state legislature must decide whether decreasing speed limits will help prevent accidents; the admissions officer

1

of a college must predict the college performance of an incoming freshman; a biologist is interested in estimating the clutch size for a particular type of bird; an economist desires to estimate the unemployment rate; an environmentalist tests whether new controls have resulted in a reduction in pollution.

In reviewing the preceding, relatively short list of possible areas of applications of statistics, the reader should recognize that good statistics is closely associated with careful thinking in many investigations. For illustration, students should appreciate how statistics is used in the endless cycle of the scientific method. We observe Nature and ask questions, we run experiments and collect data that shed light on these questions, we analyze the data and compare the results of the analysis to what we previously thought, we raise new questions, and on and on. Or if you like, statistics is clearly part of the important "plan–do–study–act" cycle: Questions are raised and investigations planned and carried out. The resulting data are studied and analyzed and then acted upon, often raising new questions.

There are many aspects of statistics. Some people get interested in the subject by collecting data and trying to make sense out of these observations. In some cases the answers are obvious and little training in statistical methods is necessary. But if a person goes very far in many investigations, he or she soon realizes that there is a need for some theory to help describe the error structure associated with the various estimates of the patterns. That is, at some point appropriate probability and mathematical models are required to make sense out of complicated data sets. Statistics and the probabilistic foundation on which statistical methods are based can provide the models to help people make decisions such as these. So in this book, we are more concerned about the mathematical, rather than the applied, aspects of statistics, although we give enough real examples so that the reader can get a good sense of a number of important applications of statistical methods.

In the study of probability we consider experiments for which the outcome cannot be predicted with certainty. Such experiments are called **random experiments**. Each experiment ends in an outcome that cannot be determined with certainty before the experiment is performed. However, the experiment is such that the collection of every possible outcome can be described and perhaps listed. This collection of all outcomes is called the outcome space or, more frequently, the **sample space** **S**. The following examples will help illustrate what we mean by random experiments, outcomes, and their associated sample spaces.

> **Example 1.1-1** A rat is selected at random from a cage, and its sex is determined. The set of possible outcomes is female and male. Thus the sample space is $S = \{\text{female, male}\} = \{F, M\}$.

> **Example 1.1-2** Each of six students selects an integer at random from the first 52 positive integers. (Or each of the six students selects at random a card from a well-shuffled deck of playing cards.) We are interested in whether at least two of these six integers match (M) or whether they are different (D). Thus $S = \{M, D\}$.

Example 1.1-3   A box of breakfast cereal contains one of four different prizes. The purchase of one box of cereal yields one of the prizes as the outcome, and the sample space is the set of four different prizes.

Example 1.1-4   A state selects a three-digit integer at random for one of its daily lottery games. Each three-digit integer is a possible outcome, and the sample space is $S = \{000, 001, 002, \ldots, 998, 999\}$.

Example 1.1-5   A fair coin is flipped successively at random until the first head is observed. If we let x denote the number of flips of the coin that are required, then $S = \{x : x = 1, 2, 3, 4, \ldots\}$.

Example 1.1-6   A fair coin is flipped successively at random until **heads** is observed on **two** successive flips. If we let y denote the number of flips of the coin that are required, then $S = \{y : y = 2, 3, 4, \ldots\}$.

Example 1.1-7   A biologist is studying certain birds called gallinules that live in a marsh. An adult bird is captured and weighed. If w denotes the weight of the bird, then the sample space would be the set of possible weights. From the biologist's knowledge of gallinules, we could let $S = \{w : 200 \leq w \leq 450\}$, where w is the weight in grams.

Example 1.1-8   In Example 1.1-7 the biologist could classify a captured bird by sex and weight. In such a case the sample space becomes $S = \{(c, w) : c = \text{F or M}, 200 \leq w \leq 450\}$, an example of a two-dimensional sample space.

Note that the outcomes of a random experiment can be numerical, as in Examples 1.1-4, 1.1-5, 1.1-6, and 1.1-7, but they do not have to be, as shown by Examples 1.1-1, 1.1-2, and 1.1-3. Often we "mathematize" those latter outcomes by assigning numbers to them; for instance, in Example 1.1-1, we could define a function, say X, such that $X(F) = 0$ and $X(M) = 1$. Such functions, defined on sample spaces, are called **random variables**.

Note the numbers of outcomes in the sample spaces in these examples. In the first four examples each set of possible outcomes is finite. The numbers of outcomes are 2, 2, 4, and 1000, respectively. In Example 1.1-5 the number of possible outcomes is infinite but countable. That is, there are as many outcomes as there are counting numbers (i.e., positive integers). The sample space for Example 1.1-7 is different from the other examples in that the set of possible outcomes is an interval of numbers. Theoretically, the weight could be any one of an infinite number of possible weights; here the number of possible outcomes is not countable. However, from a practical point of view, reported weights are selected from a finite number of possibilities. Many times it is better to conceptualize the sample space as an interval of outcomes and Example 1.1-7 is an example of a sample space of the continuous type.

If we consider a random experiment and its sample space, we note that under repeated performances of the experiment, some outcomes occur more frequently than others. For illustration, in Example 1.1-5, if this coin-flipping experiment is repeated over and over, the first head is observed on the first flip more often than on the second flip. If we can somehow determine the fractions of times a random experiment ends in the respective outcomes, we have described a *distribution* or *population*. Often we cannot determine this distribution through theoretical reasoning but must actually perform the random experiment a number of times to obtain guesses or *estimates* of these fractions. The collection of the observations that are obtained from these repeated trials is often called a *sample*. The making of a conjecture about a distribution based on the sample is called a *statistical inference*. That is, in statistics, we try to argue from the sample to the population. To understand the background behind statistical inferences that are made from the sample, we need a knowledge of some probability, basic distributions, and sampling distribution theory; these topics are considered in the early part of this book. We begin by introducing some terms needed to understand probability.

Given a sample space S, let A be a part of the collection of outcomes in S, that is, $A \subset S$. Then A is called an **event**. When the random experiment is performed and the outcome of the experiment is in A, we say that **event A has occurred**.

We are interested in defining what is meant by the probability of A, denoted by $P(A)$, and often called the chance of A occurring. Sometimes the nature of an experiment is such that the probability of A can be assigned easily. For example, when the state lottery in Example 1.1-4 selects a three-digit integer, we would expect each of the 1000 possible three-digit numbers to have the same chance of being selected, namely 1/1000. If we let $A = \{233, 323, 332\}$, then it makes sense to let $P(A) = 3/1000$. Or if we let $B = \{234, 243, 324, 342, 423, 432\}$, then we would let $P(B) = 6/1000$, the probability of the event B.

Probabilities of events associated with the other examples are perhaps not quite as obvious and straightforward. In Example 1.1-1, the probability of selecting a female rat from the cage depends on the number of female and male rats in the cage. In Example 1.1-2, the probability that at least two students select the same integer is dependent on how randomly the students make their selections.

To help us understand what is meant by the probability of A, $P(A)$, consider repeating the experiment a large number of times, say n times. Count the number of times that event A actually occurred throughout these n performances; this number is called the frequency of event A and is denoted by $\#(A)$. The ratio $\#(A)/n$ is called the **relative frequency** of event A in these n repetitions of the experiment. A relative frequency is usually very unstable for small values of n, but it tends to stabilize as n increases. (You might check this by performing the experiment described in Example 1.1-2, computing the relative frequency of M.) This suggests that we associate with event A a number, say p, that is equal to or approximately equal to the number about which the relative frequency tends to stabilize. This number p can then be taken as the number that the relative frequency of event A will be near in future performances of the experiment. Thus, although we cannot predict the outcome of a random experiment with certainty, we can, for a large value of n, predict fairly accurately the relative frequency associated with event A. The number p assigned to event A is called the **probability** of event A, and it is denoted by $P(A)$. That is, $P(A)$ represents the proportion of outcomes of a random experiment that terminate in the event A in a *large number* of trials of that experiment.

The following two examples will help to illustrate some of the ideas just presented.

Example 1.1-9 Consider the simple experiment of rolling a fair six-sided die (one of a pair of dice) and observing the outcome. The sample space is $S = \{1, 2, 3, 4, 5, 6\}$. If $A = \{1, 2\}$, we would probably let $P(A) = 2/6 = 1/3$. This experiment was simulated 500 times on a computer. We observed the following combinations of the number of trials, n, the frequency of A, $\#(A)$, and the relative frequency of A after n trials, $\#(A)/n$:

n	$\#(A)$	$\#(A)/n$
50	16	0.32
100	34	0.34
250	80	0.32
500	163	0.326

More complete results of the simulation are depicted in Figure 1.1-1. Note that our assignment of $1/3$ to $P(A)$ seems to be supported by the simulation.

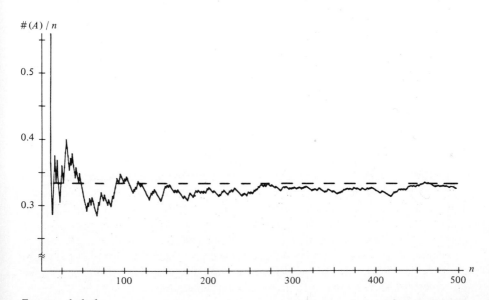

FIGURE 1.1-1

Example 1.1-10 A 12-sided die, called a dodecahedron, has 12 faces that are regular pentagons. These faces are numbered 1, 2, 3, 4, 5, 6, 7, 8, 9, 10, 11, 12. An experiment consists of rolling this die 12 times. If the face numbered k is the outcome on roll k for $k = 1$ to 12, we say that a match has occurred. The experiment is called a success if at least one match occurs during the 12 trials. Otherwise, the experiment is called a failure. The sample space is $S = \{\text{success, failure}\}$. Let $A = \{\text{success}\}$. We would like to assign a value to $P(A)$. Accordingly, this experiment was also simulated 500 times on a computer. Figure 1.1-2 depicts the results of this simulation, and the following table summarizes a few of those results:

n	$\#(A)$	$\#(A)/n$
50	37	0.74
100	65	0.65
250	166	0.664
500	318	0.636

The probability of event A is not intuitively obvious, but it will be shown in Example 2.4-6 that $P(A) = 1 - (1 - 1/12)^{12} = 0.648$. This assignment is certainly supported by the simulation (although not proved by it).

Examples 1.1-9 and 1.1-10 show that at times intuition can be used to assign probabilities correctly, although at other times probabilities may have to be assigned

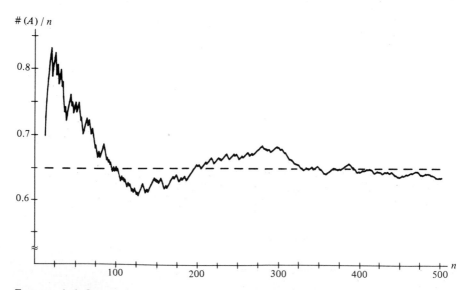

FIGURE 1.1-2

empirically. The following example gives one more illustration where intuition can help in assigning a probability to an event.

> **Example 1.1-11** A disk 2 inches in diameter is thrown at random on a tiled floor, where each tile is a square with sides 4 inches in length. Let C be the event that the disk will land entirely on one tile. In order to assign a value to $P(C)$, consider the center of the disk. In what region must the center lie to assure that the disk lies entirely on one tile? If you draw a picture, it should be clear that the center must lie within a square having sides of length 2 and lying in the center of a tile. Since the area of this square is 4 and the area of a tile is 16, it makes sense to let $P(C) = 4/16$.

Exercises

1.1-1 Describe the sample space for each of the following experiments:
   (a) A student is selected at random from a statistics class, and the student's ACT score in mathematics is determined.

   HINT: ACT test scores in mathematics are integers between 1 and 36, inclusive.

   (b) A candy bar is selected at random from a production line and is weighed.
   (c) A coin is tossed three times, and the sequence of heads and tails is observed.
   (d) Some biology students are interested in studying grackles. The sex and wing length of a trapped bird are determined.

1.1-2 Consider families that have three children and select one such family at random. List the outcomes in the following:
   (a) The sample space S as 3-tuples, agreeing for example, "gbb" would indicate that the youngest is a girl and the two oldest are boys.
   (b) The event $A = \{$at least two girls$\}$.
   (c) $B = \{$exactly one boy$\}$.
   (d) $C = \{$two youngest are girls$\}$.
   (e) Both A and B.
   (f) Either B or C (agreeing that this includes being in both B and C).

1.1-3 In each of the following random experiments describe the sample space S. Use your intuition or any experience you may have had to assign a value to the probability p of each of the events A.
   (a) The toss of an unbiased coin where the event A is heads.
   (b) The cast of an honest die where the event A occurs if we observe a 3, 4, 5, or 6.
   (c) The draw of a card from an ordinary deck of playing cards where the event A is a club.
   (d) The choice of a point from a square with opposite vertices $(0, 0)$ and $(1, 1)$, where the event A occurs if the sum of the coordinates of the point is less than 3/4.

1.1-4   A typical roulette wheel used in a casino has 38 slots that are numbered 1, 2, 3, . . . , 36, 0, 00, respectively. The 0 and 00 slots are colored green. Half of the remaining slots are red and half are black. Also half of the integers between 1 and 36 inclusive are odd, half are even, and 0 and 00 are defined to be neither odd nor even. A ball is rolled around the wheel and ends up in one of the slots; we assume each slot has equal probability of 1/38. Assume that we are interested in the number of the slot in which the ball falls.
(a) Define the sample space S.
(b) Let $A = \{0, 00\}$. Give the value of $P(A)$.
(c) Let $B = \{14, 15, 17, 18\}$. Give the value of $P(B)$.
(d) Let $D = \{odd\}$. Give the value of $P(D)$.

1.1-5   The five numbers 1, 2, 3, 4, and 5 are written on five respective disks and placed in a hat. Two disks are drawn without replacement from the hat, and the numbers written on them are observed.
(a) List the 10 possible outcomes for this experiment as pairs of numbers (order not important).
(b) If each of the 10 outcomes has probability 1/10, assign a value to the probability that the sum of the two drawn numbers is (i) 3; (ii) between 6 and 8, inclusive.

1.1-6   A box contains five computer memory chips, two of which are defective. The chips are tested one at a time, recording D for a defective chip and G for a good one. Let $A = \{$second defective is the second chip tested$\}$, $B = \{$second defective is the fourth chip tested$\}$, and $C = \{$at least four chips must be tested to find the second defective$\}$.
(a) Describe the sample space. That is, list the 10 possible outcomes, like GDGGD or GDDGG.
(b) Assuming that each of the 10 outcomes is equally likely, find $P(A)$, $P(B)$, and $P(C)$.

1.1-7   An icosahedron has 20 faces, each of which is an equilateral triangle. The faces are numbered from 0 to 9, with each number on two faces. You may think of this as a 20-sided die. This icosahedron is rolled once, and the number on the side facing up is read.
(a) Define the sample space S.
(b) Let $A = \{0, 1, 2\}$. Give the value of $P(A)$.

1.1-8   Roll a fair six-sided die six times. If the face numbered k is the outcome on roll k, for $k = 1$ to 6, a match has occurred. If at least one match occurs during the six trials, the experiment is called a success; otherwise, it is a failure. Let $A = \{$success$\}$ and estimate the value of $P(A)$ by performing this experiment a large number of times.

1.1-9   Divide a line segment into two parts by selecting a point at random. Assign probability to the event that the larger segment is at least two times longer than the shorter segment.

1.1-10 Let the interval $[-r, r]$ be the base of a semicircle. If a point is selected at random from this interval, assign a probability to the event that the length of the perpendicular segment from this point to the semicircle is less than $r/2$.

1.1-11 In Example 1.1-5, a fair coin is flipped at random until the first head is observed. Let A_x equal the event that exactly x flips are required for $x = 1, 2, 3, 4, \ldots$. Estimate the values of $P(A_x)$ for $x = 1, 2, 3, 4,$ and 5 by performing this experiment a large number of times. (You could combine your experimental results with those of your classmates.)

1.1-12 In Example 1.1-6, a fair coin is flipped at random until heads is observed on two successive flips. Let B_y equal the event that exactly y flips are required for $y = 2, 3, 4, \ldots$. Estimate the values of $P(B_y)$ for $y = 2, 3, 4,$ and 5 by performing this experiment a large number of times. (You could combine your experimental results with those of your classmates.)

1.2   Samples, Histograms, and Ogives

In Section 1.1, we considered a random experiment with outcomes falling in a sample space S. There we noted that a function determined by the outcome could be defined; it is usually denoted by some capital letter, like X, and called a random variable. In a sense, the random variable X could be thought of as some characteristic or measurement associated with the outcome, like the number of spots ''up'' on a die that is to be tossed or the weight of a bird that is to be captured. Often we are then interested in the distribution (or population) of the random variable X; that is, the fraction of times X is in a certain subset of the possible space of X-values. Suppose that an interesting event is $\{a < X \le b\}$, where a and b are real constants. Then we really want to know its probability, namely, $P(a < X \le b)$. In particular, if x is a real number, we would like to know the probability $P(X \le x)$. If this latter probability is considered for all values of x, it is called the **distribution function of the random variable** X and denoted by $F(x) = P(X \le x)$. It is called a distribution function because once it is known, we can determine other probabilities about X. For example, our intuition tells us that

$$P(a < X \le b) = F(b) - F(a),$$

because $F(b)$ represents the probability of X being less than or equal to b, and if we subtract the probability $F(a)$ of being less than or equal to a from $F(b)$, we obtain $P(a < X \le b)$.

Many times we cannot determine the distribution or distribution function of X from theoretical arguments. Thus we consider performing the experiment a number of times to determine something about the distribution of X. That is, it is helpful to repeat a random experiment several times to elicit some information about the unknown distribution. Each time the outcome occurs, the function X is computed and the collection of resulting observations, denoted by $x_1, x_2, \ldots, x_n$, is called a sample from the associated distribution. Often these observations are collected so that, in a sense that is explained later, they are independent of each other. That is, we do not want one observation to influence the others. For illustration, the outcome on one roll of a die does not affect the results of other rolls. If this type of independence exists, we often say that $x_1, x_2, \ldots, x_n$ are observations of a **random sample** of **size** n. The distribution of X from which the sample arises is sometimes called the **population**. These observed sample values, $x_1, x_2, \ldots, x_n$, will be used to elicit information about the unknown population (or distribution).

If we let $\#(\{x_i: x_i \le x\})$ equal the number of these observed values that are less than or equal to x, then the function

$$F_n(x) = \frac{\#(\{x_i: x_i \le x\})}{n},$$

defined for each real number x, is called the **empirical distribution function**.

Observe that $\#(\{x_i: x_i \le x\})$ is the frequency in n trials of the event that the X value is less than or equal to x. Hence the empirical distribution function $F_n(x)$ is the relative frequency of that event; and, for large n, we expect $F_n(x)$ to be close to the probability of the event that $\{X \le x\}$, namely $F(x) = P(X \le x)$. Thus, in some sense, $F_n(x)$ is an estimate of the distribution function $F(x)$ of X.

The probability of events other than $\{X \le x\}$ can be estimated by the observations of a random sample, $x_1, x_2, \ldots, x_n$. For example, if the event A is defined by $A = \{x: a < x \le b\}$, then

$$\frac{\#(\{x_i: a < x_i \le b\})}{n}$$

is an estimate of $P(A) = P(a < X \le b)$. Of course, this estimate equals $F_n(b) - F_n(a)$.

Say the random variable X is such that it can only take on values in a discrete set; that is, its space consists of a finite or countable number of points. Then X is said to be a **random variable of the discrete type**. For instance, in Example 1.1-4 concerning the lottery the space of the observed number X is $\{000, 001, 002, \ldots, 998, 999\}$; and in Example 1.1-6 concerning the flips of a coin, the space of Y, the number of flips until consecutive heads is observed, is $\{2, 3, 4, \ldots\}$. In Example 1.1-1 with $X(M) = 0$ and $X(F) = 1$, the space of X is $R = \{0, 1\}$.

We shall now consider experiments for which the theoretical set of possible outcomes of X forms an interval, as might occur in considerations of the heights of first-grade children, the length of life of a light bulb, or the time required to run 100 meters. Note that such observations are often rounded off so that the set of observations may seem to come from a finite set. However, we shall consider for such experiments conceptual sample spaces that are intervals of finite or infinite length. A random variable X whose space is an interval (or union of intervals) is called a **random variable of the continuous type**.

Suppose that a random experiment is such that the space of the random variable X associated with this experiment is $R = \{x: a \le x \le b\}$. If this experiment is repeated n independent times with observations $x_1, x_2, \ldots, x_n$, an estimate of $P(c \leq X \leq d)$, where $a \leq c \leq d \leq b$, is the relative frequency

$$\frac{\#(\{x_i: c \leq x_i \leq d\})}{n}.$$

That is, an estimate is given by the proportion of outcomes that fall in the interval $[c, d]$. Again, the empirical distribution function $F_n(x)$ is an estimate of the distribution function $F(x)$.

There are many characteristics associated with a random sample, $x_1, x_2, \ldots, x_n$, other than the preceding relative frequencies. Two of the main ones are the **sample mean** and the **sample variance** given by the respective formulas:

$$\bar{x} = \frac{1}{n} \sum_{i=1}^{n} x_i \quad \text{and} \quad s^2 = \frac{1}{n-1} \sum_{i=1}^{n} (x_i - \bar{x})^2.$$

For example, rolling a fair eight-sided die five times could result in the sample of $n = 5$ observations

$$x_1 = 3, \; x_2 = 7, \; x_3 = 2, \; x_4 = 5, \; x_5 = 3.$$

Then

$$\bar{x} = \frac{3 + 7 + 2 + 5 + 3}{5} = 4$$

and

$$s^2 = \frac{(3-4)^2 + (7-4)^2 + (2-4)^2 + (5-4)^2 + (3-4)^2}{4} = \frac{16}{4} = 4.$$

Also, $s = \sqrt{s^2} \geq 0$ is called the **sample standard deviation** and, in this example, $s = \sqrt{4} = 2$. Roughly, s can be thought of as the average distance that the x-values are away from the sample mean $\bar{x}$. Clearly, this is not exact because in this example the distances from $\bar{x} = 4$ are 1, 3, 2, 1, 1, with an average of 1.6. Usually, s will be somewhat larger than this average distance, but this approximate statement should give the reader some idea about the meaning of a standard deviation.

There is an alternative way of computing s^2 because

$$\sum_{i=1}^{n} (x_i - \bar{x})^2 = \sum_{i=1}^{n} (x_i^2 - 2\bar{x}x_i + \bar{x}^2) = \sum_{i=1}^{n} x_i^2 - 2\bar{x} \sum_{i=1}^{n} x_i + \sum_{i=1}^{n} \bar{x}^2$$

$$= \sum_{i=1}^{n} x_i^2 - 2\bar{x}(n\bar{x}) + n\bar{x}^2 = \sum_{i=1}^{n} x_i^2 - n\bar{x}^2,$$

since $\sum_{i=1}^{n} x_i = n\bar{x}$. Thus

$$s^2 = \frac{\Sigma x_i^2 - n\bar{x}^2}{n-1} = \frac{\Sigma x_i^2 - (\Sigma x_i)^2/n}{n-1}.$$

In our simple example,

$$s^2 = \frac{3^2 + 7^2 + 2^2 + 5^2 + 3^2 - (5)(4)^2}{4} = \frac{16}{4} = 4.$$

Of course, in this computation, $5(4)^2$ could be replaced by $(20)^2/5 = 80$.

We would like to extend this notation and obtain estimates of various probabilities associated with the distribution of the random variable X of the continuous type. In particular, we would like a reasonably simple descriptive estimate of the corresponding distribution function $F(x)$. One solution to this problem for distributions that have intervals of outcomes is now given.

Let us first concentrate on the spread of the sample by considering the smallest and the largest sample items. We shall partition the interval determined by these two values into k intervals, where in practice k is often about 10 and the intervals are of equal length. We do not always use $k = 10$; with a large sample size, the number k is frequently larger than 10 (and more than likely smaller with a small sample size). Practice with this type of procedure helps determine an appropriate value of k. As an example, consider a study of the weights of male college students. Suppose $n = 200$ and the sample values were recorded to the nearest pound so that 126 was the smallest weight and 242 the largest. Thus the **range** of the sample is $242 - 126 = 116$. Accordingly, $k = 10$ intervals, each of length 12, would cover all the sample items. But so would $k = 12$ intervals, each of length 10. Another possibility would be $k = 9$ intervals, each of length 13. We could list other possibilities; thus we see that this problem has no unique solution, and experience will help determine an appropriate k value.

For illustration, suppose we take $k = 10$ intervals, each of length 12. Even that selection does not determine the 10 intervals exactly: The first one could begin at 126 and the last one end at 246, or the first begin at 124 and the last end at 244, and so on. What is done in practice? Usually, these intervals are selected as follows:

1. Each interval begins and ends halfway between two possible values of the measurements.
2. The first interval begins about as much below the smallest value as the last interval ends above the largest.

Thus a good selection for these 10 intervals would be given by the following boundaries:

$$(124.5, 136.5) \qquad (136.5, 148.5)$$
$$(148.5, 160.5) \qquad (160.5, 172.5)$$
$$(172.5, 184.5) \qquad (184.5, 196.5)$$
$$(196.5, 208.5) \qquad (208.5, 220.5)$$
$$(220.5, 232.5) \qquad (232.5, 244.5)$$

These intervals are called **class intervals**, and the boundaries are **class boundaries**. In general, we shall denote these k class intervals by

$$(c_0, c_1), (c_1, c_2), \ldots, (c_{k-1}, c_k).$$

A frequency table is constructed that lists the class boundaries, a tabulation of the measurements in the various classes, the frequency f_i of each class, and the cumulative frequency. The probability estimates for the k classes are conveniently graphed in the form of a **relative frequency histogram**. Such a histogram is constructed by drawing a rectangle for each class having as its base the interval bounded by the class boundaries and an area equal to the relative frequency f_i/n of the observations for the class. That is, the function defined by

$$h(x) = \frac{f_i}{(n)(c_i - c_{i-1})} \quad \text{for } c_{i-1} < x \leq c_i, \, i = 1, 2, \ldots, k,$$

is called a **relative frequency histogram**, where f_i is the frequency of the ith class and n is the total number of observations.

An example will help clarify some of these terms.

Example 1.2-1   Let us consider the study of the weights of male college students with $n = 200$. Say we record each weight in the appropriate class to obtain Table 1.2-1.

For $124.5 < x \leq 136.5$, the relative frequency histogram is defined by

$$h(x) = \frac{7}{(200)(136.5 - 124.5)} = \frac{7}{(200)(12)} = 0.0029.$$

Continuing in this manner, we obtain the histogram given in Figure 1.2-1; the number within each rectangle is the area of that rectangle.

Note that the relative frequency histogram $h(x)$ satisfies the following properties:

(a) $h(x) \geq 0$ for all x.

TABLE 1.2-1

Class Interval	Tabulation	Frequency (f_i)	Cumulative Frequency
(124.5, 136.5)	++++ ||	7	7
(136.5, 148.5)	++++ ++++ |||	13	20
(148.5, 160.5)	++++ ++++ ++++ ++++	20	40
(160.5, 172.5)	++++ ++++ ++++ ++++ ++++ ++++ ++++ ||||	39	79
(172.5, 184.5)	++++ ++++ ++++ ++++ ++++ ++++ ++++ |	36	115
(184.5, 196.5)	++++ ++++ ++++ ++++ ++++ ++++ ++++ ||	37	152
(196.5, 208.5)	++++ ++++ ++++ |||	18	170
(208.5, 220.5)	++++ ++++ ++++ ||||	19	189
(220.5, 232.5)	++++ |||	8	197
(232.5, 244.5)	|||	3	200

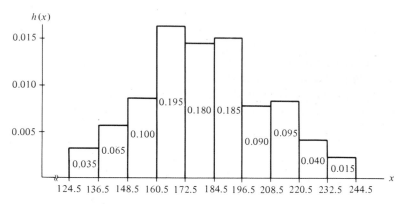

FIGURE 1.2-1

(b) The total area bounded by the x axis and below $h(x)$ equals 1; that is,

$$\int_{c_0}^{c_k} h(x) \, dx = 1.$$

(c) The probability for an event A, which is composed of a union of class intervals, can be estimated by the area above A bounded by $h(x)$; that is,

$$P(A) \approx \int_A h(x) \, dx.$$

We now return to an estimate of the distribution function $F(x)$ of X. With 200 observations, the empirical distribution function $F_{200}(x)$ is tedious to describe and graph, unless you have a computer. For all practical purposes, it amounts to listing 200 observations in order. So let us begin with the information given in Table 1.2-1, which certainly does not provide us with the individual observations. We then proceed as follows.

Let $F_n(x)$ denote the empirical distribution function. When a set of data has been classified in a frequency table like Table 1.2-1, we know the values of $F_n(x)$ at each of the class boundaries $c_0, c_1, \ldots, c_k$, namely

$$F_n(c_i) = \frac{\text{cumulative frequency up to and including } c_i}{n}.$$

Graph the points

$$[c_0, F_n(c_0) = 0], [c_1, F_n(c_1)], \ldots, [c_{k-1}, F_n(c_{k-1})], [c_k, F_n(c_k) = 1].$$

Now draw a line segment between each pair of adjacent points; that is, a line segment is drawn between $[c_{i-1}, F_n(c_{i-1})]$ and $[c_i, F_n(c_i)]$, $i = 1, 2, \ldots, k$. The data given in Table 1.2-1 provide the plot shown in Figure 1.2-2, which is commonly called an **ogive**. We shall denote this function by $H(x)$.

We make three observations about the ogive $H(x)$.

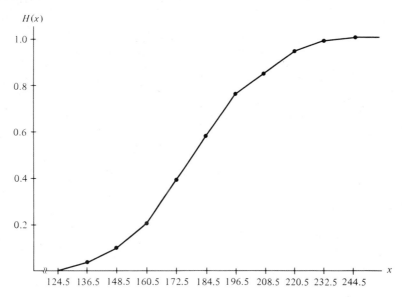

FIGURE 1.2-2

(a) Note that the ogive gives the cumulative area under the relative frequency histogram; that is,

$$H(x) = \int_{-\infty}^{x} h(t) \; dt = \int_{c_0}^{x} h(t) \; dt.$$

(b) $H(x)$ is a nondecreasing function with $0 \le H(x) \le 1$.

(c) The slope of the line segment that joins

$$[c_{i-1}, F_n(c_{i-1})] = [c_{i-1}, H(c_{i-1})] \qquad \text{and} \qquad [c_i, F_n(c_i)] = [c_i, H(c_i)]$$

is

$$\frac{H(c_i) - H(c_{i-1})}{c_i - c_{i-1}} = \frac{F_n(c_i) - F_n(c_{i-1})}{c_i - c_{i-1}} = \frac{f_i/n}{c_i - c_{i-1}}$$

$$= h(x), \qquad c_{i-1} < x \le c_i.$$

That is, for $c_{i-1} < x < c_i$, $h(x) = H'(x)$, the derivative of the ogive function.

Example 1.2-2. Suppose a six-sided die is rolled $n = 100$ times resulting in the 100 numbers:

3	6	5	1	4	5	4	3	5	6	5	4	4	5	5	2
4	6	2	2	4	3	2	3	4	6	1	5	6	5	1	4
3	4	6	4	2	6	4	3	6	4	4	4	5	5	1	5
3	1	1	4	6	4	6	1	5	1	3	5	1	1	5	2
4	2	3	1	2	3	3	2	4	6	3	5	5	4	2	4
2	2	2	2	3	4	6	6	5	2	1	6	2	2	1	5
6	4	2	2												

For these data, $\bar{x} = 358/100 = 3.58$, $s^2 = [1548 - (358)^2/100]/99 = 2.6905$, and $s = 1.64$. Since there are only six possible outcomes, $\{1, 2, 3, 4, 5, 6\}$, each occurring with a certain frequency, we can modify our computations of the mean and the variance taking advantage of repeated measurements. Denote the outcomes by u_i and the frequencies by f_i. Then we can calculate the following sums.

Outcomes (u_i)	Frequencies (f_i)	$f_i u_i$	$f_i u_i^2$
1	13	13	13
2	19	38	76
3	13	39	117
4	22	88	352
5	18	90	450
6	15	90	540
Totals	100	358	1548

Here $\Sigma_{i=1}^{100} x_i = \Sigma_{i=1}^{6} f_i u_i = 358$ and $\Sigma_{i=1}^{100} x_i^2 = \Sigma_{i=1}^{6} f_i u_i^2 = 1548$, so

$$\bar{x} = \frac{358}{100} = 3.58 \quad \text{and} \quad s^2 = \frac{1548 - (358)^2/100}{99} = 2.6905.$$

The above example can be generalized in case the data, $x_1, x_2, \ldots, x_n$ have k possible outcomes $u_1, u_2, \ldots, u_k$ with respective frequencies $f_1, f_2, \ldots, f_k$. The sample mean is

$$\bar{x} = \frac{1}{n} \sum_{i=1}^{n} x_i = \frac{1}{n} \sum_{i=1}^{k} f_i u_i = \bar{u}$$

and the sample variance is

$$s_x^2 = \frac{\sum_{i=1}^{n} x_i^2 - \left(\sum_{i=1}^{n} x_i\right)^2 / n}{n - 1} = \frac{\sum_{i=1}^{k} f_i u_i^2 - \left(\sum_{i=1}^{k} f_i u_i\right)^2 / n}{n - 1} = s_u^2.$$

Now suppose that data for an experiment have been grouped in a frequency table with k classes. In such a case we could assume that the possible outcomes are approximately equal to the midpoints of the classes, denoted by u_i and called the **class marks**. Say these outcomes occur with respective frequencies f_i, $i = 1, 2, \ldots, k$ and $n = \Sigma_{i=1}^{k} f_i$. In these cases we have k points $u_1, u_2, \ldots, u_k$ with respective frequencies $f_1, f_2, \ldots, f_k$. The sample mean and the sample variance of these grouped data are, respectively,

$$\bar{u} = \frac{1}{n} \sum_{i=1}^{k} f_i u_i$$

and

$$s_u^2 = \frac{\sum_{i=1}^{k} f_i(u_i - \bar{u})^2}{n - 1} = \frac{\sum_{i=1}^{k} f_i u_i^2 - \left(\sum_{i=1}^{k} f_i u_i\right)^2 \bigg/ n}{n - 1}$$

$$= \frac{\sum_{i=1}^{k} f_i u_i^2 - n\bar{u}^2}{n - 1}.$$

For grouped "continuous data," these statistics can be used as respective approximations of $\bar{x}$ and s_x^2, which are computed before grouping. But in the discrete case we have noted that $\bar{x} = \bar{u}$ and $s_x^2 = s_u^2$.

Example 1.2-3 To find the sample mean and sample variance of the data in Table 1.2-1, we constructed Table 1.2-2. The mean and variance of these grouped data are

$$\bar{u} = \frac{36,072.0}{200} = 180.36 \quad \text{and} \quad s_u^2 = \frac{118,972.08}{199} = 597.85.$$

Note that the variance could have been found using the formula

$$s_u^2 = \frac{118,972.08}{199} = \frac{6,624,918 - (200)(180.36)^2}{199} = 597.85.$$

TABLE 1.2-2

Class Mark (u_i)	Frequency (f_i)	$f_i u_i$	$f_i(u_i - \bar{u})^2$
130.5	7	913.5	17,402.13
142.5	13	1,852.5	18,633.93
154.5	20	3,090.0	13,374.79
166.5	39	6,493.5	7,491.88
178.5	36	6,426.0	124.55
190.5	37	7,048.5	3,804.33
202.5	18	3,645.0	8,823.23
214.5	19	4,075.5	22,145.25
226.5	8	1,812.0	17,031.20
238.5	3	715.5	10,140.78
	200	36,072.0	118,972.08

Here $\sqrt{597.85} = 24.45 = s_u$ is the standard deviation of these grouped data. In Section 4.4 we shall learn that if these data come from a normal population, then approximately 68% of the observations can be expected to fall within one standard deviation of the mean and approximately 95% within two standard deviations of the mean. Is that true for these data?

In Table 1.2-2, we note that the class interval from 160.5 to 172.5 had the largest frequency, namely 39. Often the interval with the largest frequency is called the **modal class**, and the respective class mark is often called the **mode**.

In some instances it is not always desirable to use equally spaced class boundaries in the construction of the frequency distribution and histogram. This is particularly true if the data are skewed with a very long tail. This is illustrated in Example 1.2-4.

Example 1.2-4   The following 40 losses, due to wind-related catastrophes, were recorded to the nearest $1,000,000. These data include only those losses of $2,000,000 or more; and, for convenience, they have been ordered and recorded in millions.

2	2	2	2	2	2	2	2	2	2
2	2	3	3	3	3	4	4	4	5
5	5	5	6	6	6	6	8	8	9
15	17	22	23	24	24	25	27	32	43

The sample mean and the sample standard deviation are $\bar{x} = 9.225$ and $s = 10.237$, respectively.

The empirical distribution function $F_{40}(x)$, for the data in Example 1.2-4, can be depicted by the step function in Figure 1.2-3.

One way to determine the class intervals for a frequency distribution is by constructing an ogive on the graph of the empirical distribution function. For example,

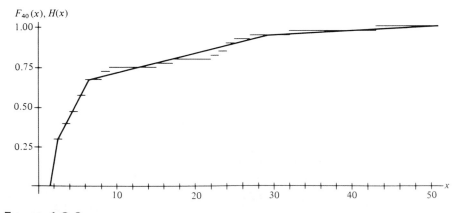

FIGURE 1.2-3

if the ogive $H(x)$ is given by a combination of four line segments joining five points on the empirical distribution function, we could use the points

$$(1.5, 0), (2.5, 0.3), (6.5, 0.675), (29.5, 0.95), \text{ and } (49.5, 1).$$

Thus we are taking as class boundaries the points

$$c_0 = 1.5, \ c_1 = 2.5, \ c_2 = 6.5, \ c_3 = 29.5, \text{ and } c_4 = 49.5.$$

This ogive is also drawn on Figure 1.2-3. The worst part of this fit is around the $x = 22$. To fit here any better would require the ogive to be steeper around $x = 24$ and flatter around $x = 20$; this would result in a histogram that differs from our intuition. Hence we believe the given ogive smooths $F_{40}(x)$ reasonably well and in a manner consistent with our prior notions. Clearly, there is a great subjective element in this fit and other persons would have different ones. It did, however, seem natural to choose the lefthand endpoint of 1.5 (because only values of 2,000,000 or more were recorded), but the right-hand endpoint of 49.5 was selected somewhat arbitrarily, although it seemed to provide a reasonable fit. Table 1.2-3 gives the resulting frequency distribution.

The relative frequency histogram corresponding to $H(x)$ is given by

$$h(x) = \begin{cases} \dfrac{12}{40}, & 1.5 < x \le 2.5, \\[2mm] \dfrac{15}{(40)(4)}, & 2.5 < x \le 6.5, \\[2mm] \dfrac{11}{(40)(23)}, & 6.5 < x \le 29.5, \\[2mm] \dfrac{2}{(40)(20)}, & 29.5 < x \le 49.5; \end{cases}$$

it is displayed in Figure 1.2-4.

The numbers in the rectangles are the respective relative frequencies. It is important to note in the case of unequal lengths among class intervals that the areas, not the heights, of the rectangles are proportional to the frequencies. In particular, the first and second classes have frequencies $f_1 = 12$ and $f_2 = 15$, and yet the height of the first is greater than the height of the second while here $f_1 < f_2$. We call 2, rather

TABLE 1.2-3

Class Interval	Frequency (f_i)
(1.5, 2.5)	12
(2.5, 6.5)	15
(6.5, 29.5)	11
(29.5, 49.5)	2
	40

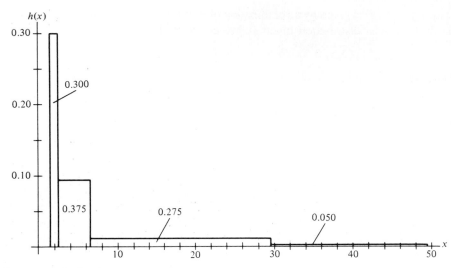

FIGURE 1.2-4

than 4.5, the mode because the relative frequency histogram is higher at 2 than it is at 4.5. Of course, if we have equal lengths among the class intervals, then the heights are proportional to the frequencies.

Exercises

1.2-1   One characteristic of a car's storage console that is checked by the manufacturer is the time in seconds that it takes for the lower storage compartment door to open completely. A random sample of size $n = 5$ yielded the following times:

$$1.1 \quad 0.9 \quad 1.4 \quad 1.1 \quad 1.0$$

(a) Find the sample mean, $\bar{x}$.
(b) Find the sample variance, s^2.
(c) Find the sample standard deviation, s.

1.2-2   The *National Geographic* reported the following litter sizes for 20 lions:

$$2 \quad 2 \quad 1 \quad 3 \quad 2 \quad 2 \quad 3 \quad 1 \quad 4 \quad 3$$
$$2 \quad 1 \quad 3 \quad 2 \quad 2 \quad 2 \quad 2 \quad 1 \quad 2 \quad 1$$

(a) Here we have only 4 possible outcomes; so proceed as in Example 1.2-2 and determine the frequency of each. Which value is the mode?
(b) Find the sample mean, $\bar{x}$.
(c) Find the sample variance, s^2.
(d) Find the sample standard deviation, s.

1.2-3 A student trainer worked in the training room in the physical education building. The number of students coming in to be taped or for treatment of other injuries was recorded each half hour yielding the following data:

3	3	2	0	4	5	6	4	4	3	2	1	2	3	0
5	5	3	2	3	5	4	1	2	0	3	2	4	2	6

(a) Assign the frequency to each of the values 0, 1, . . . , 6.
(b) Find the sample mean, $\bar{x}$.
(c) Find the sample variance, s^2.
(d) Find the sample standard deviation, s.

1.2-4 Denote a sample of n numbers by $x_1, x_2, \ldots, x_n$. Suppose that we transform each observation x_i into a new observation z_i using the transformation $z_i = ax_i + b$, $i = 1, 2, \ldots, n$. Show that the relationships between the means and the variances of $x_1, x_2, \ldots, x_n$ and $z_1, z_2, \ldots, z_n$ are

$$\bar{z} = a\bar{x} + b \qquad \text{and} \qquad s_z^2 = a^2 s_x^2.$$

1.2-5 Using the following sample: 1.0, 2.3, 3.0, 4.2, and 1.5, and the results of Exercise 1.2-4, find five observations with a mean three units larger and a variance four times as large as these.

1.2-6 The following numbers are a random sample of size 10 from some distribution:

$$-0.49 \qquad 0.90 \qquad 0.76 \qquad -0.97 \qquad -0.73$$
$$0.93 \qquad -0.88 \qquad -0.75 \qquad 0.88 \qquad 0.96.$$

(a) Graph the empirical distribution function.
(b) Use the empirical distribution function to estimate $P(X \le -0.50)$, $P(X > 0)$, and $P(-0.50 \le X \le 0.50)$.

1.2-7 The costs of 20 "300 level" textbooks for the fall were as follows:

38.95	49.95	55.50	44.00	43.30	41.00	39.95
12.95	42.00	14.95	25.95	15.95	17.95	29.00
52.00	42.65	57.00	37.35	33.95	37.00	

(a) Group these data into six classes using as class intervals (9.995, 17.995), (17.995, 25.995), and so on.
(b) Construct a relative frequency histogram and interpret the histogram.
(c) Calculate the values of the sample mean and standard deviation using the original data.
(d) Calculate the values of the sample mean and sample standard deviation from the grouped data. Are the respective answers in parts (c) and (d) close to each other?

1.2-8 Five measurements in "ohms per square" of the electronic conductive coating (that allows light to pass through it) on a thin, clear piece of glass were made and the average was calculated. This was repeated 150 times, yielding the following data:

83.86	75.86	79.65	90.57	95.37	97.97	77.00	80.80
83.53	83.17	80.20	81.42	89.14	88.68	85.15	90.11
89.03	85.00	82.57	79.46	81.20	82.80	81.28	74.64
69.85	75.60	76.60	78.30	88.51	86.32	84.03	95.27
90.05	77.50	75.14	81.33	86.12	78.30	80.30	84.96
79.30	88.96	82.76	83.13	79.60	86.20	85.16	87.86
91.00	91.10	84.04	92.10	79.20	83.76	87.87	86.71
81.89	85.72	75.84	74.27	93.08	81.75	75.66	75.35
76.55	84.86	90.68	91.02	90.97	98.30	91.84	97.41
73.60	90.65	80.20	74.75	90.35	79.66	86.88	83.00
86.24	80.50	74.25	91.20	70.16	78.40	85.60	80.82
75.95	80.75	81.86	82.18	82.98	84.00	76.85	85.00
79.50	86.56	83.30	72.40	79.20	86.20	82.36	84.08
86.11	88.25	88.93	93.12	78.30	77.24	82.52	81.37
83.72	86.90	84.37	92.60	95.01	78.95	81.40	88.40
76.10	85.33	82.95	80.20	88.21	83.49	81.00	82.72
81.12	83.62	91.18	85.90	79.01	77.56	81.13	80.60
81.65	70.70	69.36	79.09	71.35	67.20	67.43	69.95
66.76	76.35	69.45	80.13	84.26	88.13		

(a) Construct a frequency table for these 150 observations using 10 intervals of equal length, and using (66.495, 69.695) for the first class interval. Which is the modal class?

(b) Construct a relative frequency histogram and an ogive for the grouped data.

(c) Find the mean $\bar{u}$ and variance s_u^2 for the grouped data. Compare $\bar{u}$ and s_u^2 with $\bar{x} = 82.661$ and $s_x^2 = 42.228$, the mean and variance for the ungrouped data.

1.2-9 Decks have become very popular items for American homeowners. Lumber yards sell many styles of decks during the summer months, but they all use the same type of nail. The question that always arises is how many nails to buy. If 800 nails are needed, how many pounds of nails should be purchased? The weights (in grams) of the following 50 nails should help to answer the question.

9.42	8.69	8.93	8.27	8.82	8.66	8.90	8.31	9.15	9.63
9.41	8.56	8.82	8.58	8.43	8.05	8.56	8.55	8.88	8.73
8.29	8.79	8.51	8.85	9.34	9.21	8.38	8.51	8.41	8.98
8.58	9.21	8.27	8.76	9.26	8.59	8.36	8.71	8.51	8.88
9.20	8.24	8.57	8.85	8.69	8.85	9.08	9.40	9.25	8.79

(a) Group these data into seven classes using as class intervals (7.995, 8.245), (8.245, 8.495), and so on.

(b) Construct a relative frequency histogram.

(c) Construct the ogive curve.

    (d) Calculate the sample mean and sample standard deviation for the grouped data.

    (e) How many pounds of nails would you recommend? (Note that the nail weights are given in grams.)

1.2-10 (a) Construct a relative frequency histogram and an ogive for the following 90 measurements, which give the amounts of butterfat (in pounds) produced by 90 cows during a 305-day milk production period following their first calf. You may use 11 classes with the first class interval (264.5, 304.5).

486	537	513	583	453	510	570	500	458	555
618	327	350	643	500	497	421	505	637	599
392	574	492	635	460	696	593	422	499	524
539	339	472	427	532	470	417	437	388	481
537	489	418	434	466	464	544	475	608	444
573	611	586	613	645	540	494	532	691	478
513	583	457	612	628	516	452	501	453	643
541	439	627	619	617	394	607	502	395	470
531	526	496	561	491	380	345	274	672	509

    (b) Find the mean $\bar{u}$ and variance s_u^2 for the grouped data.

1.2-11 In order to test a new golf ball, each of 50 professional golfers hit the new ball three times off the tee and then hit a competing brand three times off the tee. Of course, half of the golfers hit the competing brand first, and half hit the new ball first. The total distances (in yards) for the new brand are

785	680	810	740	697	787	830	845	866	782
773	754	685	753	807	835	742	766	709	724
746	783	786	796	816	739	781	823	789	764
775	828	821	821	808	732	636	705	764	692
703	740	714	746	804	690	740	751	807	773

The respective total distances for the competing brand are

700	643	803	706	654	728	769	783	817	766
741	741	681	760	765	783	717	714	668	728
729	699	713	780	805	699	714	788	715	728
703	804	784	783	782	747	656	688	703	668
721	720	695	721	774	680	726	739	798	746

For each set of distances:

(a) Construct a frequency distribution.

(b) Draw a relative frequency histogram.

(c) Draw the ogive.

(d) Find the sample mean, sample variance, and sample standard deviation for the grouped data.

1.2-12 A small part for an automobile rear view mirror was produced on two different punch presses. In order to describe the distribution of the weights of those parts, a random sample was selected, and each piece was weighed in grams, resulting in the following data set:

3.968	3.534	4.032	3.912	3.572	4.014	3.682	3.608
3.669	3.705	4.023	3.588	3.945	3.871	3.744	3.711
3.645	3.977	3.888	3.948	3.551	3.796	3.657	3.667
3.799	4.010	3.704	3.642	3.681	3.554	4.025	4.079
3.621	3.575	3.714	4.017	4.082	3.660	3.692	3.905
3.977	3.961	3.948	3.994	3.958	3.860	3.965	3.592
3.681	3.861	3.662	3.995	4.010	3.999	3.993	4.004
3.700	4.008	3.627	3.970	3.647	3.847	3.628	3.646
3.674	3.601	4.029	3.603	3.619	4.009	4.015	3.615
3.672	3.898	3.959	3.607	3.707	3.978	3.656	4.027
3.645	3.643	3.898	3.635	3.865	3.631	3.929	3.635
3.511	3.539	3.830	3.925	3.971	3.646	3.669	3.931
4.028	3.665	3.681	3.984	3.664	3.893	3.606	3.699
3.997	3.936	3.976	3.627	3.536	3.695	3.981	3.587
3.680	3.888	3.921	3.953	3.847	3.645	4.042	3.692
3.910	3.672	3.957	3.961	3.950	3.904	3.928	3.984
3.721	3.927	3.621	4.038	3.047	3.627	3.774	3.983
3.658	4.034	3.778					

(a) Construct a frequency distribution using about 10 (say, 8 to 12) classes.
(b) Draw a relative frequency histogram.
(c) Find the sample mean for the grouped data.
(d) Find the sample variance and sample standard deviation for the grouped data.

1.2-13 An insurance company experienced the following mobile home losses in 10,000's of dollars for 50 catastrophic events:

1	2	2	3	3	4	4	5	5	5
5	6	7	7	9	9	9	10	11	12
22	24	28	29	31	33	36	38	38	38
39	41	48	49	53	55	74	82	117	134
192	207	224	225	236	280	301	308	351	527

(a) For these 50 losses, construct the empirical distribution function.
(b) Construct an ogive on the graph of the empirical distribution function.
(c) Construct a frequency distribution.
(d) Draw the resulting relative frequency histogram and determine the modal class.

1.2-14 In the casino game roulette, if a player bets $1 on red, the probability of winning $1 is 18/38 and the probability of losing $1 is 20/38. Let X equal the number of successive $1 bets that a player makes before losing $5. One hundred observations of X were simulated on a computer, yielding the following data:

23	127	877	65	101	45	61	95	21	43
53	49	89	9	75	93	71	39	25	91
15	131	63	63	41	7	37	13	19	413
65	43	35	23	135	703	83	7	17	65
49	177	61	21	9	27	507	7	5	87
13	213	85	83	75	95	247	1815	7	13
71	67	19	615	11	15	7	131	47	25
25	5	471	11	5	13	75	19	307	33
57	65	9	57	35	19	9	33	11	51
27	9	19	63	109	515	443	11	63	9

For these data:
(a) Construct the empirical distribution function.
(b) Construct an ogive on the graph of the empirical distribution function.
(c) Construct a frequency distribution.
(d) Draw the resulting relative frequency distribution.

1.3   Exploratory Data Analysis

In Section 1.2 we studied methods, like the histogram, of describing data. In recent years statisticians have become more concerned about better descriptive statistical methods. One such person is John W. Tukey. Here we give a few of his simpler, but very useful, exploratory methods. For more details, see the books by Tukey (1977) and Velleman and Hoaglin (1981).

Possibly the easiest way to begin is with an example to which all of us can relate. Say we have the following 50 test scores on a statistics examination:

93	77	67	72	52	83	66	84	59	63
75	97	84	73	81	42	61	51	91	87
34	54	71	47	79	70	65	57	90	83
58	69	82	76	71	60	38	81	74	69
68	76	85	58	45	73	75	42	93	65

Noting that 34 and 97 are the smallest and largest scores, respectively, we can make a conventional tally like that in Table 1.3-1. Along with the corresponding histogram, this tabulation describes the sample fairly well. However, it is important to note that in this summary the original test scores are lost.

We can do much the same thing, but keep those original values, through a **stem-and-leaf display**. For this particular data set, we could use the following procedure. The first number in the set, 93, is recorded as follows: the 9 (in the "tens" place) is treated as the stem, and the 3 (in the "units" place) is the corresponding leaf. Note that this leaf of 3 is the first digit after the stem of 9 in Table 1.3-2. The second number, 77, is given by the leaf of 7 after the stem of 7; the third number, 67, by the leaf of 7 after the stem of 6; the fourth number, 72, as the leaf of 2 after the stem

TABLE 1.3-1

Class Interval	Tabulation	Frequency (f_i)
(29.5, 39.5)	||	2
(39.5, 49.5)	||||	4
(49.5, 59.5)	ⷜ ||	7
(59.5, 69.5)	ⷜ ⷜ	10
(69.5, 79.5)	ⷜ ⷜ |||	13
(79.5, 89.5)	ⷜ ||||	9
(89.5, 99.5)	ⷜ	5

TABLE 1.3-2

Stem	Leaf	Frequency
3	4 8	2
4	2 7 5 2	4
5	2 9 1 4 7 8 8	7
6	7 6 3 1 5 9 0 9 8 5	10
7	7 2 5 3 1 9 0 6 1 4 6 3 5	13
8	3 4 4 1 7 3 2 1 5	9
9	3 7 1 0 3	5

of 7 (note this is the second leaf on the 7 stem); and so on. Table 1.3-2 is called a stem-and-leaf display. If the leaves are carefully aligned vertically, this table has the same effect as a histogram, but the original numbers are not lost.

There are times when it is helpful to modify the stem-and-leaf display by ordering the leaves in each row from smallest to largest. The resulting stem-and-leaf diagram is called an **ordered stem-and-leaf display**. Table 1.3-3 gives an ordered stem-and-leaf display using the data in Table 1.3-2.

TABLE 1.3-3

Stem	Leaf	Frequency
3	4 8	2
4	2 2 5 7	4
5	1 2 4 7 8 8 9	7
6	0 1 3 5 5 6 7 8 9 9	10
7	0 1 1 2 3 3 4 5 5 6 6 7 9	13
8	1 1 2 3 3 4 4 5 7	9
9	0 1 3 3 7	5

TABLE **1.3-4**

Stem	Leaf	Frequency
3*	4	1
3•	8	1
4*	2 2	2
4•	5 7	2
5*	1 2 4	3
5•	7 8 8 9	4
6*	0 1 3	3
6•	5 5 6 7 8 9 9	7
7*	0 1 1 2 3 3 4	7
7•	5 5 6 6 7 9	6
8*	1 1 2 3 3 4 4	7
8•	5 7	2
9*	0 1 3 3	4
9•	7	1

There is another modification that can also be helpful. Suppose that we want two rows of leaves with each original stem. We can do this by recording leaves 0, 1, 2, 3, and 4 with a stem adjoined with an asterisk (*) and leaves 5, 6, 7, 8, and 9 with a stem adjoined with a dot (•). Of course, in our example, by going from seven original classes to 14 classes, we lose a certain amount of smoothness with this particular data set, as illustrated in Table 1.3-4, which is also ordered.

Clearly, some imagination must be used if class intervals of different lengths are required. Tukey suggested the scheme used in the next example.

Example 1.3-1   The following numbers represent ACT composite scores for 60 entering freshman at a certain college:

26	19	22	28	31	29	25	23	20	33	23	26
30	27	26	29	20	23	18	24	29	27	32	24
25	26	22	29	21	24	20	28	23	26	30	19
27	21	32	28	29	23	25	21	28	22	25	24
19	24	35	26	25	20	31	27	23	26	30	29

An ordered stem-and-leaf display is given in Table 1.3-5. Here leaves are recorded as zeros and ones with a stem adjoined with an asterisk (*), twos and threes with a stem adjoined with t, fours and fives with a stem adjoined with f, sixes and sevens with a stem adjoined with s, and eights and nines with a stem adjoined with a dot (•).

There is a reason for constructing ordered stem-and-leaf diagrams. For a sample of n observations, $x_1, x_2, \ldots, x_n$, when the observations are ordered from small to large, the resulting ordered data are called the **order statistics** of the sample. For

TABLE 1.3-5

Stem	Leaf	Frequency
1•	8 9 9 9	4
2∗	0 0 0 0 1 1 1	7
2t	2 2 2 3 3 3 3 3 3	9
2f	4 4 4 4 4 5 5 5 5 5	10
2s	6 6 6 6 6 6 6 7 7 7 7	11
2•	8 8 8 8 9 9 9 9 9 9	10
3∗	0 0 0 1 1	5
3t	2 2 3	3
3f	5	1

years statisticians have found that order statistics and certain functions of the order statistics are extremely valuable. It is very easy to determine the values of the sample in order from an ordered stem-and-leaf display. For illustration, consider the values in Table 1.3-3 or Table 1.3-4. The order statistics of the 50 test scores are

34	38	42	42	45	47	51	52	54	57
58	58	59	60	61	63	65	65	66	67
68	69	69	70	71	71	72	73	73	74
75	75	76	76	77	79	81	81	82	83
83	84	84	85	87	90	91	93	93	97

Sometimes we give ranks to these order statistics and use the rank as the subscript on y. The first-order statistic $y_1 = 34$ has rank 1; the second-order statistic $y_2 = 38$ has rank 2; the third-order statistic $y_3 = 42$ has rank 3; the fourth-order statistic $y_4 = 42$ has rank 4, . . . ; and the 50th-order statistic $y_{50} = 97$ has rank 50. It is also about as easy to determine these values from the ordered stem-and-leaf display. We see that $y_1 \leq y_2 \leq \cdots \leq y_{50}$.

From either these order statistics or the corresponding ordered stem-and-leaf display, it is rather easy to find the **sample percentiles**. If $0 < p < 1$, the $(100p)$th sample percentile has *approximately np* sample observations less than it and also $n(1 - p)$ sample observations greater than it. One way of achieving this is to take the $(100p)$th sample percentile as the $(n + 1)p$th order statistic, provided that $(n + 1)p$ is an integer. If $(n + 1)p$ is not an integer but is equal to r plus some proper fraction, say a/b, use a weighted average of the rth and the $(r + 1)$st order statistics. That is, define the $(100p)$th sample percentile as

$$\tilde{\pi}_p = y_r + (a/b)(y_{r+1} - y_r)$$
$$= (1 - a/b)y_r + (a/b)y_{r+1}.$$

(If $p < 1/(n + 1)$ or $p > n/(n + 1)$, that sample percentile is not defined.)

For illustration, consider the 50 ordered test scores. With $p = 1/2$, we find the 50th percentile by averaging the 25th and 26th order statistics, since $(n + 1)p = (51)(1/2) = 25.5$. Thus the 50th percentile is

$$\tilde{\pi}_{0.50} = (1/2)y_{25} + (1/2)y_{26}$$
$$= (71 + 71)/2 = 71.$$

With $p = 1/4$, we have $(n + 1)p = (51)(1/4) = 12.75$; and thus the 25th sample percentile is

$$\tilde{\pi}_{0.25} = (1 - 0.75)y_{12} + (0.75)y_{13}$$
$$= (0.25)(58) + (0.75)(59) = 58.75.$$

With $p = 3/4$, so that $(n + 1)p = (51)(3/4) = 38.25$, the 75th sample percentile is

$$\tilde{\pi}_{0.75} = (1 - 0.25)y_{38} + (0.25)y_{39}$$
$$= (0.75)(81) + (0.25)(82) = 81.25.$$

Note that *approximately* 50%, 25%, and 75% of the sample observations are less than 71, 58.75, and 81.25, respectively.

Special names are given to certain percentiles. The 50th percentile is the **median** of the sample. The 25th, 50th, and 75th percentiles are the **first, second**, and **third quartiles** of the sample. For notation we let $\tilde{q}_1 = \tilde{\pi}_{0.25}$, $\tilde{q}_2 = \tilde{m} = \tilde{\pi}_{0.50}$, and $\tilde{q}_3 = \tilde{\pi}_{0.75}$. The 10th, 20th, . . . , and 90th percentiles are the **deciles** of the sample. So note that the 50th percentile is also the median, the second quartile, and the fifth decile. By using the set of 50 test scores, since $(51)(2/10) = 10.2$ and $(51)(9/10) = 45.9$, the second and ninth deciles are

$$\tilde{\pi}_{0.20} = (0.8)y_{10} + (0.2)y_{11} = (0.8)(57) + (0.2)(58) = 57.2$$

and

$$\tilde{\pi}_{0.90} = (0.1)y_{45} + (0.9)y_{46} = (0.1)(87) + (0.9)(90) = 89.7,$$

commonly called the 20th and 90th percentiles, respectively.

It is frequently instructive to note, along with the three quartiles, the extremes of the sample, namely the first and nth order statistics. (In the test scores data, they are 34 and 97, respectively.) These five numbers provide a good summary of a set of data and have a special name. The **five-number summary** of a set of data consists of the minimum, the first quartile, the median, the third quartile, and the maximum, written in that order. Furthermore, the difference between the third and first quartiles is called the **interquartile range, IQR**. That is, $IQR = \tilde{q}_3 - \tilde{q}_1$, which equals $81.25 - 58.75 = 22.50$ with the test scores data.

There is graphical means for displaying the five-number summary of a set of data that is called a **box-and-whisker diagram**. To construct a horizontal box-and-whisker diagram, or more simply, a **box plot**, draw a horizontal axis that is scaled to the data. Above the axis draw a rectangular box with the left and right sides drawn at $\tilde{q}_1$ and $\tilde{q}_3$ with a vertical line segment drawn at the median, $\tilde{q}_2 = \tilde{m}$. A left whisker is drawn as a horizontal line segment from the minimum to the midpoint of the left side of the box and a right whisker is drawn as a horizontal line segment from the midpoint of the right side of the box to the maximum. Note that the length of the box is equal to the IQR. The left and right whiskers contain the first and the fourth quarters of the data while the two middle quarters of the data are contained, respectively, in the two sections in the box, one to the left and one to the right of the median line.

> **Example 1.3-2** Using the test score data, we find that the five-number summary is given by
>
> $$y_1 = 34, \quad \tilde{q}_1 = 58.75, \quad \tilde{q}_2 = \tilde{m} = 71, \quad \tilde{q}_3 = 81.25, \quad y_{50} = 97.$$
>
> All of these were calculated above. The box-and-whisker diagram or box plot of these data is given in Figure 1.3-1. The facts that the long whisker is to the left and the box from 58.75 to 71 is slightly longer than that from 71 to 81.25 lead us to say that these data are slightly *skewed to the left*.

The next example illustrates how the box plot depicts data that are *skewed to the right*.

> **Example 1.3-3** Using the 40 losses due to wind damage that are given in Example 1.2-4, the minimum and maximum are 2 and 43. Since $(40 + 1)(1/4) = 10.25$, $(40 + 1)(1/2) = 20.5$, and $(40 + 1)(3/4) = 30.75$, we have that the three quartiles are 2, 5, and 13.5. Thus the five-number summary is given by
>
> $$y_1 = 2, \quad \tilde{q}_1 = 2, \quad \tilde{q}_2 = \tilde{m} = 5, \quad \tilde{q}_3 = 13.5, \quad y_{40} = 43.$$
>
> The box plot associated with these data is given in Figure 1.3-2. Note that the box plot indicates that the data are very skewed to the right. In fact, because $y_1 = \tilde{q}_1 = 2$, the length of the left whisker equals zero.

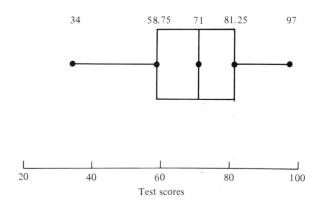

FIGURE 1.3-1

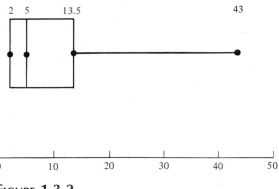

FIGURE 1.3-2

Sometimes we are interested in picking out observations that seem to be much larger or much smaller than most of the other observations. That is, we are looking for outliers. Tukey suggested a method for defining outliers that is resistant to the effect of one or two extreme values and makes use of the IQR. In a box-and-whisker diagram, construct **inner fences** to the left and right of the box at a distance of 1.5 times the IQR. **Outer fences** are constructed in the same way at a distance of 3 times the IQR. Observations that lie between the inner and the outer fences are called **suspected outliers**. Observations that lie beyond the outer fences are called **outliers**. When you are analyzing a set of data, suspected outliers deserve a closer look and outliers should be looked at very carefully. It does not follow that suspected outliers or outliers should be removed from the data unless some error (such as a recording error) has been made. Moreover, it sometimes helps to determine the cause of extreme values, because outliers can often provide useful insights to the situation under consideration (such as better ways of doing things).

Example 1.3-4 Continuing with Example 1.3-3, we find that the interquartile range is $IQR = 13.5 - 2 = 11.5$. Thus the inner fences would be constructed at a distance of $1.5(11.5) = 17.25$ to the left and right of the box, and the outer fences would be constructed at a distance of $3(11.5) = 34.5$ to the left and right of the box. Figure 1.3-3 shows a box plot with the fences. Of course, since no data fall to the left of the box, there are no fences to the left. From this box plot we see that there are two suspected outliers. However, because of the nature of the data, they are consistent with the type of losses that the insurance industry can expect to see.

Some functions of two or more order statistics are quite important in modern statistics. We mention and illustrate three more, along with the IQR, using the 50 test scores:

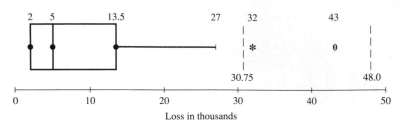

FIGURE 1.3-3

(a) **Midrange** = average of the extremes

$$= \frac{y_1 + y_n}{2}$$

$$= \frac{34 + 97}{2} = 65.5,$$

(b) **Trimean** $= \dfrac{(\text{1st quartile}) + 2(\text{2nd quartile}) + (\text{3rd quartile})}{4}$

$$= \frac{\tilde{q}_1 + 2\tilde{q}_2 + \tilde{q}_3}{4}$$

$$= \frac{58.75 + 2(71) + 81.25}{4} = 70.5,$$

(c) **Range** = difference of the extremes

$$= y_n - y_1$$
$$= 97 - 34 = 63,$$

(d) **Interquartile range** = difference of third and first quartiles

$$= \tilde{q}_3 - \tilde{q}_1$$
$$= 81.25 - 58.75 = 22.5.$$

Thus we see that the mean, the median, the midrange, and the trimean are measures of the middle of the sample. In some sense, the standard deviation, the range, and the interquartile range provide measures of spread of the sample.

Exercises

1.3-1 Let X denote the concentration of $CaCO_3$ in milligrams per liter. Twenty observations of X are

130.8	129.9	131.5	131.2	129.5
132.7	131.5	127.8	133.7	132.2
134.8	131.7	133.9	129.8	131.4
128.8	132.7	132.8	131.4	131.3

(a) Construct an ordered stem-and-leaf display, using stems of 127, 128, . . . , 134.

(b) Find the (i) midrange, (ii) trimean, (iii) range, (iv) interquartile range, (v) median, (vi) sample mean, and (vii) sample variance.

(c) Draw a box-and-whisker diagram.

1.3-2   A company manufactures mints that have a label weight of 20.4 grams. It regularly samples from the production line and weighs the selected mints. During two mornings of production it sampled 81 mints with the following weights:

21.8	21.7	21.7	21.6	21.3	21.6	21.5	21.3	21.2
21.0	21.6	21.6	21.6	21.5	21.4	21.8	21.7	21.6
21.6	21.3	21.9	21.9	21.6	21.0	20.7	21.8	21.7
21.7	21.4	20.9	22.0	21.3	21.2	21.0	21.0	21.9
21.6	21.5	21.5	21.1	21.3	21.3	21.2	21.0	20.8
21.6	21.6	21.5	21.5	21.2	21.5	21.4	21.4	21.3
21.2	21.8	21.7	21.7	21.6	20.5	21.8	21.7	21.5
21.4	21.4	21.9	21.8	21.7	21.4	21.3	20.9	20.9
20.7	21.1	20.8	20.6	20.6	22.0	22.0	21.7	21.6

(a) Construct an ordered stem-and-leaf display.

(b) Find (i) the three quartiles, (ii) the 60th percentile, and (iii) the 15th percentile.

(c) Construct a box-and-whisker display.

1.3-3   For each of 32 female grackles that were at least one year old, the wing chord was measured (in centimeters), yielding the following data:

12.8	12.7	13.0	12.5	12.8	12.9	12.2	12.8
12.6	12.4	12.3	12.7	12.7	13.5	12.5	12.5
12.8	12.2	12.7	12.3	12.4	12.8	12.4	11.2
12.6	12.8	13.8	12.0	11.7	12.1	13.6	12.4

(a) Construct an ordered stem-and-leaf display.

(b) Draw a box-and-whisker diagram.

1.3-4   For each of 25 male grackles that were at least one year old, the wing chord was measured (in centimeters), yielding the following data:

14.1	14.0	13.9	14.4	13.0
13.6	14.3	13.6	13.8	14.1
13.7	14.3	14.2	13.4	14.1
13.5	14.8	14.4	13.8	14.0
13.5	14.5	13.5	13.8	14.4

(a) Construct an ordered stem-and-leaf display.

(b) Draw a box-and-whisker diagram.

1.3-5   For the roulette data in Exercise 1.2-14:

(a) Construct an ordered stem-and-leaf display.

(b) Find (i) y_1, the smallest order statistic; (ii) y_{100}, the largest order statistic; (iii) $\tilde{q}_1$ and $\tilde{q}_3$, the first and third quartiles; and (iv) $\tilde{m}$, the median.

(c) Draw a box-and-whisker diagram.

(d) Find the locations of the inner and outer fences and draw a box plot that shows the fences, the suspected outliers, and the outliers.

(e) Find $\bar{x}$, the sample mean.

1.3-6 In the casino game roulette, if a player bets $1 on red (or on black or on odd or on even), the probability of winning $1 is 18/38, and the probability of losing $1 is 20/38. Suppose that a player begins with $5 and makes successive $1 bets. Let Y equal the player's maximum capital before losing the $5. One hundred observations of Y were simulated on a computer, yielding the following data:

25	9	5	5	5	9	6	5	15	45
55	6	5	6	24	21	16	5	8	7
7	5	5	35	13	9	5	18	6	10
19	16	21	8	13	5	9	10	10	6
23	8	5	10	15	7	5	5	24	9
11	34	12	11	17	11	16	5	15	5
12	6	5	5	7	6	17	20	7	8
8	6	10	11	6	7	5	12	11	18
6	21	6	5	24	7	16	21	23	15
11	8	6	8	14	11	6	9	6	10

(a) Construct an ordered stem-and-leaf display.

(b) Find the five-number summary of the data and draw a box-and-whisker diagram.

(c) Calculate the *IQR* and the locations of the inner and outer fences.

(d) Draw a box plot that shows the fences, suspected outliers, and outliers.

(e) Find the 90th percentile.

1.3-7 The weights (in grams) of 25 indicator housings used on gauges are as follows:

102.0	106.3	106.6	108.8	107.7
106.1	105.9	106.7	106.8	110.2
101.7	106.6	106.3	110.2	109.9
102.0	105.8	109.1	106.7	107.3
102.0	106.8	110.0	107.9	109.3

(a) Construct an ordered stem-and-leaf display using integers as the stems and tenths as the leaves.

(b) Find the five-number summary of the data and draw a box plot.

(c) Are there any suspected outliers or outliers?

1.3-8 For the insurance data in Exercise 1.2-13:

(a) Construct an ordered stem-and-leaf display.

    (b) Find the five-number summary of the data and draw a box-and-whisker diagram.

    (c) Calculate the *IQR* and the locations of the inner and outer fences.

    (d) Draw a box plot that shows the fences, suspected outliers, and outliers.

1.3-9 Using the costs of the ''300 level'' textbooks in Exercise 1.2-7:

    (a) Construct an ordered stem-and-leaf diagram using as stems 1∗, 1•, 2∗, 2•, etc., and three digit leaves. (A leaf would consist of the ones place and two decimal places.)

    (b) Calculate the five-number summary and draw a box plot.

    (c) Are there any suspected outliers?

1.4   Graphical Comparisons of Data Sets

    Graphical methods are extremely valuable in showing how two or more data sets are related to each other. In this section we extend two graphical techniques from exploratory data analysis, namely, stem-and-leaf displays and box-and-whisker diagrams, as well as give a third graphical technique. We begin with an extension of ordered stem-and-leaf diagrams, which are then used to help us find the order statistics for the other graphical techniques.

    To compare the characteristics of two treatments (two populations), take a random sample of size n from the first, $x_1, x_2, \ldots, x_n$, and a random sample of size m from the second, $y_1, y_2, \ldots, y_m$. Assuming that these measurements are about the same order of magnitude, we construct a **back-to-back stem-and-leaf display** or a **two-sided stem-and-leaf display**. In doing this, place the stems that are appropriate for both sets of data down the center of the page, placing the leaves of the x-values on the left and those for the y-values on the right. It is helpful, for further applications, to order the leaves from small to large, beginning at the stem and proceeding outward.

    Example 1.4-1 At a heat treating company, iron castings and steel forgings are heat treated to achieve desired mechanical properties and machinability. One steel forging is ''annealed'' to soften the part for easy machining. Two lots of this part, made of 1020 steel, are heat treated in two different furnaces. The parts are heat treated to a hardness specification, in this case the Rockwell G scale. The specification for this part is 36–66 (Rockwell G), and it is desirable for the parts to be close to the middle of the hardness specification. The hardness results for furnace no. 10 were

49	47	48	49	51	48	47	48	49	51	47	46	48
49	50	46	46	47	48	49	46	46	47	48	47	

The hardness results for furnace no. 14 were

43	43	47	48	36	43	43	47	47	51	41	43	46
43	49	41	38	44	40	46	39	38	41	38	41	

TABLE 1.4-1

Frequency	Furnace No. 10 Leaves	Stem	Furnace No. 14 Leaves	Frequency
0		3s	6	1
0		3•	8 8 8 9	4
0		4*	0 1 1 1 1	5
0		4t	3 3 3 3 3	6
0		4f	4	1
11	7 7 7 7 7 7 6 6 6 6 6	4s	6 6 7 7 7	5
11	9 9 9 9 9 8 8 8 8 8 8	4•	8 9	2
3	1 1 0	5*	1	1

The back-to-back stem-and-leaf display in Table 1.4-1 gives an excellent comparison of the results of the two furnaces.

This back-to-back stem-and-leaf display shows quite dramatically that the two furnaces are not operating similarly and that furnace no. 10 has not only less variability, but also produces hardness values much closer to the middle of the hardness specification.

Recall that the box-and-whisker diagram was useful in showing where the middle half of the observations fall and whether there are any outliers, and also gave an indication of the skewness of the data. Box plots can also be used to compare two or more sets of data by drawing the box plots on the same graph. The box plots may be drawn either horizontally or vertically.

We continue with the preceding example.

Example 1.4-2 If we use the hardness results for furnaces no. 10 and no. 14 in the last example, since Table 1.4-1 gives an ordered back-to-back stem-and-leaf display, the quartiles can be found easily. The locations of the first and third quartiles are the averages of the 6th and 7th ordered observations and the 19th and 20th ordered observations, respectively, and the median is the 13th observation. So for furnace no. 10, the five-number summary is 46, 47, 48, 49, and 51. The five-number summary for furnace no. 14 is 36, 40.5, 43, 46.5, and 51. The box plots for the two furnaces are given in Figure 1.4-1. Note that this figure confirms what we had noted from the back-to-back stem-and-leaf display.

Clearly, more than two box-and-whisker displays can be used to compare three or more samples and thus their corresponding distributions. More formal methods of testing the equality of several means is considered in Section 8.1.

Example 1.4-3 Hogg and Ledolter [see R. V. Hogg and J. Ledolter, *Applied Statistics for Engineers and Physical Scientists,* 2nd ed. (New York:

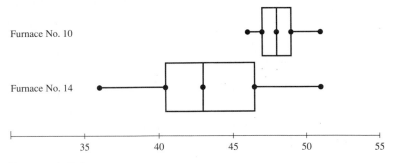

FIGURE **1.4-1**

Macmillan, 1992)] report that a civil engineer wishes to compare the strengths of three different types of beams, one (A) made of steel and two (B and C) made of different and more expensive alloys. A certain deflection (in units of 0.001 inch) was measured for each beam when submitted to a given force; thus a small deflection would indicate a beam of great strength. The order statistics for the three samples of sizes $n_1 = 8$, $n_2 = 6$, and $n_3 = 6$ are the following:

A:	79	82	83	84	85	86	86	87
B:	74	75	76	77	78	82		
C:	77	78	79	79	79	82		

The corresponding box-and-whisker displays are given in Figure 1.4-2. The deflection seems to be greatest in A, while B and C are about the same. However, with such small sample sizes, it is difficult to say for certain.

We now introduce a third graphical technique for comparing two sets of data. To begin, suppose that each of two sets of data contains n numbers. Furthermore, order each set so that $x_1 \leq x_2 \leq x_3 \leq \cdots \leq x_n$ and $y_1 \leq y_2 \leq y_3 \leq \cdots \leq y_n$. Thus we have the order statistics for each random sample. Each of x_r and y_r is called the **quantile of order** $r/(n+1)$ for the respective samples. Note that x_r and y_r are also called the $100[r/(n + 1)]$th percentiles.

In a **quantile–quantile plot** or a **q–q plot**, the quantiles of one sample are plotted against the corresponding quantiles of the other sample. That is, with equal sample sizes, we graph the points (x_1, y_1), (x_2, y_2), . . . , (x_n, y_n). If both samples were exactly the same, each pair of corresponding quantiles would plot on a straight line with slope 1 and intercept 0. If, on the other hand, the mean of the first sample (which is plotted as the first coordinate) is shifted over, say d units, but otherwise the samples are the same, pairs would plot on a straight line with slope 1 but the intercept is now $-d$. A slope less than 1 would indicate that the variability of the first sample (plotted as the x-coordinate of the pair) is greater than that of the second (plotted as the y-coordinate of the pair). Of course, a slope greater than 1 would also indicate that the variability of the second sample is greater than that of the first.

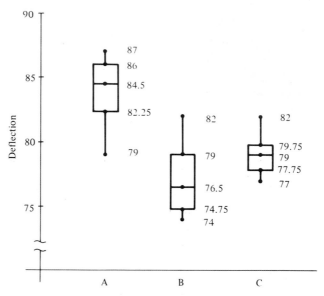

FIGURE **1.4-2**

With a little thought, other graphs suggest other situations. For illustration, suppose that the q–q plot is on a curve whose slope increases with higher quantiles. This suggests that the variability of the first sample decreases relative to that of the second with higher values of each. For instance, the first sample might be skewed to the left and the second skewed to the right.

Usually with real data sets, the q–q plot is not exactly on a straight line. Let us consider a case study by J. H. Sheesley. [See R. D. Snee, L. B. Hare, and J. R. Trout, *Experiments in Industry* (Milwaukee, Wis.: American Society for Quality Control, 1985).]

Example 1.4-4 In Sheesley's study, the performance of two different types of lead wires were compared in the production of ordinary household light bulbs. Due to the fact that some of the lead wires made by the old method would misfeed, the company changed the way it produced lead wires in hopes of improving the situation. To compare the old method with the new method, the results of 12 production runs with the old method were compared with 12 runs with the new. The two sets of order statistics (the observations ordered from small to large) of the average hourly number of misfeeding leads were the following:

Old:	10.8	17.6	18.0	18.3	19.2	19.6
	19.9	21.4	22.7	23.2	23.7	39.4
New:	7.8	11.2	11.5	12.4	16.7	16.8
	18.9	21.5	23.7	25.6	26.9	28.1

Since the ith order statistic of each sample is the $[i/(n + 1)]$th quantile of the sample, $i = 1, 2, \ldots, 12$, the 12 pairs

(quantile from old, quantile from new)

are plotted in Figure 1.4-3. From the figure we note, except for the extreme quantiles, the sample values of the old wires are closer together than those of the new wires. There are more points below the 45° line through the origin; this would suggest that the old distribution might have a larger mean than that of the new distribution, but the improvement has not been outstanding.

It is interesting to note that two box-and-whisker diagrams alongside each other would have provided about the same information. Since $(0.25)(13) = 3.25$, $(0.5)(13) = 6.5$, and $(0.75)(13) = 9.75$, the 25th, 50th, and 75th percentiles are the respective weighted averages of the 3rd and 4th, 6th and 7th, and the 9th and 10th order statistics. The two box-and-whisker diagrams are depicted in Figure 1.4-4. Thus we see that the interquartile range is much smaller for the old than the new and the whiskers longer for the old method. In addition, the comparison of the two box-and-whisker diagrams indicates that, based on these data, the new method produces somewhat fewer misfeeds than the old. It seems to us, however, as if even more improvements could be made and then more data collected.

We will not always be so fortunate as to have equal sample sizes. If the sample sizes are small but unequal, the order statistics of the smaller sample can be used to determine which quantiles are to be used.

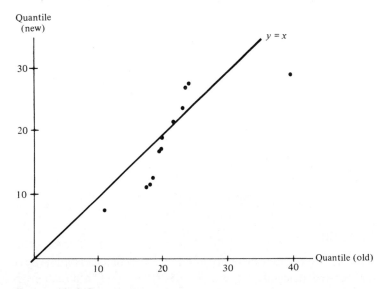

FIGURE 1.4-3

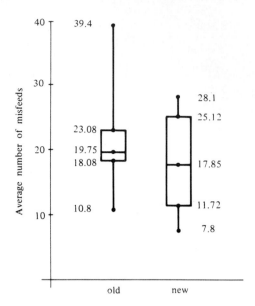

FIGURE 1.4-4

In general, suppose that we would like to find the $(100p)$th percentile (pth quantile) of a sample. If $r = (n + 1)p$ is an integer, then x_r is that percentile; for example, if $n = 9$ and $p = 0.30$, then $r = (9 + 1)(0.30) = 3$, so x_3 is the 30th percentile or quantile of order $p = 3/10$. If $(n + 1)p$ is not an integer, we use a weighted average of the two adjacent order statistics, x_r and x_{r+1}, where r is the greatest integer $[(n + 1)p]$ in $(n + 1)p$. For example, if $n = 11$ and we want to find the 30th percentile, we calculate $r = [(11 + 1)(0.30)] = [3.6] = 3$. We then use the 3rd order statistic plus 0.6 of the difference between the 4th and 3rd order statistics or, equivalently, a weighted average of the 3rd and the 4th order statistics:

$$x_3 + (0.6)(x_4 - x_3) = (0.4)x_3 + (0.6)x_4.$$

Suppose that we want to make a q–q plot using $n = 11$ observations of X and $m = 13$ observations of Y. We shall use the quantiles of the smaller sample to select the corresponding quantiles of the larger sample. That is, for the smaller sample the $n = 11$ order statistics $x_1 < x_2 < \cdots < x_{11}$ are the respective $(100p)$th percentiles (pth quantiles) of the sample for $p = 1/12, 2/12, 3/12, \ldots , 11/12$. To find the corresponding percentiles (quantiles) of the larger sample of size $m = 13$ using the order statistics $y_1 < y_2 < \cdots < y_{13}$, we can use Table 1.4-2 (see Exercise 1.4-5):

Exercises

1.4-1 Some nurses in Ottawa County Public Health conducted a survey of women who received inadequate prenatal care. They used information from birth certificates to select mothers for the survey. The mothers that were selected

TABLE 1.4-2

p	$p(n + 1)$	pth Quantile of x's	$p(m + 1)$	pth Quantile of y's
1/12	1	x_1	1.17	$(0.83)y_1 + (0.17)y_2$
2/12	2	x_2	2.33	$(0.67)y_2 + (0.33)y_3$
3/12	3	x_3	3.50	$(0.50)y_3 + (0.50)y_4$
4/12	4	x_4	4.67	$(0.33)y_4 + (0.67)y_5$
5/12	5	x_5	5.83	$(0.17)y_5 + (0.83)y_6$
6/12	6	x_6	7.00	y_7
7/12	7	x_7	8.17	$(0.83)y_8 + (0.17)y_9$
8/12	8	x_8	9.33	$(0.67)y_9 + (0.33)y_{10}$
9/12	9	x_9	10.50	$(0.50)y_{10} + (0.50)y_{11}$
10/12	10	x_{10}	11.67	$(0.33)y_{11} + (0.67)y_{12}$
11/12	11	x_{11}	12.83	$(0.17)y_{12} + (0.83)y_{13}$

were divided into two groups: 14 mothers who said they had 5 or fewer prenatal visits and 14 mothers who said they had 6 or more prenatal visits. The birthweights (in ounces) of their babies were

Mothers with 5 or fewer visits

49	108	110	82	93	114	134
114	96	52	101	114	120	116

Mothers with 6 or more visits

113	108	93	119	119	98	106
87	153	116	129	97	110	131

(a) Construct an ordered back-to-back stem-and-leaf display using integer stems 4, 5, . . . , 13, 14, 15.
(b) Calculate the five-number summary for each set of weights and draw box plots for the two sets of weights on the same figure.
(c) Draw a q–q plot of the paired order statistics.
(d) What conclusions can you draw from these limited data?

1.4-2 Construct an ordered back-to-back stem-and-leaf display for the data in Example 1.4-4. Does this confirm the conclusions given in that example?

1.4-3 At the beginning of the semester in Health Dynamics, some college students were weighed under water in order to determine their percentage of body fat. The weights in kilograms for 25 females were

1.2	1.5	1.2	1.5	1.6	2.0	1.1
2.3	1.0	2.0	1.8	1.5	2.1	1.9
2.1	1.9	1.9	1.4	1.8	2.7	0.8
1.6	1.4	1.7	0.9			

The weights in kilograms for 25 males were

5.2	3.7	5.2	2.7	4.7	5.2	3.8
4.2	4.0	4.1	4.5	3.7	4.2	4.6
4.3	3.6	4.2	3.2	3.4	3.1	4.2
2.8	3.7	4.9	3.9			

(a) Construct an ordered back-to-back stem-and-leaf display.
(b) Find the five-number summary for each set of weights.
(c) Draw box-and-whisker diagrams on the same graph.
(d) Use the paired order statistics to draw a $q-q$ plot.
(e) What conclusions can you draw from your graphical displays?

1.4-4 At the beginning of the semester in Health Dynamics, the Forced Vital Capacity (FVC) (the volume of air a person can expel from the lungs) was one measurement taken for some college students. The FVC in liters for 25 females were

2.9	2.8	3.4	3.2	3.7	3.8	4.1
3.1	4.2	3.8	3.7	3.4	3.5	3.5
2.7	4.2	3.4	4.0	3.5	3.7	3.4
3.4	3.1	3.1	3.6			

The FVC in liters for 25 males were

5.1	5.4	5.8	4.3	6.0	5.4	5.7
5.6	5.3	4.7	6.1	6.3	4.6	5.5
6.7	5.3	4.3	4.1	4.8	4.5	5.8
4.1	4.8	5.3	6.7			

(a) Construct an ordered back-to-back stem-and-leaf display.
(b) Find the five-number summary for each set of volumes.
(c) Draw box-and-whisker diagrams on the same graph.
(d) Use the paired order statistics to draw a $q-q$ plot.
(e) What conclusions can you draw from your graphical displays?

1.4-5 A botanist is interested in comparing the growth response of dwarf pea stems to two different levels of the hormone indoleacetic acid (IAA). Using 16-day-old pea plants, the botanist obtains 5-millimeter sections and floats these sections on solutions with different hormone concentrations to observe the effect of the hormone. Let X and Y denote, respectively, the growth in millimeters that can be attributed to the hormone during the first 26 hours after sectioning for $(0.5)10^{-4}$ and 10^{-4} levels of concentration of IAA. Given $n = 11$ observations of X

0.8	1.8	1.0	0.1	0.9	1.7	1.0	1.4	0.9	1.2	0.5

and $m = 13$ observations of Y

1.0	0.8	1.6	2.6	1.3	1.1	2.4
1.8	2.5	1.4	1.9	2.0	1.2,	

(a) Construct an ordered back-to-back stem-and-leaf display.

(b) Find the five-number summary for each set of lengths.

(c) Draw box-and-whisker diagrams on the same graph.

(d) To construct a $q-q$ plot, use Table 1.4-2 to find the quantiles of the observations of Y.

(e) What conclusions can you draw from your graphical displays?

1.4-6 In Exercise 1.2-11, data were given comparing a new golf ball with a competing brand. Using these data,

(a) Construct an ordered back-to-back stem-and-leaf display.

(b) Find the five-number summary for each set of distances.

(c) Draw box-and-whisker diagrams on the same graph.

(d) Use about every fifth pair of order statistics to construct a $q-q$ plot.

(e) Does it seem as if the new ball is better than the competing brand?

1.4-7 Using the wing chord data in Exercises 1.3-3 and 1.3-4,

(a) Construct an ordered back-to-back stem-and-leaf display.

(b) Find the five-number summary for each set of lengths.

(c) Draw box-and-whisker diagrams on the same graph.

(d) Use about 9 to 12 quantiles to draw a $q-q$ plot (e.g., $p = 0.1, 0.2, \ldots,$ 0.9).

(e) What conclusions can you draw from your graphical displays?

1.4-8 The precipitation of platinum sulfide by three different methods led to the following results (already ordered in magnitude):

Method	Platinum Recovered from 10.12 Milligrams Pt						
1	10.28	10.29	10.31	10.33	10.33	10.33	10.34
2	10.22	10.23	10.23	10.24	10.27	10.28	10.29
3	10.22	10.25	10.30	10.33	10.36	10.38	10.38

(a) Draw three box-and-whisker diagrams.

(b) Construct three $q-q$ plots: method 1 against method 2, method 2 against method 3, method 3 against method 1.

(c) Make some appropriate comments about the graphical displays.

1.4-9 For an aerosol product, there are three weights: the tare weight (container weight), the concentrate weight, and the propellant weight. Let $X_1, X_2,$ and X_3 denote the weights in grams of the propellant for products produced on three different days. Nine observations of each random variable were as follows:

$$X_1: 43.06 \quad 43.32 \quad 42.63 \quad 42.86 \quad 43.05$$
$$42.87 \quad 42.94 \quad 42.80 \quad 42.36$$
$$X_2: 42.33 \quad 42.81 \quad 42.13 \quad 42.41 \quad 42.39$$
$$42.10 \quad 42.42 \quad 41.42 \quad 42.52$$

X_3:   42.83   42.57   42.96   43.16   42.25
      42.24   42.20   41.97   42.61

(a) Find the five-number summary for each set of weights.
(b) Construct three box-and-whisker diagrams on the same graph.
(c) What conclusions can you draw?

1.5   Time Sequences and Digidots

Thus far we have emphasized the importance of good summaries of data sets, in particular visual displays such as histograms, stem-and-leaf displays, $q–q$ plots, and box-and-whisker diagrams. Here we want to consider another important variable, namely time, that often contributes to variability. In a **time sequence**, observations are recorded in the order in which they were collected as ordered pairs where the x-coordinate denotes the time (in order of collection, which could be days, weeks, years, etc.) and the y-coordinate records the observation. If measurements are taken sequentially and plotted in this time sequence, trends, cycles, or major changes in processes (like government interventions) are often observed. For illustration, consider the three time sequences in Figure 1.5-1.

Figure 1.5-1(a) might be a recording of sales of a company for the last 12 years. Despite occasional drops in sales from one year to the next, there has been an upward trend. Suppose, however, that only the last two points had been recorded for the stockholders. Since sales went down, they might believe that there are some difficulties and that changes should be made. If they look at the entire sequence, however, they clearly see the long-term improvement and only variability about that upward trend accounts for the last drop. On the other hand, if several successive drops appear, then some action might be needed.

The points in Figure 1.5-1(b) might also represent sales, but now plotted by quarters. Clearly, in this business under consideration, there is a cyclic effect with the second and third quarters doing better than the first and fourth. Again we note that there seems to be an overall upward trend.

Figure 1.5-1(c) could be the plot of some measure of success (possibly sales) of the company. Clearly, something happened after the tenth week. Possibly management did not like the past level of performance and instituted a new program (possibly one based on statistical quality control). On the other hand, there might be some outside intervention (like war) that increased the sales of this particular product.

We have listed three possible scenarios associated with these three time series. There might be others; but, in any case, we clearly see that plotting the points in a time sequence often helps explore the variability of the outcomes associated with a particular system. That is, time is often an important factor in dealing with variability.

The following example to which many of you can relate illustrates a time sequence with a cycle effect.

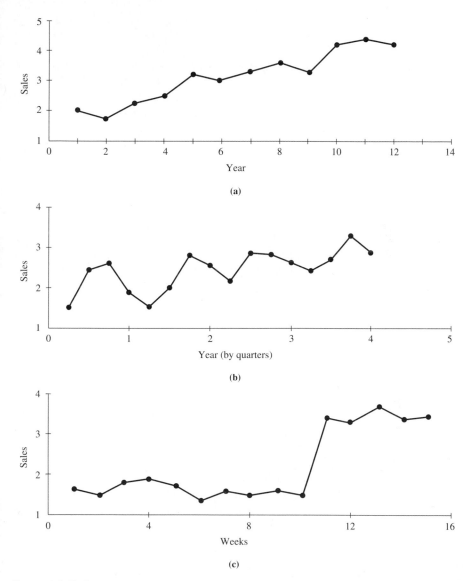

FIGURE 1.5-1

Example 1.5-1 One of the VAX computers on a college campus is primarily available for students to use. The following CPU times in days for January through December for two consecutive years were reported:

4.6	7.9	5.9	7.4	11.1	1.8	4.4	2.5
5.9	9.3	9.6	4.7	7.1	8.5	9.9	13.8
2.2	2.4	2.3	1.0	3.9	4.6	5.3	5.0

A time sequence of these data is given in Figure 1.5-2.

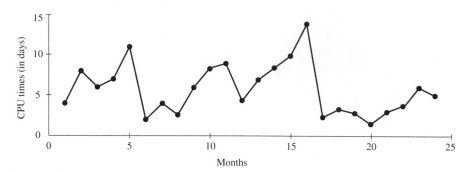

FIGURE 1.5-2

Sometimes, however, time is not an important factor in the response variable and yet time sequence plots still help us understand the concept of variation. Suppose that a process or individual performs at some average level, sometimes doing better and sometimes worse than that average. For example, suppose a certain worker's output is plotted by the month in Figure 1.5-3.

Let us say the manager of the company decides to do the following. On good months (like the second month) the worker is given a "pat on the back" (maybe a bonus or even being named "worker of the month"). What happens? Usually the performance goes down the next month. Recall that the worker is averaging below the performance of that high month; that is, she was performing above average when she was made worker of the month. So there is a tendency to do worse the month following a very good performance. (Statisticians sometimes call this "regression toward the mean.")

This explains what is often called the "sophomore jinx" in sports. A player might have an unusually good first year, actually one far above his average. So the next year, he does worse. Fans will say it's the sophomore jinx; he probably got overconfident and didn't work as hard. Although there might be some truth to this, it could also be that he worked just as hard as before but still went down because his

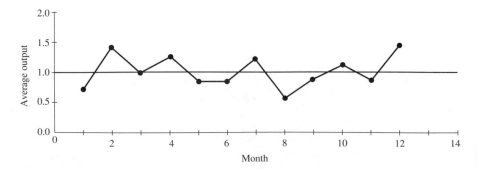

FIGURE 1.5-3

first year was over his average. So, often the "rookie of the year" will do worse the second year because not many players win that award with an average year. That is, he was performing far above his average when he won the award.

Back to the worker described by Figure 1.5-3. On bad months (like the eighth month), the manager could call in the worker and have a serious talk with her or, in some way, penalize her. What happens? Usually, she does better the next month (again regression toward the mean). So the boss gets the idea that it is better to get tough because the workers seem to do better after these little talks. Of course, like the "pat on the back" technique, this does not really work because the workers keep varying about their average. They usually do better after a very poor performance and worse after an excellent performance.

If the manager does not like the average level of performance, then he or she must do something to change the system. Most of the time, the individual workers can do very little to change the process, for major decisions for improvements must be made by the manager or even in the boardroom. After these decisions, the workers must be given "roadmaps directing them to these improvements." Most workers (as well as students) want to do better and would like to take pride in their performance. Managers (as well as teachers) should provide for the opportunity to improve the self esteem of those working for them.

We have seen that a time sequence gives a picture over time of the magnitudes of the observations as they were collected. Earlier we saw that a stem-and-leaf display gives a picture of the complete set of observations. We now look at a combination of these two techniques. Recently, J. Stuart Hunter (*The American Statistician* 42, 1988, p. 54) has enhanced Tukey's stem-and-leaf display by combining it with a time-sequence plot. This is illustrated in the next example.

Example 1.5-2   A combination of a stem-and-leaf display and a time-sequence plot is illustrated using the following 20 quiz scores of an individual in a two-semester calculus course. Each quiz was graded on a 20-point scale, and they are given in time sequence.

18	20	16	12	15	15	10	13	14	9
12	10	8	11	12	11	10	11	8	6

As each score became available, it was recorded in two ways: as a **digit** in the stem-and-leaf display on the left in Figure 1.5-4 (note that the stems decrease in magnitude) and as a **dot** on the time-sequence plot on the right. Hence Hunter called such a display a **digidot plot**.

With only the stem-and-leaf display (i.e., without the time sequence) an observer would see a distribution of scores that had a median of 11.5 and also that 9.5–11.5 is the **modal interval**, that is, the interval (or class) with the highest frequency. However, since the time history is also recorded, it is clear that the student was not making a great effort throughout the year in this calculus class. If interested, students can correct trends like this, but they must make changes in their behavior to do so. By continuing the past pattern

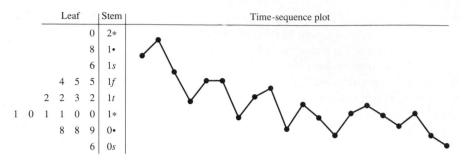

Leaf				Stem	Time-sequence plot
			0	2*	
			8	1•	
			6	1s	
	4	5	5	1f	
	2	2	3	2	1t
1	0	1	1	0 0	1*
		8	8	9	0•
			6	0s	

FIGURE 1.5-4

nothing will change and the downward trend will continue. What improvements could be made to change this? Each student must answer this in his or her own way.

Example 1.5-2 illustrates that a digidot plot takes each observation in the order in which it is observed, records the final digit(s) as leaves on a stem-and-leaf display, and also plots a dot for each observation on a time-sequence plot. This one display provides the marginal frequencies and the time order of the observations.

Example 1.5-3   Containers of a prewash concentrate are selected regularly from the production line and the concentrate is weighed. The target weight is 330.5–336.5 grams. We now construct a digidot for the following 60 observations that were observed in the order left to right, line by line.

337.9	338.1	337.3	338.3	337.3	338.1	337.3	338.0
337.9	338.9	337.6	338.2	338.1	338.6	338.0	338.4
334.4	334.3	334.5	333.9	335.1	334.0	335.6	334.1
335.0	334.8	335.2	335.2	335.2	334.0	334.5	334.6
335.4	335.9	335.6	335.8	334.5	335.4	335.0	335.5
335.1	335.0	334.1	334.6	334.4	334.7	333.8	335.0
334.0	334.3	334.0	333.7	334.7	333.5	334.8	333.9
335.7	335.1	335.1	334.6				

The digidot plot for these data is given in Figure 1.5-5. It is pretty clear that during the beginning of production, the fill weights were slightly above the target weight. This is good news for the customer as long as the containers do not overflow. However, if the trend had continued, this would have resulted in a loss of profit for the company. After adjustments were made, the process seemed to come under control, with the weights reflecting the normal variability for a process like this.

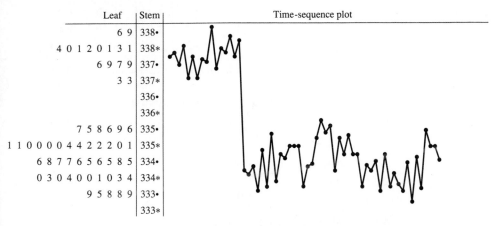

Leaf	Stem	Time-sequence plot
6 9	338•	
4 0 1 2 0 1 3 1	338*	
6 9 7 9	337•	
3 3	337*	
	336•	
	336*	
7 5 8 6 9 6	335•	
1 1 0 0 0 0 4 4 2 2 2 0 1	335*	
6 8 7 7 6 5 6 5 8 5	334•	
0 3 0 4 0 0 1 0 3 4	334*	
9 5 8 8 9	333•	
	333*	

Prewash concentrate fill weights

FIGURE 1.5-5

Exercises

1.5-1 A VAX computer that was used by the administration on a college campus reported the following CPU times (in days) for January through December for two consecutive years.

11.0	10.2	9.8	8.8	7.1	8.8	13.4	15.2
13.1	17.7	14.2	14.7	16.2	12.4	14.8	11.7
9.5	10.0	12.2	11.8	12.0	13.2	10.7	12.0

Construct and interpret a time sequence of these data.

1.5-2 The following data give the percentage of solids in a product. The specifications call for the percentage to be between 60.0% and 72.0%. The data were observed in the order left to right, line by line.

69.8	71.3	65.6	66.3	70.1	71.9	69.6	71.9
71.1	71.7	71.9	69.8	66.8	68.3	64.4	64.2
65.1	63.7	66.2	61.9	66.1	62.9	66.9	67.3
63.3	63.4	67.2	67.4	65.5	66.2	67.5	67.3
66.9	66.5	65.5	63.9	64.6	62.3	66.2	67.2
66.0	69.8	69.7	71.0	69.8	66.0	70.3	65.5
67.0	66.8	67.6	68.6	66.5	66.2	70.4	68.1
64.3	65.2	68.0	65.1				

Construct and interpret a digidot plot using as stems 61, 62, . . . , 71.

1.5-3 Twenty-four "3-pound" bags of apples were filled and weighed automatically in the following order by scale no. 5, yielding the weights:

3.26	3.62	3.39	3.12	3.53	3.30	3.10	3.26
3.19	3.22	3.14	3.39	3.31	3.21	3.49	3.41
3.02	3.17	3.20	3.12	3.42	3.36	3.21	3.26

(a) Construct and interpret a digidot plot using as stems 3.0∗, 3.0•, 3.1∗, 3.1•, . . . , 3.6∗.

(b) Compare scale no. 5 with scale no. 6 in Exercise 1.5-4.

1.5-4 Twenty-four "3-pound" bags of apples were filled and weighed automatically in the following order by scale no. 6, yielding the weights:

3.22	2.97	3.00	3.08	3.04	3.09	3.04	3.21
3.00	3.05	3.11	3.00	3.10	3.13	3.08	3.22
3.12	3.04	3.03	3.02	3.10	3.09	3.18	3.00

(a) Construct and interpret a digidot plot using as stems 2.9•, 3.0∗, 3.0•, . . . 3.2∗.

(b) Compare scale no. 6 with scale no. 5 in Exercise 1.5-3.

1.5-5 The year, x, and the birth rate, y, for 1960–1987 in the United States were as follows:

(1960, 23.7)	(1961, 23.3)	(1962, 22.4)	(1963, 21.7)
(1964, 21.1)	(1965, 19.4)	(1966, 18.4)	(1967, 17.8)
(1968, 17.6)	(1969, 17.9)	(1970, 18.4)	(1971, 17.2)
(1972, 15.6)	(1973, 14.8)	(1974, 14.8)	(1975, 14.6)
(1976, 14.6)	(1977, 15.1)	(1978, 15.0)	(1979, 15.6)
(1980, 15.9)	(1981, 15.8)	(1982, 15.9)	(1983, 15.5)
(1984, 15.5)	(1985, 15.8)	(1986, 15.6)	(1987, 15.7)

Construct and interpret a digidot plot of these data listing the years along the x-axis and using for stems 14, 15, . . . , 23.

1.5-6 The year, x, and the birthrate, y, for 1960–1987 in Michigan were as follows:

(1960, 24.9)	(1961, 24.4)	(1962, 23.0)	(1963, 22.3)
(1964, 21.6)	(1965, 20.3)	(1966, 19.9)	(1967, 18.9)
(1968, 18.3)	(1969, 19.0)	(1970, 19.3)	(1971, 18.1)
(1972, 16.3)	(1973, 15.6)	(1974, 15.1)	(1975, 14.7)
(1976, 14.4)	(1977, 15.1)	(1978, 15.1)	(1979, 15.6)
(1980, 15.7)	(1981, 15.2)	(1982, 15.0)	(1983, 14.5)
(1984, 15.0)	(1985, 15.2)	(1986, 15.1)	(1987, 15.3)

Construct and interpret a digidot plot of these data listing the years along the x-axis and using for stems 14, 15, . . . , 24.

1.5-7 Dish drop viscosity has a specification of 120–230 cps. One hundred observations were selected in the following order and measured, yielding the following viscosities (read across, line by line):

158	147	158	159	169	151	166	151	143	169
153	174	151	164	185	168	140	180	176	154
160	187	145	164	158	169	153	149	144	157
156	183	157	140	162	158	160	180	154	160
164	168	154	158	164	159	153	170	158	170
150	161	169	166	154	157	138	155	134	165
161	172	156	145	153	143	152	152	156	163
179	157	135	172	143	154	165	145	152	145
171	189	144	154	147	187	147	159	167	151
153	168	148	188	152	165	155	140	157	176

Construct and interpret a digidot plot of these measurements.

1.5-8 A manufacturer of car windows has studs in the windows for attaching them to the car. A standard "stud pull-out test" measures the force required to pull a stud out of a window. The target mean is 123. Seventy observations, in order, were as follows:

140	159	138	102	84	126	147	126	103	92
149	155	135	120	94	149	143	109	101	86
144	154	120	105	97	149	151	140	103	99
157	140	120	96	87	146	137	120	93	89
149	154	139	100	84	148	142	130	98	81
112	135	109	84	87	112	135	109	84	87
118	135	90	77	125	126	131	78	75	126

Construct and interpret a digidot plot of these data. Use as stems 7, 8, . . . , 15.

1.5-9 When fruit-stripe gum is rolled out for sticks, the target thickness is 6.7/100 of an inch. The measurements of 30 sticks of gum, selected in the given order, were

6.70	6.80	7.00	6.95	6.85	6.70	6.60	6.75
6.70	6.60	6.70	6.60	6.95	7.00	7.00	6.95
6.80	6.85	6.75	6.65	6.55	6.70	6.90	7.00
6.75	6.85	6.60	6.60	6.60	6.55		

Construct and interpret a digidot plot using as stems 65, 66, . . . , 78.

1.6 Scatter Plots, Least Squares, and Correlation

In Section 1.5, we plotted a response variable, like sales, against another variable, which in that section was time. By doing so, we often get valuable insights about the situation under consideration. Moreover, we can make similar bivariate plots when one of the variables is not time, but some other type of explanatory variable. For example, a student's ACT or SAT score can be of some help in predicting that student's grade point average in college. Also, the weight of a car frequently suggests the number of gallons of gasoline needed to drive that car 100 miles. In each of these examples, the role of the time variable is taken by another explanatory variable: the scholastic aptitude score and the weight of the car, respectively. We denote this "independent" variable by x and the response variable by y. In our examples, y equals college GPA and number of gallons needed to drive the car 100 miles, respectively.

Suppose that we observe n points with coordinates (x_i, y_i), $i = 1, 2, \ldots, n$, where x_i is a preliminary exam score and y_i is a final exam score. These might plot as in Figure 1.6-1; here $n = 10$.

Such a diagram is called a **scatter plot**. Here it looks as if a straight line $y = \alpha + \beta x$ would fit these data fairly well. Clearly, not all the points lie on any straight line, but one line could serve as the pattern about which the points fall. One such line is also depicted in Figure 1.6-1.

How do we select values of the parameters, α and β, of the line so that, in some sense, the line "best" fits the data? One way of doing this is as follows. Look at a typical point, which we have denoted by (x_i, y_i). The point on the line that is either directly below or above that point (x_i, y_i) is the point $(x_i, \alpha + \beta x_i)$. Accordingly, the vertical distance between these two points is $|y_i - \alpha - \beta x_i|$. So, in some way, we want to fit the line to minimize these distances $|y_i - \alpha - \beta x_i|$, $i = 1, 2, \ldots, n$; that is, we want the line $y = \alpha + \beta x$ to be as close as possible to the observed points. One popular way to do this is to square each of the n distances and to add these squared values together to obtain

$$H(\alpha, \beta) = \sum_{i=1}^{n} (y_i - \alpha - \beta x_i)^2.$$

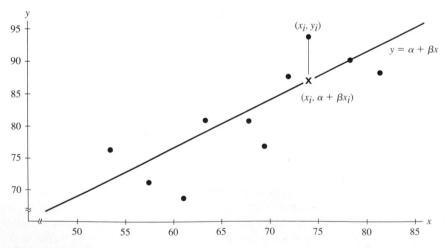

FIGURE 1.6-1

We then want to select α and β to minimize $H(\alpha, \beta)$. Since we are *minimizing* the *sum of squared values*, we call this the **method of least squares**. We now have a calculus problem: We want to find the values of α and β that minimize the function $H(\alpha, \beta)$.

REMARK If you have not studied multivariate calculus, we simply note here that the rule for finding a minimum is about the same as that in the calculus of a single variable. Recall that there one takes the first derivative and equates it to zero. For functions of two variables, we take two first derivatives: Treat β as a constant and take the derivative with respect to α, and then treat α as a constant and take the derivative with respect to β. These two derivatives are denoted, respectively, by

$$\frac{\partial H(\alpha, \beta)}{\partial \alpha} \quad \text{and} \quad \frac{\partial H(\alpha, \beta)}{\partial \beta}$$

and are called first partial derivatives. These are equated to zero in our search for the values of α and β that minimize $H(\alpha, \beta)$. Just as in one-variable calculus, there are certain properties that the second derivatives must satisfy to be certain that we have obtained a relative minimum, but we will not worry about those here for, in this case, the solution does in fact provide the absolute minimum.

We set the two first partial derivatives equal to zero, namely

$$\frac{\partial H(\alpha, \beta)}{\partial \alpha} = \sum_{i=1}^{n} 2(y_i - \alpha - \beta x_i)(-1) = 0,$$

$$\frac{\partial H(\alpha, \beta)}{\partial \beta} = \sum_{i=1}^{n} 2(y_i - \alpha - \beta x_i)(-x_i) = 0.$$

Rewriting these derivatives, we obtain (because $\sum_{i=1}^{n} \alpha = n\alpha$)

$$n\alpha + \left(\sum_{i=1}^{n} x_i\right)\beta = \sum_{i=1}^{n} y_i,$$

$$\left(\sum_{i=1}^{n} x_i\right)\alpha + \left(\sum_{i=1}^{n} x_i^2\right)\beta = \sum_{i=1}^{n} x_i y_i.$$

If the first of these equations is multiplied by $(\sum_{i=1}^{n} x_i)$ and the second is multiplied by n, the difference of the two equations eliminates the α term and gives us

$$\left[n\sum_{i=1}^{n} x_i^2 - \left(\sum_{i=1}^{n} x_i\right)^2\right]\beta = n\sum_{i=1}^{n} x_i y_i - \left(\sum_{i=1}^{n} x_i\right)\left(\sum_{i=1}^{n} y_i\right).$$

That is, if we denote the β value of the solution by $\hat{\beta}$, it is given by

$$\hat{\beta} = \frac{n\sum_{i=1}^{n} x_i y_i - \left(\sum_{i=1}^{n} x_i\right)\left(\sum_{i=1}^{n} y_i\right)}{n\sum_{i=1}^{n} x_i^2 - \left(\sum_{i=1}^{n} x_i\right)^2} = \frac{\sum_{i=1}^{n} x_i y_i - \left(\sum_{i=1}^{n} x_i\right)\left(\sum_{i=1}^{n} y_i\right)\Big/ n}{\sum_{i=1}^{n} x_i^2 - \left(\sum_{i=1}^{n} x_i\right)^2\Big/ n}.$$

Substituting $\hat{\beta}$ in an earlier equation, we see that the α value of the solution is

$$\hat{\alpha} = \frac{\sum_{i=1}^{n} y_i - \left(\sum_{i=1}^{n} x_i\right)\hat{\beta}}{n} = \bar{y} - \hat{\beta}\bar{x},$$

where $\bar{x} = (\sum_{i=1}^{n} x_i)/n$ and $\bar{y} = (\sum_{i=1}^{n} y_i)/n$. That is, the best fitting straight line, in the sense of least squares, is

$$\hat{y} = \hat{\alpha} + \hat{\beta}x = \bar{y} - \hat{\beta}\bar{x} + \hat{\beta}x = \bar{y} + \hat{\beta}(x - \bar{x}).$$

If we study the solution $\hat{\beta}$, its denominator equals $\sum_{i=1}^{n}(x_i - \bar{x})^2$ because

$$\sum_{i=1}^{n} (x_i - \bar{x})^2 = \sum_{i=1}^{n} x_i^2 - 2\bar{x}\sum_{i=1}^{n} x_i + n\bar{x}^2 = \sum_{i=1}^{n} x_i^2 - \left(\sum_{i=1}^{n} x_i\right)^2 \Big/ n.$$

Moreover, the numerator is equal to $\sum_{i=1}^{n}(x_i - \bar{x})(y_i - \bar{y})$ because

$$\sum_{i=1}^{n} (x_i - \bar{x})(y_i - \bar{y}) = \sum_{i=1}^{n} (x_i y_i - \bar{x}y_i - \bar{y}x_i + \bar{x}\bar{y})$$

$$= \sum_{i=1}^{n} x_i y_i - \bar{x}\sum_{i=1}^{n} y_i - \bar{y}\sum_{i=1}^{n} x_i + n\bar{x}\bar{y}$$

$$= \sum_{i=1}^{n} x_i y_i - \frac{1}{n}\left(\sum_{i=1}^{n} x_i\right)\left(\sum_{i=1}^{n} y_i\right) - \frac{1}{n}\left(\sum_{i=1}^{n} y_i\right)\left(\sum_{i=1}^{n} x_i\right) + \frac{1}{n}\left(\sum_{i=1}^{n} x_i\right)\left(\sum_{i=1}^{n} y_i\right)$$

$$= \sum_{i=1}^{n} x_i y_i - \frac{1}{n}\left(\sum_{i=1}^{n} x_i\right)\left(\sum_{i=1}^{n} y_i\right).$$

That is,

$$\hat{\beta} = \frac{\sum_{i=1}^{n} (x_i - \bar{x})(y_i - \bar{y})}{\sum_{i=1}^{n} (x_i - \bar{x})^2}.$$

Recall that $s_x^2 = [1/(n-1)]\sum_{i=1}^{n}(x_i - \bar{x})^2$ and $s_y^2 = [1/(n-1)]\sum_{i=1}^{n}(y_i - \bar{y})^2$ are called the sample variances of the x-values and y-values, respectively. In this last expression for $\hat{\beta}$, the numerator divided by $n - 1$, is called the **covariance** of x and y and is given by

$$c_{xy} = \frac{1}{n - 1} \sum_{i=1}^{n} (x_i - \bar{x})(y_i - \bar{y}).$$

If we consider the covariance of the *standardized* x- and y-values, namely $(x_i - \bar{x})/s_x$ and $(y_i - \bar{y})/s_y$, $i = 1, 2, \ldots, n$, we obtain another popular statistic called the **correlation coefficient,**

$$r = \frac{1}{n - 1} \sum_{i=1}^{n} \left(\frac{x_i - \bar{x}}{s_x} \right)\left(\frac{y_i - \bar{y}}{s_y} \right) = \frac{\dfrac{1}{n - 1} \sum_{i=1}^{n} (x_i - \bar{x})(y_i - \bar{y})}{s_x s_y} = \frac{c_{xy}}{s_x s_y}.$$

Thus we now see that

$$\hat{\beta} = \frac{c_{xy}}{s_x^2} = \left(\frac{c_{xy}}{s_x s_y} \right)\left(\frac{s_y}{s_x} \right) = r\frac{s_y}{s_x}$$

and the best-fitting straight line is

$$\hat{y} = \bar{y} + r\frac{s_y}{s_x}(x - \bar{x}).$$

Again it must be emphasized that all the points (x_i, y_i), $i = 1, 2, \ldots, n$, do *not* lie on this line, but will be reasonably close to it.

To calculate the value of r, it is sometimes useful to note that

$$r = \frac{\dfrac{1}{n - 1} \sum_{i=1}^{n} (x_i - \bar{x})(y_i - \bar{y})}{\sqrt{\dfrac{1}{n - 1} \sum_{i=1}^{n} (x_i - \bar{x})^2} \sqrt{\dfrac{1}{n - 1} \sum_{i=1}^{n} (y_i - \bar{y})^2}}$$

$$= \frac{\sum_{i=1}^{n} x_i y_i - \left(\sum_{i=1}^{n} x_i \right)\left(\sum_{i=1}^{n} y_i \right)\bigg/ n}{\sqrt{\sum_{i=1}^{n} x_i^2 - \left(\sum_{i=1}^{n} x_i \right)^2 \bigg/ n} \sqrt{\sum_{i=1}^{n} y_i^2 - \left(\sum_{i=1}^{n} y_i \right)^2 \bigg/ n}}.$$

An example using data from H. V. Henderson and P. F. Velleman, ''Building Multiple Regression Models Interactively'' [*Biometrics* 37:391–411 (1981)], will help to illustrate these characteristics.

Example 1.6-1. Consider Table 1.6-1, in which x equals the weight of a car in thousands of pounds and y equals the number of gallons of gasoline needed to drive 100 miles. The scatter plot of these $n = 10$ data points is

TABLE 1.6-1

Car	x	y	$(x-\bar{x})(y-\bar{y})$	$(x-\bar{x})^2$	$(y-\bar{y})^2$
AMC Concord	3.4	5.5	0.555	0.25	1.2321
Chevrolet Caprice	3.8	5.9	1.359	0.81	2.2801
Ford Country Squire Wagon	4.1	6.5	2.532	1.44	4.4521
Chevrolet Chevette	2.2	3.3	0.763	0.49	1.1881
Toyota Corona	2.6	3.6	0.237	0.09	0.6241
Ford Mustang Ghia	2.9	4.6	0.000	0.00	0.0441
Mazda GLC	2.0	2.9	1.341	0.81	2.2201
AMC Sprint	2.7	3.6	0.158	0.04	0.6241
VW Rabbit	1.9	3.1	1.290	1.00	1.6641
Buick Century	3.4	4.9	0.255	0.25	0.2601
Sum	29.0	43.9	8.490	5.18	14.5890

given in Figure 1.6-2. We find that the means are $\bar{x} = 29.0/10 = 2.90$, $\bar{y} = 43.9/10 = 4.39$, and the standard deviations are $s_x = \sqrt{5.18/9} = 0.759$, $s_y = \sqrt{14.589/9} = 1.273$. Since the covariance equals

$$c_{xy} = \frac{8.490}{9} = 0.943,$$

we find that

$$r = \frac{8.490/9}{\sqrt{5.18/9}\sqrt{14.589/9}} = 0.977.$$

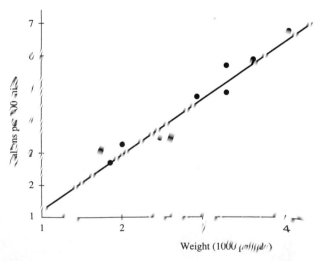

FIGURE 1.6-2

last expression for $\hat{\beta}$, the numerator divided by $n - 1$, is called the **covariance** of x and y and is given by

$$c_{xy} = \frac{1}{n-1} \sum_{i=1}^{n} (x_i - \bar{x})(y_i - \bar{y}).$$

If we consider the covariance of the *standardized* x- and y-values, namely $(x_i - \bar{x})/s_x$ and $(y_i - \bar{y})/s_y$, $i = 1, 2, \ldots , n$, we obtain another popular statistic called the **correlation coefficient**,

$$r = \frac{1}{n-1} \sum_{i=1}^{n} \left(\frac{x_i - \bar{x}}{s_x} \right) \left(\frac{y_i - \bar{y}}{s_y} \right) = \frac{\dfrac{1}{n-1} \sum\limits_{i=1}^{n} (x_i - \bar{x})(y_i - \bar{y})}{s_x s_y} = \frac{c_{xy}}{s_x s_y}.$$

Thus we now see that

$$\hat{\beta} = \frac{c_{xy}}{s_x^2} = \left(\frac{c_{xy}}{s_x s_y} \right) \left(\frac{s_y}{s_x} \right) = r \frac{s_y}{s_x}$$

and the best-fitting straight line is

$$\hat{y} = \bar{y} + r \frac{s_y}{s_x} (x - \bar{x}).$$

Again it must be emphasized that all the points (x_i, y_i), $i = 1, 2, \ldots , n$, do *not* lie on this line, but will be reasonably close to it.

To calculate the value of r, it is sometimes useful to note that

$$r = \frac{\dfrac{1}{n-1} \sum\limits_{i=1}^{n} (x_i - \bar{x})(y_i - \bar{y})}{\sqrt{\dfrac{1}{n-1} \sum\limits_{i=1}^{n} (x_i - \bar{x})^2} \sqrt{\dfrac{1}{n-1} \sum\limits_{i=1}^{n} (y_i - \bar{y})^2}}$$

$$= \frac{\sum\limits_{i=1}^{n} x_i y_i - \left(\sum\limits_{i=1}^{n} x_i \right) \left(\sum\limits_{i=1}^{n} y_i \right) \Big/ n}{\sqrt{\sum\limits_{i=1}^{n} x_i^2 - \left(\sum\limits_{i=1}^{n} x_i \right)^2 \Big/ n} \sqrt{\sum\limits_{i=1}^{n} y_i^2 - \left(\sum\limits_{i=1}^{n} y_i \right)^2 \Big/ n}}.$$

An example using data from H. V. Henderson and P. F. Velleman, "Building Multiple Regression Models Interactively" [*Biometrics* 37:391–411 (1981)], will help to illustrate these characteristics.

> **Example 1.6-1.**　Consider Table 1.6-1, in which x equals the weight of a car in thousands of pounds and y equals the number of gallons of gasoline needed to drive 100 miles. The scatter plot of these $n = 10$ data points is

TABLE 1.6-1

Car	x	y	$(x - \bar{x})(y - \bar{y})$	$(x - \bar{x})^2$	$(y - \bar{y})^2$
AMC Concord	3.4	5.5	0.555	0.25	1.2321
Chevrolet Caprice	3.8	5.9	1.359	0.81	2.2801
Ford Country Squire Wagon	4.1	6.5	2.532	1.44	4.4521
Chevrolet Chevette	2.2	3.3	0.763	0.49	1.1881
Toyota Corona	2.6	3.6	0.237	0.09	0.6241
Ford Mustang Ghia	2.9	4.6	0.000	0.00	0.0441
Mazda GLC	2.0	2.9	1.341	0.81	2.2201
AMC Sprint	2.7	3.6	0.158	0.04	0.6241
VW Rabbit	1.9	3.1	1.290	1.00	1.6641
Buick Century	3.4	4.9	0.255	0.25	0.2601
Sums	29.0	43.9	8.490	5.18	14.5890

given in Figure 1.6-2. We find that the means are $\bar{x} = 29.0/10 = 2.90$, $\bar{y} = 43.9/10 = 4.39$, and the standard deviations are $s_x = \sqrt{5.18/9} = 0.759$, $s_y = \sqrt{14.589/9} = 1.273$. Since the covariance equals

$$c_{xy} = \frac{8.490}{9} = 0.943,$$

we find that

$$r = \frac{8.490/9}{\sqrt{5.18/9}\sqrt{14.589/9}} = 0.977.$$

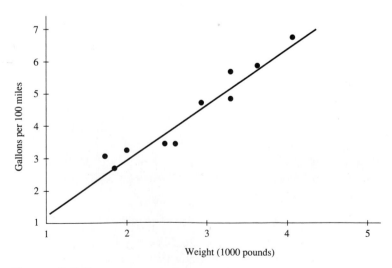

FIGURE 1.6-2

The best-fitting straight line is

$$\hat{y} = 4.39 + (0.977)\frac{1.273}{0.759}(x - 2.90) = 1.639x - 0.363$$

and this is actually the line depicted in Figure 1.6-2. Note that we could also have found the slope of the line using

$$\hat{\beta} = \frac{\displaystyle\sum_{i=1}^{n}(x_i - \bar{x})(y_i - \bar{y})}{\displaystyle\sum_{i=1}^{n}(x_i - \bar{x})^2} = \frac{8.490}{5.18} = 1.639.$$

The values of r and $\hat{\beta}$ can sometimes be found more easily by constructing a table like Table 1.6-2.

Using the sums in Table 1.6-2, we obtain

$$r = \frac{\displaystyle\sum_{i=1}^{n}x_i y_i - \left(\sum_{i=1}^{n}x_i\right)\left(\sum_{i=1}^{n}y_i\right)\Big/n}{\sqrt{\displaystyle\sum_{i=1}^{n}x_i^2 - \left(\sum_{i=1}^{n}x_i\right)^2\Big/n}\ \sqrt{\displaystyle\sum_{i=1}^{n}y_i^2 - \left(\sum_{i=1}^{n}y_i\right)^2\Big/n}}$$

$$= \frac{135.80 - (29.0)(43.9)/10}{\sqrt{89.28 - (29.0)^2/10}\ \sqrt{207.31 - (43.9)^2/10}}$$

$$= \frac{8.49}{\sqrt{5.18}\sqrt{14.589}}$$

$$= 0.977.$$

TABLE 1.6-2

x	y	x^2	xy	y^2
3.4	5.5	11.56	18.70	30.25
3.8	5.9	14.44	22.42	34.81
4.1	6.5	16.81	26.65	42.25
2.2	3.3	4.84	7.26	10.89
2.6	3.6	6.76	9.36	12.96
2.9	4.6	8.41	13.34	21.16
2.0	2.9	4.00	5.80	8.41
2.7	3.6	7.29	9.72	12.96
1.9	3.1	3.61	5.89	9.61
3.4	4.9	11.56	16.66	24.01
29.0	43.9	89.28	135.80	207.31

$$\hat{\beta} = \frac{\sum\limits_{i=1}^{n} x_i y_i - \left(\sum\limits_{i=1}^{n} x_i\right)\left(\sum\limits_{i=1}^{n} y_i\right)\Big/n}{\sum\limits_{i=1}^{n} x_i^2 - \left(\sum\limits_{i=1}^{n} x_i\right)^2\Big/n}$$

$$= \frac{135.80 - (29.0)(43.9)/10}{89.28 - (29.0)^2/10}$$

$$= \frac{8.490}{5.180}$$

$$= 1.639.$$

Of course, $\hat{\alpha} = \bar{y} - \hat{\beta}\bar{x} = 4.39 - (1.639)(2.90) = -0.363$.

A few remarks should be made about the correlation coefficient r.

(a) If we think of the horizontal and vertical lines through the point $(\bar{x}, \bar{y})$ as dividing the plane into four quadrants, note that points in the first and third quadrants give positive products $(x_i - \bar{x})(y_i - \bar{y})$ and those in the second and fourth quadrants produce negative products. Thus, if there are more and larger deviations from $(\bar{x}, \bar{y})$ in the first and third quadrants than in the second and fourth, the correlation coefficient is positive. That is, if larger x-values tend to produce larger y-values and smaller x-values tend to produce smaller y-values, then $r > 0$. On the other hand, $r < 0$ if the points tend to favor the second and fourth quadrants; this happens when large x-values tend to produce small y-values and small x-values tend to produce large y-values.

(b) Note that

$$\sum_{i=1}^{n} [(y_i - \bar{y}) - t(x_i - \bar{x})]^2 \geq 0$$

for every real t. That is, the quadratic expression in t,

$$\left[\sum_{i=1}^{n} (x_i - \bar{x})^2\right]t^2 - 2\left[\sum_{i=1}^{n} (x_i - \bar{x})(y_i - \bar{y})\right]t + \left[\sum_{i=1}^{n} (y_i - \bar{y})^2\right]$$

is greater than or equal to zero for all real t. Hence it's discriminant, $b^2 - 4ac$, must be nonpositive; that is,

$$\left[2\sum_{i=1}^{n} (x_i - \bar{x})(y_i - \bar{y})\right]^2 - 4\left[\sum_{i=1}^{n} (x_i - \bar{x})^2\right]\left[\sum_{i=1}^{n} (y_i - \bar{y})^2\right] \leq 0.$$

Thus

$$\frac{\left[\displaystyle\sum_{i=1}^{n} (x_i - \bar{x})(y_i - \bar{y})\right]^2}{\left[\displaystyle\sum_{i=1}^{n} (x_i - \bar{x})^2\right]\left[\displaystyle\sum_{i=1}^{n} (y_i - \bar{y})^2\right]} = r^2 \le 1.$$

It follows that $-1 \le r \le 1$, since $r^2 \le 1$.

(c) The correlation coefficient r quantifies the *linear* association between x and y. If $r = 1$, then all the points are on a straight line with positive slope, s_y/s_x. If, on the other hand, $r = -1$, then all the points are on a straight line with negative slope, $-s_y/s_x$. Moreover, r could equal zero, indicating a weak linear relationship between x and y; yet there could be some other strong nonlinear relationship, like a quadratic one (see Exercise 1.6-12). In any case, one statistic, here r, might miss certain aspects of the data that are discernable to the eye. That is the reason we encourage you to plot the data in looking for the patterns.

From Example 1.6-1, with $r = 0.977$, we see that there is a strong linear relationship. This certainly agrees with our observation of the scatter plot. Thus, for these data, the weight of the car is highly correlated with the number of gallons needed to travel a specified distance. In this case, weight is the major explanation of mileage performance. However, two variables could be highly correlated, and yet one does not cause the other. For example, before air conditioning and modern hospitals, it was true that infant mortality was highly correlated to the softness of asphalt streets. Clearly, one was not causing the other, but a third variable (temperature) made both of the other variables increase or decrease together.

To visualize better the relationship between r and a plot of n observed points $(x_1, y_1), \ldots, (x_n, y_n)$, we have generated three different sets of 50 pairs of observations from three bivariate distributions. In the next example we list the corresponding values of $\bar{x}$, $\bar{y}$, s_x^2, s_y^2, r, and the observed best-fitting line. Each set of points and corresponding line are plotted on the same graph.

Example 1.6-2   Three random samples, each of size $n = 50$, were taken from three different bivariate distributions. The corresponding sample characteristics are

(a) $\bar{x} = 11.905$, $s_x^2 = 14.095$, $\bar{y} = 8.271$, $s_y^2 = 6.851$, and $r = 0.799$, so the best-fitting line is

$$\hat{y} = 8.271 + 0.799\sqrt{\frac{6.851}{14.095}} \, (x - 11.905) = 0.557x + 1.639.$$

(b) $\bar{x} = 12.038$, $s_x^2 = 15.011$, $\bar{y} = 7.790$, $s_y^2 = 7.931$, and $r = 0.169$, so the best-fitting line is

$$\hat{y} = 7.790 + 0.169\sqrt{\frac{7.931}{15.011}}\,(x - 12.038) = 0.123x + 6.311.$$

(c) $\bar{x} = 12.095$, $s_x^2 = 16.762$, $\bar{y} = 8.040$, $s_y^2 = 7.655$, and $r = -0.689$, so the best-fitting line is

$$\hat{y} = 8.040 - 0.689\sqrt{\frac{7.655}{16.762}}\,(x - 12.095) = -0.466x + 13.672.$$

In Figure 1.6-3, these respective lines and the corresponding sample points are plotted. Note the effect that the value of r has on the slope of the line and the variability of the points about that line.

The next example shows that two paired variables may be clearly related (dependent), yet have a correlation coefficient r close to zero. This, however, is not unexpected, since we recall that r does, in a sense, measure the *linear* relationship between two random variables. That is, the linear relationship between the variables could be zero, whereas higher order ones could be quite strong.

Example 1.6-3   Twenty-five observations of paired variables were simulated, yielding the following data:

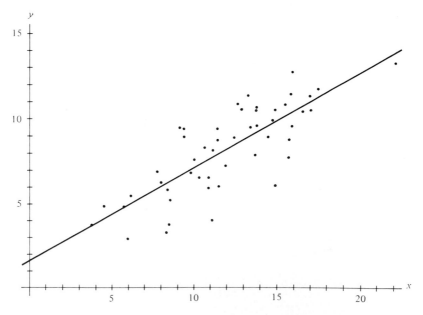

FIGURE 1.6-3(a)

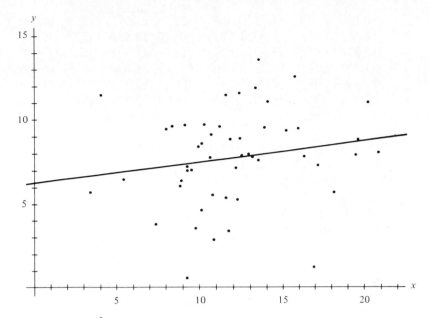

FIGURE 1.6-3(b)

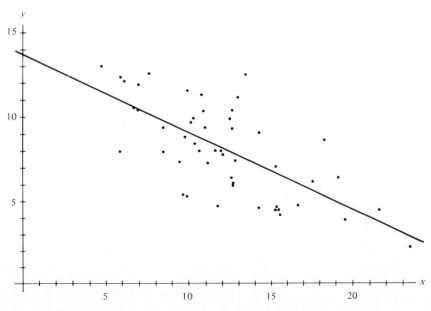

FIGURE 1.6-3(c)

(6.91, 17.52)	(4.32, 22.69)	(2.38, 17.61)	(7.98, 14.29)
(8.26, 10.77)	(2.00, 12.87)	(3.10, 18.63)	(7.69, 16.77)
(2.21, 14.97)	(3.42, 19.16)	(8.18, 11.15)	(5.39, 22.41)
(1.19, 7.50)	(3.21, 19.06)	(5.47, 23.89)	(7.35, 16.63)
(2.32, 15.09)	(7.54, 14.75)	(1.27, 10.75)	(7.33, 17.42)
(8.41, 9.40)	(8.72, 9.83)	(6.09, 22.33)	(5.30, 21.37)
(7.30, 17.36)			

For this set of observations $\bar{x} = 5.33$, $\bar{y} = 16.17$, $s_x^2 = 6.521$, $s_y^2 = 20.865$, and $r = -0.06$. Note that r is very close to zero even though the variables seem very dependent; that is, it seems that a quadratic expression would fit the data very well. In Exercise 1.6-11 the reader is asked to fit $y = a + bx + cx^2$ to these 25 points by the method of least squares. See Figure 1.6-4 for a plot of the 25 points.

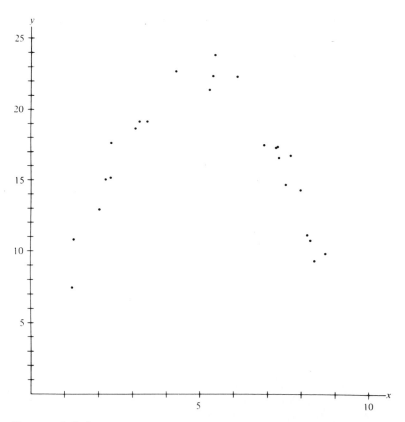

FIGURE 1.6-4

Exercises

1.6-1   A reader survey for a birding magazine asked respondents to give the number of binoculars (x) and the number of bird reference books (y) that they owned. The following data were returned for 10 respondents:

$$(1, 0) \quad (3, 3) \quad (6, 7) \quad (2, 6) \quad (2, 1)$$
$$(5, 9) \quad (1, 1) \quad (3, 1) \quad (3, 10) \quad (0, 1)$$

(a) Find the equation of the least squares regression line.
(b) Find the value of the correlation coefficient r.
(c) Plot the data and the least squares regression line on the same graph.

1.6-2   There was interest in discovering whether some college students learned to use the library more effectively during the academic year. A library knowledge test was given at the beginning and the end of the semester with the scores denoted by x and y, respectively. The scores for $n = 14$ students were as follows:

$$(15, 21) \quad (19, 21) \quad (22, 25) \quad (17, 16) \quad (13, 19)$$
$$(24, 23) \quad (20, 23) \quad (18, 22) \quad (22, 22) \quad (19, 19)$$
$$(13, 18) \quad (17, 12) \quad (15, 20) \quad (17, 21)$$

(a) Find the equation of the least squares regression line.
(b) Find the value of the correlation coefficient r.
(c) Plot the data and the least squares regression line on the same graph.

1.6-3   From the *Environmental Council Report to the President for 1988,* we read that the carbon monoxide outputs (y) into the atmosphere in millions of tons from transportation sources (cars, buses, airplanes, etc.) for seven consecutive years (x) yielded the following paired data (x, y):

$$(1, 52.6) \quad (2, 51.6) \quad (3, 48.1) \quad (4, 48.3)$$
$$(5, 48.4) \quad (6, 45.2) \quad (7, 42.6)$$

(a) Find the equation of the least squares regression line.
(b) Find the value of the correlation coefficient r.
(c) Plot the data and the least squares regression line on the same graph.
(d) Has there been improvement in decreasing carbon monoxide outputs?

1.6-4   Ten college students took the Undergraduate Record Exam (URE) when they were juniors and the Graduate Record Exam (GRE) when they were seniors. The Quantitative URE score (x) and the Quantitative GRE score (y) for each of these 10 students is given in the following list of ordered pairs (x, y):

$$(550, 570) \quad (670, 730) \quad (490, 450) \quad (410, 540) \quad (570, 560)$$
$$(490, 400) \quad (450, 420) \quad (490, 520) \quad (780, 710) \quad (520, 620)$$

(a) Verify that $\bar{x} = 542.0$, $\bar{y} = 552.0$, $s_x^2 = 12{,}040.0$, $s_y^2 = 12{,}640.0$, and $r = 0.79$.

(b) Find the equation of the best-fitting line.

(c) Plot the 10 points and the line on the same graph.

1.6-5 The respective high school and college GPAs for 20 college seniors as ordered pairs (x, y) are

(3.75, 3.19)	(3.45, 3.34)	(2.87, 2.23)	(3.60, 3.46)
(3.42, 2.97)	(4.00, 3.79)	(2.65, 2.55)	(3.10, 2.50)
(3.47, 3.15)	(2.60, 2.26)	(4.00, 3.76)	(2.30, 2.11)
(2.47, 2.11)	(3.36, 3.01)	(3.60, 2.92)	(3.65, 3.09)
(3.30, 3.05)	(2.58, 2.63)	(3.80, 3.22)	(3.79, 3.27)

(a) Verify that $\bar{x} = 3.29$, $\bar{y} = 2.93$, $s_x^2 = 0.283$, $s_y^2 = 0.260$, and $r = 0.92$.

(b) Find the equation of the best-fitting line.

(c) Plot the 20 points and the line on the same graph.

1.6-6 The respective high school GPA and the SAT mathematics score for 25 college students as ordered pairs (x, y) are

(4.00, 577)	(2.53, 453)	(3.45, 407)	(2.48, 539)
(2.69, 534)	(2.82, 584)	(2.33, 464)	(2.21, 525)
(2.59, 545)	(3.37, 499)	(3.00, 446)	(2.93, 466)
(3.25, 491)	(2.90, 433)	(3.64, 556)	(3.23, 394)
(2.46, 497)	(2.62, 460)	(2.75, 413)	(2.82, 440)
(3.51, 608)	(4.00, 657)	(3.72, 449)	(2.78, 323)
(3.33, 413)			

(a) Verify that $\bar{x} = 3.02$, $\bar{y} = 486.92$, $s_x^2 = 0.258$, $s_y^2 = 5875.74$, and $r = 0.275$.

(b) Find the equation of the best-fitting line.

(c) Plot the 25 points and the line on the same graph.

1.6-7 Twenty 1990 domestic automobiles were selected from the 1990 *Car Buyer's Guide* magazine (Volume 10, Issue 1). Engine displacement in liters (x) and torque (y) is given for each of these cars in the following list of paired data:

(3.8, 210)	(4.5, 270)	(5.7, 330)	(3.0, 171)	(2.2, 122)
(3.8, 315)	(1.6, 101)	(3.8, 220)	(4.5, 245)	(5.7, 340)
(2.2, 220)	(2.0, 203)	(5.0, 300)	(1.6, 87)	(3.3, 185)
(4.3, 235)	(3.1, 170)	(3.3, 183)	(3.0, 200)	(1.0, 58)

(a) Find the equation of the least squares regression line.

(b) Find the value of the correlation coefficient r.

(c) Plot the data and the least squares regression line on the same graph.

(d) Is there a linear relationship between engine displacement and torque?

1.6-8 The following data give the ACT Math and ACT Verbal scores for 15 students:

$$\begin{array}{lllll}
(16, 19) & (18, 17) & (22, 18) & (20, 23) & (17, 20) \\
(25, 21) & (21, 24) & (23, 18) & (24, 18) & (31, 25) \\
(27, 29) & (28, 24) & (30, 24) & (27, 23) & (28, 24)
\end{array}$$

(a) Verify that $\bar{x} = 23.8$, $\bar{y} = 21.8$, $s_x^2 = 22.457$, $s_y^2 = 11.600$, and $r = 0.626$.

(b) Find the equation of the best-fitting line.

(c) Plot the 15 points and the line on the same graph.

1.6-9 Each of 16 professional golfers hits off the tee a brand A golf ball and a brand B golf ball, eight hitting ball A before ball B and eight hitting ball B before ball A. Let X and Y equal the distances traveled in yards for ball A and for ball B, respectively. The following data, (x, y), were observed:

$$\begin{array}{llll}
(265, 252) & (272, 276) & (246, 243) & (260, 246) \\
(274, 275) & (263, 246) & (255, 244) & (258, 245) \\
(276, 259) & (274, 267) & (274, 260) & (269, 267) \\
(244, 251) & (212, 222) & (254, 255) & (224, 231)
\end{array}$$

(a) Verify that $\bar{x} = 257.50$, $\bar{y} = 252.44$, $s_x^2 = 341.333$, $s_y^2 = 218.796$, and $r = 0.867$.

(b) Find the equation of the best-fitting line.

(c) Plot the 16 points and the line on the same graph.

1.6-10 Fourteen pairs of gallinules were captured and weighed. Let X equal the male weight and Y the female weight. The following weights in grams were observed:

$$\begin{array}{lllll}
(405, 321) & (396, 378) & (457, 351) & (450, 320) & (415, 365) \\
(403, 328) & (370, 372) & (435, 314) & (425, 375) & (425, 355) \\
(415, 355) & (400, 340) & (425, 398) & (420, 330)
\end{array}$$

(a) Verify that $\bar{x} = 417.2$, $s_x = 22.36$, $\bar{y} = 350.1$, $s_y = 25.56$, and $r = -0.252$.

(b) Find the equation of the best-fitting line.

(c) Plot the points and the line on the same graph.

1.6-11 We would like to fit the quadratic curve $y = a + bx + cx^2$ to a set of points $(x_1, y_1), (x_2, y_2), \ldots, (x_n, y_n)$ by the method of least squares. To do this, let

$$h(a, b, c) = \sum_{i=1}^{n} (y_i - a - bx_i - cx_i^2)^2.$$

(a) By setting the three first partial derivatives of h with respect to a, b, and c equal to zero, show that a, b, and c satisfy the following set of equations, all sums going from 1 to n:

$$an + b \sum x_i + c \sum x_i^2 = \sum y_i;$$
$$a \sum x_i + b \sum x_i^2 + c \sum x_i^3 = \sum x_i y_i;$$
$$a \sum x_i^2 + b \sum x_i^3 + c \sum x_i^4 = \sum x_i^2 y_i.$$

(b) For the data given in Example 1.6-4, $\sum x_i = 133.34$, $\sum x_i^2 = 867.75$, $\sum x_i^3 = 6197.21$, $\sum x_i^4 = 46,318.88$, $\sum y_i = 404.22$, $\sum x_i y_i = 2138.38$, $\sum x_i^2 y_i = 13,380,30$. Show that $a = -1.88$, $b = 9.86$, and $c = -0.995$.

(c) Plot the points and this least squares quadratic regression curve on the same graph.

1.6-12 Let a random number X be selected from the interval $(1, 9)$. For each observed value of $X = x$, let a random number Y be selected from the interval $(x^2 - 10x + 26, x^2 - 10x + 30)$. Twenty-five observations of X and Y generated on a computer are

(4.16, 2.66)	(2.88, 8.60)	(4.97, 3.76)	(2.02, 12.81)
(2.69, 10.14)	(2.54, 8.69)	(1.49, 14.07)	(2.13, 10.36)
(2.44, 8.04)	(3.20, 4.41)	(4.20, 3.01)	(8.74, 17.15)
(3.17, 6.79)	(5.39, 1.63)	(8.43, 14.23)	(6.10, 4.75)
(5.47, 1.82)	(8.17, 14.55)	(2.18, 12.76)	(3.18, 6.56)
(8.26, 12.89)	(6.62, 6.72)	(2.68, 9.53)	(8.06, 11.63)
(6.87, 5.96)			

(a) For these data, $\sum x_i = 116.04$, $\sum x_i^2 = 675.35$, $\sum x_i^3 = 4551.52$, $\sum x_i^4 = 33,331.38$, $\sum y_i = 213.52$, $\sum x_i y_i = 1036.97$, $\sum x_i^2 y_i = 6661.79$. Show that the equation of the least squares quadratic regression curve is equal to $1.026x^2 - 10.296x + 28.612$.

(b) Plot the points and the least squares quadratic regression curve on the same graph.

1.6-13 After keeping records for 6 years, a telephone company is interested in predicting the number of telephones that will be in service in year 7. The following data are available. The number of telephones in service is given in thousands.

Year	Number of Telephones
1	91
2	93
3	95
4	99
5	102
6	105

(a) Find the quadratic curve of best fit, $y = a + bx + cx^2$, that could be used for this prediction.

(b) Plot the points and the curve on the same graph.

(c) Predict the number of telephones that will be in service in year 7.

1.6-14 For male freshmen in a health fitness program, let X equal a participant's percentage of body fat at the beginning of the semester and let Y equal the change in this percentage (percentage at the end of the semester minus percentage at the beginning of the semester, so that a negative y indicates a loss). Twelve observations of (X, Y) are

(13.1, 1.1)	(16.8, 0.5)	(17.9, −1.3)	(10.6, −2.2)
(8.2, −1.1)	(10.4, −0.2)	(17.4, −2.0)	(10.5, −1.4)
(5.4, 0.5)	(14.3, −4.6)	(11.1, 1.0)	(5.3, 1.7)

(a) Verify that $\bar{x} = 11.750$, $s_x = 4.298$, $\bar{y} = -0.667$, $s_y = 1.788$, and $r = -0.395$.

(b) Find the equation of the best-fitting line (i.e., the least squares regression line).

(c) Plot the points and the line on the same graph.

1.6-15 Let X equal the number of milligrams of tar and Y the number of milligrams of carbon monoxide per filtered cigarette (100 millimeters in length) measured by the Federal Trade Commission. A sample of 12 brands yielded the following data:

(5, 7)	(17, 16)	(9, 11)	(8, 9)	(11, 9)	(12, 11)
(15, 15)	(20, 20)	(11, 10)	(13, 13)	(13, 11)	(11, 14)

(a) Verify that $\bar{x} = 12.08$, $s_x = 4.010$, $\bar{y} = 12.17$, $s_y = 3.614$, and $r = 0.915$.

(b) Find the equation of the best-fitting line.

(c) Plot the points and the line on the same graph.

1.6-16 For each of 20 statistics students, let X and Y equal the mother's and father's ages, respectively. The observed data are

(50, 52)	(51, 50)	(50, 53)	(52, 51)	(47, 50)
(48, 51)	(51, 52)	(48, 50)	(48, 48)	(64, 65)
(40, 41)	(52, 62)	(46, 48)	(49, 52)	(51, 56)
(54, 58)	(44, 46)	(53, 56)	(50, 49)	(51, 55)

(a) Verify that $\bar{x} = 49.95$, $s_x = 4.628$, $\bar{y} = 52.25$, $s_y = 5.418$, and $r = 0.888$.

(b) Find the equation of the best-fitting line.

(c) Plot the points and the line on the same graph.

1.7 Design of Experiments

Many observations that are taken occur without active participation of an experimenter. The analysis of such data represents a passive use of statistics that is often extremely helpful in making future decisions. However, in trying to create better products or services, there are frequently a number of factors (independent variables or "parameters") that we can control, and we want to set these at levels which will produce the best products.

For example, in making a cake mix, a company must decide how much sugar, flour, vanilla, salt, butter, egg mix, and milk to use, and then also recommend what time and temperature should be used in baking the mixture. Each of these variables (from amount of sugar to temperature) is called a factor, and the experimenters must decide which level is best for each factor. If we can list nine or so factors for making a cake, think of how many factors would be associated with a more complicated product.

Much can be learned about a process by considering the most influential factors and changing them according to a specified plan to try to find the best level of each. How does a cake mix company know that a certain mixture should be baked at 350° for 35 minutes? Why not use 325° for 40 minutes? Or 375° for 30 minutes? The company must experiment with these levels of time and temperature, as well as with those of the other factors, to determine the optimal. Only by experimenting can the company determine the best level for each factor.

Such investigations involve collecting and analyzing data from carefully designed experiments. Since a poorly designed experiment may not shed light on a particular question that we want answered, it is important to plan an experiment before starting to perform it. Trying to rescue a poorly designed experiment after it has been run and the data have been collected usually leads to less-than-satisfactory results. Thus it is important to design the experiment ahead of time so that the validity of the experimental results is assured. The *design of experiments* is a very important area of statistics that teaches us how to perform experiments to maximize our knowledge of the situation under consideration with a relatively few number of experimental runs.

As an example of such experimentation, let us return to the time and temperature needed to bake the best cake. Suppose that this is a completely new cake mix and the company has no idea about the right time and temperature, except that they will probably be in the ranges of 25–50 minutes and 250°–400°, respectively. Suppose that we have some measure of the "goodness" of a cake, possibly scores of professional tasters who grade each cake on a 100-point scale.

Unknown to anyone, suppose that the real goodness response to the cakes made at different times and temperatures is given in Figure 1.7-1. That is, this response surface is depicted by contour curves which are like "isotherm curves." For example, a cake baked at 300° for 35 minutes would get a score of 59 from the tasters. (Although there would be variation among tasters, let us say they average 59 for such a cake.) Clearly, if we knew this response surface ahead of time, we should bake this cake mix at about 363° for about 31 minutes to obtain a score above 90. However, we do *not* know this surface. It is our job to find out something about it so that we can determine the optimal time and temperature at which to bake this cake. We do this by baking the cake mixture at different times and temperatures.

One procedure, called "changing one factor at a time" is sometimes used; but, as we see in Figure 1.7-1, that scheme often misses the optimal levels of the factors. In

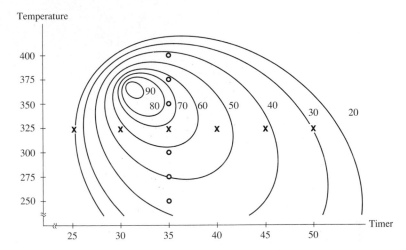

FIGURE 1.7-1

it, we set each of the factors except the first at some middle value. In our case, we only have two factors; so we set the second at 325°, say. Now let us run the experiment 6 times: at 25, 30, 35, 40, 45, and 50 minutes. The respective scores are *about* 20, 58, 68, 53, 41, and 30 by interpolating between contour curves at the x points. That is, with the temperature set at 325°, the maximum score occurs at 35 minutes. Thus we now set the time at 35 minutes and perform six more experiments with respective temperatures of 250°, 275°, 300°, 350°, 375°, and 400°, recalling that 35 minutes and 325° has already produced a score of about 68. The seven scores, with 68 inserted in the correct position, are about 43, 50, 59, 68, 76, 63, and 42. Thus it looks as if the maximum occurs at 350° for 35 minutes; also, the maximum value is 76. Clearly, we have missed the true optimum levels of these two factors by using the "changing one factor at a time" scheme.

Wouldn't it have been much better to run at first four experiments changing the levels of each factor? That is, take two middle values of each factor: say 35 and 40 minutes, 300° and 325°. Then run four experiments at (35, 300°), (35, 325°), (40, 300°), and (40, 325°), obtaining scores of about 59, 68, 53, and 53, respectively. These values are plotted in Figure 1.7-2.

It seems fairly clear that the surface is increasing as we move to the left and upward. Thus we might run three more experiments at (30, 325°), (30, 350°), and (35, 350°) obtaining scores of about 58, 69, and 76, respectively; these are also noted in Figure 1.7-2. Clearly, we must continue upward and we perform two more experiments at (30, 375°) and (35, 375°) obtaining 64 and 63. At this point, it seems as if we should investigate points around times of 30 to 35 minutes and temperatures in the neighborhood of 350°, which will just about give us an optimal solution. But note how changing both factors in experimenting is better than changing one at a time. This is even more true as the number of factors increases.

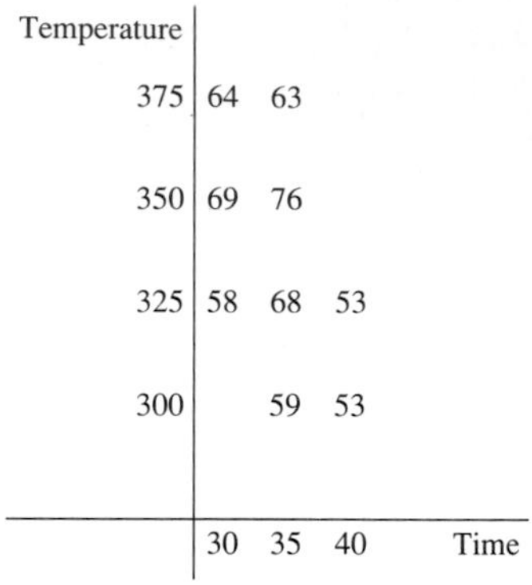

Temperature

375	64	63	
350	69	76	
325	58	68	53
300		59	53

                    30   35   40   Time

FIGURE 1.7-2

In the preceding discussion, in which we first ran four experiments, with two factors at two levels each, is often called a 2^2 factorial experiment. If we had three factors involved and ran each at two levels, that would give us $2 \times 2 \times 2 = 8$ experimental runs, which is called a 2^3 factorial experiment. In general, with k factors at 2 levels each, we have a 2^k factorial experiment. Note that with $k = 10$ factors and only two levels each, we would need $2^{10} = 1024$ experimental runs. Often this is too many, and in advanced courses on design of experiments we learn how to run only a fraction of this number without losing too much information. (See also Section 8.3.)

Let us say that we now wish to compare two different methods of doing something (say teaching statistics). In essence we have only one factor, teaching methods, and we have only two levels: Method 1 (M_1) and Method 2 (M_2). If we had 50 students who were participating in this study, it would be desirable to place 25 in each group. The first group uses M_1 and the second M_2. Clearly, if the investigators favor M_2 over M_1, it would not be fair to let them place the better students in the second group. The assignments should be made at random so that each combination of 50 students taken 25 at a time has the same chance of being the first group. One way of doing this is placing the 50 names on equal size slips of paper and then drawing 25 from a bowl at random.

Suppose that we knew a little more about these students and were able to pair them up by mathematical ability, say. Student S_1 is about the same as S_2, S_3 about the same as S_4, and so on. Then let us select one from each pair at random (like with the flip of a coin) for the first group. We would say that the pair S_1, S_2 is a block; pair S_3, S_4 is a block; and so on. This procedure of assigning students to the respective groups is called **blocking**.

To illustrate blocking, suppose that we wish to test the wearing ability of two different types of soles for tennis shoes. Say we have 30 grade-school children who would participate. One way to select 15 for each group would be at random, and the first group would wear shoes with the first sole and the second group the second sole. After one month, say, we would compare the amount of wear for each group. A better way would be to block as follows: Let each student use a pair of shoes that has both types of soles, one on the right foot and the other on the left, as we recognize that each foot must take essentially the same number of steps (to be safe, we might let 15 students, selected at random, wear one type of sole on the right foot and the other 15 wear the other type on the right foot). In this experiment each student is a block. Sometimes in experiments, siblings or twins or two rats from the same litter could be used as a block.

Actually, a lot of good design of experiments is just plain common sense, but sometimes it is helpful for statisticians to note a few things about careful designs for those interested in running important experiments and studies. Our simple recommendation is to block what you can and randomize what you cannot block. However, if a person is involved in a complicated experiment or study, he or she should study advanced methods in the design of experiments or see a friendly statistician.

Exercises

These and similar problems could be assigned as projects if time permits.

1.7-1 Design an experiment to find out how the distance to the basket and the angle from the perpendicular to the backboard affects your shooting percentage in basketball. Make certain you randomize the order of the positions from which you shoot to avoid "warm up" and "fatigue" effects. Display your results. (Essentially, you are estimating a response surface and you might draw contours.)

1.7-2 Suppose that we want to test the effects of two factors in making popcorn. One factor is the type of popper with two levels: oil-based or air-based. The other factor, type of popcorn, has three levels: gourmet, national brand, generic brand. Clearly, one response measurement would be the amount of popcorn that is obtained from $\frac{1}{2}$ cup of popcorn. But you could also try some sort of taste measurement, the amount of popcorn obtained from 10 cents' worth of each of popcorn, or some combination of taste and cost. What are your conclusions?

1.7-3 Suppose that we plan to compare two detergents, say A and B. We will wash equally dirty white clothes with each detergent, also changing two other factors: hot or cold water, soft or hard water. Thus, we have a three-factor experiment in which each factor has two levels; a 2^3 factorial experiment is under consideration and, without replication, we needed 8 runs. Replications might be desirable using 16 or 24 runs. What is the response variable?

1.7-4 We want to investigate the effects of five factors on the growth of pinto beans. They are

  • Soaking fluid: water or regular beer.
  • Salinity: no salt or salt.

- Acidity: no vinegar or vinegar.
- Temperature: refrigerator or room.
- Soaking time: 2 hours or 4 hours.

Use 3 tablespoons of soaking fluid. For salt add $\frac{1}{4}$ teaspoon to soaking liquid. For vinegar, add 1 teaspoon to the soaking fluid. Note that this is a 2^5 factorial experiment and, without replications, 32 runs are needed to cover all possible combinations.

2

Probability

2.1 Properties of Probability

In Section 1.1, the collection of all possible outcomes (the *universal set*) of a random experiment is denoted by S and is called the *sample space*. A subset A of S is a partial collection of elements of S and is called by probabilists an **event** A. That is, we say the event has occurred if the outcome of the experiment is an element in A.

Since, in studying probability, the words *set* and *event* are interchangeable, the reader might want to review **algebra of sets**, found in the Appendix. For convenience, however, we remind the reader of a little of that terminology:

- ϕ denotes the **null** or **empty** set;
- $A \subset B$ means A is a **subset** of B;
- $A \cup B$ is the **union** of A and B;
- $A \cap B$ is the **intersection** of A and B;
- A' is the **complement** of A (i.e., all elements in S not in A).

Some of these sets are depicted in Figure 2.1-1.

Special terminology associated with events that is often used by probabilists includes the following:

1. $A_1, A_2, \ldots, A_k$ are **mutually exclusive events**, which means that $A_i \cap A_j = \phi$, $i \neq j$, that is, if $A_1, A_2, \ldots, A_k$ are disjoint sets;
2. $A_1, A_2, \ldots, A_k$ are **exhaustive events** means $A_1 \cup A_2 \cup \cdots \cup A_k = S$.

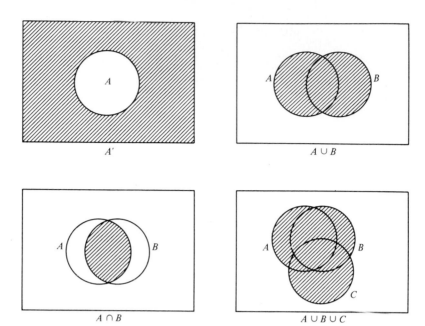

FIGURE 2.1-1

So if $A_1, A_2, \ldots, A_k$ are **mutually exclusive and exhaustive events**, we know that $A_i \cap A_j = \phi$, $i \neq j$, and $A_1 \cup A_2 \cup \cdots \cup A_k = S$.

Let the event A be a subset of the sample space S. We now wish to associate with A a number $P(A)$ about which the relative frequency $\#(A)/n$ of the event A tends to stabilize with large n. A function such as $P(A)$, which is evaluated for a set A, is called a **set function**. In this section we consider the probability set function $P(A)$ and discuss some of its properties. In succeeding sections we shall describe how the probability set function is defined for particular experiments.

To help decide what properties the probability set function should satisfy, consider properties possessed by the relative frequency $\#(A)/n$. For example, $\#(A)/n$ is always nonnegative. If $A = S$, the sample space, then the outcome of the experiment will always belong to S, and thus $\#(S)/n = 1$. Also, if A and B are two mutually exclusive events, then $\#(A \cup B)/n = \#(A)/n + \#(B)/n$. Hopefully, these remarks will help to motivate the following definition.

DEFINITION 2.1-1   **Probability**   *is a set function P that assigns to each event A in the sample space S a number $P(A)$, called the probability of the event A, such that the following properties are satisfied:*

(a) *$P(A) \geq 0$,*
(b) *$P(S) = 1$,*
(c) *If $A_1, A_2, A_3, \ldots$ are events and $A_i \cap A_j = \phi$, $i \neq j$, then*

$$P(A_1 \cup A_2 \cup \cdots \cup A_k) = P(A_1) + P(A_2) + \cdots + P(A_k)$$

for each positive integer k, and

$$P(A_1 \cup A_2 \cup A_3 \cup \cdots) = P(A_1) + P(A_2) + P(A_3) + \cdots.$$

REMARK Thus far our interpretation of probability has been the **relative frequency approach**. For example, we motivated Definition 2.1-1 by considering the relative frequencies of the events S, A, and B. However, many persons extend probabilities to other situations that cannot be repeated a large number of times. If probability is thought of as a rational measure of belief, then the probability $P(A) = 2/3$ would mean that their personal or **subjective** probability of the event A is equal to 2/3. If such persons are not opposed to gambling, this could be interpreted as a willingness on their part to bet on the outcome A so that the possible payoffs are in the ratio (called **odds**) $P(A)/[1 - P(A)] = (2/3)/(1/3) = 2$; that is, here the odds are 2 to 1 in favor of A. Moreover, if they truly believe that $P(A) = 2/3$ is correct, they would be willing to accept either side of the bet: (1) win one unit if A occurs and lose two units if it does not, or (2) win two units if A does not occur and lose one unit if it does. Thus assigning subjective probability to an event is much like the old problem of dividing a candy bar between two children: one cuts the bar in two parts and the other chooses a part first. For example, if a person truly believes that 0.6 is the chance that Michigan will beat Ohio State on a certain Saturday in the fall, then the odds are $0.6/(1 - 0.6) = 3/2$. Such a person would be willing to bet on Michigan, receiving two units if Michigan won and losing three units if Ohio State won; or the other way around, betting on Ohio State and receiving three units if Ohio State won and losing two units if Michigan won. It would be worthwhile to select some upcoming athletic event in which the class is interested and ask the students to assign subjective probabilities to the outcomes. In any case, the mathematical properties of probability given in Definition 2.1-1 and the following theorems are consistent with either interpretation, and the development of the subject does not depend on which approach is used.

The following theorems give some other important properties of the probability set function. When one considers these theorems, it is important to understand the theoretical concepts and proofs. However, if the reader keeps the relative frequency concept in mind, the theorems should also have some intuitive appeal.

THEOREM 2.1-1 *For each event A,*

$$P(A) = 1 - P(A').$$

PROOF: We have

$$S = A \cup A' \qquad \text{and} \qquad A \cap A' = \phi.$$

Thus, from properties (b) and (c), it follows that

$$1 = P(A) + P(A').$$

Hence

$$P(A) = 1 - P(A'). \qquad \square$$

Example 2.1-1   A fair coin is flipped successively until the same face is observed on successive flips. Let $A = \{x: x = 3, 4, 5, \ldots\}$; that is, A is the event that it will take three or more flips of the coin to observe the same face on two consecutive flips. To find $P(A)$, we first find the probability of $A' = \{x: x = 2\}$, the complement of A. In two flips of a coin, the possible outcomes are $\{HH, HT, TH, TT\}$, and we assume that each of these four points has the same chance of being observed. Thus

$$P(A') = P(\{HH, TT\}) = \tfrac{2}{4}.$$

It follows from Theorem 2.1-1 that

$$P(A) = 1 - P(A') = 1 - \tfrac{2}{4} = \tfrac{2}{4}.$$

THEOREM 2.1-2 $P(\phi) = 0$.

PROOF: In Theorem 2.2-1, take $A = \phi$ so that $A' = S$. Thus

$$P(\phi) = 1 - P(S) = 1 - 1 = 0. \qquad \square$$

THEOREM 2.1-3   *If events A and B are such that $A \subset B$, then $P(A) \leq P(B)$.*

PROOF: Now

$$B = A \cup (B \cap A') \qquad \text{and} \qquad A \cap (B \cap A') = \phi.$$

Hence, from property (c),

$$P(B) = P(A) + P(B \cap A') \geq P(A)$$

because from property (a),

$$P(B \cap A') \geq 0. \qquad \square$$

THEOREM 2.1-4   *For each event A, $P(A) \leq 1$.*

PROOF: Since $A \subset S$, we have by Theorem 2.1-3 and property (b) that

$$P(A) \leq P(S) = 1,$$

which gives the desired result. $\square$

Property (a) along with Theorem 2.1-4 shows that, for each event A,

$$0 \leq P(A) \leq 1.$$

THEOREM 2.1-5   *If A and B are any two events, then*

$$P(A \cup B) = P(A) + P(B) - P(A \cap B).$$

PROOF: The event $A \cup B$ can be represented as a union of mutually exclusive events, namely,

$$A \cup B = A \cup (A' \cap B).$$

Hence, by property (c),

$$P(A \cup B) = P(A) + P(A' \cap B).$$

However,

$$B = (A \cap B) \cup (A' \cap B),$$

which is a union of mutually exclusive events. Thus

$$P(B) = P(A \cap B) + P(A' \cap B)$$

and

$$P(A' \cap B) = P(B) - P(A \cap B).$$

If this result is substituted in the equation involving $P(A \cup B)$, we obtain

$$P(A \cup B) = P(A) + P(B) - P(A \cap B),$$

which is the desired result. ☐

Example 2.1-2   A faculty leader was meeting two students in Paris, one arriving by train from Amsterdam and the other arriving by train from Brussels at approximately the same time. Let A and B be the events that the trains are on time, respectively. If $P(A) = 0.93$, $P(B) = 0.89$, and $P(A \cap B) = 0.87$, the probability that at least one train is on time is

$$P(A \cup B) = P(A) + P(B) - P(A \cap B)$$
$$= 0.93 + 0.89 - 0.87 = 0.95.$$

THEOREM 2.1-6   *If A, B, and C are any three events, then*

$$P(A \cup B \cup C) = P(A) + P(B) + P(C) - P(A \cap B)$$
$$- P(A \cap C) - P(B \cap C) + P(A \cap B \cap C).$$

PROOF: Write

$$A \cup B \cup C = A \cup (B \cup C)$$

and then apply Theorem 2.1-5. The details are left as an exercise. ☐

Example 2.1-3   Continuing with Example 2.1-2, say that a third student is arriving from Cologne. Let C be the event that this train is on time with $P(C) = 0.91$, $P(B \cap C) = 0.85$, $P(A \cap C) = 0.86$, and $P(A \cap B \cap C) = 0.81$. The probability that at least one of these trains is on time is

$$P(A \cup B \cup C) = P(A) + P(B) + P(C) - P(A \cap B) - P(A \cap C)$$
$$- P(B \cap C) + P(A \cap B \cap C)$$
$$= 0.93 + 0.89 + 0.91 - 0.87 - 0.85 - 0.86 + 0.81$$
$$= 0.96.$$

Let a probability set function be defined on a sample space S. Let $S = \{e_1, e_2, \ldots, e_m\}$, where each e_i is a possible outcome of the experiment. We call e_i a **simple event**. The integer m is called the total number of ways in which the random experiment can terminate. If each of these outcomes has the same probability of occurring, we say that the m outcomes are **equally likely**. That is,

$$P(\{e_i\}) = \frac{1}{m}, \qquad i = 1, 2, \ldots, m.$$

If the number of simple outcomes in an event A is h, the integer h is called the number of ways that are favorable to the event A. In this case $P(A)$ is equal to the number of ways favorable to the event A divided by the total number of ways in which the experiment can terminate. That is, under this assumption of equally likely outcomes, we have that

$$P(A) = \frac{h}{m} = \frac{N(A)}{N(S)},$$

where $h = N(A)$ is the number of ways A can occur and $m = N(S)$ is the number of ways S can occur. It should be emphasized that in order to assign the probability h/m to the event A, we must assume that each of the events $e_1, e_2, \ldots, e_m$ has the same probability $1/m$. This assumption is then an important part of our probability model; if it is not realistic in an application, the probability of the event A cannot be computed this way.

Example 2.1-4   Let a card be drawn at random from an ordinary deck of 52 playing cards. The sample space S is the set of $m = 52$ different cards, and it is reasonable to assume that each of these cards has the same probability for selection, $1/52$. Accordingly, if A is the set of outcomes that are kings, $P(A) = 4/52 = 1/13$ because there are $h = 4$ kings in the deck. That is, $1/13$ is the probability of drawing a card that is a king provided that each of the 52 cards has the same probability.

In Example 2.1-4, the computations are very easy because there is no difficulty in the determination of the appropriate values of h and m. However, instead of drawing only one card, suppose that 13 are taken at random and without replacement. We can think of each possible 13-card hand as being an outcome in a sample space, and it is reasonable to assume that each of these outcomes has the same probability. To use the above method to assign the probability of a hand, consisting of seven spades and six hearts, for illustration, we must be able to count the number h of all such hands as well as the number m of possible 13-card hands. In these more complicated situations, we need better methods of determining h and m. We discuss some of these counting techniques in Section 2.2.

Exercises

2.1-1 Draw one card at random from a standard deck of cards. The sample space S is the collection of the 52 cards. Assume that the probability set function assigns 1/52 to each of these 52 outcomes. Let

$$A = \{x: x \text{ is a jack, queen, or king}\},$$
$$B = \{x: x \text{ is a 9, 10, or jack and } x \text{ is red}\},$$
$$C = \{x: x \text{ is a club}\},$$
$$D = \{x: x \text{ is a diamond, a heart, or a spade}\}.$$

Find
(a) $P(A)$.   (b) $P(A \cap B)$.   (c) $P(A \cup B)$.
(d) $P(C \cup D)$.   (e) $P(C \cap D)$.

2.1-2 A coin is tossed four times, and the sequence of heads and tails is observed.
(a) List each of the 16 sequences in the sample space S.
(b) Let events A, B, C, and D be given by $A = \{$at least 3 heads$\}$, $B = \{$at most 2 heads$\}$, $C = \{$heads on the third toss$\}$, and $D = \{$1 head and 3 tails$\}$. If the probability set function assigns 1/16 to each outcome in the sample space, find (i) $P(A)$, (ii) $P(A \cap B)$, (iii) $P(B)$, (iv) $P(A \cap C)$, (v) $P(D)$, (vi) $P(A \cup C)$, and (vii) $P(B \cap D)$.

2.1-3 A field of beans is planted with three seeds per hill. For each hill of beans, let A_i be the event that i seeds germinate, $i = 0, 1, 2, 3$. Suppose that $P(A_0) = 1/64$, $P(A_1) = 9/64$, and $P(A_2) = 27/64$. Give the value of $P(A_3)$.

2.1-4 Consider the trial on which a 3 is first observed in successive rolls of a four-sided die. Let A be the event that 3 is observed on the first trial. Let B be the event that at least two trials are required to observe a 3. Assuming that each side has probability 1/4, find
(a) $P(A)$.
(b) $P(B)$.
(c) $P(A \cup B)$.

2.1-5 A fair eight-sided die is rolled once. Let $A = \{2, 4, 6, 8\}$, $B = \{3, 6\}$, $C = \{2, 5, 7\}$, and $D = \{1, 3, 5, 7\}$. Assume that each face has the same probability.
(a) Give the values of (i) $P(A)$, (ii) $P(B)$, (iii) $P(C)$, and (iv) $P(D)$.
(b) For the events A, B, C, and D, list all pairs of events that are mutually exclusive.
(c) Give the values of (i) $P(A \cap B)$, (ii) $P(B \cap C)$, and (iii) $P(C \cap D)$.
(d) Give the values of (i) $P(A \cup B)$, (ii) $P(B \cup C)$, and (iii) $P(C \cup D)$ using Theorem 2.1-5.

2.1-6 If $P(A) = 0.4$, $P(B) = 0.5$, and $P(A \cap B) = 0.3$, find
(a) $P(A \cup B)$.
(b) $P(A \cap B')$.
(c) $P(A' \cup B')$.

2.1-7 Continuing with Examples 2.1-1 and 1.1-6 and Exercise 2.1-2,
  (a) If B is the event that it will take four or more flips of a coin to observe the same face on consecutive flips, first find $P(B')$ and then find $P(B)$.
  (b) If C is the event that it will take four or more flips of a coin to observe heads on consecutive flips, first find $P(C')$ and then find $P(C)$.

2.1-8 If $S = A \cup B$, $P(A) = 0.7$, and $P(B) = 0.9$, find $P(A \cap B)$.

2.1-9 If $P(A \cup B) = 0.7$, $P(A) = 0.5$, and $P(B) = 0.3$, find $P(A \cap B)$.

2.1-10 If $P(A) = 0.4$, $P(B) = 0.5$, and $P(A \cup B) = 0.7$, find
  (a) $P(A \cap B)$.
  (b) $P(A' \cup B')$.

2.1-11 Roll a fair six-sided die three times. Let $A_1 = \{1 \text{ or } 2 \text{ on the first roll}\}$, $A_2 = \{3 \text{ or } 4 \text{ on the second roll}\}$, and $A_3 = \{5 \text{ or } 6 \text{ on the third roll}\}$. It is given that $P(A_i) = 1/3$, $i = 1, 2, 3$; $P(A_i \cap A_j) = (1/3)^2$, $i \neq j$; and $P(A_1 \cap A_2 \cap A_3) = (1/3)^3$.
  (a) Use Theorem 2.1-6 to find $P(A_1 \cup A_2 \cup A_3)$.
  (b) Show that $P(A_1 \cup A_2 \cup A_3) = 1 - (1 - 1/3)^3$.

2.1-12 Prove Theorem 2.1-6.

2.1-13 For each positive integer n, let $P(\{n\}) = (1/2)^n$. Let $A = \{n: 1 \leq n \leq 10\}$, $B = \{n: 1 \leq n \leq 20\}$ and $C = \{n: 11 \leq n \leq 20\}$. Find
  (a) $P(A)$. (b) $P(B)$. (c) $P(A \cup B)$.
  (d) $P(A \cap B)$. (e) $P(C)$. (f) $P(B')$.

2.1-14 Let x equal a number that is selected randomly from the closed interval from zero to one, that is $[0, 1]$. Use your intuition to assign values to
  (a) $P(\{x: 0 \leq x \leq 1/3\})$.
  (b) $P(\{x: 1/3 \leq x \leq 1\})$.
  (c) $P(\{x: x = 1/3\})$.
  (d) $P(\{x: 1/2 < x < 5\})$.

2.2 Methods of Enumeration

In this section we develop counting techniques that are useful in determining the number of outcomes associated with the events of certain random experiments. We begin with a consideration of the multiplication principle.

Multiplication Principle

Suppose that an experiment (or procedure) E_1 has n_1 outcomes and for each of these possible outcomes an experiment (procedure) E_2 has n_2 possible outcomes. The composite experiment (procedure) E_1E_2 that consists of performing first E_1 and then E_2 has n_1n_2 possible outcomes.

Example 2.2-1 Let E_1 denote the selection of a rat from a cage containing female (F) and male (M) rats. Let E_2 denote the administering of either drug A (A), drug B (B), or a placebo (P) to the selected rat. The outcome for the composite experiment can be denoted by an ordered pair, such as (F, P). In fact, the set of all possible outcomes can be denoted by the following rectangular array:

$$
\begin{array}{ccc}
(\text{F, A}) & (\text{F, B}) & (\text{F, P}) \\
(\text{M, A}) & (\text{M, B}) & (\text{M, P})
\end{array}
$$

Another way of illustrating the multiplication principle is with a tree diagram like that in Figure 2.2-1. The diagram shows that there are $n_1 = 2$ possibilities (branches) for the sex of the rat and that for each of these outcomes there are $n_2 = 3$ possibilities (branches) for the drug.

Clearly, the multiplication principle can be extended to a sequence of more than two experiments or procedures. Suppose that the experiment E_i has n_i, $i = 1, 2, \ldots, m$, possible outcomes after previous experiments have been performed. The composite experiment $E_1 E_2 \cdots E_m$, which consists of performing E_1, then $E_2, \ldots$, and finally E_m, has $n_1 n_2 \cdots n_m$ possible outcomes.

> **Example 2.2-2** A certain food service gives the following choices for dinner: E_1, soup or tomato juice; E_2, steak or shrimp; E_3, french fried potatoes or baked potatoes; E_4, corn or peas; E_5, Jello, tossed salad, cottage cheese, or cole slaw; E_6, cake, cookies, pudding, brownie, vanilla ice cream, chocolate ice cream, or orange sherbet; E_7, coffee, tea, milk, or punch. How many different dinner selections are possible if one of the listed choices is made for each of $E_1, E_2, \ldots$, and E_7? By the multiplication principle there are
>
> $$(2)(2)(2)(2)(4)(7)(4) = 1792 \text{ different combinations.}$$

Although the multiplication principle is fairly simple and easy to understand, it will be extremely useful in developing various counting techniques. Suppose that a set contains n objects. Consider the problem of drawing r objects from this set. The order in which the objects are drawn may or may not be important. In addition, it is possible that a drawn object is replaced before the next object is drawn. Accordingly, we give some definitions and show how the multiplication principle can be used to count the number of possibilities.

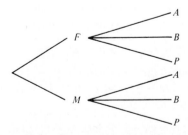

FIGURE 2.2-1

DEFINITION 2.2-1 *If r objects are selected from a set of n objects, and if the order of selection is noted, the selected set of r objects is called an* **ordered sample of size** *r.*

DEFINITION 2.2-2 **Sampling with replacement** *occurs when an object is selected and then replaced before the next object is selected.*

By the multiplication principle, the number of possible ordered samples of size r taken from a set of n objects is n^r when sampling with replacement.

Example 2.2-3 A die is rolled five times. The number of possible ordered samples is $6^5 = 7776$. Note that rolling a die is equivalent to sampling with replacement.

Example 2.2-4 An urn contains 10 balls numbered 0, 1, 2, . . . , 9. If four balls are selected, one at a time and with replacement, the number of possible ordered samples is $10^4 = 10,000$. Note that this is the number of four-digit integers between 0000 and 9999, inclusive.

Suppose that n positions are to be filled with n different objects. There are n choices for filling the first position, $n - 1$ for the second, . . . , and 1 choice for the last position. So by the multiplication principle there are

$$n(n - 1) \cdots (2)(1) = n!$$

possible arrangements. The symbol $n!$ is read n factorial. By definition, we take $0! = 1$; that is, we say that zero positions can be filled with zero objects in one way.

DEFINITION 2.2-3 *Each of the n! arrangements (in a row) of n objects is called a* **permutation** *of the n objects.*

Example 2.2-5 The number of permutations of the four letters a, b, c, and d is clearly $4! = 24$. However, the number of possible four-letter code words using the four letters a, b, c, and d if letters may be repeated is $4^4 = 256$.

If only r positions are to be filled with objects selected from n different objects, $r \le n$, the number of possible ordered arrangements is

$$P(n, r) = n(n - 1)(n - 2) \cdots (n - r + 1).$$

That is, there are n ways to fill the first position, $(n - 1)$ ways to fill the second, and so on until there are $[n - (r - 1)] = (n - r + 1)$ ways to fill the rth position.
In terms of factorials we have that

$$P(n, r) = \frac{n(n - 1) \cdots (n - r + 1)(n - r) \cdots (3)(2)(1)}{(n - r) \cdots (3)(2)(1)} = \frac{n!}{(n - r)!}.$$

DEFINITION 2.2-4 *Each of the P(n, r) arrangements is called a* **permutation of** n **objects taken** r **at a time**.

Example 2.2-6 The number of possible four-letter code words, selecting from the 26 letters in the alphabet, in which all four letters are different is

$$P(26, 4) = (26)(25)(24)(23) = \frac{26!}{22!} = 358,800.$$

Example 2.2-7 The number of four-digit integers, using different values for digits and allowing a leading zero, is

$$P(10, 4) = 10 \cdot 9 \cdot 8 \cdot 7 = \frac{10!}{6!} = 5040.$$

DEFINITION 2.2-5 **Sampling without replacement** *occurs when an object is not replaced after it has been selected.*

By the multiplication principle, the number of possible ordered samples of size r taken from a set of n objects, sampling without replacement, is

$$n(n - 1) \cdots (n - r + 1) = \frac{n!}{(n - r)!},$$

which is equivalent to $P(n, r)$, the number of permutations of n objects taken r at a time.

Example 2.2-8 The number of ordered samples of five cards that can be drawn without replacement from a standard deck of 52 playing cards is

$$(52)(51)(50)(49)(48) = \frac{52!}{47!} = 311,875,200.$$

Often the order of selection is not important and interest centers only on the selected set of r objects. That is, we are interested in the number of subsets of size r that can be selected from a set of n different objects. In order to find the number of (unordered) subsets of size r, we count, in two different ways, the number of ordered subsets of size r that can be taken from the n distinguishable objects. By equating these two answers, we are able to count the number of (unordered) subsets of size r.

Let C denote the number of (unordered) subsets of size r that can be selected from n different objects. We can obtain each of the $P(n, r)$ ordered subsets by first selecting one of the C unordered subsets of r objects and then ordering these r objects. Since the latter can be carried out in $r!$ ways, the multiplication principle yields $(C)(r!)$ ordered subsets; so $(C)(r!)$ must equal $P(n, r)$. Thus we have

$$(C)(r!) = \frac{n!}{(n-r)!}$$

or

$$C = \frac{n!}{r!(n-r)!} = \binom{n}{r} = C(n, r).$$

Accordingly, a set of n different objects possesses

$$\binom{n}{r} = \frac{n!}{r!(n-r)!}$$

unordered subsets of size $r \leq n$.

We could also say that the number of ways in which r objects can be selected without replacement from n objects, when the order of selection is disregarded, is $\binom{n}{r}$. This motivates the following definition.

> **DEFINITION 2.2-6** *Each of the $C(n, r)$ unordered subsets is called a* **combination of n objects taken r at a time** *where*
>
> $$C(n, r) = \binom{n}{r}.$$

Example 2.2-9 The number of possible five-card hands (hands in five-card poker) drawn from a deck of 52 playing cards is

$$C(52, 5) = \binom{52}{5} = \frac{52!}{5!47!} = 2,598,960.$$

Example 2.2-10 The number of possible 13-card hands (hands in bridge) that can be selected from a deck of 52 playing cards is

$$C(52, 13) = \binom{52}{13} = \frac{52!}{13!39!} = 635,013,559,600.$$

The numbers $\binom{n}{r}$ are frequently called **binomial coefficients**, since they arise in the expansion of a binomial. We shall illustrate this by giving a justification of the binomial expansion

$$(a + b)^n = \sum_{r=0}^{n} \binom{n}{r} b^r a^{n-r}. \tag{2.2-1}$$

In the expansion of

$$(a + b)^n = (a + b)(a + b) \cdots (a + b),$$

either an a or a b is selected from each of the n factors. One possible product is then $b^r a^{n-r}$; this occurs when b is selected from each of r factors and a from each of the remaining $n - r$ factors. But this latter operation can be completed in $\binom{n}{r}$ ways, which then must be the coefficient of $b^r a^{n-r}$, as shown in equation 2.2-1.

The binomial coefficients are given in Table I in the Appendix for selected values of n and r. Note that for some combinations of n and r, the table uses the fact that

$$\binom{n}{r} = \frac{n!}{r!(n-r)!} = \frac{n!}{(n-r)!r!} = \binom{n}{n-r}.$$

That is, the number of ways in which r objects can be selected out of n objects is equal to the number of ways in which the $n - r$ objects can be selected out of n objects.

Example 2.2-11 Assume that each of the $\binom{52}{5} = 2,598,960$ five-card hands drawn from a deck of 52 playing cards has the same probability for being selected. The number of possible five-card hands that are all spades (event A) is

$$N(A) = \binom{13}{5}\binom{39}{0} = \frac{13!}{5!8!} = 1287$$

from Table I in the Appendix. Thus the probability of an all-spade five-card hand is

$$P(A) = \frac{N(A)}{N(S)} = \frac{1287}{2,598,960} = 0.000495.$$

Suppose now that the event B is the set of outcomes in which exactly three cards are kings and exactly two cards are queens. We can select the three kings in any one of

$$\binom{4}{3}$$ ways and the two queens in any one of $\binom{4}{2}$ ways.

By the multiplication principle, the number of outcomes in B is

$$N(B) = \binom{4}{3}\binom{4}{2}\binom{44}{0},$$

where $\binom{44}{0}$ gives the number of ways in which 0 cards are selected out of the non-kings and non-queens and of course is equal to one. Thus

$$P(B) = \frac{N(B)}{N(S)} = \frac{\binom{4}{3}\binom{4}{2}\binom{44}{0}}{\binom{52}{5}} = \frac{24}{2,598,960} = 0.0000092.$$

Finally, let C be the set of outcomes in which there are exactly two kings, two queens, and one jack. Then

$$P(C) = \frac{N(C)}{N(S)} = \frac{\binom{4}{2}\binom{4}{2}\binom{4}{1}\binom{40}{0}}{\binom{52}{5}} = \frac{144}{2,598,960} = 0.000055$$

because the numerator of this fraction is the number of outcomes in C.

Now suppose that a set contains n objects of two types, r of one type and $n - r$ of the other type. The number of permutations of these n objects is $n!$. However, they are not all distinguishable. To count the number of distinguishable arrangements, first select r out of the n positions for the objects of the first type. This can be done in $\binom{n}{r}$ ways. Then fill in the remaining positions with the objects of the second type. Thus the number of distinguishable arrangements is

$$C(n, r) = \binom{n}{r} = \frac{n!}{r!(n - r)!}.$$

DEFINITION 2.2-7 *Each of the $C(n, r)$ permutations of n objects, r of one type and $n - r$ of another type, is called a **distinguishable permutation**.*

Example 2.2-12 A coin is flipped 10 times and the sequence of heads and tails is observed. The number of possible 10-tuplets that result in four heads and six tails is

$$\binom{10}{4} = \frac{10!}{4!6!} = \frac{10!}{6!4!} = \binom{10}{6} = 210.$$

Example 2.2-13 Students on a boat send signals back to shore by arranging seven colored flags on a vertical flagpole. If they have four orange and three blue flags, they can send

$$\binom{7}{4} = \frac{7!}{4!3!} = 35$$

different signals. Note that if they had seven flags of different colors, they could send $7! = 5040$ different signals.

The foregoing results can be extended. Suppose that in a set of n objects, n_1 are similar, n_2 are similar, . . . , n_s are similar, where $n_1 + n_2 + \cdots + n_s = n$. The number of distinguishable permutations of the n objects is (see Exercise 2.2-16)

$$\binom{n}{n_1, n_2, \ldots, n_s} = \frac{n!}{n_1!n_2!\cdots n_s!}. \tag{2.2-2}$$

Example 2.2-14 If the students on the boat have three red flags, four yellow flags, and two blue flags to arrange on a vertical pole, the number of possible signals is

$$\binom{9}{3,\,4,\,2} = \frac{9!}{3!4!2!} = 1260.$$

The argument used in determining the binomial coefficients in the expansion of $(a + b)^n$ can be extended to find the expansion of $(a_1 + a_2 + \cdots + a_s)^n$. The coefficient of $a_1^{n_1} a_2^{n_2} \cdots a_s^{n_s}$ is

$$\binom{n}{n_1,\, n_2,\, \ldots,\, n_s} = \frac{n!}{n_1! n_2! \cdots n_s!}.$$

This is sometimes called a **multinomial coefficient**.

When r objects are selected out of n objects, we are often interested in the number of possible outcomes. We have seen that for ordered samples, there are n^r possible outcomes when sampling with replacement and $P(n, r)$ outcomes when sampling without replacement. For unordered samples, there are $C(n, r)$ outcomes when sampling without replacement. Each of the outcomes above is equally likely provided the experiment is performed in a fair manner.

We shall now count the number of possible samples of size r that can be selected out of n objects when the order is irrelevant and when sampling with replacement. For example, if a six-sided die is rolled 10 times (or 10 six-sided dice are rolled once), how many possible unordered outcomes are there? To count the number of possible outcomes, think of listing r 0's for the r objects that are to be selected. Then insert $(n - 1)$ slashes to partition the r objects into n sets, the first set giving objects of the first kind, and so on. So if $n = 6$ and $r = 10$ in the die illustration, a possible outcome is

$$0 \ \ 0 \ \ / \ \ / \ \ 0 \ \ 0 \ \ 0 \ \ / \ \ 0 \ \ / \ \ 0 \ \ 0 \ \ 0 \ \ / \ \ 0,$$

which says to take two 1's, zero 2's, three 3's, one 4, three 5's, and one 6. In general, each outcome is a permutation of r 0's and $(n - 1)$ /'s. Each distinguishable permutation is equivalent to an unordered sample. The number of distinguishable permutations, and hence the number of unordered samples of size r that can be selected out of n objects when sampling with replacement, is

$$C(n - 1 + r, r) = \frac{(n - 1 + r)!}{r!(n - 1)!}.$$

Example 2.2-15 Roll a pair of six-sided dice. The number of distinguishable outcomes is

$$\frac{(6 - 1 + 2)!}{2!(6 - 1)!} = \frac{7!}{2!5!} = 21.$$

You should be able to list these 21 outcomes. Note that these outcomes are *not* equally likely under the usual probabilistic assumptions about dice.

The number of samples of size r that can be selected out of n objects is summarized in the following table.

Number of Samples of r Out of n Objects

	With Replacement	Without Replacement
Ordered	n^r	$\dfrac{n!}{(n-r)!}$
Unordered	$\dfrac{(n-1+r)!}{(n-1)!r!}$	$\dfrac{n!}{r!(n-r)!}$

Exercises

2.2-1 A boy found a bicycle lock for which the combination was unknown. The correct combination is a four-digit number, $d_1 d_2 d_3 d_4$, where d_i, $i = 1, 2, 3, 4$, is selected from 1, 2, 3, 4, 5, 6, 7, 8. How many different lock combinations are possible with such a lock?

2.2-2 How many different signals can be made using four flags of different colors on a vertical flagpole if exactly three flags are used for each signal?

2.2-3 How many different license plates are possible if a state uses
(a) Two letters followed by a four-digit integer (leading zeros permissible)?
(b) Three letters followed by a three-digit integer?

2.2-4 From a collection of nine paintings, four are to be selected to hang side by side on a gallery wall in positions 1, 2, 3, and 4. In how many ways can this be done?

2.2-5 How many four-letter code words are possible using the letters in HOPE if
(a) The letters may not be repeated?
(b) The letters may be repeated?

2.2-6 Some albatrosses return to the world's only mainland colony of royal albatrosses on Otago Peninsula near Dunedin, New Zealand, every two years to nest and raise their young. In order to learn more about the albatross, colored plastic bands are placed on their legs so that they can be identified from a distance. Suppose that three bands are placed on one leg, selecting from the colors red, yellow, green, white, and blue. Find the number of different color codes that are possible for banding an albatross if
(a) The three bands are different colors.
(b) Repeated colors are permissible.

2.2-7   In a state lottery four digits are drawn at random one at a time with replacement from 0 to 9. Suppose that you win if any permutation of your selected integers is drawn. Give the probability of winning if you select
(a) 6, 7, 8, 9.
(b) 6, 7, 8, 8.
(c) 7, 7, 8, 8.
(d) 7, 8, 8, 8.

2.2-8   Suppose that Chicago and Toronto are playing in a preliminary round of the Stanley Cup hockey playoffs in which they play until one team wins three games. That is, it is a best-of-five series that ends when one team has three victories. Considering the possible orderings for the winning team, in how many ways could this series end?

2.2-9   The World Series in baseball continues until either the American League team or the National League team wins four games. How many different orders are possible if the series goes
(a) Four games?
(b) Five games?
(c) Six games?
(d) Seven games?

2.2-10   How many different varieties of pizza can be made if you have the following choices: size—small, medium, or large; crust—thin 'n crispy, hand-tossed, or pan; toppings—(cheese is automatic) there are 12 toppings from which you may select from 0 to 12?

2.2-11   A cafe lets you order a deli sandwich your way. There are 6 choices for bread, 4 choices for meat, 4 choices for cheese, and 12 different garnishes (condiments). How many different sandwich possibilities are there if you choose
(a) One bread, one meat, and one cheese?
(b) One bread, one meat, one cheese, and from 0 to 12 garnishes?
(c) One bread; 0, 1, or 2 meats; 0, 1, or 2 cheeses; and from 0 to 12 garnishes?

2.2-12   Pascal's triangle gives a method for calculating the binomial coefficients; it begins as follows:

$$
\begin{array}{ccccccccc}
 & & & & 1 & & & & \\
 & & & 1 & & 1 & & & \\
 & & 1 & & 2 & & 1 & & \\
 & 1 & & 3 & & 3 & & 1 & \\
1 & & 4 & & 6 & & 4 & & 1 \\
\end{array}
$$

1   5   10   10   5   1
:   :   :   :   :   :

The nth row of this triangle gives the coefficients for $(a + b)^{n-1}$. To find an entry in the table other than a 1 on the boundary, add the two nearest numbers in the row directly above. The equation, called **Pascal's equation**,

$$\binom{n}{r} = \binom{n-1}{r} + \binom{n-1}{r-1}$$

explains why Pascal's triangle works. Prove that this equation is correct.

2.2-13 Among nine presidential candidates at a debate, three are Republicans and six are Democrats.
   (a) In how many different ways can the nine candidates be lined up?
   (b) How many lineups by party are possible if each candidate is labeled either R or D?
   (c) For each of the nine candidates, you are to decide whether the candidate did a good job or a poor job; that is, give each of the nine candidates a grade of G or P. How many different "score cards" are possible?

2.2-14 Prove:

$$\sum_{r=0}^{n} (-1)^r \binom{n}{r} = 0 \qquad \text{and} \qquad \sum_{r=0}^{n} \binom{n}{r} = 2^n.$$

HINT: Consider $(1 - 1)^n$ and $(1 + 1)^n$ or use Pascal's equation and proof by induction.

2.2-15 A poker hand is defined as drawing five cards at random without replacement from a deck of 52 playing cards. Find the probability of each of the following poker hands:
   (a) Four of a kind (four cards of equal face values).
   (b) Full house (one pair and one triple of cards with equal face value).
   (c) Three of a kind (three equal face values plus two different cards).
   (d) Two pairs (two pairs of equal face value plus one other card).
   (e) One pair (one pair of equal face value plus three different cards).

2.2-16 A college plans to place on a long table 15 computers of which 6 are IBM PCs, 5 are Swan 386SX microcomputers, and 4 are Dell 316SX microcomputers (all IBM compatible).
   (a) Show that the number of distinguishable arrangements is 15!/6!5!4! by first selecting positions for the IBMs, then selecting positions for the Swans, and finally selecting positions for the Dells.
   (b) Generalize your argument to prove Equation 2.2-2.

2.2-17 A box of candy hearts contains 52 hearts of which 19 are white, 10 are tan, 7 are pink, 3 are purple, 5 are yellow, 2 are orange, and 6 are green. If you select 9 pieces of candy randomly from the box, without replacement, give the probability that
   (a) Three of the hearts are white.
   (b) Three are white, 2 are tan, 1 is pink, 1 is yellow, and 2 are green.

2.2-18   An office furniture manufacturer makes modular storage files. It offers its customers two choices for the base and four choices for the top, and the modular storage files come in five different heights. The customer may choose any combination of the five different-sized modules so that the finished file has a base, a top, and 1, 2, 3, 4, 5, or 6 storage modules.

    (a) How many choices does the customer have if the completed file has four storage modules, a top, and a base? The order in which the four storage modules are stacked is irrelevant.

    (b) The manufacturer would like to use in its advertising the number of different files that are possible—selecting one of the two bases, one of the four tops, and then either 1 or 2 or 3 or 4 or 5 or 6 storage modules, selecting any combination of the five different sizes with the order of stacking irrelevant. What is the number of possibilities?

2.2-19   The "eating club" is hosting a make-your-own sundae at which the following are provided:

Ice Cream Flavors	Toppings
Chocolate	Caramel
Cookies-n-cream	Hot fudge
Strawberry	Marshmallow
Vanilla	M&M's
	Nuts
	Strawberries

    (a) How many sundaes are possible using 1 flavor of ice cream and 3 different toppings?

    (b) How many sundaes are possible using 1 flavor of ice cream and from 0 to 6 toppings?

    (c) How many different combinations of flavors of 3 scoops of ice cream are possible if it is permissible to make all 3 scoops the same flavor?

2.2-20   A bag of 36 dum-dum pops (suckers) contains up to 10 flavors. That is, there are from 0 to 36 suckers of each of 10 flavors in the bag. How many different flavor combinations are possible?

2.3   Conditional Probability

We introduce the idea of conditional probability by means of an example.

Example 2.3-1   Suppose that we are given 20 tulip bulbs that are very similar in appearance and told that 8 tulips will bloom early, 12 will bloom late, 13 will be red, and 7 will be yellow, in accordance with the various

TABLE 2.3-1

	Early (E)	Late (L)	Totals
Red (R)	5	8	13
Yellow (Y)	3	4	7
Totals	8	12	20

combinations of Table 2.3-1. If one bulb is selected at random, the probability that it will produce a red tulip (R) is given by $P(R) = 13/20$, under the assumption that each bulb is "equally likely." Suppose, however, that close examination of the bulb will reveal whether it will bloom early (E) or late (L). If we consider an outcome only if it results in a tulip bulb that will bloom early, only eight outcomes in the sample space are now of interest. Thus it is natural to assign, under this limitation, the probability of 5/8 to R; that is, $P(R \mid E) = 5/8$, where $P(R \mid E)$ is read as the probability of R given that E has occurred. Note that

$$P(R \mid E) = \frac{5}{8} = \frac{N(R \cap E)}{N(E)} = \frac{N(R \cap E)/20}{N(E)/20} = \frac{P(R \cap E)}{P(E)},$$

where $N(R \cap E)$ and $N(E)$ are the numbers of outcomes in events $R \cap E$ and E, respectively.

This example is illustrative of a number of common situations. That is, in some random experiments, we are interested only in those outcomes that are elements of a subset B of the sample space S. This means, for our purposes, that the sample space is effectively the subset B. We are now confronted with the problem of defining a probability set function with B as the "new" sample space. That is, for a given event A we want to define $P(A \mid B)$, the probability of A considering only those outcomes of the random experiment that are elements of B. The previous example gives us the clue to that definition. That is, for experiments in which each outcome is equally likely, it makes sense to define $P(A \mid B)$ by

$$P(A \mid B) = \frac{N(A \cap B)}{N(B)},$$

where $N(A \cap B)$ and $N(B)$ are the numbers of outcomes in $A \cap B$ and B, respectively. If we divide the numerator and the denominator of this fraction by $N(S)$, the number of outcomes in the sample space, we have

$$P(A \mid B) = \frac{N(A \cap B)/N(S)}{N(B)/N(S)} = \frac{P(A \cap B)}{P(B)}.$$

We are thus led to the following definition.

DEFINITION 2.3-1 *The* **conditional probability** *of an event A, given that event B has occurred, is defined by*

$$P(A|B) = \frac{P(A \cap B)}{P(B)},$$

provided that $P(B) > 0$.

A formal use of the definition is given in the following example.

Example 2.3-2 If $P(A) = 0.4$, $P(B) = 0.5$, and $P(A \cap B) = 0.3$, then $P(A|B) = 0.3/0.5 = 0.6$ and $P(B|A) = P(A \cap B)/P(A) = 0.3/0.4 = 0.75$.

We can think of the "given B" as specifying the new sample space for which we now want to calculate the probability of that part of A that is contained in B to determine $P(A|B)$. The following two examples illustrate this idea.

Example 2.3-3 Suppose that $P(A) = 0.7$, $P(B) = 0.3$, and $P(A \cap B) = 0.2$. These probabilities are listed on the Venn diagram in Figure 2.3-1. Given that the outcome of the experiment belongs to B, what then is the probability of A? We are effectively restricting the sample space to B; of the probability $P(B) = 0.3$, 0.2 corresponds to $P(A \cap B)$ and hence to A. That is, $0.2/0.3 = 2/3$ of the probability of B corresponds to A. Of course, formally by definition, we also obtain

$$P(A|B) = \frac{P(A \cap B)}{P(B)} = \frac{0.2}{0.3} = \frac{2}{3}.$$

Example 2.3-4 A pair of four-sided dice is rolled and the sum is determined. Let A be the event that a sum of 3 is rolled and let B be the event that a sum of 3 or a sum of 5 is rolled. In a sequence of rolls the probability that a sum of 3 is rolled before a sum of 5 is rolled can be thought of as the conditional probability of a sum of 3 given that a sum of 3 or 5 has occurred; that is, the conditional probability of A, given B,

$$P(A|B) = \frac{P(A \cap B)}{P(B)} = \frac{P(A)}{P(B)} = \frac{2/16}{6/16} = \frac{2}{6}.$$

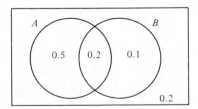

FIGURE 2.3-1

Note that for this example, the only outcomes of interest are those having a sum of 3 or a sum of 5, and of these six equally likely outcomes, two have a sum of 3. See Figure 2.3-2. (See also Exercise 2.3-13.)

It is interesting to note that conditional probability satisfies the axioms for a probability function, namely, with $P(B) > 0$,

(a) $P(A|B) \geq 0$.
(b) $P(B|B) = 1$.
(c) If $A_1, A_2, A_3, \ldots$ are mutually exclusive events, then

$$P(A_1 \cup A_2 \cup \cdots \cup A_k|B) = P(A_1|B) + P(A_2|B) + \cdots + P(A_k|B),$$

for each positive integer k, and

$$P(A_1 \cup A_2 \cup \cdots|B) = P(A_1|B) + P(A_2|B) + \cdots.$$

Properties (a) and (b) are evident because

$$P(A|B) = \frac{P(A \cap B)}{P(B)} \geq 0,$$

since $P(A \cap B) \geq 0$ and $P(B) > 0$, and

$$P(B|B) = \frac{P(B \cap B)}{P(B)} = \frac{P(B)}{P(B)} = 1.$$

Property (c) holds because, for the second part of (c),

$$P(A_1 \cup A_2 \cup \cdots|B) = \frac{P[(A_1 \cup A_2 \cup \cdots) \cap B]}{P(B)}$$

$$= \frac{P[(A_1 \cap B) \cup (A_2 \cap B) \cup \cdots]}{P(B)}.$$

But $(A_1 \cap B), (A_2 \cap B), \ldots$ are also mutually exclusive events; so

$$P(A_1 \cup A_2 \cup \cdots|B) = \frac{P(A_1 \cap B) + P(A_2 \cap B) + \cdots}{P(B)}$$

$$= \frac{P(A_1 \cap B)}{P(B)} + \frac{P(A_2 \cap B)}{P(B)} + \cdots$$

$$= P(A_1|B) + P(A_2|B) + \cdots.$$

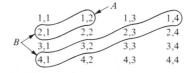

FIGURE 2.3-2

The first part of property (c) is proved in a similar manner.

Many times the conditional probability of an event is clear because of the nature of an experiment. The following example illustrates this.

Example 2.3-5   At a fair a vendor has 25 helium balloons on strings: 10 balloons are yellow, 8 are red, and 7 are green. A balloon is selected at random and sold. Given that the balloon sold is yellow, what is the probability that the next balloon selected at random is also yellow? Of the 24 remaining balloons, 9 are yellow; so a natural value to assign to this conditional probability is 9/24.

In Example 2.3-5, let A be the event that the first balloon selected is yellow, and let B be the event that the second balloon selected is yellow. Suppose that we are interested in the probability that both selected balloons are yellow. That is, we are interested in finding $P(A \cap B)$. We noted in Example 2.3-5 that

$$P(B|A) = \frac{P(A \cap B)}{P(A)} = \frac{9}{24}.$$

Thus, multiplying through by $P(A)$, we have

$$P(A \cap B) = P(A)P(B|A) = P(A)\left(\frac{9}{24}\right) \qquad (2.3\text{-}1)$$

or

$$P(A \cap B) = \left(\frac{10}{25}\right)\left(\frac{9}{24}\right).$$

That is, equation (2.3-1) gives us a general rule for the probability of the intersection of two events once we know the conditional probability $P(B|A)$.

DEFINITION 2.3-2   *The probability that two events, A and B, both occur is given by the* **multiplication rule**

$$P(A \cap B) = P(A)P(B|A)$$

or by

$$P(A \cap B) = P(B)P(A|B).$$

Sometimes, after considering the nature of the random experiment, one can make reasonable assumptions so that it is easier to assign $P(B)$ and $P(A|B)$ rather than $P(A \cap B)$. Then $P(A \cap B)$ can be computed with these assignments. This will be illustrated in Examples 2.3-6 and 2.3-7.

Example 2.3-6   A bowl contains seven blue chips and three red chips. Two chips are to be drawn successively at random and without replacement. We want to compute the probability that the first draw results in a red chip (A) and the second draw results in a blue chip (B). It is reasonable to assign the following probabilities:

$$P(A) = \frac{3}{10} \quad \text{and} \quad P(B|A) = \frac{7}{9}.$$

The probability of red on the first draw and blue on the second draw is

$$P(A \cap B) = \left(\frac{3}{10}\right)\left(\frac{7}{9}\right) = \frac{7}{30}.$$

It should be noted that in many instances, it is possible to compute a probability by two seemingly different methods. For illustration, consider Example 2.3-6 but find the probability of drawing a red chip on each of the two draws. Following that example, it is

$$\left(\frac{3}{10}\right)\left(\frac{2}{9}\right) = \frac{1}{15}.$$

However, we can also find this probability using combinations as follows:

$$\frac{\binom{3}{2}\binom{7}{0}}{\binom{10}{2}} = \frac{\frac{(3)(2)}{(1)(2)}}{\frac{(10)(9)}{(1)(2)}} = \frac{1}{15}.$$

Thus we obtain the same answer, as we should, provided that our reasoning is consistent with the underlying assumptions.

Example 2.3-7 From an ordinary deck of playing cards, cards are to be drawn successively at random and without replacement. The probability that the third spade appears on the sixth draw is computed as follows: Let A be the event of two spades in the first five cards drawn, and let B be the event of a spade on the sixth draw. Thus the probability that we wish to compute is $P(A \cap B)$. It is reasonable to take

$$P(A) = \frac{\binom{13}{2}\binom{39}{3}}{\binom{52}{5}} = 0.274 \quad \text{and} \quad P(B|A) = \frac{11}{47} = 0.234.$$

The desired probability $P(A \cap B)$ is then the product of these two numbers, $(0.274)(0.234) = 0.064$.

Example 2.3-8 Continuing with Example 2.3-4, in which a pair of four-sided dice is rolled, the probability for rolling a sum of 3 on the first roll and then, continuing the sequence of rolls, rolling a sum of 3 before rolling a sum of 5 is

$$\frac{2}{16} \cdot \frac{2}{6} = \frac{4}{96} = \frac{1}{24}.$$

The multiplication rule can be extended to three or more events. In the case of three events, we have, by using the multiplication rule for two events,

$$P(A \cap B \cap C) = P[(A \cap B) \cap C]$$
$$= P(A \cap B)P(C|A \cap B).$$

But

$$P(A \cap B) = P(A)P(B|A).$$

Hence

$$P(A \cap B \cap C) = P(A)P(B|A)P(C|A \cap B).$$

This type of argument can be used to extend the multiplication rule to more than three events, and the general formula for k events can be officially proved by mathematical induction.

Example 2.3-9   Four cards are to be dealt successively at random and without replacement from an ordinary deck of playing cards. The probability of receiving in order a spade, a heart, a diamond, and a club is

$$\left(\frac{13}{52}\right)\left(\frac{13}{51}\right)\left(\frac{13}{50}\right)\left(\frac{13}{49}\right),$$

a result that follows from the extension of the multiplication rule and reasonable assignments to the probabilities involved.

We close this section with a different type of example.

Example 2.3-10   A grade school boy has five blue and four white marbles in his left pocket and four blue and five white marbles in his right pocket. If he transfers one marble at random from his left to his right pocket, what is the probability of his then drawing a blue marble from his right pocket? For notation let BL, BR, and WL denote drawing blue from left pocket, blue from right pocket, and white from left pocket, respectively. Then

$$P(BR) = P(BL \cap BR) + P(WL \cap BR)$$
$$= P(BL)P(BR|BL) + P(WL)P(BR|WL)$$
$$= \left(\frac{5}{9}\right)\left(\frac{5}{10}\right) + \left(\frac{4}{9}\right)\left(\frac{4}{10}\right) = \frac{41}{90}.$$

Exercises

In solving certain of these exercises, make the usual and rather natural assumptions.

2.3-1   A common test for AIDS is called the ELISA test. Among 1,000,000 people who are given the ELISA test, we can expect results similar to those given in the table.

	B_1: Carry AIDS Virus	B_2: Do Not Carry AIDS Virus	Totals
A_1: Test positive	4,885	73,630	78,515
A_2: Test negative	115	921,370	921,485
Totals	5,000	995,000	1,000,000

If one of these 1,000,000 people is selected randomly, find the following probabilities:
(a) $P(B_1)$.
(b) $P(A_1)$.
(c) $P(A_1|B_2)$.
(d) $P(B_1|A_1)$.

2.3-2   The following table classifies 1456 people by their sex and by whether or not they favor a gun law.

	Male (S_1)	Female (S_2)	Totals
Favor (A_1)	392	649	1041
Oppose (A_2)	241	174	415
Totals	633	823	1456

Compute the following probabilities if one of these 1456 persons is selected randomly:
(a) $P(A_1)$.
(b) $P(A_1|S_1)$.
(c) $P(A_1|S_2)$.

2.3-3   When persons fold their arms, let A_1 and A_2 be the events that their left arm or right arm, respectively, is on top. When persons fold their hands, let B_1 and B_2 be the events that their left thumb and right thumb, respectively, are on top. A survey in one statistics class yielded the following table.

	B_1	B_2	Totals
A_1	26	10	36
A_2	12	10	22
Totals	38	20	58

If a student is selected randomly, find the following probabilities:
(a) $P(A_1 \cap B_1)$.
(b) $P(A_1 \cup B_1)$.
(c) $P(A_1 | B_1)$.
(d) $P(B_2 | A_2)$.

2.3-4 Two cards are drawn successively and without replacement from an ordinary deck of playing cards. Compute the probability of drawing
(a) Two hearts.
(b) A heart on the first draw, a club on the second draw.
(c) A heart on the first draw, an ace on the second draw.

HINT: In part (c), note that a heart can be drawn by getting the ace of hearts or one of the other 12 hearts.

2.3-5 Suppose that $P(A) = 0.7$, $P(B) = 0.5$, and $P([A \cup B]') = 0.1$.
(a) Find $P(A \cap B)$.
(b) Give $P(A | B)$.
(c) Give $P(B | A)$.

2.3-6 A hand of 13 cards is to be dealt at random and without replacement from an ordinary deck of playing cards. Find the conditional probability that there are at least three kings in the hand given that the hand contains at least two kings.

2.3-7 Suppose that the genes for eye color for a certain male fruit fly are (R, W) and the genes for eye color for the mating female fruit fly are (R, W), where R and W represent red and white, respectively. Their offspring receive one gene for eye color from each parent.
(a) Define the sample space for the genes for eye color for the offspring.
(b) Assume that each of the four possible outcomes has equal probability. If an offspring ends up with either two red genes or one red and one white gene for eye color, its eyes will look red. Given that an offspring's eyes look red, what is the conditional probability that it has two red genes for eye color?

2.3-8 Suppose that there are 14 songs on a compact disk (CD) and you like 8 of them. When using the random button selector on a CD player, each of the 14 songs is played once in a random order. Find the probability that among the first 2 songs that are played,
(a) You like both of them.

    (b) You like neither of them.

    (c) You like exactly one of them.

2.3-9   Suppose that there are 12 songs on a compact disk (CD) of which two are your favorites. When using the random button selector on a CD player, each of the 12 selections is played once in a random order. Find the probability that the second of your 2 favorites is

    (a) The third song that is played.

    (b) The sixth song that is played.

2.3-10   An urn contains 17 balls marked LOSE and 3 balls marked WIN. You and an opponent take turns selecting at random a single ball from the urn without replacement. The person who selects the third WIN ball wins the game. It does not matter who selected the first two WIN balls.

    (a) If you draw first, find the probability that you win the game on your fourth draw.

    (b) If you draw first, what is the probability that you win?

    HINT: You could win on your second, third, fourth, . . . , or tenth draw, not on your first.

2.3-11   In a string of 12 Christmas tree light bulbs, 3 are defective. The bulbs are selected at random and tested, one at a time, until the third defective bulb is found. Compute the probability that the third defective bulb is the

    (a) Third bulb tested.

    (b) Fifth bulb tested.

    (c) Tenth bulb tested.

2.3-12   A small grocery store had 10 cartons of milk, 2 of which were sour. If you are going to buy the sixth carton of milk sold that day at random, compute the probability of selecting a carton of sour milk.

2.3-13   In the gambling game "craps" a pair of dice is rolled and the outcome of the experiment is the sum of the dice. The bettor wins on the first roll if the sum is 7 or 11. The bettor loses on the first roll if the sum is 2, 3, or 12. If the sum is 4, 5, 6, 8, 9, or 10, that number is called the bettor's "point." Once the point is established, the rule is: If the bettor rolls a 7 before the "point," the bettor loses; but if the "point" is rolled before a 7, the bettor wins.

    (a) List the 36 outcomes in the sample space for the roll of a pair of dice. Assume that each of them has a probability of 1/36.

    (b) Find the probability that the bettor wins on the first roll. That is, find the probability of rolling a 7 or 11, $P(7 \text{ or } 11)$.

    (c) Given that 8 is the outcome on the first roll, find the probability that the bettor now rolls the point 8 before rolling a 7 and thus wins. Note that at this stage in the game the only outcomes of interest are 7 and 8. Thus find $P(8 \mid 7 \text{ or } 8)$.

(d) The probability that a bettor rolls an 8 on the first roll and then wins is given by $P(8)P(8|7 \text{ or } 8)$. Show that the value of this probability is (5/36)(5/11).

(e) Show that the total probability that a bettor wins in the game of craps is 0.49293.

HINT: Note that the bettor can win in one of several mutually exclusive ways: by rolling a 7 or 11 on the first roll or by establishing one of the points 4, 5, 6, 8, 9, or 10 on the first roll and then obtaining that point before a 7 on successive rolls.

2.3-14 A single card is drawn at random from each of six well-shuffled decks of playing cards. Let A be the event that all six cards drawn are different.
(a) Find $P(A)$.
(b) Find the probability that at least two of the drawn cards match.

2.3-15 Consider the birthdays of the students in a class of size r. Assume that the year consists of 365 days.
(a) How many different ordered samples of birthdays are possible (r in sample) allowing repetitions (with replacement)?
(b) The same as part (a) except requiring that all the students have different birthdays (without replacement)?
(c) If we can assume that each ordered outcome in part (a) has the same probability, what is the probability that no two students have the same birthday?
(d) For what value of r is the probability in part (c) about equal to 1/2? Is this number surprisingly small?

HINT: Use a calculator or computer to find r.

2.3-16 You are a member of a class of 18 students. A bowl contains 18 chips, 1 blue and 17 red. Each student is to take 1 chip from the bowl without replacement. The student who draws the blue chip is guaranteed an A for the course.
(a) If you have a choice of drawing first, fifth, or last, which position would you choose? Justify your choice using probability.
(b) Suppose the bowl contains 2 blue and 16 red chips. What position would you now choose?

2.3-17 A drawer contains four black, six brown, and eight olive socks. Two socks are selected at random from the drawer.
(a) Compute the probability that both socks are the same color.
(b) Compute the probability that both socks are olive if it is known that they are the same color.

2.3-18 Bowl C contains 6 red chips and 4 blue chips. Five of these 10 chips are selected at random and without replacement and put in bowl D, which was originally empty. One chip is then drawn at random from bowl D. Given that this chip is blue, find the conditional probability that 2 red chips and 3 blue chips were transferred from bowl C to bowl D.

2.3-19 Bowl A contains three red and two white chips, and bowl B contains four red and three white chips. A chip is drawn at random from bowl A and transferred to bowl B. Compute the probability of then drawing a red chip from bowl B.

2.3-20 An urn contains four balls numbered 1 through 4. The balls are selected one at a time without replacement. A match occurs if ball numbered m is the mth ball selected. Let the event A_i denote a match on the ith draw, $i = 1, 2, 3, 4$. Show that

(a) $P(A_i) = \dfrac{3!}{4!}$.

(b) $P(A_i \cap A_j) = \dfrac{2!}{4!}$.

(c) $P(A_i \cap A_j \cap A_k) = \dfrac{1!}{4!}$.

(d) The probability of at least one match is

$$P(A_1 \cup A_2 \cup A_3 \cup A_4) = 1 - \frac{1}{2!} + \frac{1}{3!} - \frac{1}{4!}.$$

(e) Extend this exercise so that there are n balls in the urn. Show that the probability of at least one match is

$$P(A_1 \cup A_2 \cup \cdots \cup A_n) = 1 - \frac{1}{2!} + \frac{1}{3!} - \frac{1}{4!} + \cdots + \frac{(-1)^{n+1}}{n!}$$

$$= 1 - \left(1 - \frac{1}{1!} + \frac{1}{2!} - \frac{1}{3!} + \cdots + \frac{(-1)^n}{n!}\right).$$

(f) What is the limit of this probability as n increases without bound?

2.4   Independent Events

For certain pairs of events, the occurrence of one of them may or may not change the probability of the occurrence of the other. In the latter case they are said to be **independent events**. However, before giving the formal definition of independence, let us consider an example.

Example 2.4-1 Flip a coin twice and observe the sequence of heads and tails. The sample space is then

$$S = \{HH, HT, TH, TT\}.$$

It is reasonable to assign a probability of 1/4 to each of these four outcomes. Let

$$A = \{\text{heads on the first flip}\} = \{HH, HT\},$$
$$B = \{\text{tails on the second flip}\} = \{HT, TT\},$$
$$C = \{\text{tails on both flips}\} = \{TT\}.$$

Now $P(B) = 2/4 = 1/2$. However, if we are given that C has occurred, then $P(B|C) = 1$ because $C \subset B$. That is, the knowledge of the occurrence of C has changed the probability of B. On the other hand, if we are given that A has occurred,

$$P(B|A) = \frac{P(A \cap B)}{P(A)} = \frac{1/4}{2/4} = \frac{1}{2} = P(B).$$

So the occurrence of A has not changed the probability of B. Hence the probability of B does not depend upon knowledge about event A, so we say that A and B are independent events. That is, events A and B are independent if the occurrence of one of them does not affect the probability of the occurrence of the other. A more mathematical way of saying this is

$$P(B|A) = P(B) \qquad \text{or} \qquad P(A|B) = P(A),$$

provided that $P(A) > 0$ or, in the latter case, $P(B) > 0$. With the first of these equalities and the multiplication rule (Definition 2.3-2), we have

$$P(A \cap B) = P(A)P(B|A) = P(A)P(B).$$

The second of these equalities, namely $P(A|B) = P(A)$, gives us the same result:

$$P(A \cap B) = P(B)P(A|B) = P(B)P(A).$$

This example motivates the following definition of independent events.

DEFINITION 2.4-1 *Events A and B are* **independent** *if and only if*

$$P(A \cap B) = P(A)P(B).$$

Otherwise A and B are called **dependent** *events.*

Events that are independent are sometimes called **statistically independent, stochastically independent**, or **independent in a probability sense**, but in most instances we use independent without a modifier if there is no possibility of misunderstanding. It is interesting to note that the definition always holds if $P(A) = 0$ or $P(B) = 0$ because then $P(A \cap B) = 0$, since $(A \cap B) \subset A$ and $(A \cap B) \subset B$. Thus the left-hand and right-hand member of $P(A \cap B) = P(A)P(B)$ are both equal zero and thus are equal to each other.

Example 2.4-2 A red die and a white die are rolled. Let event $A = \{4$ on the red die$\}$ and event $B = \{$sum of dice is odd$\}$. Of the 36 equally likely outcomes, 6 are favorable to A, 18 are favorable to B, and 3 are favorable to $A \cap B$. Thus

$$P(A)P(B) = \left(\frac{6}{36}\right)\left(\frac{18}{36}\right) = \frac{3}{36} = P(A \cap B).$$

Hence A and B are independent by Definition 2.4-1.

Example 2.4-3   A red die and a white die are rolled. Let event $C = \{5$ on red die$\}$ and event $D = \{$sum of dice is 11$\}$. Of the 36 equally likely outcomes, 6 are favorable to C, 2 are favorable to D, and 1 is favorable to $C \cap D$. Thus

$$P(C)P(D) = \left(\frac{6}{36}\right)\left(\frac{2}{36}\right) = \frac{1}{108} \neq \frac{1}{36} = P(C \cap D).$$

Hence C and D are dependent events by Definition 2.4-1.

THEOREM 2.4-1   *If A and B are independent events, then the following pairs of events are also independent:*
(a) A *and* B'.
(b) A' *and* B.
(c) A' *and* B'.

PROOF OF (a):   We know that conditional probability satisfies the axioms for a probability function. Hence, if $P(A) > 0$, then $P(B'|A) = 1 - P(B|A)$. Thus

$$\begin{aligned}
P(A \cap B') &= P(A)P(B'|A) = P(A)[1 - P(B|A)]\\
&= P(A)[1 - P(B)]\\
&= P(A)P(B'),
\end{aligned}$$

since $P(B|A) = P(B)$ by hypothesis. Thus A and B' are independent events. The proofs for parts (b) and (c) are left as exercises. $\qquad\square$

Before extending the definition of independent events to more than two events, we present the following example.

Example 2.4-4   An urn contains four balls numbered 1, 2, 3, and 4. One ball is to be drawn at random from the urn. Let the events A, B, and C be defined by $A = \{1, 2\}$, $B = \{1, 3\}$, $C = \{1, 4\}$. Then $P(A) = P(B) = P(C) = 1/2$. Furthermore,

$$P(A \cap B) = \frac{1}{4} = P(A)P(B),$$

$$P(A \cap C) = \frac{1}{4} = P(A)P(C),$$

$$P(B \cap C) = \frac{1}{4} = P(B)P(C),$$

which implies that A, B, and C are independent in pairs (called **pairwise independence**). However, since $A \cap B \cap C = \{1\}$, we have

$$P(A \cap B \cap C) = \frac{1}{4} \neq \frac{1}{8} = P(A)P(B)P(C).$$

That is, something seems to be lacking for the complete independence of A, B, and C.

This example illustrates the reason for the second condition in Definition 2.4-2.

DEFINITION 2.4-2 *Events A, B, and C are* **mutually independent** *if and only if the following two conditions hold:*

(a) *They are pairwise independent; that is,*

$$P(A \cap B) = P(A)P(B), \ P(A \cap C) = P(A)P(C),$$

   and

$$P(B \cap C) = P(B)P(C).$$

(b) $P(A \cap B \cap C) = P(A)P(B)P(C).$

Definition 2.6-2 can be extended to mutual independence of four or more events. In this extension, each pair, triple, quartet, and so on, must satisfy this type of multiplication rule. If there is no possibility of misunderstanding, *independent* is often used without the modifier *mutually* when considering several events.

Example 2.4-5   A rocket has a built-in redundant system. In this system, if component K_1 fails, it is bypassed and component K_2 is used. If component K_2 fails, it is bypassed and component K_3 is used. (An example of such components is computer systems.) Suppose that the probability of failure of any one of these components is 0.15 and assume that the failures of these components are mutually independent events. Let A_i denote the event that component K_i fails for $i = 1, 2, 3$. Because the system fails if K_1 fails and K_2 fails and K_3 fails, the probability that the system does not fail is given by

$$\begin{aligned} P[(A_1 \cap A_2 \cap A_3)'] &= 1 - P(A_1 \cap A_2 \cap A_3) \\ &= 1 - P(A_1)P(A_2)P(A_3) \\ &= 1 - (0.15)^3 \\ &= 0.9966. \end{aligned}$$

One way to increase the reliability of such a system is to add more components (realizing that this also adds weight and takes up space). For example, if a fourth component K_4 were added to this system, the probability that the system does not fail is

$$P[(A_1 \cap A_2 \cap A_3 \cap A_4)'] = 1 - (0.15)^4 = 0.9995.$$

The proof and illustration of the following results are left as exercises. If A, B, and C are mutually independent events, then the following events are also independent:

(a) A and $(B \cap C)$;
(b) A and $(B \cup C)$;
(c) A' and $(B \cap C')$.

In addition, A', B', and C' are mutually independent.

Many experiments consist of a sequence of n trials that are mutually independent. If the outcomes of the trials, in fact, do not have anything to do with one another, then events, such that each is associated with a different trial, should be independent in the probability sense. That is, if the event A_i is associated with the ith trial, $i = 1, 2, \ldots, n$, then

$$P(A_1 \cap A_2 \cap \cdots \cap A_n) = P(A_1)P(A_2)\cdots P(A_n).$$

Example 2.4-6   A fair 12-sided die is rolled 12 independent times. Let A_i be the event that side i is observed on the ith roll, called a match on the ith trial, $i = 1, 2, \ldots, 12$. Thus $P(A_i) = 1/12$, and $P(A_i') = 1 - 1/12 = 11/12$. If we let B denote the event that at least one match occurs, then B' is the event that no matches occur. Thus

$$P(B) = 1 - P(B') = 1 - P(A_1' \cap A_2' \cap \cdots \cap A_{12}')$$
$$= 1 - \left(\frac{11}{12}\right)\left(\frac{11}{12}\right)\cdots\left(\frac{11}{12}\right) = 1 - \left(\frac{11}{12}\right)^{12}.$$

The sample space for an experiment of n trials is a set of n-tuples, where the ith component denotes the outcome on the ith trial. For example, if a six-sided die is rolled five times,

$$S = \{(O_1, O_2, O_3, O_4, O_5): O_i = 1, 2, 3, 4, 5, \text{ or } 6, \text{ for } i = 1, 2, 3, 4, 5\}.$$

That is, S is a set of five-tuples, where each component is one of the first six positive integers.

If a coin is tossed two times,

$$S = \{(O_1, O_2): O_i = \text{H or T}, i = 1, 2\}.$$

We often drop the commas and parentheses and let, for illustration, (H, T) = HT, as in Example 2.4-1.

Example 2.4-7   Urn I contains two red balls and three white balls; urn II contains two red balls and one white ball; urn III contains one red ball and three white balls. At the ith trial a ball is drawn from urn i, $i = $ I, II, III. Assume that the trials are independent. Under reasonable assumptions, the probabilities of some of the possible outcomes are (in an obvious notation)

$$P(\{(R, R, R)\}) = P(RRR) = \left(\frac{2}{5}\right)\left(\frac{2}{3}\right)\left(\frac{1}{4}\right),$$

$$P(\{(W, R, W)\}) = P(WRW) = \left(\frac{3}{5}\right)\left(\frac{2}{3}\right)\left(\frac{3}{4}\right),$$

$$P(\{(R, W, W)\}) = P(RWW) = \left(\frac{2}{5}\right)\left(\frac{1}{3}\right)\left(\frac{3}{4}\right).$$

Example 2.4-8 An urn contains three red, two white, and four yellow balls. An ordered sample of size 3 is drawn from the urn. If the balls are drawn with replacement so that one outcome does not change the probabilities of the others, the trials are independent. Under reasonable assumptions, the probabilities of the two given outcomes (in an obvious notation) are

$$P(\text{RWY}) = \left(\frac{3}{9}\right)\left(\frac{2}{9}\right)\left(\frac{4}{9}\right) = \frac{8}{243}$$

and

$$P(\{\text{YYR or RWW}\}) = \left(\frac{4}{9}\right)\left(\frac{4}{9}\right)\left(\frac{3}{9}\right) + \left(\frac{3}{9}\right)\left(\frac{2}{9}\right)\left(\frac{2}{9}\right) = \frac{20}{243}.$$

If the balls are drawn without replacement, the trials are dependent. The probabilities, again under reasonable assumptions, of the two outcomes are

$$P(\text{RWY}) = \left(\frac{3}{9}\right)\left(\frac{2}{8}\right)\left(\frac{4}{7}\right) = \frac{1}{21}$$

and

$$P(\{\text{YYR or RWW}\}) = \left(\frac{4}{9}\right)\left(\frac{3}{8}\right)\left(\frac{3}{7}\right) + \left(\frac{3}{9}\right)\left(\frac{2}{8}\right)\left(\frac{1}{7}\right) = \frac{7}{84}.$$

Example 2.4-9 Suppose that on five consecutive days an "instant winner" lottery ticket is purchased and the probability of winning is 1/5 on each day. Assuming independent trials,

$$P(\text{WWLLL}) = \left(\frac{1}{5}\right)^2\left(\frac{4}{5}\right)^3,$$

$$P(\text{LWLWL}) = \left(\frac{4}{5}\right)\left(\frac{1}{5}\right)\left(\frac{4}{5}\right)\left(\frac{1}{5}\right)\left(\frac{4}{5}\right) = \left(\frac{1}{5}\right)^2\left(\frac{4}{5}\right)^3.$$

In general, the probability of purchasing two winning tickets and three losing tickets is

$$\binom{5}{2}\left(\frac{1}{5}\right)^2\left(\frac{4}{5}\right)^3 = \frac{5!}{2!3!}\left(\frac{1}{5}\right)^2\left(\frac{4}{5}\right)^3 = 0.2048$$

because there are $\binom{5}{2}$ ways to select the positions (or days) for the winning tickets and each of these $\binom{5}{2}$ ways has the probability $(1/5)^2(4/5)^3$.

Exercises

2.4-1 Let A and B be independent events with $P(A) = 0.7$ and $P(B) = 0.2$. Compute
(a) $P(A \cap B)$.
(b) $P(A \cup B)$.
(c) $P(A' \cup B')$.

2.4-2 Let $P(A) = 0.3$ and $P(B) = 0.6$.
(a) Find $P(A \cup B)$ when A and B are independent.
(b) Find $P(A|B)$ when A and B are mutually exclusive.

2.4-3 Let A and B be independent events with $P(A) = 1/4$ and $P(B) = 2/3$. Compute
(a) $P(A \cap B)$.   (b) $P(A \cap B')$.   (c) $P(A' \cap B')$.
(d) $P[(A \cup B)']$.   (e) $P(A' \cap B)$.

2.4-4 Prove parts (b) and (c) of Theorem 2.4-1.

2.4-5 If $P(A) = 0.8$, $P(B) = 0.5$, and $P(A \cup B) = 0.9$, are A and B independent events? Why?

2.4-6 If A, B, and C are mutually independent, show that the following pairs of events are independent: A and $(B \cap C)$, A and $(B \cup C)$, A' and $(B \cap C')$. Also show that A', B', and C' are mutually independent.

2.4-7 Each of three persons fires one shot at a target. Let A_i denote the event that the target is hit by person i, $i = 1, 2, 3$. If we assume A_1, A_2, A_3 are mutually independent and if $P(A_1) = 0.7$, $P(A_2) = 0.9$, $P(A_3) = 0.8$, compute the probability that exactly two people hit the target (i.e., one misses).

2.4-8 Suppose that A, B, and C are mutually independent events and that $P(A) = 0.5$, $P(B) = 0.8$, and $P(C) = 0.9$. In the Venn diagram in Figure 2.4-1, there are eight disjoint subsets of the sample space.
(a) Draw a Venn diagram like Figure 2.4-1 and fill in the probabilities for each of these eight subsets.
(b) Give the probabilities that (i) all three events occur, (ii) exactly two of the three events occur, and (iii) none of the events occur.

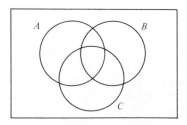

Figure 2.4-1

2.4-9   Let D_1, D_2, D_3 be three four-sided dice whose sides have been labeled as follows:

$$D_1: 0\ 3\ 3\ 3 \qquad D_2: 2\ 2\ 2\ 5 \qquad D_3: 1\ 1\ 4\ 6$$

These dice are rolled at random. Let A, B, and C be the events that the outcome on die D_1 is larger than the outcome on D_2, the outcome on D_2 is larger than the outcome on D_3, the outcome on D_3 is larger than the outcome on D_1, respectively. Show that

(a) $P(A) = \frac{9}{16}$.

(b) $P(B) = \frac{9}{16}$.

(c) $P(C) = \frac{10}{16}$.

Do you find it interesting that each of the probabilities that D_1 ''beats'' D_2, D_2 ''beats'' D_3, and D_3 ''beats'' D_1 is greater than $\frac{1}{2}$? Thus it is difficult to determine the ''best'' die.

2.4-10   (a) If the events A and B are mutually exclusive, are A and B always independent? If the answer is no, can they ever be independent? Explain.
(b) If $A \subset B$, can A and B ever be independent events? Explain.

2.4-11   Flip an unbiased coin five independent times. Compute the probability of
(a) HHTHT.
(b) THHHT.
(c) HTHTH.
(d) Three heads occurring in the five trials.

2.4-12   An urn contains two red balls and four white balls. Sample successively five times at random and with replacement so that the trials are independent. Compute the probability of
(a) WWRWR.
(b) RWWWR.

2.4-13   In Example 2.4-5, suppose that the probability of failure of a component is $p = 0.4$. Find the probability that the system does not fail if the number of redundant components is
(a) 3.
(b) 8.

2.4-14   An urn contains 10 red and 10 white balls. The balls are drawn from the urn at random, one at a time. Find the probability that the fourth white ball is the sixth ball drawn if the sampling is done
(a) With replacement.
(b) Without replacement.
(c) In the World Series the American League (red) and National League (white) teams play until one team wins four games. Do you think that this urn model could be used to describe the probabilities of a 4-, 5-, 6-, or 7-game series? If your answer is yes, would you choose sampling with or without replacement in your model?

2.4-15 An urn contains five balls, one marked WIN and four marked LOSE. You and another player take turns selecting a ball from the urn, one at a time. The first person to select the WIN ball is the winner. If you draw first, find the probability that you will win if the sampling is done
(a) With replacement.
(b) Without replacement.

2.4-16 If you buy a Lotto ticket for the Michigan Lottery, you may say yes to **Zinger**. The word *Yes* then appears next to a six-digit integer (selected randomly from 000000 to 999999, inclusive) on your Lotto ticket. To win something playing the Zinger, your numbers must match, in exact order drawn from left to right, the first 2, 3, 4, 5, or 6 digits of the selected Zinger number. Find the probability that you match
(a) The first 2 but not the first 3 digits.
(b) The first 3 but not the first 4 digits.
(c) The first 4 but not the first 5 digits.
(d) The first 5 but not the first 6 digits.
(e) All 6 digits.

2.4-17 Extend Example 2.4-6 to an n-sided die. That is, suppose that a fair n-sided die is rolled n independent times. A match occurs if side i is observed on the ith trial, $i = 1, 2, \ldots, n$.
(a) Show that the probability of at least one match is

$$1 - \left(\frac{n-1}{n}\right)^n = 1 - \left(1 - \frac{1}{n}\right)^n.$$

(b) Find the limit of this probability as n increases without bound.

2.4-18 An urn contains n balls numbered from 1 through n. A random sample of n balls is selected from the urn, one at a time. A match occurs if ball numbered i is selected on the ith draw. For $n = 1$ to 15, find the probability of at least one match if the sampling is done
(a) With replacement (see Exercise 2.4-17).
(b) Without replacement (see Exercise 2.3-20).
(c) How much does n affect these probabilities?
(d) How does sampling with and without replacement affect these probabilities?
(e) Illustrate these probabilities empirically, either performing the experiments physically or simulating them on a computer.

★2.5 Bayes' Theorem

In this section we consider the well-known Bayes' theorem. We begin with an example.

Example 2.5-1 Bowl B_1 contains two red and four white chips; bowl B_2 contains one red and two white chips; and bowl B_3 contains five red and four white chips. Say that the probabilities for selecting the bowls are not the same but are given by $P(B_1) = 1/3$, $P(B_2) = 1/6$, and $P(B_3) = 1/2$, where B_1, B_2, and B_3 are the events that bowls B_1, B_2, and B_3 are chosen, respectively. The experiment consists of selecting a bowl with these probabilities and then drawing a chip at random from that bowl. Let us compute the probability of event R, drawing a red chip, say $P(R)$. Note that $P(R)$ is dependent first of all on which bowl is selected and then on the probability of drawing a red chip from the selected bowl. That is, the event R is the union of the mutually exclusive events $B_1 \cap R$, $B_2 \cap R$, and $B_3 \cap R$. Thus

$$P(R) = P(B_1 \cap R) + P(B_2 \cap R) + P(B_3 \cap R)$$
$$= P(B_1)P(R|B_1) + P(B_2)P(R|B_2) + P(B_3)P(R|B_3)$$
$$= \left(\frac{1}{3}\right)\left(\frac{2}{6}\right) + \left(\frac{1}{6}\right)\left(\frac{1}{3}\right) + \left(\frac{1}{2}\right)\left(\frac{5}{9}\right) = \frac{4}{9}.$$

Suppose now that the outcome of the experiment is a red chip, but we do not know from which bowl it was drawn. Accordingly, we compute the conditional probability that the chip was drawn from bowl B_1, namely, $P(B_1|R)$. From the definition of conditional probability and the result above, we have that

$$P(B_1|R) = \frac{P(B_1 \cap R)}{P(R)}$$

$$= \frac{P(B_1)P(R|B_1)}{P(B_1)P(R|B_1) + P(B_2)P(R|B_2) + P(B_3)P(R|B_3)}$$

$$= \frac{(1/3)(2/6)}{(1/3)(2/6) + (1/6)(1/3) + (1/2)(5/9)} = \frac{2}{8}.$$

Similarly, we have that

$$P(B_2|R) = \frac{P(B_2 \cap R)}{P(R)} = \frac{(1/6)(1/3)}{4/9} = \frac{1}{8}$$

and

$$P(B_3|R) = \frac{P(B_3 \cap R)}{P(R)} = \frac{(1/2)(5/9)}{4/9} = \frac{5}{8}.$$

Note that the conditional probabilities $P(B_1|R)$, $P(B_2|R)$, and $P(B_3|R)$ have changed from the original probabilities $P(B_1)$, $P(B_2)$, and $P(B_3)$ in a way that agrees with your intuition. Namely, once the red chip has been observed, the probability concerning B_3 seems more favorable than originally because B_3 has a larger percentage of red chips than do B_1 and B_2. The conditional probabilities of B_1 and B_2 decrease from their original ones once the red chip is observed. Frequently, the original probabilities are called *prior probabilities*, and the conditional probabilities are the *posterior probabilities*.

We generalize the result of Example 2.5-1. Let $B_1, B_2, \ldots, B_m$ constitute a *partition* of the sample space S. That is,

$$S = B_1 \cup B_2 \cup \cdots \cup B_m \quad \text{and} \quad B_i \cap B_j = \phi, \quad i \neq j.$$

Of course, the events $B_1, B_2, \ldots, B_m$ are mutually exclusive and exhaustive (since the union of the disjoint sets equals the sample space S). Furthermore, suppose the **prior probability** of the event B_i is positive; that is $P(B_i) > 0, i = 1, \ldots, m$. If A is an event, then A is the union of m mutually exclusive events, namely,

$$A = (B_1 \cap A) \cup (B_2 \cap A) \cup \cdots \cup (B_m \cap A).$$

Thus

$$P(A) = \sum_{i=1}^{m} P(B_i \cap A)$$

$$= \sum_{i=1}^{m} P(B_i)P(A|B_i). \tag{2.5-1}$$

If $P(A) > 0$, we have that

$$P(B_k|A) = \frac{P(B_k \cap A)}{P(A)}, \quad k = 1, 2, \ldots, m. \tag{2.5-2}$$

Using equation (2.5-1) and replacing $P(A)$ in equation (2.5-2), we have **Bayes' formula**:

$$P(B_k|A) = \frac{P(B_k)P(A|B_k)}{\sum\limits_{i=1}^{m} P(B_i)P(A|B_i)}, \quad k = 1, 2, \ldots, m.$$

The conditional probability $P(B_k|A)$ is often called the **posterior probability** of B_k. The following example illustrates one application of Bayes' result.

Example 2.5-2 In a certain factory, machines I, II, and III are all producing springs of the same length. Of their production, machines I, II, and III produce 2%, 1%, and 3% defective springs, respectively. Of the total production of springs in the factory, machine I produces 35%, machine II produces 25%, and machine III produces 40%. If one spring is selected at random from the total springs produced in a day, the probability that it is defective, in an obvious notation, equals

$$P(D) = P(I)P(D|I) + P(II)P(D|II) + P(III)P(D|III)$$

$$= \left(\frac{35}{100}\right)\left(\frac{2}{100}\right) + \left(\frac{25}{100}\right)\left(\frac{1}{100}\right) + \left(\frac{40}{100}\right)\left(\frac{3}{100}\right) = \frac{215}{10,000}.$$

If the selected spring is defective, the conditional probability that it was produced by machine III is, by Bayes' formula,

$$P(\text{III}|D) = \frac{P(\text{III})P(D|\text{III})}{P(D)} = \frac{(40/100)(3/100)}{215/10,000} = \frac{120}{215}.$$

Note how the posterior probability of III increased from the prior probability of III after the defective spring was observed because III produces a larger percentage of defectives than do I and II.

Exercises

2.5-1   Bowl B_1 contains 2 white chips, bowl B_2 contains 2 red chips, bowl B_3 contains 2 white and 2 red chips, and bowl B_4 contains 3 white chips and 1 red chip. The probabilities of selecting bowl B_1, B_2, B_3, or B_4 are 1/2, 1/4, 1/8, and 1/8, respectively. A bowl is selected using these probabilities, and a chip is then drawn at random. Find
(a) $P(W)$, the probability of drawing a white chip.
(b) $P(B_1|W)$, the conditional probability that bowl B_1 had been selected, given that a white chip was drawn.

2.5-2   Bean seeds from supplier A have an 85% germination rate and those from supplier B have a 75% germination rate. A seed packaging company purchases 40% of their bean seeds from supplier A and 60% from supplier B and mixes these seeds together.
(a) Find the probability that a seed selected at random from the mixed seeds will germinate, say $P(G)$.
(b) Given that a seed germinates, find the probability that the seed was purchased from supplier A.

2.5-3   The Belgium 20 franc coin (B20), the Italian 500 lire coin (I500), and the Hong Kong 5 dollar coin (HK5) are approximately the same size. Coin purse 1 (C1) contains 6 of each of these coins. Coin purse 2 (C2) contains 9 B20's, 6 I500's, and 3 HK5's. A fair four-sided die is rolled. If the outcome is {1}, a coin is selected randomly from C1. If the outcome belongs to {2, 3, 4}, a coin is selected randomly from C2. Find
(a) $P(B20)$, the probability of selecting a Belgian coin.
(b) $P(C1|B20)$, the probability that the coin was selected from C1, given that it was a Belgian coin.

2.5-4   Bowl A contains two red chips; bowl B contains two white chips; and bowl C contains one red chip and one white chip. A bowl is selected at random (with equal probabilities), and one chip is taken at random from that bowl.
(a) Compute the probability of selecting a white chip, say $P(W)$.
(b) If the selected chip is white, compute the conditional probability that the other chip in the bowl is red.
HINT: Compute $P(C|W)$.

2.5-5   On the desk of a mathematics professor there are two boxes of computer disks, each containing 10 disks. Call the boxes B_1 and B_2. Box B_1 contains six Verbatim (V) disks and four Control Data (CD) disks. Box B_2 contains two Verbatim disks and eight Control Data disks. A box is selected with probabilities $P(B_1) = 3/4$ and $P(B_2) = 1/4$. A disk is then selected at random from the box. Find
(a) $P(V)$.
(b) $P(B_1|V)$.

2.5-6   A package, say A, of 24 crocus bulbs contains 8 yellow, 8 white, and 8 purple crocus bulbs. A package, say B, of 24 crocus bulbs contains 6 yellow, 6 white, and 12 purple crocus bulbs. One of the two packages is selected at random.
(a) If 3 bulbs from this package were planted and all 3 yielded purple flowers, compute the conditional probability that package B was selected.
(b) If the 3 bulbs yielded 1 yellow flower, 1 white flower, and 1 purple flower, compute the conditional probability that package A was selected.

2.5-7   Each bag in a large box contains 25 tulip bulbs. Three-fourths of the bags contain bulbs for 5 red and 20 yellow tulips; one-fourth of the bags contain bulbs for 15 red and 10 yellow tulips. A bag is selected at random and one bulb is planted. Give the probability that it will produce a
(a) Red tulip, say $P(R)$.
(b) Yellow tulip, say $P(Y)$.
(c) If the tulip is red, find the conditional probability that a bag having 15 red and 10 yellow tulips was selected.

3

Discrete Distributions

3.1 Random Variables of the Discrete Type

A sample space S may be difficult to describe if the elements of S are not numbers. We shall now discuss how we can use a rule by which each simple outcome of a random experiment, an element s of S, may be associated with a real number x. We begin the discussion with an example.

Example 3.1-1 A rat is selected at random from a cage and its sex is determined. The set of possible outcomes is female and male. Thus the sample space is $S = \{$female, male$\} = \{$F, M$\}$. Let X be a function defined on S such that $X(F) = 0$ and $X(M) = 1$. Thus X is a real-valued function that has the sample space S as its domain and the set of real numbers $\{x: x = 0, 1\}$ as its range. We call X a random variable and, in this example, the space associated with X is $\{x: x = 0, 1\}$.

We now formulate the definition of a random variable.

DEFINITION 3.1-1 *Given a random experiment with a sample space S, a function X that assigns to each element s in S one and only one real number $X(s) = x$ is called a* **random variable**. *The* **space** *of X is the set of real numbers $\{x: x = X(s), s \in S\}$, where $s \in S$ means the element s belongs to the set S.*

115

REMARK As we give examples of random variables and their probability distributions, the reader soon recognizes that in treating an outcome of a random experiment the experimenter must take some type of measurement (or measurements). This measurement can be thought of as the outcome of a random variable. We would simply like to know the probability of a measurement ending in A, a subset of the space of X. If this is known for all subsets A, then we know the probability distribution of the random variable. Obviously, in practice, we do not very often know this distribution exactly. Hence statisticians make conjectures about these distributions; that is, we construct probabilistic models for random variables. The ability of a statistician to model a real situation appropriately is a valuable trait; some of us are better than others. In this chapter we introduce some probability models in which the spaces of the random variables consist of sets of integers.

It may be that the set S has elements that are themselves real numbers. In such an instance we could write $X(s) = s$ so that X is the identity function and the space of X is also S. This is illustrated in Example 3.1-2.

Example 3.1-2 Let the random experiment be the cast of a die, observing the number of spots on the side facing up. The sample space associated with this experiment is $S = \{1, 2, 3, 4, 5, 6\}$. For each $s \in S$, let $X(s) = s$. The space of the random variable X is then $\{1, 2, 3, 4, 5, 6\}$.

If we associate a probability of $1/6$ with each outcome, then, for example, $P(X = 5) = 1/6$, $P(2 \leq X \leq 5) = 4/6$, and $P(X \leq 2) = 2/6$ seem to be reasonable assignments, where $\{2 \leq X \leq 5\}$ means

$$\{X = 2, 3, 4, \text{ or } 5\}$$

and $\{X \leq 2\}$ means $\{X = 1 \text{ or } 2\}$, in this example.

The student will no doubt recognize two major difficulties here:

1. In many practical situations the probabilities assigned to the events are unknown.
2. Since there are many ways of defining a function X on S, which function do we want to use?

As a matter of fact, the solutions to these problems in particular cases are major concerns in applied statistics. In considering (2), statisticians try to determine what *measurement* (or measurements) should be taken on an outcome; that is, how best do we "mathematize" the outcome (which, for the anthropologist, might be a skull)? These measurement problems are most difficult and can be answered only by getting involved in a practical project. For (1), we often need, through repeated observations (called sampling), to estimate these probabilities or percentages. For example, what percentage of newborn girls in the University of Iowa Hospital weigh less than 7 pounds? Here a newborn baby girl is the outcome, and we have measured her one way (by weight); but obviously there are many other ways of measuring her. If we let X be the weight in pounds, we are interested in the probability $P(X < 7)$, and we can only estimate this by repeated observations. One obvious way of estimating this is by use of the relative frequency of $\{X < 7\}$ after a number of observations. If additional assumptions can be made, we will study, in this text, other ways of estimating this probability. It is this latter aspect with which mathematical statistics is concerned. That is, if we assume certain models, we find that the theory of statistics can explain how best to draw conclusions or make predictions.

In many instances, it is clear exactly what function X the experimenter wants to define on the sample space. For example, the caster in the dice game ''craps'' is concerned about the sum of the spots, say X, that are up on the pair of dice. Hence we go directly to the space of X, which is usually denoted by R. After all, in the dice game the caster is directly concerned only with the probabilities associated with X. Hence for convenience, the reader can, in many instances, think of the space of X as being the sample space.

Let X denote a random variable with space R. Suppose that we know how the probability is distributed over the various subsets A of R; that is, we can compute $P(X \in A)$. In this sense, we speak of the distribution of the random variable X, meaning, of course, the distribution of probability associated with the space R of X.

Let X denote a random variable with one-dimensional space R, a subset of the real numbers. Suppose that the space R contains a countable number of points; that is, R contains either a finite number of points or the points of R can be put into a one-to-one correspondence with the positive integers. Such a set R is called a set of discrete points or simply a discrete sample space. Furthermore, the random variable X is called a random variable of the **discrete type**, and X is said to have a distribution of the discrete type.

For a random variable X of the discrete type, the probability $P(X = x)$ is frequently denoted by $f(x)$, and this function $f(x)$ is called the **probability density function**. Note that some authors refer to $f(x)$ as the probability function, the frequency function, or the probability mass function. We prefer ''probability density function,'' and it is hereafter abbreviated p.d.f.

Let $f(x)$ be the p.d.f. of the random variable X of the discrete type, and let R be the space of X. Since $f(x) = P(X = x)$, $x \in R$, $f(x)$ must be positive for $x \in R$ and we want all these probabilities to add to 1 because each $P(X = x)$ represents the fraction of times x can be expected to occur. Moreover, to determine the probability associated with the event $A \subset R$, we would sum the probabilities of the x values in A. That is, we want $f(x)$ to satisfy the properties

(a) $f(x) > 0$, $\quad x \in R$;

(b) $\displaystyle\sum_{x \in R} f(x) = 1$;

(c) $P(X \in A) = \displaystyle\sum_{x \in A} f(x)$, $\quad$ where $\quad A \subset R$.

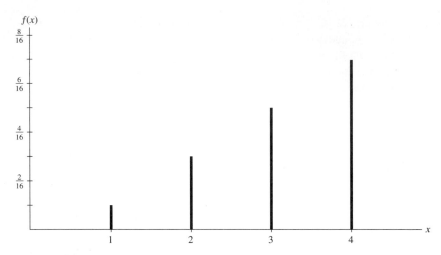

FIGURE 3.1-1

We usually let $f(x) = 0$ when $x \notin R$ and thus the domain of $f(x)$ is the set of real numbers. When we define the p.d.f. $f(x)$ and do not say zero elsewhere, then we tacitly mean that $f(x)$ has been defined at all x's in the space R, and it is assumed that $f(x) = 0$ elsewhere, namely, $f(x) = 0$, $x \notin R$. Since the probability $P(X = x) = f(x) > 0$ when $x \in R$ and since R contains all the probability associated with X, R is sometimes referred to as the **support** of X as well as the space of X.

> **Example 3.1-3** Roll a four-sided die twice and let X equal the larger of the two outcomes if they are different and the common value if they are the same. The sample space for this experiment is $S = \{(d_1, d_2): d_1 = 1, 2, 3, 4; d_2 = 1, 2, 3, 4\}$, where each of these 16 points has probability 1/16. Then $P(X = 1) = P[(1, 1)] = 1/16$, $P(X = 2) = P[\{(1, 2), (2, 1), (2, 2)\}] = 3/16$, and similarly $P(X = 3) = 5/16$ and $P(X = 4) = 7/16$. That is, the p.d.f. of X can be written simply as
>
> $$f(x) = P(X = x) = \frac{2x - 1}{16}, \quad x = 1, 2, 3, 4.$$
>
> We could add that $f(x) = 0$ elsewhere; but if we do not, the reader should take $f(x)$ to equal zero when $x \notin R$.

A better understanding of a particular probability distribution can often be obtained with a graph that depicts the p.d.f. of X. Note that the graph of the p.d.f., when $f(x) > 0$, would be simply the set of points $\{[x, f(x)]: x \in R\}$, where R is the space of X. Two types of graphs can be used to give a better visual appreciation of the p.d.f., namely, a bar graph and a probability histogram. A **bar graph** of the p.d.f. $f(x)$ of the random variable X is a graph having a vertical line segment drawn

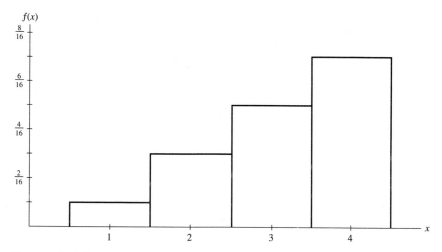

FIGURE 3.1-2

from $(x, 0)$ to $[x, f(x)]$ at each x in R, the space of X. If X can only assume integer values, a **probability histogram** of the p.d.f. $f(x)$ is a graphical representation that has a rectangle of height $f(x)$ and a base of length 1, centered at x for each $x \in R$, the space of X.

Figures 3.1-1 and 3.1-2 display a bar graph and a probability histogram, respectively, for the p.d.f. $f(x)$ defined in Example 3.1-3.

Instead of considering events such as $\{X = x\}$ and the corresponding probabilities $f(x) = P(X = x)$, let us now consider events of the form $\{X \leq x\}$ and the probabilities $P(X \leq x)$. The following example will help the reader follow the more general discussion.

Example 3.1-4 Let the random variable X of the discrete type have the p.d.f. $f(x) = x/6$, $x = 1, 2, 3$. Then, for example,

$$P(X \leq 1) = f(1) = \frac{1}{6},$$

$$P(X \leq 3) = f(1) + f(2) + f(3) = \frac{1}{6} + \frac{2}{6} + \frac{3}{6} = 1,$$

$$P(X \leq 0) = 0,$$

$$P\left(X \leq \frac{3}{2}\right) = \frac{1}{6},$$

and

$$P\left(X \leq \frac{7}{3}\right) = \frac{1}{6} + \frac{2}{6} = \frac{1}{2}.$$

In general, let $F(x) = P(X \le x)$ be defined for each real number x. Then

$$F(x) = \begin{cases} 0, & -\infty < x < 1, \\ 1/6, & 1 \le x < 2, \\ 3/6, & 2 \le x < 3, \\ 1, & 3 \le x < \infty. \end{cases}$$

The graph of $y = F(x)$ is depicted in Figure 3.1-3. Note that $F(x)$ is a nonde-creasing and right-hand continuous function that has a minimum value of zero and a maximum value of one. The height of each jump at x corresponds to $f(x) = P(X = x)$, $x = 1, 2, 3$. The p.d.f., $f(x)$, is displayed as a bar graph in Figure 3.1-4. Note the relationship between the graphs of $f(x)$ and $F(x)$; $f(x)$ represents the probability at $x \in R$, and $F(x)$ cumulates all the probability from points that are less than or equal to x.

Let X be a random variable of the discrete type with space R and p.d.f. $f(x) = P(X = x)$, $x \in R$. Now take x to be a real number and consider the set A of all points in R that are less than or equal to x. That is,

$$A = \{t: t \le x \text{ and } t \in R\}.$$

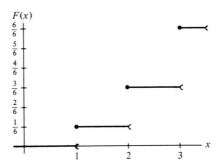

FIGURE 3.1-3

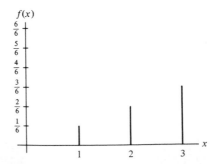

FIGURE 3.1-4

Define the function $F(x)$ by

$$F(x) = P(X \le x) = \sum_{t \in A} f(t).$$

The function $F(x)$ is called the **distribution function** (sometimes **cumulative distribution function**) of the discrete-type random variable X.

Several properties of a distribution function $F(x)$ can be listed as a consequence of the fact that probability must be a value between 0 and 1, inclusive:

(a) $0 \le F(x) \le 1$ because $F(x)$ is a probability.
(b) $F(x)$ is a nondecreasing function of x.
(c) $F(y) = 1$, where y is any value greater than or equal to the largest value in R; and $F(z) = 0$, where z is any value less than the smallest value in R.
(d) If X is a random variable of the discrete type, then $F(x)$ is a step function, and the height of a step at x, $x \in R$, equals the probability $P(X = x)$.

It is clear that the probability distribution associated with the random variable X can be described by either the distribution function $F(x)$ or by the probability density function $f(x)$. The function used is a matter of convenience; in most instances, $f(x)$ is easier to use than is $F(x)$.

In Section 2.1 we discussed the relationship between the probability $P(A)$ of an event A and the relative frequency $\#(A)/n$ of occurrences of event A in n repetitions of an experiment. We shall now extend those ideas.

Suppose that a random experiment is repeated n independent times. Let $A = \{X = x\}$, the event that x is the outcome of the experiment and let $B = \{X \le x\}$, the event that the outcome of the experiment is less than or equal to x. Then we would expect the relative frequency $\#(A)/n$ to be close to $f(x)$ and $\#(B)/n$ to be close to $F(x)$. The following example illustrates this.

Example 3.1-5 A tetrahedron (four-sided die with outcomes 1, 2, 3, 4) is rolled twice. Let X equal the sum of the two outcomes. Then the possible values that X can equal are 2, 3, 4, 5, 6, 7, and 8. The p.d.f. of X is given by $f(x) = (4 - |x - 5|)/16$, for $x = 2, 3, 4, 5, 6, 7, 8$. That is, $f(2) = 1/16$, $f(3) = 2/16, f(4) = 3/16, f(5) = 4/16, f(6) = 3/16, f(7) = 2/16$, and $f(8) = 1/16$. Intuitively, these probabilities seem correct if we think of the 16 points (result on first roll, result on second roll), and assume that each has probability $1/16$. Then note that $X = 2$ only for the point $(1, 1)$; $X = 3$ for the two points $(2, 1)$ and $(1, 2)$; and so on. This experiment was simulated 1000 times on a computer; Table 3.1-1 lists the results and compares the relative frequencies with the corresponding probabilities as well as the cumulative relative frequencies with the corresponding cumulative probabilities.

Two graphs can be used to display the results given in Table 3.1-1. The probability histogram of the p.d.f. $f(x)$ of X is given in Figure 3.1-5. The heights of the dashed lines represent the observed relative frequencies of the

TABLE 3.1-1

x	Number of Observations of x	Relative Frequency of x	Probability of $\{X = x\}$, $f(x)$	Cumulative Relative Frequency	Cumulative Probability $F(x)$
2	71	0.071	0.0625	0.071	0.0625
3	124	0.124	0.1250	0.195	0.1875
4	194	0.194	0.1875	0.389	0.3750
5	258	0.258	0.2500	0.647	0.6250
6	177	0.177	0.1875	0.824	0.8125
7	122	0.122	0.1250	0.946	0.9375
8	54	0.054	0.0625	1.000	1.0000

corresponding x values. The histogram associated with the dashed lines is the relative frequency histogram. For random experiments of the discrete type, this relative frequency histogram of a set of data gives an estimate of the probability histogram of the associated random variable when the latter is unknown.

The values of the distribution function $F(x)$ at the integers 2, 3, . . . , 8 are given in the last column in Table 3.1-1. The graph of $y = F(x)$ is given in Figure 3.1-6. The heights of the dashed lines represent the values of the cumulative relative frequencies, namely, values of the empirical distribution function. Note that the empirical distribution function is a characteristic of the set of data that can be used to estimate the distribution function of the random variable X when this latter function is unknown. Estimation is considered in detail later in this book.

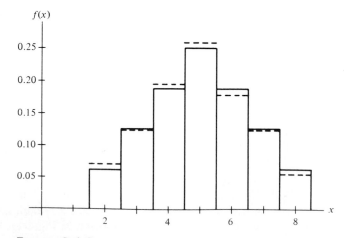

FIGURE 3.1-5

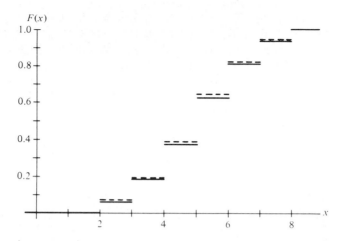

FIGURE 3.1-6

Exercises

3.1-1 Let the p.d.f. of X be defined by $f(x) = x/9$, $x = 2, 3, 4$.
(a) Sketch a bar graph for this p.d.f.
(b) Determine the distribution function and sketch its graph.

3.1-2 Let a chip be taken at random from a bowl that contains six white chips, three red chips, and one blue chip. Let the random variable $X = 1$ if the outcome is a white chip; let $X = 5$ if the outcome is a red chip; and let $X = 10$ if the outcome is a blue chip.
(a) Find the p.d.f. of X.
(b) Graph the p.d.f. as a bar graph.
(c) Find the distribution function of X.
(d) Graph the distribution function.

3.1-3 For each of the following, determine the constant c so that $f(x)$ satisfies the conditions of being a p.d.f. for a random variable X. Depict each p.d.f. as a bar graph.
(a) $f(x) = x/c$, $x = 1, 2, 3, 4$.
(b) $f(x) = cx$, $x = 1, 2, 3, \ldots, 10$.
(c) $f(x) = c(1/4)^x$, $x = 1, 2, 3, \ldots$.
(d) $f(x) = c(x + 1)^2$, $x = 0, 1, 2, 3$.
(e) $f(x) = x/c$, $x = 1, 2, 3, \ldots, n$.

3.1-4 The State of Michigan generates a three-digit number at random six days a week for their daily lottery. The numbers are generated one digit at a time. Consider the following set of 50 three-digit numbers as 150 one-digit integers that were generated at random:

```
169  938  506  757  594  656  444  809  321  545
732  146  713  448  861  612  881  782  209  752
571  701  852  924  766  633  696  023  601  789
137  098  534  826  642  750  827  689  979  000
933  451  945  464  876  866  236  617  418  988
```

Let X denote the outcome when a single digit is generated.
(a) With true random numbers, what is the p.d.f. of X? Draw the probability histogram.
(b) For the 150 observations, determine the relative frequencies of 0, 1, 2, 3, 4, 5, 6, 7, 8, and 9, respectively.
(c) Draw the relative frequency histogram of the observations on the same graph paper as that of the probability histogram. Use a colored or dashed line for the relative frequency histogram.
(d) For the 150 observations, determine the cumulative relative frequencies for 0, 1, 2, 3, 4, 5, 6, 7, 8, and 9, respectively.
(e) Draw both the theoretical and empirical distribution functions on the same graph paper.

3.1-5   The p.d.f. of X is $f(x) = (5 - x)/10$, $x = 1, 2, 3, 4$.
(a) Graph the p.d.f. as a bar graph.
(b) Define and graph the distribution function of X.
(c) Use the following independent observations of X, simulated on a computer, to construct a table like Table 3.1-1.

```
3  1  2  2  3  2  2  2  1  3  3  2  3  2  4  4  2  1  1  3
3  1  2  2  1  1  4  2  3  1  1  1  2  1  3  1  1  3  3  1
1  1  1  1  1  4  1  3  1  2  4  1  1  2  3  4  3  1  4  2
2  1  3  2  1  4  1  1  1  2  1  3  4  3  2  1  4  4  1  3
2  2  2  1  2  3  1  1  4  2  1  4  2  1  2  3  1  4  2  3
```

(d) Construct a probability histogram and a relative frequency histogram like Figure 3.1-5.
(e) Graph the theoretical and empirical distribution functions, as in Figure 3.1-6.

3.1-6   Let a random experiment be the cast of a pair of unbiased dice, each having six faces, and let the random variable X denote the sum of the dice.
(a) Determine the p.d.f. $f(x)$ of X and draw the bar graph for $f(x)$.

HINT: Picture the sample space consisting of the 36 points, (result on first die, result on second die), and assume that each has probability 1/36. Find the probability for each possible outcome of X, namely, $x = 2, 3, 4, \ldots, 12$.

(b) Find $P(X \leq 1)$, $P(X \leq 3)$, $P(X \leq 7/2)$.
(c) Plot the distribution function $F(x)$ of X.

3.1-7   Let a random experiment be the cast of a pair of unbiased six-sided dice and let X equal the larger of the outcomes if they are different and the common value if they are equal.
(a) Define the p.d.f. of X and draw a probability histogram.
(b) Define the distribution function of X and draw its graph.

3.1-8   Cast a pair of six-sided dice 50 or 100 times. For each cast, record the value of the larger outcome, X.
(a) Compare your observations of X with the distribution of X (see Exercise 3.1-7) by drawing the probability histogram and the relative frequency histogram on the same graph.
(b) On the same graph, draw the empirical distribution function of the observations of X along with the distribution function of X.

3.1-9   Let the p.d.f. of X be defined by $f(x) = (1 + |x - 3|)/11$, $x = 1, 2, 3, 4, 5$.
(a) Graph the p.d.f. of X as a bar graph.
(b) Define and graph the distribution function of X.

3.1-10   Five cards are selected at random without replacement from a standard, thoroughly shuffled 52-card deck of playing cards. Let X equal the number of face cards (kings, queens, jacks) in the hand. Forty observations of X yielded the following data:

```
2 1 2 1 0 0 1 0 1 1 0 2 0 2 3 0 1 1 0 3
1 2 0 2 0 2 0 1 0 1 1 2 1 0 1 1 2 1 1 0
```

(a) Argue that the p.d.f. of X is

$$f(x) = \frac{\binom{12}{x}\binom{40}{5-x}}{\binom{52}{5}}, \qquad x = 0, 1, 2, 3, 4, 5$$

and thus $f(0) = 0.253$, $f(1) = 0.422$, $f(2) = 0.251$, $f(3) = 0.066$, $f(4) = 0.0076$, $f(5) = 0.0003$.
(b) Draw a probability histogram for this distribution.
(c) Determine the relative frequencies of 0, 1, 2, 3, and superimpose the relative frequency histogram on your probability histogram.
(d) Define the distribution function of X.
(e) Draw both the theoretical and the empirical distribution functions on the same graph.
(f) Generate your own data and compare your data with the theoretical model.

3.1-11   Let $f(x)$ be the p.d.f. of a random variable X. Find the distribution function $F(x)$ of X and sketch its graph for each of the following:
(a) $f(x) = 1$, $x = 3$.
(b) $f(x) = 1/3$, $x = 1, 2, 3$.
(c) $f(x) = x/15$, $x = 1, 2, 3, 4, 5$.
(d) $f(x) = (1/4)(3/4)^x$, $x = 0, 1, 2, \ldots$.

3.2   Mathematical Expectation

An extremely important concept in summarizing important characteristics of distributions of probability is that of mathematical expectation, which is introduced by an example.

Example 3.2-1   An enterprising young man who needs a little extra money devises a game of chance in which some of his friends might wish to participate. The game that he proposes is to let the participant cast an unbiased die and then receive a payment according to the following schedule: If the event $A = \{1, 2, 3\}$ occurs, he receives 1¢; if $B = \{4, 5\}$ occurs, he receives 5¢; and if $C = \{6\}$ occurs, he receives 35¢. The probabilities of the respective events are assumed to be 3/6, 2/6, and 1/6, since the die is unbiased. The problem that now faces the young man is the determination of the amount that should be charged for the opportunity of playing the game. He reasons, correctly, that if the game is played a large number of times about 3/6 of the trials will require a payment of 1¢, about 2/6 of them will require one of 5¢, and about 1/6 of them will require one of 35¢. Thus the approximate average payment is

$$(1)\left(\frac{3}{6}\right) + (5)\left(\frac{2}{6}\right) + (35)\left(\frac{1}{6}\right) = 8.$$

That is, he expects to pay 8¢ "on the average." Note that he never pays exactly 8¢; the payment is either 1¢, 5¢, or 35¢. However, the "weighted average" of 1, 5, and 35, in which the weights are the respective probabilities 3/6, 2/6, and 1/6, equals eight. Such a weighted average is called the **mathematical expectation** of payment. Thus, if the young man decides to charge 10¢ per play, he would make 2¢ per play "on the average." Since the most that a player would lose at the charge of 10¢ per play is 9¢, the young man might find that several players are attracted by the possible gain of 25¢.

A more mathematical way of formulating the preceding example would be to let X be the random variable defined by the outcome of the cast of the die. Thus the p.d.f. of X is

$$f(x) = \frac{1}{6}, \qquad x = 1, 2, 3, 4, 5, 6.$$

In terms of the observed value x, the payment is given by the function

$$u(x) = \begin{cases} 1, & x = 1, 2, 3, \\ 5, & x = 4, 5, \\ 35, & x = 6. \end{cases}$$

The mathematical expectation of payment is then equal to

$$\sum_{x=1}^{6} u(x)f(x)$$

$$= (1)\left(\frac{1}{6}\right) + (1)\left(\frac{1}{6}\right) + (1)\left(\frac{1}{6}\right) + (5)\left(\frac{1}{6}\right) + (5)\left(\frac{1}{6}\right) + (35)\left(\frac{1}{6}\right)$$

$$= (1)\left(\frac{3}{6}\right) + (5)\left(\frac{2}{6}\right) + (35)\left(\frac{1}{6}\right)$$

$$= 8.$$

This discussion suggests the more general definition of mathematical expectation of a function of X.

> **DEFINITION 3.2-1** *If $f(x)$ is the p.d.f. of the random variable X of the discrete type with space R and if the summation*
>
> $$\sum_{R} u(x)f(x) = \sum_{x \in R} u(x)f(x)$$
>
> *exists, then the sum is called the* **mathematical expectation** *or the* **expected value** *of the function $u(X)$, and it is denoted by $E[u(X)]$. That is,*
>
> $$E[u(X)] = \sum_{R} u(x)f(x).$$

We can think of the expected value $E[u(X)]$ as a weighted mean of $u(x)$, $x \in R$, where the weights are the probabilities $f(x) = P(X = x)$, $x \in R$.

REMARK The usual definition of mathematical expectation of $u(X)$ requires that the sum converge absolutely; that is,

$$\sum_{x \in R} |u(x)|f(x)$$

exists. However, in this book, each $u(x)$ is such that the convergence is absolute, and we do not burden the student with this additional requirement.

There is another important observation that must be made about the consistency of this definition. Certainly, this function $u(X)$ of the random variable X is itself a random variable, say Y. Suppose that we find the p.d.f. of Y to be $g(y)$ on the support R_1. Then $E(Y)$ is given by the summation

$$\sum_{y \in R_1} yg(y).$$

In general, it is true that

$$\sum_{x \in R} u(x)f(x) = \sum_{y \in R_1} yg(y);$$

that is, the same expectation is obtained by either method. We do not prove this general result but only illustrate it in the following example.

Example 3.2-2 Let the random variable X have the p.d.f.

$$f(x) = \frac{1}{3}, \qquad x \in R,$$

where $R = \{-1, 0, 1\}$. Let $u(X) = X^2$. Then

$$\sum_{x \in R} x^2 f(x) = (-1)^2 \left(\frac{1}{3}\right) + (0)^2 \left(\frac{1}{3}\right) + (1)^2 \left(\frac{1}{3}\right) = \frac{2}{3}.$$

However, the support of the random variable $Y = X^2$ is $R_1 = \{0, 1\}$ and

$$P(Y = 0) = P(X = 0) = \frac{1}{3},$$

$$P(Y = 1) = P(X = -1) + P(X = 1) = \frac{1}{3} + \frac{1}{3} = \frac{2}{3}.$$

That is,

$$g(y) = \begin{cases} \dfrac{1}{3}, & y = 0, \\ \dfrac{2}{3}, & y = 1; \end{cases}$$

and $R_1 = \{0, 1\}$. Hence

$$\sum_{y \in R_1} yg(y) = (0)\left(\frac{1}{3}\right) + (1)\left(\frac{2}{3}\right) = \frac{2}{3},$$

which illustrates the preceding observation.

Before presenting additional examples, we list some useful facts about mathematical expectation in the following theorem.

THEOREM 3.2-1 *When it exists, mathematical expectation E satisfies the following properties:*
(a) *If c is a constant, $E(c) = c$.*
(b) *If c is a constant and u is a function,*

$$E[cu(X)] = cE[u(X)].$$

(c) *If c_1 and c_2 are constants and u_1 and u_2 are functions, then*

$$E[c_1 u_1(X) + c_2 u_2(X)] = c_1 E[u_1(X)] + c_2 E[u_2(X)].$$

PROOF: First, we have for the proof of (a) that

$$E(c) = \sum_R cf(x) = c \sum_R f(x) = c$$

because

$$\sum_R f(x) = 1.$$

Next, to prove (b), we see that

$$E[cu(X)] = \sum_R cu(x)f(x)$$

$$= c \sum_R u(x)f(x)$$

$$= cE[u(X)].$$

Finally, the proof of (c) is given by

$$E[c_1 u_1(X) + c_2 u_2(X)] = \sum_R [c_1 u_1(x) + c_2 u_2(x)]f(x)$$

$$= \sum_R c_1 u_1(x)f(x) + \sum_R c_2 u_2(x)f(x).$$

By applying (b), we obtain

$$E[c_1 u_1(X) + c_2 u_2(X)] = c_1 E[u_1(X)] + c_2 E[u_2(X)]. \qquad \square$$

Property (c) can be extended to more than two terms by mathematical induction; that is, we have

(c)′ $\quad E\left[\displaystyle\sum_{i=1}^{k} c_i u_i(X) \right] = \displaystyle\sum_{i=1}^{k} c_i E[u_i(X)].$

Because of property (c)′, mathematical expectation E is called a **linear** or **distributive** operator.

Example 3.2-3   Let X have the p.d.f.

$$f(x) = \frac{x}{10}, \qquad x = 1, 2, 3, 4.$$

Then

$$E(X) = \sum_{x=1}^{4} x\left(\frac{x}{10}\right) = (1)\left(\frac{1}{10}\right) + (2)\left(\frac{2}{10}\right) + (3)\left(\frac{3}{10}\right) + (4)\left(\frac{4}{10}\right) = 3,$$

$$E(X^2) =$$

$$\sum_{x=1}^{4} x^2\left(\frac{x}{10}\right) = (1)^2\left(\frac{1}{10}\right) + (2)^2\left(\frac{2}{10}\right) + (3)^2\left(\frac{3}{10}\right) + (4)^2\left(\frac{4}{10}\right) = 10,$$

and

$$E[X(5 - X)] = 5E(X) - E(X^2) = (5)(3) - 10 = 5.$$

Example 3.2-4　Let $u(x) = (x - b)^2$, where b is not a function of X, and suppose $E[(X - b)^2]$ exists. To find that value of b for which $E[(X - b)^2]$ is a minimum, we write

$$E[(X - b)^2] = E[X^2 - 2bX + b^2]$$
$$= E(X^2) - 2bE(X) + b^2$$

because $E(b^2) = b^2$. Thus the derivative of $E[(X - b)^2]$ with respect to b is $-2E(X) + 2b$. This derivative is equal to zero when $b = E(X)$ and $E(X)$ is the value of b that minimizes $E[(X - b)^2]$. In the next section we learn that $E(X)$ and $E[(X - b)^2]$, when $b = E(X)$, have special names.

Exercises

3.2-1　The number 3 appears on each of two chips in a bowl and the number 9 on a third chip in the bowl. Consider the following game. A player draws one chip at random from the bowl and receives either $3 or $9, depending on the number on that chip. If he or she plays this game a large number of times, how many dollars can the player expect to receive on the average per play?

3.2-2　Let X be the number selected at random from the first 10 positive integers. Assuming equal probabilities on these 10 integers, compute $E[X(11 - X)]$.

3.2-3　Find $E(X)$ for each of the distributions given in Exercise 3.1-11.

HINT:　Part (d). Note that the difference $E(X) - (3/4)E(X)$ is equal to the sum of a geometric series.

3.2-4　Let the random variable X have the p.d.f.

$$f(x) = \frac{(|x| + 1)^2}{9}, \qquad x = -1, 0, 1.$$

Compute $E(X)$, $E(X^2)$, and $E(3X^2 - 2X + 4)$.

3.2-5　In a particular lottery 3,000,000 tickets are sold each week for 50¢ apiece. Out of the 3,000,000 tickets, 12,006 are drawn and awarded prizes: twelve thousand $25 prizes, four $10,000 prizes, one $50,000 prize, and one

$200,000 prize. If you purchased a single ticket each week, what is the expected value of this game to you?

3.2-6   In a state lottery a three-digit integer is selected at random. If a player bets $1 on a particular number, the payoff, if that number is selected, is $500 minus the $1 paid for the ticket. Let X equal the payoff to the bettor $-$1, or $499, and find $E(X)$.

3.2-7   In the gambling game chuck-a-luck, for a $1 bet it is possible to win $1, $2, or $3 with respective probabilities 75/216, 15/216, and 1/216. One dollar is *lost* with probability 125/216. Let X equal the payoff for this game and find $E(X)$. Note that when a bet is won, the $1 that was bet is returned to the bettor in addition to the $1, $2, or $3 that is won.

3.2-8   Let the p.d.f. of X be defined by $f(x) = 6/(\pi^2 x^2)$, $x = 1, 2, 3, \ldots$. Show that $E(X)$ does not exist in this case.

3.2-9   Let us select at random a number from the first n positive integers. If the payment is equal to the reciprocal of the number, find an expression for the expected payment. Evaluate this number when $n = 5$. Approximate this number when $n = 100$.

HINT: Use a modification of the integral test for testing convergence of series.

3.2-10   Let X be a random variable with support $\{1, 2, 3, 5, 15, 25, 50\}$, each point of which has the same probability 1/7. Argue that $c = 5$ is the value that minimizes $E(|X - c|)$. Compare this to the value of b that minimizes $E[(X - b)^2]$.

3.2-11   If you purchase a Lotto ticket for the Michigan Lottery, you may say yes to Zinger. The word *Yes* then appears next to a six-digit integer on your Lotto ticket. To win something playing Zinger, your number must match, *in exact order drawn from left to right,* the first 2, 3, 4, 5, or 6 digits of the selected Zinger number. If you match the first 2 digits, you win $20. Matching the first 3 digits wins $100; matching the first 4 digits wins $500; matching the first 5 digits wins $5,000; and matching all 6 digits wins $100,000. Note that you may win at most one of these prizes at a time.
(a) What is the expected payoff of this game?
(b) If it costs $1 to play, what is your expected loss each play?

3.2-12   A roulette wheel used in a United States casino has 38 slots of which 18 are red, 18 are black, and 2 are green. A roulette wheel used in a French casino has 37 slots of which 18 are red, 18 are black, and 1 is green. A ball is rolled around the wheel and ends up in one of the slots with equal probability. Suppose that a player bets on red. If a $1 bet is placed, the player wins $1 if the ball ends up in a red slot (the player's $1 bet is returned). If the ball ends up in a black or green slot, the player loses $1. Find the expected value of this game to the player in
(a) The United States.
(b) France.

3.2-13   In the gambling game craps (see Exercise 2.3-13), the player wins \$1 with probability 0.49293 and loses \$1 with probability 0.50707 for each \$1 bet. What is the expected value of the game to the player?

3.3   The Mean, Variance, and Standard Deviation

Certain mathematical expectations are so important that they have special names. In this section we consider two of them: the mean and the variance.

If X is a random variable with p.d.f. $f(x)$ of the discrete type and space

$$R = \{b_1, b_2, b_3, \ldots\},$$

then

$$E(X) = \sum_R xf(x)$$
$$= b_1 f(b_1) + b_2 f(b_2) + b_3 f(b_3) + \cdots$$

is the weighted average of the numbers belonging to R, where the weights are given by the p.d.f. $f(x)$. We call $E(X)$ the **mean** of X (or the mean of the distribution) and denote it by μ. That is, $\mu = E(X)$.

REMARK   In mechanics, the weighted average of the points $b_1, b_2, b_3, \ldots$ in one-dimensional space is called the centroid of the system. Those readers without a mechanics background can think of the centroid as being the point of balance for the system in which the weights $f(b_1), f(b_2), f(b_3), \ldots$ are placed upon the points $b_1, b_2, b_3, \ldots$.

Example 3.3-1   Let X have the p.d.f.

$$f(x) = \begin{cases} \dfrac{1}{8}, & x = 0, 3, \\[2mm] \dfrac{3}{8}, & x = 1, 2. \end{cases}$$

The mean of X is

$$\mu = E(X) = 0\left(\frac{1}{8}\right) + 1\left(\frac{3}{8}\right) + 2\left(\frac{3}{8}\right) + 3\left(\frac{1}{8}\right) = \frac{3}{2}.$$

The next example shows that if the outcomes of X are equally likely (i.e., each of the outcomes has the same probability), then the mean of X is the arithmetic average of these outcomes.

Example 3.3-2 Roll a fair die and let X denote the outcome. Thus X has the p.d.f. $f(x) = 1/6$, $x = 1, 2, 3, 4, 5, 6$. Then

$$E(X) = \sum_{x=1}^{6} x\left(\frac{1}{6}\right) = \frac{1 + 2 + 3 + 4 + 5 + 6}{6} = \frac{7}{2},$$

which is the arithmetic average of the first six positive integers.

We have noted that the mean $\mu = E(X)$ is the centroid of a system of weights or a measure of the central location of the probability distribution of X. A measure of the dispersion or spread of a distribution is defined as follows. If $u(x) = (x - \mu)^2$ and $E[(X - \mu)^2]$ exists, the **variance**, frequently denoted by σ^2 or Var(X), of a random variable X of the discrete type (or variance of the distribution) is defined by

$$\sigma^2 = E[(X - \mu)^2] = \sum_{R} (x - \mu)^2 f(x).$$

The positive square root of the variance is called the **standard deviation** of X and is denoted by

$$\sigma = \sqrt{\text{Var}(X)} = \sqrt{E[(X - \mu)^2]}.$$

Example 3.3-3 Let X denote the outcome when rolling a fair die. From Example 3.3-2 we know that $\mu = 7/2 = 3.5$. Thus

$$\sigma^2 = E[(X - 3.5)^2] = \sum_{x=1}^{6} (x - 3.5)^2 \frac{1}{6}$$

$$= [(1 - 3.5)^2 + (2 - 3.5)^2 + \cdots + (6 - 3.5)^2]\left(\frac{1}{6}\right)$$

$$= \frac{35}{12}.$$

The standard deviation of X is $\sigma = \sqrt{35/12} = 1.708$, approximately.

It is worthwhile to note that the variance can be computed in another manner. We have

$$\sigma^2 = E[(X - \mu)^2] = E(X^2 - 2\mu X + \mu^2),$$

which, by the distributive property of E, is

$$\sigma^2 = E(X^2) - 2\mu E(X) + \mu^2 = E(X^2) - 2\mu^2 + \mu^2 = E(X^2) - \mu^2.$$

Sometimes $\sigma^2 = E(X^2) - \mu^2$ provides an easier way of computing Var(X) than does $\sigma^2 = E[(X - \mu)^2]$. Thus, in Example 3.3-3 we could have first computed

$$E(X^2) = \sum_{x=1}^{6} x^2\left(\frac{1}{6}\right) = \frac{1^2 + 2^2 + \cdots + 6^2}{6} = \frac{91}{6}$$

and then

$$\sigma^2 = \text{Var}(X) = \frac{91}{6} - \left(\frac{7}{2}\right)^2 = \frac{35}{12}.$$

Example 3.3-4 Let the p.d.f. of X be defined by $f(x) = x/6$, $x = 1, 2, 3$. The mean of X is

$$\mu = E(X) = 1\left(\frac{1}{6}\right) + 2\left(\frac{2}{6}\right) + 3\left(\frac{3}{6}\right) = \frac{7}{3}.$$

To find the variance and standard deviation of X we first find

$$E(X^2) = 1^2\left(\frac{1}{6}\right) + 2^2\left(\frac{2}{6}\right) + 3^2\left(\frac{3}{6}\right) = \frac{36}{6} = 6.$$

Thus the variance of X is

$$\sigma^2 = E(X^2) - \mu^2 = 6 - \left(\frac{7}{3}\right)^2 = \frac{5}{9}$$

and the standard deviation of X is

$$\sigma = \sqrt{5/9} = 0.745.$$

Although most students understand that $\mu = E(X)$ is, in some sense, a measure of the middle of the distribution of X, it is difficult to get much of a feeling for the variance and the standard deviation. The following example illustrates that the standard deviation is a measure of dispersion or spread of the points belonging to the space R.

Example 3.3-5 Let X have the p.d.f. $f(x) = 1/3$, $x = -1, 0, 1$. Here the mean is

$$\mu = \sum_{x=-1}^{1} xf(x) = (-1)\left(\frac{1}{3}\right) + (0)\left(\frac{1}{3}\right) + (1)\left(\frac{1}{3}\right) = 0.$$

Accordingly, the variance, denoted by σ_X^2, is

$$\sigma_X^2 = E[(X - 0)^2]$$

$$= \sum_{x=-1}^{1} x^2 f(x)$$

$$= (-1)^2 \left(\frac{1}{3}\right) + (0)^2 \left(\frac{1}{3}\right) + (1)^2 \left(\frac{1}{3}\right)$$

$$= \frac{2}{3},$$

so the standard deviation is $\sigma_X = \sqrt{2/3}$. Next let another random variable Y have the p.d.f. $g(y) = 1/3$, $y = -2, 0, 2$. Its mean is also zero, and it is easy to show that $\mathrm{Var}(Y) = 8/3$, so the standard deviation of Y is $\sigma_Y = 2\sqrt{2/3}$. Here the standard deviation of Y is twice that of X, reflecting the fact that the probability of Y is spread out twice as much as that of X.

Now let X be a random variable with mean μ_X and variance σ_X^2. Of course, $Y = aX + b$, where a and b are constants, is a random variable, too. The mean of Y is

$$\mu_Y = E(Y) = E(aX + b) = aE(X) + b = a\mu_X + b.$$

Moreover, the variance of Y is

$$\sigma_Y^2 = E[(Y - \mu_Y)^2] = E[(aX + b - a\mu_X - b)^2] = E[a^2(X - \mu_X)^2] = a^2\sigma_X^2.$$

Thus $\sigma_Y = |a|\sigma_X$. For illustration, in Example 3.3-5 we note that the relationship between the two distributions could be explained by $Y = 2X$ so that $\sigma_Y^2 = 4\sigma_X^2$ and thus $\sigma_Y = 2\sigma_X$, which we had observed there. In addition, we see that adding or subtracting a constant from X does not change the variance. For illustration, $\mathrm{Var}(X - 1) = \mathrm{Var}(X)$, because $a = 1$ and $b = -1$.

Let r be a positive integer. If

$$E(X^r) = \sum_R x^r f(x)$$

exists, it is called the rth **moment** of the distribution about the origin. The expression **moment** has its origin in the study of mechanics. In addition, the expectation

$$E[(X - b)^r] = \sum_R (x - b)^r f(x)$$

is called the rth moment of the distribution about b.

For a given positive integer r,

$$E[(X)_r] = E[X(X - 1)(X - 2) \cdots (X - r + 1)]$$

is called the rth **factorial moment**. We note that the second factorial moment is equal to the difference of the second and first moments:

$$E[X(X - 1)] = E(X^2) - E(X).$$

There is another formula that can be used for computing the variance that uses the second factorial moment and sometimes simplifies the calculations. First find the values of $E(X)$ and $E[X(X - 1)]$. Then

$$\sigma^2 = E[X(X - 1)] + E(X) - [E(X)]^2,$$

since, using the distributive property of E, this becomes

$$\sigma^2 = E(X^2) - E(X) + E(X) - [E(X)]^2 = E(X^2) - \mu^2.$$

Example 3.3-6 Continuing with Example 3.3-4, we find that

$$E[X(X-1)] = 1(0)\left(\frac{1}{6}\right) + 2(1)\left(\frac{2}{6}\right) + 3(2)\left(\frac{3}{6}\right) = \frac{22}{6}.$$

Thus

$$\sigma^2 = E[X(X-1)] + E(X) - [E(X)]^2$$
$$= \frac{22}{6} + \frac{7}{3} - \left(\frac{7}{3}\right)^2 = \frac{5}{9}.$$

REMARK Recall that the empirical distribution is defined by placing the weight (probability) of $1/n$ on each of n observations $x_1, x_2, \ldots, x_n$. Then the mean of this empirical distribution is

$$\sum_{i=1}^{n} x_i \frac{1}{n} = \frac{\sum_{i=1}^{n} x_i}{n} = \bar{x}.$$

The symbol $\bar{x}$ represents the **mean of the empirical distribution**. We shall see that $\bar{x}$ is usually close in value to $\mu = E(X)$; thus, when μ is unknown, $\bar{x}$ will be used to estimate μ.

Similarly, the **variance of the empirical distribution** can be computed. Let v denote this variance so that it is equal to

$$v = \sum_{i=1}^{n} (x_i - \bar{x})^2 \frac{1}{n} = \sum_{i=1}^{n} x_i^2 \frac{1}{n} - \bar{x}^2 = \frac{1}{n} \sum_{i=1}^{n} x_i^2 - \bar{x}^2.$$

This last statement is true because, in general,

$$\sigma^2 = E(X^2) - \mu^2.$$

Note that there is a relationship between the sample variance s^2 and variance v of the empirical distribution, namely $s^2 = nv/(n-1)$. Of course, with large n, the difference between s^2 and v is very small. Usually, we use s^2 to estimate σ^2 when σ^2 is unknown.

Exercises

3.3-1 Let the p.d.f. of X be given by $f(0) = 3/10, f(1) = 3/10, f(2) = 1/10$, and $f(3) = 3/10$. Compute the mean, variance, and standard deviation of X.

3.3-2 If the p.d.f. of X is given by $f(x)$, (i) depict the p.d.f. as a probability histogram and find the values of (ii) μ, (iii) σ^2, and (iv) σ. (v) Mark the five points μ, $\mu \pm \sigma$, $\mu \pm 2\sigma$ on the graph.

(a) $f(x) = \dfrac{2 - |x - 4|}{4}$, $\quad x = 3, 4, 5.$

(b) $f(x) = \dfrac{4 - |x - 4|}{16}$, $\quad x = 1, 2, 3, 4, 5, 6, 7.$

3.3-3 Find the mean and variance for the following discrete distributions:

(a) $f(x) = \dfrac{1}{5}$, $\quad x = 5, 10, 15, 20, 25.$

(b) $f(x) = 1$, $\quad x = 5.$

(c) $f(x) = \dfrac{4 - x}{6}$, $\quad x = 1, 2, 3.$

3.3-4 Find the mean and the variance of the distribution that has the distribution function

$$F(x) = \begin{cases} 0, & x < 10, \\ \dfrac{1}{4}, & 10 \le x < 15, \\ \dfrac{3}{4}, & 15 \le x < 20, \\ 1, & 20 \le x. \end{cases}$$

HINT: First find the p.d.f. of X.

3.3-5 For each of the following distributions, find $\mu = E(X)$ and then find $\sigma^2 = E[(X - \mu)^2]$:

(a) $f(x) = \dfrac{1 + |x - 3|}{11}$, $\quad x = 1, 2, 3, 4, 5.$

(b) $f(0) = 8/27$, $f(1) = 12/27$, $f(2) = 6/27$, $f(3) = 1/27.$

3.3-6 For each of the following distributions, find $\mu = E(X)$, $E[X(X - 1)]$, and $\sigma^2 = E[X(X - 1)] + E(X) - \mu^2$:

(a) $f(x) = \dfrac{3!}{x!(3 - x)!} \left(\dfrac{1}{4}\right)^x \left(\dfrac{3}{4}\right)^{3-x}$, $\quad x = 0, 1, 2, 3.$

(b) $f(x) = \dfrac{4!}{x!(4 - x)!} \left(\dfrac{1}{2}\right)^4$, $\quad x = 0, 1, 2, 3, 4.$

3.3-7 Given $E(X + 4) = 10$ and $E[(X + 4)^2] = 116$, determine (a) $\text{Var}(X + 4)$, (b) μ, and (c) σ^2.

3.3-8 Let μ and σ^2 denote the mean and variance of the random variable X. Determine $E[(X - \mu)/\sigma]$ and $E\{[(X - \mu)/\sigma]^2\}$.

3.3-9 A measure of **skewness** is defined by

$$\frac{E[(X - \mu)^3]}{\{E[(X - \mu)^2]\}^{3/2}} = \frac{E[(X - \mu)^3]}{(\sigma^2)^{3/2}}.$$

When a distribution is symmetrical about the mean, the skewness is equal to zero. If the probability histogram has a longer "tail" to the right than to the left, the measure of skewness is positive, and we say that the distribution is skewed positively or to the right. If the probability histogram has a longer tail to the left than to the right, the measure of skewness is negative, and we say that the distribution is skewed negatively or to the left. If the p.d.f. of X is given by $f(x)$, (i) depict the p.d.f. as a probability histogram and find the values of (ii) the mean, (iii) the standard deviation, and (iv) skewness.

(a) $f(x) = \begin{cases} \dfrac{2^{6-x}}{64}, & x = 1, 2, 3, 4, 5, 6, \\[2ex] \dfrac{1}{64}, & x = 7. \end{cases}$

(b) $f(x) = \begin{cases} \dfrac{1}{64}, & x = 1, \\[2ex] \dfrac{2^{x-2}}{64}, & x = 2, 3, 4, 5, 6, 7. \end{cases}$

3.3-10   Let the p.d.f. of X be defined by $f(x) = (2 - |x - 2|)/4$, $x = 1, 2, 3$.
(a) Calculate (i) μ, (ii) σ^2, and (iii) σ.
(b) Depict the p.d.f. as a probability histogram.
(c) Use the 50 following observations of X and calculate the following characteristics of the empirical distribution: (i) the mean $\bar{x}$, (ii) the sample variance s^2, and (iii) the sample standard deviation s.

2	2	2	3	2	2	2	2	2	3
1	2	1	2	3	1	1	2	2	3
1	1	2	3	2	1	2	2	3	3
3	2	2	2	2	2	3	3	1	3
2	2	2	2	2	3	2	3	2	2

(d) Superimpose the relative frequency histogram of these observations on the probability histogram.

3.3-11   Let X equal the larger outcome when a pair of four-sided dice is rolled. The p.d.f. of X is

$$f(x) = \frac{2x - 1}{16}, \qquad x = 1, 2, 3, 4.$$

(a) Find the mean, variance, and standard deviation of X.
(b) Calculate the sample mean, sample variance, and sample standard deviation of the following 100 observations of X:

4	4	4	4	2	2	2	3	1	4	3	3	2	3	2	4	4	2	3	4	4	3	4	3	4
3	3	2	4	4	3	2	3	3	3	2	4	4	3	4	1	4	3	4	4	4	3	2	4	4
4	3	1	3	2	4	4	4	4	1	3	4	3	2	4	4	3	3	1	3	3	3	3	2	2
2	3	4	3	3	2	4	2	3	3	2	4	4	3	4	4	4	4	3	4	4	4	4	4	4

(c) Draw the graphs of the probability histogram and the relative frequency histogram on the same figure.

(d) Draw the graphs of the theoretical distribution function and the empirical distribution function on the same figure.

(e) Generate your own data, either rolling dice (four-sided or six-sided) or simulating this experiment using a computer.

3.3-12   Let X be the smaller outcome when a pair of four-sided dice is rolled. The p.d.f. of X is

$$f(x) = \frac{9 - 2x}{16}, \qquad x = 1, 2, 3, 4.$$

(a) Find the mean, variance, and standard deviation of X.

(b) Calculate the sample mean, sample variance, and sample standard deviation of the following 100 observations of X:

```
1 2 3 1 2 2 2 2 4 1 1 2 2 3 1 2 1 3 1 1 1 1 2 3 1
2 1 2 2 2 3 1 3 1 1 1 3 2 1 2 3 2 1 1 2 1 2 1 1 1
1 1 3 2 1 4 2 3 1 3 1 2 3 1 1 1 3 2 1 1 1 1 2 2 2
3 1 3 1 1 1 1 3 2 1 2 1 1 1 4 1 1 4 1 2 3 2 2 4 1
```

(c) Draw the graphs of the probability histogram and the relative frequency histogram on the same figure.

(d) Draw the graphs of the theoretical distribution function and the empirical distribution function on the same figure.

(e) Generate your own data, either rolling dice (four-sided or six-sided) or simulating this experiment using a computer.

3.3-13   A Bingo card has 25 squares with numbers on 24 of them, the center being a free square. The integers that are placed on the Bingo card are selected randomly and without replacement from 1 to 75, inclusive. When a game called "cover-up" is played, balls numbered from 1 to 75, inclusive, are selected randomly and without replacement until a player covers each of the numbers on his card. Let X equal the number of balls that must be drawn to cover all the numbers on a card.

(a) Show that the p.d.f. of X, for $x = 24, 25, \ldots, 75$, is

$$f(x) = \frac{\binom{24}{23}\binom{51}{x-24}}{\binom{75}{x-1}} \cdot \frac{1}{75 - (x-1)} = \frac{\binom{24}{24}\binom{51}{x-24}}{\binom{75}{x}} \cdot \frac{24}{x}.$$

(b) What value of X is most likely to occur? In other words, what is the mode of this distribution?

(c) To show that the mean of X is $(24)(76)/25 = 72.96$, use the combinatorial identity

$$\binom{k+n+1}{k+1} = \sum_{x=k}^{n+k} \binom{x}{k}.$$

(d) Show that $E[X(X + 1)] = \dfrac{24 \cdot 77 \cdot 76}{26} = 5401.8462.$

(e) Show that $\text{Var}(X) = E[X(X + 1)] - E[X] - [E(X)]^2 = \dfrac{46,512}{8125} =$

5.7246 and $\sigma = 2.39$.

(f) Show that the distribution function of X, for $x = 24, 25, \ldots, 75$, is

$$F(x) = \frac{\dbinom{x}{24}}{\dbinom{75}{24}}.$$

(g) The following 100 observations of X were simulated on the computer.

```
75 73 74 74 73 75 75 71 71 74 75 74 74 73 75 75 75 71 68 75
74 72 73 75 74 67 73 71 74 74 73 75 71 73 71 68 74 75 66 75
75 74 75 71 75 72 75 74 75 75 72 75 75 74 73 75 62 64 75 72
74 74 75 73 75 75 73 74 75 68 75 69 74 75 61 73 73 73 72 74
68 74 71 73 75 73 74 72 74 73 75 73 71 70 62 74 74 72 75 74
```

Compare these data with the probability model. In particular, make the following comparisons: (i) $\bar{x}$ with μ, (ii) s^2 with σ^2, (iii) s with σ, (iv) $(1/100)\Sigma_{i=1}^{100} x_i(x_i + 1)$ with $E[X(X + 1)]$.

(h) Compare (i) the probability histogram with the relative frequency histogram and (ii) the theoretical distribution function with the empirical distribution function.

(i) Simulate similar data using a computer and make the same comparisons using your data.

3.4   The Discrete Uniform and Hypergeometric Distributions

We now begin the task of developing discrete probability models that are appropriate for a variety of random experiments. For each of them we will give some examples and applications as well as the p.d.f., mean, variance, and standard deviation. Note that we usually use letters at the end of the alphabet, like W, Y, Z as well as X, as random variables.

Perhaps the easiest probability model is the one that puts equal probability on each of its points. We begin with an example.

Example 3.4-1   Several states select a three- or four-digit integer randomly for their daily lottery game. Each of these games depends on their ability to select the digits from 0 to 9, inclusive, randomly. That is, if Y is equal to the outcome of one of the digits, then Y must be an integer that is selected randomly from 0 to 9, inclusive. It seems appropriate to define the p.d.f. of Y by

$$g(y) = P(Y = y) = \tfrac{1}{10}, \qquad y = 0, 1, 2, 3, 4, 5, 6, 7, 8, 9.$$

Upon selecting from positive integers and generalizing this example, let X equal an integer that is selected randomly from the first m positive integers. We say that X has a **discrete uniform distribution on the integers** $1, 2, \ldots, m$. The p.d.f. of X is defined by

$$f(x) = \frac{1}{m}, \qquad x = 1, 2, \ldots, m.$$

The mean of X is given by

$$\mu = E(X) = \sum_{x=1}^{m} x\left(\frac{1}{m}\right) = \frac{1}{m} \sum_{x=1}^{m} x$$

$$= \left(\frac{1}{m}\right) \frac{m(m+1)}{2} = \frac{m+1}{2}.$$

To find the variance of X, we first find

$$E(X^2) = \sum_{x=1}^{m} x^2\left(\frac{1}{m}\right) = \frac{1}{m} \sum_{x=1}^{m} x^2$$

$$= \left(\frac{1}{m}\right) \frac{m(m+1)(2m+1)}{6} = \frac{(m+1)(2m+1)}{6}.$$

Thus the variance of X is

$$\sigma^2 = \text{Var}(X) = E[(X - \mu)^2]$$

$$= E(X^2) - \mu^2 = \frac{(m+1)(2m+1)}{6} - \left(\frac{m+1}{2}\right)^2$$

$$= \frac{m^2 - 1}{12}.$$

For example, we find that if X equals the outcome when rolling a fair six-sided die, the p.d.f. of X is

$$f(x) = \frac{1}{6}, \qquad x = 1, 2, 3, 4, 5, 6;$$

the respective mean and variance of X (see Examples 3.3-2 and 3.3-3) are

$$\mu = \frac{6+1}{2} = 3.5 \qquad \text{and} \qquad \sigma^2 = \frac{6^2 - 1}{12} = \frac{35}{12}.$$

If we let $Y = X - 1$, then Y is equal to an integer selected randomly from the first m nonnegative integers: $0, 1, 2, \ldots, m - 1$. The p.d.f. of Y is

$$g(y) = P(Y = y) = P(X - 1 = y) = P(X = y + 1) = \frac{1}{m}, y = 0, 1, \ldots, m - 1.$$

Furthermore, the mean and variance of Y are

$$\mu_Y = E(Y) = E(X - 1) = E(X) - 1 = \frac{m + 1}{2} - 1 = \frac{m - 1}{2};$$

$$\text{Var}(Y) = \text{Var}(X - 1) = \text{Var}(X) = \frac{m^2 - 1}{12}.$$

Thus, for Example 3.4-1, the mean and variance of Y are

$$\mu_Y = \frac{10 - 1}{2} = \frac{9}{2}$$

and

$$\text{Var}(Y) = \frac{10^2 - 1}{12} = \frac{99}{12}.$$

Our next probability model uses the material in Section 2.2 on methods of enumeration. Consider a collection of $n = n_1 + n_2$ similar objects, n_1 of them belonging to one of two dichotomous classes (red chips, say) and n_2 of them belonging to the second class (blue chips, say). A collection of r objects is selected from these n objects at random and without replacement. Find the probability that exactly x (where the integer x satisfies $x \le r$, $x \le n_1$, and $r - x \le n_2$) of these r objects are red (i.e., x belong to the first class and $r - x$ belong to the second). Of course, we can select x red chips in any one of

$$\binom{n_1}{x} \text{ ways and } r - x \text{ blue chips in any one of } \binom{n_2}{r - x}$$

ways. By the multiplication principle, the product

$$\binom{n_1}{x}\binom{n_2}{r - x}$$

equals the number of ways the joint operation can be performed. If we assume that each of the $\binom{n}{r}$ ways of selecting r objects from $n = n_1 + n_2$ objects has the same probability, we have that the probability of selecting exactly x red chips is

$$P(X = x) = \frac{\binom{n_1}{x}\binom{n_2}{r - x}}{\binom{n}{r}},$$

where $x \le r$, $x \le n_1$, and $r - x \le n_2$. We say that the random variable X has a **hypergeometric distribution**.

Example 3.4-2   In a small pond there are 50 fish, 10 of which have been tagged. If a fisherman's catch consists of 7 fish, selected at random and without replacement, and X denotes the number of tagged fish, the probability that exactly 2 tagged fish are caught is

$$P(X = 2) = \frac{\binom{10}{2}\binom{40}{5}}{\binom{50}{7}} = \frac{(45)(658{,}008)}{99{,}884{,}400} = 0.296,$$

approximately.

Example 3.4-3   A lot, consisting of 100 fuses, is inspected by the following procedure. Five fuses are chosen at random and tested; if all 5 blow at the correct amperage, the lot is accepted. Suppose that the lot contains 20 defective fuses. If X is a random variable equal to the number of defective fuses in the sample of 5, the probability of accepting the lot is

$$P(X = 0) = \frac{\binom{20}{0}\binom{80}{5}}{\binom{100}{5}} = 0.32,$$

approximately. More generally, if x is one of the numbers 0, 1, 2, 3, 4, or 5, then

$$P(X = x) = \frac{\binom{20}{x}\binom{80}{5-x}}{\binom{100}{5}}, \qquad x = 0, 1, 2, 3, 4, 5.$$

Example 3.4-4   Some examples of hypergeometric probability histograms are given in Figure 3.4-1. The values of n_1, n_2, and r are given with each figure.

It should be true that $\sum_{x=0}^{r} f(x) = 1$. To show this, and also to find the mean and variance for a hypergeometric random variable, we need some mathematical tools that are given in Appendix A. For example, Theorem A-1 that is proved there is

$$\binom{n}{r} = \sum_{x=0}^{r} \binom{n_1}{x}\binom{n_2}{r-x}, \tag{3.4-1}$$

where $n = n_1 + n_2$ and it is understood that $\binom{k}{j} = 0$ if $j > k$. From this result, dividing both sides of the equation by $\binom{n}{r}$, it follows that $\sum_{x=0}^{r} f(x) = 1$.

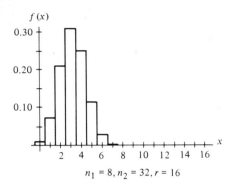

$n_1 = 8, n_2 = 32, r = 16$

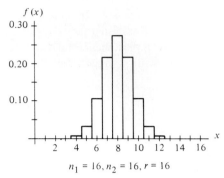

$n_1 = 16, n_2 = 16, r = 16$

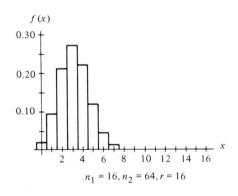

$n_1 = 16, n_2 = 64, r = 16$

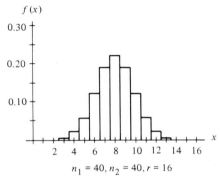

$n_1 = 40, n_2 = 40, r = 16$

FIGURE 3.4-1

It requires a little tricky calculation to show that

$$\mu = r\left(\frac{n_1}{n}\right)$$

and

$$\sigma^2 = r\left(\frac{n_1}{n}\right)\left(\frac{n_2}{n}\right)\left(\frac{n-r}{n-1}\right).$$

Hints for these calculations are given in Exercises 3.4-17 and 3.4-18.

Example 3.4-5 For Example 3.4-2, $n_1 = 10$, $n_2 = 40$, $r = 7$. The mean of X is

$$\mu = r\left(\frac{n_1}{n}\right) = 7\left(\frac{10}{50}\right) = 1.4,$$

the variance of X is

$$\sigma^2 = r\left(\frac{n_1}{n}\right)\left(\frac{n_2}{n}\right)\left(\frac{n-r}{n-1}\right)$$

$$= 7\left(\frac{10}{50}\right)\left(\frac{40}{50}\right)\left(\frac{43}{49}\right) = 0.983,$$

and the standard deviation of X is

$$\sigma = \sqrt{0.983} = 0.99.$$

Example 3.4-6   A supplier ships parts to another company in lots of 25 parts, some of which could be defective. The receiving company has an *acceptance sampling plan* which states that $r = 5$ parts are to be taken at random and without replacement from each lot. If there are zero defectives among those 5 parts, the entire lot is accepted; otherwise, the lot is rejected. That is, if one or more defectives are found among the 5 in the sample, the lot is rejected. If X is the number of defectives among the 5 sampled parts, X must have a hypergeometric distribution; but here we do not know the values of n_1, the number of defectives in the lot, and n_2, the number of good parts in the lot, but only that $n_1 + n_2 = 25$. Clearly, we want the probability of accepting the lot, namely $P(X = 0)$, to be large if n_1 is very small. This probability is usually called the *operating characteristic curve* when treated as a function of n_1 or equivalently, the fraction defective, $p = n_1/25$, in the lot. That is, the operating characteristic curve is

$$OC(p) = P(X = 0) = \frac{\binom{n_1}{0}\binom{25 - n_1}{5}}{\binom{25}{5}},$$

where $p = n_1/25$. Letting $n_1 = 0, 1, 2, \ldots$ so that $p = 0.00,\ 0.04,\ 0.08, \ldots$, we find using Table I in the Appendix that

$$OC(0.00) = \frac{\binom{0}{0}\binom{25}{5}}{\binom{25}{5}} = 1.00,$$

$$OC(0.04) = \frac{\binom{1}{0}\binom{24}{5}}{\binom{25}{5}} = 0.800,$$

$$OC(0.08) = \frac{\binom{2}{0}\binom{23}{5}}{\binom{25}{5}} = 0.633,$$

$$OC(0.12) = \frac{\binom{3}{0}\binom{22}{5}}{\binom{25}{5}} = 0.496,$$

$$OC(0.16) = \frac{\binom{4}{0}\binom{21}{5}}{\binom{25}{5}} = 0.383,$$

and so on. Upon observing these probabilities, the company might find this acceptance sampling plan unsatisfactory because possibly, with $n_1 = (25)(0.04) = 1$, $OC(0.04) = P(X = 0) = 0.8$ is too low and, with $n_1 = (25)(0.16) = 4$, $OC(0.16) = P(X = 0) = 0.383$ is too high. Accordingly, if this is so, the plan must be changed. In practice in industry, usually lot sizes are much larger than 25, sometimes running into the thousands, and the sample sizes might be in the hundreds rather than $r = 5$. Here we kept the numbers small to use Table I, but we illustrate what is actually done in practice with those larger values in Section 3.7 on the Poisson distribution.

Exercises

3.4-1 In a lottery, a three-digit integer is selected at random from 000 to 999, inclusive. Let X equal the integer that is selected on a particular day. (i) Define the p.d.f. of X and find the (ii) mean, (iii) variance, and (iv) standard deviation of X.

3.4-2 Let X equal the outcome when rolling a fair 12-sided die. Define the p.d.f. of X and find the mean, variance, and standard deviation of X.

3.4-3 Let X equal the number of the month of the year in which a person is born; that is, record 1 for January, 2 for February, and so on. Does X have the same distribution, approximately, as the outcome when rolling a fair 12-sided die in Exercise 3.4-2? Use the following birth months of 60 students of one of the authors to help answer this question.

```
 1   1  11   4   2  11   5  10   8   3   4   1  10   7   8   9   7   1   9   1
11   4   8   1   7   8  12   2   3  12   7   5   4   6   8  12   6  12   9   9
 4  10   7   1  10   9  11   8  12   7   1   4   6   5   5   6   4   5  10  10
```

In particular, find the
(a) Sample mean, $\bar{x}$.

(b) Sample variance, s^2.

(c) Sample standard deviation, s.

Compare these sample characteristics with the distribution characteristics from Exercise 3.4-2. Also,

(d) Construct a relative frequency histogram and superimpose the probability histogram.

(e) What is your conclusion using these data?

(f) Collect your own data and answer the same questions.

3.4-4 Let X have a uniform distribution on the integers from 1 to 365, inclusive.

(a) Define the p.d.f. of X.

(b) Find the mean, variance, and standard deviation of X.

(c) Let Y equal the day of the year on which a person is born. Does Y have the same distribution as X, approximately? To help answer this question, use the following data which give the day of the year on which 50 beginning math students were born.

230	283	273	47	340	8	183	27	82	174
277	51	193	39	157	280	301	243	317	148
160	206	352	163	56	139	295	28	120	193
313	145	221	55	92	268	193	319	61	98
335	126	104	360	34	69	220	4	298	120

For these data, find (i) the sample mean, (ii) the sample variance, and (iii) the sample standard deviation, comparing these sample characteristics of Y with the distribution characteristics of X. What is your conclusion with these limited data?

(d) Collect additional data and make the same comparisons.

3.4-5 Let a and b be integers with $a < b$. Let Y have a discrete uniform distribution on the integers between a and b, inclusive.

(a) Define the p.d.f. of Y.

(b) Find the mean, variance, and standard deviation of Y.

3.4-6 A bag contains 24 pieces of candy, of which 12 are peppermints and 12 are butterscotch. Let X equal the number of peppermints in a sample of five pieces of candy selected at random and without replacement from the bag. Find

(a) $P(X = 2)$.

(b) $P(X \leq 2)$.

(c) $\mu = E(X)$.

(d) $\sigma^2 = \text{Var}(X)$.

3.4-7 Say that there are 3 defective items in a lot of 50 items. A sample of size 10 is taken at random and without replacement. Let X denote the number of defective items in the sample. Find the probability that the sample contains

(a) Exactly 1 defective item.

(b) At most 1 defective item.

(c) Find the mean and variance of X.

3.4-8   A jar contains 25 pieces of candy, of which 11 are yogurt-covered nuts and 14 are yogurt-covered raisins. Let X equal the number of nuts in a random sample of 7 pieces of candy that are selected without replacement. Find
(a) $P(X = 3)$.
(b) $P(X = 6)$.
(c) The mean of X.
(d) The variance of X.

3.4-9   In a lot of 100 light bulbs, there are 5 bad bulbs. An inspector inspects 10 bulbs selected at random. Find the probability of finding at least one defective bulb.

HINT: First compute the probability of finding no defectives in the sample.

3.4-10   In LOTTO 47, Michigan's lottery game, a player selects 6 integers out of the first 47 positive integers. The state then randomly selects 6 out of the first 47 integers. Cash prizes are given to a player who matches 4, 5, or 6 integers. Let X equal the number of integers selected by a player that match integers selected by the state.
(a) Define the p.d.f. of X.
(b) Calculate the mean, variance, and standard deviation of X.
(c) How many standard deviations above the mean are 4, 5, and 6, the numbers of matches that yield prizes?
(d) What values of X are most likely to occur?
(e) On February 2, 1991, Michigan's third largest prize was offered, so there were a lot of bets placed. Of the 10,200,000 bets that were placed, 2 people matched all 6 numbers with each winning $12,034,747 (most of this paid by losers during preceding games), 206 matched 5 numbers to win $2500, and 10,460 matched 4 numbers to win $100. Are these numbers of winners consistent with the probability model? Do you think that players select their 6 numbers randomly?

3.4-11   In Florida's lottery game, a player selects 6 integers out of the first 49 positive integers. The state then randomly selects 6 out of the first 49 integers. Cash prizes are given to a player who matches 3, 4, 5, or 6 integers. Let X equal the number of integers selected by a player that match integers selected by the state.
(a) Define the p.d.f. of X.
(b) Calculate the mean, variance, and standard deviation of X.
(c) How many standard deviations above the mean are 3, 4, 5, and 6, the numbers of matches that yield prizes?
(d) What value of X is most likely to occur?
(e) On September 15, 1990, the second largest jackpot in the United States caused Lotto fever and approximately 105,000,000 tickets were purchased for $1 each. Among these tickets, 1,829,677 matched 3 numbers and won $5.50; 96,036 matched 4 numbers and won $108; 1,581 matched 5 numbers and won $4488; while 6 matched all 6 numbers and won

$17,500,000. (Much of the prize money was provided by losing tickets during previous weeks.) Are these values consistent with the probability model? (Lotto officials reported that the most popular combination of numbers selected by the players was 1, 2, 3, 4, 5, 6.)

3.4-12 In Michigan's lottery game, KENO, the player picks 10 integers between 1 and 80, inclusive. The state selects 22 integers randomly from 1 to 80, inclusive. Let X equal the number of integers selected by the player that match integers selected by the state. If $X = 10, 9, 8, 7, 6$, the player wins $250,000, $2500, $250, $25, and $7, respectively. If $X = 0$, the player receives a free $1 Instant Ticket.
(a) Define the p.d.f. of X.
(b) Calculate the mean, variance, and standard deviation of X.
(c) How many standard deviations above the mean are 6, 7, 8, 9, and 10, the numbers of matches that yield cash prizes?
(d) How many standard deviations below the mean is 0, the number of matches for which a $1 ticket is the prize?
(e) What is the expected value of this game to the player, assuming that the prize for 0 matches is worth $1 and the cost for one ticket is $1?
(f) Of course, in part (e), a $1 ticket is not really worth $1, but rather the expected value of the game, say E. Hence we have the equation

$$E = (E)P(X = 0) + (7)P(X = 6) + (25)P(X = 7) + (250)P(X = 8)$$
$$+ (2500)P(X = 9) + (250,000)P(X = 10).$$

Solve for E.

3.4-13 A bridge hand is defined as 13 cards selected at random without replacement from a deck of 52 playing cards.
(a) What is the probability of being dealt at random a bridge hand that does not contain a spade?
(b) What is the probability that a bridge hand contains exactly five hearts?
(c) Find the probability that a bridge hand contains at most one ace.

3.4-14 For the distributions depicted in Figure 3.4-1, find μ, σ^2, and σ when
(a) $n_1 = 8$, $n_2 = 32$, and $r = 16$.
(b) $n_1 = 16$, $n_2 = 16$, and $r = 16$.
(c) $n_1 = 16$, $n_2 = 64$, and $r = 16$.
(d) $n_1 = 40$, $n_2 = 40$, and $r = 16$.

3.4-15 A lot of $n_1 + n_2 = 25$ items is accepted if the number of defectives, X, among $r = 5$ items taken at random and without replacement from the lot is less than or equal to 1. The operating characteristic curve is

$$OC(p) = P(X \le 1) = P(X = 0) + P(X = 1),$$

where $p = n_1/25$. Determine the probabilities OC(0.04), OC(0.08), OC(0.12), and OC(0.16) and compare them to those in Example 3.4-6.

3.4-16 A lot of $n_1 + n_2 = 100$ items is accepted if the number of defectives, X, among $r = 10$ items taken at random and without replacement from the lot is equal to zero. The operating characteristic curve is $OC(p) = P(X = 0)$, where $p = n_1/100$. Evaluate $OC(p)$ where $p = 0.00, 0.01, 0.02$, and 0.05. Here Table I cannot be used, but these probabilities are not difficult to evaluate with a small calculator.

3.4-17 To find the mean of a hypergeometric random variable,
   (a) First show that
$$\binom{n}{r} = \frac{n}{r}\binom{n-1}{r-1}.$$

   (b) In the summation for the mean, make the change of variables $k = x - 1$ and replace the denominator using the expression in part (a).
   (c) In the resulting expression, use equation (3.4-1) with n and r replaced with $n - 1$ and $r - 1$, respectively, to show that
$$\mu = r\left(\frac{n_1}{n}\right).$$

3.4-18 To find the variance of a hypergeometric random variable,
   (a) First show that
$$E[X(X - 1)] = \frac{n_1(n_1 - 1)(r)(r - 1)}{n(n - 1)},$$

   making a change of variables $k = x - 2$ and noting that
$$\binom{n}{r} = \frac{n(n-1)}{r(r-1)}\binom{n-2}{r-2}.$$

   (b) Now show that
$$\sigma^2 = r\left(\frac{n_1}{n}\right)\left(\frac{n_2}{n}\right)\left(\frac{n-r}{n-1}\right).$$

3.5 Bernoulli Trials and the Binomial Distribution

The probability models for random experiments that will be described in this section occur frequently in applications.

A **Bernoulli experiment** is a random experiment, the outcome of which can be classified in but one of two mutually exclusive and exhaustive ways, say, success or failure (e.g., female or male, life or death, nondefective or defective). A sequence of **Bernoulli trials** occurs when a Bernoulli experiment is performed several *independent* times so that the probability of success, say p, remains the *same* from trial to trial. That is, in such a sequence we let p denote the probability of success on each trial. In addition, we shall frequently let $q = 1 - p$ denote the probability of failure; that is, we shall use q and $1 - p$ interchangeably.

> **Example 3.5-1**   Suppose that the probability of germination of a beet seed is 0.8 and the germination of a seed is called a success. If 10 seeds are planted and independence can be assumed, this corresponds to 10 Bernoulli trials with $p = 0.8$.

> **Example 3.5-2**   In the Michigan daily lottery the probability of winning when placing a six-way boxed bet is 0.006. A bet placed on each of 12 successive days would correspond to 12 Bernoulli trials with $p = 0.006$.

Let X be a random variable associated with a Bernoulli trial by defining it as follows:

$$X(\text{success}) = 1 \quad \text{and} \quad X(\text{failure}) = 0.$$

That is, the two outcomes, success and failure, are denoted by one and zero, respectively. The p.d.f. of X can be written as

$$f(x) = p^x(1 - p)^{1-x}, \quad x = 0, 1,$$

and we say that X has a **Bernoulli distribution**. The expected value of X is

$$\mu = E(X) = \sum_{x=0}^{1} xp^x(1 - p)^{1-x} = (0)(1 - p) + (1)(p) = p,$$

and the variance of X is

$$\sigma^2 = \text{Var}(X) = \sum_{x=0}^{1} (x - p)^2 p^x(1 - p)^{1-x}$$
$$= p^2(1 - p) + (1 - p)^2 p = p(1 - p) = pq.$$

It follows that the standard deviation of X is

$$\sigma = \sqrt{p(1 - p)} = \sqrt{pq}.$$

In a sequence of n Bernoulli trials, we shall let X_i denote the Bernoulli random variable associated with the ith trial. An observed sequence of n Bernoulli trials will then be an n-tuple of zeros and ones.

> **Example 3.5-3**   Out of millions of instant lottery tickets, suppose that 20% are winners. If five such tickets are purchased, a possible observed sequence is (0, 0, 0, 1, 0), in which the fourth ticket is a winner and the other four are losers. Assuming independence among winning and losing tickets, the probability of this outcome is

> $$(0.8)(0.8)(0.8)(0.2)(0.8) = (0.2)(0.8)^4.$$

Example 3.5-4   If five beet seeds are planted in a row, a possible observed sequence would be (1, 0, 1, 0, 1) in which the first, third, and fifth seeds germinated and the other two did not. If the probability of germination is $p = 0.8$, the probability of this outcome is, assuming independence,

$$(0.8)(0.2)(0.8)(0.2)(0.8) = (0.8)^3(0.2)^2.$$

In a sequence of Bernoulli trials, we are often interested in the total number of successes and not in the order of their occurrence. If we let the random variable X equal the number of observed successes in n Bernoulli trials, the possible values of X are $0, 1, 2, \ldots, n$. If x successes occur, where $x = 0, 1, 2, \ldots, n$, then $n - x$ failures occur. The number of ways of selecting x positions for the x successes in the n trials is

$$\binom{n}{x} = \frac{n!}{x!(n - x)!}.$$

Since the trials are independent and since the probabilities of success and failure on each trial are, respectively, p and $q = 1 - p$, the probability of each of these ways is $p^x(1 - p)^{n-x}$. Thus the p.d.f. of X, say $f(x)$, is the sum of the probabilities of these $\binom{n}{x}$ mutually exclusive events; that is,

$$f(x) = \binom{n}{x}p^x(1 - p)^{n-x}, \qquad x = 0, 1, 2, \ldots, n.$$

These probabilities are called binomial probabilities, and the random variable X is said to have a **binomial distribution**.

Summarizing, a binomial experiment satisfies the following properties:

1. A Bernoulli (success–failure) experiment is performed n times.
2. The trials are independent.
3. The probability of success on each trial is a constant p; the probability of failure is $q = 1 - p$.
4. The random variable X counts the number of successes in the n trials.

A binomial distribution will be denoted by the symbol $b(n, p)$ and we say that the distribution of X is $b(n, p)$. The constants n and p are called the **parameters** of the binomial distribution; they correspond to the number n of independent trials and the probability p of success on each trial. Thus, if we say that the distribution of X is $b(12, 1/4)$, we mean that X is the number of successes in $n = 12$ Bernoulli trials with probability $p = 1/4$ of success on each trial.

Example 3.5-5   In the instant lottery with 20% winning tickets, if X is equal to the number of winning tickets among $n = 8$ that are purchased, the probability of purchasing 2 winning tickets is

$$f(2) = P(X = 2) = \binom{8}{2}(0.2)^2(0.8)^6 = 0.2936.$$

The distribution of the random variable X is $b(8, 0.2)$.

Example 3.5-6 In Example 3.5-1, the number X of seeds that germinate in $n = 10$ independent trials is $b(10, 0.8)$; that is,

$$f(x) = \binom{10}{x}(0.8)^x(0.2)^{10-x}, \qquad x = 0, 1, 2, \ldots, 10.$$

In particular,

$$P(X \le 8) = 1 - P(X = 9) - P(X = 10)$$
$$= 1 - 10(0.8)^9(0.2) - (0.8)^{10} = 0.6242.$$

Example 3.5-7 In order to obtain a better feeling for the effect of the parameters n and p on the distribution of probabilities, four probability histograms are displayed in Figure 3.5-1.

Values of the distribution function of a random variable X that is $b(n, p)$ are given in Table II in the Appendix for selected values of n and p. The use of this table is illustrated in the next example.

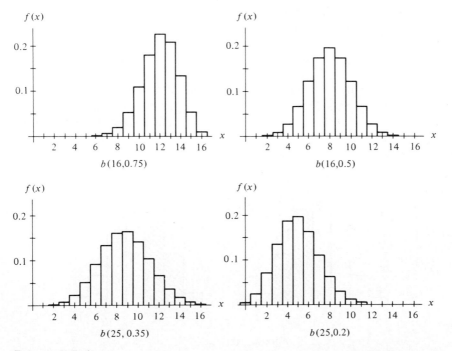

FIGURE 3.5-1

Example 3.5-8   Leghorn chickens are raised for laying eggs. If $p = 0.5$ is the probability of a female chick hatching, assuming independence, the probability that there are exactly 6 females out of 10 newly hatched chicks selected at random is

$$\binom{10}{6}\left(\frac{1}{2}\right)^6\left(\frac{1}{2}\right)^4 = P(X \le 6) - P(X \le 5)$$

$$= 0.8281 - 0.6230 = 0.2051,$$

since $P(X \le 6) = 0.8281$ and $P(X \le 5) = 0.6230$ from Table II in the Appendix. The probability of at least 6 female chicks is

$$\sum_{x=6}^{10} \binom{10}{x}\left(\frac{1}{2}\right)^x\left(\frac{1}{2}\right)^{10-x} = 1 - P(X \le 5) = 1 - 0.6230 = 0.3770.$$

Although probabilities for the binomial distribution $b(n, p)$ are given in Table II in the Appendix for selected p values less than or equal to 0.5, the next example demonstrates that this table can also be used for p values greater than 0.5. In later sections we learn how to approximate certain binomial probabilities with those of other distributions.

Example 3.5-9   Suppose that we are in one of those rare times when 65% of the American public approve of the way the President of the United States is handling his job. Take a random sample of $n = 8$ Americans and let Y equal the number who give approval. Then the distribution of Y is $b(8, 0.65)$. To find $P(Y \ge 6)$, note that

$$P(Y \ge 6) = P(8 - Y \le 8 - 6) = P(X \le 2),$$

where $X = 8 - Y$ counts the number who disapprove. Since $q = 1 - p = 0.35$ equals the probability of disapproval by each person selected, the distribution of X is $b(8, 0.35)$ (see Figure 3.5-2). From Table II in the Appendix, since $P(X \le 2) = 0.4278$, it follows that $P(Y \ge 6) = 0.4278$.

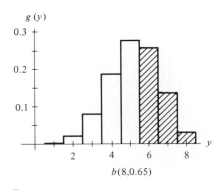

$b(8,0.65)$

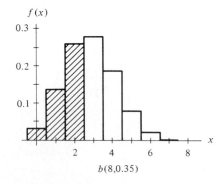

$b(8,0.35)$

Figure 3.5-2

Similarly,

$$P(Y \le 5) = P(8 - Y \ge 8 - 5)$$
$$= P(X \ge 3) = 1 - P(X \le 2)$$
$$= 1 - 0.4278 = 0.5722$$

and

$$P(Y = 5) = P(8 - Y = 8 - 5)$$
$$= P(X = 3) = P(X \le 3) - P(X \le 2)$$
$$= 0.7064 - 0.4278 = 0.2786.$$

Recall that if n is a positive integer, then

$$(a + b)^n = \sum_{x=0}^{n} \binom{n}{x} b^x a^{n-x}.$$

Thus the sum of the binomial probabilities, if we use the above binomial expansion with $b = p$ and $a = 1 - p$, is

$$\sum_{x=0}^{n} \binom{n}{x} p^x (1 - p)^{n-x} = [(1 - p) + p]^n = 1,$$

a result that had to follow from the fact that $f(x)$ is a p.d.f.

We use the binomial expansion to find the mean and the variance of the binomial random variable X that is $b(n, p)$. The mean is given by

$$\mu = E(X) = \sum_{x=0}^{n} x \frac{n!}{x!(n - x)!} p^x (1 - p)^{n-x}.$$

Since the first term of this sum is equal to zero, this can be written as

$$\mu = \sum_{x=1}^{n} \frac{n!}{(x - 1)!(n - x)!} p^x (1 - p)^{n-x}$$

because $x/x! = 1/(x - 1)!$ when $x > 0$.

If we let $k = x - 1$ in the latter sum, we obtain

$$\mu = \sum_{k=0}^{n-1} \frac{n!}{k!(n - k - 1)!} p^{k+1} (1 - p)^{n-k-1}$$

$$= np \sum_{k=0}^{n-1} \frac{(n - 1)!}{k!(n - 1 - k)!} p^k (1 - p)^{n-1-k} = np$$

because the summand in this expression is that of the binomial p.d.f. $b(n - 1, p)$.

To find the variance, we first determine the second factorial moment $E[X(X - 1)]$:

$$E[X(X - 1)] = \sum_{x=0}^{n} x(x - 1) \frac{n!}{x!(n - x)!} p^x (1 - p)^{n-x}.$$

The first two terms in this summation equal zero; thus we find that

$$E[X(X-1)] = \sum_{x=2}^{n} \frac{n!}{(x-2)!(n-x)!} p^x (1-p)^{n-x}$$

after observing that $x(x-1)/x! = 1/(x-2)!$ when $x > 1$. Letting $k = x - 2$, we obtain

$$E[X(X-1)] = \sum_{k=0}^{n-2} \frac{n!}{k!(n-k-2)!} p^{k+2}(1-p)^{n-k-2}$$

$$= n(n-1)p^2 \sum_{k=0}^{n-2} \frac{(n-2)!}{k!(n-2-k)!} p^k (1-p)^{n-2-k}.$$

Since the last summand is that of the binomial p.d.f. $b(n-2, p)$, we obtain

$$E[X(X-1)] = n(n-1)p^2.$$

Thus

$$\sigma^2 = \text{Var}(X) = E(X^2) - [E(X)]^2 = E[X(X-1)] + E(X) - [E(X)]^2$$
$$= n(n-1)p^2 + np - (np)^2 = -np^2 + np = np(1-p).$$

Summarizing, if X is $b(n, p)$, we obtain

$$\mu = np, \qquad \sigma^2 = np(1-p) = npq, \qquad \text{and} \qquad \sigma = \sqrt{np(1-p)}.$$

Note that when p is the probability of success on each trial, the expected number of successes in n trials is np, a result that agrees with most of our intuitions.

Example 3.5-10 Suppose that observation over a long period of time has disclosed that, on the average, one out of 10 items produced by a process is defective. Select five items independently from the production line and test them. Let X denote the number of defective items among the $n = 5$ items. Then X is $b(5, 0.1)$. Furthermore,

$$E(X) = 5(0.1) = 0.5, \qquad \text{Var}(X) = 5(0.1)(0.9) = 0.45.$$

For example, the probability of observing at most one defective item is

$$P(X \le 1) = \binom{5}{0}(0.1)^0(0.9)^5 + \binom{5}{1}(0.1)^1(0.9)^4 = 0.9185.$$

The next example shows the relationship between the binomial probability model and a set of observed data. Repetitions of this example would of course yield different results, in which the fits could be better or worse than the given one.

Example 3.5-11 Consider the simple experiment of flipping a fair coin five independent times. If X equals the number of heads that are observed,

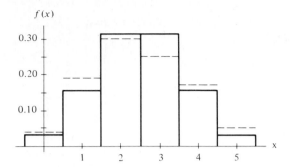

FIGURE 3.5-3

then X is $b(5, 0.5)$, and $\mu = 2.5$, $\sigma^2 = 1.25$, and $\sigma = 1.118$. This experiment was simulated 100 times, yielding the following data:

```
2 3 2 4 1 2 1 1 4 2 4 2 0 4 4 2 4 4 3 4
2 2 4 4 1 1 3 3 1 4 2 3 1 2 4 1 2 5 3 2
4 3 2 2 2 3 5 2 0 3 2 1 3 4 2 2 4 0 2 1
3 3 2 3 2 1 3 2 2 2 1 1 3 3 1 1 4 2 1 5
3 2 3 0 3 5 3 2 4 3 3 5 2 3 3 1 3 2 1 1
```

For these data $\bar{x} = 2.47$, $s^2 = 1.5243$, and $s = 1.235$. In Figure 3.5-3 the probability histogram and the relative frequency histogram (as dashed lines) are given.

Suppose that an urn contains n_1 success balls and n_2 failure balls and we let $p = n_1/(n_1 + n_2)$. Let X equal the number of success balls in a random sample of size r that is taken from this urn. If the sampling is done one at a time with replacement, the distribution of X is $b(r, p)$; and if the sampling is done without replacement, X has a hypergeometric distribution with p.d.f.

$$f(x) = \frac{\binom{n_1}{x}\binom{n_2}{r-x}}{\binom{n_1 + n_2}{r}}, \qquad x = 0, 1, \ldots, n_1.$$

When $n_1 + n_2$ is large and r is relatively small, it makes little difference if the sampling is done with or without replacement. In Figure 3.5-4 the probability histograms are compared for different combinations of r, n_1, and n_2. You are asked to compute some of these probabilities in Exercise 3.5-19.

Exercises

3.5-1 An urn contains 7 red and 11 white balls. Draw one ball at random from the urn. Let $X = 1$ if a red ball is drawn, and let $X = 0$ if a white ball is drawn. Give the p.d.f., mean, and variance of X.

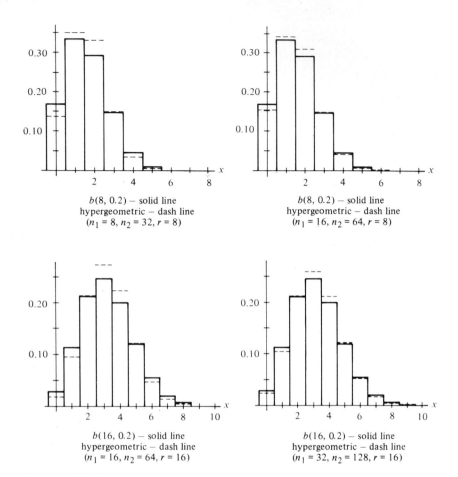

$b(8, 0.2)$ — solid line
hypergeometric — dash line
$(n_1 = 8, n_2 = 32, r = 8)$

$b(8, 0.2)$ — solid line
hypergeometric — dash line
$(n_1 = 16, n_2 = 64, r = 8)$

$b(16, 0.2)$ — solid line
hypergeometric — dash line
$(n_1 = 16, n_2 = 64, r = 16)$

$b(16, 0.2)$ — solid line
hypergeometric — dash line
$(n_1 = 32, n_2 = 128, r = 16)$

Figure 3.5-4

3.5-2 Suppose that in Exercise 3.5-1, $X = 1$ if a red ball is drawn, and $X = -1$ if a white ball is drawn. Give the p.d.f., mean, and variance of X.

3.5-3 On a five-question multiple-choice test there are five possible answers, of which one is correct (C) and four are incorrect (I). If a student guesses randomly and independently, find the probability of
(a) Being correct only on questions 1 and 4 (i.e., scoring C, I, I, C, I).
(b) Being correct on two questions.

3.5-4 Suppose that 40% of American homes have a microwave oven. Let X equal the number of American homes in a random sample of $n = 25$ homes that have a microwave oven. Find the probability that
(a) X is at most 11.
(b) X is at least 7.

(c) X is equal to 8.

(d) Give the mean, variance, and standard deviation of X.

3.5-5 Suppose that 70% of all residents of the Wellsburg, W. Va., area are at risk for heart disease [*Time,* December 12, 1988]. Let X equal the number out of $n = 16$ of these residents who are selected randomly that are at risk for heart disease. Find the probability that

(a) X is at least 13.

(b) X is at most 11.

(c) X is equal to 12.

(d) Give the mean, variance, and standard deviation of X.

3.5-6 It is claimed that 15% of the ducks in a particular region have patent schistosome infection. Suppose that seven ducks are selected at random. Let X equal the number of ducks that are infected.

(a) Assuming independence, how is X distributed?

(b) Find (i) $P(X \geq 2)$, (ii) $P(X = 1)$, and (iii) $P(X \leq 3)$.

3.5-7 Suppose that 2000 points are selected independently and at random from the unit square $S = \{(x, y): 0 \leq x < 1, 0 \leq y < 1\}$. Let W equal the number of points that fall in $A = \{(x, y): x^2 + y^2 < 1\}$.

(a) How is W distributed?

(b) Give the mean, variance, and standard deviation of W.

(c) What is the expected value of $W/500$?

(d) By selecting n pairs of random numbers from Table IX in the Appendix (or using a random number generator on a computer), generate n observations of W and use these to give an estimate for π. (More will be said about estimation later on in this text.)

(e) By using n triplets of random numbers, could you extend part (d) to estimate the volume of a ball of radius 1 in 3-space, namely $(4/3)\pi$?

(f) If you use a computer, could you extend these techniques to estimate the ''volume'' of a ball of radius 1 in n-space?

3.5-8 It is believed that 20% of Americans do not have any health insurance. Suppose this is true and let X equal the number with no health insurance in a random sample of $n = 15$ Americans.

(a) How is X distributed?

(b) Give the mean, variance, and standard deviation of X.

(c) Find $P(X \geq 5)$.

3.5-9 Let p equal the proportion of all college and university students who would say *yes* to the question, ''Would you drink from the same glass as your friend if you suspected that this friend were an AIDS virus carrier?'' Assume that $p = 0.10$. Let X equal the number of students out of a random sample of size $n = 9$ who would say *yes* to this question.

(a) How is X distributed?

(b) Give the values of the mean, variance, and standard deviation of X.

(c) Find (i) $P(X = 2)$ and (ii) $P(X \geq 2)$.

3.5-10 Suppose that, in manufacturing, the probability of a certain item being defective is $p = 0.05$. Suppose further that the quality of an item is independent of the quality of the other manufactured items. An inspector selects six items at random. Let X equal the number of defective items in the sample.
(a) How is X distributed?
(b) Give the values of $E(X)$ and $\text{Var}(X)$.
(c) Find (i) $P(X = 0)$, (ii) $P(X \le 1)$, and (iii) $P(X \ge 2)$.

3.5-11 According to a *Des Moines Register* poll, 40% of Iowa farmers support an independent Palestinian state. Let X equal the number of Iowa farmers out of a random sample of $n = 20$ who support an independent Palestinian state. Find
(a) $P(X \le 10)$.
(b) $P(X \ge 8)$.
(c) $P(X = 9)$.

3.5-12 One of the 12 courses in a Chinese vegetarian banquet at a Buddhist temple was a platter of mushrooms. When attempting to pick up a round, slippery mushroom with chopsticks, an American tourist was successful 55% of the time on the first try. Let X equal the number of times that the tourist was successful on the first try at picking up 14 mushrooms. Assume that the "first tries" are independent trials.
(a) Give the mean and variance of X.
(b) Find $P(X < 8)$ and $P(X > 6)$.

3.5-13 A random variable X has a binomial distribution with mean 6 and variance 3.6. Find $P(X = 4)$.

3.5-14 A certain type of mint has a label weight of 20.4 grams. Suppose that the probability is 0.90 that a mint weighs more than 20.7 grams. Let X equal the number of mints that weigh more than 20.7 grams in a sample of 8 mints selected at random.
(a) How is X distributed?
(b) Find (i) $P(X = 8)$, (ii) $P(X \le 6)$, and (iii) $P(X \ge 6)$.

3.5-15 It is given that the probability of germination of each beet seed is 0.8. Thus if X denotes the number of seeds that germinate in a set of nine seeds, then X is $b(9, 0.8)$, provided that the assumption of independence is valid.
(a) Give the values of μ, σ^2, and σ.
(b) Draw the probability histogram for the p.d.f. of X.
(c) If 100 sets of nine beet seeds were planted, calculate $\bar{x}$, s^2, and s for the following 100 observations.

7	9	7	7	9	7	9	8	5	8	8	6	8	7	8	8	9	7	8	8
8	7	8	7	7	7	6	7	9	7	6	7	6	7	6	5	7	8	7	4
8	8	7	6	9	7	8	7	7	6	7	9	6	6	8	8	8	7	9	5
7	4	7	8	8	9	6	7	6	8	7	5	7	7	9	7	7	8	7	9
7	5	9	8	7	7	8	9	8	5	9	9	8	6	9	8	8	8	9	7

(d) Superimpose the relative frequency histogram of these observations on the graph of the probability histogram.

3.5-16 In the casino game chuck-a-luck, three unbiased six-sided dice are rolled. One possible bet is $1 on fives, and the payoff is equal to $1 for each five on that roll. In addition, the dollar bet is returned if at least one five is rolled. The dollar that was bet is lost only if no fives are rolled. Let X denote the payoff for this game. Then X can equal -1, 1, 2, or 3.

(a) Define the p.d.f. $f(x)$.

(b) Calculate μ, σ^2, and σ.

(c) Depict the p.d.f. as the probability histogram.

(d) Use the 100 following observations of X to calculate $\bar{x}$, s^2, and s:

-1	1	-1	-1	1	-1	-1	-1	-1	1
-1	1	-1	2	1	-1	-1	-1	1	-1
1	-1	-1	1	1	-1	-1	-1	-1	1
-1	-1	2	1	1	-1	-1	1	-1	1
-1	-1	1	2	-1	-1	1	-1	-1	-1
-1	1	-1	-1	-1	1	-1	1	-1	2
-1	1	-1	1	-1	2	1	-1	-1	1
-1	1	1	-1	1	1	-1	3	1	-1
-1	-1	1	1	1	1	-1	-1	2	-1
1	-1	1	-1	1	2	-1	1	1	-1

(e) Superimpose the relative frequency histogram of these observations on the probability histogram.

3.5-17 It is claimed that for a particular lottery, 1/10 of the 50,000,000 tickets will win a prize. What is the probability of winning at least one prize if you purchase

(a) 10 tickets?

(b) 15 tickets?

3.5-18 For the lottery described in Exercise 3.5-17, find the smallest number of tickets that must be purchased so that the probability of winning at least one prize is greater than

(a) 0.50.

(b) 0.95.

3.5-19 Construct a table that gives the probabilities for the $b(8, 0.2)$ distribution, the hypergeometric distribution with $n_1 = 8$, $n_2 = 32$, and $r = 8$, and the hypergeometric distribution with $n_1 = 16$, $n_2 = 64$, and $r = 8$ (see Figure 3.5-4).

3.6 Geometric and Negative Binomial Distributions

To obtain a binomial random variable, we observed a sequence of n Bernoulli trials and counted the number of successes. Suppose now that we do not fix the number of Bernoulli trials in advance but instead continue to observe the sequence of Bernoulli trials until a certain number, say r, of successes occurs. The random variable of interest is the number of trials needed to observe the rth success.

We first discuss this problem when $r = 1$. That is, consider a sequence of Bernoulli trials with probability p of success. This sequence is observed until the first success occurs. Let X denote the trial number on which this first success occurs. For example, if F and S represent failure and success, respectively, and the sequence starts with F, F, F, S, . . . , then $X = 4$. Moreover, because the trials are independent, the probability of such a sequence is

$$P(X = 4) = (q)(q)(q)(p) = q^3p = (1 - p)^3p.$$

In general, the p.d.f., $f(x) = P(X = x)$, of X is given by

$$f(x) = (1 - p)^{x-1}p, \qquad x = 1, 2, \ldots$$

because there must be $x - 1$ failures before the first success that occurs on trial x. We say that X has a **geometric distribution**.

Recall that for a geometric series (see Appendix A for a review), the sum is given by

$$\sum_{k=0}^{\infty} ar^k = \sum_{k=1}^{\infty} ar^{k-1} = \frac{a}{1 - r}$$

when $|r| < 1$. Thus

$$\sum_{x=1}^{\infty} f(x) = \sum_{x=1}^{\infty} (1 - p)^{x-1}p = \frac{p}{1 - (1 - p)} = 1$$

so that $f(x)$ does satisfy the properties of a p.d.f.

From the sum of a geometric series we also note that, when k is an integer,

$$P(X > k) = \sum_{x=k+1}^{\infty} (1 - p)^{x-1}p = \frac{(1 - p)^k p}{1 - (1 - p)} = (1 - p)^k = q^k,$$

and thus the value of the distribution function at a positive integer k is

$$P(X \le k) = \sum_{x=1}^{k} (1 - p)^{x-1}p = 1 - P(X > k) = 1 - (1 - p)^k = 1 - q^k.$$

Example 3.6-1 Some biology students were checking the eye color for a large number of fruit flies. For the individual fly, suppose that the probability of white eyes is 1/4 and the probability of red eyes is 3/4, and that we may treat these flies as independent Bernoulli trials. The probability that at least four flies have to be checked for eye color to observe a white-eyed fly is given by

$$P(X \ge 4) = P(X > 3) = q^3 = \left(\frac{3}{4}\right)^3 = 0.422.$$

The probability that at most four flies have to be checked for eye color to observe a white-eyed fly is given by

$$P(X \le 4) = 1 - q^4 = 1 - \left(\frac{3}{4}\right)^4 = 0.684.$$

The probability that the first fly with white eyes is the fourth fly that is checked is

$$P(X = 4) = q^{4-1}p = \left(\frac{3}{4}\right)^3 \left(\frac{1}{4}\right) = 0.105.$$

It is also true that

$$\begin{aligned}
P(X = 4) &= P(X \le 4) - P(X \le 3) \\
&= [1 - (3/4)^4] - [1 - (3/4)^3] \\
&= \left(\frac{3}{4}\right)^3 \left(\frac{1}{4}\right).
\end{aligned}$$

In general,

$$f(x) = P(X = x) = \left(\frac{3}{4}\right)^{x-1} \left(\frac{1}{4}\right), \qquad x = 1, 2, 3, \ldots .$$

A probability histogram is shown in Figure 3.6-1 with $r = 1$, $p = 0.25$.

To find the mean and the variance for the geometric distribution, we will use the following results about the sum and the first and second derivatives of a geometric series. For $-1 < r < 1$, let

$$g(r) = \sum_{k=0}^{\infty} ar^k = \frac{a}{1 - r}. \tag{3.6-1}$$

Then

$$g'(r) = \sum_{k=1}^{\infty} akr^{k-1} = \frac{a}{(1 - r)^2}, \tag{3.6-2}$$

and

$$g''(r) = \sum_{k=2}^{\infty} ak(k - 1)r^{k-2} = \frac{2a}{(1 - r)^3}. \tag{3.6-3}$$

If X has a geometric distribution and $0 < p < 1$, then the mean of X is given by

$$E(X) = \sum_{x=1}^{\infty} xq^{x-1}p = \frac{p}{(1 - q)^2} = \frac{1}{p}$$

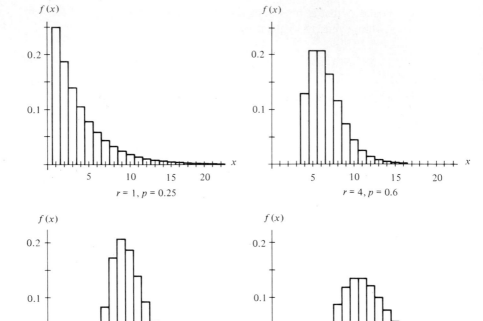

FIGURE 3.6-1

using formula (3.6-2) with $a = p$ and $r = q$. Note, for example, that if $p = 1/4$ is the probability of success, then $E(X) = 1/(1/4) = 4$ trials are needed on the average to observe a success.

To find the variance of X, we first find the second factorial moment $E[X(X - 1)]$. We have

$$E[X(X - 1)] = \sum_{x=1}^{\infty} x(x - 1)q^{x-1}p = \sum_{x=2}^{\infty} pqx(x - 1)q^{x-2} = \frac{2pq}{(1 - q)^3} = \frac{2q}{p^2}$$

using formula (3.6-3) with $a = pq$ and $r = q$. Thus the variance of X is

$$\mathrm{Var}(X) = E(X^2) - [E(X)]^2 = \{E[X(X - 1)] + E(X)\} - [E(X)]^2$$

$$= \frac{2q}{p^2} + \frac{1}{p} - \frac{1}{p^2}$$

$$= \frac{2q + p - 1}{p^2} = \frac{q}{p^2} = \frac{1 - p}{p^2}.$$

The standard deviation of X is

$$\sigma = \sqrt{(1 - p)/p^2}.$$

Example 3.6-2 Continuing with Example 3.6-1, with $p = 1/4$, we obtain

$$\mu = \frac{1}{1/4} = 4,$$

$$\sigma^2 = \frac{3/4}{(1/4)^2} = 12,$$

and

$$\sigma = \sqrt{12} = 3.464.$$

Locate μ, $\mu \pm \sigma$, and $\mu \pm 2\sigma$ on the graph in Figure 3.6-1 ($r = 1$, $p = 0.25$). Note that

$$P(X \leq 4) = 0.578.$$

The probability that X is more than one standard deviation below the mean is

$$P(X < 4 - 3.464) = P(X \leq 0) = 0.$$

However the probability that X is more than one standard deviation above the mean is

$$P(X > 4 + 3.464) = P(X > 7) = (0.75)^7 = 0.1335.$$

Example 3.6-3 Let X equal the number of people that you have to ask, selecting randomly, to find someone who was born in the same month as you were. Assuming that each month is equally likely, the p.d.f. of X is $f(x) = (11/12)^{x-1}(1/12)$, $x = 1, 2, 3, \ldots$. The mean, variance, and standard deviation of X are

$$\mu = \frac{1}{1/12} = 12, \qquad \sigma^2 = \frac{11/12}{(1/12)^2} = (11)(12) = 132, \qquad \text{and} \qquad \sigma = 11.489.$$

Also,

$$P(X > 23) = (11/12)^{23} = 0.135,$$
$$P(X < 10) = 1 - (11/12)^9 = 0.543.$$

Consider the problem of obtaining a random ordering of the first n positive integers. For example, suppose that we would like a random ordering (permutation) of the first six positive integers. We could obtain this random ordering by rolling a fair six-sided die. The first cast of the die would give the first outcome in the random ordering. To obtain the second number in the ordering, only five of the six possible outcomes are eligible. After the first $k - 1$ positions have been filled with unique integers, the number of candidates for position k is $6 - k + 1$ for $k = 1, 2, 3, 4, 5, 6$. The probability of selecting one of these eligible integers is $p_k = (6 - k + 1)/6$. If X_k denotes the number of trials (casts of the die) needed to observe the first success (an integer that has not yet been selected), then X_k has a geometric distribution with $p = p_k$.

An interesting problem is to determine the average number of casts of the die that are needed to obtain a random ordering of 1, 2, 3, 4, 5, 6. Let X_k equal the number of casts required to fill position k. If we let $W = X_1 + X_2 + \cdots + X_6$, then W denotes the total number of casts required. So the average number of casts required is

$$
\begin{aligned}
E(W) &= E(X_1 + X_2 + \cdots + X_6) \\
&= E(X_1) + E(X_2) + \cdots + E(X_6) \\
&= 1 + \frac{1}{5/6} + \frac{1}{4/6} + \cdots + \frac{1}{1/6} = 14.7.
\end{aligned}
$$

REMARK The fact that $E(X_1 + X_2 + \cdots + X_6) = E(X_1) + E(X_2) + \cdots + E(\dot{X}_6)$ is considered in Section 5.2.

Example 3.6-4 A computer science student was interested in simulating on the computer the shuffling of a deck of 52 playing cards. Essentially, this problem requires the random ordering of the first 52 positive integers. A random number generator can be used to obtain this ordering. If we think of this random number generator as a 52-sided die, what is the average or expected number of times that this "die" would have to be cast to obtain the random ordering? Letting W denote the total number of casts, we obtain

$$
E(W) = \sum_{k=1}^{52} \frac{1}{(52 - k + 1)/52} = \sum_{k=1}^{52} \frac{52}{53 - k} = 235.978.
$$

We turn now to the more general problem of observing a sequence of Bernoulli trials until exactly r successes occur, where r is a fixed positive integer. Let the random variable X denote the number of trials needed to observe the rth success. That is, X is the trial number on which the rth success is observed. By the multiplication rule of probabilities, the p.d.f. of X, say $g(x)$, equals the product of the probability

$$
\binom{x - 1}{r - 1} p^{r-1}(1 - p)^{x-r} = \binom{x - 1}{r - 1} p^{r-1} q^{x-r}
$$

of obtaining exactly $r - 1$ successes in the first $x - 1$ trials and the probability p of a success on the rth trial. Thus the p.d.f. of X is

$$
g(x) = \binom{x - 1}{r - 1} p^r (1 - p)^{x-r} = \binom{x - 1}{r - 1} p^r q^{x-r}, \qquad x = r, r + 1, \ldots .
$$

We say that X has a **negative binomial distribution**.

Example 3.6-5 Suppose that the biology students in Example 3.6-1 check the eye color of fruit flies until the third white-eyed fruit fly is observed. Then, with $p = 1/4$, the probability of observing the third white-eyed fruit fly on the thirteenth trial is

$$\binom{13 - 1}{3 - 1}\left(\frac{1}{4}\right)^3\left(\frac{3}{4}\right)^{13-3} = 0.058.$$

The probability that at most 12 flies have to be checked to find the third white-eyed fly is

$$\sum_{x=3}^{12}\binom{x - 1}{3 - 1}\left(\frac{1}{4}\right)^3\left(\frac{3}{4}\right)^{x-3} = 0.609.$$

The reason for calling this the negative binomial distribution is the following. Consider $h(w) = (1 - w)^{-r}$, the binomial $(1 - w)$ with the negative exponent $-r$. Using Maclaurin's series expansion (see Appendix A.4), we have

$$(1 - w)^{-r} = \sum_{k=0}^{\infty}\frac{h^{(k)}(0)}{k!}w^k = \sum_{k=0}^{\infty}\binom{r + k - 1}{r - 1}w^k.$$

If we let $x = k + r$ in the summation, then $k = x - r$ and

$$(1 - w)^{-r} = \sum_{x=r}^{\infty}\binom{r + x - r - 1}{r - 1}w^{x-r} = \sum_{x=r}^{\infty}\binom{x - 1}{r - 1}w^{x-r},$$

the summand of which is, except for the factor p^r, the negative binomial probability when $w = q$. In particular, we see that the sum of the probabilities for the negative binomial distribution is 1 because

$$\sum_{x=r}^{\infty}g(x) = \sum_{x=r}^{\infty}\binom{x - 1}{r - 1}p^r q^{x-r} = p^r(1 - q)^{-r} = 1.$$

In Exercise 3.6-19 the reader is asked to show that the mean of X is

$$\mu = E(X) = \frac{r}{p}.$$

Note that this mean is r times the mean of a geometric random variable.
In Exercise 3.6-20 the reader is asked to show that the variance of X is

$$\sigma^2 = \frac{rq}{p^2} = \frac{r(1 - p)}{p^2}.$$

Example 3.6-7 Suppose that during practice, a basketball player can make a free throw 80% of the time. Furthermore, assume that a sequence of free-throw shooting can be thought of as independent Bernoulli trials. Let X equal the minimum number of free throws that this player must attempt to make a total of 10 shots. The p.d.f. of X is

$$g(x) = \binom{x - 1}{10 - 1}(0.80)^{10}(0.20)^{x-10}, \qquad x = 10, 11, 12, \ldots$$

The mean, variance, and standard deviation of X are

$$\mu = 10\left(\frac{1}{0.80}\right) = 12.5, \qquad \sigma^2 = \frac{10(0.20)}{0.80^2} = 3.125, \qquad \text{and} \qquad \sigma = 1.768.$$

And we have, for example,

$$P(X = 12) = g(12) = \binom{11}{9}(0.80)^{10}(0.20)^2 = 0.236.$$

Example 3.6-8   In order to consider the effect of p and r on the negative binomial distribution, Figure 3.6-1 gives the probability histograms for four combinations of p and r. Note that since $r = 1$ in the first of these, it represents a geometric p.d.f.

Exercises

3.6-1   If a student selects answers randomly and independently on a true–false examination, determine the probability that
(a) The first correct answer is that of question 3.
(b) At most three questions must be answered to obtain the first correct answer.

3.6-2   Let X equal the number of people that you must ask, selecting randomly, in order to find someone with the same birthday as yours. Assuming each day of the year is equally likely (and ignoring February 29),
(a) Define the p.d.f. of X.
(b) Give the values of the mean, variance, and standard deviation of X.
(c) Find $P(X > 400)$ and $P(X < 300)$.

3.6-3   For each question on a multiple-choice test, there are five possible answers of which exactly one is correct. If a student selects answers at random, give the probability that the first question answered correctly is question 5.

3.6-4   The probability that a machine produces a defective item is 0.01. Each item is checked as it is produced. Assume that these are independent trials and compute the probability that at least 100 items must be checked to find one that is defective.

3.6-5   Apples are packaged automatically in 3-pound bags. Suppose that 4% of the time the bag of apples weighs less than 3 pounds. If you select bags randomly and weigh them in order to find one underweight bag of apples, find the probability that the number of bags that must be selected is

(a) At least 20.

(b) At most 20.

(c) Exactly 20.

3.6-6 Suppose that the probability of manufacturing a defective rear view mirror for a car is 0.08. Suppose further that the quality of any one mirror is independent of the quality of any other mirror. If an inspector selects mirrors at random from the production line, give the probability that the first defective mirror is the eighth mirror selected.

3.6-7 According to a representative for an automobile manufacturer, the company uses 3000 lock-and-key combinations on its vehicles. Suppose that you find a key for one of these cars.

(a) Give the expected number of vehicles that you would have to check to find one that your key fit.

(b) Give the probability that you would have to check at least 3000 vehicles to find one that your key fit.

(c) Give the probability that at most 2000 vehicles would have to be checked to find one that your key fit.

3.6-8 For a particular instant lottery the bettor can determine immediately whether a purchased ticket is a winning ticket. Suppose that the probability of a winning ticket is $p = 1/10$.

(a) What is the expected number of tickets that must be purchased in order to find a winning ticket?

(b) Give the probability that more than 10 tickets must be purchased to find a winning ticket.

(c) Give the probability that exactly 20 tickets must be purchased to find two winning tickets.

3.6-9 Flip an unbiased coin in a sequence of independent trials. Compute the probability that the first head is observed on the fifth trial, given that tails are observed on each of the first three trials.

3.6-10 Let X have a geometric distribution. Show that

$$P(X > k + j \mid X > k) = P(X > j),$$

where k and j are nonnegative integers.

REMARK We sometimes say that in this situation there has been loss of memory.

3.6-11 Let X equal the number of flips of a fair coin that are required to observe the same face on consecutive flips.

(a) Define the p.d.f. of X.

(b) Give the values of the mean, variance, and standard deviation of X.

(c) Find the values of (i) $P(X \le 3)$, (ii) $P(X \ge 5)$, and (iii) $P(X = 3)$.

3.6-12 Let X equal the number of flips of a fair coin that are required to observe heads on consecutive flips. Let f_n equal the nth Fibonacci number where $f_1 = 1, f_2 = 1$, and $f_n = f_{n-1} + f_{n-2}$, $n = 3, 4, 5, \ldots$.

(a) Show that the p.d.f. of X is

$$f(x) = \frac{f_{x-1}}{2^x}, \qquad x = 2, 3, 4, \ldots .$$

(b) Use the fact that

$$f_x = \frac{1}{\sqrt{5}} \left[\left(\frac{1 + \sqrt{5}}{2} \right)^x - \left(\frac{1 - \sqrt{5}}{2} \right)^x \right]$$

to show that $\sum_{x=2}^{\infty} f(x) = 1$.

(c) Show that $\mu = E(X) = 6$.

(d) Show that $E[X(X - 1)] = 52$ so that the variance of X is $\sigma^2 = 22$ and the standard deviation is $\sigma = 4.690$.

(e) Either simulate this experiment on the computer or physically toss a coin to support the theoretical characteristics of this distribution.

3.6-13 One of four different prizes was randomly put into each box of a cereal. If a family decided to buy this cereal until it obtains at least one of each of the four different prizes, what is the expected number of boxes of cereal that must be purchased?

3.6-14 A tea distributor randomly placed one of 15 English porcelain miniature animals in a 100-bag box.

(a) On the average, how many boxes of tea must be purchased by a customer to obtain a complete collection?

(b) If the customer uses one tea bag per day, how long will that take, approximately?

3.6-15 An urn contains 11 red balls. A ball is drawn at random from the urn and replaced by a black ball. What is the expected number of draws needed so that all of the balls in the urn are black?

NOTE: This exercise is based on an experiment in biology in which the students studied how food was "eaten" by paramecia.

3.6-16 Show that 63/512 is the probability that the fifth head is observed on the tenth independent flip of an unbiased coin.

3.6-17 A CD player has a magazine that holds six CDs. The machine is capable of randomly selecting a disk and then selecting a song randomly from that disk. Suppose that five disks are albums by Paul McCartney and one is by Billy Joel. The player selects songs until a song by Billy Joel is played after which the machine is turned off. Give the probability that the machine is turned off after

(a) The sixth song.

(b) At least five songs have been played.

(c) Suppose now that the songs continue to be played until the second song by Billy Joel has been played. Find the probability that at most four songs are played.

3.6-18 Let X have a negative binomial distribution. As illustrated in Figure 3.6-1, verify that
(a) $P(X = 5) = P(X = 6)$ when $r = 4$ and $p = 0.6$.
(b) $P(X = 20) = P(X = 21)$ when $r = 15$ and $p = 0.7$.

3.6-19 To find the mean of a negative binomial random variable, X, use Maclaurin's series expansion for $h(w) = (1 - w)^{-r}$, formula (3.6-4).
(a) Show that

$$h'(w) = r(1 - w)^{-r-1} = \sum_{k=1}^{\infty} \binom{r + k - 1}{r - 1} kw^{k-1}.$$

(b) Using the series expansion for $h'(w)$, show that $E(X - r) = p^r qr(1 - q)^{-r-1}$.

(c) Show that $E(X) = \dfrac{r}{p}$.

3.6-20 To find the variance of a negative binomial random variable, X, use the series expansion for $h'(w)$ in Exercise 3.6-19(a).
(a) Show that

$$h''(w) = r(r + 1)(1 - w)^{-r-2} = \sum_{k=2}^{\infty} \binom{r + k - 1}{r - 1} k(k - 1)w^{k-2}.$$

(b) Show that $E[(X - r)(X - r - 1)] = (q^2/p^2)(r)(r + 1)$.
(c) Using Exercise 3.6-19(b) and part (b) of this problem, show that $\mathrm{Var}(X - r) = rq/p^2$.
(d) Find the variance of X.

3.6-21 Suppose that the basketball player in Example 3.6-7 can make a free throw 60% of the time. Let X equal the minimum number of free throws that this player must attempt to make a total of 10 shots.
(a) Give the mean, variance, and standard deviation of X.
(b) Find $P(X = 16)$.

3.7 The Poisson Distribution

Some experiments result in counting the number of times particular events occur in given times or on given physical objects. For example, we could count the number of phone calls arriving at a switchboard between 9 and 10 A.M., the number of flaws in 100 feet of wire, the number of customers that arrive at a ticket window between 12 noon and 2 P.M., or the number of defects in a 100-foot roll of aluminum screen that is 2 feet wide. Each count can be looked upon as a random variable associated with an approximate Poisson process provided the conditions in Definition 3.7-1 are satisfied.

DEFINITION 3.7-1 *Let the number of changes that occur in a given continuous interval be counted. We have an* **approximate Poisson process** *with parameter* $\lambda > 0$ *if the following are satisfied:*

(a) *The numbers of changes occurring in nonoverlapping intervals are independent.*

(b) *The probability of exactly one change in a sufficiently short interval of length h is approximately λh.*

(c) *The probability of two or more changes in a sufficiently short interval is essentially zero.*

Suppose that an experiment satisfies the three points of an approximate Poisson process. Let X denote the number of changes in an interval of "length 1" (where "length 1" represents one unit of the quantity under consideration). We would like to find an approximation for $P(X = x)$, where x is a nonnegative integer. To achieve this, we partition the unit interval into n subintervals of equal length $1/n$. If n is sufficiently large (i.e., much larger than x), we shall approximate the probability that x changes occur in this unit interval by finding the probability that one change occurs in each of exactly x of these n subintervals. The probability of one change occurring in any one subinterval of length $1/n$ is approximately $\lambda(1/n)$ by condition (b). The probability of two or more changes in any one subinterval is essentially zero by condition (c). So for each subinterval, exactly one change occurs with a probability of approximately $\lambda(1/n)$. Consider the occurrence or nonoccurrence of a change in each subinterval as a Bernoulli trial. By condition (a) we have a sequence of n Bernoulli trials with probability p approximately equal to $\lambda(1/n)$. Thus an approximation for $P(X = x)$ is given by the binomial probability

$$\frac{n!}{x!(n-x)!}\left(\frac{\lambda}{n}\right)^{x}\left(1-\frac{\lambda}{n}\right)^{n-x}.$$

In order to obtain a better approximation, choose a larger value for n. If n increases without bound, we have that

$$\lim_{n\to\infty}\frac{n!}{x!(n-x)!}\left(\frac{\lambda}{n}\right)^{x}\left(1-\frac{\lambda}{n}\right)^{n-x}$$

$$= \lim_{n\to\infty}\frac{n(n-1)\cdots(n-x+1)}{n^{x}}\frac{\lambda^{x}}{x!}\left(1-\frac{\lambda}{n}\right)^{n}\left(1-\frac{\lambda}{n}\right)^{-x}.$$

Now, for fixed x, we have (see Appendix A.3)

$$\lim_{n\to\infty}\frac{n(n-1)\cdots(n-x+1)}{n^{x}} = \lim_{n\to\infty}\left[1\left(1-\frac{1}{n}\right)\cdots\left(1-\frac{x-1}{n}\right)\right] = 1,$$

$$\lim_{n\to\infty}\left(1-\frac{\lambda}{n}\right)^{n} = e^{-\lambda},$$

and

$$\lim_{n\to\infty}\left(1-\frac{\lambda}{n}\right)^{-x} = 1.$$

Thus

$$\lim_{n\to\infty}\frac{n!}{x!(n-x)!}\left(\frac{\lambda}{n}\right)^x\left(1-\frac{\lambda}{n}\right)^{n-x} = \frac{\lambda^x e^{-\lambda}}{x!} = P(X = x),$$

approximately. The distribution of probability associated with this process has a special name.

We say that the random variable X has a **Poisson distribution** if its p.d.f. is of the form

$$f(x) = \frac{\lambda^x e^{-\lambda}}{x!}, \qquad x = 0, 1, 2, \ldots ,$$

where $\lambda > 0$.

It is easy to see that $f(x)$ enjoys the properties of a p.d.f. because clearly $f(x) \geq 0$ and, from the Maclaurin's series expansion of e^λ (see Appendix A.4), we have

$$\sum_{x=0}^{\infty}\frac{\lambda^x e^{-\lambda}}{x!} = e^{-\lambda}\sum_{x=0}^{\infty}\frac{\lambda^x}{x!} = e^{-\lambda}e^{\lambda} = 1.$$

To discover the exact role of the parameter $\lambda > 0$, let us find some of the characteristics of the Poisson distribution.

The mean for the Poisson distribution is given by

$$E(X) = \sum_{x=0}^{\infty} x \frac{\lambda^x e^{-\lambda}}{x!} = e^{-\lambda}\sum_{x=1}^{\infty}\frac{\lambda^x}{(x-1)!}$$

because $(0)f(0) = 0$ and $x/x! = 1/(x-1)!$ when $x > 0$.

If we let $k = x - 1$, then

$$E(X) = e^{-\lambda}\sum_{k=0}^{\infty}\frac{\lambda^{k+1}}{k!} = \lambda e^{-\lambda}\sum_{k=0}^{\infty}\frac{\lambda^k}{k!} = \lambda e^{-\lambda}e^{\lambda} = \lambda.$$

That is, the parameter λ is the mean of the Poisson distribution.

To find the variance, we first determine the second factorial moment $E[X(X-1)]$. We have

$$E[X(X-1)] = \sum_{x=0}^{\infty} x(x-1)\frac{\lambda^x e^{-\lambda}}{x!} = e^{-\lambda}\sum_{x=2}^{\infty}\frac{\lambda^x}{(x-2)!}$$

because $(0)(0 - 1)f(0) = 0$, $(1)(1 - 1)f(1) = 0$, and $x(x - 1)/x! = 1/(x - 2)!$ when $x > 1$. If we let $k = x - 2$, then

$$E[X(X - 1)] = e^{-\lambda} \sum_{k=0}^{\infty} \frac{\lambda^{k+2}}{k!} = \lambda^2 e^{-\lambda} \sum_{k=0}^{\infty} \frac{\lambda^k}{k!} = \lambda^2 e^{-\lambda} e^{\lambda} = \lambda^2.$$

Thus

$$\mathrm{Var}(X) = E(X^2) - [E(X)]^2 = E[X(X - 1)] + E(X) - [E(X)]^2$$
$$= \lambda^2 + \lambda - \lambda^2 = \lambda.$$

That is, for the Poisson distribution, $\mu = \sigma^2 = \lambda$.

Table III in the Appendix gives values of the distribution function of a Poisson random variable for selected values of λ. This table is illustrated in the next example.

Example 3.7-1   Let X have a Poisson distribution with a mean of $\lambda = 5$. Then using Table III in the Appendix,

$$P(X \le 6) = \sum_{x=0}^{6} \frac{5^x e^{-5}}{x!} = 0.762,$$

$$P(X > 5) = 1 - P(X \le 5) = 1 - 0.616 = 0.384,$$

and

$$P(X = 6) = P(X \le 6) - P(X \le 5) = 0.762 - 0.616 = 0.146.$$

Example 3.7-2   In order to consider the effect of λ on the p.d.f. $f(x)$ of X, Figure 3.7-1 contains the probability histograms of $f(x)$ for four different values of λ.

Example 3.7-3   Let X denote the number of alpha particles emitted by barium-133 in 1/10 of a second and counted by a Geiger counter. One hundred observations of X produced the data in Table 3.7-1. For these data, $\bar{x} = 559/100 = 5.59$, and $s^2 = 3{,}619/99 - (559)^2/9900 = 4.992$. Thus we see that the sample mean and sample variance are fairly close to each other. In Figure 3.7-2 the probability histogram for a Poisson distribution with $\lambda = 5.59$ is drawn. The dashed lines represent the relative frequency histogram of the data. Accordingly, it looks as if the Poisson distribution provides a reasonable probability model for this random variable X. Later in the text, consideration is given to formal tests of the goodness of fit of various models.

If events in a Poisson process occur at a mean rate of λ per unit, then the expected number of occurrences in an interval of length t is λt. For example, if phone calls arrive at a switchboard following a Poisson process at a mean rate of three per minute, then the expected number of phone calls in a 5-minute period is $(3)(5) =$

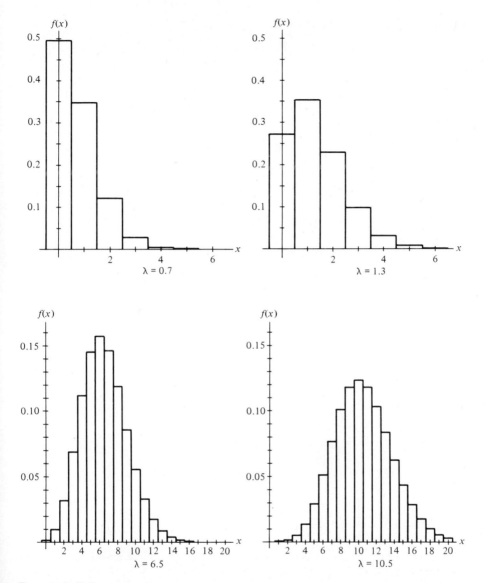

FIGURE 3.7-1

15. Or if calls arrive at a mean rate of 22 in a 5-minute period, the expected number of calls per minute is $\lambda = 22(1/5) = 4.4$. Moreover, the number of occurrences, say X, in the interval of length t has the Poisson p.d.f.

$$f(x) = \frac{(\lambda t)^x e^{-\lambda t}}{x!}, \qquad x = 0, 1, 2, \ldots .$$

TABLE 3.7-1

Outcome (x)	Frequency (f)	fx	fx²
1	1	1	1
2	4	8	16
3	13	39	117
4	19	76	304
5	16	80	400
6	15	90	540
7	9	63	441
8	12	96	768
9	7	63	567
10	2	20	200
11	1	11	121
12	1	12	144
	100	559	3619

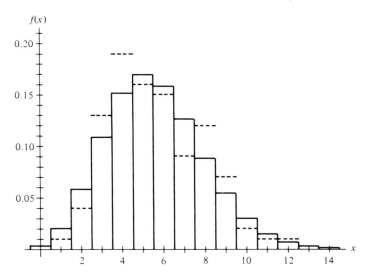

FIGURE 3.7-2

This follows by treating the interval of length t as if it were the "unit interval" with mean λt instead of λ.

Example 3.7-4 Flaws (bad records) on a used computer tape occur on the average of one flaw per 1200 feet. If one assumes a Poisson distribution, what is the distribution of X, the number of flaws in a 4800-foot roll? The expected number of flaws in $4800 = 4(1200)$ feet is 4; that is, $E(X) = 4$. Thus the p.d.f. of X is

$$f(x) = \frac{4^x e^{-4}}{x!}, \qquad x = 0, 1, 2, \ldots,$$

and, in particular,

$$P(X = 0) = \frac{4^0 e^{-4}}{0!} = e^{-4} = 0.018,$$

$$P(X \le 4) = 0.629,$$

by Table III in the Appendix.

Example 3.7-5 Telephone calls enter a college switchboard on the average of two every 3 minutes. If one assumes an approximate Poisson process, what is the probability of five or more calls arriving in a 9-minute period? Let X denote the number of calls in a 9-minute period. We see that $E(X) = 6$; that is, on the average, six calls will arrive during a 9-minute period. Thus, using Table III in the Appendix,

$$P(X \ge 5) = 1 - P(X \le 4) = 1 - \sum_{x=0}^{4} \frac{6^x e^{-6}}{x!}$$

$$= 1 - 0.285 = 0.715.$$

Not only is the Poisson distribution important in its own right, but it can also be used to approximate probabilities for a binomial distribution. If X has a Poisson distribution with parameter λ, we saw that with n large,

$$P(X = x) \approx \binom{n}{x}\left(\frac{\lambda}{n}\right)^x\left(1 - \frac{\lambda}{n}\right)^{n-x},$$

where $p = \lambda/n$ so that $\lambda = np$ in the above binomial probability. That is, if X has the binomial distribution $b(n, p)$ with large n, then

$$\frac{(np)^x e^{-np}}{x!} \approx \binom{n}{x}p^x(1 - p)^{n-x}.$$

This approximation is reasonably good if n is large. But since λ was a fixed constant in that earlier argument, p should be small since $np = \lambda$. In particular, the approximation is quite accurate if $n \ge 20$ and $p \le 0.05$, and it is very good if $n \ge 100$ and $np \le 10$.

Example 3.7-6 A manufacturer of Christmas tree light bulbs knows that 2% of its bulbs are defective. Approximate the probability that a box of 100 of these bulbs contains at most three defective bulbs. Assuming independence, we have a binomial distribution with parameters $p = 0.02$ and $n = 100$. The Poisson distribution with $\lambda = 100(0.02) = 2$ gives

$$\sum_{x=0}^{3} \frac{2^x e^{-2}}{x!} = 0.857,$$

from Table III in the Appendix. Using the binomial distribution, we obtain, after some tedious calculations,

$$\sum_{x=0}^{3} \binom{100}{x} (0.02)^x (0.98)^{100-x} = 0.859.$$

Hence, in this case, the Poisson approximation is extremely close to the true value, but much easier to find.

Example 3.7-7 In Figure 3.7-3 Poisson probability histograms have been drawn along with dashed lines representing the respective binomial probabilities so that we can see whether or not these are close to each other. If the

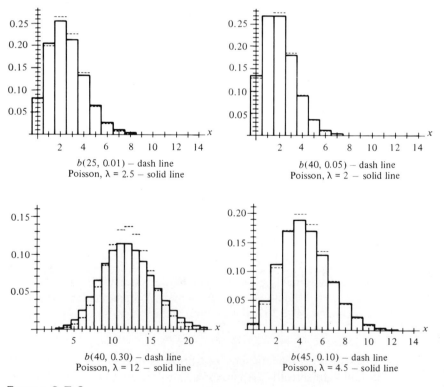

$b(25, 0.01)$ − dash line
Poisson, $\lambda = 2.5$ − solid line

$b(40, 0.05)$ − dash line
Poisson, $\lambda = 2$ − solid line

$b(40, 0.30)$ − dash line
Poisson, $\lambda = 12$ − solid line

$b(45, 0.10)$ − dash line
Poisson, $\lambda = 4.5$ − solid line

FIGURE 3.7-3

distribution of X is $b(n, p)$, the approximating Poisson distribution has a mean of $\lambda = np$. Note that the approximation is not good when p is large (e.g., $p = 0.30$).

Example 3.7-8   A lot of 1000 parts is shipped to a company. A sampling plan dictates that $n = 100$ parts are to be taken at random and without replacement and the lot accepted if no more than two of these 100 parts are defective. Here Ac $= 2$ is usually called the **acceptance number**. The operating characteristic curve

$$OC(p) = P(X \leq 2),$$

where p is the fraction defective in the lot, is really the sum of the three hypergeometric probabilities

$$P(X = x) = \frac{\binom{n_1}{x}\binom{1000 - n_1}{100 - x}}{\binom{1000}{100}}, \qquad x = 0, 1, 2,$$

where $n_1 = (1000)(p)$. However, we have seen that the hypergeometric distribution can be approximated by the binomial distribution, which in turn can be approximated by the Poisson distribution when n is large and p is small. This is exactly our situation since $n = 100$ and since we are interested in values of p in the range 0.00 to 0.10. Thus

$$OC(p) = P(X \leq 2) \approx \sum_{x=0}^{2} \frac{(100p)^x e^{-100p}}{x!}.$$

Using Table III in the Appendix, we find that some approximate values of the operating characteristic curve are $OC(0.01) = 0.920$, $OC(0.02) = 0.677$, $OC(0.03) = 0.423$, $OC(0.05) = 0.125$, and $OC(0.10) = 0.003$ and it is plotted in Figure 3.7-4. If this is not a satisfactory operating characteristic curve for the type of parts involved, the sample size n and the acceptance number Ac must be changed.

Exercises

3.7-1   Let X have a Poisson distribution with a mean of 4. Find
(a) $P(2 \leq X \leq 5)$.
(b) $P(X \geq 3)$.
(c) $P(X \leq 3)$.

3.7-2   Let X have a Poisson distribution with a variance of 3. Find $P(X = 2)$.

3.7-3   Customers arrive at a travel agency at a mean rate of 11 per hour. Assuming that the number of arrivals per hour has a Poisson distribution, give the probability that more than 10 customers arrive in a given hour.

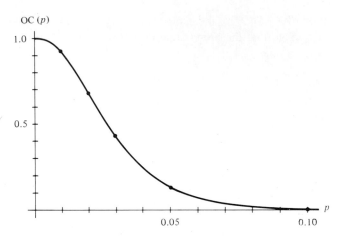

FIGURE 3.7-4

3.7-4 If X has a Poisson distribution so that $3P(X = 1) = P(X = 2)$, find $P(X = 4)$.

3.7-5 Let X equal the number of students who arrive at the card catalog in a college library during a 15-minute time period. Assume that X has a Poisson distribution with an average of 4.2 arrivals every 15 minutes. Find
(a) $P(2 < X < 6)$.
(b) $P(X > 4)$.

3.7-6 Flaws in a certain type of drapery material appear on the average of one in 150 square feet. If we assume the Poisson distribution, find the probability of at most one flaw in 225 square feet.

3.7-7 A certain type of aluminum screen that is 2 feet wide has on the average one flaw in a 100-foot roll. Find the probability that a 50-foot roll has no flaws.

3.7-8 With probability 0.001, a prize of $499 is won in the Michigan Daily Lottery when a $1 bet is placed. Let Y equal the number of $499 prizes won by a gambler after placing n straight bets. Note that Y is $b(n, 0.001)$. After placing $n = 2000$ $1 bets, a gambler is behind if $Y \leq 4$. Use the Poisson distribution to approximate $P(Y \leq 4)$ when $n = 2000$.

3.7-9 Suppose that the probability of suffering a side effect from a certain flu vaccine is 0.005. If 1000 persons are innoculated, find approximately the probability that
(a) At most 1 person suffers.
(b) 4, 5, or 6 persons suffer.

3.7-10 A roll of a biased die results in a two only 1/10 of the time. Let X denote the number of two's in 100 rolls of this die. Approximate.
(a) $P(3 \leq X \leq 7)$.
(b) $P(X \geq 5)$.

3.7-11 The mean of a Poisson random variable X is $\mu = 9$. Compute

$$P(\mu - 2\sigma < X < \mu + 2\sigma).$$

3.7-12 A Geiger counter was set up in the physics laboratory to record the number of alpha particle emissions of carbon-14 in 0.5 second. The following are 50 observations.

4	6	6	12	11	11	10	5	10	7
9	6	9	11	9	6	4	9	11	8
10	11	7	4	5	8	6	5	8	7
6	3	4	6	4	12	7	14	5	9
7	10	9	6	6	12	10	7	12	13

(a) Calculate the sample mean and sample variance for these data. Are these two measurements close to each other?

(b) On the same graph depict the probability histogram for a Poisson distribution with $\lambda = 8$ (we use 8 only because it is the closest value of λ to $\bar{x}$ in Table III in the Appendix), with the relative frequency histogram of the data superimposed.

3.7-13 For determining the half-lives of radioactive isotopes, it is important to know what the background radiation is in a given detector over a period of time. Data taken in a γ-ray detection experiment over 300 one-second intervals yielded the following data:

0	2	4	6	6	1	7	4	6	1	1	2	3	6	4	2	7	4	4	2	2	5	4	4	4
1	2	4	3	2	2	5	0	3	1	1	0	0	5	2	7	1	3	3	3	2	3	1	4	1
3	5	3	5	1	3	3	0	3	2	6	1	1	4	6	3	6	4	4	2	2	4	3	3	6
1	6	2	5	0	6	3	4	3	1	1	4	6	1	5	1	1	4	1	4	1	1	1	3	3
4	3	3	2	5	2	1	3	5	3	2	7	0	4	2	3	3	5	6	1	4	2	6	4	2
0	4	4	7	3	5	2	2	3	1	3	1	3	6	5	4	8	2	2	4	2	2	1	4	7
5	2	1	1	4	1	4	3	6	2	1	1	2	2	2	2	3	5	4	3	2	2	3	3	2
4	4	3	2	2	3	6	1	1	3	3	2	1	4	5	5	1	2	3	3	1	3	7	2	5
4	2	0	6	2	3	2	3	0	4	4	5	2	5	3	0	4	6	2	2	2	2	2	5	2
2	3	4	2	3	7	1	1	7	1	3	6	0	5	3	0	0	3	3	0	2	4	3	1	2
3	3	3	4	3	2	2	7	5	3	5	1	1	2	2	6	1	3	1	4	4	2	3	4	5
1	3	4	3	1	0	3	7	4	0	5	2	5	4	4	2	2	3	2	4	6	5	5	3	4

Do these look like observations of a Poisson random variable with mean $\lambda = 3$? To help answer this question, do the following.

(a) Group the data; that is, find the frequencies of $0, 1, 2, \ldots, 8$.

(b) Calculate the sample mean and sample variance. Are they approximately equal to each other?

(c) Construct a probability histogram and a relative frequency histogram on the same graph.

(d) What is your conclusion? (Later in this book we give tests of hypotheses to help give a statistical answer.)

3.7-14   Let X equal the number of telephone calls that are received at the college switchboard from on-campus during a 15-minute period. The following numbers of calls were received during each of 26 time periods:

$$
\begin{array}{cccccccccccc}
4 & 8 & 5 & 3 & 1 & 3 & 2 & 5 & 6 & 7 & 4 & 4 & 5 \\
2 & 3 & 6 & 4 & 1 & 2 & 5 & 6 & 7 & 5 & 7 & 5 & 1
\end{array}
$$

(a) Calculate the sample mean and sample variance for these data. Are they approximately equal to each other?

(b) Assume that $\lambda = 4.2$. Compare $P(X \le 3)$ with the proportion of observations that are less than or equal to 3.

(c) Compare $P(X > 5)$ with the proportion of observations that are greater than 5.

(d) Graph the empirical and theoretical distribution functions on the same figure.

(e) Does it look like the Poisson distribution with $\lambda = 4.2$ could be the correct probability model based on these limited data?

3.7-15   Let X equal the number of green peanut m&m's in packages of size 22. Forty-five observations of X yielded the following frequencies for the possible outcomes of X:

Outcome (x):	0	1	2	3	4	5	6	7	8	9
Frequency:	0	2	4	5	7	9	8	5	3	2

(a) Use these data to construct a relative frequency histogram and superimpose over it a Poisson p.d.f. with a mean of $\lambda = 5$.

(b) Do these data appear to be observations of a Poisson random variable? (In a later chapter we will give a statistical test to determine whether the answer is yes or no.)

3.7-16   Let X equal the number of chocolate chips in a chocolate-chip cookie. Sixty-two observations of X yielded the following frequencies for the possible outcomes of X:

Outcome (x):	0	1	2	3	4	5	6	7	8	9	10
Frequency:	0	0	2	8	7	13	13	10	4	4	1

(a) Use these data to graph the empirical distribution function and superimpose on your graph the distribution function for a Poisson distribution with a mean of $\lambda = 5.6$.

(b) Do these data seem to be observations of a Poisson random variable with a mean of $\lambda = 5.6$? (In a later chapter we will give a test to help answer similar questions.)

3.7-17   While testing a used computer tape for bad records, a computer operator counted the number of flaws per 100 feet of tape. Let X equal this random variable. Given that 40 observations of X yielded 5 zeros, 7 ones, 12 twos, 9 threes, 5 fours, 1 five, and 1 six,

(a) Construct a relative frequency histogram and superimpose a Poisson p.d.f. with $\lambda = 2.2$.

(b) Does the Poisson distribution seem to be a possible probability model for this experiment?

3.7-18   A lot of 10,000 items is accepted if a sample of $n = 400$ has no more than Ac = 3 defective items. Determine the approximate values of OC(0.002), OC(0.004), OC(0.006), OC(0.01), and OC(0.02) and plot the operating characteristic curve.

3.7-19   Consider a lot of 5000 items. Design a sampling plan (i.e., find n and Ac) so that OC(0.02) = 0.95 and OC(0.08) = 0.10 approximately.

HINT: Start with $n = 50$ and Ac = 2 and modify as necessary, noting that increasing n creates a steeper OC curve.

3.8   Moment-Generating Functions

In the preceding sections we have seen the importance of the mean, variance, and standard deviation of a random variable X. For some distributions it was fairly difficult to find directly $E(X)$ and $E(X^2)$, the first and second moments, or $E[X(X-1)]$, the second factorial moment which was sometimes used in finding the variance. Accordingly, we define a function of a real variable t that can be used to find $E(X)$ and $E(X^2)$ as well as the other moments of X. We also define a function that can be used to find the factorial moments of X.

DEFINITION 3.8-1   *Let X be a random variable of the discrete type with p.d.f. $f(x)$ and space R. If there is a positive number h such that*

$$E(e^{tX}) = \sum_{x \in R} e^{tx} f(x)$$

exists for $-h < t < h$, then the function of t defined by

$$M(t) = E(e^{tX})$$

is called the **moment-generating function** *of X.*

From the theory of mathematical analysis, it can be shown that the existence of $M(t)$, for $-h < t < h$, implies that derivatives of $M(t)$ of all orders exist at $t = 0$; moreover, it is permissible to interchange differentiation and summation. Thus

$$M'(t) = \sum_{x \in R} x e^{tx} f(x),$$

$$M''(t) = \sum_{x \in R} x^2 e^{tx} f(x),$$

and, for each positive integer r,

$$M^{(r)}(t) = \sum_{x \in R} x^r e^{tx} f(x).$$

Setting $t = 0$, we see that

$$M'(0) = \sum_{x \in R} x f(x) = E(X),$$

$$M''(0) = \sum_{x \in R} x^2 f(x) = E(X^2),$$

and, in general,

$$M^{(r)}(0) = \sum_{x \in R} x^r f(x) = E(X^r).$$

In particular, if the moment-generating function exists,

$$\mu = M'(0) \quad \text{and} \quad \sigma^2 = M''(0) - [M'(0)]^2.$$

The above argument shows that we can find the moments of X by differentiating $M(t)$. It must be emphasized that in use, we generally first evaluate the summation representing $M(t)$, obtaining a closed form solution, and then differentiate that solution to obtain the moments of X. The following examples illustrate the use of the moment-generating function for finding the first and second moments and then the mean and variance.

Example 3.8-1 Let X have a binomial distribution $b(n, p)$ with p.d.f.

$$f(x) = \binom{n}{x} p^x (1 - p)^{n-x}, \qquad x = 0, 1, 2, \ldots, n.$$

The moment-generating function of X is

$$M(t) = E(e^{tX}) = \sum_{x=0}^{n} e^{tx} \binom{n}{x} p^x (1 - p)^{n-x}$$

$$= \sum_{x=0}^{n} \binom{n}{x} (pe^t)^x (1 - p)^{n-x}.$$

Using the formula for the binomial expansion with $a = 1 - p$ and $b = pe^t$, we see that

$$M(t) = [(1 - p) + pe^t]^n$$

for all real values of t. The first two derivatives of $M(t)$ are

$$M'(t) = n[(1 - p) + pe^t]^{n-1}(pe^t)$$

and

$$M''(t) = n(n - 1)[(1 - p) + pe^t]^{n-2}(pe^t)^2 + n[(1 - p) + pe^t]^{n-1}(pe^t).$$

Thus

$$\mu = E(X) = M'(0) = np$$

and

$$\sigma^2 = E(X^2) - [E(X)]^2 = M''(0) - [M'(0)]^2$$
$$= n(n - 1)p^2 + np - (np)^2 = np(1 - p),$$

as was shown in Section 3.5.

In the special case when $n = 1$, X has a Bernoulli distribution and

$$M(t) = (1 - p) + pe^t$$

for all real values of t, $\mu = p$, and $\sigma^2 = p(1 - p)$.

Example 3.8-2   Let X have a Poisson distribution with p.d.f.

$$f(x) = \frac{\lambda^x e^{-\lambda}}{x!}, \qquad x = 0, 1, 2, \ldots.$$

The moment-generating function of X is

$$M(t) = E(e^{tX}) = \sum_{x=0}^{\infty} e^{tx} \frac{\lambda^x e^{-\lambda}}{x!} = e^{-\lambda} \sum_{x=0}^{\infty} \frac{(\lambda e^t)^x}{x!}.$$

From the series representation of the exponential function, we have that

$$M(t) = e^{-\lambda} e^{\lambda e^t} = e^{\lambda(e^t - 1)}$$

for all real values of t. Now

$$M'(t) = \lambda e^t e^{\lambda(e^t - 1)},$$

and

$$M''(t) = (\lambda e^t)^2 e^{\lambda(e^t - 1)} + \lambda e^t e^{\lambda(e^t - 1)}.$$

The values of the mean and variance of X are

$$\mu = M'(0) = \lambda$$

and

$$\sigma^2 = M''(0) - [M'(0)]^2 = (\lambda^2 + \lambda) - \lambda^2 = \lambda.$$

Although the moment-generating function, when it exists, is a useful tool for determining moments, its major importance is the fact that it uniquely determines the distribution. Another way of stating this is that if two random variables X and Y have the same moment-generating function, then X and Y are identically distributed. Although the rigorous proof of this property is based on the theory of transforms in analysis, it seems fairly evident for distributions of the discrete type. First note clearly the form of the moment-generating function of a discrete random variable. If X has a p.d.f. $f(x)$ with support $\{b_1, b_2, \ldots\}$, then

$$M(t) = \sum_R e^{tx} f(x)$$

$$= f(b_1)e^{tb_1} + f(b_2)e^{tb_2} + \cdots.$$

Hence the coefficient of e^{tb_i} is $f(b_i) = P(X = b_i)$. That is, if we write a moment-generating function of a discrete-type random variable X in the above form, the probability of any value of X, say b_i, is the coefficient of e^{tb_i}.

Example 3.8-3 Let the moment-generating function of X be defined by

$$M(t) = \frac{1}{15}e^t + \frac{2}{15}e^{2t} + \frac{3}{15}e^{3t} + \frac{4}{15}e^{4t} + \frac{5}{15}e^{5t}.$$

Then, for example, the coefficient of e^{2t} is 2/15. Thus $f(2) = P(X = 2) = 2/15$. In general, we see that the p.d.f. of X is $f(x) = x/15$, $x = 1, 2, 3, 4, 5$.

When the moment-generating function exists, derivatives of all orders exist at $t = 0$. Thus it is possible to represent $M(t)$ as a Maclaurin's series, namely,

$$M(t) = M(0) + M'(0)\left(\frac{t}{1!}\right) + M''(0)\left(\frac{t^2}{2!}\right) + M'''(0)\left(\frac{t^3}{3!}\right) + \cdots.$$

If the Maclaurin's series expansion of $M(t)$ exists and the moments are given, we can frequently sum the Maclaurin's series to obtain the closed form of $M(t)$. This is illustrated in the next example.

Example 3.8-4 Let the moments of X be defined by

$$E(X^r) = 0.8, \qquad r = 1, 2, 3, \ldots.$$

The moment-generating function of X is then

$$M(t) = M(0) + \sum_{r=1}^{\infty} 0.8\left(\frac{t^r}{r!}\right) = 1 + 0.8 \sum_{r=1}^{\infty} \frac{t^r}{r!}$$

$$= 0.2 + 0.8 \sum_{r=0}^{\infty} \frac{t^r}{r!} = 0.2e^{0t} + 0.8e^{1t}.$$

Thus

$$P(X = 0) = 0.2 \quad \text{and} \quad P(X = 1) = 0.8.$$

Recall that we found the second factorial moment to help us find the variance of some of the distributions before we had the moment-generating function. There is another generating function that can help us find the factorial moments.

DEFINITION 3.8-2 *Let X be a random variable of the discrete type with p.d.f. $f(x)$, space R, and moment-generating function $M(t)$. Then*

$$\eta(t) = E(t^X) = E[e^{X(\ln t)}] = M(\ln t)$$

is called the **factorial moment-generating function**.

To see how $\eta(t)$ can generate the factorial moments, we proceed as we did with the moment-generating function. The derivatives of $\eta(t)$ are

$$\eta(t) = E(t^X) = \sum_{x \in R} t^x f(x),$$

$$\eta'(t) = \sum_{x \in R} xt^{x-1} f(x),$$

$$\eta''(t) = \sum_{x \in R} x(x-1)t^{x-2} f(x),$$

and, for each positive integer r,

$$\eta^{(r)}(t) = \sum_{x \in R} x(x-1) \cdots (x-r+1)t^{x-r} f(x).$$

Setting $t = 1$ gives us

$$\eta(1) = \sum_{x \in R} 1 f(x) = 1,$$

$$\eta'(1) = \sum_{x \in R} xf(x) = E(X),$$

$$\eta''(1) = \sum_{x \in R} x(x-1)f(x) = E[X(X-1)],$$

and, for each positive integer r,

$$\eta^{(r)}(1) = \sum_{x \in R} x(x-1) \cdots (x-r+1)f(x)$$

$$= E[X(X-1) \cdots (X-r+1)] = E[(X)_r],$$

the rth factorial moment of X.

It follows that the mean and variance of X are given by

$$\mu = E(X) = \eta'(1),$$
$$\sigma^2 = E[X(X - 1)] + E(X) - [E(X)]^2 = \eta''(1) + \eta'(1) - [\eta'(1)]^2.$$

Note that given $M(t)$, the moment-generating function of X, the factorial moment-generating function of X is $\eta(t) = M(\ln t)$. Also, given the factorial moment-generating function of X, the moment-generating function of X is $M(t) = \eta(e^t)$. The next two examples illustrate these ideas.

Example 3.8-5 Let X have a geometric distribution with p.d.f. $f(x) = (1 - p)^{x-1}p$, $x = 1, 2, 3, \ldots$. The factorial moment-generating function of X is

$$\eta(t) = E(t^X) = \sum_{x=1}^{\infty} t^x (1 - p)^{x-1} p$$

$$= \frac{p}{1 - p} \sum_{x=1}^{\infty} [(1 - p)t]^x = \left(\frac{p}{1 - p}\right) \frac{(1 - p)t}{1 - (1 - p)t}$$

$$= \frac{pt}{1 - (1 - p)t}, \qquad |t| < \frac{1}{1 - p}.$$

To find the factorial moments we differentiate $\eta(t)$ as follows:

$$\eta'(t) = \frac{[1 - (1 - p)t]p + pt(1 - p)}{[1 - (1 - p)t]^2} = \frac{p}{[1 - (1 - p)t]^2},$$

$$\eta''(t) = \frac{2p(1 - p)}{[1 - (1 - p)t]^3}.$$

It follows that

$$\mu = \eta'(1) = \frac{p}{p^2} = \frac{1}{p}$$

and

$$\sigma^2 = \eta''(1) + \eta'(1) - [\eta'(1)]^2$$
$$= \frac{2p(1 - p)}{p^3} + \frac{1}{p} - \frac{1}{p^2} = \frac{1 - p}{p^2}.$$

The moment-generating function of X is

$$M(t) = \eta(e^t) = \frac{pe^t}{1 - (1 - p)e^t}, \qquad |e^t| < \frac{1}{1 - p}.$$

Example 3.8-6 Let X have a negative binomial distribution with p.d.f.

$$f(x) = \binom{x - 1}{r - 1} p^r (1 - p)^{x-r}, \qquad x = r, r + 1, r + 2, \ldots$$

The factorial moment-generating function of X is

$$\eta(t) = \sum_{x=r}^{\infty} t^x \binom{x-1}{r-1} p^r (1-p)^{x-r}$$

$$= (pt)^r \sum_{x=r}^{\infty} \binom{x-1}{r-1} [(1-p)t]^{x-r}$$

$$= \frac{(pt)^r}{[1-(1-p)t]^r}, \qquad |t| < \frac{1}{1-p}.$$

The moment-generating function of X is

$$M(t) = \eta(e^t)$$

$$= \frac{(pe^t)^r}{[1-(1-p)e^t]^r}, \qquad t < -\ln(1-p).$$

Either of these generating functions could be used to show that

$$\mu = \frac{r}{p}$$

and

$$\sigma^2 = \frac{r(1-p)}{p^2}.$$

However, the calculations get a little messy and an easier way to find these characteristics is given in Exercise 3.8-10.

Exercises

3.8-1 Find the moment-generating function when the p.d.f. of X is defined by:

(a) $f(x) = \dfrac{1}{3}, \qquad x = 1, 2, 3.$

(b) $f(x) = 1, \qquad x = 5.$

(c) $f(x) = \dfrac{5-x}{10}, \qquad x = 1, 2, 3, 4.$

(d) $f(x) = (0.3)^x(0.7)^{1-x}, \qquad x = 0, 1.$

3.8-2 Define the p.d.f. and give the values of μ, σ^2, and σ when the moment-generating function of X is defined by

(a) $M(t) = 1/3 + (2/3)e^t.$

(b) $M(t) = (0.25 + 0.75e^t)^{12}.$

(c) $M(t) = e^{4.6(e^t-1)}.$

3.8-3 Let X have a binomial distribution $b(n, p)$. Show that the factorial moment-generating function of X is $\eta(t) = [(1-p) + pt]^n$ directly by finding $E(t^X)$.

3.8-4 Let X have a Poisson distribution with mean λ. Show that the factorial moment-generating function of X is $\eta(t) = e^{\lambda(t-1)}$ directly by finding $E(t^X)$.

3.8-5 Find the moment-generating function, mean, and variance of X if the p.d.f. of X is $f(x) = (1/2)(2/3)^x$, $x = 1, 2, 3, 4, \ldots$.

3.8-6 If the moment-generating function of X is

$$M(t) = \frac{2}{5}e^t + \frac{1}{5}e^{2t} + \frac{2}{5}e^{3t},$$

find the mean, variance, and p.d.f. of X.

3.8-7 (i) Give the name of the distribution of X (if it has a name); (ii) find the values of μ and σ^2; and (iii) calculate $P(1 \le X \le 2)$ when the moment-generating function of X is given by
   (a) $M(t) = (0.3 + 0.7e^t)^5$.
   (b) $M(t) = e^{4(e^t-1)}$.
   (c) $M(t) = \dfrac{0.3e^t}{1 - 0.7e^t}$, $t < -\ln(0.7)$.
   (d) $M(t) = 0.45 + 0.55e^t$.
   (e) $M(t) = 0.3e^t + 0.4e^{2t} + 0.2e^{3t} + 0.1e^{4t}$.
   (f) $M(t) = (0.6e^t)^2(1 - 0.4e^t)^{-2}$, $t < -\ln(0.4)$.
   (g) $M(t) = \sum_{x=1}^{10}(0.1)e^{tx}$.

3.8-8 Find $P(X = 3)$ if the moment-generating function of X is
   (a) $M(t) = (0.65 + 0.35e^t)^{14}$.
   (b) $M(t) = e^{5.6(e^t-1)}$.
   (c) $M(t) = 0.25e^t + 0.35e^{3t} + 0.40e^{5t}$.
   (d) $M(t) = \dfrac{0.35e^t}{1 - 0.65e^t}$, $t < -\ln(0.65)$.
   (e) $M(t) = \sum_{x=1}^{15}(1/15)e^{tx}$.

3.8-9 If $E(X^r) = 5^r$, $r = 1, 2, 3, \ldots$, find $M(t)$, the moment-generating function of X, and the p.d.f. of X.

3.8-10 Let the moment-generating function $M(t)$ of X exist for $-h < t < h$. Consider the function $R(t) = \ln M(t)$. The first two derivatives of $R(t)$ are, respectively,

$$R'(t) = \frac{M'(t)}{M(t)} \quad \text{and} \quad R''(t) = \frac{M(t)M''(t) - [M'(t)]^2}{[M(t)]^2}.$$

Setting $t = 0$, show that
   (a) $\mu = R'(0)$.
   (b) $\sigma^2 = R''(0)$.

3.8-11 Use the result of Exercise 3.8-10 to find the mean and variance for the
   (a) Bernoulli distribution.
   (b) Binomial distribution.
   (c) Poisson distribution.
   (d) Geometric distribution.
   (e) Negative binomial distribution.

4

Continuous Distributions

4.1   Random Variables of the Continuous Type

Random variables whose spaces are not composed of a countable number of points but are intervals or a union of intervals are said to be of the **continuous type**. Recall that the relative frequency histogram $h(x)$ associated with n observations of a random variable of that type is a nonnegative function defined so that the total area between its graph and the x axis equals one. In addition, $h(x)$ is constructed so that the integral

$$\int_a^b h(x) \, dx$$

is an estimate of the probability $P(a < X < b)$, where the interval (a, b) is a subset of the space R of the random variable X (see Section 1.2).

Let us now consider what happens to the function $h(x)$ in the limit, as n increases without bound and as the lengths of the class intervals decrease to zero. It is to be hoped that $h(x)$ will become closer and closer to some function, say $f(x)$, that gives the true probabilities, such as $P(a < X < b)$, through the integral

$$P(a < X < b) = \int_a^b f(x) \, dx.$$

That is, $f(x)$ should be a nonnegative function such that the total area between its graph and the x axis equals one. Moreover, the probability $P(a < X < b)$ is the area bounded by the graph of $f(x)$, the x axis, and the lines $x = a$ and $x = b$. Thus we say that the **probability density function (p.d.f.)** of a random variable X of the **continuous type**, with space R that is an interval or union of intervals, is an integrable function $f(x)$ satisfying the following conditions:

(a) $f(x) > 0$, $x \in R$.
(b) $\int_R f(x)\ dx = 1$.
(c) The probability of the event $X \in A$ is

$$P(X \in A) = \int_A f(x)\ dx.$$

The corresponding distribution of probability is said to be one of the continuous type.

Example 4.1-1   Let the random variable X be the distance in feet between bad records on a used computer tape. Suppose that a reasonable probability model for X is given by the p.d.f.

$$f(x) = \frac{1}{40}e^{-x/40}, \qquad 0 \le x < \infty.$$

Note that $R = \{x: 0 \le x < \infty\}$ and $f(x) > 0$ for $x \in R$. Also,

$$\int_R f(x)\ dx = \int_0^\infty \frac{1}{40}e^{-x/40}\ dx$$
$$= \lim_{b \to \infty}\ [-e^{-x/40}]_0^b$$
$$= 1 - \lim_{b \to \infty}\ e^{-b/40} = 1.$$

The probability that the distance between bad records is greater than 40 feet is

$$P(X > 40) = \int_{40}^\infty \frac{1}{40}e^{-x/40}\ dx = e^{-1} = 0.368.$$

The p.d.f. and the probability of interest are depicted in Figure 4.1-1.

So that we can avoid repeated references to the space (or support) R of the random variable X, we shall adopt the same convention when describing probability density functions of the continuous type as we did in the discrete case. We extend the definition of the p.d.f. $f(x)$ to the entire set of real numbers by letting it equal zero when $x \notin R$. For example,

$$f(x) = \begin{cases} \dfrac{1}{40}e^{-x/40}, & 0 \le x < \infty, \\ 0, & \text{elsewhere,} \end{cases}$$

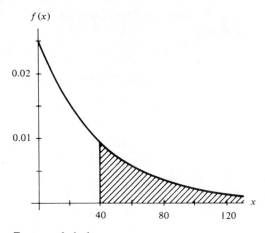

FIGURE **4.1-1**

has the properties of a p.d.f. of a continuous-type random variable X having support $\{x: 0 \leq x < \infty\}$. It will always be understood that $f(x) = 0$ when $x \notin R$, even when this is not explicitly written out.

The **distribution function** of a random variable X of the continuous type, defined in terms of the p.d.f. of X, is given by

$$F(x) = P(X \leq x) = \int_{-\infty}^{x} f(t) \, dt.$$

From the fundamental theorem of calculus we have, for x values for which the derivative $F'(x)$ exists, that $F'(x) = f(x)$.

Example 4.1-2   Continuing with Example 4.1-1, if the p.d.f. of X is

$$f(x) = \begin{cases} 0, & -\infty < x < 0, \\ \dfrac{1}{40} e^{-x/40}, & 0 \leq x < \infty, \end{cases}$$

the distribution function of X is $F(x) = 0$ for $x \leq 0$ and, for $x > 0$,

$$F(x) = \int_{-\infty}^{x} f(t) \, dt = \int_{0}^{x} \frac{1}{40} e^{-t/40} \, dt$$

$$= -e^{-t/40} \Big|_{0}^{x} = 1 - e^{-x/40}.$$

Note that

$$F'(x) = \begin{cases} 0, & -\infty < x < 0, \\ \dfrac{1}{40} e^{-x/40}, & 0 < x < \infty. \end{cases}$$

Also $F'(0)$ does not exist. (Sketch a graph of $y = F(x)$ to see why this is true.)

Since there are no steps or jumps in a distribution function, $F(x)$, of the continuous type, it must be true that

$$P(X = b) = 0$$

for all real values of b. This agrees with the fact that the integral $\int_b^b f(x)\,dx$ is taken to be zero in calculus. Thus we see that

$$P(a \leq X \leq b) = P(a < X < b) = P(a \leq X < b) = P(a < X \leq b) = F(b) - F(a),$$

provided that X is a random variable of the continuous type. Moreover, we can change the definition of a p.d.f. of a random variable of the continuous type at a finite (actually countable) number of points without altering the distribution of probability. For illustration,

$$f(x) = \begin{cases} 0, & -\infty < x < 0, \\ \dfrac{1}{40} e^{-x/40}, & 0 \leq x < \infty, \end{cases}$$

and

$$f(x) = \begin{cases} 0, & -\infty < x \leq 0, \\ \dfrac{1}{40} e^{-x/40}, & 0 < x < \infty, \end{cases}$$

are equivalent in the computation of probabilities involving this random variable.

Example 4.1-3   Let Y be a continuous random variable with p.d.f. $g(y) = 2y$, $0 < y < 1$. The distribution function of Y is defined by

$$G(y) = \begin{cases} 0, & y < 0, \\ \displaystyle\int_0^y 2t\,dt = y^2, & 0 \leq y < 1, \\ 1, & 1 \leq y. \end{cases}$$

Figure 4.1-2 gives the graph of the p.d.f. $g(y)$ and the graph of the distribution function $G(y)$. For examples of computations of probabilities, consider

$$P\left(\frac{1}{2} < Y \leq \frac{3}{4}\right) = G\left(\frac{3}{4}\right) - G\left(\frac{1}{2}\right) = \left(\frac{3}{4}\right)^2 - \left(\frac{1}{2}\right)^2 = \frac{5}{16}$$

and

$$P\left(\frac{1}{4} \leq Y < 2\right) = G(2) - G\left(\frac{1}{4}\right) = 1 - \left(\frac{1}{4}\right)^2 = \frac{15}{16}.$$

Recall that the p.d.f. $f(x)$ of a random variable of the discrete type is bounded by one because $f(x)$ gives a probability, namely

$$f(x) = P(X = x).$$

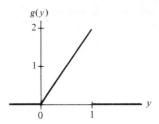

FIGURE 4.1-2

For random variables of the continuous type, the p.d.f. does not have to be bounded (see Exercises 4.1-2(c) and 4.1-3(c)). The restriction is that the area between the p.d.f. and the x axis must equal one. Furthermore, it should be noted that the p.d.f. of a random variable X of the continuous type does not need to be a continuous function. For example,

$$f(x) = \begin{cases} \dfrac{1}{2}, & 0 < x < 1 \quad \text{or} \quad 2 < x < 3, \\ 0, & \text{elsewhere,} \end{cases}$$

enjoys the properties of a p.d.f. of a distribution of the continuous type, and yet $f(x)$ has discontinuities at $x = 0, 1, 2,$ and 3. However, the distribution function associated with a distribution of the continuous type is always a continuous function.

For continuous-type random variables, the definitions associated with mathematical expectation are the same as those in the discrete case except that integrals replace summations. For illustration, let X be a continuous random variable with a p.d.f. $f(x)$.

The **expected value of X** or **mean of X** is

$$\mu = E(X) = \int_{-\infty}^{\infty} xf(x) \, dx.$$

The **variance of X** is

$$\sigma^2 = \text{Var}(X) = \int_{-\infty}^{\infty} (x - \mu)^2 f(x) \, dx.$$

The **standard deviation of X** is

$$\sigma = \sqrt{\text{Var}(X)}.$$

The **moment-generating function**, if it exists, is

$$M(t) = \int_{-\infty}^{\infty} e^{tx} f(x) \, dx, \qquad -h < t < h.$$

Moreover, results such as $\sigma^2 = E(X^2) - \mu^2$ and $\mu = M'(0)$, $\sigma^2 = M''(0) - [M'(0)]^2$ are still valid. Again it is important to note that the moment-generating function completely determines the distribution.

Example 4.1-4   For the random variable Y in Example 4.1-3,

$$\mu = E(Y) = \int_0^1 y(2y)\, dy = \left[\left(\frac{2}{3}\right)y^3\right]_0^1 = \frac{2}{3}$$

and

$$\sigma^2 = \text{Var}(Y) = E(Y^2) - \mu^2$$

$$= \int_0^1 y^2(2y)\, dy - \left(\frac{2}{3}\right)^2 = \left[\left(\frac{1}{2}\right)y^4\right]_0^1 - \frac{4}{9} = \frac{1}{18}.$$

Example 4.1-5   Let X have the p.d.f.

$$f(x) = \begin{cases} xe^{-x}, & 0 \le x < \infty, \\ 0, & \text{elsewhere.} \end{cases}$$

Then

$$M(t) = \int_0^\infty e^{tx} xe^{-x}\, dx = \lim_{b \to \infty} \int_0^b xe^{-(1-t)x}\, dx$$

$$= \lim_{b \to \infty} \left[-\frac{xe^{-(1-t)x}}{1-t} - \frac{e^{-(1-t)x}}{(1-t)^2} \right]_0^b$$

$$= \lim_{b \to \infty} \left[-\frac{be^{-(1-t)b}}{1-t} - \frac{e^{-(1-t)b}}{(1-t)^2} \right] + \frac{1}{(1-t)^2}$$

$$= \frac{1}{(1-t)^2},$$

provided that $t < 1$. Note that $M(0) = 1$, which is true for every moment-generating function. Now

$$M'(t) = \frac{2}{(1-t)^3} \quad \text{and} \quad M''(t) = \frac{6}{(1-t)^4}.$$

Thus

$$\mu = M'(0) = 2$$

and

$$\sigma^2 = M''(0) - [M'(0)]^2 = 6 - 2^2 = 2.$$

Let X be a continuous-type random variable with p.d.f. $f(x)$ and distribution function $F(x)$. The $(100p)$th **percentile** is a number π_p such that the area under $f(x)$ to the left of π_p is p. That is,

$$p = \int_{-\infty}^{\pi_p} f(x)\, dx = F(\pi_p).$$

The 50th percentile is called the **median**. We let $m = \pi_{0.50}$. The 25th and 75th percentiles are called the **first** and **third quartiles**, respectively, denoted by $q_1 = \pi_{0.25}$ and $q_3 = \pi_{0.75}$. Of course, the median $m = \pi_{0.50} = q_2$ is also called the second quartile.

Example 4.1-6 Let the p.d.f. of X be $f(x) = 1 - |x - 1|, 0 \le x \le 2$. Then the distribution function of X is defined by

$$F(x) = \begin{cases} 0, & x < 0, \\ \dfrac{x^2}{2}, & 0 \le x < 1, \\ 1 - \dfrac{(2-x)^2}{2}, & 1 \le x < 2, \\ 1, & 2 \le x. \end{cases}$$

To find, for example, the 32nd percentile, $\pi_{0.32}$, we solve

$$F(\pi_{0.32}) = 0.32.$$

Since $\pi_{0.32}$ is between 0 and 1 for this distribution because $0.32 < F(1) = 1/2$, we have

$$F(\pi_{0.32}) = \frac{\pi_{0.32}^2}{2} = 0.32$$

or

$$\pi_{0.32} = \sqrt{0.64} = 0.8.$$

To find the 92nd percentile, $\pi_{0.92}$, we first note that the 92nd percentile is between 1 and 2. Thus we solve

$$F(\pi_{0.92}) = 1 - \frac{(2 - \pi_{0.92})^2}{2} = 0.92,$$

which yields

$$\pi_{0.92} = 1.6.$$

The 32nd and 92nd percentiles are shown on the graph of the p.d.f. in Figure 4.1-3 and on the graph of the distribution function in Figure 4.1-4.

Exercises

4.1-1 Let the random variable X have the p.d.f. $f(x) = 2(1 - x), 0 \le x \le 1$, zero elsewhere.
(a) Sketch the graph of this p.d.f.
(b) Determine and sketch the graph of the distribution function of X.
(c) Find (i) $P(0 \le X \le 1/2)$, (ii) $P(1/4 \le X \le 3/4)$, (iii) $P(X = 3/4)$, and (iv) $P(X \ge 3/4)$.

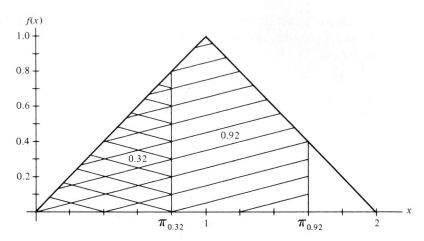

FIGURE 4.1-3

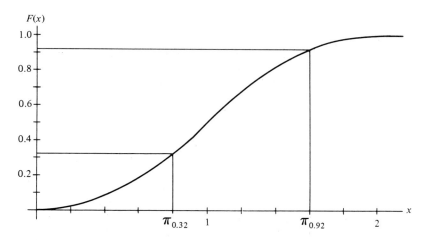

FIGURE 4.1-4

4.1-2 For each of the following functions, (i) find the constant c so that $f(x)$ is a p.d.f. of a random variable X, (ii) find the distribution function, $F(x) = P(X \le x)$, and (iii) sketch graphs of $y = f(x)$ and $y = F(x)$.
  (a) $f(x) = x^3/4$, $0 < x < c$;
  (b) $f(x) = (3/16)x^2$, $-c < x < c$;
  (c) $f(x) = c/\sqrt{x}$, $0 < x < 1$. Is this p.d.f. bounded?

4.1-3 For each of the following functions, (i) find the constant c so that $f(x)$ is a p.d.f. of a random variable X, (ii) find the distribution function, $F(x) = P(X \le x)$, and (iii) sketch graphs of $y = f(x)$ and $y = F(x)$.
  (a) $f(x) = 4x^c$, $0 \le x \le 1$,

(b) $f(x) = c\sqrt{x}, \quad 0 \le x \le 4,$
(c) $f(x) = c/x^{3/4}, \quad 0 < x < 1.$

4.1-4 For each of the distributions in Exercise 4.1-2, find μ, σ^2, and σ.

4.1-5 For each of the distributions in Exercise 4.1-3, find μ, σ^2, and σ.

4.1-6 Let $f(x) = \cos x$, $0 \le x \le \pi/2$, be the p.d.f. of X.
(a) Sketch the graph of this p.d.f.
(b) Determine and sketch the graph of the distribution function of X.

4.1-7 Let $f(x) = (1/2) \sin x$, $0 \le x \le \pi$, be the p.d.f. of X. Find
(a) μ.
(b) σ^2.
(c) Sketch the graph of the p.d.f. of X.
(d) Determine and sketch the graph of the distribution function of X.

4.1-8 The p.d.f. of X is $f(x) = c/x^2$, $1 < x < \infty$, zero elsewhere.
(a) Calculate the value of c so that $f(x)$ is a p.d.f.
(b) Show that $E(X)$ does not exist.

4.1-9 The p.d.f. of Y is $g(y) = d/y^3$, $1 < y < \infty$, zero elsewhere.
(a) Calculate the value of d so that $g(y)$ is a p.d.f.
(b) Find $E(Y)$.
(c) Show that $\mathrm{Var}(Y)$ does not exist.

4.1-10 Sketch the graphs of the following probability density functions. Also find and sketch the graphs of the distribution functions associated with these distributions. Note carefully the relationship between the shape of the graph of the p.d.f. and the concavity of the graph of the distribution function.

(a) $f(x) = \left(\dfrac{3}{2}\right)x^2, \quad -1 < x < 1.$

(b) $f(x) = \dfrac{1}{2}, \quad -1 < x < 1.$

(c) $f(x) = \begin{cases} x + 1, & -1 < x < 0, \\ 1 - x, & 0 \le x < 1. \end{cases}$

4.1-11 Find the mean and variance for each of the distributions in Exercise 4.1-10.

4.1-12 Let $R(t) = \ln M(t)$, where $M(t)$ is the moment-generating function of a random variable of the continuous type. Show that
(a) $\mu = R'(0)$.
(b) $\sigma^2 = R''(0)$.

4.1-13 If $M(t) = (1 - t)^{-2}$, $t < 1$, use $R(t) = \ln M(t)$ and the result in Exercise 4.1-12 to find
(a) μ.
(b) σ^2.

4.1-14 Find the moment-generating function $M(t)$ of the distribution with p.d.f. $f(x) = (1/10)e^{-x/10}$, $0 < x < \infty$. Use $M(t)$ or $R(t) = \ln M(t)$ to determine the mean μ and the variance σ^2.

4.1-15 The logistic distribution is associated with the distribution function $F(x) = (1 + e^{-x})^{-1}$, $-\infty < x < \infty$. Find the p.d.f. of the logistic distribution and show that its graph is symmetric about $x = 0$.

4.1-16 Let $f(x) = 1/2$, $0 < x < 1$ or $2 < x < 3$, zero elsewhere, be the p.d.f. of X.
(a) Sketch the graph of this p.d.f.
(b) Define the distribution function of X and sketch its graph.
(c) Find $q_1 = \pi_{0.25}$.
(d) Find $m = \pi_{0.50}$. Is it unique?
(e) Find $q_3 = \pi_{0.75}$.

4.1-17 Let $f(x) = 1/2$, $-1 < x < 1$. Find
(a) $m = \pi_{0.5}$.
. (b) $q_1 = \pi_{0.25}$.
(c) $\pi_{0.90}$.

4.1-18 Let $f(x) = (x + 1)/2$, $-1 < x < 1$. Find
(a) $\pi_{0.64}$.
(b) $q_1 = \pi_{0.25}$.
(c) $\pi_{0.81}$.

4.1-19 Let the random variable X_n have the p.d.f. $f(x_n) = n$, $0 < x_n < 1/n$.
(a) Define the distribution function of X_n, say $F_n(x_n)$.
(b) Graph the p.d.f. and the distribution function for $n = 1, 2, 3, \ldots, 10$.

4.1-20 One hundred observations of X having a certain distribution were taken and then ordered, yielding

0.19	0.25	0.26	0.26	0.28	0.31	0.39	0.42	0.44	0.45
0.48	0.53	0.54	0.57	0.66	0.69	0.70	0.72	0.73	0.79
0.83	0.85	0.87	0.90	0.91	0.94	0.96	0.99	1.01	1.02
1.02	1.03	1.05	1.07	1.13	1.13	1.14	1.20	1.22	1.27
1.28	1.30	1.34	1.34	1.37	1.39	1.40	1.43	1.43	1.46
1.49	1.50	1.51	1.53	1.54	1.56	1.56	1.57	1.60	1.60
1.61	1.62	1.62	1.62	1.63	1.64	1.66	1.67	1.70	1.71
1.71	1.71	1.72	1.73	1.73	1.74	1.75	1.76	1.77	1.80
1.81	1.84	1.85	1.85	1.86	1.87	1.87	1.88	1.90	1.91
1.91	1.92	1.92	1.93	1.93	1.93	1.94	1.95	1.98	1.99

(a) Calculate the sample mean $\bar{x}$.
(b) Calculate the sample standard deviation s.
(c) Group these data into 10 classes: $(0.00, 0.20]$, $(0.20, 0.40]$, $\ldots$, $(1.80, 2.00]$.
(d) Construct a relative frequency histogram with the p.d.f. $f(x) = x/2$, $0 < x < 2$, superimposed. Is there a good fit?
(e) Construct an ogive curve with the distribution function $F(x) = x^2/4$, $0 < x < 2$, superimposed. Is there a good fit?
(f) Find μ and σ, assuming that the p.d.f. of X is $f(x) = x/2$, $0 < x \le 2$. Is $\bar{x}$ close to μ? Is s close to σ?

4.2 The Uniform and Exponential Distributions

Let the random variable X denote the outcome when a point is selected at random from an interval $[a, b]$, $-\infty < a < b < \infty$. If the experiment is performed in a fair manner, it is reasonable to assume that the probability that the point is selected from the interval $[a, x]$, $a \le x < b$ is $(x - a)/(b - a)$. That is, the probability is proportional to the length of the interval so that the distribution function of X is

$$F(x) = \begin{cases} 0, & x < a, \\ \dfrac{x - a}{b - a}, & a \le x < b, \\ 1, & b \le x. \end{cases}$$

Because X is a continuous-type random variable, $F'(x)$ is equal to the p.d.f. of X whenever $F'(x)$ exists; thus when $a < x < b$, we have $f(x) = F'(x) = 1/(b - a)$.

The random variable X has a **uniform distribution** if its p.d.f. is equal to a constant on its support. In particular, if the support is the interval $[a, b]$, then

$$f(x) = \frac{1}{b - a}, \qquad a \le x \le b.$$

Moreover, we shall say that X is $U(a, b)$. This distribution is also referred to as **rectangular** because the graph of $f(x)$ suggests that name. See Figure 4.2-1 for the graph of $f(x)$ and the distribution function $F(x)$. Note that we could have taken $f(a) = 0$ or $f(b) = 0$ without altering the probabilities, since this is a continuous-type distribution, and we shall do this in some cases.

The mean, variance, and moment-generating function of X are not difficult to calculate (see Exercise 4.2-1). They are

$$\mu = \frac{a + b}{2}, \qquad \sigma^2 = \frac{(b - a)^2}{12},$$

$$M(t) = \begin{cases} \dfrac{e^{tb} - e^{ta}}{t(b - a)}, & t \neq 0, \\ 1, & t = 0. \end{cases}$$

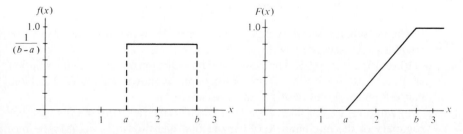

FIGURE 4.2-1

An important uniform distribution is that for which $a = 0$ and $b = 1$, namely $U(0, 1)$. If X is $U(0, 1)$, approximate values of X can be simulated on most computers using a random number generator. In fact, it should be called a **pseudo-random number generator** because the programs that produce the random numbers are usually such that if the starting number (the seed number) is known, all subsequent numbers in the sequence may be determined by simple arithmetical operations. Yet, despite their deterministic origin, these computer-produced numbers do behave as if they were truly randomly generated, and we shall not encumber our terminology by adding *pseudo*. Examples of computer-produced random numbers are given in the Appendix in Table IX. Treat each of these numbers as a decimal—that is, divide each of them by 10^4.

The next example looks at some of the numbers in Table IX.

Example 4.2-1   In this example we use some statistical techniques that were studied earlier and that can be used to help determine whether the uniform distribution is the appropriate probability model. In particular, to help decide whether or not the data in Appendix Table IX are observations from a uniform distribution, $U(0, 1)$, we could construct a histogram and empirical distribution function of all of the data, comparing these with the p.d.f. and distribution function of the uniform distribution, $U(0, 1)$. Instead, the techniques we illustrate in this example use fewer numbers of observations. Suppose that we select the first 19 numbers in the third column of Table IX. They are

0.6960	0.4658	0.4930	0.7843	0.1233	0.4779	0.4871
0.0315	0.6814	0.2581	0.7244	0.9705	0.0460	0.8287
0.2055	0.2906	0.5334	0.8071	0.3384		

For these 19 numbers, the sample mean is $\bar{x} = 0.4865$ and the sample standard deviation is $s = 0.2797$. These should be compared with the mean, $\mu = 0.5000$, and standard deviation, $\sigma = \sqrt{1/12} = 0.2887$, of the uniform distribution, $U(0, 1)$. Graphically, a technique that can sometimes yield insight into the data is the q–q plot of the data quantiles versus the corresponding $U(0, 1)$ distribution quantiles. We first order the data as follows:

0.0315	0.0460	0.1233	0.2055	0.2581	0.2906	0.3384
0.4658	0.4779	0.4871	0.4930	0.5334	0.6814	0.6960
0.7244	0.7843	0.8071	0.8287	0.9705		

The reason for selecting a sample of size 19 is because the respective quantiles are associated with $i/(n + 1)$, $i = 1, 2, \ldots, 19$, namely 0.05, 0.10, 0.15, ..., 0.95. The respective quantiles of the $U(0, 1)$ distribution are 0.05, 0.10, 0.15, ..., 0.95. The q–q plot is shown in Figure 4.2-2. Note that the points do fall close to the line $y = x$.

In addition, we consider a box-and-whisker diagram with the five-number summary of the minimum, 0.0315; the first quartile, $q_1 = 0.2581$; the median, $m = 0.4871$; the third quartile, $q_3 = 0.7244$; and the maximum,

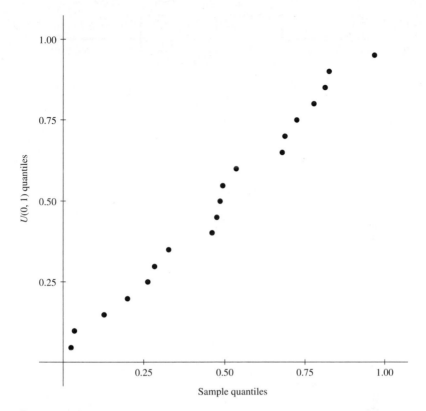

FIGURE 4.2-2

0.9705. The box plot is shown in Figure 4.2-3. Note especially the symmetry of this box plot.

Since the data are ordered, it is also quite easy to construct the empirical distribution function and draw the theoretical distribution function on the same graph. This is illustrated in Figure 4.2-4. In Section 10.7 a statistical test is given that is based on the greatest vertical distance between the empirical and theoretical distribution functions. This test is called the Kolmogorov-Smirnov goodness of fit test, but for now we simply use an "eyeball" test.

We must point out that every set of 19 observations from Table IX or from a $U(0, 1)$ distribution will not yield characteristics and figures exactly like these, and in some cases the fit will perhaps not look very good. In addition, larger samples are preferable to samples of size 19, especially if a computer can help with the computations and the graphics.

We turn now to a continuous distribution that is related to the Poisson distribution. When previously observing a process of the (approximate) Poisson type, we counted the number of changes occurring in a given interval. This number was a

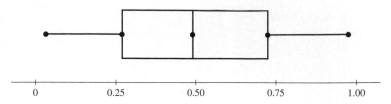

FIGURE 4.2-3

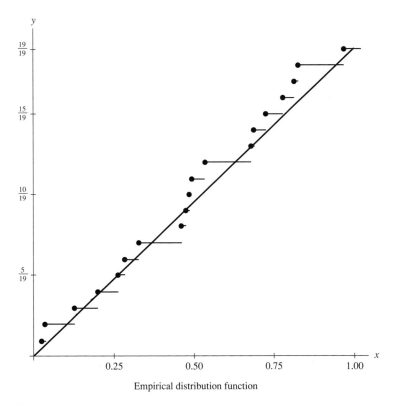

Empirical distribution function

FIGURE 4.2-4

discrete-type random variable with a Poisson distribution. But not only is the number of changes a random variable; the waiting times between successive changes are also random variables. However, the latter are of the continuous type, since each of them can assume any positive value. In particular, let W denote the waiting time until the first change occurs when observing a Poisson process in which the mean number of changes in the unit interval is λ. Then W is a continuous-type random variable, and we proceed to find its distribution function.

Because this waiting time is nonnegative, the distribution function $F(w) = 0$, $w < 0$. For $w \geq 0$,

$$
\begin{aligned}
F(w) = P(W \leq w) &= 1 - P(W > w) \\
&= 1 - P \text{ (no changes in } [0, w]) \\
&= 1 - e^{-\lambda w},
\end{aligned}
$$

since we previously discovered that $e^{-\lambda w}$ equals the probability of no changes in an interval of length w. That is, if the mean number of changes per unit interval is λ, then the mean number of changes in an interval of length w is proportional to w, namely, λw. Thus, when $w > 0$, the p.d.f. of W is given by

$$
\begin{aligned}
F'(w) &= \lambda e^{-\lambda w} \\
&= f(w).
\end{aligned}
$$

We often let $\lambda = 1/\theta$ and say that the random variable X has an **exponential distribution** if its p.d.f. is defined by

$$
f(x) = \frac{1}{\theta} e^{-x/\theta}, \qquad 0 \leq x < \infty,
$$

where the parameter $\theta > 0$. Accordingly, the waiting time W until the first change in a Poisson process has an exponential distribution with $\theta = 1/\lambda$. To determine the exact meaning of the parameter θ, we first find the moment-generating function. It is

$$
\begin{aligned}
M(t) &= \int_0^\infty e^{tx} \left(\frac{1}{\theta}\right) e^{-x/\theta} \, dx = \lim_{b \to \infty} \int_0^b \left(\frac{1}{\theta}\right) e^{-(1 - \theta t)x/\theta} \, dx \\
&= \lim_{b \to \infty} \left[-\frac{e^{-(1 - \theta t)x/\theta}}{1 - \theta t} \right]_0^b = \frac{1}{1 - \theta t}, \qquad t < \frac{1}{\theta}.
\end{aligned}
$$

Thus

$$
M'(t) = \frac{\theta}{(1 - \theta t)^2}
$$

and

$$
M''(t) = \frac{2\theta^2}{(1 - \theta t)^3}.
$$

Hence, for an exponential distribution, we have

$$
\mu = M'(0) = \theta \qquad \text{and} \qquad \sigma^2 = M''(0) - [M'(0)]^2 = \theta^2.
$$

So if λ is the mean number of changes in the unit interval, then $\theta = 1/\lambda$ is the mean waiting time for the first change. In particular, suppose that $\lambda = 7$ is the mean number of changes per minute; then the mean waiting time for the first change is 1/7 of a minute.

Example 4.2-2 Let X have an exponential distribution with a mean of 40. The p.d.f. of X is

$$f(x) = \frac{1}{40} e^{-x/40}, \qquad 0 \leq x < \infty.$$

See Figure 4.1-1 for the graph of this p.d.f. The probability that X is less than 36 is

$$P(X < 36) = \int_0^{36} \frac{1}{40} e^{-x/40} \, dx = 1 - e^{-36/40} = 0.593.$$

Let X have an exponential distribution with mean $\mu = \theta$. Then the distribution function of X is

$$F(x) = \begin{cases} 0, & -\infty < x < 0, \\ 1 - e^{-x/\theta}, & 0 \leq x < \infty. \end{cases}$$

The p.d.f. and distribution function are graphed in Figure 4.2-5 for $\theta = 5$. The median, m, is found by solving $F(m) = 0.5$. That is,

$$1 - e^{-m/\theta} = 0.5.$$

Thus

$$m = -\theta \ln (0.5).$$

The median $m = -5 \ln (0.5) = 3.466$ and mean $\theta = 5$ are both indicated on the graphs.

It is useful to note that for an exponential random variable, X, we have that

$$P(X > x) = 1 - F(x) = 1 - (1 - e^{-x/\theta})$$
$$= e^{-x/\theta}.$$

Example 4.2-3 Customers arrive in a certain shop according to an approximate Poisson process at a mean rate of 20 per hour. What is the probability that the shopkeeper will have to wait more than 5 minutes for the arrival of the first customer? Let X denote the waiting time *in minutes* until the first customer arrives and note that $\lambda = 1/3$ is the expected number of arrivals per minute. Thus

$$\theta = \frac{1}{\lambda} = 3$$

and

$$f(x) = \frac{1}{3} e^{-(1/3)x}, \qquad 0 \leq x < \infty.$$

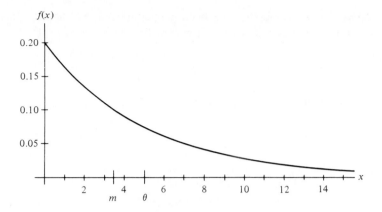

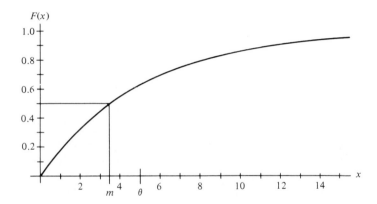

FIGURE 4.2-5

Hence

$$P(X > 5) = \int_{5}^{\infty} \frac{1}{3} e^{-(1/3)x} \, dx = e^{-5/3} = 0.1889.$$

The median time until the first arrival is

$$m = -3 \ln(0.5) = 2.0794.$$

Example 4.2-4 Suppose that the life of a certain type of electronic component has an exponential distribution with a mean life of 500 hours. If X denotes the life of this component (or the time to failure of this component), then

$$P(X > x) = \int_{x}^{\infty} \frac{1}{500} e^{-t/500} \, dt = e^{-x/500}.$$

Suppose that the component has been in operation for 300 hours. The conditional probability that it will last for another 600 hours is

$$P(X > 900 | X > 300) = \frac{P(X > 900)}{P(X > 300)} = \frac{e^{-900/500}}{e^{-300/500}} = e^{-6/5}.$$

It is important to note that this conditional probability is exactly equal to $P(X > 600) = e^{-6/5}$. That is, the probability that it will last an additional 600 hours, given that it has operated 300 hours, is the same as the probability that it would last 600 hours when first put into operation. Thus, for such components, an old component is as good as a new one, and we say that the failure rate is constant. Certainly, with constant failure rate, there is no advantage in replacing components that are operating satisfactorily. Obviously, this is not true in practice because most would have an increasing failure rate; hence the exponential distribution is probably not the best model for the probability distribution of such a life.

REMARK In Exercise 4.2-11, the result of Example 4.2-4 is generalized; namely, if the length of life has an exponential distribution, the probability that it will last a time of at least $x + y$ units, given that it has lasted at least x units, is exactly the same as the probability that it will last at least y units when first put into operation. In effect, this states that the exponential distribution has a forgetfulness (or no memory) property. It is also interesting to observe that, for continuous random variables whose support is $(0, \infty)$, the exponential distribution is the only distribution with this forgetfulness property. Recall, however, that when we considered distributions of the discrete type, we noted that the geometric distribution has this property (see Exercise 3.6-10).

Returning to the uniform distribution, we let the distribution of X be $U(0, 1)$. Given n independent observations of X—for example, n random numbers—the following theorem enables us to simulate observations of a continuous random variable that has as its distribution function $y = F(x)$.

THEOREM 4.2-1 *Let Y have a distribution that is $U(0, 1)$. Let $F(x)$ have the properties of a distribution function of the continuous type with $F(a) = 0, F(b) = 1$, and suppose that $F(x)$ is strictly increasing for $a < x < b$, where a and b could be $-\infty$ and ∞, respectively. Then the random variable X defined by $X = F^{-1}(Y)$ is a continuous-type random variable with distribution function $F(x)$.*

PROOF: The distribution function of X is

$$P(X \leq x) = P[F^{-1}(Y) \leq x].$$

However, since $F(x)$ is strictly increasing, $F^{-1}(Y) \leq x$ is equivalent to $Y \leq F(x)$ and hence

$$P(X \leq x) = P[Y \leq F(x)].$$

But Y is $U(0, 1)$; so $P(Y \leq y) = y$ for $0 < y < 1$, and accordingly

$$P(X \leq x) = P[Y \leq F(x)] = F(x), \qquad 0 < F(x) < 1.$$

That is, the distribution function of X is $F(x)$. □

We now illustrate how Theorem 4.2-1 can be used to simulate observations from a given distribution.

Example 4.2-5 To see how we can simulate observations from an exponential distribution with a mean of $\theta = 10$, note that the distribution function of X is $F(x) = 1 - e^{-x/10}$ when $0 \leq x < \infty$. Solving $y = F(x)$ for x yields $x = F^{-1}(y) = -10 \ln (1 - y)$. So given a sequence of random numbers y_1, $y_2, \ldots, y_n$, we would expect $x_i = -10 \ln (1 - y_i)$, $i = 1, 2, \ldots, n$, to represent n observations of an exponential random variable X with mean $\theta = 10$. Table 4.2-1 gives the values of 15 random numbers, y_i, along with the values of $x_i = -10 \ln (1 - y_i)$.

As a rough check on whether these 15 observations seem to come from an exponential distribution with mean 10, let us construct a q–q plot. We must order these 15 observations and plot them against the corresponding quantiles q_i of the exponential distribution with mean $\theta = 10$. These are the solution of

$$F(q_i) = 1 - e^{-q_i/10} = \frac{i}{16}$$

TABLE 4.2-1

y	$x = -10 \ln (1 - y)$
0.1514	1.6417
0.6697	11.0775
0.0527	0.5414
0.4749	6.4417
0.2900	3.4249
0.2354	2.6840
0.9662	33.8729
0.0043	0.0431
0.1003	1.0569
0.9192	25.1578
0.4971	6.8736
0.7293	13.0674
0.9118	24.2815
0.8225	17.2878
0.5915	8.9526

or

$$q_i = -10 \ln \left(1 - \frac{i}{16} \right).$$

The ordered x_i's and the respective quantiles, q_i, are given in Table 4.2-2.

Figure 4.2-6 shows the q–q plot for these data. The linearity of this q–q plot confirms the possibility of the exponential distribution being the correct model. The box plot, Figure 4.2-7, shows the skewness of the data as it should be for an exponential distribution. In addition, the sample mean is $\bar{x} = 10.4270$ and the sample standard deviation is $s = 10.4311$, both of which are close to $\theta = 10$. We again caution the reader that in similar simulations, the fit is sometimes better and sometimes not as good.

The following probability integral transformation theorem is the converse of Theorem 4.2-1.

THEOREM 4.2-2   *Let X have the distribution function $F(x)$ of the continuous type that is strictly increasing. Then the random variable Y, defined by $Y = F(X)$, has a distribution that is $U(0, 1)$.*

PROOF: The distribution function of Y is

$$P(Y \le y) = P[F(X) \le y], \qquad 0 < y < 1.$$

However, $F(X) \le y$ is equivalent to $X \le F^{-1}(y)$; thus

$$P(Y \le y) = P[X \le F^{-1}(y)].$$

TABLE 4.2-2

Ordered x's	Quantiles
0.0431	0.6454
0.5414	1.3353
1.0569	2.0764
1.6417	2.8768
2.6840	3.7469
3.4249	4.7000
6.4417	5.7536
6.8736	6.9315
8.9526	8.2668
11.0775	9.8083
13.0674	11.6315
17.2878	13.8629
24.2815	16.7398
25.1578	20.7944
33.8729	27.7259

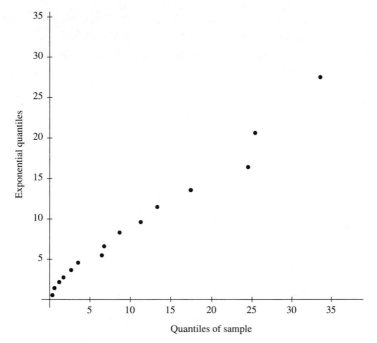

FIGURE 4.2-6

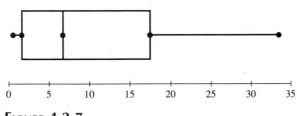

FIGURE 4.2-7

Since $P(X \leq x) = F(x)$, we have that

$$P(Y \leq y) = P[X \leq F^{-1}(y)] = F[F^{-1}(y)] = y, \qquad 0 < y < 1,$$

which is the distribution function of a $U(0, 1)$ random variable. $\square$

Although in our statements and proofs of Theorems 4.2-1 and 4.2-2 we required $F(x)$ to be strictly increasing, this restriction can be dropped, and both theorems are still true. In our exposition we did not want to trouble students with certain difficulties that are experienced if $F(x)$ is not strictly increasing.

Exercises

4.2-1 Show that the mean, variance, and moment-generating function of the uniform distribution are as given in this section.

4.2-2 Let $f(x) = 1/2$, $-1 \le x \le 1$, be the p.d.f. of X. Graph the p.d.f. and distribution function, and record the mean and variance of X.

4.2-3 Customers arrive randomly at a bank teller's window. Given that one customer arrived during a particular 10-minute period, let X equal the time within the 10 minutes that the customer arrived. If X is $U(0, 10)$, find
(a) The p.d.f. of X.
(b) $P(X \ge 8)$.
(c) $P(2 \le X < 8)$.
(d) $E(X)$.
(e) $\text{Var}(X)$.

4.2-4 If the moment-generating function of X is

$$M(t) = \frac{e^{5t} - e^{4t}}{t}, \qquad t \neq 0, \qquad \text{and} \qquad M(0) = 1,$$

find
(a) $E(X)$.
(b) $\text{Var}(X)$.
(c) $P(4.2 < X \le 4.7)$.

4.2-5 Let Y have a uniform distribution $U(0, 1)$, and let
$$W = a + (b - a)Y, \ a < b.$$
(a) Find the distribution function of W.

HINT: Find $P[a + (b - a)Y \le w]$.

(b) How is W distributed?

4.2-6 Let X have an exponential distribution with a mean of $\theta = 20$. Compute
(a) $P(10 < X < 30)$.
(b) $P(X > 30)$.
(c) $P(X > 40 | X > 10)$.

4.2-7 Let the p.d.f. of X be $f(x) = (1/2)e^{-x/2}$, $0 \le x < \infty$.
(a) What are the mean, variance, and moment-generating function of X?
(b) Calculate $P(X > 3)$.
(c) Calculate $P(X > 5 | X > 2)$.

4.2-8 Telephone calls enter a college switchboard according to a Poisson process on the average of two every 3 minutes. Let X denote the waiting time until the first call that arrives after 10 A.M.
(a) What is the p.d.f. of X?
(b) Find $P(X > 2)$.

4.2-9 What are the p.d.f., the mean, and the variance of X if the moment-generating function of X is given by the following?
(a) $M(t) = \dfrac{1}{1 - 3t}, \qquad t < \dfrac{1}{3}.$

(b) $M(t) = \dfrac{3}{3 - t}, \qquad t < 3.$

4.2-10 Let X equal the number of students who use a college card catalog every 15 minutes. Assume that X has a Poisson distribution with mean 5. Let W equal the time in minutes between two student arrivals. Then W has an exponential distribution with mean $\theta = 3$. Find
(a) $P(W \geq 6)$.
(b) $P(W > 12 | W > 6)$.

4.2-11 Let X have an exponential distribution with mean $\theta > 0$. Show that

$$P(X > x + y | X > x) = P(X > y).$$

4.2-12 Let $F(x)$ be the distribution function of the continuous-type random variable X, and assume that $F(x) = 0$ for $x \leq 0$ and $0 < F(x) < 1$ for $0 < x$. Prove that if

$$P(X > x + y | X > x) = P(X > y),$$

then

$$F(x) = 1 - e^{-\lambda x}, \qquad 0 < x.$$

HINT: Show that $g(x) = 1 - F(x)$ satisfies the functional equation

$$g(x + y) = g(x)g(y),$$

which implies that $g(x) = a^{cx}$.

4.2-13 Let X equal the number of bad records in each 100 feet of a used computer tape. Assume that X has a Poisson distribution with mean 2.5. Let W equal the number of feet before the first bad record is found.
(a) Give the mean number of flaws per foot.
(b) How is W distributed?
(c) Give the mean and variance of W.
(d) Find (i) $P(W \leq 20)$, (ii) $P(W > 40)$, and (iii) $P(W > 60 | W > 20)$.

4.2-14 Let W equal the number of feet between bad records on the computer tape in Exercise 4.2-13. Ten observations of W are

$$67 \quad 16 \quad 7 \quad 35 \quad 97 \quad 28 \quad 74 \quad 5 \quad 9 \quad 37$$

(a) Calculate the sample mean $\overline{w}$ and sample variance s^2 of these data.
(b) Graph the empirical distribution function of these data along with the distribution function for the exponential distribution with mean $\theta = 40$.
(c) Do these data look like observations from an exponential distribution with a mean of 40?

4.2-15 Let X equal the time between calls that are made over the public safety radio. On four different days (February 14, 21, and 28, and March 6) and during a period of one hour on each day, the following observations of X were made:

$$5 \quad 7 \quad 8 \quad 20 \quad 17 \quad 2 \quad 24 \quad 8 \quad 8 \quad 6 \quad 4$$
$$3 \quad 42 \quad 10 \quad 18 \quad 5 \quad 7 \quad 8 \quad 4 \quad 5 \quad 10$$

If calls arrive randomly in accordance with an approximate Poisson process, then the distribution of X should be approximately exponential.

(a) Calculate the values of the sample mean and sample standard deviation. Are they close to each other in value?

(b) Construct a q–q plot of the ordered observations versus the respective quantiles of the exponential distribution with a mean of $\theta = 1$. If this is approximately linear, the exponential model is supported. Since the mean of these data is not close to 1, this line will not have slope 1, but a linear fit will still indicate an exponential model. What is your conclusion?

(c) Construct a box-and-whisker diagram. Does it indicate that the data are skewed, as would be true for observations from an exponential distribution?

4.2-16   Some biology students were interested in analyzing the amount of time that bees spend in flower patches gathering nectar. In particular, they were wondering whether it made a difference whether there was a low density or a high density of flowers. Thirty-seven bees visited the low-density flower patch and spent the following times (in seconds):

10	40	20	85	105	85	270	40	60	90
360	30	520	210	45	16	6	420	90	515
71	448	353	526	279	279	505	157	240	840
240	180	300	450	225	60	300			

(a) Calculate the sample mean and sample standard deviation. Are they close in value to each other?

(b) Construct an ordered stem-and-leaf diagram.

(c) Construct a box-and-whisker diagram.

(d) Does the exponential probability model look like it could be used?

Thirty-nine bees visited the high-density flower patch and spent the following times (in seconds):

235	210	95	146	195	840	185	610	680	990
146	404	119	47	9	4	10	169	270	95
329	151	211	127	154	35	225	140	158	116
46	113	149	420	120	45	10	18	105	

(e) Calculate the sample mean and sample standard deviation. Are they close in value to each other?

(f) Construct an ordered stem-and-leaf diagram.

(g) Construct a box-and-whisker diagram.

(h) Does the exponential probability model look like it could be used?

Compare the data for the two groups by

(i) Constructing back-to-back stem-and-leaf diagrams.

(j) Constructing the two box plots on the same diagram.

(k) How would you compare the times in the low- and high-density flower patches?

4.2-17 There are times when a **shifted exponential model** is appropriate. Let the p.d.f. of X be $f(x) = \dfrac{1}{\theta} e^{-(x-\alpha)/\theta}$, $\alpha < x < \infty$.

(a) Define the distribution function of X.

(b) Calculate the mean and variance of X.

(c) Do you think that this would be a better model for any of the examples and exercises for which data are given? (Also see Exercise 10.7-13.)

4.2-18 The p.d.f. of X is $f(x) = 2x$, $0 < x < 1$.

(a) Find the distribution function of X.

(b) Describe how an observation of X can be simulated.

(c) Simulate 10 observations of X.

(d) Graph the empirical and theoretical distribution functions associated with X.

4.2-19 (a) Simulate a random sample of size $n = 10$ from an exponential distribution with mean $\theta = 3$.

(b) Sketch the empirical and theoretical distribution functions on the same set of axes.

4.2-20 A certain type of aluminum screen 2 feet in width has on the average three flaws in a 100-foot roll.

(a) What is the probability that the first 40 feet in a roll contain no flaws?

(b) What assumption did you make to solve part (a)?

4.2-21 Find the third quartile, q_3, for the exponential distribution with mean θ. If a random sample of size $n = 5$ is taken from this distribution, determine the probability that at least three of the sample items are less than q_3. Does the latter probability depend upon the fact that the distribution is exponential? Or would it hold for every distribution of the continuous type?

4.2-22 Let X have a **logistic distribution** with p.d.f.

$$f(x) = \frac{e^{-x}}{(1 + e^{-x})^2}, \quad -\infty < x < \infty.$$

Show that

$$Y = \frac{1}{1 + e^{-X}}$$

has a $U(0, 1)$ distribution.

4.3 The Gamma and Chi-Square Distributions

In the (approximate) Poisson process with mean λ, we have seen that the waiting time until the first change has an exponential distribution. We now let W denote the waiting time until the αth change occurs and find the distribution of W.

The distribution function of W, when $w \geq 0$, is given by

$$F(w) = P(W \leq w) = 1 - P(W > w)$$
$$= 1 - P \text{ (fewer than } \alpha \text{ changes occur in } [0, w])$$
$$= 1 - \sum_{k=0}^{\alpha-1} \frac{(\lambda w)^k e^{-\lambda w}}{k!}, \qquad (4.3\text{-}1)$$

since the number of changes in the interval $[0, w]$ has a Poisson distribution with mean λw. Because W is a continuous-type random variable, $F'(w)$ is equal to the p.d.f. of W whenever this derivative exists. We have, provided $w > 0$, that

$$F'(w) = \lambda e^{-\lambda w} - e^{-\lambda w} \sum_{k=1}^{\alpha-1} \left[\frac{k(\lambda w)^{k-1}\lambda}{k!} - \frac{(\lambda w)^k \lambda}{k!} \right]$$

$$= \lambda e^{-\lambda w} - e^{-\lambda w} \left[\lambda - \frac{\lambda(\lambda w)^{\alpha-1}}{(\alpha - 1)!} \right]$$

$$= \frac{\lambda(\lambda w)^{\alpha-1}}{(\alpha - 1)!} e^{-\lambda w}.$$

If $w < 0$, then $F(w) = 0$ and $F'(w) = 0$. A p.d.f. of this form is said to be one of the gamma type, and the random variable W is said to have a gamma distribution.

Before determining the characteristics of the gamma distribution, let us consider the gamma function for which the distribution is named. The **gamma function** is defined by

$$\Gamma(t) = \int_0^\infty y^{t-1} e^{-y} \, dy, \qquad 0 < t.$$

This integral is positive for $0 < t$ because the integrand is positive. Values of it are often given in a table of integrals. If $t > 1$, integration of the gamma function of t by parts yields

$$\Gamma(t) = [-y^{t-1} e^{-y}]_0^\infty + \int_0^\infty (t - 1) y^{t-2} e^{-y} \, dy$$

$$= (t - 1) \int_0^\infty y^{t-2} e^{-y} \, dy = (t - 1)\Gamma(t - 1).$$

For example, $\Gamma(6) = 5\Gamma(5)$ and $\Gamma(3) = 2\Gamma(2) = (2)(1)\Gamma(1)$. Whenever $t = n$, a positive integer, we have, by repeated application of $\Gamma(t) = (t - 1)\Gamma(t - 1)$, that

$$\Gamma(n) = (n - 1)\Gamma(n - 1) = (n - 1)(n - 2) \cdots (2)(1)\Gamma(1).$$

However,

$$\Gamma(1) = \int_0^\infty e^{-y} \, dy = 1.$$

Thus, when n is a positive integer, we have that

$$\Gamma(n) = (n - 1)!;$$

and, for this reason, the gamma function is called the generalized factorial. Incidentally, $\Gamma(1)$ corresponds to $0!$, and we have noted that $\Gamma(1) = 1$, which is consistent with earlier discussions.

Let us now formally define the p.d.f. of the gamma distribution and find its characteristics. The random variable X has a **gamma distribution** if its p.d.f. is defined by

$$f(x) = \frac{1}{\Gamma(\alpha)\theta^{\alpha}} x^{\alpha-1} e^{-x/\theta}, \qquad 0 \le x < \infty.$$

Hence, W, the waiting time until the αth change in a Poisson process, has a gamma distribution with parameters α and $\theta = 1/\lambda$. To see that $f(x)$ actually has the properties of a p.d.f., note that $f(x) \ge 0$ and

$$\int_{-\infty}^{\infty} f(x)\, dx = \int_{0}^{\infty} \frac{x^{\alpha-1} e^{-x/\theta}}{\Gamma(\alpha)\theta^{\alpha}}\, dx,$$

which, by the change of variables $y = x/\theta$, equals

$$\int_{0}^{\infty} \frac{(\theta y)^{\alpha-1} e^{-y}}{\Gamma(\alpha)\theta^{\alpha}}\, \theta\, dy = \frac{1}{\Gamma(\alpha)} \int_{0}^{\infty} y^{\alpha-1} e^{-y}\, dy = \frac{\Gamma(\alpha)}{\Gamma(\alpha)} = 1.$$

The moment-generating function of X is (Exercise 4.3-3)

$$M(t) = \frac{1}{(1 - \theta t)^{\alpha}}, \qquad t < \frac{1}{\theta}.$$

The mean and variance are (Exercise 4.3-4)

$$\mu = \alpha\theta \qquad \text{and} \qquad \sigma^2 = \alpha\theta^2.$$

Example 4.3-1 Suppose that an average of 30 customers per hour arrive at a shop in accordance with a Poisson process. That is, if a minute is our unit, then $\lambda = 1/2$. What is the probability that the shopkeeper will wait more than 5 minutes before both of the first two customers arrive? If X denotes the waiting time in minutes until the second customer arrives, then X has a gamma distribution with $\alpha = 2$, $\theta = 1/\lambda = 2$. Hence

$$P(X > 5) = \int_{5}^{\infty} \frac{x^{2-1} e^{-x/2}}{\Gamma(2)2^2}\, dx = \int_{5}^{\infty} \frac{x e^{-x/2}}{4}\, dx$$

$$= \frac{1}{4}[(-2)xe^{-x/2} - 4e^{-x/2}]_{5}^{\infty}$$

$$= \frac{7}{2} e^{-5/2} = 0.287.$$

We could also have used equation (4.3-1) with $\lambda = 1/\theta$ because α is an integer. From equation (4.3-1) we have

$$P(X > x) = \sum_{k=0}^{\alpha-1} \frac{(x/\theta)^k e^{-x/\theta}}{k!}.$$

Thus, with $x = 5$, $\alpha = 2$, and $\theta = 2$, this is equal to

$$P(X > x) = \sum_{k=0}^{2-1} \frac{(5/2)^k e^{-5/2}}{k!}$$

$$= e^{-5/2}\left(1 + \frac{5}{2}\right) = \left(\frac{7}{2}\right)e^{-5/2}.$$

Example 4.3-2 Telephone calls arrive at a switchboard at a mean rate of $\lambda = 2$ per minute according to a Poisson process. Let X denote the waiting time in minutes until the fifth call arrives. The p.d.f. of X, with $\alpha = 5$ and $\theta = 1/\lambda = 1/2$, is

$$f(x) = \frac{2^5 x^4}{4!} e^{-2x}, \qquad 0 \le x < \infty.$$

The mean and the variance of X are, respectively, $\mu = 5/2$ and $\sigma^2 = 5/4$.

In order to see the effect of the parameters on the shape of the p.d.f., several combinations of α and θ have been used for graphs that are displayed in Figure 4.3-1. Note that for a fixed θ, as α increases, the probability moves to the right. The same is true for increasing θ, with fixed α. Since $\theta = 1/\lambda$, as θ increases, λ decreases. That is, if $\theta_2 > \theta_1$, then $\lambda_2 = 1/\theta_2 < \lambda_1 = 1/\theta_1$. So if the mean number of changes per unit decreases, the waiting time to observe α changes can be expected to increase.

We now consider a special case of the gamma distribution that plays an important role in statistics. Let X have a gamma distribution with $\theta = 2$ and $\alpha = r/2$, where r is a positive integer. The p.d.f. of X is

$$f(x) = \frac{1}{\Gamma(r/2)2^{r/2}} x^{r/2-1} e^{-x/2}, \qquad 0 \le x < \infty.$$

We say that X has a **chi-square distribution with r degrees of freedom**, which we abbreviate by saying X is $\chi^2(r)$. The mean and the variance of this chi-square distribution are

$$\mu = \alpha\theta = \left(\frac{r}{2}\right)2 = r \qquad \text{and} \qquad \sigma^2 = \alpha\theta^2 = \left(\frac{r}{2}\right)2^2 = 2r.$$

That is, the mean equals the number of degrees of freedom, and the variance equals twice the number of degrees of freedom. An explanation of "number of degrees of

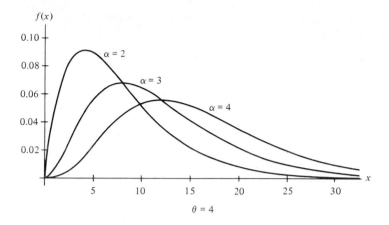

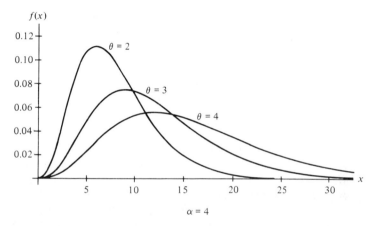

FIGURE 4.3-1

freedom'' is given later. From the results concerning the more general gamma distribution, we see that its moment-generating function is

$$M(t) = (1 - 2t)^{-r/2}, \qquad t < \frac{1}{2}.$$

In Figure 4.3-2 the graphs of chi-square p.d.f.'s for $r = 2, 3, 5$, and 8 are given. Note the relationship between the mean, $\mu = r$, and the point at which the p.d.f. obtains its maximum (see Exercise 4.3-13).

Because the chi-square distribution is so important in applications, tables have been prepared giving the values of the distribution function

$$F(x) = \int_0^x \frac{1}{\Gamma(r/2)2^{r/2}} w^{r/2-1} e^{-w/2} \, dw$$

for selected values of r and x. For an example, see Table IV in the Appendix.

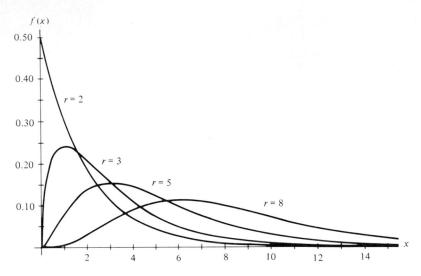

FIGURE 4.3-2

Example 4.3-3 Let X have a chi-square distribution with $r = 5$ degrees of freedom. Then, using Table IV in the Appendix,

$$P(1.145 \leq X \leq 12.83) = F(12.83) - F(1.145) = 0.975 - 0.050 = 0.925$$

and

$$P(X > 15.09) = 1 - F(15.09) = 1 - 0.99 = 0.01.$$

Example 4.3-4 If X is $\chi^2(7)$, two constants, a and b, such that

$$P(a < X < b) = 0.95$$

are $a = 1.690$ and $b = 16.01$. Other constants a and b can be found, and we are only restricted in our choices by the limited table.

Probabilities like that of Example 4.3-4 are so important in statistical applications that we use special symbols for a and b. Let α be a positive probability (that is usually less than 0.5) and let X have a chi-square distribution with r degrees of freedom. Then $\chi_\alpha^2(r)$ is a number such that

$$P[X \geq \chi_\alpha^2(r)] = \alpha.$$

That is, $\chi_\alpha^2(r)$ is the $100(1 - \alpha)$ percentile (or upper 100α percent point) of the chi-square distribution with r degrees of freedom. Then the 100α percentile is the number $\chi_{1-\alpha}^2(r)$ such that

$$P[X \leq \chi_{1-\alpha}^2(r)] = \alpha.$$

That is, the probability to the right of $\chi_{1-\alpha}^2(r)$ is $1 - \alpha$ (see Figure 4.3-3).

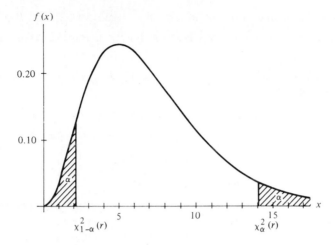

FIGURE 4.3-3

Example 4.3-5 Let X have a chi-square distribution with seven degrees of freedom. Then, using Table IV in the Appendix, $\chi^2_{0.05}(7) = 14.07$ and $\chi^2_{0.95}(7) = 2.167$. These are the points that are indicated on Figure 4.3-3.

Example 4.3-6 If customers arrive at a shop on the average of 30 per hour in accordance with a Poisson process, what is the probability that the shopkeeper will have to wait longer than 9.390 minutes for the first nine customers to arrive? Note that the mean rate of arrivals per minute is $\lambda = 1/2$. Thus, $\theta = 2$ and $\alpha = r/2 = 9$. If X denotes the waiting time until the ninth arrival, X is $\chi^2(18)$. Hence

$$P(X > 9.390) = 1 - 0.05 = 0.95.$$

Example 4.3-7 If X has an exponential distribution with a mean of 2, the p.d.f. of X is

$$f(x) = \frac{1}{2}e^{-x/2} = \frac{x^{2/2-1}e^{-x/2}}{\Gamma(2/2)2^{2/2}}, \qquad 0 \le x < \infty.$$

That is, X is $\chi^2(2)$. Thus, for illustration,

$$P(0.051 < X < 7.378) = 0.975 - 0.025 = 0.95.$$

Exercises

4.3-1 Telephone calls enter a college switchboard at a mean rate of 2/3 call per minute according to a Poisson process. Let X denote the waiting time until the tenth call arrives.
(a) What is the p.d.f. of X?
(b) What are the moment-generating function, mean, and variance of X?

4.3-2 If X has a gamma distribution with $\theta = 4$ and $\alpha = 2$, find $P(X < 5)$.

4.3-3 Find the moment-generating function for the gamma distribution with parameters α and θ.

HINT: In the integral representing $E(e^{tX})$, change variables by letting $y = (1 - \theta t)x/\theta$, where $1 - \theta t > 0$.

4.3-4 If X has a gamma distribution, show that $E(X) = \alpha\theta$ and $\text{Var}(X) = \alpha\theta^2$.

4.3-5 If the moment-generating function of a random variable W is

$$M(t) = (1 - 7t)^{-20},$$

find the p.d.f., mean, and variance of W.

4.3-6 Let X denote the number of alpha particles emitted by barium-133 and observed by a Gieger counter in a fixed position. Assume that $\lambda = 14.7$ is the mean number of counts per second. Let W denote the waiting time to observe 100 counts. Twenty-five independent observations of W are

6.9	7.3	6.7	6.4	6.3
5.9	7.0	7.1	6.5	7.6
7.2	7.1	6.1	7.3	7.6
7.6	6.7	6.3	5.7	6.7
7.5	5.3	5.4	7.4	6.9

(a) Give the p.d.f., mean, and variance of W.
(b) Calculate the sample mean and sample variance of the 25 observations of W.
(c) Use the relative frequency of event $\{W \le 6.6\}$ to approximate $P(W \le 6.6)$.

4.3-7 Consider the computer tape described in Exercise 4.2-13 but now let W equal the number of feet before the third bad record is encountered.
(a) How is W distributed?
(b) Give the mean and variance of W.
(c) Find $P(W \le 100)$.

HINT: Use formula (4.3-1) with $\lambda w = 2.5$.

4.3-8 Let X equal the number of alpha particle emissions of carbon-14 that are counted by a Geiger counter per second. Assume that the distribution of X is Poisson with mean 16. Let W equal the time in seconds before the seventh count is made.
(a) Give the distribution of W.
(b) Find $P(W \le 0.5)$.

4.3-9 If X is $\chi^2(17)$, find
(a) $P(X < 7.564)$.
(b) $P(X > 27.59)$.
(c) $P(6.408 < X < 27.59)$.
(d) $\chi^2_{0.95}(17)$.
(e) $\chi^2_{0.025}(17)$.

4.3-10 If X is $\chi^2(12)$, find constants a and b such that
$$P(a < X < b) = 0.90 \quad \text{and} \quad P(X < a) = 0.05.$$

4.3-11 If X is $\chi^2(23)$, find the following:
(a) $P(14.85 < X < 32.01)$.
(b) Constants a and b such that $P(a < X < b) = 0.95$ and $P(X < a) = 0.025$.
(c) The mean and variance of X.
(d) $\chi^2_{0.05}(23)$ and $\chi^2_{0.95}(23)$.

4.3-12 If the moment-generating function of X is $M(t) = (1 - 2t)^{-12}$, $t < 1/2$, find
(a) $E(X)$.
(b) $\text{Var}(X)$.
(c) $P(15.66 < X < 42.98)$.

4.3-13 Find the point at which a chi-square p.d.f. obtains its maximum when $r > 2$.

4.3-14 Cars arrive at a toll booth at a mean rate of five cars every 10 minutes according to a Poisson process. Find the probability that the toll collector will have to wait longer than 26.30 minutes before collecting the eighth toll.

4.3-15 If 15 observations are taken independently from a chi-square distribution with four degrees of freedom, find the probability that at most three of the sample items exceed 7.779.

4.3-16 If 10 observations are taken independently from a chi-square distribution with 19 degrees of freedom, find the probability that exactly 2 of the 10 sample items exceed 30.14.

4.3-17 Suppose that 100 random numbers from the interval $[0, 1)$ are generated on a computer. Then let y_i equal the number of random numbers, x, such that $(i - 1)/10 \le x < i/10$, $i = 1, 2, \ldots, 10$. It will be shown in Section 9.5 that

$$Q = \sum_{i=1}^{10} \frac{(y_i - 10)^2}{10}$$

has a distribution that is approximately $\chi^2(9)$. The following data give 100 observations of Q:

12.6	6.8	4.6	9.2	12.6	17.4	7.8	12.4	5.4	1.8
14.6	7.0	2.8	11.2	4.8	12.6	9.0	8.6	18.2	12.6
5.2	17.6	6.2	19.8	9.8	4.4	8.6	6.2	4.0	17.0
7.6	6.2	15.4	2.0	11.2	12.8	5.4	10.6	12.2	5.2
1.8	16.2	7.4	9.4	10.2	13.4	14.0	6.0	9.2	9.0
10.6	16.2	5.6	7.2	14.2	7.2	9.4	11.4	6.6	15.2
7.8	10.6	2.6	12.2	7.6	9.8	18.6	15.0	6.4	10.0
8.2	7.0	4.8	7.2	5.4	5.0	10.4	10.0	14.2	11.8
3.0	23.0	8.2	7.6	4.4	10.4	6.2	14.2	4.0	6.8
4.4	12.0	8.8	4.0	11.4	4.0	12.2	3.8	13.0	10.6

(a) Construct a frequency distribution.
(b) Draw the relative frequency histogram. Superimpose the $\chi^2(9)$ p.d.f.
(c) Calculate the sample mean and sample variance. Are they close to $E(Q) = 9$ and $\mathrm{Var}(Q) = 18$, respectively?
(d) Use the data to approximate $P(Q \le 3.325)$ and $P(Q \ge 14.68)$. Compare the approximations with the probabilities selected from Table IV in the Appendix for the $\chi^2(9)$ distribution.

4.4   The Normal Distribution

The normal distribution is perhaps the most important distribution in statistical applications since many measurements have (approximate) normal distributions. One explanation of this fact is the role of the normal distribution in the Central Limit Theorem. One form of this theorem will be considered in Section 5.4.

We give the definition of the p.d.f. for the normal distribution, verify that it is a p.d.f., and then justify the use of μ and σ^2 in this formula. That is, we will show that μ and σ^2 are actually the mean and the variance of this distribution. The random variable X has a **normal distribution** if its p.d.f. is defined by

$$f(x) = \frac{1}{\sigma\sqrt{2\pi}} \exp\left[-\frac{(x-\mu)^2}{2\sigma^2}\right], \qquad -\infty < x < \infty,$$

where μ and σ are parameters satisfying $-\infty < \mu < \infty$, $0 < \sigma < \infty$, and also where $\exp[v]$ means e^v. Briefly, we say that X is $N(\mu, \sigma^2)$.

Clearly, $f(x) > 0$. We now evaluate the integral

$$I = \int_{-\infty}^{\infty} \frac{1}{\sigma\sqrt{2\pi}} \exp\left[-\frac{(x-\mu)^2}{2\sigma^2}\right] dx,$$

showing that it is equal to 1. In I, change variables of integration by letting $z = (x - \mu)/\sigma$. Then

$$I = \int_{-\infty}^{\infty} \frac{1}{\sqrt{2\pi}} e^{-z^2/2}\, dz.$$

Since $I > 0$, if $I^2 = 1$, then $I = 1$. Now

$$I^2 = \frac{1}{2\pi}\left[\int_{-\infty}^{\infty} e^{-x^2/2}\, dx\right]\left[\int_{-\infty}^{\infty} e^{-y^2/2}\, dy\right],$$

or, equivalently,

$$I^2 = \frac{1}{2\pi}\int_{-\infty}^{\infty}\int_{-\infty}^{\infty} \exp\left(-\frac{x^2+y^2}{2}\right) dx\, dy.$$

Letting $x = r \cos \theta$, $y = r \sin \theta$ (i.e., using polar coordinates), we have

$$I^2 = \frac{1}{2\pi} \int_0^{2\pi} \int_0^{\infty} e^{-r^2/2} r \, dr \, d\theta$$

$$= \frac{1}{2\pi} \int_0^{2\pi} d\theta = \frac{1}{2\pi} \, 2\pi = 1.$$

Thus $I = 1$, and we have shown that $f(x)$ has the properties of a p.d.f. The moment-generating function of X is

$$M(t) = \int_{-\infty}^{\infty} \frac{e^{tx}}{\sigma\sqrt{2\pi}} \exp\left[-\frac{(x - \mu)^2}{2\sigma^2}\right] dx$$

$$= \int_{-\infty}^{\infty} \frac{1}{\sigma\sqrt{2\pi}} \exp\left\{-\frac{1}{2\sigma^2}[x^2 - 2(\mu + \sigma^2 t)x + \mu^2]\right\} dx.$$

To evaluate this integral, we complete the square in the exponent

$$x^2 - 2(\mu + \sigma^2 t)x + \mu^2 = [x - (\mu + \sigma^2 t)]^2 - 2\mu\sigma^2 t - \sigma^4 t^2.$$

Thus

$$M(t) = \exp\left(\frac{2\mu\sigma^2 t + \sigma^4 t^2}{2\sigma^2}\right) \int_{-\infty}^{\infty} \frac{1}{\sigma\sqrt{2\pi}} \exp\left\{-\frac{1}{2\sigma^2}[x - (\mu + \sigma^2 t)]^2\right\} dx.$$

Note that the integrand in the last integral is like the p.d.f. of a normal distribution with μ replaced by $\mu + \sigma^2 t$. However, the normal p.d.f. integrates to one for all real μ, in particular when it equals $\mu + \sigma^2 t$. Thus

$$M(t) = \exp\left(\frac{2\mu\sigma^2 t + \sigma^4 t^2}{2\sigma^2}\right) = \exp\left(\mu t + \frac{\sigma^2 t^2}{2}\right).$$

Now

$$M'(t) = (\mu + \sigma^2 t) \exp\left(\mu t + \frac{\sigma^2 t^2}{2}\right)$$

and

$$M''(t) = [(\mu + \sigma^2 t)^2 + \sigma^2] \exp\left(\mu t + \frac{\sigma^2 t^2}{2}\right).$$

Thus

$$E(X) = M'(0) = \mu,$$
$$\text{Var}(X) = M''(0) - [M'(0)]^2 = \mu^2 + \sigma^2 - \mu^2 = \sigma^2.$$

That is, the parameters μ and σ^2 in the p.d.f. of X are the mean and the variance of X.

Example 4.4-1   If the p.d.f. of X is

$$f(x) = \frac{1}{\sqrt{32\pi}} \exp \left[-\frac{(x + 7)^2}{32} \right], \qquad -\infty < x < \infty,$$

then X is $N(-7, 16)$. That is, X has a normal distribution with a mean $\mu = -7$, variance $\sigma^2 = 16$, and the moment-generating function

$$M(t) = \exp (-7t + 8t^2).$$

Example 4.4-2   If the moment-generating function of X is

$$M(t) = \exp (5t + 12t^2),$$

then X is $N(5, 24)$, and its p.d.f. is

$$f(x) = \frac{1}{\sqrt{48\pi}} \exp \left[-\frac{(x - 5)^2}{48} \right], \qquad -\infty < x < \infty.$$

If Z is $N(0, 1)$, we shall say that Z has a standard normal distribution. Moreover, the distribution function of Z is

$$\Phi(z) = P(Z \le z) = \int_{-\infty}^{z} \frac{1}{\sqrt{2\pi}} e^{-w^2/2} \, dw.$$

It is not possible to evaluate this integral by finding an antiderivative that can be expressed as an elementary function. However, numerical approximations for integrals of this type have been tabulated and are given in Tables Va and Vb in the Appendix. The bell-shaped curved in Figure 4.4-1 represents the graph of the p.d.f. of Z, and the shaded area equals $\Phi(z_0)$.

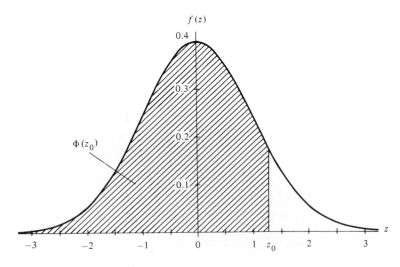

FIGURE 4.4-1

Values of $\Phi(z)$ for $z \geq 0$ are given in Appendix Table Va. Because of the symmetry of the standard normal p.d.f., it is true that $\Phi(-z) = 1 - \Phi(z)$ for all real z. Thus Appendix Table Va is sufficient. However, it is sometimes convenient to be able to read $\Phi(-z)$, for $z > 0$, directly from a table. This can be done using values in Appendix Table Vb which lists right tail probabilities. Again, because of the symmetry of the standard normal p.d.f., when $z > 0$, $\Phi(-z) = P(Z \leq -z) = P(Z > z)$ can be read directly from Appendix Table Vb.

Example 4.4-3 If Z is $N(0, 1)$, then, using Appendix Table Va, we obtain

$$P(Z \leq 1.24) = \Phi(1.24) = 0.8925,$$
$$P(1.24 \leq Z \leq 2.37) = \Phi(2.37) - \Phi(1.24) = 0.9911 - 0.8925 = 0.0986,$$
$$P(-2.37 \leq Z \leq -1.24) = P(1.24 \leq Z \leq 2.37) = 0.9086.$$

Using Appendix Table Vb, we find that

$$P(Z > 1.24) = 0.1075,$$
$$P(Z \leq -2.14) = P(Z \geq 2.14) = 0.0162,$$

and using both tables, we obtain

$$P(-2.14 \leq Z \leq 0.77) = P(Z \leq 0.77) - P(Z \leq -2.14)$$
$$= 0.7794 - 0.0162 = 0.7632.$$

There are times when we want to read the normal probability table in the opposite way, essentially finding the inverse of the standard normal distribution function. That is, given a probability p, find a constant a so that $P(Z \leq a) = p$. This is illustrated in the next example.

Example 4.4-4 If the distribution of Z is $N(0, 1)$, to find constants a and b such that

$$P(Z \leq a) = 0.9147 \qquad \text{and} \qquad P(Z \geq b) = 0.0526,$$

find the given probabilities in Appendix Tables Va and Vb, respectively, and read off the corresponding values of z. From Appendix Table Va we see that $a = 1.37$ and from Appendix Table Vb we see that $b = 1.62$.

In statistical applications we are often interested in finding a number z_α such that

$$P(Z \geq z_\alpha) = \alpha,$$

where Z is $N(0, 1)$ and α is usually less than 0.5. That is, z_α is the $100(1 - \alpha)$ percentile (sometimes called the upper 100α percent point) for the standard normal distribution (see Figure 4.4-2). The value of z_α is given in Appendix Table Va for selected values of α. For other values of α, z_α can be found in Appendix Table Vb.

Because of the symmetry of the normal p.d.f.,

$$P(Z \leq -z_\alpha) = P(Z \geq z_\alpha) = \alpha.$$

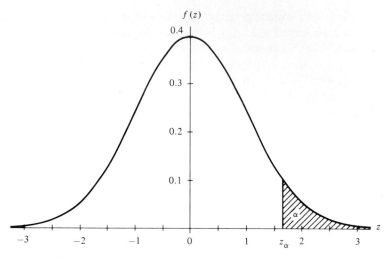

FIGURE 4.4-2

Also, since the subscript is the "right-tail" probability,

$$z_{1-\alpha} = -z_\alpha.$$

For example,

$$z_{0.95} = z_{1-0.05} = -z_{0.05}.$$

Example 4.4-5   To find $z_{0.0125}$, note that

$$P(Z \geq z_{0.0125}) = 0.0125.$$

Thus

$$z_{0.0125} = 2.24$$

from Appendix Table Vb. Also,

$$z_{0.05} = 1.645 \qquad \text{and} \qquad z_{0.025} = 1.960$$

using the last rows of Appendix Table Va.

REMARK   Recall that the $(100p)$th percentile, π_p, for a random variable X is a number such that $P(X \leq \pi_p) = p$. If Z is $N(0, 1)$, since

$$P(Z \geq z_\alpha) = \alpha,$$

it follows that

$$P(Z < z_\alpha) = 1 - \alpha.$$

Thus z_α is the $100(1 - \alpha)$th percentile for the standard normal distribution, $N(0, 1)$. For example, $z_{0.05} = 1.645$ is the $100(1 - 0.05) = 95$th percentile and $z_{0.95} = -1.645$ is the $100(1 - 0.95) = 5$th percentile.

If X is $N(\mu, \sigma^2)$, the next theorem shows that the random variable $(X - \mu)/\sigma$ is $N(0, 1)$. Thus Tables Va and Vb in the Appendix can be used to find probabilities concerning X.

THEOREM 4.4-1 *If X is $N(\mu, \sigma^2)$, then $Z = (X - \mu)/\sigma$ is $N(0, 1)$.*

PROOF: The distribution function of Z is

$$P(Z \leq z) = P\left(\frac{X - \mu}{\sigma} \leq z\right) = P(X \leq z\sigma + \mu)$$

$$= \int_{-\infty}^{z\sigma + \mu} \frac{1}{\sigma\sqrt{2\pi}} \exp\left[-\frac{(x - \mu)^2}{2\sigma^2}\right] dx.$$

In the integral representing $P(Z \leq z)$, use the change of variable of integration given by $w = (x - \mu)/\sigma$ (i.e., $x = w\sigma + \mu$) to obtain

$$P(Z \leq z) = \int_{-\infty}^{z} \frac{1}{\sqrt{2\pi}} e^{-w^2/2} \, dw.$$

But this is the expression for $\Phi(z)$, the distribution function of a standardized normal random variable. Hence Z is $N(0, 1)$. $\quad\square$

This theorem can be used to find probabilities about X, which is $N(\mu, \sigma^2)$, as follows:

$$P(a \leq X \leq b) = P\left(\frac{a - \mu}{\sigma} \leq \frac{X - \mu}{\sigma} \leq \frac{b - \mu}{\sigma}\right) = \Phi\left(\frac{b - \mu}{\sigma}\right) - \Phi\left(\frac{a - \mu}{\sigma}\right),$$

since $(X - \mu)/\sigma$ is $N(0, 1)$.

Example 4.4-6 If X is $N(3, 16)$, then

$$P(4 \leq X \leq 8) = P\left(\frac{4 - 3}{4} \leq \frac{X - 3}{4} \leq \frac{8 - 3}{4}\right)$$

$$= \Phi(1.25) - \Phi(0.25) = 0.8944 - 0.5987 = 0.2957,$$

$$P(0 \leq X \leq 5) = P\left(\frac{0 - 3}{4} \leq Z \leq \frac{5 - 3}{4}\right)$$

$$= \Phi(0.5) - \Phi(-0.75) = 0.6915 - 0.2266 = 0.4649,$$

and

$$P(-2 \leq X \leq 1) = P\left(\frac{-2 - 3}{4} \leq Z \leq \frac{1 - 3}{4}\right)$$

$$= \Phi(-0.5) - \Phi(-1.25) = 0.3085 - 0.1056 = 0.2029.$$

Example 4.4-7 If X is $N(25, 36)$, we find a constant c such that

$$P(|X - 25| \leq c) = 0.9544.$$

We want

$$P\left(\frac{-c}{6} \le \frac{X - 25}{6} \le \frac{c}{6}\right) = 0.9544.$$

Thus

$$\Phi\left(\frac{c}{6}\right) - \left[1 - \Phi\left(\frac{c}{6}\right)\right] = 0.9544$$

and

$$\Phi\left(\frac{c}{6}\right) = 0.9772.$$

Hence $c/6 = 2$ and $c = 12$. That is, the probability that X falls within two standard deviations of its mean is the same as the probability that the standard normal variable Z falls within two units (standard deviations) of zero.

In the next theorem we give a relationship between the chi-square and normal distributions.

THEOREM 4.4-2   *If the random variable X is $N(\mu, \sigma^2)$, $\sigma^2 > 0$, then the random variable $V = (X - \mu)^2/\sigma^2 = Z^2$ is $\chi^2(1)$.*

PROOF: Because $V = Z^2$, where $Z = (X - \mu)/\sigma$ is $N(0, 1)$, the distribution function $G(v)$ of V is, for $v \ge 0$,

$$G(v) = P(Z^2 \le v) = P(-\sqrt{v} \le Z \le \sqrt{v}).$$

That is, with $v \ge 0$,

$$G(v) = \int_{-\sqrt{v}}^{\sqrt{v}} \frac{1}{\sqrt{2\pi}} e^{-z^2/2} \, dx = 2 \int_{0}^{\sqrt{v}} \frac{1}{\sqrt{2\pi}} e^{-z^2/2} \, dz.$$

If we change the variable of integration by writing $z = \sqrt{y}$, then, since $D_y(z) = 1/(2\sqrt{y})$, we have

$$G(v) = \int_{0}^{v} \frac{1}{\sqrt{2\pi y}} e^{-y/2} \, dy, \qquad 0 \le v.$$

Of course, $G(v) = 0$, when $v < 0$. Hence the p.d.f. $g(v) = G'(v)$ of the continuous-type random variable V is, by one form of the fundamental theorem of calculus,

$$g(v) = \frac{1}{\sqrt{\pi}\sqrt{2}} v^{1/2-1} e^{-v/2}, \qquad 0 < v < \infty.$$

Since $g(v)$ is a p.d.f., it must be true that

$$\int_{0}^{\infty} \frac{1}{\sqrt{\pi}\sqrt{2}} v^{1/2-1} e^{-v/2} \, dv = 1.$$

The change of variables $x = v/2$ yields

$$1 = \frac{1}{\sqrt{\pi}} \int_0^\infty x^{1/2-1} e^{-x} \, dx = \frac{1}{\sqrt{\pi}} \Gamma\left(\frac{1}{2}\right).$$

Hence $\Gamma(1/2) = \sqrt{\pi}$, and thus V is $\chi^2(1)$.   □

Example 4.4-8 If Z is $N(0, 1)$, then

$$P(|Z| < 1.96 = \sqrt{3.842}) = 0.95$$

and, of course,

$$P(Z^2 < 3.842) = 0.95$$

from the chi-square table with $r = 1$.

Given a set of observations of a random variable X, it is a challenge to determine the distribution of X. In particular, how can we decide whether or not X has an approximate normal distribution? If we have a large number of observations of X, a histogram of the observations can often be helpful. (See Exercises 4.4-11 and 4.4-12.) For small samples, a comparison of the empirical distribution function with a normal distribution function can be used. Or a q–q plot can be used to check on whether the sample arises from a normal distribution. For example, suppose the quantiles of a sample were plotted against the corresponding quantiles of a certain normal distribution and these pairs of points were on a straight line with slope 1 and intercept 0. Of course, we would then believe that we have an ideal sample from that normal distribution with that certain mean and standard deviation. This plot, however, requires that we know the mean and the standard deviation of this normal distribution, and we usually do not. However, since the quantile, q_p, of $N(\mu, \sigma^2)$ is related to the corresponding one, z_{1-p}, of $N(0, 1)$ by $q_p = \mu + \sigma z_{1-p}$, we can always plot the quantiles of the sample against the corresponding ones of $N(0, 1)$ and get the needed information. That is, if the sample quantiles are plotted as the x-coordinate of the pair and the $N(0, 1)$ quantiles as the y-coordinate and if the graph is almost a straight line, then it is reasonable to assume that the sample arises from a normal distribution. Moreover, the reciprocal of the slope of that straight line is a good estimate of the standard deviation σ because $z_{1-p} = (q_p - \mu)/\sigma$.

Example 4.4-9 Have you ever wondered about the weights of carrots in a prepackaged "1-pound" bag? Let X equal the weight of such a bag of carrots. Would you expect X to have a normal distribution? One of the authors frequently has his carrots weighed at the checkout. Say $n = 12$ observations of X were

1.12	1.13	1.19	1.25	1.06	1.31
1.12	1.23	1.29	1.17	1.20	1.11

With 12 observations, the standard normal quantiles that we need correspond to $p = 1/13, 2/13, \ldots, 12/13$.

k	Weights (x)	$p = k/13$	z_{1-p}
1	1.06	0.0769	-1.43
2	1.11	0.1538	-1.02
3	1.12	0.2308	-0.74
4	1.12	0.3077	-0.50
5	1.13	0.3846	-0.29
6	1.17	0.4615	-0.10
7	1.19	0.5385	0.10
8	1.20	0.6154	0.29
9	1.23	0.6923	0.50
10	1.25	0.7692	0.74
11	1.29	0.8462	1.02
12	1.31	0.9231	1.43

A q–q plot of these data is shown in Figure 4.4-3. Note that the points do fall close to a straight line so the normal probability model seems to be appropriate based on these few data.

This is an experiment that you could easily replicate to obtain more data and perhaps confirm these results or show that they are not true.

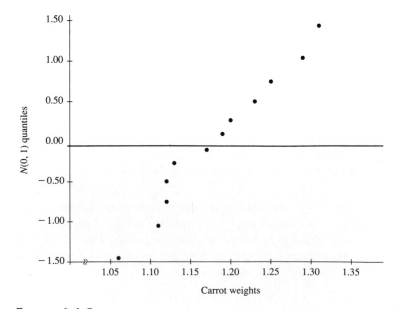

FIGURE 4.4-3

Exercises

4.4-1 If Z is $N(0, 1)$, find
(a) $P(0.53 < Z \leq 2.06)$.   (b) $P(-0.79 \leq Z < 1.52)$.
(c) $P(-2.63 < Z \leq -0.51)$.   (d) $P(Z > -1.77)$.
(e) $P(Z > 2.89)$.   (f) $P(|Z| < 1.96)$.
(g) $P(|Z| < 1)$.   (h) $P(|Z| < 2)$.
(i) $P(|Z| < 3)$.

4.4-2 If Z is $N(0, 1)$, find
(a) $P(0 \leq Z \leq 0.87)$.   (b) $P(-2.64 \leq Z \leq 0)$.
(c) $P(-2.13 \leq Z \leq -0.56)$.   (d) $P(|Z| > 1.39)$.
(e) $P(Z < -1.62)$.   (f) $P(|Z| > 1)$.
(g) $P(|Z| > 2)$.   (h) $P(|Z| > 3)$.

4.4-3 Find the values of
(a) $z_{0.01}$.   (b) $-z_{0.005}$.
(c) $z_{0.0475}$.   (d) $z_{0.985}$.

4.4-4 Find the values of
(a) $z_{0.10}$.   (b) $-z_{0.05}$.
(c) $-z_{0.0485}$.   (d) $z_{0.9656}$.

4.4-5 If Z is $N(0, 1)$, find values of c such that
(a) $P(Z \geq c) = 0.025$.   (b) $P(|Z| \leq c) = 0.95$.
(c) $P(Z > c) = 0.05$.   (d) $P(|Z| \leq c) = 0.90$.

4.4-6 If the moment-generating function of X is $M(t) = \exp(166t + 200t^2)$, find
(a) The mean of X.   (b) The variance of X.
(c) $P(170 < X < 200)$.   (d) $P(148 \leq X \leq 172)$.

4.4-7 If X is normally distributed with a mean of 6 and a variance of 25, find
(a) $P(6 \leq X \leq 12)$.   (b) $P(0 \leq X \leq 8)$.
(c) $P(-2 < X \leq 0)$.   (d) $P(X > 21)$.
(e) $P(|X - 6| < 5)$.   (f) $P(|X - 6| < 10)$.
(g) $P(|X - 6| < 15)$.

4.4-8 If the moment-generating function of X is $M(t) = \exp(-6t + 32t^2)$, find
(a) $P(-4 \leq X < 16)$.   (b) $P(-10 < X \leq 0)$.

4.4-9 If X is $N(650, 625)$, find
(a) $P(600 \leq X < 660)$.
(b) A constant $c > 0$ such that $P(|X - 650| \leq c) = 0.9544$.

4.4-10 If X is $N(\mu, \sigma^2)$, show that $Y = aX + b$ is $N(a\mu + b, a^2\sigma^2)$, $a \neq 0$.

HINT: Find the distribution function $P(Y \leq y)$ of Y and, in the resulting integral, let $w = ax + b$ or, equivalently, $x = (w - b)/a$.

4.4-11 A packaged product has a label weight of 450 grams. The company goal is to fill each package with at least 450 grams but at most 458 grams. To check these goals, a random sample of 100 packages was selected and weighed, yielding the following weights rounded to the nearest gram:

457	457	455	457	454	454	457	455	456	459
457	458	456	456	461	457	458	452	457	460
453	458	452	454	454	456	455	456	451	454
456	457	457	453	455	459	458	457	458	457
461	457	455	458	458	455	457	458	456	463
455	455	455	456	456	456	455	456	460	456
456	457	458	454	455	456	459	457	457	451
450	453	453	459	450	453	452	458	456	457
451	458	456	460	455	455	456	460	457	456
457	456	460	459	457	455	461	455	457	457

(a) Group these data into 14 classes and draw a relative frequency histogram.
(b) Do these data seem to come from a normal distribution with mean μ about equal to $\bar{x} = 456.2$ and variance about equal to $s^2 = 5.93$? Sketch the p.d.f. for the normal distribution $N(456.2, 5.93)$ on your histogram.

4.4-12   A company manufactures windows that are then inserted into an automobile. Each window has 5 studs for attaching it. A pull-out test is used to determine the force required to pull a stud out of a window. (Note that this is an example of destructive testing.) Let X equal the force required for pulling studs out of position 4. Sixty observations of X were as follows:

159	150	147	160	155	142	143	151	154	133
151	146	140	146	137	148	154	157	142	153
135	144	135	165	118	158	126	147	123	140
125	151	153	158	144	163	150	150	137	164
137	156	139	134	171	144	160	147	155	175
162	160	149	149	158	152	165	131	150	120

(a) Construct an ordered stem-and-leaf diagram using as stems 11•, 12∗, 12•, and so on.
(b) Construct a q–q plot using about every 5th observation in the ordered array and the corresponding quantiles of $N(0, 1)$.
(c) Does it look like X has a normal distribution?

4.4-13   Whenever you fly, the favorite snack (because it is free) is a bag of peanuts. Let X equal the weight of peanuts in a bag that has a 14-gram label weight. Sixteen observations of X, that have been ordered, are

13.9	14.4	14.6	14.7	14.7	15.2	15.2	15.2
15.3	15.4	15.4	15.5	15.6	15.6	15.9	16.4

Determine whether X could have a normal distribution by constructing a q–q plot. What is your conclusion?

4.4-14   A candy maker produces mints that have a label weight of 20.4 grams. Assume that the distribution of the weights of these mints is $N(21.37, 0.16)$.

(a) Let X denote the weight of a single mint selected at random from the production line. Find $P(X > 22.07)$.

(b) Suppose that 15 mints are selected at random and weighed. Let Y equal the number of these mints that weigh less than 20.857 grams. Then find $P(Y \leq 2)$.

4.4-15 Let X equal the birth weight (in grams) of babies in the United States. Assuming that the distribution of X is $N(3315, 575^2)$, find
(a) $P(2584.75 \leq X \leq 4390.25)$.
(b) $P(2619.25 \leq X \leq 3642.75)$.
(c) $P(3119.50 \leq X \leq 3579.50)$.
(d) Let Y equal the number of babies that weigh less than 2719 grams at birth among 25 of these babies selected independently. Find $P(Y \leq 4)$.

4.4-16 Let X equal the birth weight (in grams) of babies in Singapore. Assuming that the distribution of X is $N(3135, 459^2)$, find
(a) $P(2546.56 \leq X \leq 3723.44)$.
(b) $P(2379.94 \leq X \leq 3890.06)$.
(c) $P(2067.37 \leq X \leq 4202.63)$.
(d) Let Y equal the number of babies that weigh less than 2546.56 grams at birth among 20 of these babies selected independently. Find $P(Y \leq 3)$.

4.4-17 Let the distribution of X be $N(\mu, \sigma^2)$. Show that the points of inflection of the graph of the p.d.f. of X occur at $x = \mu \pm \sigma$.

4.4-18 If X is $N(7, 4)$, find $P[15.364 \leq (X - 7)^2 \leq 20.096]$.

4.4-19 If the moment-generating function of X is $M(t) = e^{500t + 5000t^2}$, find $P[27,060 \leq (X - 500)^2 \leq 50,240]$.

★4.5 Other Models

The binomial, Poisson, gamma, chi-square, and normal models are frequently used models in statistics. However, many other interesting and very useful models can be found. We begin with a modification of one of the postulates of an approximate Poisson process as given in Section 3.7. In that definition, the numbers of changes occurring in nonoverlapping intervals are independent, and the probability of at least two changes in a sufficiently small interval is essentially zero. We continue to use these postulates, but now we say that the probability of exactly one change in a sufficiently short interval of length h is approximately λh, *where λ is a nonnegative function of the position of this interval*. To be explicit, say $p(x, w)$ is the probability of x changes in the interval $(0, w)$, $0 \leq w$. Then, the last postulate, in more formal terms, becomes

$$p(x + 1, w + h) - p(x, w) \approx \lambda(w)h,$$

where $\lambda(w)$ is a nonnegative function of w. Thus, if h equals one small unit, $\lambda(w)$ approximates the probability of one change (frequently failure) in the unit interval from w to $w + 1$ and is called the **failure rate**. Accordingly, if we want the approximate probability of zero changes in the interval $(0, w + h)$ we could take, from the independence, the probability of zero changes in the interval $(0, w)$ times that of zero changes in the interval $(w, w + h)$. That is,

$$p(0, w + h) \approx p(0, w)[1 - \lambda(w)h]$$

because the probability of one or more changes in $(w, w + h)$ is about equal to $\lambda(w)h$. Equivalently,

$$\frac{p(0, w + h) - p(0, w)}{h} \approx -\lambda(w)p(0, w).$$

Taking limits as $h \to 0$, we have

$$D_w[p(0, w)] = -\lambda(w)p(0, w).$$

That is, the resulting differential equation is

$$\frac{D_w[p(0, w)]}{p(0, w)} = -\lambda(w);$$

and thus

$$\ln p(0, w) = -\int_0^w \lambda(t) \, dt + c_1.$$

Therefore,

$$p(0, w) = \exp\left[-\int_0^w \lambda(t) \, dt + c_1\right] = c_2 \exp\left[-\int_0^w \lambda(t) \, dt\right],$$

where $c_2 = e^{c_1}$. However, the boundary condition of the probability of zero changes in an interval of length zero must be one; that is, $p(0, 0) = 1$. So if we select

$$H(w) = \int_0^w \lambda(t) \, dt$$

to be such that $H(0) = 0$, then $c_2 = 1$. That is,

$$p(0, w) = e^{-H(w)},$$

where $H'(w) = \lambda(w)$ and $H(0) = 0$.

   Suppose that we now let the continuous-type random variable W be the interval necessary to produce the first change; then the distribution function of W is

$$G(w) = P(W \le w) = 1 - P(W > w), \qquad 0 \le w.$$

Because zero changes in the interval $(0, w)$ are the same as $W > w$, then

$$G(w) = 1 - p(0, w) = 1 - e^{-H(w)}, \qquad 0 \le w.$$

The p.d.f. of W is

$$g(w) = G'(w) = H'(w)e^{-H(w)} = \lambda(w) \exp\left[-\int_0^w \lambda(t) \, dt\right], \qquad 0 \le w.$$

From this, we see that the failure rate in terms of $g(w)$ and $G(w)$ is

$$\lambda(w) = \frac{g(w)}{1 - G(w)}.$$

In many applications of this result, W can be thought of as a random time interval. For example, if one change means "death" or "failure" of the item under consideration, then W is actually the length of life of the item. Usually $\lambda(w)$, which is commonly called the **failure rate** or **force of mortality**, is an increasing function of w. That is, the larger w (the older the item) is, the better is the chance of failure within a short interval of length h, namely $\lambda(w)h$. As we review the exponential distribution of Section 4.2, we note that there $\lambda(w)$ is a constant; that is, the failure rate or force of mortality does not increase as the item gets older. If this were true in human populations, it would mean that a person 80 years old would have as much chance of living another year as would a person 20 years old (sort of a mathematical "fountain of youth"). However, a constant failure rate (force of mortality) is not the case in most human populations nor in most populations of manufactured items. That is, the failure rate $\lambda(w)$ is usually an increasing function of w. We give two important examples of useful probabilistic models.

Example 4.5-1   Let

$$H(w) = \left(\frac{w}{\beta}\right)^{\alpha}, \qquad 0 \le w,$$

so that the failure rate is

$$\lambda(w) = H'(w) = \frac{\alpha w^{\alpha - 1}}{\beta^{\alpha}},$$

where $\alpha > 0$, $\beta > 0$. Then the p.d.f. of W is

$$g(w) = \frac{\alpha w^{\alpha - 1}}{\beta^{\alpha}} \exp\left[-\left(\frac{w}{\beta}\right)^{\alpha}\right], \qquad 0 \le w.$$

Frequently, in engineering, this distribution, with appropriate values of α and β, is excellent for describing the life of a manufactured item. Often α is greater than one but less than five. This p.d.f. is frequently called that of the **Weibull distribution** and, in model fitting, is a strong competitor to the gamma p.d.f. See Figure 4.5-1 for some examples of Weibull p.d.f.'s.

Example 4.5-2   Persons are often shocked to learn that human mortality increases almost exponentially once a person reaches 25 years of age. Depending on which mortality table is used, one finds that this increase is about 10% each year, which means that the rate of mortality will double about every 7 years. Although this fact can be shocking, we can be thankful that the force of mortality starts very low. The probability that a man in reasonably good

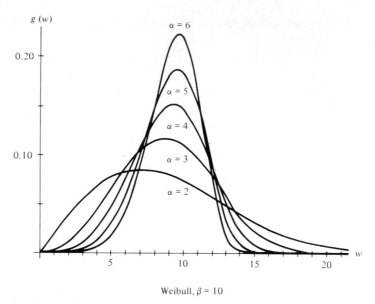

FIGURE 4.5-1

health at age 55 dies within the next year is only about 1% (it is not this large for a woman). Now, assuming an exponential force of mortality, we have

$$\lambda(w) = H'(w) = ae^{bw}, \qquad a > 0, \qquad b > 0.$$

Thus

$$H(w) = \frac{a}{b}e^{bw} + c.$$

However, $H(0) = 0$; so $c = -a/b$. Thus

$$G(w) = 1 - \exp\left[-\frac{a}{b}e^{bw} + \frac{a}{b}\right], \qquad 0 \leq w,$$

and

$$g(w) = ae^{bw}\exp\left[-\frac{a}{b}e^{bw} + \frac{a}{b}\right], \qquad 0 \leq w,$$

are the distribution function and p.d.f. associated with the famous **Gompertz law** found in actuarial science.

Both the gamma and Weibull distributions are skewed. In many studies (life testing, response times, incomes, etc.) these are valuable distributions for model selection. Another attractive one is called the **lognormal distribution**. This involves transforming a random variable, here normal. (See Example 4.5-3 and Figure 4.5-2).

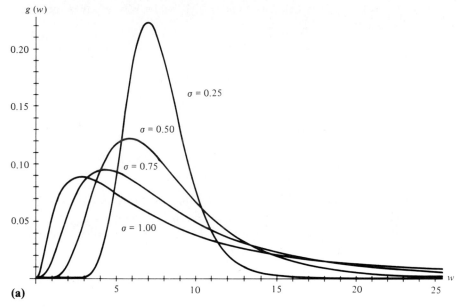

(a)

lognormal, $\mu = 2$

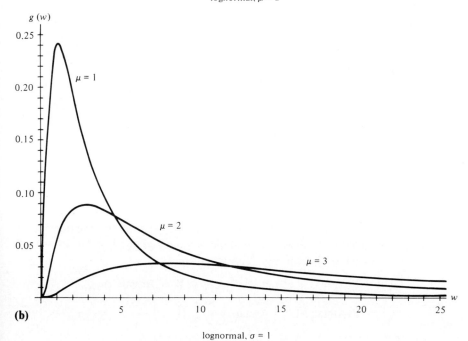

(b)

lognormal, $\sigma = 1$

FIGURE 4.5-2

Example 4.5-3 Say that X is $N(\mu, \sigma^2)$. If we let $W = e^X$, then the distribution function of W is

$$G(w) = P(W \leq w) = P(e^X \leq w) = P(X \leq \ln w), \qquad 0 < w.$$

That is,

$$G(w) = \int_{-\infty}^{\ln w} \frac{1}{\sigma\sqrt{2\pi}} \exp\left[-\frac{(x-\mu)^2}{2\sigma^2}\right] dx, \qquad 0 < w,$$

and thus the p.d.f. of W is

$$g(w) = G'(w) = \frac{1}{\sigma w \sqrt{2\pi}} \exp\left[-\frac{(\ln w - \mu)^2}{2\sigma^2}\right], \qquad 0 < w.$$

It is most important to observe that μ and σ^2 are not the mean and the variance of W, but those of $X = \ln W$, which has a normal distribution. Incidentally, it is easy to see why this distribution has been called the lognormal. See Figure 4.5-2 for some examples of lognormal p.d.f.'s.

There is another interesting distribution, this one involving a transformation of a uniform random variable.

Example 4.5-4 A spinner is mounted at the point $(0, 1)$. Let w be the smallest angle between the y-axis and the spinner (see Figure 4.5-3) Assume that w is the value of a random variable W that has a uniform distribution on the interval $(-\pi/2, \pi/2)$. That is, W is $U(-\pi/2, \pi/2)$, and the distribution function of W is

$$P(W \leq w) = F(w) = \begin{cases} 0, & -\infty < w < -\dfrac{\pi}{2}, \\[2mm] \left(w + \dfrac{\pi}{2}\right)\left(\dfrac{1}{\pi}\right), & -\dfrac{\pi}{2} \leq w < \dfrac{\pi}{2}, \\[2mm] 1, & \dfrac{\pi}{2} \leq w < \infty. \end{cases}$$

The relationship between x and w is given by $x = \tan w$; that is, x is the point on the x-axis which is the intersection of that axis and the linear extension of the spinner. To find the distribution of the random variable $X = \tan W$, we see that the distribution function of X is given by

$$G(x) = P(X \leq x) = P(\tan W \leq x) = P(W \leq \text{Arctan } x)$$

$$= F(\text{Arctan } x) = \left(\text{Arctan } x + \frac{\pi}{2}\right)\left(\frac{1}{\pi}\right), \qquad -\infty < x < \infty.$$

The last equality follows because $-\pi/2 < w = \text{Arctan } x < \pi/2$. The p.d.f. of X is given by

$$g(x) = G'(x) = \frac{1}{\pi(1 + x^2)}, \qquad -\infty < x < \infty.$$

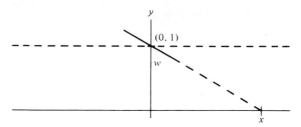

FIGURE 4.5-3

In Figure 4.5-4 the graph of this **Cauchy p.d.f.** is given. Although it looks similar to the graph for the standard normal distribution, it has striking differences. For example, in Exercise 4.5-10 you will be asked to show that $E(X)$ does not exist because the *tails* of the Cauchy p.d.f. contain too much probability for this p.d.f. to "balance" at $x = 0$.

To help appreciate the large probability in the tails of the Cauchy distribution, it is useful to simulate some observations of a Cauchy random variable. To simulate an observation of W, we can first begin with a random number, Y, that is an observation from the $U(0, 1)$ distribution. Given $Y = y$, $w = \pi y - \pi/2$ is an observation of W (see Exercise 4.2-5). Thus an observation of X is given by

$$x = \tan w = \tan \left(\pi y - \frac{\pi}{2} \right).$$
(4.5-1)

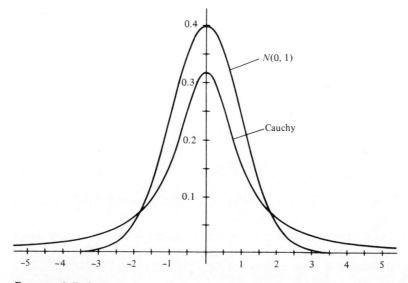

FIGURE 4.5-4

TABLE 4.5-1

y	x
0.1514	−1.9415
0.6697	0.5901
0.0527	−5.9847
0.4749	−0.0790
0.2900	−0.7757
0.2354	−1.0962
0.9662	9.3820
0.0043	−74.0211
0.1003	−3.0678
0.9192	3.8545

In Table 4.5-1 the values of y are the first 10 random numbers in the last column of Table IX in the Appendix. The corresponding values of x are given by equation (4.5-1). Although most of these observations from the Cauchy distribution are relatively small in magnitude, we see that a very large value (in magnitude) occurs occasionally. Another way of looking at this situation is by considering sightings (or firing of a gun) from an observation tower, here with coordinates $(0, 1)$, at independent random angles, each with the uniform distribution $U(-\pi/2, \pi/2)$; the target points would then be at Cauchy observations.

Exercises

4.5-1 (a) Show that a Weibull p.d.f. has a maximum when

$$w = \beta \left(\frac{\alpha - 1}{\alpha} \right)^{1/\alpha} \qquad \text{for } \alpha > 1.$$

(b) To what is this value close when α is large? That is, find

$$\lim_{\alpha \to \infty} \beta \left(\frac{\alpha - 1}{\alpha} \right)^{1/\alpha}.$$

(c) Find the value of the maximum when (see Figure 4.5-1) (i) $\alpha = 2$, (ii) $\alpha = 4$, and (iii) $\alpha = 6$.

4.5-2 Graph the Weibull p.d.f. with $\alpha = 1$. What is another name for this distribution?

4.5-3 Let the life W (in years) of the usual family car have a Weibull distribution with $\alpha = 2$.
(a) Show that β must equal 10 for $P(W > 5) = e^{-1/4} \approx 0.7788$.

HINT: $P(W > 5) = e^{-H(5)}$.

(b) Take eight observations from the uniform distribution $U(0, 1)$ independently and simulate eight values from the Weibull ($\alpha = 2, \beta = 10$) distribution.

HINT: You may use the random number table. Also recall that if U has that uniform distribution, then $F^{-1}(U)$ has a distribution with distribution function F.

4.5-4 (a) Show that the median m of the Weibull distribution is $m = \beta(\ln 2)^{1/\alpha}$.
(b) To what value is the median close when α is large? That is, find

$$\lim_{\alpha \to \infty} \beta(\ln 2)^{1/\alpha}.$$

4.5-5 Suppose that the length W of a man's life does follow the Gompertz distribution with $\lambda(w) = a(1.1)^w = ae^{(\ln 1.1)w}$, $P(W \le 53 \mid 52 < W) = 0.01$. Determine the constant a and $P(W \le 71 \mid 70 < W)$.

4.5-6 Let Y_1 be the smallest item of three independent observations W_1, W_2, W_3, from a Weibull distribution with parameters α and β. Show that Y_1 has a Weibull distribution. What are the parameters of this latter distribution?

HINT: $G(y) = P(Y_1 \le y) = 1 - P(y < W_i, i = 1, 2, 3) = 1 - [P(y < W_1)]^3$.

4.5-7 A frequent force of mortality used in actuarial science is $\lambda(w) = ae^{bw} + c$. Find the distribution function and p.d.f. associated with this **Makeham's law**.

4.5-8 Let W have a lognormal distribution with parameters μ and σ^2.
(a) Show that the median of W is e^μ.
(b) Show that the point at which the p.d.f. is a maximum (the **mode** of W) is $e^{\mu - \sigma^2}$.

4.5-9 Let W have a lognormal distribution. Find the median of W, the mode of W, and the value of the p.d.f. at the mode when
(a) $\mu = 2, \sigma = 0.25$.
(b) $\mu = 2, \sigma = 0.50$.
(c) $\mu = 2, \sigma = 0.75$.
(d) $\mu = 2, \sigma = 1.00$.
(e) $\mu = 1, \sigma = 1.00$.
(f) $\mu = 3, \sigma = 1.00$.
Check these results with Figure 4.5-2.

4.5-10 Let $f(x) = 1/[\pi(1 + x^2)]$, $-\infty < x < \infty$, be the p.d.f. of the Cauchy random variable X. Show that $E(X)$ does not exist.

4.5-11 Let X have a Cauchy distribution. Find
(a) $P(X > 1)$.
(b) $P(X > 5)$.
(c) $P(X > 10)$.

4.5-12 (a) Use Table IX in the Appendix or a computer to simulate nine observations of a Cauchy random variable.

(b) Find the sample mean and the sample median for your nine observations. Which of these statistics seems to give the better estimate of the center (here zero) of the distribution? Compare your results with those of other students.

★4.6   Mixed Distributions and Censoring

Thus far we have considered random variables that are either discrete or continuous. In most applications these are the types that are encountered. However, on some occasions, combinations of the two types of random variables are found. That is, in some experiments, positive probability is assigned to each of certain points and also is spread over an interval of outcomes, each point of which has zero probability. An illustration will help clarify these remarks.

Example 4.6-1 A bulb for a slide projector is tested by turning it on, letting it burn for 1 hour, and then turning it off. Let X equal the length of time that the bulb performs satisfactorily during this test. There is positive probability that the bulb will burn out when it is turned on; hence

$$P(X = 0) > 0.$$

It could also burn out during the 1-hour time period; thus

$$P(0 < X < 1) > 0$$

with $P(X = x) = 0$ when $x \in (0, 1)$. In addition, $P(X = 1) > 0$. The act of turning it off after 1 hour so that the actual failure time beyond 1 hour is not observed is called censoring, which is considered later in this section.

The distribution function for a distribution of the mixed type will be a combination of those for the discrete and continuous types. That is, at each point of positive probability the distribution function will be discontinuous so that the height of the step there equals the corresponding probability; at all other points the distribution function will be continuous.

Example 4.6-2 Let X have a distribution function $F(x)$ defined by

$$F(x) = \begin{cases} 0, & x < 0, \\ \dfrac{x^2}{4}, & 0 \le x < 1, \\ \dfrac{1}{2}, & 1 \le x < 2, \\ \dfrac{x}{3}, & 2 \le x < 3, \\ 1, & 3 \le x. \end{cases}$$

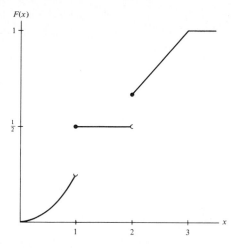

FIGURE 4.6-1

This distribution function is depicted in Figure 4.6-1. Probabilities can be computed using $F(x)$; for illustration consider

$$P(0 < X < 1) = \frac{1}{4},$$

$$P(0 < X \le 1) = \frac{1}{2},$$

$$P(X = 1) = \frac{1}{4},$$

and

$$P(1 \le X \le 2) = \frac{2}{3} - \frac{1}{4} = \frac{5}{12}.$$

Example 4.6-3   Consider the following game. An unbiased coin is tossed. If the outcome is heads, the player receives \$2. If the outcome is tails, the player spins a balanced spinner that has a scale from 0 to 1 and receives that fraction of a dollar associated with the point selected by the spinner. If X denotes the amount received, the space of X is $R = [0, 1) \cup \{2\}$. The distribution function of X is defined by

$$F(x) = \begin{cases} 0, & x < 0, \\[2mm] \dfrac{x}{2}, & 0 \le x < 1, \\[2mm] \dfrac{1}{2}, & 1 \le x < 2, \\[2mm] 1, & 2 \le x. \end{cases}$$

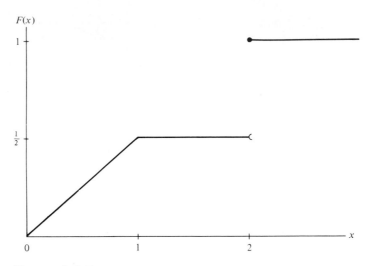

F(x)

FIGURE **4.6-2**

The graph of the distribution function $F(x)$ is given in Figure 4.6-2.

Suppose that the random variable X has a distribution of the mixed type. To find the expectation of the function $u(X)$ of X, a combination of a sum and a Riemann integral is used, as shown in Example 4.6-4.

Example 4.6-4   We shall find the mean and variance for the random variable given in Example 4.6-2. Note that there $F'(x) = x/2$ when $0 < x < 1$, and $F'(x) = 1/3$ when $2 < x < 3$; also $P(X = 1) = 1/4$ and $P(X = 2) = 1/6$. Accordingly, we have

$$\mu = E(X) = \int_0^1 x\left(\frac{x}{2}\right) dx + 1\left(\frac{1}{4}\right) + 2\left(\frac{1}{6}\right) + \int_2^3 x\left(\frac{1}{3}\right) dx$$

$$= \left[\frac{x^3}{6}\right]_0^1 + \frac{1}{4} + \frac{1}{3} + \left[\frac{x^2}{6}\right]_2^3 = \frac{19}{12}$$

and

$$\sigma^2 = E(X^2) - [E(X)]^2$$

$$= \int_0^1 x^2\left(\frac{x}{2}\right) dx + 1^2\left(\frac{1}{4}\right) + 2^2\left(\frac{1}{6}\right) + \int_2^3 x^2\left(\frac{1}{3}\right) dx - \left(\frac{19}{12}\right)^2 = \frac{31}{48}.$$

Frequently, in life testing, we know that the length of life, say X, exceeds the number b, but the exact value is unknown. This is called **censoring**. For instance, this can happen when a subject in a cancer study simply disappears; the investigator knows that the subject has lived a certain number of months, but the exact length of life is unknown. Or it might happen when an investigator does not have enough time in an investigation to observe the moments of deaths of all the animals, say rats, in some study. Censoring can also occur in the insurance industry; in particular, consider a loss with a limited-pay policy in which the top amount is exceeded but it is not known by how much.

> **Example 4.6-5** Reinsurance companies are concerned with large losses because they might agree, for illustration, to cover losses due to wind damages that are between \$2,000,000 and \$10,000,000. Say that X equals the size of a wind loss in millions of dollars, and suppose that it has the distribution function
>
> $$F(x) = 1 - \left(\frac{10}{10 + x}\right)^3, \qquad 0 \le x < \infty.$$
>
> If losses beyond \$10,000,000 are reported only as 10, then the distribution function of this censored distribution is
>
> $$F(x) = \begin{cases} 1 - \left(\dfrac{10}{10 + x}\right)^3, & 0 \le x < 10, \\ 1, & 10 \le x < \infty, \end{cases}$$
>
> which has a jump of $[10/(10 + 10)]^3 = 1/8$ at $x = 10$.

Exercises

4.6-1 From the graph of the distribution function determine the indicated probabilities.

(a) $P(X < 0)$.   (b) $P(X < -1)$.   (c) $P(X \le -1)$.

(d) $P(X < 1)$.   (e) $P\left(-1 \le X < \dfrac{1}{2}\right)$.

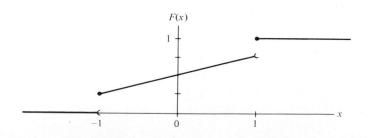

4.6-2 From the graph of the distribution function determine the indicated probabilities.

(a) $P\left(-\dfrac{1}{2} \le X \le \dfrac{1}{2}\right)$.   (b) $P\left(\dfrac{1}{2} < X < 1\right)$.   (c) $P\left(\dfrac{3}{4} < X < 2\right)$.

(d) $P(X > 1)$.   (e) $P(2 < X < 3)$.

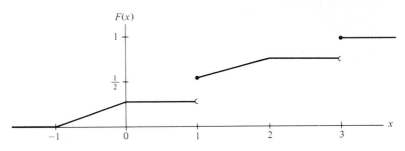

4.6-3 Let X be a random variable of the mixed type having the distribution function

$$F(x) = \begin{cases} 0, & x < 0, \\[2mm] \dfrac{x^2}{4}, & 0 \le x < 1, \\[2mm] \dfrac{x + 1}{4}, & 1 \le x < 2, \\[2mm] 1, & 2 \le x. \end{cases}$$

(a) Carefully sketch the graph of $F(x)$.

(b) Find the mean and the variance of X.

(c) Find $P(1/4 < X < 1)$, $P(X = 1)$, $P(X = 1/2)$, and $P(1/2 \le X < 2)$.

4.6-4 Find the mean and variance of X if the distribution function of X is

$$F(x) = \begin{cases} 0, & x < 0, \\[2mm] 1 - \left(\dfrac{2}{3}\right)e^{-x}, & 0 \le x. \end{cases}$$

4.6-5 Consider the following game. A fair die is rolled. If the outcome is even, the player receives a number of dollars equal to the outcome on the die. If the outcome is odd, a number is selected at random from the interval $[0, 1)$ with a balanced spinner, and the player receives that fraction of a dollar associated with the point selected.

(a) Define and sketch the distribution function of X, the amount received.

(b) Find the expected value of X.

(c) Simulate 10 plays of this game.

(d) Is the average of your 10 plays close to the expected value calculated in part (b)?

4.6-6 Compute the means of both the uncensored and the censored distributions in Example 4.6-5.

HINT: Use integration by parts.

4.6-7 The distribution function of X is defined by

$$F(x) = \begin{cases} 0, & -\infty < x < 0, \\ \frac{1}{4}(x^2 + 1), & 0 \le x < 1, \\ \frac{1}{4}(x + 2), & 1 \le x < 2, \\ 1, & 2 \le x < \infty. \end{cases}$$

(a) Sketch a graph of $y = F(x)$.
(b) Find $P(0 < X < 1)$.
(c) Find $P(1 \le X < 1.5)$.
(d) Find $\mu = E(X)$.
(e) Find $\sigma^2 = \text{Var}(X)$.

5

Sampling Distribution Theory

5.1 Multivariate Distributions

There are many random experiments or situations that involve more than one random variable. For example, a college admissions department might be interested in the ACT mathematics score X and the ACT verbal score Y of prospective students. Or manufactured items might be classified into three or more categories: Here X might represent the number of good items among n items; Y would be the number of "seconds," and the number of defectives would then be $n - X - Y$. Or to illustrate the Mendelian theory of inheritance in a biology laboratory, 400 kernels of corn could be classified into four categories: smooth and yellow, smooth and purple, wrinkled and yellow, and wrinkled and purple. The numbers in these four categories could be denoted by X_1, X_2, X_3, and $X_4 = 400 - X_1 - X_2 - X_3$, respectively.

In order to deal with situations such as these, it will be necessary to extend certain definitions as well as give new ones. We begin with the discrete case.

DEFINITION 5.1-1 *Let X and Y be two random variables defined on a discrete probability space. Let R denote the corresponding two-dimensional space of X and Y, the two random variables of the discrete type. The probability that X = x and Y = y is denoted by f(x, y) = P(X = x, Y = y). The func-*

tion $f(x, y)$ *is called the* **joint probability density function** *(joint p.d.f.) of X and Y and has the following properties:*

(a) $0 \leq f(x, y) \leq 1$.

(b) $\displaystyle\sum\sum_{(x, y) \in R} f(x, y) = 1$.

(c) $P[(X, Y) \in A] = \displaystyle\sum\sum_{(x, y) \in A} f(x, y)$, *where A is a subset of the space R.*

The following example will make this definition more meaningful.

Example 5.1-1   Roll a pair of unbiased dice. For each of the 36 sample points with probability 1/36, let X denote the smaller and Y the larger outcome on the dice. For example, if the outcome is (3, 2) then the observed values are $X = 2$, $Y = 3$. Incidentally, the event $\{X = 2, Y = 3\}$ could occur in two ways: (3, 2) or (2, 3); so its probability is

$$\frac{1}{36} + \frac{1}{36} = \frac{2}{36}.$$

If the outcome is (2, 2), then the observed values are $X = 2$, $Y = 2$ and $P(X = 2, Y = 2) = \frac{1}{36}$ since the event $\{X = 2, Y = 2\}$ can occur in only one way. The joint p.d.f. of X and Y is given by the probabilities

$$f(x, y) = \begin{cases} \dfrac{1}{36}, & 1 \leq x = y \leq 6, \\[2mm] \dfrac{2}{36} & 1 \leq x < y \leq 6, \end{cases}$$

when x and y are integers. Figure 5.1-1 depicts the probabilities of the various points of the space R.

Notice that certain numbers have been recorded in the bottom and lefthand margins of Figure 5.1-1. These numbers are the respective column and row totals of the probabilities. The column totals are the respective probabilities that X will assume the values in the x space $R_1 = \{1, 2, 3, 4, 5, 6\}$, and the row totals are the respective probabilities that Y will assume the values in the y space $R_2 = \{1, 2, 3, 4, 5, 6\}$. That is, the totals describe probability density functions of X and Y, respectively. Since each collection of these probabilities is frequently recorded in the margins and satisfies the properties of a p.d.f. of one random variable, each is called a marginal p.d.f.

DEFINITION 5.1-2   *Let X and Y have the joint probability density function $f(x, y)$ with space R. The probability density function of X alone, called the* **marginal probability density function of** X, *is defined by*

$$f_1(x) = \sum_y f(x, y), \qquad x \in R_1,$$

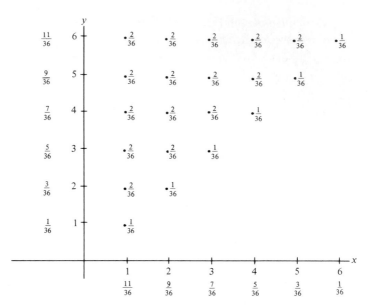

FIGURE 5.1-1

where the summation is taken over all possible y values for each given x in the x space R_1. That is, the summation is over all (x, y) in R with a given x value. Similarly, the **marginal probability density function of** *Y is defined by*

$$f_2(y) = \sum_x f(x, y), \qquad y \in R_2,$$

where the summation is taken over all possible x values for each given y in the y space R_2. The random variables X and Y are **independent** *if and only if*

$$f(x, y) \equiv f_1(x)f_2(y), \qquad x \in R_1, y \in R_2;$$

otherwise X and Y are said to be **dependent**.

Example 5.1-2   Let the joint p.d.f. of X and Y be defined by

$$f(x, y) = \frac{x + y}{21}, \qquad x = 1, 2, 3, \quad y = 1, 2.$$

Then

$$f_1(x) = \sum_y f(x, y) = \sum_{y=1}^{2} \frac{x + y}{21}$$
$$= \frac{x + 1}{21} + \frac{x + 2}{21} = \frac{2x + 3}{21}, \qquad x = 1, 2, 3;$$

and

$$f_2(y) = \sum_x f(x, y) = \sum_{x=1}^{3} \frac{x+y}{21} = \frac{6+3y}{21}, \qquad y = 1, 2.$$

Note that both $f_1(x)$ and $f_2(y)$ satisfy the properties of a probability density function. Since $f(x, y) \neq f_1(x)f_2(y)$, X and Y are dependent.

Example 5.1-3 Let the joint p.d.f. of X and Y be

$$f(x, y) = \frac{xy^2}{30}, \qquad x = 1, 2, 3, \quad y = 1, 2.$$

The marginal probability density functions are

$$f_1(x) = \sum_{y=1}^{2} \frac{xy^2}{30} = \frac{x}{6}, \qquad x = 1, 2, 3,$$

and

$$f_2(y) = \sum_{x=1}^{3} \frac{xy^2}{30} = \frac{y^2}{5}, \qquad y = 1, 2.$$

Then $f(x, y) \equiv f_1(x)f_2(y)$ for $x = 1, 2, 3$ and $y = 1, 2$; thus X and Y are independent.

Example 5.1-4 Let the joint p.d.f. of X and Y be

$$f(x, y) = \frac{xy^2}{13}, \qquad (x, y) = (1, 1), (1, 2), (2, 2).$$

Then the p.d.f. of X is

$$f_1(x) = \begin{cases} \dfrac{5}{13}, & x = 1, \\[2mm] \dfrac{8}{13}, & x = 2, \end{cases}$$

and that of Y is

$$f_2(y) = \begin{cases} \dfrac{1}{13}, & y = 1, \\[2mm] \dfrac{12}{13}, & y = 2. \end{cases}$$

Thus $f(x, y) \neq f_1(x)f_2(y)$ for $x = 1, 2$ and $y = 1, 2$, and X and Y are dependent.

Note that in Example 5.1-4 the support R of X and Y is "triangular." Whenever this support R is not "rectangular," the random variables must be dependent because R cannot then equal the product set $\{(x, y): x \in R_1, y \in R_2\}$. That is, if we observe that the support R of X and Y is not a product set, then X and Y must be dependent. For illustration, in Example 5.1-4, X and Y are dependent because $R = \{(1, 1), (1, 2), (2, 2)\}$ is not a product set. On the other hand, if R equals the product set $\{(x, y): x \in R_1, y \in R_2\}$ and if the formula for $f(x, y)$ is the product of an expression in x alone and an expression in y alone, then X and Y are independent, as illustrated in Example 5.1-3. Example 5.1-2 illustrates the fact that the support can be rectangular but the formula for $f(x, y)$ is not such a product and thus X and Y are dependent.

The notion of a joint p.d.f. of two discrete random variables can be extended to a joint p.d.f. of n random variables of the discrete type. Briefly, the joint p.d.f. of the n random variables $X_1, X_2, \ldots, X_n$ is defined by

$$f(x_1, x_2, \ldots, x_n) = P(X_1 = x_1, X_2 = x_2, \ldots, X_n = x_n)$$

over an appropriate space R. Furthermore, $f(x_1, x_2, \ldots, x_n)$ satisfies properties similar to those given in Definition 5.1-1. In addition, the **marginal probability density function** of one of n discrete random variables, say X_k, is found by summing $f(x_1, x_2, \ldots, x_n)$ over all x_i's except x_k; that is,

$$f_k(x_k) = \sum_{x_1} \cdots \sum_{x_{k-1}} \sum_{x_{k+1}} \cdots \sum_{x_n} f(x_1, x_2, \ldots, x_n), \qquad x_k \in R_k.$$

The random variables $X_1, X_2, \ldots, X_n$ are **mutually independent** if and only if

$$f(x_1, x_2, \ldots, x_n) \equiv f_1(x_1)f_2(x_2) \cdots f_n(x_n), \qquad x_1 \in R_1, x_2 \in R_2, \ldots, x_n \in R_n.$$

If $X_1, X_2, \ldots, X_n$ are not independent, they are said to be **dependent**.

We are now in a position to examine more formally the concept of a random sample from a distribution. Recall that when we collected the n observations, $x_1, x_2, \ldots, x_n$, of X we wanted them in some sense to be independent, which we now observe is actually mutual independence. That is, *before* the sample is actually taken, we want the corresponding random variables $X_1, X_2, \ldots, X_n$ to be mutually independent and each to have the same distribution and, of course, the same p.d.f., say $f(x)$. That is, the numbers $X_1, X_2, \ldots, X_n$ that are to be observed should be mutually independent and identically distributed random variables with joint p.d.f. $f(x_1)f(x_2) \cdots f(x_n)$.

Example 5.1-5 Let X_1, X_2, X_3, X_4 be four mutually independent and identically distributed random variables with the common Poisson p.d.f.

$$f(x) = \frac{2^x e^{-2}}{x!}, \qquad x = 0, 1, 2, \ldots.$$

The joint p.d.f. of X_1, X_2, X_3, X_4, is

$$f(x_1)f(x_2)f(x_3)f(x_4) = \left(\frac{2^{x_1}e^{-2}}{x_1!}\right)\left(\frac{2^{x_2}e^{-2}}{x_2!}\right)\left(\frac{2^{x_3}e^{-2}}{x_3!}\right)\left(\frac{2^{x_4}e^{-2}}{x_4!}\right)$$

$$= \frac{2^{x_1+x_2+x_3+x_4}e^{-8}}{x_1!x_2!x_3!x_4!},$$

for $x_i = 0, 1, 2, \ldots$, and $i = 1, 2, 3, 4$. Then, for illustration,

$$P(X_1 = 3, X_2 = 1, X_3 = 2, X_4 = 1) = f(3)f(1)f(2)f(1)$$

$$= \frac{2^{3+1+2+1}e^{-8}}{3!1!2!1!}$$

$$= \frac{2^7 e^{-8}}{12} = \frac{32}{3}e^{-8}.$$

Let us also compute the probability that exactly one of the X's equals zero. First we treat zero as "success." If W equals the number of successes, then the distribution of W is $b(4, e^{-2})$ because

$$P(X_i = 0) = e^{-2}, \qquad i = 1, 2, 3, 4.$$

Thus the probability of one success and three failures is

$$P(W = 1) = \frac{4!}{1!3!}\,(e^{-2})(1 - e^{-2})^3.$$

We are now prepared to define officially a random sample and other related terms. Say a random experiment that results in a random variable X having p.d.f. $f(x)$ is repeated n independent times. Let X_1, X_2, $\ldots$, X_n denote the n random variables associated with these outcomes. The collection of these random variables, which are mutually independent and identically distributed, is called a **random sample** from a distribution with p.d.f. $f(x)$. The number n is called the **sample size**. The common distribution of the random variables in a random sample is sometimes called the **population** from which the sample is taken.

Let X_1, X_2, $\ldots$, X_n be random variables of the discrete type having a joint distribution. We consider the mathematical expectation of functions of these random variables. If $u(X_1, X_2, \ldots, X_n)$ is a function of n random variables of the discrete type that have a joint p.d.f. $f(x_1, x_2, \ldots, x_n)$ and space R, then

$$E[u(X_1, X_2, \ldots, X_n)] = \sum \cdots \sum_{(x_1,\ldots,x_n)} u(x_1, x_2, \ldots, x_n)f(x_1, x_2, \ldots, x_n),$$

if it exists, is called the **mathematical expectation** (or **expected value**) of

$$u(X_1, X_2, \ldots, X_n).$$

Example 5.1-6 There are eight similar chips in a bowl: three marked $(0, 0)$, two marked $(1, 0)$, two marked $(0, 1)$, and one marked $(1, 1)$. A player selects a chip at random and is given the sum of the two coordinates in dollars. If X_1 and X_2 represent those two coordinates, respectively, their joint p.d.f. is

$$f(x_1, x_2) = \frac{3 - x_1 - x_2}{8}, \qquad x_1 = 0, 1 \text{ and } x_2 = 0, 1.$$

Thus

$$E(X_1 + X_2) = \sum_{x_2=0}^{1} \sum_{x_1=0}^{1} (x_1 + x_2) \frac{3 - x_1 - x_2}{8}$$

$$= (0)\left(\frac{3}{8}\right) + (1)\left(\frac{2}{8}\right) + (1)\left(\frac{2}{8}\right) + (2)\left(\frac{1}{8}\right) = \frac{3}{4}.$$

That is, the expected payoff is 75¢.

The following mathematical expectations, subject to their existence, have special names:

(a) If $u_1(X_1, X_2, \ldots, X_n) = X_i$, then

$$E[u_1(X_1, X_2, \ldots, X_n)] = E(X_i) = \mu_i$$

is called the **mean** of X_i, $i = 1, 2, \ldots, n$.

(b) If $u_2(X_1, X_2, \ldots, X_n) = (X_i - \mu_i)^2$, then

$$E[u_2(X_1, X_2, \ldots, X_n)] = E[(X_i - \mu_i)^2] = \sigma_i^2 = \text{Var}(X_i)$$

is called the **variance** of X_i, $i = 1, 2, \ldots, n$.

The idea of joint distributions of two or more random variables of the discrete type can be extended to that of two or more random variables of the continuous type. The definitions are really the same except that integrals replace summations. We begin with two random variables of the continuous type.

The **joint probability density function** of two continuous-type random variables is an integrable function $f(x, y)$ with the following properties:

(a) $f(x, y) \geq 0$.
(b) $\int_{-\infty}^{\infty} \int_{-\infty}^{\infty} f(x, y) \, dx \, dy = 1$.
(c) $P[(X, Y) \in A] = \iint_{A} f(x, y) \, dx \, dy$,

where $\{(X, Y) \in A\}$ is an event defined in the plane.

Property (c) implies that $P[(X, Y) \in A]$ is the volume of the solid over the region A in the xy-plane and bounded by the surface $z = f(x, y)$.

Example 5.1-7   Let X and Y have the joint p.d.f. $f(x, y) = e^{-x-y}, 0 < x < \infty, 0 < y < \infty$. The graph of $z = f(x, y)$ is given in Figure 5.1-2; in this figure

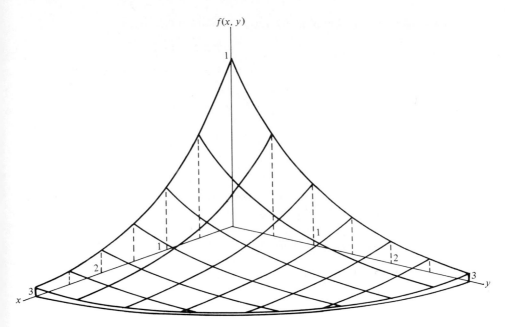

FIGURE 5.1-2

$f(x, y)$ is shown only when $x^2 + y^2 \leq 9$. Let $A = \{(x, y): 0 < x < \infty, 0 < y < x/3\}$. The probability that (X, Y) falls in A is given by

$$P[(X, Y) \in A] = \int_0^\infty \int_0^{x/3} e^{-x-y} \, dy \, dx = \int_0^\infty e^{-x}[-e^{-y}]_0^{x/3} \, dx$$

$$= \int_0^\infty [e^{-x} - e^{-4x/3}] \, dx = \left[-e^{-x} + \frac{3}{4}e^{-4x/3}\right]_0^\infty$$

$$= \frac{1}{4}.$$

The respective **marginal p.d.f.**'s of continuous-type random variables X and Y are given by

$$f_1(x) = \int_{-\infty}^\infty f(x, y) \, dy, \quad x \in R_1,$$

and

$$f_2(y) = \int_{-\infty}^\infty f(x, y) \, dx, \quad y \in R_2,$$

where R_1 and R_2 are the spaces of X and Y. The definitions associated with mathematical expectations are the same as those associated with the discrete case after replacing the summations with integrations.

Example 5.1-8 Let X and Y have the joint p.d.f.

$$f(x, y) = 2, \qquad 0 \leq x \leq y \leq 1.$$

Then $R = \{(x, y): 0 \leq x \leq y \leq 1\}$ is the support and, for illustration,

$$P\left(0 \leq X \leq \frac{1}{2}, 0 \leq Y \leq \frac{1}{2}\right) = P\left(0 \leq X \leq Y, 0 \leq Y \leq \frac{1}{2}\right)$$

$$= \int_0^{1/2} \int_0^y 2 \; dx \; dy = \int_0^{1/2} 2y \; dy = \frac{1}{4}.$$

The shaded region in Figure 5.1-3 is the region of integration that is a subset of R, and the given probability is the volume above that region under the surface $z = 2$. The marginal p.d.f.'s are given by

$$f_1(x) = \int_x^1 2 \; dy = 2(1 - x), \qquad 0 \leq x \leq 1,$$

and

$$f_2(y) = \int_0^y 2 \; dx = 2y, \qquad 0 \leq y \leq 1.$$

Three illustrations of expected values are

$$E(X) = \int_0^1 \int_x^1 2x \; dy \; dx = \int_0^1 2x(1 - x) \; dx = \frac{1}{3},$$

$$E(Y) = \int_0^1 \int_0^y 2y \; dx \; dy = \int_0^1 2y^2 \; dy = \frac{2}{3},$$

$$E(Y^2) = \int_0^1 \int_0^y 2y^2 \; dx \; dy = \int_0^1 2y^3 \; dy = \frac{1}{2}.$$

From these calculations we see that $E(X)$, $E(Y)$, and $E(Y^2)$ could be calculated using the marginal p.d.f.'s as well as the joint one.

The definition of independent random variables of the continuous type carries over naturally from the discrete case. That is, X and Y are **independent** if and only if the joint p.d.f. factors into the product of their marginal p.d.f.'s; namely,

$$f(x, y) \equiv f_1(x) f_2(y), \; x \in R_1, \; y \in R_2.$$

Thus the random variables X and Y in Example 5.1-7 are independent. In addition, the rules that allow us to easily determine dependent and independent random variables are also valid here. For illustration, X and Y in Example 5.1-8 are dependent because the support R is not a product space, since it is bounded by the diagonal line $y = x$.

If there are n random variables, $X_1, X_2, \ldots, X_n$, with joint p.d.f. $f(x_1, x_2, \ldots, x_n)$, the **marginal p.d.f.** of any one of them, say X_k, is given by an $(n - 1)$-fold integral in which x_k is fixed; namely,

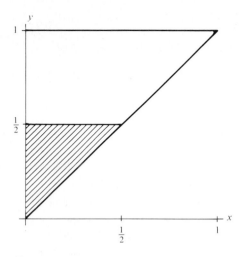

FIGURE 5.1-3

$$f_k(x_k) = \int_{-\infty}^{\infty} \cdots \int_{-\infty}^{\infty} f(x_1, x_2, \ldots, x_n) dx_1 \cdots dx_{k-1} dx_{k+1} \cdots dx_n, \ x_k \in R_k.$$

These n random variables are mutually independent if and only if

$$f(x_1, x_2, \ldots, x_n) \equiv f_1(x_1) f_2(x_2) \cdots f_n(x_n), \ x_1 \in R_1, \ x_2 \in R_2, \cdots, x_n \in R_n.$$

It follows that, if $X_1, X_2, \ldots, X_n$ are observations of a random sample of size n from a distribution with p.d.f. $f(x)$, then the joint p.d.f. of $X_1, X_2, \ldots, X_n$ is

$$f(x_1) f(x_2) \cdots f(x_n).$$

Example 5.1-9 Let X_1, X_2, X_3 be a random sample from a chi-square distribution with 4 degrees of freedom. This is a distribution of the continuous type with p.d.f.

$$f(x) = \frac{1}{\Gamma(\frac{4}{2}) 2^{4/2}} x^{4/2-1} e^{-x/2} = \frac{1}{2^2} x e^{-x/2}, \ 0 < x < \infty.$$

Thus the joint p.d.f. of X_1, X_2, X_3 is

$$f(x_1) f(x_2) f(x_3) = \left(\frac{1}{4} x_1 e^{-x_1/2} \right) \left(\frac{1}{4} x_2 e^{-x_2/2} \right) \left(\frac{1}{4} x_3 e^{-x_3/2} \right)$$

$$= \frac{1}{64} x_1 x_2 x_3 e^{-(x_1 + x_2 + x_3)/2},$$

where $R = \{(x_1, x_2, x_3): 0 < x_i < \infty, \ i = 1, 2, 3\}$.

Exercises

5.1-1 Let the joint p.d.f. of X and Y be defined by

$$f(x, y) = \frac{x + y}{32}, \qquad x = 1, 2, \quad y = 1, 2, 3, 4.$$

Find

(a) $f_1(x)$, the marginal p.d.f. of X.   (b) $f_2(y)$, the marginal p.d.f. of Y.
(c) $P(X > Y)$.   (d) $P(Y = 2X)$.
(e) $P(X + Y = 3)$.   (f) $P(X \le 3 - Y)$.
(g) Are X and Y independent or dependent?

5.1-2 Roll a red and a black four-sided die. Let X equal the outcome on the red die, and let Y equal the outcome on the black die.
(a) On graph paper, show the space of X and Y.
(b) Define the joint p.d.f. on the space (similar to Figure 5.1-1).
(c) Give the marginal p.d.f. of X in the margin.
(d) Give the marginal p.d.f. of Y in the margin.
(e) Are X and Y dependent or independent? Why?

5.1-3 Roll a red and a black four-sided die. Let X equal the outcome on the red die and let Y equal the sum of the two dice.
(a) On graph paper, describe the space of X and Y.
(b) Define the joint p.d.f. on the space (similar to Figure 5.1-1).
(c) Give the marginal p.d.f. of X in the margin.
(d) Give the marginal p.d.f. of Y in the margin.
(e) Are X and Y dependent or independent? Why?

5.1-4 Roll a pair of six-sided dice; one red and one black. Let X equal the outcome on the red die and let Y equal the absolute value of the difference of the two outcomes.
(a) On a graph, describe the space of X and Y.
(b) Define the joint p.d.f. on the space (similar to Figure 5.1-1).
(c) Give the marginal p.d.f.'s in the margins.
(d) Calculate the values of $\mu_Y = E(Y)$ and $\sigma_Y^2 = \text{Var}(Y)$.
(e) Roll a pair of dice 50 times to generate 50 observations of Y. Find the values $\bar{y}$ and s_y^2, comparing them with μ_Y and σ_Y^2, respectively.

5.1-5 A box contains a mixture of tea bags—15 spice, 5 orange, 10 mint, and 20 green. Select 4 tea bags at random and without replacement. Find the probability that
(a) One of each kind of tea is selected.
(b) All 4 tea bags are green tea.

5.1-6 Let X_1, X_2, X_3 denote a random sample of size $n = 3$ from a distribution with the geometric p.d.f.

$$f(x) = \left(\frac{3}{4}\right)\left(\frac{1}{4}\right)^{x-1}, \qquad x = 1, 2, 3, \ldots .$$

That is, X_1, X_2, and X_3 are mutually independent and each has this geometric distribution.

(a) Compute $P(X_1 = 1, X_2 = 3, X_3 = 1)$.

(b) Determine $P(X_1 + X_2 + X_3 = 5)$.

(c) If Y equals the maximum of X_1, X_2, X_3, find

$$P(Y \le 2) = P(X_1 \le 2)P(X_2 \le 2)P(X_3 \le 2).$$

5.1-7 In a biology laboratory, corn is used to illustrate the Mendelian theory of inheritance. It is claimed that the four categories for the kernels of corn, smooth and yellow, wrinkled and yellow, smooth and purple, and wrinkled and purple, will occur in the ratio 9:3:3:1. Out of 208 kernels of corn, let X_1, X_2, X_3 and $X_4 = 208 - X_1 - X_2 - X_3$ denote the numbers in the four categories, respectively, and assume the theory is true.

(a) Give the joint p.d.f. of X_1, X_2, X_3, and X_4, and describe the support in 3-space.

(b) Give the marginal p.d.f. of X_1.

5.1-8 Toss a fair die 12 independent times. Let X_i denote the number of times i occurs, $i = 1, 2, 3, 4, 5, 6$.

(a) What is the joint p.d.f. of X_1, X_2, . . . , X_6?

(b) Find the probability that each outcome occurs two times.

(c) Find $P(X_i = 2)$.

(d) Are X_1, X_2, . . . , X_6 mutually independent?

5.1-9 Let $f(x, y) = 2e^{-x-y}$, $0 \le x \le y < \infty$, be the joint p.d.f. of X and Y. Find $f_1(x)$ and $f_2(y)$, the marginal p.d.f.'s of X and Y, respectively. Are X and Y independent?

5.1-10 Let $f(x, y) = 3/2$, $x^2 \le y \le 1$, $0 \le x \le 1$, be the joint p.d.f. of X and Y. Find

(a) $P(0 \le X \le 1/2)$.   (b) $P(1/2 \le Y \le 1)$.

(c) $P(1/2 \le X \le 1, 1/2 \le Y \le 1)$.   (d) $P(X \ge 1/2, Y \ge 1/2)$.

(e) Are X and Y independent?

5.1-11 Let $f(x, y) = 1/4$, $0 \le x \le 2$, $0 \le y \le 2$, be the joint p.d.f. of X and Y. Find $f_1(x)$ and $f_2(y)$, the marginal probability density functions. Are the two random variables independent?

5.1-12 Let X and Y have the joint p.d.f. $f(x, y) = x + y$, $0 \le x \le 1$, $0 \le y \le 1$.

(a) Find the marginal p.d.f.'s $f_1(x)$ and $f_2(y)$ and show that $f(x, y) \ne f_1(x)f_2(y)$. Thus X and Y are dependent. Compute

(b) μ_X.   (c) μ_Y.   (d) σ_X^2.   (e) σ_Y^2.

5.1-13 Let $f(x, y) = e^{-x-y}$, $0 < x < \infty$, $0 < y < \infty$, be the joint p.d.f. of X and Y. Argue that X and Y are independent and compute

(a) $P(X < Y)$.

(b) $P(X > 1, Y > 1)$.

(c) $P(X = Y)$.

(d) $P(X < 2)$.

(e) $P(0 < X < \infty, X/3 < Y < 3X)$.

(f) $P(0 < X < \infty, 3X < Y < \infty)$.

5.1-14 Let X_1, X_2 be independent and have distributions that are $U(0, 1)$. The joint p.d.f. of X_1 and X_2 is $f(x_1, x_2) = 1$, $0 \le x_1 \le 1$, $0 \le x_2 \le 1$.

(a) Show that $P(X_1^2 + X_2^2 \le 1) = \pi/4$.

(b) Using pairs of random numbers, find an approximation of $\pi/4$.

HINT: For n pairs of random numbers, the relative frequency

$$\frac{\#[\{(x_1, x_2): x_1^2 + x_2^2 \le 1\}]}{n}$$

is an approximation of $\pi/4$.

(c) Let Y equal the $\#[\{(x_1, x_2): x_1^2 + x_2^2 \le 1\}]$ for n independent pairs of random numbers. How is Y distributed?

5.1-15 Let $X_1, X_2, \ldots, X_n$ be independent and have distributions that are $U(0, 1)$. The joint p.d.f. of $X_1, X_2, \ldots, X_n$ is $f(x_1, x_2, \ldots, x_n) = 1$, $0 \le x_1 \le 1$, $0 \le x_2 \le 1$, $\ldots$, $0 \le x_n \le 1$. Show that

(a) $P(X_1 + X_2 \le 1) = 1/2!$.

(b) $P(X_1 + X_2 + X_3 \le 1) = 1/3!$.

(c) $P(X_1 + X_2 + \cdots + X_n \le 1) = 1/n!$.

HINT: Draw figures for parts (a) and (b).

5.2 Distributions of Sums of Independent Random Variables

Functions of random variables are usually of interest in statistical applications; and, of course, we have already considered some functions, called **statistics**, of the observations of a random sample. Two important statistics are the sample mean $\bar{X}$ and the sample variance S^2. Of course, in a particular sample, say $x_1, x_2, \ldots, x_n$, we observed definite values of these statistics, $\bar{x}$ and s^2; however, we should recognize that each value is only one observation of the respective random variables, $\bar{X}$ and S^2. That is, each $\bar{X}$ or S^2 (or, more generally, any statistic) is also a random variable with its own distribution. In this chapter we determine the distributions of some of the important statistics and, more generally, distributions of functions of random variables. The derivations of distributions of these functions fall in that area of statistics usually referred to as **sampling distribution theory**.

We begin with an example that illustrates a different way of finding the distribution of a sum of random variables in a random sample.

Example 5.2-1 Let us cast an unbiased four-sided die two independent times and observe the outcomes on these trials, say X_1 and X_2. This is equivalent to saying that X_1 and X_2 are the observations of a random sample of size $n = 2$ from a distribution with p.d.f.

$$f(x) = \frac{1}{4}, \qquad x = 1, 2, 3, 4.$$

We shall give another way to find the p.d.f. of $Y = X_1 + X_2$ (see Exercise 5.1-3). Note that the space of Y is $R = \{2, 3, 4, 5, 6, 7, 8\}$. To determine the p.d.f. of Y, say $g(y) = P(Y = y)$, $y \in R$, we first look at some special cases. Let $y = 3$. We note that the event $\{Y = 3\}$ can occur in two mutually exclusive ways: $\{X_1 = 1, X_2 = 2\}$ and $\{X_1 = 2, X_2 = 1\}$. Thus recalling that X_1 and X_2 are independent, we obtain

$$g(3) = P(Y = 3) = P(X_1 = 1, X_2 = 2) + P(X_1 = 2, X_2 = 1)$$

$$= \left(\frac{1}{4}\right)\left(\frac{1}{4}\right) + \left(\frac{1}{4}\right)\left(\frac{1}{4}\right) = \frac{2}{16}.$$

Similarly,

$$g(5) = P(Y = 5) = P(X_1 = 1, X_2 = 4) + P(X_1 = 2, X_2 = 3)$$
$$+ P(X_1 = 3, X_2 = 2) + P(X_1 = 4, X_2 = 1)$$

$$= \left(\frac{1}{4}\right)\left(\frac{1}{4}\right) + \left(\frac{1}{4}\right)\left(\frac{1}{4}\right) + \left(\frac{1}{4}\right)\left(\frac{1}{4}\right) + \left(\frac{1}{4}\right)\left(\frac{1}{4}\right) = \frac{4}{16}.$$

In general, we see that the p.d.f. in Example 5.2-1 is given by the following formula, called a **convolution formula**,

$$g(y) = P(Y = y) = \sum_{k=1}^{y-1} f(k)f(y - k),$$

where some of the summands could equal zero.

We can give the p.d.f. either by Table 5.2-1 or by the formula

$$g(y) = \frac{4 - |y - 5|}{16}, \qquad y = 2, 3, 4, 5, 6, 7, 8.$$

TABLE 5.2-1

y	$g(y)$
2	1/16
3	2/16
4	3/16
5	4/16
6	3/16
7	2/16
8	1/16

Using the p.d.f. of Y, its mean is

$$E(Y) = \sum_{y=2}^{8} yg(y)$$

$$= 2\left(\frac{1}{16}\right) + 3\left(\frac{2}{16}\right) + 4\left(\frac{3}{16}\right) + 5\left(\frac{4}{16}\right) + 6\left(\frac{3}{16}\right) + 7\left(\frac{2}{16}\right) + 8\left(\frac{1}{16}\right)$$

$$= 5.$$

In Example 5.2-1, since $Y = X_1 + X_2$, we anticipate that $E(Y) = 5$, which is computed using the p.d.f. of Y, would be equal to $E(X_1 + X_2)$, which is computed using the joint p.d.f. of X_1 and X_2, namely

$$E(X_1 + X_2) = \sum_{x_2=1}^{4} \sum_{x_1=1}^{4} (x_1 + x_2)\left(\frac{1}{4}\right)\left(\frac{1}{4}\right).$$

Of course, this is true because

$$E(X_1 + X_2) = E(X_1) + E(X_2) = \frac{5}{2} + \frac{5}{2} = 5.$$

This special result with a linear function of two random variables extends to more general functions of several random variables. We accept the following without proof.

THEOREM 5.2-1 *Let $X_1, X_2, \ldots, X_n$ be n random variables with joint p.d.f. $f(x_1, x_2, \ldots, x_n)$. Let the random variable $Y = u(X_1, X_2, \ldots, X_n)$ have the p.d.f. $g(y)$. Then, in the discrete case,*

$$E(Y) = \sum_{y} yg(y) = \sum\sum\cdots\sum_{(x_1,x_2,\ldots,x_n)} u(x_1, x_2, \ldots, x_n)f(x_1, x_2, \ldots, x_n),$$

provided that these summations exist. For random variables of the continuous type, integrals replace the summations.

The next theorem proves that the expected value of the product of functions of n mutually independent random variables is the product of their expected values.

THEOREM 5.2-2 *If $X_1, X_2, \ldots, X_n$ are mutually independent random variables having p.d.f.'s $f_1(x_1), f_2(x_2), \ldots, f_n(x_n)$ and $E[u_i(X_i)]$, $i = 1, 2, \ldots, n$, exist, then*

$$E[u_1(X_1)u_2(X_2)\cdots u_n(X_n)] = E[u_1(X_1)]E[u_2(X_2)]\cdots E[u_n(X_n)].$$

PROOF: In the discrete case, we have that

$$E[u_1(X_1)u_2(X_2) \cdots u_n(X_n)]$$

$$= \sum \sum \cdots \sum_{(x_1, x_2, \ldots, x_n)} u_1(x_1)u_2(x_2) \cdots u_n(x_n)f_1(x_1)f_2(x_2) \cdots f_n(x_n)$$

$$= \sum_{x_1} u_1(x_1)f_1(x_1) \sum_{x_2} u_2(x_2)f_2(x_2) \cdots \sum_{x_n} u_n(x_n)f_n(x_n)$$

$$= E[u_1(X_1)]E[u_2(X_2)] \cdots E[u_n(X_n)].$$

In the proof for the continuous case, integrals replace summations. $\square$

REMARK Sometimes students recognize that $X^2 = X \cdot X$ and thus believe that $E(X^2)$ is equal to $[E(X)][E(X)] = [E(X)]^2$ because the above theorem states that the expected value of the product is the product of the expected values. However, note the hypothesis of independence in the theorem, and certainly X is not independent of itself. Incidentally, if $E(X^2)$ did equal $[E(X)]^2$, then the variance of X

$$\sigma^2 = E(X^2) - [E(X)]^2$$

would always equal zero. This really happens only in the case of degenerate (one point) distributions.

Example 5.2-2 It is interesting to note that these two theorems allow us to determine the mean, the variance, and the moment-generating function of a function such as $Y = X_1 + X_2$, where X_1 and X_2 have been defined in Example 5.1-1. We have already seen that $\mu_Y = 5$.
The variance of Y is

$$\sigma_Y^2 = E[(Y - \mu_Y)^2] = E[(X_1 + X_2 - \mu_1 - \mu_2)^2],$$

where $\mu_i = E(X_i) = 5/2$, $i = 1, 2$. Thus

$$\sigma_Y^2 = E[(X_1 - \mu_1) + (X_2 - \mu_2)]^2$$
$$= E[(X_1 - \mu_1)^2 + 2(X_1 - \mu_1)(X_2 - \mu_2) + (X_2 - \mu_2)^2].$$

But, from an earlier result about the expected value being a linear operator, we have

$$\sigma_Y^2 = E[(X_1 - \mu_1)^2] + 2E[(X_1 - \mu_1)(X_2 - \mu_2)] + E[(X_2 - \mu_2)^2].$$

However, since X_1 and X_2 are independent, then

$$E[(X_1 - \mu_1)(X_2 - \mu_2)] = [E(X_1 - \mu_1)][E(X_2 - \mu_2)]$$
$$= (\mu_1 - \mu_1)(\mu_2 - \mu_2) = 0.$$

Thus

$$\sigma_Y^2 = \sigma_1^2 + \sigma_2^2,$$

where

$$\sigma_i^2 = E[(X_i - \mu_i)^2], \qquad i = 1, 2.$$

In the case in which X_1 and X_2 are the outcomes on two independent casts of a four-sided die, we have $\sigma_i^2 = 5/4$, $i = 1, 2$, and, hence

$$\sigma_Y^2 = \frac{5}{4} + \frac{5}{4} = \frac{5}{2}.$$

Finally, the moment-generating function is

$$M_Y(t) = E(e^{tY}) = E[e^{t(X_1+X_2)}] = E(e^{tX_1}e^{tX_2}).$$

The independence of X_1 and X_2 implies that

$$M_Y(t) = E(e^{tX_1})E(e^{tX_2}).$$

For our example in which X_1 and X_2 have the same p.d.f.

$$f(x) = \frac{1}{4} \qquad x = 1, 2, 3, 4,$$

and thus the same moment-generating function

$$M_X(t) = \frac{1}{4}\, e^t + \frac{1}{4}\, e^{2t} + \frac{1}{4}\, e^{3t} + \frac{1}{4}\, e^{4t},$$

we have that $M_Y(t) = [M_X(t)]^2$ equals

$$\frac{1}{16}\, e^{2t} + \frac{2}{16}\, e^{3t} + \frac{3}{16}\, e^{4t} + \frac{4}{16}\, e^{5t} + \frac{3}{16}\, e^{6t} + \frac{2}{16}\, e^{7t} + \frac{1}{16}\, e^{8t}.$$

Note that the coefficient of e^{bt} is equal to the probability $P(Y = b)$; for illustration, $4/16 = P(Y = 5)$. This agrees with the result found in Example 5.2-1, and thus we see that we could find the distribution of Y by determining its moment-generating function.

In this section we are restricting attention to those functions that are linear combinations of independent random variables. We shall first prove an important theorem about the mean and the variance of such a linear combination, extending the results in the last example.

THEOREM 5.2-3 *If $X_1, X_2, \ldots, X_n$ are n independent random variables with respective means $\mu_1, \mu_2, \ldots, \mu_n$ and variances $\sigma_1^2, \sigma_2^2, \ldots, \sigma_n^2$, then the mean and the variance of $Y = \sum_{i=1}^{n} a_i X_i$, where $a_1, a_2, \ldots, a_n$ are real constants, are*

$$\mu_Y = \sum_{i=1}^{n} a_i \mu_i \qquad and \qquad \sigma_Y^2 = \sum_{i=1}^{n} a_i^2 \sigma_i^2,$$

respectively.

PROOF: We have that

$$\mu_Y = E(Y) = E\left(\sum_{i=1}^{n} a_i X_i\right) = \sum_{i=1}^{n} a_i E(X_i) = \sum_{i=1}^{n} a_i \mu_i$$

because the expected value of the sum is the sum of the expected values (i.e., E is a linear operator). Also

$$\sigma_Y^2 = E[(Y - \mu_Y)^2] = E\left[\left(\sum_{i=1}^{n} a_i X_i - \sum_{i=1}^{n} a_i \mu_i\right)^2\right]$$

$$= E\left\{\left[\sum_{i=1}^{n} a_i(X_i - \mu_i)\right]^2\right\} = E\left[\sum_{i=1}^{n}\sum_{j=1}^{n} a_i a_j (X_i - \mu_i)(X_j - \mu_j)\right].$$

Again using the fact that E is a linear operator, we obtain

$$\sigma_Y^2 = \sum_{i=1}^{n}\sum_{j=1}^{n} a_i a_j E[(X_i - \mu_i)(X_j - \mu_j)].$$

However, if $i \neq j$, then from the independence of X_i and X_j we have

$$E[(X_i - \mu_i)(X_j - \mu_j)] = E(X_i - \mu_i)E(X_j - \mu_j) = (\mu_i - \mu_i)(\mu_j - \mu_j) = 0.$$

Thus the variance can be written as

$$\sigma_Y^2 = \sum_{i=1}^{n} a_i^2 E[(X_i - \mu_i)^2] = \sum_{i=1}^{n} a_i^2 \sigma_i^2. \qquad \square$$

We give two illustrations of the theorem.

Example 5.2-3 Let the independent random variables X_1 and X_2 have respectively means $\mu_1 = -4$ and $\mu_2 = 3$ and variances $\sigma_1^2 = 4$ and $\sigma_2^2 = 9$. The mean and the variance of $Y = 3X_1 - 2X_2$ are, respectively,

$$\mu_Y = (3)(-4) + (-2)(3) = -18$$

and

$$\sigma_Y^2 = (3)^2(4) + (-2)^2(9) = 72.$$

Example 5.2-4 Let $X_1, X_2, \ldots, X_n$ be a random sample of size n from a distribution with mean μ and variance σ^2. First let $Y = X_1 - X_2$; then

$$\mu_Y = \mu - \mu = 0 \qquad \text{and} \qquad \sigma_Y^2 = (1)^2\sigma^2 + (-1)^2\sigma^2 = 2\sigma^2.$$

Now consider the sample mean

$$\bar{X} = \frac{X_1 + X_2 + \cdots + X_n}{n},$$

which is a linear function with each $a_i = 1/n$. Then

$$\mu_{\bar{X}} = \sum_{i=1}^{n}\left(\frac{1}{n}\right)\mu = \mu \qquad \text{and} \qquad \sigma_{\bar{X}}^2 = \sum_{i=1}^{n}\left(\frac{1}{n}\right)^2 \sigma^2 = \frac{\sigma^2}{n}.$$

That is, the mean of $\overline{X}$ is that of the distribution from which the sample arose, but the variance of $\overline{X}$ is that of the underlying distribution divided by n.

In some applications it is sufficient to know the mean and variance of a linear combination of random variables, say Y. However, it is often helpful to know exactly how Y is distributed. The next theorem can frequently be used to find the distribution of a linear combination of independent random variables.

THEOREM 5.2-4 *If $X_1, X_2, \ldots, X_n$ are independent random variables with respective moment-generating functions $M_{X_i}(t)$, $i = 1, 2, 3, \ldots, n$, then the moment-generating function of $Y = \sum_{i=1}^{n} a_i X_i$ is*

$$M_Y(t) = \prod_{i=1}^{n} M_{X_i}(a_i t).$$

PROOF: The moment-generating function of Y is given by

$$M_Y(t) = E[e^{tY}] = E[e^{t(a_1 X_1 + a_2 X_2 + \cdots + a_n X_n)}]$$
$$= E[e^{a_1 t X_1} e^{a_2 t X_2} \cdots e^{a_n t X_n}]$$
$$= E[e^{a_1 t X_1}] E[e^{a_2 t X_2}] \cdots E[e^{a_n t X_n}]$$

using Theorem 5.2-2. However, since

$$E(e^{tX_i}) = M_{X_i}(t),$$

then

$$E(e^{a_i t X_i}) = M_{X_i}(a_i t).$$

Thus we have that

$$M_Y(t) = M_{X_1}(a_1 t) M_{X_2}(a_2 t) \cdots M_{X_n}(a_n t)$$

$$= \prod_{i=1}^{n} M_{X_i}(a_i t). \qquad \Box$$

A corollary follows immediately, and it will be used in some important examples.

COROLLARY 5.2-1 *If $X_1, X_2, \ldots, X_n$ are observations of a random sample from a distribution with moment-generating function $M(t)$, then*
(a) *the moment-generating function of $Y = \sum_{i=1}^{n} X_i$ is*

$$M_Y(t) = \prod_{i=1}^{n} M(t) = [M(t)]^n;$$

(b) *the moment-generating function of $\overline{X} = \sum_{i=1}^{n} (1/n) X_i$ is*

$$M_{\overline{X}}(t) = \prod_{i=1}^{n} M\left(\frac{t}{n}\right) = \left[M\left(\frac{t}{n}\right)\right]^n.$$

PROOF: For (a), let $a_i = 1$, $i = 1, 2, \ldots, n$, in Theorem 5.2-4. For (b), take $a_i = 1/n$, $i = 1, 2, \ldots, n$. $\square$

The following examples and the exercises give some important applications of Theorem 5.2-4 and its corollary.

Example 5.2-5 Let $X_1, X_2, \ldots, X_n$ denote the outcomes on n Bernoulli trials. The moment-generating function of X_i, $i = 1, 2, \ldots, n$, is

$$M(t) = q + pe^t.$$

If

$$Y = \sum_{i=1}^{n} X_i,$$

then

$$M_Y(t) = \prod_{i=1}^{n}(q + pe^t) = (q + pe^t)^n.$$

Thus we again see that Y is $b(n, p)$.

Example 5.2-6 Let X_1, X_2, X_3 be the observations of a random sample of size $n = 3$ from the exponential distribution having mean θ and, of course, moment-generating function $M(t) = 1/(1 - \theta t)$, $t < 1/\theta$. The moment-generating function of $Y = X_1 + X_2 + X_3$ is

$$M_Y(t) = [(1 - \theta t)^{-1}]^3 = (1 - \theta t)^{-3}, \qquad t < 1/\theta,$$

which is that of a gamma distribution with parameters $\alpha = 3$ and θ. Thus Y has this distribution. On the other hand, the moment-generating function of $\overline{X}$ is

$$M_{\overline{X}}(t) = \left[\left(1 - \frac{\theta t}{3}\right)^{-1}\right]^3 = \left(1 - \frac{\theta t}{3}\right)^{-3}, \qquad t < 3/\theta;$$

and hence the distribution of $\overline{X}$ is gamma with parameters $\alpha = 3$ and $\theta/3$, respectively.

Exercises

5.2-1   Let X_1 and X_2 be observations of a random sample of size $n = 2$ from a distribution with p.d.f. $f(x) = x/6$, $x = 1, 2, 3$. Then find the p.d.f. of $Y = X_1 + X_2$. Determine the mean and the variance of the sum in two ways.

5.2-2   Let X_1 and X_2 be a random sample of size $n = 2$ from a distribution with p.d.f. $f(x) = 6x(1 - x)$, $0 < x < 1$. Find the mean and the variance of $Y = X_1 + X_2$.

5.2-3   Let X_1, X_2, X_3 be a random sample of size 3 from the distribution with p.d.f. $f(x) = 1/4$, $x = 1, 2, 3, 4$. For example, observe three independent rolls of a fair four-sided die.

(a) Find the p.d.f. of $Y = X_1 + X_2 + X_3$.

(b) Sketch a bar graph of the p.d.f. of Y.

5.2-4   Let X_1 and X_2 be two independent random variables with respective means μ_1 and μ_2 and variances σ_1^2 and σ_2^2. Show that the mean and the variance of $Y = X_1 X_2$ are $\mu_1 \mu_2$ and $\sigma_1^2 \sigma_2^2 + \mu_1^2 \sigma_2^2 + \mu_2^2 \sigma_1^2$, respectively.

HINT: Note that $E(Y) = E(X_1)E(X_2)$ and $E(Y^2) = E(X_1^2)E(X_2^2)$.

5.2-5   Let X_1 and X_2 be two independent random variables with respective means 3 and 7 and variances 9 and 25. Compute the mean and the variance of $Y = -2X_1 + X_2$.

5.2-6   Let X_1 and X_2 have independent distributions $b(n_1, p)$ and $b(n_2, p)$. Find the moment-generating function of $Y = X_1 + X_2$. How is Y distributed?

5.2-7   Let X_1, X_2, X_3 be mutually independent random variables with Poisson distributions having means 2, 1, 4, respectively.

(a) Find the moment-generating function of the sum $Y = X_1 + X_2 + X_3$.

(b) How is Y distributed?

(c) Compute $P(3 \leq Y \leq 9)$.

5.2-8   Generalize Exercise 5.2-7 by showing that the sum of n independent Poisson random variables with respective means $\mu_1, \mu_2, \ldots, \mu_n$ is Poisson with mean

$$\mu_1 + \mu_2 + \cdots + \mu_n.$$

5.2-9   Let X_1, X_2, X_3, X_4, X_5 be a random sample of size 5 from a geometric distribution with $p = 1/3$.

(a) Find the moment-generating function of $Y = X_1 + X_2 + X_3 + X_4 + X_5$.

(b) How is Y distributed?

5.2-10   Let $W = X_1 + X_2 + \cdots + X_h$, a sum of h mutually independent and identically distributed exponential random variables with mean θ. Show that W has a gamma distribution with mean $h\theta$.

5.2-11   Let X_1, X_2, X_3 denote a random sample of size 3 from a gamma distribution with $\alpha = 7$ and $\theta = 5$.

(a) Find the moment-generating function of $Y = X_1 + X_2 + X_3$.

(b) How is Y distributed?

5.2-12   The moment-generating function of X is

$$M_X(t) = \left(\frac{1}{4}\right)(e^t + e^{2t} + e^{3t} + e^{4t});$$

the moment-generating function of Y is

$$M_Y(t) = \left(\frac{1}{3}\right)(e^t + e^{2t} + e^{3t});$$

X and Y are independent random variables. Let $W = X + Y$.
(a) Find the moment-generating function of W.
(b) Give the p.d.f. of W; that is, determine $P(W = w)$, $w = 2, 3, \ldots, 7$, from the moment-generating function of W.

5.2-13 Let X equal the outcome when a four-sided die is rolled. Let Y equal the outcome when a six-sided die is rolled. Let $W = X + Y$.
(a) Find the moment-generating function of W.
(b) Give the p.d.f. of W.

5.2-14 Let X and Y, with respective p.d.f.'s $f(x)$ and $g(y)$, be independent discrete random variables, each of whose support is a subset of the nonnegative integers $0, 1, 2, \ldots$. Show that the p.d.f. of $W = X + Y$ is given by the convolution formula

$$h(w) = \sum_{x=0}^{w} f(x)g(w - x), \quad w = 0, 1, 2, \ldots .$$

HINT: Argue that $h(w) = P(W = w)$ is the probability of the $w + 1$ mutually exclusive events $(x, w - x)$, $x = 0, 1, \ldots, w$, or find the moment-generating function of W.

5.2-15 Let X_1, X_2, X_3, X_4 be a random sample from a distribution having p.d.f. $f(x) = (x + 1)/6$, $x = 0, 1, 2$.
(a) Use Exercise 5.2-14 to find the p.d.f. of $W_1 = X_1 + X_2$.
(b) What is the p.d.f. of $W_2 = X_3 + X_4$?
(c) Now find the p.d.f. of $W = W_1 + W_2 = X_1 + X_2 + X_3 + X_4$.

5.2-16 Roll a fair four-sided die eight times and denote the outcomes by $X_1, X_2, \ldots, X_8$. Use Exercise 5.2-14 to find the p.d.f.'s of
(a) $X_1 + X_2$.

(b) $\displaystyle\sum_{i=1}^{4} X_i$.

(c) $\displaystyle\sum_{i=1}^{8} X_i$.

5.3 Random Functions Associated with Normal Distributions

In statistical applications, it is often assumed that the population from which a sample is taken is normally distributed, $N(\mu, \sigma^2)$. There is then interest in estimating the parameters μ and σ^2 or in testing conjectures about these parameters. The usual statistics that are used in these activities are the sample mean $\overline{X}$ and the sample variance S^2; thus we need to know something about the distribution of these statistics or functions of these statistics.

THEOREM 5.3-1 *If $X_1, X_2, \ldots, X_n$ are observations of a random sample of size n from the normal distribution $N(\mu, \sigma^2)$, then the distribution of the sample mean $\overline{X} = (1/n) \sum_{i=1}^{n} X_i$ is $N(\mu, \sigma^2/n)$.*

PROOF: Since the moment-generating function of each X is

$$M_X(t) = \exp\left(\mu t + \frac{\sigma^2 t^2}{2} \right),$$

the moment-generating function of

$$\overline{X} = \frac{1}{n} \sum_{i=1}^{n} X_i$$

is, from the corollary of Theorem 5.2-4, equal to

$$M_{\overline{X}}(t) = \left\{ \exp\left[\mu\left(\frac{t}{n}\right) + \frac{\sigma^2 (t/n)^2}{2} \right] \right\}^n$$

$$= \exp\left[\mu t + \frac{(\sigma^2/n) t^2}{2} \right].$$

However, the moment-generating function uniquely determines the distribution of the random variable. Since this one is that associated with the normal distribution $N(\mu, \sigma^2/n)$, the sample mean $\overline{X}$ is $N(\mu, \sigma^2/n)$. $\square$

Theorem 5.3-1 shows that if $X_1, X_2, \ldots, X_n$ is a random sample from the normal distribution, $N(\mu, \sigma^2)$, then the probability distribution of $\overline{X}$ is also normal with the same mean μ but a variance σ^2/n. This means that $\overline{X}$ has a greater probability of falling in an interval containing μ than does a single observation, say X_1. For example, if $\mu = 50$, $\sigma^2 = 16$, $n = 64$, then $P(49 < \overline{X} < 51) = 0.9544$, whereas $P(49 < X_1 < 51) = 0.1974$. This is illustrated again in the next example.

> **Example 5.3-1** Let $X_1, X_2, \ldots, X_n$ be a random sample from the $N(50, 16)$ distribution. We know that the distribution of $\overline{X}$ is $N(50, 16/n)$. To illustrate the effect of n, the graph of the p.d.f. of $\overline{X}$ is given in Figure 5.3-1 for $n = 1, 4, 16$, and 64. When $n = 64$, compare the areas that represent $P(49 < \overline{X} < 51)$ and $P(49 < X_1 < 51)$.

Before we can find the distribution of S^2, or $(n-1)S^2/\sigma^2$, we give two preliminary theorems that are important in their own right. The first is a direct result of the material in Section 5.2 and gives an additive property for independent chi-square random variables.

THEOREM 5.3-2 *Let the distributions of $X_1, X_2, \ldots, X_k$ be $\chi^2(r_1), \chi^2(r_2), \ldots, \chi^2(r_k)$, respectively. If $X_1, X_2, \ldots, X_k$ are independent, then $Y = X_1 + X_2 + \cdots + X_k$ is $\chi^2(r_1 + r_2 + \cdots + r_k)$.*

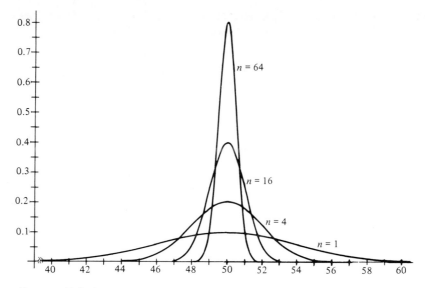

FIGURE 5.3-1

PROOF: We give the proof for $k = 2$ but it is similar for a general k. The moment-generating function of Y is given by

$$M_Y(t) = E[e^{tY}] = E[e^{t(X_1 + X_2)}] = E[e^{tX_1}e^{tX_2}].$$

Because X_1 and X_2 are independent, this last expectation can be factored so that

$$M_Y(t) = E[e^{tX_1}]E[e^{tX_2}] = (1 - 2t)^{-r_1/2}(1 - 2t)^{-r_2/2}, \qquad t < \frac{1}{2},$$

since X_1 and X_2 have chi-square distributions. Thus

$$M_Y(t) = (1 - 2t)^{-(r_1 + r_2)/2}, \qquad t < \frac{1}{2},$$

the moment-generating function for a chi-square distribution with $r = r_1 + r_2$ degrees of freedom. The uniqueness of the moment-generating function implies that Y is $\chi^2(r_1 + r_2)$. For the general k, Y is $\chi^2(r_1 + r_2 + \cdots + r_k)$. $\qquad\square$

The next theorem combines and extends the results of Theorems 4.4-2 and 5.3-2 and gives one interpretation of degrees of freedom.

THEOREM 5.3-3 *Let $Z_1, Z_2, \ldots, Z_r$ have standard normal distributions, $N(0, 1)$. If these random variables are mutually independent, then $W = Z_1^2 + Z_2^2 + \cdots + Z_r^2$ has a distribution that is $\chi^2(r)$.*

PROOF: By Theorem 4.4-2, Z_i^2 is $\chi^2(1)$ for $i = 1, 2, \ldots, r$. From Theorem 5.3-2, with $k = r$, $Y = W$, and $r_i = 1$, we see that W is $\chi^2(r)$. $\square$

COROLLARY 5.3-1 *If $X_1, X_2, \ldots, X_r$ have mutually independent normal distributions $N(\mu_i, \sigma_i^2)$, $i = 1, 2, \ldots, r$, respectively, then the distribution of*

$$W = \sum_{i=1}^{r} \frac{(X_i - \mu_i)^2}{\sigma_i^2}$$

is $\chi^2(r)$.

PROOF: This follows since $Z_i = (X_i - \mu_i)/\sigma_i$ is $N(0, 1)$, $i = 1, 2, \ldots, r$. $\square$

Note that the number of terms in the summation and the number of degrees of freedom are equal in Theorem 5.2-3 and the corollary.

The following theorem gives an important result that will be used in statistical applications. In connection with those, we will use the sample variance S^2 to estimate the variance, σ^2, when sampling from the normal distribution, $N(\mu, \sigma^2)$. More will be said about S^2 at the time of its use.

THEOREM 5.3-4 *If $X_1, X_2, \ldots, X_n$ are observations of a random sample of size n from the normal distribution $N(\mu, \sigma^2)$,*

$$\overline{X} = \frac{1}{n} \sum_{i=1}^{n} X_i$$

and

$$S^2 = \frac{1}{n-1} \sum_{i=1}^{n} (X_i - \overline{X})^2,$$

then

(a) $\overline{X}$ and S^2 are independent.

(b) $\dfrac{(n-1)S^2}{\sigma^2} = \dfrac{\sum_{i=1}^{n}(X_i - \overline{X})^2}{\sigma^2}$ *is $\chi^2(n-1)$.*

PROOF: We are not prepared to prove (a) in this book; so we accept it without proof here. To prove (b), note that

$$\sum_{i=1}^{n} \left(\frac{X_i - \mu}{\sigma} \right)^2 = \sum_{i=1}^{n} \left[\frac{(X_i - \overline{X}) + (\overline{X} - \mu)}{\sigma} \right]^2$$

$$= \sum_{i=1}^{n} \left(\frac{X_i - \overline{X}}{\sigma} \right)^2 + \frac{n(\overline{X} - \mu)^2}{\sigma^2} \qquad (5.3\text{-}1)$$

because the cross-product term is equal to

$$2\sum_{i=1}^{n} \frac{(\overline{X} - \mu)(X_i - \overline{X})}{\sigma^2} = \frac{2(\overline{X} - \mu)}{\sigma^2} \sum_{i=1}^{n} (X_i - \overline{X}) = 0.$$

But $Y_i = (X_i - \mu)/\sigma$, $i = 1, 2, \ldots, n$, are mutually independent standardized normal variables. Hence $W = \sum_{i=1}^{n} Y_i^2$ is $\chi^2(n)$ by Corollary 5.3-1. Moreover, since $\overline{X}$ is $N(\mu, \sigma^2/n)$, then

$$Z^2 = \left(\frac{\overline{X} - \mu}{\sigma/\sqrt{n}}\right)^2 = \frac{n(\overline{X} - \mu)^2}{\sigma^2}$$

is $\chi^2(1)$ by Theorem 4.4-2. In this notation, equation (5.3-1) becomes

$$W = \frac{(n-1)S^2}{\sigma^2} + Z^2.$$

However, from (a), $\overline{X}$ and S^2 are independent; thus Z^2 and S^2 are also independent. In the moment-generating function of W, this independence permits us to write

$$E[e^{t\{(n-1)S^2/\sigma^2 + Z^2\}}] = E[e^{t(n-1)S^2/\sigma^2} e^{tZ^2}]$$
$$= E[e^{t(n-1)S^2/\sigma^2}]E[e^{tZ^2}].$$

Since W and Z^2 have chi-square distributions, we can substitute their moment-generating functions to obtain

$$(1 - 2t)^{-n/2} = E[e^{t(n-1)S^2/\sigma^2}](1 - 2t)^{-1/2}.$$

Equivalently, we have

$$E[e^{t(n-1)S^2/\sigma^2}] = (1 - 2t)^{-(n-1)/2}, \qquad t < \frac{1}{2}.$$

This, of course, is the moment-generating function of a $\chi^2(n-1)$ variable, and accordingly $(n-1)S^2/\sigma^2$ has this distribution. □

Combining the results of Corollary 5.3-1 and Theorem 5.3-4, we see that when sampling from a normal distribution,

$$U = \sum_{i=1}^{n} \frac{(X_i - \mu)^2}{\sigma^2}$$

is $\chi^2(n)$, and

$$W = \sum_{i=1}^{n} \frac{(X_i - \overline{X})^2}{\sigma^2}$$

is $\chi^2(n-1)$. That is, when the population mean, μ, in $\sum (X_i - \mu)^2$ is replaced by the sample mean, $\overline{X}$, one degree of freedom is lost. There are more general situations in which a degree of freedom is lost for each parameter estimated in certain chi-square random variables (e.g., see Section 9.5).

Example 5.3-2   Let X_1, X_2, X_3, X_4 be a random sample of size 4 from the normal distribution, $N(76.4, 383)$. Then

$$U = \sum_{i=1}^{4} \frac{(X_i - 76.4)^2}{383}$$

is $\chi^2(4)$,

$$W = \sum_{i=1}^{4} \frac{(X_i - \overline{X})^2}{383}$$

is $\chi^2(3)$, and, for examples,

$$P(0.711 \leq U \leq 7.779) = 0.90 - 0.05 = 0.85,$$
$$P(0.352 \leq W \leq 6.251) = 0.90 - 0.05 = 0.85.$$

In later sections we shall illustrate the importance of the chi-square distribution in applications.

We now prove another theorem which deals with linear functions of independent normally distributed random variables.

THEOREM 5.3-5   *If $X_1, X_2, \ldots, X_n$ are n mutually independent normal variables with means $\mu_1, \mu_2, \ldots, \mu_n$ and variances $\sigma_1^2, \sigma_2^2, \ldots, \sigma_n^2$, respectively, then the linear function*

$$Y = \sum_{i=1}^{n} c_i X_i$$

has the normal distribution

$$N\left(\sum_{i=1}^{n} c_i \mu_i, \sum_{i=1}^{n} c_i^2 \sigma_i^2 \right).$$

PROOF: By Theorem 5.2-4, we have

$$M_Y(t) = \prod_{i=1}^{n} M_{X_i}(c_i t) = \prod_{i=1}^{n} \exp(\mu_i c_i t + \sigma_i^2 c_i^2 t^2 / 2)$$

because $M_{X_i}(t) = \exp(\mu_i t + \sigma_i^2 t^2 / 2)$, $i = 1, 2, \ldots, n$. Thus

$$M_Y(t) = \exp\left[\left(\sum_{i=1}^{n} c_i \mu_i \right) t + \left(\sum_{i=1}^{n} c_i^2 \sigma_i^2 \right) \left(\frac{t^2}{2} \right) \right].$$

This is the moment-generating function of a distribution which is

$$N\left(\sum_{i=1}^{n} c_i \mu_i, \sum_{i=1}^{n} c_i^2 \sigma_i^2 \right)$$

and thus Y has this normal distribution.   □

From Theorem 5.3-5 we can make the observation that the difference of two independent normally distributed random variables, say $Y = X_1 - X_2$, has the normal distribution $N(\mu_1 - \mu_2, \sigma_1^2 + \sigma_2^2)$.

> **Example 5.3-3**   Let X_1 and X_2 equal the number of pounds of butterfat produced by holstein cows on the Koopman and the Vliestra farms, respectively, during the 305-day lactation period following the births of calves. Assume that the distribution of X_1 is $N(693.2, 22820)$ and the distribution of X_2 is $N(631.7, 19205)$. If a cow is selected randomly from each herd and the butterfat productions are compared, we shall find $P(X_1 > X_2)$. That is, we shall find the probability that the butterfat produced by the Koopman farm cow exceeds that produced by the Vliestra farm cow. (Sketch p.d.f.'s on the same graph for these two normal distributions.) If we let $Y = X_1 - X_2$, then the distribution of Y is $N(693.2 - 631.7, 22820 + 19205)$. Thus,
>
> $$P(X_1 > X_2) = P(Y > 0) = P\left(\frac{Y - 61.5}{\sqrt{42025}} > \frac{0 - 61.5}{205} \right)$$
>
> $$= P(Z > -0.30) = 0.6179.$$

Exercises

5.3-1   Let $X_1, X_2, \ldots, X_{16}$ be a random sample from a normal distribution $N(77, 25)$. Compute
(a) $P(77 < \overline{X} < 79.5)$.
(b) $P(74.2 < \overline{X} < 78.4)$.

5.3-2   Let X be $N(50, 36)$. Using the same set of axes, sketch the graphs of the probability density functions of
(a) X.
(b) $\overline{X}$, the mean of a random sample of size 9 from this distribution.
(c) $\overline{X}$, the mean of a random sample of size 36 from this distribution.

5.3-3   Let X equal the widest diameter (in millimeters) of the fetal head measured between the 16th and 25th weeks of pregnancy. Assume that the distribution of X is $N(46.58, 40.96)$. Let $\overline{X}$ be the sample mean of a random sample of $n = 16$ observations of X.
(a) Give the value of $E(\overline{X})$.
(b) Give the value of $\text{Var}(\overline{X})$.
(c) Find $P(44.42 \le \overline{X} \le 48.98)$.

5.3-4   Let X equal the weight of the soap in a ''6-pound'' box. Assume that the distribution of X is $N(6.05, 0.0004)$.
(a) Find $P(X < 6.0171)$.
(b) If nine boxes of soap are selected at random from the production line, find the probability that at most two boxes weigh less than 6.0171 each.
(c) Let $\overline{X}$ be the sample mean of the nine boxes. Find $P(\overline{X} \le 6.035)$.

5.3-5 Let $Z_1, Z_2, \ldots, Z_7$ be a random sample from the standard normal distribution $N(0, 1)$. Let $W = Z_1^2 + Z_2^2 + \cdots + Z_7^2$. Find $P(1.69 < W < 14.07)$.

5.3-6 If $X_1, X_2, \ldots, X_{16}$ is a random sample of size $n = 16$ from the normal distribution $N(50, 100)$, determine

(a) $P\left(796.2 \leq \sum_{i=1}^{16} (X_i - 50)^2 \leq 2630 \right)$.

(b) $P\left(726.1 \leq \sum_{i=1}^{16} (X_i - \overline{X})^2 \leq 2500 \right)$.

5.3-7 Let X equal the weight (in grams) of a nail of the type that is used for making decks. Assume that the distribution of X is $N(8.78, 0.16)$. Let $\overline{X}$ be the mean of a random sample of the weights of $n = 9$ nails.
(a) Sketch, on the same set of axes, the graphs of the p.d.f.'s of X and of $\overline{X}$.
(b) Let S^2 be the sample variance of the 9 weights. Find constants a and b so that $P(a \leq S^2 \leq b) = 0.90$.

HINT: That probability is equivalent to $P(8a/0.16 \leq 8S^2/0.16 \leq 8b/0.16)$ and $8S^2/0.16$ is $\chi^2(8)$. Find $8a/0.16$ and $8b/0.16$ from the tables.

5.3-8 At a heat-treating company, iron castings and steel forgings are heat-treated to achieve desired mechanical properties and machinability. One steel forging is annealed to soften the part for each machining. Two lots of this part, made of 1020 steel, are heat-treated in two different furnaces. The specification for this part is 36–66 on the Rockwell G scale. Let X_1 and X_2 equal the respective hardness measurements for parts selected randomly from furnaces 1 and 2. Assume that the distributions of X_1 and X_2 are $N(47.88, 2.19)$ and $N(43.04, 14.89)$, respectively.
(a) Sketch the p.d.f.'s of X_1 and X_2 on the same graph.
(b) Compute $P(X_1 > X_2)$.

5.3-9 Let X equal the force required to pull a stud out of a window that is to be inserted into an automobile. Assume that the distribution of X is $N(147.8, 12.3^2)$.
(a) Find $P(X < 163.3)$.
(b) If $\overline{X}$ is the mean and S^2 is the variance of a random sample of size $n = 25$ from this distribution of X, determine $P(\overline{X} \leq 150.9)$.
(c) Find constants a and b so that $P(a \leq S^2 \leq b) = 0.90$. (See the Hint in Exercise 5.3-7.)

5.3-10 Suppose that the distribution of the weight of a prepackaged "1-pound bag" of carrots is $N(1.18, 0.07^2)$ and the distribution of the weight of a prepackaged "3-pound bag" of carrots is $N(3.22, 0.09^2)$. Selecting bags at random, find the probability that the sum of three 1-pound bags exceeds the weight of one 3-pound bag.

HINT: First determine the distribution of Y, the sum of the three, and then compute $P(Y > W)$, where W is the weight of the 3-pound bag.

5.3-11 Let X denote the wing length in millimeters of a male gallinule and Y the wing length in millimeters of a female gallinule. Assume that X is $N(184.09, 39.37)$ and Y is $N(171.93, 50.88)$, and that X and Y are independent. If a male and a female gallinule are captured, what is the probability that X is greater than Y?

5.3-12 Suppose that for a particular population of students SAT mathematics scores are $N(529, 5732)$ and SAT verbal scores are $N(474, 6368)$. Select two students at random, and let X equal the first student's math score and Y the second student's verbal score. Find $P(X > Y)$.

5.3-13 Suppose that the length of life in hours, say X, of a light bulb manufactured by company A is $N(800, 14{,}400)$ and the length of life in hours, say Y, of a light bulb manufactured by company B is $N(850, 2500)$. One bulb is selected from each company and is burned until "death."
   (a) Find the probability that the length of life of the bulb from company A exceeds the length of life of the bulb from company B by at least 15 hours.
   (b) Find the probability that at least one of the bulbs "lives" for at least 920 hours.

5.4   The Central Limit Theorem

In Section 5.2 we found that the mean $\overline{X}$ of a random sample of size n from a distribution with mean μ and variance $\sigma^2 > 0$ is a random variable with the properties that

$$E(\overline{X}) = \mu \quad \text{and} \quad \text{Var}(\overline{X}) = \frac{\sigma^2}{n}.$$

Thus, as n increases, the variance of $\overline{X}$ decreases. Consequently, the distribution of $\overline{X}$ clearly depends on n, and we see that we are dealing with sequences of distributions.

In Theorem 5.3-1 we considered the p.d.f. of $\overline{X}$ when sampling from the normal distribution $N(\mu, \sigma^2)$. We showed that the distribution of $\overline{X}$ is $N(\mu, \sigma^2/n)$, and in Figure 5.3-1, by graphing the p.d.f.'s for several values of n, we illustrated that as n increases, the probability becomes concentrated in a small interval centered at μ. That is, as n increases, $\overline{X}$ tends to converge to μ, or $(\overline{X} - \mu)$ tends to converge to 0 in some probability sense.

In general, if we let

$$W = \frac{\sqrt{n}}{\sigma}(\overline{X} - \mu) = \frac{\overline{X} - \mu}{\sigma/\sqrt{n}},$$

where $\overline{X}$ is the mean of a random sample of size n from some distribution with mean μ and variance σ^2, then for each positive integer n,

$$E(W) = E\left[\frac{\overline{X} - \mu}{\sigma/\sqrt{n}}\right] = \frac{E(\overline{X}) - \mu}{\sigma/\sqrt{n}} = \frac{\mu - \mu}{\sigma/\sqrt{n}} = 0$$

and

$$\text{Var}(W) = E(W^2) = E\left[\frac{(\overline{X} - \mu)^2}{\sigma^2/n}\right] = \frac{E[(\overline{X} - \mu)^2]}{\sigma^2/n} = \frac{\sigma^2/n}{\sigma^2/n} = 1.$$

Thus, while $\overline{X} - \mu$ "degenerates" to zero, the factor $\sqrt{n}/\sigma$ in $\sqrt{n}(\overline{X} - \mu)/\sigma$ "spreads out" the probability enough to prevent this degeneration. What then is the distribution of W as n increases? One observation that might shed some light on the answer to this question can be made immediately. If the sample arises from a normal distribution then, from Theorem 5.3-1, we know that $\overline{X}$ is $N(\mu, \sigma^2/n)$ and hence W is $N(0, 1)$ for each positive n. Thus, in the limit, the distribution of W must be $N(0, 1)$. So if the solution of the question does not depend on the underlying distribution (i.e., it is unique), the answer must be $N(0, 1)$. As we will see, this is exactly the case, and this result is so important it is called the Central Limit Theorem, the proof of which is given in Section 5.6.

THEOREM 5.4-1   (**Central Limit Theorem**) *If $\overline{X}$ is the mean of a random sample $X_1, X_2, \ldots, X_n$ of size n from a distribution with a finite mean μ and a finite positive variance σ^2, then the distribution of*

$$W = \frac{\overline{X} - \mu}{\sigma/\sqrt{n}} = \frac{\displaystyle\sum_{i=1}^{n} X_i - n\mu}{\sqrt{n}\sigma}$$

is $N(0, 1)$ in the limit as $n \to \infty$.

A practical use of the Central Limit Theorem is approximating, when n is "sufficiently large," the distribution function of W, namely

$$P(W \le w) \approx \int_{-\infty}^{w} \frac{1}{\sqrt{2\pi}} e^{-z^2/2} \, dz = \Phi(w).$$

We present some illustrations of this application, discuss "sufficiently large," and try to give an intuitive feeling for the Central Limit Theorem.

    Example 5.4-1   Let $\overline{X}$ denote the mean of a random sample of size $n = 15$ from the distribution whose p.d.f. is $f(x) = (3/2)x^2$, $-1 < x < 1$. Here $\mu = 0$ and $\sigma^2 = 3/5$. Thus

$$P(0.03 \le \overline{X} \le 0.15) = P\left(\frac{0.03 - 0}{\sqrt{3/5}/\sqrt{15}} \le \frac{\overline{X} - 0}{\sqrt{3/5}/\sqrt{15}} \le \frac{0.15 - 0}{\sqrt{3/5}/\sqrt{15}}\right)$$

$$= P(0.15 \le W \le 0.75)$$

$$\approx \Phi(0.75) - \Phi(0.15)$$

$$= 0.7734 - 0.5596 = 0.2138.$$

    Example 5.4-2   Let $X_1, X_2, \ldots, X_{20}$ denote a random sample of size 20 from the uniform distribution $U(0, 1)$. Here $E(X_i) = 1/2$ and $\text{Var}(X_i) = 1/12$, $i = 1, 2, \ldots, 20$. If $Y = X_1 + X_2 + \cdots + X_{20}$, then

$$P(Y \le 9.1) = P\left(\frac{Y - 20(1/2)}{\sqrt{20/12}} \le \frac{9.1 - 10}{\sqrt{20/12}}\right) = P(W \le -0.697)$$
$$\approx \Phi(-0.697)$$
$$= 0.2423.$$

Also,

$$P(8.5 \le Y \le 11.7) = P\left(\frac{8.5 - 10}{\sqrt{5/3}} \le \frac{Y - 10}{\sqrt{5/3}} \le \frac{11.7 - 10}{\sqrt{5/3}}\right)$$
$$= P(-1.162 \le W \le 1.317)$$
$$\approx \Phi(1.317) - \Phi(-1.162)$$
$$= 0.9061 - 0.1226 = 0.7835.$$

Example 5.4-3 Let $\overline{X}$ denote the mean of a random sample of size 25 from the distribution whose p.d.f. is $f(x) = x^3/4$, $0 < x < 2$. It is easy to show that $\mu = 8/5 = 1.6$ and $\sigma^2 = 8/75$. Thus

$$P(1.5 \le \overline{X} \le 1.65) = P\left(\frac{1.5 - 1.6}{\sqrt{8/75}/\sqrt{25}} \le \frac{\overline{X} - 1.6}{\sqrt{8/75}/\sqrt{25}} \le \frac{1.65 - 1.6}{\sqrt{8/75}/\sqrt{25}}\right)$$
$$= P(-1.531 \le W \le 0.765)$$
$$\approx \Phi(0.765) - \Phi(-1.531)$$
$$= 0.7779 - 0.0629 = 0.7150.$$

These examples have shown how the Central Limit Theorem can be used for approximating certain probabilities concerning the mean $\overline{X}$ or the sum $Y = \sum_{i=1}^{n} X_i$ of a random sample. That is, $\overline{X}$ is approximately $N(\mu, \sigma^2/n)$, and Y is approximately $N(n\mu, n\sigma^2)$ when n is "sufficiently large," where μ and σ^2 are the mean and the variance of the underlying distribution from which the sample arose. Generally, if n is greater than 25 or 30, these approximations will be good. However, if the underlying distribution is symmetric, unimodal, and of the continuous type, a value of n as small as 4 or 5 can yield a very adequate approximation. Moreover, if the original distribution is approximately normal, $\overline{X}$ would have a distribution very close to normal when n equals 2 or 3. In fact, we know that if the sample is taken from $N(\mu, \sigma^2)$, $\overline{X}$ is exactly $N(\mu, \sigma^2/n)$ for every $n = 1, 2, 3, \ldots$.

The following examples will help to illustrate the previous remarks and will give the reader a better intuitive feeling about the Central Limit Theorem. In particular, we shall see how the size of n affects the distribution of $\overline{X}$ and $Y = \sum X_i$ for samples from several underlying distributions.

Example 5.4-4 Let X_1, X_2, X_3, X_4 be a random sample of size 4 from the uniform distribution $U(0, 1)$. Then $\mu = 1/2$ and $\sigma^2 = 1/12$. We shall compare the graph of the p.d.f. of

$$Y = \sum_{i=1}^{n} X_i$$

with the graph of the $N[n(1/2), n(1/12)]$ p.d.f. for $n = 2$ and 4, respectively.

To find the p.d.f. of $Y = X_1 + X_2$, we first find the distribution function of Y. The joint p.d.f. of X_1 and X_2 is

$$f(x_1, x_2) = 1, \quad 0 \le x_1 \le 1, 0 \le x_2 \le 1.$$

The distribution function of Y is

$$G(y) = P(Y \le y) = P(X_1 + X_2 \le y).$$

If $0 < y < 1$,

$$G(y) = \int_0^y \int_0^{y-x_1} 1 \, dx_2 \, dx_1 = \frac{y^2}{2},$$

and if $1 < y < 2$,

$$G(y) = 1 - \int_{y-1}^1 \int_{y-x_1}^1 1 \, dx_2 \, dx_1 = 1 - \frac{(2-y)^2}{2}.$$

Thus the p.d.f. of Y is

$$g(y) = G'(y) = \begin{cases} y, & 0 < y < 1, \\ 2 - y, & 1 < y < 2. \end{cases}$$

Adding the point $g(1) = 1$, we obtain the triangular p.d.f. that is graphed in Figure 5.4-1(a). In this figure the $N[2(1/2), 2(1/12)]$ p.d.f. is also graphed.

The p.d.f. of $Y = X_1 + X_2 + X_3 + X_4$ could be found in a similar manner. While this is more difficult than when $n = 2$, it is

$$g(y) = \begin{cases} \dfrac{y^3}{6}, & 0 \le y < 1, \\[2ex] \dfrac{-3y^3 + 12y^2 - 12y + 4}{6}, & 1 \le y < 2, \\[2ex] \dfrac{3y^3 - 24y^2 + 60y - 44}{6}, & 2 \le y < 3, \\[2ex] \dfrac{-y^3 + 12y^2 - 48y + 64}{6}, & 3 \le y \le 4. \end{cases}$$

This p.d.f. is graphed in Figure 5.4-1(b) along with the $N[4(1/2), 4(1/12)]$ p.d.f. If we are interested in finding $P(1.7 \le Y \le 3.2)$, this could be done by evaluating

$$\int_{1.7}^{3.2} g(y) \, dy,$$

which is tedious (see Exercise 5.4-10). It is much easier to use a normal approximation, which results in a number very close to the exact value.

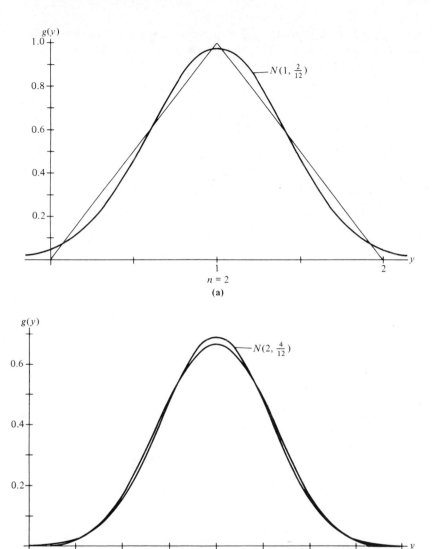

FIGURE 5.4-1

In Example 5.4-4 and Exercise 5.4-10 we show that even for a small value of n, like $n = 4$, the sum of the sample items has an approximate normal distribution. The following example illustrates that for some underlying distributions (particularly skewed ones) n must be quite large to obtain a satisfactory approximation. In order to keep the scale on the horizontal axis the same for each value of n, we will use the following result.

Let $f(x)$ and $F(x)$ be the p.d.f. and distribution function of a random variable, X, of the continuous type having mean μ and variance σ^2. Let $W = (X - \mu)/\sigma$. The distribution function of W is given by

$$G(w) = P(W \le w) = P\left(\frac{X - \mu}{\sigma} \le w\right)$$

$$= P(X \le \sigma w + \mu) = F(\sigma w + \mu).$$

Thus the p.d.f. of W is given by

$$g(w) = F'(\sigma w + \mu) = \sigma f(\sigma w + \mu).$$

Example 5.4-5   Let $X_1, X_2, \ldots, X_{100}$ be a random sample of size 100 from a chi-square distribution with one degree of freedom. If

$$Y = \sum_{i=1}^{n} X_i,$$

then Y is $\chi^2(n)$, and $E(Y) = n$, $\mathrm{Var}(Y) = 2n$. Let

$$W = \frac{Y - n}{\sqrt{2n}}.$$

The p.d.f. of W is given by

$$g(w) = \sqrt{2n}\frac{(\sqrt{2n}\,w + n)^{n/2-1}}{\Gamma\left(\dfrac{n}{2}\right)2^{n/2}}e^{-(\sqrt{2n}w+n)/2}, \qquad -n/\sqrt{2n} < w < \infty.$$

Note that $w > -n/\sqrt{2n}$ corresponds to $y > 0$. In Figure 5.4-2(a) and (b), the graph of W is given along with the $N(0, 1)$ p.d.f. for $n = 20$ and 100, respectively.

In order to have an intuitive feeling about how the sample size n affects the distribution of $W = (\overline{X} - \mu)/(\sigma/\sqrt{n})$, it is helpful to simulate values of W on a computer using different values of n and different underlying distributions. The next example illustrates this.

REMARK   We simulate observations from a distribution of X having a continuous-type distribution function $F(x)$ as follows. Suppose $F(a) = 0$, $F(b) = 1$, and $F(x)$ is strictly increasing for $a < x < b$. Let $Y = F(x)$ and let the distribution of Y be $U(0, 1)$. If y is an observed value of Y, then $x = F^{-1}(y)$ is an observed value of X (see Exercise 5.4-11). Thus, if y is the value of a computer-generated random number, then $x = F^1(y)$ is the simulated value of X.

Example 5.4-6   It is often difficult to find the exact distribution of $W = (\overline{X} - \mu)/(\sigma/\sqrt{n})$. However, let us consider some empirical evidence about

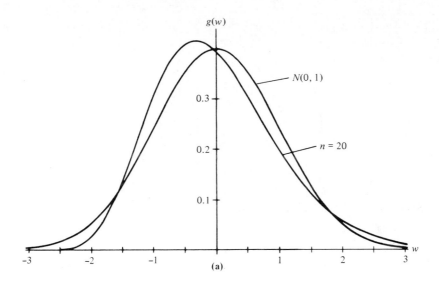

(a)

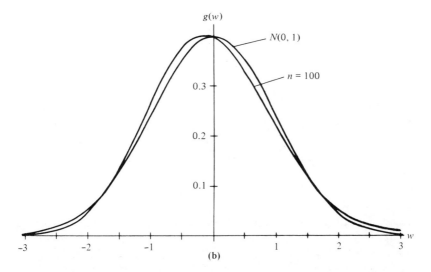

(b)

FIGURE 5.4-2

the distribution of W by simulating random samples on the computer. Let X_1, $X_2, \ldots, X_n$ denote a random sample of size n from the distribution with p.d.f. $f(x)$, distribution function $F(x)$, mean μ, and variance σ^2. We shall generate 1000 random samples of size n from this distribution and compute a value of W for each sample, thus obtaining 1000 observed values of W. A histogram of these 1000 values is constructed using 21 intervals of equal length. We depict the results of this experiment as follows:

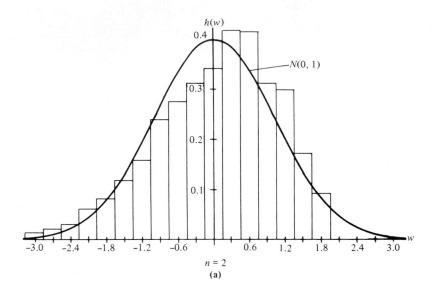

$n = 2$

(a)

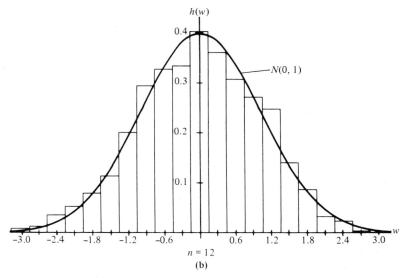

$n = 12$

(b)

FIGURE 5.4-3

(a) In Figure 5.4-3, $f(x) = (x + 1)/2$ and $F(x) = (x + 1)^2/4$ for $-1 < x < 1$; $\mu = 1/3$, $\sigma^2 = 2/9$; and $n = 2$ and 12 (see Exercise 9.5-18).

(b) In Figure 5.4-4, $f(x) = (3/2)x^2$ and $F(x) = (x^3 + 1)/2$ for $-1 < x < 1$; $\mu = 0$, $\sigma^2 = 3/5$; and $n = 2$ and 12 (see Exercise 9.5-19). (Sketch the graph of $y = f(x)$ and give an intuitive argument as to why the histogram for $n = 2$ looks the way it does.)

The $N(0, 1)$ p.d.f. has been superimposed on each histogram.

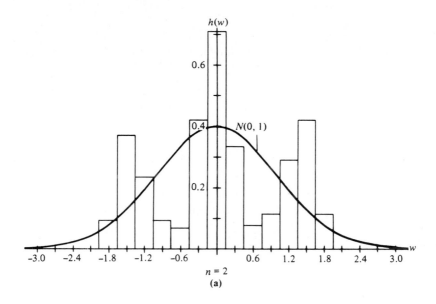

$n = 2$

(a)

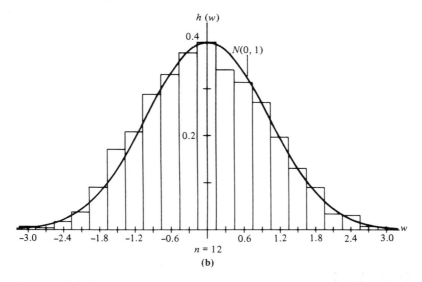

$n = 12$

(b)

FIGURE 5.4-4

Note very clearly that these examples have not proved anything. They are presented to give evidence of the truth of the Central Limit Theorem. So far all the illustrations have concerned distributions of the continuous type. However, the hypotheses for the Central Limit Theorem do not require the distribution to be continuous. We shall consider applications of the Central Limit Theorem for discrete-type distributions in the next section.

Exercises

5.4-1 Let $\overline{X}$ be the mean of a random sample of size 12 from the uniform distribution on the interval (0, 1). Approximate $P(1/2 \leq \overline{X} \leq 2/3)$.

5.4-2 Let $Y = X_1 + X_2 + \cdots + X_{15}$ be the sum of a random sample of size 15 from the distribution whose p.d.f. is $f(x) = (3/2)x^2$, $-1 < x < 1$. Approximate

$$P(-0.3 \leq Y \leq 1.5).$$

5.4-3 Let $\overline{X}$ be the mean of a random sample of size 36 from an exponential distribution with mean 3. Approximate $P(2.5 \leq \overline{X} \leq 4)$.

5.4-4 Approximate $P(39.75 \leq \overline{X} \leq 41.25)$, where $\overline{X}$ is the mean of a random sample of size 32 from a distribution with mean $\mu = 40$ and variance $\sigma^2 = 8$.

5.4-5 Let $X_1, X_2, \ldots, X_{18}$ be a random sample of size 18 from a chi-square distribution with $r = 1$. Recall that $\mu = 1$, $\sigma^2 = 2$.

(a) How is $Y = \sum_{i=1}^{18} X_i$ distributed?

(b) Using the result of part (a), we see from Table IV in the Appendix that

$$P(Y \leq 9.390) = 0.05 \qquad \text{and} \qquad P(Y \leq 34.80) = 0.99.$$

Compare these two probabilities with the approximations found using the Central Limit Theorem.

5.4-6 A random sample of size $n = 18$ is taken from the distribution with p.d.f. $f(x) = 1 - x/2$, $0 \leq x \leq 2$.
(a) Find μ and σ^2.
(b) Find, approximately, $P(2/3 \leq \overline{X} \leq 5/6)$.

5.4-7 Let X equal the maximal oxygen intake of a human on a treadmill, where the measurements are in milliliters of oxygen per minute per kilogram of weight. Assume that for a particular population the mean of X is $\mu = 54.030$ and the standard deviation is $\sigma = 5.8$. Let $\overline{X}$ be the sample mean of a random sample of size $n = 47$. Find $P(52.761 \leq \overline{X} \leq 54.453)$, approximately.

5.4-8 Let X equal the weight in grams of a miniature candy bar. Assume that $\mu = E(X) = 24.43$ and $\sigma^2 = \text{Var}(X) = 2.20$. Let $\overline{X}$ be the sample mean of a random sample of $n = 30$ candy bars. Find
(a) $E(\overline{X})$.
(b) $\text{Var}(\overline{X})$.
(c) $P(24.17 \leq \overline{X} \leq 24.82)$, approximately.

5.4-9 Let X equal the birth weight in grams of a baby born in the Sudan. Assume that $E(X) = 3320$ and $\text{Var}(X) = 660^2$. Let $\overline{X}$ be the sample mean of a random sample of size $n = 225$. Find $P(3233.76 \leq \overline{X} \leq 3406.24)$, approximately.

5.4-10 In Example 5.4-4, compute $P(1.7 \leq Y \leq 3.2)$ and compare this answer with the normal approximation of this probability.

5.4-11 Under the conditions given in the Remark of this section, show that $X = F^{-1}(Y)$ has a distribution function $F(x)$.

HINT: $P(X \le x) = P[F^{-1}(Y) \le x] = P[Y \le F(x)]$ and recall that Y is $U(0, 1)$ with distribution function $P(Y \le y) = y$, $0 \le y < 1$.

5.5 Approximations for Discrete Distributions

In this section we illustrate how the normal distribution can be used to approximate probabilities for certain discrete-type distributions. One of the most important discrete distributions is the binomial distribution. To see how the Central Limit Theorem can be applied, recall that a binomial random variable can be described as the sum of Bernoulli random variables. That is, let $X_1, X_2, \ldots , X_n$ be a random sample from a Bernoulli distribution with a mean $\mu = p$ and a variance $\sigma^2 = p(1 - p)$, where $0 < p < 1$. Then $Y = \sum_{i=1}^{n} X_i$ is $b(n, p)$. The Central Limit Theorem states that the distribution of

$$W = \frac{Y - np}{\sqrt{np(1 - p)}} = \frac{\overline{X} - p}{\sqrt{p(1 - p)/n}}$$

is $N(0, 1)$ in the limit as $n \to \infty$. Thus, if n is sufficiently large, the distribution of Y is approximately $N[np, np(1 - p)]$, and probabilities for the binomial distribution $b(n, p)$ can be approximated using this normal distribution. A rule often stated is that n is "sufficiently large" if $np \ge 5$ and $n(1 - p) \ge 5$, and it can be used as a guide.

Note that we shall be approximating probabilities for a discrete distribution with probabilities for a continuous distribution. Let us discuss a reasonable procedure in this situation. If V is $N(\mu, \sigma^2)$, $P(a < V < b)$ is equivalent to the area bounded by the p.d.f. of V, the v axis, $v = a$, and $v = b$. If Y is $b(n, p)$, recall that the probability histogram for Y was defined as follows. For each y such that $k - 1/2 < y < k + 1/2$, let

$$f(y) = \frac{n!}{k!(n - k)!}p^k(1 - p)^{n-k}, \qquad k = 0, 1, 2, \ldots , n.$$

Then $P(Y = k)$ can be represented by the area of the rectangle with a height of $P(Y = k)$ and a base of length 1 centered at k. Figure 5.5-1 shows the graph of the

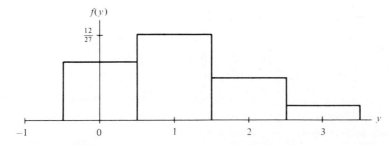

FIGURE 5.5-1

probability histogram for the binomial distribution $b(3, 1/3)$. When using the normal distribution to approximate probabilities for the binomial distribution, areas under the p.d.f. for the normal distribution will be used to approximate areas of rectangles in the probability histogram for the binomial distribution.

Example 5.5-1 Let Y be $b(10, 1/2)$. Then, by the Central Limit Theorem, $P(a < Y < b)$ can be approximated using the normal distribution with mean $10(1/2) = 5$ and variance $10(1/2)(1/2) = 5/2$. Figure 5.5-2 shows the graph of the probability histogram for $b(10, 1/2)$ and the graph of the p.d.f. of $N(5, 5/2)$. Note that the area of the rectangle whose base is

$$\left(k - \frac{1}{2}, k + \frac{1}{2} \right)$$

and the area under the normal curve between $k - 1/2$ and $k + 1/2$ are approximately equal for each integer k.

Example 5.5-2 Let Y be $b(18, 1/6)$. Because $np = 18(1/6) = 3 < 5$, the normal approximation is not as good here. Figure 5.5-3 illustrates this by depicting the skewed probability histogram for $b(18, 1/6)$ and the symmetric p.d.f. of $N(3, 5/2)$.

Example 5.5-3 Let Y have the binomial distribution of Example 5.5-1 and Figure 5.5-2, namely $b(10, 1/2)$. Then

$$P(3 \le Y < 6) = P(2.5 \le Y \le 5.5)$$

because $P(Y = 6)$ is not in the desired answer. But the latter equals

$$P\left(\frac{2.5 - 5}{\sqrt{10/4}} \le \frac{Y - 5}{\sqrt{10/4}} \le \frac{5.5 - 5}{\sqrt{10/4}} \right) \approx \Phi(0.316) - \Phi(-1.581)$$

$$= 0.6240 - 0.0570 = 0.5670$$

Using Appendix Table II, we find that $P(3 \le Y < 6) = 0.5683$.

Example 5.5-4 Let Y be $b(36, 1/2)$. Then

$$P(12 < Y \le 18) = P(12.5 \le Y \le 18.5)$$

$$= P\left(\frac{12.5 - 18}{\sqrt{9}} \le \frac{Y - 18}{\sqrt{9}} \le \frac{18.5 - 18}{\sqrt{9}} \right)$$

$$\approx \Phi(0.167) - \Phi(-1.833)$$

$$= 0.5329.$$

Note that 12 was increased to 12.5 because $P(Y = 12)$ is not included in the desired probability. Using the binomial formula, we find that (you may verify this answer using your calculator)

$$P(12 < Y \le 18) = P(13 \le Y \le 18) = 0.5334.$$

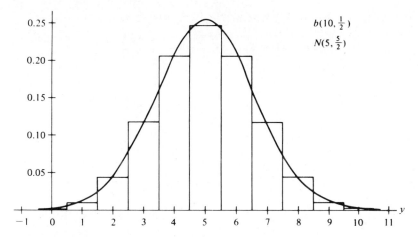

FIGURE 5.5-2

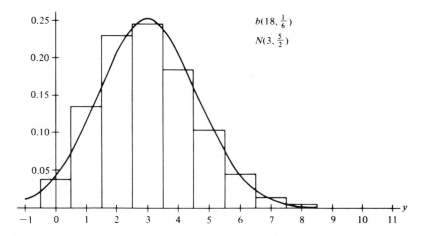

FIGURE 5.5-3

Also,

$$P(Y = 20) = P(19.5 \leq Y \leq 20.5)$$

$$= P\left(\frac{19.5 - 18}{\sqrt{9}} \leq \frac{Y - 18}{\sqrt{9}} \leq \frac{20.5 - 18}{\sqrt{9}}\right)$$

$$\approx \Phi(0.833) - \Phi(0.5)$$

$$= 0.1060.$$

Using the binomial formula, we have $P(Y = 20) = 0.1063$. So, in this situation, the approximation is extremely good.

Note that, in general, if Y is $b(n, p)$,

$$P(Y \le k) \approx \Phi\left(\frac{k + 1/2 - np}{\sqrt{npq}}\right)$$

and

$$P(Y < k) \approx \Phi\left(\frac{k - 1/2 - np}{\sqrt{npq}}\right),$$

because in the first case k is included and in the second it is not.

We now show how the Poisson distribution with large enough mean can be approximated using a normal distribution.

Example 5.5-5 A random variable having a Poisson distribution with mean 20 can be thought of as the sum Y of the items of a random sample of size 20 from a Poisson distribution with mean 1. Thus

$$W = \frac{Y - 20}{\sqrt{20}}$$

has a distribution that is approximately $N(0, 1)$, and the distribution of Y is approximately $N(20, 20)$ (see Figure 5.5-4). So, for illustration,

$$P(16 < Y \le 21) = P(16.5 \le Y \le 21.5)$$
$$= P\left(\frac{16.5 - 20}{\sqrt{20}} \le \frac{Y - 20}{\sqrt{20}} \le \frac{21.5 - 20}{\sqrt{20}}\right)$$
$$\approx \Phi(0.335) - \Phi(-0.783)$$
$$= 0.4142.$$

Note that 16 is increased to 16.5 because $Y = 16$ is not included in the event $16 < Y \le 21$. The answer using the Poisson formula is 0.4226.

In general, if Y has a Poisson distribution with mean λ, then the distribution of

$$W = \frac{Y - \lambda}{\sqrt{\lambda}}$$

is approximately $N(0, 1)$ when λ is sufficiently large.

Exercises

5.5-1 Let the distribution of Y be $b(25, 1/2)$. Find the following probabilities in two ways: exactly using Appendix Table II and approximately using the Central Limit Theorem. Compare the two results in each of the three cases.
(a) $P(10 < Y \le 12)$.
(b) $P(12 \le Y < 15)$.
(c) $P(Y = 12)$.

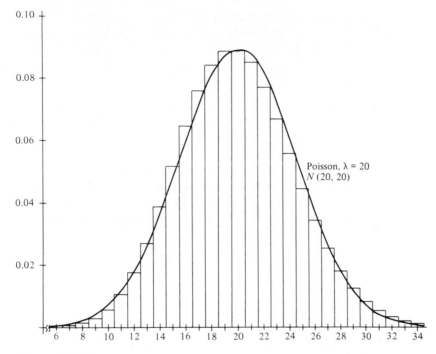

FIGURE 5.5-4

5.5-2 Among the gifted 7th-graders who score very high on a mathematics exam, approximately 20% are left-handed or ambidextrous. Let X equal the number of left-handed or ambidextrous students among a random sample of $n = 25$ gifted 7th-graders. Find $P(2 < X < 9)$
   (a) Using Appendix Table II.
   (b) Approximately, using the Central Limit Theorem.

REMARK Since X has a skewed distribution, the approximation is not as good as for the symmetrical distribution where $p = 0.50$. Note that $np \geq 5$ is barely satisfied.

5.5-3 A public opinion poll in Southern California was conducted to determine whether Southern Californians are prepared for the big earthquake that experts predict will devastate the region sometime in the next 50 years. It was learned that "60% have not secured objects in their homes that might fall and cause injury and damage during a temblor." In a random sample of $n = 864$ Southern Californians, let X equal the number who "have not secured objects in their homes." Find $P(496 \leq X \leq 548)$, approximately.

5.5-4 Let X equal the number out of $n = 48$ mature aster seeds that will germinate when $p = 0.75$ is the probability that a particular seed germinates. Determine $P(35 \leq X \leq 40)$, approximately.

5.5-5   Let p equal the proportion of all college students who would say *yes* to the question, "Would you drink from the same glass as your friend if you suspected that this friend were an AIDS virus carrier?" Assume that $p = 0.10$. Let X equal the number of students out of a random sample of size $n = 100$ who would say *yes* to this question. Compute $P(X \leq 11)$, approximately.

REMARK   The value used for p was based on a poll conducted in a class at San Diego State University. It would be interesting for you to conduct a survey at your college to estimate the value of p.

5.5-6   Many things once regarded as luxuries are now regarded by most Americans as necessities. For example, the automobile is regarded as a necessity by 90% of Americans (*Public Opinion*, 1984). Let X equal the number of Americans in a random sample of $n = 100$ who regard the automobile as a necessity. Assuming that it is still true that $p = 0.90$, determine $P(89 \leq X \leq 94)$, approximately.

5.5-7   Let $X_1, X_2, \ldots, X_{48}$ be a random sample of size 48 from the distribution with p.d.f. $f(x) = 1/x^2$, $1 < x < \infty$. Approximate the probability that at most 10 of these random variables have values greater than 4.

HINT: Let the ith trial be a success if $X_i > 4$, $i = 1, 2, \ldots, 48$.

5.5-8   A candy maker produces mints that have a label weight of 20.4 grams. Assume that the distribution of the weights of these mints is $N(21.37, 0.16)$.
   (a) Let X denote the weight of a single mint selected at random from the production line. Find $P(X < 20.857)$.
   (b) During a particular shift 100 mints are selected at random and weighed. Let Y equal the number of these mints that weigh less than 20.857 grams. Find approximately $P(Y \leq 5)$.
   (c) Let $\overline{X}$ equal the sample mean of the 100 mints selected and weighed on a particular shift. Find $P(21.31 \leq \overline{X} \leq 21.39)$.

5.5-9   Let X equal the number of alpha particles emitted by barium-133 per second and counted by a Geiger counter. Assume that X has a Poisson distribution with $\lambda = 49$. Approximate $P(45 < X < 60)$.

5.5-10   Let X equal the number of alpha particles counted by a Geiger counter during 30 seconds. Assume that the distribution of X is Poisson with a mean of 4829. Determine approximately $P(4776 \leq X \leq 4857)$.

5.5-11   Let $X_1, X_2, \ldots, X_{30}$ be a random sample of size 30 from a Poisson distribution with a mean of 2/3. Approximate

(a) $P\left(15 < \sum_{i=1}^{30} X_i \leq 22\right).$

(b) $P\left(21 \leq \sum_{i=1}^{30} X_i < 27\right).$

(c) $P\left(15 \leq \sum_{i=1}^{30} X_i \leq 22\right)$.

5.5-12 In the casino game roulette, the probability of winning with a bet on red is $p = 18/38$. Let Y equal the number of winning bets out of 1000 independent bets that are placed. Find $P(Y > 500)$ approximately.

5.5-13 Let X denote the payoff for \$1 bets in the game chuck-a-luck. Then X can equal -1, 1, 2, or 3, such that $E(X) = -0.0787$ and $\mathrm{Var}(X) = 1.2392$. Given 300 independent observations of X, say $X_1, X_2, \ldots, X_{300}$, let $Y = \sum_{i=1}^{300} X_i$. Then Y represents the amount won after 300 bets. Approximate
(a) $P(Y \leq -21)$.
(b) $P(Y \geq 21)$.

5.5-14 Let Y equal the number of \$499 prizes won by a gambler after placing n straight independent \$1 bets in the Michigan daily lottery, in which a prize of \$499 is won with probability 0.001. Therefore, Y is $b(n, 0.001)$. If $n = 4200$, a gambler is behind if $Y \leq 8$; use the Poisson distribution to approximate $P(Y \leq 8)$. What is the normal approximation to the same probability? Which approximation is better in your opinion?

5.5-15 If X is $b(100, 0.1)$, find the approximate value of $P(12 \leq X \leq 14)$ using
(a) The normal approximation.
(b) The Poisson approximation.
(c) The binomial p.d.f.

5.5-16 Let $X_1, X_2, \ldots, X_{36}$ be a random sample of size 36 from the geometric distribution with p.d.f. $f(x) = (1/4)^{x-1}(3/4)$, $x = 1, 2, 3, \ldots$. Approximate

(a) $P\left(46 \leq \sum_{i=1}^{36} X_i \leq 49\right)$.
(b) $P(1.25 \leq \overline{X} \leq 1.50)$.

HINT: Observe that the distribution of the sum is of the discrete type.

5.5-17 A die is rolled 24 independent times. Let Y be the sum of the 24 resulting values. Recalling that Y is a random variable of the discrete type, approximate
(a) $P(Y \geq 86)$.
(b) $P(Y < 86)$.
(c) $P(70 < Y \leq 86)$.

5.6   Limiting Moment-Generating Functions

We would like to begin this section by showing that the binomial distribution can be approximated by the Poisson distribution when n is sufficiently large and p fairly small by taking the limit of a moment-generating function. Consider the moment-generating function of Y, which is $b(n, p)$. We shall take the limit of this as

$n \to \infty$ such that $np = \lambda$ is a constant; thus $p \to 0$. The moment generating function of Y is

$$M(t) = (1 - p + pe^t)^n.$$

Because $p = \lambda/n$, we have that

$$M(t) = \left[1 - \frac{\lambda}{n} + \frac{\lambda}{n}e^t \right]^n$$

$$= \left[1 + \frac{\lambda(e^t - 1)}{n} \right]^n.$$

Since

$$\lim_{n \to \infty} \left(1 + \frac{b}{n} \right)^n = e^b,$$

we have

$$\lim_{n \to \infty} M(t) = e^{\lambda(e^t - 1)},$$

which exists for all real t. But this is the moment-generating function of a Poisson random variable with mean λ. Hence this Poisson distribution seems like a reasonable approximation to the binomial one when n is large and p is small. This approximation is usually found to be fairly successful if $n \geq 20$ and $p \leq 0.05$ and very successful if $n \geq 100$ and $np \leq 10$. Obviously, it could be used in other situations too; we only want to stress that the approximation becomes better with larger n and smaller p.

Example 5.6-1   Let Y be $b(50, 1/25)$. Then

$$P(Y \leq 1) = \left(\frac{24}{25} \right)^{50} + 50 \left(\frac{1}{25} \right) \left(\frac{24}{25} \right)^{49} = 0.400.$$

Since $\lambda = np = 2$, the Poisson approximation is

$$P(Y \leq 1) \approx 0.406,$$

from Table III in the Appendix.

The preceding result illustrates the theorem we now state: *If a sequence of moment-generating functions approaches a certain one, say $M(t)$, then the limit of the corresponding distributions must be the distribution corresponding to $M(t)$.* This statement certainly appeals to one's intution! In a more advanced course, the proof of this theorem is given and there the existence of the moment-generating function is not even needed, for we would use the characteristic function $\phi(t) = E(e^{itX})$ instead.

The preceding theorem is used to prove the Central Limit Theorem. To help in the understanding of this proof, let us first consider a different problem, that of the limiting distribution of the mean $\overline{X}$ of a random sample $X_1, X_2, \ldots, X_n$ from a distribution with mean μ. If the distribution has moment-generating function $M(t)$, the moment-generating function of $\overline{X}$ is $[M(t/n)]^n$. But, by Taylor's expansion, there exists a number t_1 between 0 and t/n such that

$$M\left(\frac{t}{n}\right) = M(0) + M'(t_1)\frac{t}{n}$$

$$= 1 + \frac{\mu t}{n} + \frac{[M'(t_1) - M'(0)]t}{n}$$

because $M(0) = 1$ and $M'(0) = \mu$. Since $M'(t)$ is continuous at $t = 0$ and since $t_1 \rightarrow 0$ as $n \rightarrow \infty$, we know that

$$\lim_{n \to \infty} [M'(t_1) - M'(0)] = 0.$$

Thus, using a result from advanced calculus, we obtain

$$\lim_{n \to \infty} \left[M\left(\frac{t}{n}\right)\right]^n = \lim_{n \to \infty} \left\{1 + \frac{\mu t}{n} + \frac{[M'(t_1) - M'(0)]t}{n}\right\}^n$$

$$= e^{\mu t},$$

for all real t. But this limit is the moment-generating function of a degenerate distribution with all of the probability on μ. Accordingly, $\overline{X}$ has this limiting distribution, indicating that $\overline{X}$ converges to μ in a certain sense. This is one form of the law of large numbers.

We have seen that, in some probability sense, $\overline{X}$ converges to μ in the limit, or, equivalently, $\overline{X} - \mu$ converges to zero. Let us multiply the difference $\overline{X} - \mu$ by some function of n so that the result will not converge to zero. In our search for such a function, it is natural to consider

$$W = \frac{\overline{X} - \mu}{\sigma/\sqrt{n}} = \frac{\sqrt{n}(\overline{X} - \mu)}{\sigma} = \frac{Y - n\mu}{\sqrt{n}\sigma},$$

where Y is the sum of the items of the random sample. The reason for this is that W is a standardized random variable and has mean 0 and variance 1 for each positive integer n.

PROOF (of the Central Limit Theorem):
We first consider

$$E[\exp(tW)] = E\left\{\exp\left[\left(\frac{t}{\sqrt{n}\sigma}\right)\left(\sum_{i=1}^{n} X_i - n\mu\right)\right]\right\}$$

$$= E\left\{\exp\left[\left(\frac{t}{\sqrt{n}}\right)\left(\frac{X_1 - \mu}{\sigma}\right)\right] \cdots \exp\left[\left(\frac{t}{\sqrt{n}}\right)\left(\frac{X_n - \mu}{\sigma}\right)\right]\right\}$$

$$= E\left\{\exp\left[\left(\frac{t}{\sqrt{n}}\right)\left(\frac{X_1 - \mu}{\sigma}\right)\right]\right\} \cdots E\left\{\exp\left(\frac{t}{\sqrt{n}}\right)\left(\frac{X_n - \mu}{\sigma}\right)\right]\right\},$$

which follows from the mutual independence of $X_1, X_2, \ldots, X_n$. Then

$$E[\exp(tW)] = \left[m\left(\frac{t}{\sqrt{n}} \right) \right]^n, \qquad -h < \frac{t}{\sqrt{n}} < h,$$

where

$$m(t) = E\left\{ \exp\left[t\left(\frac{X_i - \mu}{\sigma} \right) \right] \right\}, \qquad -h < t < h,$$

is the common moment-generating function of each

$$Y_i = \frac{X_i - \mu}{\sigma}, \qquad i = 1, 2, \ldots, n.$$

Since $E(Y_i) = 0$ and $E(Y_i^2) = 1$, it must be that

$$m(0) = 1, \qquad m'(0) = E\left(\frac{X_i - \mu}{\sigma} \right) = 0, \qquad m''(0) = E\left[\left(\frac{X_i - \mu}{\sigma} \right)^2 \right] = 1.$$

Hence, using Taylor's formula with a remainder, we can find a number t_1 between 0 and t such that

$$m(t) = m(0) + m'(0)t + \frac{m''(t_1)t^2}{2} = 1 + \frac{m''(t_1)t^2}{2}.$$

By adding and subtracting $t^2/2$, we have that

$$m(t) = 1 + \frac{t^2}{2} + \frac{[m''(t_1) - 1]t^2}{2}.$$

Using this expression of $m(t)$ in $E[\exp(tW)]$, we can represent the moment-generating function of W by

$$E[\exp(tW)] = \left\{ 1 + \frac{1}{2}\left(\frac{t}{\sqrt{n}} \right)^2 + \frac{1}{2}[m''(t_1) - 1]\left(\frac{t}{\sqrt{n}} \right)^2 \right\}^n$$

$$= \left\{ 1 + \frac{t^2}{2n} + \frac{[m''(t_1) - 1]t^2}{2n} \right\}^n, \qquad -\sqrt{n}h < t < \sqrt{n}h,$$

where now t_1 is between 0 and $t/\sqrt{n}$. Since $m''(t)$ is continuous at $t = 0$ and $t_1 \to 0$ as $n \to \infty$, we have that

$$\lim_{n \to \infty} [m''(t_1) - 1] = 1 - 1 = 0.$$

Thus, using a result from advanced calculus, we have that

$$\lim_{n \to \infty} E[\exp(tW)] = \lim_{n \to \infty} \left\{ 1 + \frac{t^2}{2n} + \frac{[m''(t_1) - 1]t^2}{2n} \right\}^n$$

$$= \lim_{n \to \infty} \left\{ 1 + \frac{t^2/2}{n} \right\}^n = e^{t^2/2},$$

for all real t. This means that the limiting distribution of

$$W = \frac{\bar{X} - \mu}{\sigma/\sqrt{n}} = \frac{\sum\limits_{i=1}^{n} X_i - n\mu}{\sqrt{n}\sigma}$$

is $N(0, 1)$. This completes the proof of the Central Limit Theorem. □

Examples of the use of the Central Limit Theorem as an approximating distribution have been given in Sections 5.4 and 5.5.

Exercises

5.6-1   Let Y be the number of defectives in a box of 50 articles taken from the output of a machine. Each article is defective with probability 0.01. What is the probability that $Y = 0, 1, 2,$ or 3
(a) Using the binomial distribution?
(b) Using the Poisson approximation?

5.6-2   The probability that a certain type of inoculation takes effect is 0.995. Use the Poisson distribution to approximate the probability that at most 2 out of 400 people given the inoculation find that it has not taken effect.

HINT: Let $p = 1 - 0.995 = 0.005$.

5.6-3   Let S^2 be the sample variance of a random sample of size n from $N(\mu, \sigma^2)$. Show that the limit, as $n \to \infty$, of the moment-generating function of S^2 is $e^{\sigma^2 t}$. Thus, in the limit, the distribution of S^2 is degenerate with probability 1 at σ^2.

5.6-4   Let Y be $\chi^2(n)$. Use the Central Limit Theorem to demonstrate that $W = (Y - n)/\sqrt{2n}$ has a limiting distribution that is $N(0, 1)$.

HINT: Think of Y as being the sum of a random sample from a certain distribution.

5.6-5   Let Y have a Poisson distribution with mean $3n$. Use the Central Limit Theorem to show that $W = (Y - 3n)/\sqrt{3n}$ has a limiting distribution that is $N(0, 1)$.

5.7   The *t* and *F* Distributions

Two distributions that play an important role in statistical applications will be introduced in this section.

THEOREM 5.7-1   *If Z is a random variable that is $N(0, 1)$, if U is a random variable that is $\chi^2(r)$, and if Z and U are independent, then*

$$T = \frac{Z}{\sqrt{U/r}}$$

has a t distribution with r degrees of freedom. Its p.d.f. is

$$g(t) = \frac{\Gamma[(r + 1)/2]}{\sqrt{\pi r}\,\Gamma(r/2)(1 + t^2/r)^{(r+1)/2}}, \qquad -\infty < t < \infty.$$

REMARK This distribution was first discovered by W. S. Gosset when he was working for an Irish brewery. Because that brewery did not want other breweries to know that statistical methods were being used, Gosset published under the pseudonym Student. Thus this distribution is often known as Student's *t* distribution.

PROOF: In the proof, we first find an expression for the distribution function of *T* and then take its derivative to find the p.d.f. of *T*. Since *Z* and *U* are independent, the joint p.d.f. of *Z* and *U* is

$$g(z, u) = \frac{1}{\sqrt{2\pi}} e^{-z^2/2} \frac{1}{\Gamma(r/2)2^{r/2}} u^{r/2-1} e^{-u/2}, \qquad -\infty < z < \infty,\ 0 < u < \infty.$$

The distribution function $F(t) = P(T \le t)$ of *T* is given by

$$F(t) = P(Z/\sqrt{U/r} \le t)$$
$$= P(Z \le \sqrt{U/r}\,t)$$
$$= \int_0^\infty \int_{-\infty}^{\sqrt{(u/r)}\,t} g(z, u)\, dz\, du.$$

That is,

$$F(t) = \frac{1}{\sqrt{\pi}\,\Gamma(r/2)} \int_0^\infty \left[\int_{-\infty}^{\sqrt{(u/r)}\,t} \frac{e^{-z^2/2}}{2^{(r+1)/2}}\, dz \right] u^{r/2-1} e^{-u/2}\, du.$$

The p.d.f. of *T* is the derivative of the distribution function; so applying the Fundamental Theorem of Calculus to the inner integral we see that

$$f(t) = F'(t) = \frac{1}{\sqrt{\pi}\,\Gamma(r/2)} \int_0^\infty \frac{e^{-(u/2)(t^2/r)}}{2^{(r+1)/2}} \sqrt{\frac{u}{r}}\, u^{r/2-1} e^{-u/2}\, du$$
$$= \frac{1}{\sqrt{\pi r}\,\Gamma(r/2)} \int_0^\infty \frac{u^{(r+1)/2-1}}{2^{(r+1)/2}} e^{-(u/2)(1+t^2/r)}\, du.$$

In the integral make the change of variables

$$y = (1 + t^2/r)u \qquad \text{so that} \qquad \frac{du}{dy} = \frac{1}{1 + t^2/r}.$$

Thus we find that

$$f(t) = \frac{\Gamma\!\left(\dfrac{r+1}{2}\right)}{\sqrt{\pi r}\,\Gamma(r/2)} \left[\frac{1}{(1 + t^2/r)^{(r+1)/2}} \right] \int_0^\infty \frac{y^{(r+1)/2-1}}{\Gamma\!\left(\dfrac{r+1}{2}\right)2^{(r+1)/2}} e^{-y/2}\, dy.$$

The integral in this last expression for $f(t)$ is equal to 1 because the integrand is like the p.d.f. of a chi-square distribution with $r + 1$ degrees of freedom. Thus the p.d.f. is as given in the theorem. □

Note that the distribution of T is completely determined by the number r. Since it is, in general, difficult to evaluate the distribution function of T, some values of $P(T \le t)$ are found in Table VI in the Appendix for $r = 1, 2, 3, \ldots, 30$. Also observe that the graph of the p.d.f. of T is symmetrical with respect to the vertical axis $t = 0$ and is very similar to the graph of the p.d.f. of the standard normal distribution $N(0, 1)$. Figure 5.7-1 shows the graphs of the probability density functions of T when $r = 1, 3,$ and 7 and of $N(0, 1)$. In this figure we see that the tails of the t distribution are heavier than those of a normal one; that is, there is more extreme probability in the t distribution than in the standardized normal one.

Because of the symmetry of the t distribution about $t = 0$, the mean (if it exists) must equal zero. That is, it can be shown that $E(T) = 0$ when $r \ge 2$. When $r = 1$, the t distribution is the Cauchy distribution, and we noted in Section 4.5 that the mean and thus the variance do not exist for the Cauchy distribution. The variance of T is

$$\text{Var}(T) = E(T^2) = \frac{r}{r - 2}, \qquad \text{when } r \ge 3.$$

The variance does not exist when $r = 1$ or 2. Although it is fairly difficult to compute these moments from the p.d.f. of T, they can be found (Exercise 5.7-4) using the definition of T and the independence of Z and U, namely

$$E(T) = E(Z)E\left(\sqrt{\frac{r}{U}}\right) \qquad \text{and} \qquad E(T^2) = E(Z^2)E\left(\frac{r}{U}\right).$$

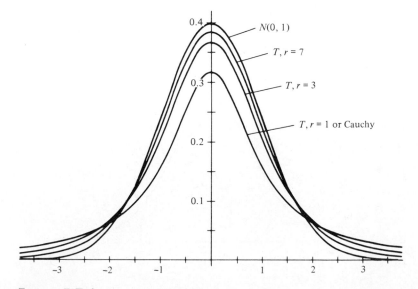

FIGURE 5.7-1

For notational purposes we shall let $t_\alpha(r)$ denote the constant for which

$$P(T \geq t_\alpha(r)) = \alpha,$$

when T has a t distribution with r degrees of freedom; that is, $t_\alpha(r)$ is the $100(1 - \alpha)$ percentile (sometimes called the upper 100α percent point) of the t distribution with r degrees of freedom (See Figure 5.7-2). In this figure, $r = 7$.

Let us consider some illustrations of the use of the t-table and this notation for right tail probabilities.

Example 5.7-1   Let T have a t distribution with seven degrees of freedom. Then, from Table VI in the Appendix, we have

$$P(T \leq 1.415) = 0.90,$$

$$P(T \leq -1.415) = 1 - P(T < 1.415) = 0.10,$$

and

$$P(-1.895 < T < 1.415) = 0.90 - 0.05 = 0.85.$$

We also have, for example, $t_{0.10}(7) = 1.415$, $t_{0.90}(7) = -t_{0.10}(7) = -1.415$, and $t_{0.025}(7) = 2.365$.

Example 5.7-2   Let T have a t distribution with a variance of 5/4. Thus $r/(r - 2) = 5/4$, and $r = 10$. Then

$$P(-1.812 \leq T \leq 1.812) = 0.90$$

and $t_{0.05}(10) = 1.812$, $t_{0.01}(10) = 2.764$, and $t_{0.99}(10) = -2.764$.

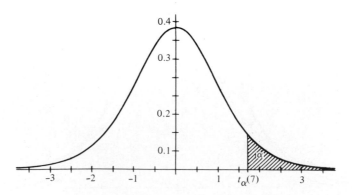

FIGURE 5.7-2

Example 5.7-3   Let T have a t distribution with 14 degrees of freedom. Find a constant c such that $P(|T| < c) = 0.90$. From Table VI in the Appendix we see that $P(T \leq 1.761) = 0.95$ and therefore $c = 1.761 = t_{0.05}(14)$.

It is often true in statistical applications that there is interest in estimating the mean μ or testing hypotheses about the mean when the variance σ^2 is unknown. Theorem 5.7-1, in which a T random variable is defined, along with Theorems 5.3-1 and 5.3-4, provide a random variable that can be used in such situations. That is, if $X_1, X_2, \ldots, X_n$ is a random sample from the normal distribution $N(\mu, \sigma^2)$, then $(\overline{X} - \mu)/(\sigma/\sqrt{n})$ is $N(0, 1)$, $(n - 1)S^2/\sigma^2$ is $\chi^2(n - 1)$, and the two are independent. Thus we have that

$$T = \frac{(\overline{X} - \mu)/(\sigma/\sqrt{n})}{\sqrt{(n - 1)S^2/\sigma^2(n - 1)}} = \frac{\overline{X} - \mu}{S/\sqrt{n}}$$

has a t distribution with $r = n - 1$ degrees of freedom. It was a random variable like this one that motivated Gosset's search for the distribution of T. This t statistic will play an important role in statistical applications.

For a second illustration of the application of Student's t, let $X_1, X_2, \ldots, X_n$ and $Y_1, Y_2, \ldots, Y_m$ be random samples from the independent normal distributions, $N(\mu_X, \sigma_X^2)$ and $N(\mu_Y, \sigma_Y^2)$, respectively. The distribution of $\overline{X}$ is $N(\mu_X, \sigma_X^2/n)$ and the distribution of $\overline{Y}$ is $N(\mu_Y, \sigma_Y^2/m)$. Since $\overline{X}$ and $\overline{Y}$ are independent, the distribution $\overline{X} - \overline{Y}$ is $N(\mu_X - \mu_Y, \sigma_X^2/n + \sigma_Y^2/m)$, and

$$Z = \frac{\overline{X} - \overline{Y} - (\mu_X - \mu_Y)}{\sqrt{\sigma_X^2/n + \sigma_Y^2/m}}$$

is $N(0, 1)$. The statistics

$$\frac{(n - 1)S_X^2}{\sigma_X^2} = \frac{\sum\limits_{i=1}^{n} (X_i - \overline{X})^2}{\sigma_X^2}$$

and

$$\frac{(m - 1)S_Y^2}{\sigma_Y^2} = \frac{\sum\limits_{i=1}^{m} (Y_i - \overline{Y})^2}{\sigma_Y^2}$$

have distributions that are $\chi^2(n-1)$ and $\chi^2(m-1)$, respectively. Because the normal distributions are independent, these chi-square statistics are independent, and thus the distribution of

$$U = \frac{(n - 1)S_X^2}{\sigma_X^2} + \frac{(m - 1)S_Y^2}{\sigma_Y^2}$$

is $\chi^2(n+m-2)$. A random variable T with the t distribution having $r = n + m - 2$ degrees of freedom is given by

$$T = \frac{Z}{\sqrt{U/(n + m - 2)}}$$
$$= \frac{[\bar{X} - \bar{Y} - (\mu_X - \mu_Y)]/\sqrt{\sigma_X^2/n + \sigma_Y^2/m}}{\sqrt{[(n - 1)S_X^2/\sigma_X^2 + (m - 1)S_Y^2/\sigma_Y^2]/[n + m - 2]}}.$$

In the statistical applications we sometimes assume that the two variances are the same, say $\sigma_X^2 = \sigma_Y^2 = \sigma^2$, in which case

$$T = \frac{\bar{X} - \bar{Y} - (\mu_X - \mu_Y)}{\sqrt{\{[(n - 1)S_X^2 + (m - 1)S_Y^2]/(n + m - 2)\}[(1/n) + (1/m)]}}$$

and neither T nor its distribution depend on σ^2.

Example 5.7-4 Let $X_1, X_2, \ldots, X_9$ and $Y_1, Y_2, \ldots, Y_{16}$ be random samples of sizes $n = 9$ and $m = 16$ from the normal independent distributions $N(\mu_X, \sigma^2)$ and $N(\mu_Y, \sigma^2)$, respectively, where σ^2 is unknown. Then

$$P\left(-2.306 \leq \frac{\bar{X} - \mu_X}{S_X/\sqrt{9}} \leq 2.306\right) = 0.95 \qquad (5.7\text{-}1)$$

because $(\bar{X} - \mu_X)/(S_X/\sqrt{9})$ has a t distribution with $r = 8$ degrees of freedom. Also because

$$T = \frac{\bar{X} - \bar{Y} - (\mu_X - \mu_Y)}{\sqrt{[(8S_X^2 + 15S_Y^2)/23][(1/9) + (1/16)]}}$$

has a t distribution with $r = 9 + 16 - 2 = 23$ degrees of freedom,

$$P(-1.714 \leq T \leq 1.714) = 0.90. \qquad (5.7\text{-}2)$$

Note that in both equation (5.7-1) and equation (5.7-2), after the data have been observed, $\bar{X}$, $\bar{Y}$, S_X^2, and S_Y^2 can be calculated so that only μ_X and μ_Y will be unknown. In the next chapter we shall use this information to construct interval estimates of unknown means and differences of means.

Another important distribution in statistical applications is introduced in the following theorem.

THEOREM 5.7-2 *If U and V are independent chi-square variables with r_1 and r_2 degrees of freedom, respectively, then*

$$F = \frac{U/r_1}{V/r_2}$$

has an F distribution with r_1 and r_2 degrees of freedom. Its p.d.f. is

$$h(w) = \frac{\Gamma[(r_1 + r_2)/2](r_1/r_2)^{r_1/2}w^{r_1/2-1}}{\Gamma(r_1/2)\Gamma(r_2/2)(1 + r_1 w/r_2)^{(r_1+r_2)/2}}, \qquad 0 < w < \infty.$$

REMARK For many years the random variable defined in Theorem 5.7-2 has been called F, a symbol first proposed by George Snedecor to honor R. A. Fisher, who used a modification of this ratio in several statistical applications.

PROOF: In this proof, we first find an expression for the distribution function of F and then take its derivative to find the p.d.f. of F. Since U and V are independent, the joint p.d.f. of U and V is

$$g(u, v) = \frac{u^{r_1/2-1}e^{-u/2}}{\Gamma(r_1/2)2^{r_1/2}} \frac{v^{r_2/2-1}e^{-v/2}}{\Gamma(r_2/2)2^{r_2/2}}, \qquad 0 < u < \infty, \quad 0 < v < \infty.$$

In this derivation we let $W = F$ to avoid using f as a symbol for a variable. The distribution function $F(w) = P(W \leq w)$ of F is

$$F(w) = P\left(\frac{U/r_1}{V/r_2} \leq w\right) = P\left(U \leq \frac{r_1}{r_2} wV\right)$$

$$= \int_0^\infty \int_0^{(r_1/r_2)wv} g(u, v) \, du \, dv.$$

That is,

$$F(w) = \frac{1}{\Gamma(r_1/2)\Gamma(r_2/2)} \int_0^\infty \left[\int_0^{(r_1/r_2)vw} \frac{u^{r_1/2-1}e^{-u/2}}{2^{(r_1+r_2)/2}} \, du\right] v^{r_2/2-1}e^{-v/2} \, dv.$$

The p.d.f. of $F = W$ is the derivative of the distribution function; so applying the Fundamental Theorem of Calculus to the inner integral, we have

$$f(w) = F'(w)$$

$$= \frac{1}{\Gamma(r_1/2)\Gamma(r_2/2)} \int_0^\infty \frac{[(r_1/r_2)vw]^{r_1/2-1}}{2^{(r_1+r_2)/2}} e^{-(r_1/2r_2)(vw)}\left(\frac{r_1}{r_2}v\right) v^{r_2/2-1}e^{-v/2} \, dv$$

$$= \frac{(r_1/r_2)^{r_1/2}w^{r_1/2-1}}{\Gamma(r_1/2)\Gamma(r_2/2)} \int_0^\infty \frac{v^{(r_1+r_2)/2-1}}{2^{(r_1+r_2)/2}} e^{-(v/2)[1+(r_1/r_2)w]} \, dv.$$

In the integral, make the change of variables

$$y = \left(1 + \frac{r_1}{r_2} w\right)v \quad \text{so that} \quad \frac{dv}{dy} = \frac{1}{1 + (r_1/r_2)w}.$$

Thus we see that

$$f(w) = \frac{(r_1/r_2)^{r_1/2}\Gamma[(r_1 + r_2)/2]w^{r_1/2-1}}{\Gamma(r_1/2)\Gamma(r_2/2)[1 + (r_1w/r_2)]^{(r_1+r_2)/2}} \int_0^\infty \frac{y^{(r_1+r_2)/2-1}e^{-y/2}}{\Gamma[(r_1 + r_2)/2]2^{(r_1+r_2)/2}} \, dy.$$

The integral in this last expression for $f(w)$ is equal to 1 because the integrand is like a p.d.f. of a chi-square distribution with $r_1 + r_2$ degrees of freedom. Thus the p.d.f. $f(w)$ is as given in the theorem. $\quad\square$

Note that the F distribution depends on two parameters, r_1 and r_2, in that order. The first parameter is the number of degrees of freedom in the numerator, and the second is the number of degrees of freedom in the denominator. See Figure 5.7-3 for graphs of the p.d.f. of the F distribution for four pairs of degrees of freedom. It can be shown that

$$E(F) = \frac{r_2}{r_2 - 2} \quad \text{and} \quad \text{Var}(F) = \frac{2r_2^2(r_1 + r_2 - 2)}{r_1(r_2 - 2)^2(r_2 - 4)},$$

provided that r_2 is large enough, namely $r_2 > 2$ and $r_2 > 4$, respectively. To verify these two expressions, we note, using the independence of U and V in the definition of F, that

$$E(F) = E\left(\frac{U}{r_1}\right)E\left(\frac{r_2}{V}\right) \quad \text{and} \quad E(F^2) = E\left[\left(\frac{U}{r_1}\right)^2\right]E\left[\left(\frac{r_2}{V}\right)^2\right].$$

In Exercise 5.7-9 the student is asked to find $E(U)$, $E(1/V)$, $E(U^2)$, and $E(1/V^2)$.

Some values of the distribution function $P(F \leq f)$ of the F distribution are given in Table VII in the Appendix. For notational purposes, if F has an F distribution with r_1 and r_2 degrees of freedom, we say that the distribution of F is $F(r_1, r_2)$. Furthermore, we will let $F_\alpha(r_1, r_2)$ denote the constant [the upper 100α percent point of $F(r_1, r_2)$] for which

$$P[F \geq F_\alpha(r_1, r_2)] = \alpha.$$

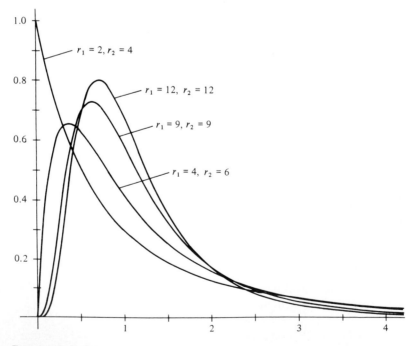

Figure 5.7-3

Example 5.7-5 If the distribution of F is $F(r_1, r_2)$, then from Table VII in the Appendix we see that

when $r_1 = 7$, $r_2 = 8$, $P(F \leq 3.50) = 0.95$, so $F_{0.05}(7, 8) = 3.50$;

when $r_1 = 9$, $r_2 = 4$, $P(F \leq 14.66) = 0.99$, so $F_{0.01}(9, 4) = 14.66$.

To use Table VII to find the F values corresponding to the 0.01, 0.025, and 0.05 cumulative probabilities we need to note the following: Since $F = (U/r_1)/(V/r_2)$, where U and V are independent and $\chi^2(r_1)$ and $\chi^2(r_2)$, respectively, then $1/F = (V/r_2)/(U/r_1)$ must have a distribution that is $F(r_2, r_1)$. Note the change in the order of the parameters in the distribution of $1/F$. Now if the distribution of F is $F(r_1, r_2)$, then

$$P[F \geq F_\alpha(r_1, r_2)] = \alpha$$

and

$$P\left[\frac{1}{F} \leq \frac{1}{F_\alpha(r_1, r_2)}\right] = \alpha.$$

The complement of $\{1/F \leq 1/F_\alpha(r_1, r_2)\}$ is $\{1/F > 1/F_\alpha(r_1, r_2)\}$. Thus

$$P\left[\frac{1}{F} > \frac{1}{F_\alpha(r_1, r_2)}\right] = 1 - \alpha. \tag{5.7-3}$$

Since the distribution of $1/F$ is $F(r_2, r_1)$, by definition of $F_{1-\alpha}(r_2, r_1)$, we have

$$P\left[\frac{1}{F} > F_{1-\alpha}(r_2, r_1)\right] = 1 - \alpha. \tag{5.7-4}$$

From equations (5.7-3) and (5.7-4) we see that

$$F_{1-\alpha}(r_2, r_1) = \frac{1}{F_\alpha(r_1, r_2)}. \tag{5.7-5}$$

The use of equation (5.7-5) is illustrated in the next example.

Example 5.7-6 If the distribution of F is $F(4, 9)$, constants c and d such that

$$P(F \leq c) = 0.01 \qquad \text{and} \qquad P(F \leq d) = 0.05$$

are given by

$$c = F_{0.99}(4, 9) = \frac{1}{F_{0.01}(9, 4)} = \frac{1}{14.66} = 0.0682;$$

$$d = F_{0.95}(4, 9) = \frac{1}{F_{0.05}(9, 4)} = \frac{1}{6.00} = 0.1667.$$

Furthermore, if F is $F(6, 9)$ then

$$P(F \le 0.2439) = P\left(\frac{1}{F} \ge \frac{1}{0.2439}\right) = P\left(\frac{1}{F} \ge 4.100\right) = 0.05$$

because the distribution of $1/F$ is $F(9, 6)$.

The F distribution also has many important applications in normal sampling theory, one of which uses the following random variable. Let $X_1, X_2, \ldots, X_n$ and $Y_1, Y_2, \ldots, Y_m$ be random samples of sizes n and m from two independent normal distributions $N(\mu_X, \sigma_X^2)$ and $N(\mu_Y, \sigma_Y^2)$, respectively. We know that $(n-1)S_X^2/\sigma_X^2$ is $\chi^2(n-1)$ and $(m-1)S_Y^2/\sigma_Y^2$ is $\chi^2(m-1)$. The independence of the distributions implies that S_X^2 and S_Y^2 are independent so that

$$F = \frac{(m-1)S_Y^2/\sigma_Y^2(m-1)}{(n-1)S_X^2/\sigma_X^2(n-1)} = \frac{S_Y^2/\sigma_Y^2}{S_X^2/\sigma_X^2}$$

has an F distribution with $r_1 = m - 1$ and $r_2 = n - 1$ degrees of freedom. This result is used in the next chapter.

Exercises

5.7-1 Let T have a t distribution with r degrees of freedom. Find
(a) $P(T \ge 2.228)$ when $r = 10$.
(b) $P(T \le 2.228)$ when $r = 10$.
(c) $P(|T| \ge 2.228)$ when $r = 10$.
(d) $P(-1.753 \le T \le 2.602)$ when $r = 15$.
(e) $P(1.330 \le T \le 2.552)$ when $r = 18$.

5.7-2 Let T have a t distribution with $r = 19$. Find c such that
(a) $P(|T| \ge c) = 0.05$.
(b) $P(|T| \ge c) = 0.01$.
(c) $P(T \ge c) = 0.025$.
(d) $P(|T| \le c) = 0.95$.

5.7-3 Find
(a) $t_{0.05}(13)$.
(b) $t_{0.01}(15)$.
(c) $t_{0.95}(17)$.
(d) $t_{0.975}(5)$.

5.7-4 *Let T have a t distribution with r degrees of freedom. Show that $E(T) = 0$, $r \ge 2$, and $\mathrm{Var}(T) = r/(r-2)$, provided that $r \ge 3$, by first finding $E(Z)$, $E(1/\sqrt{U})$, $E(Z^2)$, and $E(1/U)$.*

5.7-5 Let T have a t distribution with r degrees of freedom. Show that T^2 has an F distribution with 1 and r degrees of freedom.

HINT: Consider $T^2 = U^2/(V/r)$.

5.7-6 Let F have an F distribution with r_1 and r_2 degrees of freedom. Find
(a) $P(F \ge 3.02)$ when $r_1 = 9$, $r_2 = 10$.

(b) $P(F \leq 4.14)$ when $r_1 = 7$, $r_2 = 15$.

(c) $P(F \leq 0.1508)$ when $r_1 = 8$, $r_2 = 5$.

HINT: $0.1508 = 1/6.63$.

(d) $P(0.1323 \leq F \leq 2.79)$ when $r_1 = 6$, $r_2 = 15$.

5.7-7 Let F have an F distribution with r_1 and r_2 degrees of freedom. Find numbers a and b such that

(a) $P(a \leq F \leq b) = 0.90$ when $r_1 = 8$, $r_2 = 6$.

(b) $P(a \leq F \leq b) = 0.98$ when $r_1 = 8$, $r_2 = 6$.

5.7-8 Find

(a) $F_{0.05}(5, 9)$.

(b) $F_{0.025}(9, 7)$.

(c) $F_{0.99}(8, 5)$.

(d) $F_{0.95}(5, 7)$.

5.7-9 Find the mean and the variance of an F random variable with r_1 and r_2 degrees of freedom by first finding $E(U)$, $E(1/V)$, $E(U^2)$, and $E(1/V^2)$ as suggested in the text.

5.7-10 Let $\overline{X}$ and S^2 be the sample mean and sample variance associated with a random sample of size $n = 16$ from a normal distribution $N(\mu, 225)$.

(a) Find constants a and b so that

$$P(a \leq S^2 \leq b) = 0.95.$$

(b) Find a constant c so that

$$P\left(-c \leq \frac{\overline{X} - \mu}{S} \leq c\right) = 0.95.$$

5.7-11 Let X and Y denote the wing lengths (in millimeters) of a male and a female gallinule, respectively. Assume that the respective distributions of X and Y are $N(184.09, 39.37)$ and $N(171.93, 50.88)$ and that X and Y are independent. Let $\overline{X}$ and $\overline{Y}$ equal the sample means of random samples of sizes $n = m = 16$ birds of each sex.

(a) Find $P(\overline{X} - \overline{Y} \geq 9.12)$.

(b) Let S_X^2 and S_Y^2 be the respective sample variances. If, in fact, σ_X^2 and σ_Y^2 are unknown but equal, find a constant c so that

$$P\left(-c \leq \frac{\overline{X} - \overline{Y} - 12.16}{\sqrt{[(15S_X^2 + 15S_Y^2)/30](2/16)}} \leq c\right) = 0.95.$$

5.7-12 Let $X_1, X_2, \ldots, X_9$ be a random sample of size 9 from a normal distribution $N(54, 10)$ and let Y_1, Y_2, Y_3, Y_4 be a random sample of size 4 from an independent normal distribution $N(54, 12)$. Compute

$$P\left(\frac{1}{61.09} \leq \frac{\sum_{i=1}^{4}(Y_i - \overline{Y})^2}{9}}{\sum_{i=1}^{9}(X_i - \overline{X})^2} \leq 2.439\right).$$

5.7-13 Let X_1 and X_2 have independent gamma distributions with parameters α, θ and β, θ, respectively. Let $W = X_1/(X_1 + X_2)$. Use a method which is similar to that given in the proofs of Theorems 5.7-1 and 5.7-2 to show that the p.d.f. of W is

$$g(w) = \frac{\Gamma(\alpha + \beta)}{\Gamma(\alpha)\Gamma(\beta)} w^{\alpha-1}(1 - w)^{\beta-1}, \qquad 0 < w < 1.$$

We say that W has a **beta distribution** with parameters α and β (see Example 5.9-3).

5.7-14 Let X have a beta distribution with parameters α and β. Show that the mean and variance of X are

$$\mu = \frac{\alpha}{\alpha + \beta} \qquad \text{and} \qquad \sigma^2 = \frac{\alpha\beta}{(\alpha + \beta + 1)(\alpha + \beta)^2}.$$

HINT: In evaluating $E(X)$ and $E(X^2)$ compare the integrands to the p.d.f.'s of beta distributions with parameters $\alpha + 1$, β and $\alpha + 2$, β, respectively.

5.7-15. Let Z_1, Z_2, Z_3 be a random sample of size 3 from a standard normal distribution $N(0, 1)$.

(a) How is U distributed if

$$U = \frac{Z_3}{\sqrt{(Z_1^2 + Z_2^2)/2}} \ ?$$

(b) Let $V = Z_1/Z_2$. Show that V has a Cauchy distribution.

HINT: Use a method similar to the proof of Theorems 5.7-1 and 5.7-2. Note the quadrants in which $V > 0$, $V < 0$, $Z_2 > 0$, and $Z_2 < 0$.

(c) Let

$$W = \frac{Z_2}{\sqrt{(Z_1^2 + Z_2^2)/2}}.$$

Show that the distribution function of W is

$$F(w) = \begin{cases} 0, & w \leq -\sqrt{2}, \\ \dfrac{1}{\pi}\left(\operatorname{Arctan}\sqrt{(2 - w^2)/w^2}\right), & -\sqrt{2} < w < 0, \\ \dfrac{1}{2}, & w = 0, \\ 1 - \dfrac{1}{\pi}\left(\operatorname{Arctan}\sqrt{(2 - w^2)/w^2}\right), & 0 < w \leq \sqrt{2}, \\ 1, & \sqrt{2} \leq w. \end{cases}$$

HINT: What relationship is there between parts (b) and (c)?

(d) Show that the p.d.f. of W is

$$f(w) = \frac{1}{\pi\sqrt{2 - w^2}}, \qquad -\sqrt{2} < w < \sqrt{2}.$$

Note that this is a U-shaped distribution. Why does it differ so much from that in part (a) when the definitions for U and W are so similar?

(e) Show that the distribution function of W, for $-\sqrt{2} < w < \sqrt{2}$, can be defined by

$$F(w) = \frac{1}{2} + \frac{1}{\pi}\left(\text{Arcsin } \frac{w}{\sqrt{2}}\right) = \frac{1}{2} + \frac{1}{\pi}\left(\text{Arctan } \frac{w}{\sqrt{2 - w^2}}\right).$$

(f) Find the means and variances (if they exist) of U, V, and W.

5.8   Understanding Variability and Control Charts

As frequently happens in studying a textbook like this, we are so involved in the mathematics of probability, distributions, sampling distributions, and even descriptive statistics that we miss the real meanings of some of these concepts, particularly as far as the role they play in day-to-day activities. We note some of these in this transition section between the earlier chapters covering probability and distributions and the later ones on statistical inference.

Even the definitions and distributions of the t and F random variables say something about understanding variability. Both t and F are defined as the ratio of two independent random variables. The resulting distributions have heavier tails than the respective random variables in the numerators. For example, the t random variable has a standardized normal random variable in the numerator and the p.d.f. of t is much heavier in the tails than that of $N(0, 1)$. That is, there is a tendency to get more outliers in sampling from the t distribution than from $N(0, 1)$.

We often observe the situation in which the denominator is a random variable, creating a ratio that has more variability than we think it should have. For illustration, suppose we observe successive sample averages, say $\bar{x}_1, \bar{x}_2, \cdots, \bar{x}_k$. However, the sample sizes change from one sample to the next, say $n_1, n_2, \cdots, n_k$. If these sample sizes vary enough, the $\bar{x}$-values will look as if they arose from an approximate t-distribution. That is why in practice we often see more variation in a plot of sample means than we might expect; there are different sample sizes.

In Chapter 1, we did note a few important things about variability. For instance, it explains what is commonly called the "sophomore jinx": some freshmen (rookies) have great first years (above their averages) and then most of them do worse the second year. The same explanation can be applied to movies that have sequels; often the original was outstanding, but even with the same cast and same director, the sequel is usually worse than the original.

We had also noted that workers who do very well (above their averages) and who thus might be given some reward, usually do worse during the next work period. On the other hand, those who have bad periods (below average) and are reprimanded frequently do better the next. Yet it is wrong to think that a policy of reprimand is better than that of a pat on the back. Both situations can be explained by understanding the worker's pattern of variation. If the employer really wants to improve the outcome of the process, he or she must consider ways of improving the level (average). The workers can sometimes make small adjustments, but they need road maps to make the major ones. Thus it is really the responsibility of management to improve the process substantially. That is, management must realize that working harder with techniques that have failed will not improve the situation much. Changes must be made, and often data and the resulting statistical analysis can suggest changes for the better.

It is also disturbing to see the following situation: Suppose that the manager has created a team (say 10) of outstanding workers, and it is time to give raises. The workers are ranked from one to ten. (It is always true that 10 workers, no matter how good or how bad, will always get the ranks 1, 2, 3, . . . , 10.) Then say they get raises according to their ranks, which might have been determined simply by a random process for this particular period. That is, in the next period, their ranks might be entirely different. How does the small raise make the one with lowest rank feel if, in fact, all 10 are members of an outstanding team and there might be little or no difference among the persons ranked 1 and 10? Not good, and he or she is less likely to help the ones with higher ranks in a future period. That is, this does not promote teamwork. As a matter of fact, there is always this danger in any kind of reward, such as the ''worker of the month.'' Others are not likely to help that person the next month, because they want to be the worker of the month and receive the corresponding bonus.

A better way of rewarding a good team would be to give them essentially the same raises. Clearly, a good manager will continuously monitor performance and supply appropriate feedback to the workers. Of course, if one worker is consistently on the high side, then he or she should be considered for promotion or a substantial raise. On the other hand, if some worker is on the low end most of the time, some help for this worker is in order: maybe additional training or even a transfer to a different department for which the worker's talents are more suitable. It is important not to demean this worker because each of us needs to take pride in our accomplishments, even though they may be little. If at all possible, firing the worker should be avoided, possibly by finding some other suitable job in the company.

Some teachers will announce to a class of 30 students that there will be only so many A's, so many B's, and so on. This is wrong! If those students are competing for those given number of A's, why would anyone want to help anyone else? It completely destroys any sort of teamwork, and yet that is what students must learn to do when they are on the job later in life. Clearly, it is better to say that all of them can earn an A (they probably won't) and encourage them to work together. In that way, the teacher has a better opportunity to improve the level of the entire class which, after all, is the real objective. Too frequently mathematics and science courses discourage interaction among students, and teachers must do everything possible to break down those barriers.

Many of these preceding observations are those of W. Edwards Deming, an esteemed statistician who went to Japan after World War II and taught the Japanese how to make quality products. For his work there, he was awarded the Emperor's Medal and the Japanese established the Deming Prize to be awarded each year to the company or individual contributing the most to quality improvement. One of the things Deming continues to stress is the need for "profound knowledge," and a major item in that is understanding variation and statistical theory.

In making quality products, you want to reduce the variation as much as possible and move the level closer to the target. Deming believes that barriers between departments, between management and workers, and among workers must be broken down to improve communications and the ability to work as a team. The lines of communication, all the way from suppliers to customers, must be open to help reduce variation and improve products and services. For example, he argues that a company should not buy only on price tag but have a few reliable suppliers (possibly one) because that will reduce variation. That is, many different suppliers would obviously increase variation. Moreover, he argues that you should become partners, friends if you like, with your suppliers. You learn to trust the other; in that way, you can use methods like "just in time," in which your inventory can be kept reasonably small, to keep costs down.

If each of us thinks about these ideas, we might become obsessed by understanding variation, and that might make a big difference in our everyday lives. For illustration, once we have selected good suppliers, we continue to go to the same barber, the same service station, the same clothier, the same bank, and on and on. We like to buy items, even if a little more expensive, from places that will give us good service if something goes wrong. If this does not happen, then of course we must consider changing suppliers.

Clearly, listening to the customer can help improve the quality of our products and services. We can then meet—or even exceed—the expectations of our customers. We should continually try to improve by reducing the variation and moving the process to a better level (high is not always good, as in golf). Although the customer—as well as the supplier—must be part of the total team, more often management must continue to look for better ways to do things, ways that the customer never would have imagined. For example, in the early days of automobiles not many owners would have thought of driving on pneumatic tires. Harvey Firestone did, and in this way exceeded the expectations of those early customers.

Deming also preaches constancy of purpose. If management ideas tend to change too much, employees really do not know what to do and cannot do their best. That is, they are mixed up, increasing variation. It is better for them to receive a constant signal from their employer, a signal that changes only if research dictates ways of improving.

More training and education for everyone associated with a company also decreases variation by teaching how to make the product more uniform. Workers must know that they do not have to be afraid to make suggestions to improve the process. Often, being team members will make it easier for workers to speak up without fear of reprisal.

Many of us remember playing a game called "telephone": one whispers a message to the next person, who whispers the message to the next, and so on. The message at the end is compared to the original message, and it is usually much different from the original. This is like trying to hit a target (say at zero) with a random variable X_1. Then starting with X_1 as the center and adding on another error X_2, and so on, creating the sum of the errors, $X_1 + X_2 + \cdots + X_n$, which has an ever-increasing variance with independent errors. Deming would say, "We are off to the Milky Way." Yet we actually do this in business and industry by letting worker train worker. Once errors are introduced, they will stay there, and others will add on and on. (Incidentally, we might say the same thing about too many layers of management.) To decrease variation, wouldn't it be better to have a master instructor train each worker (or have fewer layers of management)?

Deming also notes that requiring quotas does not help the quality. A foreman who has a quota of 100 per day will often ship out 90 good and 10 bad items just to make the quota. Clearly, it would reduce the variation and satisfy the customer better if only the 90 good ones were sent on.

This leads to the point that a final inspection of products often does not really improve the quality. With such a procedure, you can only eliminate the bad ones and send on the good ones. Improvements in the design of the products and manufacturing processes are needed. If these are done well, often with the help of statistical methods, that final inspection can be eliminated. That is, improvements should be made early in the manufacturing process rather than try to correct things at an end inspection by weeding out the bad items.

The first chapter stressed that observations should be plotted in time sequence if they are taken at different times. One major application of this idea is the use of quality control charts, which W. A. Shewhart invented in the 1920s. In making products, every so often (each hour, each day, each week depending upon how many items are being produced) a sample of size n of them is taken, and they are measured resulting in the observations $x_1, x_2, \ldots, x_n$. The average $\bar{x}$ and the standard deviation s are computed. This is done k times, and the k values of $\bar{x}$ and s are averaged resulting in $\bar{\bar{x}}$ and $\bar{s}$, respectively; usually k is equal to some number between 10 and 30.

If the true mean μ and standard deviation σ of the process were known, then the Central Limit Theorem states that almost all of the $\bar{x}$ values would plot between $\mu - 3\sigma/\sqrt{n}$ and $\mu + 3\sigma/\sqrt{n}$, unless the system has actually changed. However, we know neither μ nor σ, and thus μ is estimated by $\bar{\bar{x}}$ and $3\sigma/\sqrt{n}$ by $A_3\bar{s}$, where A_3 is a factor depending upon n that can be found in books on statistical quality control. For example, a few values of A_3 (and some other constants that will be used later) are given in Table 5.8-1 for typical values of n.

The estimates of $\mu \pm 3\sigma/\sqrt{n}$ are called the upper control limit (UCL), $\bar{\bar{x}} + A_3\bar{s}$, and the lower control limit (LCL), $\bar{\bar{x}} - A_3\bar{s}$, and $\bar{\bar{x}}$ provides the estimate of the centerline. A typical plot is given in Figure 5.8-1.

Here, in the 13th sampling period, $\bar{x}$ is outside the control limits, indicating the process has changed and some investigation and action is needed to correct this change, which seems like a shift upward in the process.

It should be noted that there is a control chart for the s values, too. From sampling distribution theory, values of B_3 and B_4 have been determined and are given in Table 5.8-1 so that we know that almost all s values should be between $B_3\bar{s}$ and $B_4\bar{s}$ if there is no change in the underlying distribution. So again, if an individual s value is

TABLE 5.8-1

n	A_3	B_3	B_4	A_2	D_3	D_4
4	1.63	0	2.27	0.73	0	2.28
5	1.43	0	2.09	0.58	0	2.11
6	1.29	0.03	1.97	0.48	0	2.00
8	1.10	0.185	1.815	0.37	0.14	1.86
10	0.98	0.28	1.72	0.31	0.22	1.78
20	0.68	0.51	1.49	0.18	0.41	1.59

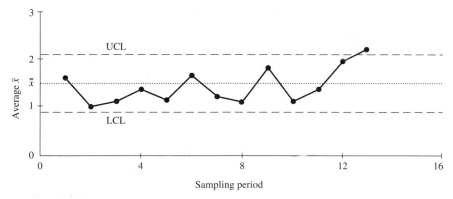

FIGURE 5.8-1

outside these control limits, some action should be taken as it seems as if there has been a change in the variation of the underlying distribution.

Often, when these charts are first constructed after $k = 10$ to 30 sampling periods, many points fall outside the control limits. A team consisting of workers, the foreman, the supervisor, an engineer, and even a statistician should try to find the reasons why this has occurred, and the situation should be corrected. After this is done and the points plot within the control limits, the process is "in statistical control." However, being in statistical control is not a guarantee of satisfaction with the products. Since $A_3 \bar{s}$ is an estimate of $3\sigma/\sqrt{n}$, then $\sqrt{n} A_3 \bar{s}$ is an estimate of 3σ, and with an underlying distribution close to a normal one, almost all items would be between $\bar{\bar{x}} \pm \sqrt{n} A_3 \bar{s}$. If these limits are too wide, then corrections must be made again.

If the variation is under control (i.e., $\bar{x}$ and s within their control limits), we say the variations seen in $\bar{x}$ and s are due to common causes. If products made under such a system with these existing common causes are satisfactory, then the production continues. If either $\bar{x}$ or s, however, is outside the control limits, that is an indication that some special causes are at work, which must be corrected. That is, a team should investigate why and some action should be taken.

TABLE 5.8-2: Console Opening Times

Group	x_1	x_2	x_3	x_4	x_5	$\bar{x}$	s	R
1	1.2	1.8	1.7	1.3	1.4	1.48	0.26	0.6
2	1.5	1.2	1.0	1.0	1.8	1.30	0.35	0.8
3	0.9	1.6	1.0	1.0	1.0	1.10	0.28	0.7
4	1.3	0.9	0.9	1.2	1.2	1.06	0.18	0.4
5	0.7	0.8	0.9	0.6	0.8	0.76	0.11	0.3
6	1.2	0.9	1.1	1.0	1.0	1.04	0.10	0.3
7	1.1	0.9	1.1	1.0	1.4	1.10	0.19	0.5
8	1.4	0.9	0.9	1.1	1.0	1.06	0.19	0.5
9	1.3	1.4	1.1	1.5	1.6	1.38	0.19	0.5
10	1.6	1.5	1.4	1.3	1.5	1.46	0.11	0.3
						$\bar{\bar{x}} = 1.17$	$\bar{s} = 0.20$	$\bar{R} = 0.49$

Example 5.8-1 A company produces a storage console. Twice a day nine critical characteristics are tested on five consoles that are selected randomly from the production line. One of these characteristics is the time it takes the lower storage component door to open completely. Table 5.8-2 lists the opening times in seconds for the consoles that were tested during one week. Also included in the table are the sample means, sample standard deviations, and the ranges.

The upper control limit (UCL) and the lower control limit (LCL) for $\bar{x}$ are found using A_3 from Table 5.8-1 and $n = 5$ as follows:

$$\text{UCL} = \bar{\bar{x}} + A_3\bar{s} = 1.17 + 1.43(0.20) = 1.46$$

and

$$\text{LCL} = \bar{\bar{x}} - A_3\bar{s} = 1.17 - 1.43(0.20) = 0.88.$$

These control limits and the sample means are plotted on the $\bar{x}$-chart in Figure 5.8-2.

The UCL and LCL for s are found using B_3 and B_4 in Table 5.8-1 and $n = 5$ as follows:

$$\text{UCL} = B_4\bar{s} = 2.09(0.20) = 0.42$$

and

$$\text{LCL} = B_3\bar{s} = 0(0.20) = 0.$$

These control limits and the sample standard deviations are plotted on the s-chart in Figure 5.8-3.

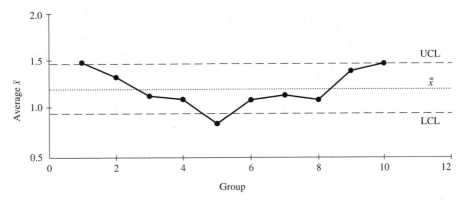

FIGURE 5.8-2: $\bar{x}$-chart

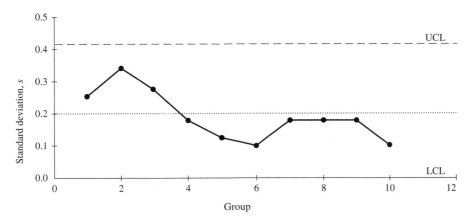

FIGURE 5.8-3: s-chart

Almost all of the observations should lie between $\bar{\bar{x}} \pm \sqrt{n}A_3\bar{s}$, namely,

$$\text{UCL} = 1.17 + \sqrt{5}(1.43)(0.20) = 1.81$$

and

$$\text{LCL} = 1.17 - \sqrt{5}(1.43)(0.20) = 0.53.$$

This is illustrated in Figure 5.8-4, in which all 50 observations do fall within these control limits.

   In most books on statistical quality control, there is an alternate way of constructing the limits on an $\bar{x}$-chart. For each sample, compute the range, R, which is defined in Section 1.3 as the absolute value of the difference of the extremes of the sample. This computation is much easier than that for calculating s. After k samples are taken, compute the average of these R-values, obtaining $\bar{R}$ as well as $\bar{\bar{x}}$. The

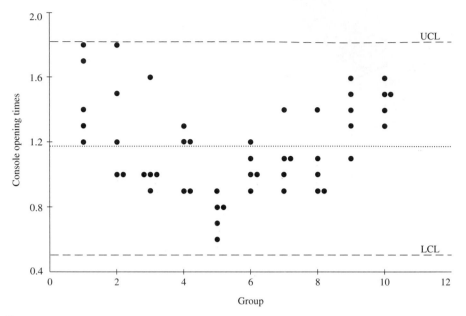

FIGURE 5.8-4

statistic $A_2 \overline{R}$ serves as an estimate of $3\sigma/\sqrt{n}$, where A_2 is found in Table 5.8-1. Thus the estimates of $\mu \pm 3\sigma\sqrt{n}$, namely $\overline{\overline{x}} \pm A_2\overline{R}$, can be used as the UCL and LCL of an $\overline{x}$-chart.

In addition, $\sqrt{n}A_2\overline{R}$ is an estimate of 3σ; so with an underlying distribution that is close to a normal one, we find that almost all observations are within the limits $\overline{\overline{x}} \pm \sqrt{n}A_2\overline{R}$.

Moreover, an R-chart can be constructed with centerline $\overline{R}$ and control limits equal to $D_3\overline{R}$ and $D_4\overline{R}$, where D_3 and D_4 are given in Table 5.8-1 and were determined so that almost all R-values should be between the control limits if there is no change in the underlying distribution. Thus an R falling outside those limits would indicate a change in the spread of the underlying distribution, and some corrective action should be considered.

The use of R, rather than s, is illustrated in the next example.

> **Example 5.8-2** Using the data in Example 5.8-1, we compute UCL and LCL for an $\overline{x}$-bar chart using $\overline{\overline{x}} \pm A_2\overline{R}$ as follows:
>
> $$\text{UCL} = \overline{\overline{x}} + A_2\overline{R} = 1.17 + 0.58(0.49) = 1.45$$
>
> and
>
> $$\text{LCL} = \overline{\overline{x}} - A_2\overline{R} = 1.17 - 0.58(0.49) = 0.89.$$

Note that these are very close to the limits that were found for the $\bar{x}$-chart in Figure 5.8-2 using $\bar{\bar{x}} \pm A_3\bar{s}$. In addition, almost all of the observations should lie within the limits $\bar{\bar{x}} \pm \sqrt{n}A_2\bar{R}$. These limits are

$$\text{UCL} = 1.17 + \sqrt{5}(0.58)(0.49) = 1.81$$

and

$$\text{LCL} = 1.17 - \sqrt{5}(0.58)(0.49) = 0.53.$$

Note that these are the same as the limits found in Example 5.8-1 and plotted in Figure 5.8-4.

An R-chart can be constructed with centerline $\bar{R} = 0.49$ and control limits given by

$$\text{UCL} = D_4\bar{R} = 2.11(0.49) = 1.03$$

and

$$\text{LCL} = D_3\bar{R} = 0(0.49) = 0.$$

Figure 5.8-5 illustrates this control chart for the range.

There are two other Shewhart control charts, the p- and c- charts. The Central Limit Theorem, which provided a justification for the three sigma limits in the $\bar{x}$-chart, also justifies the control limits in the p-chart. Suppose the number of defec-

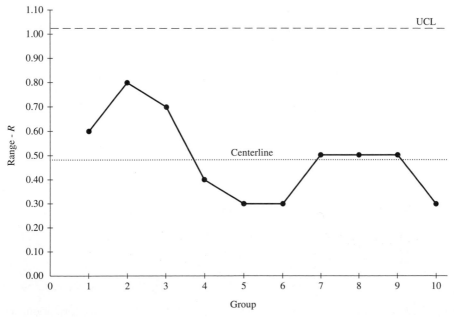

FIGURE 5.8-5: *R*-chart

tives among n items that are selected randomly, say D, has a binomial distribution $b(n, p)$. The limits $p \pm 3\sqrt{p(1-p)/n}$ should include almost all of the D/n values. However, p must be approximated by observing k values of D, say $D_1, D_2, \ldots, D_k$, and computing what is called $\bar{p}$ in the statistical quality control literature, namely,

$$\bar{p} = \frac{D_1 + D_2 + \cdots D_k}{kn}.$$

Thus the LCL and UCL are given by

$$\text{LCL} = \bar{p} - 3\sqrt{\bar{p}(1-\bar{p})/n}$$

and

$$\text{UCL} = \bar{p} + 3\sqrt{\bar{p}(1-\bar{p})/n}.$$

If the process is in control, almost all D/n values are between LCL and UCL. This may not be satisfactory and improvements might be needed to decrease $\bar{p}$. If satisfactory, however, let the process continue under these common causes of variation until a point, D/n, outside the control limits would indicate that some special cause has changed the variation. Incidentally, if D/n is below the LCL, this might very well indicate that some type of change for the better has been made, and we want to find out why. In general, outlying statistics can often suggest that good (as well as bad) breakthroughs have been made.

The following example gives the results of a simple experiment that you can easily duplicate.

Example 5.8-3   Let D_i equal the number of yellow candies in a 1.69-ounce bag. Because the number of pieces of candy varies from bag to bag, we shall use an average value for n when we construct the control limits. Table 5.8-3 lists, for 20 packages, the number of pieces of candy in the package, the number of yellow ones, and the proportion of yellow ones.

For these data,

$$\bar{p} = \frac{219}{1124} = 0.195 \quad \text{and} \quad \bar{n} = \frac{1124}{20} \approx 56.$$

Thus the UCL and LCL are given by

$$\text{LCL} = \bar{p} - 3\sqrt{\bar{p}(1-\bar{p})/56} = 0.195 - 3\sqrt{0.195(0.805)/56} = 0.036$$

and

$$\text{UCL} = \bar{p} + 3\sqrt{\bar{p}(1-\bar{p})/56} = 0.195 + 3\sqrt{0.195(0.805)/56} = 0.354.$$

The control chart for p is depicted in Figure 5.8-6. (For your information, the ''true'' value for p is 0.20.)

**TABLE 5.8-3: Data on
Yellow Candies**

Package	n_i	D_i	$\bar{p}_i$
1	56	8	0.14
2	55	13	0.24
3	58	12	0.21
4	56	13	0.23
5	57	14	0.25
6	54	5	0.09
7	56	14	0.25
8	57	15	0.26
9	54	11	0.20
10	55	13	0.24
11	57	10	0.18
12	59	8	0.14
13	54	10	0.19
14	55	11	0.20
15	56	12	0.21
16	57	11	0.19
17	54	6	0.11
18	58	7	0.12
19	58	12	0.21
20	58	14	0.24
	1124	219	

Consider the following explanation of the c-chart. Suppose the number of flaws, say C, on some product has a Poisson distribution with parameter λ. If λ is sufficiently large, as in Example 5.5-5, we considered approximating the discrete Poisson distribution by the continuous $N(\lambda, \lambda)$ distribution. Thus the interval from $\lambda - 3\sqrt{\lambda}$ to $\lambda + 3\sqrt{\lambda}$ contains virtually all of the C values. Since λ is unknown, however, it must be approximated by $\bar{c}$, the average of the k values, $c_1, c_2, \ldots,$ c_k. Hence the two control limits are computed as

$$\text{LCL} = \bar{c} - 3\sqrt{\bar{c}} \qquad \text{and} \qquad \text{UCL} = \bar{c} + 3\sqrt{\bar{c}}.$$

The remarks made about the $\bar{x}$- and $\bar{p}$-charts apply to the c-chart as well, but we must remember that each c value is the number of flaws on one manufactured item, not an average $\bar{x}$ or a fraction defective D/n.

REMARK In observing time-sequence plots, as with control charts, do not read too much into a short sequence of points. For example, after a pep talk by the coach of a women's golf team, three successive decreasing scores might indicate that the coach had some influence. But even if the process has not changed, the probability of three decreasing scores (ruling out ties) is 1/6, since there are 3! = 6 equally likely ways of arranging three points. Often business people are worse offenders because they might think two increasing sales points indicate improvement. We know that with no change in the system, the probability of that is 1/2. Now maybe 4 points that successively increase after an intervention would be cause for claiming improvement because the probability of such an event is 1/4! = 1/24 if no change has actually been made. This probability is small enough to justify a celebration.

In this section we have listed a few simple, but often overlooked, ideas about variation that the reader might find useful throughout life. We hope that persons who have taken a statistics course understand some of these basic concepts. It could change their lives.

Exercises

5.8-1   It is important to control the viscosity of liquid dishwasher soap so that it flows out of the container but does not run out too rapidly. Thus samples are taken randomly throughout the day and the viscosity is measured. Use the following 20 sets of 5 observations for this exercise.

					$\bar{x}$	s	R
158	147	158	159	169	158.20	7.79	22
151	166	151	143	169	156.00	11.05	26
153	174	151	164	185	165.40	14.33	34
168	140	180	176	154	163.60	16.52	40
160	187	145	164	158	162.80	15.29	42
169	153	149	144	157	154.40	9.48	25
156	183	157	140	162	159.60	15.47	43
158	160	180	154	160	162.40	10.14	26
164	168	154	158	164	161.60	5.55	14
159	153	170	158	170	162.00	7.65	17
150	161	169	166	154	160.00	7.97	19
157	138	155	134	165	149.80	13.22	31
161	172	156	145	153	157.40	10.01	27
143	152	152	156	163	153.20	7.26	20
179	157	135	172	143	157.20	18.63	44
154	165	145	152	145	152.20	8.23	20
171	189	144	154	147	161.00	18.83	45
187	147	159	167	151	162.20	15.85	40
153	168	148	188	152	161.80	16.50	40
165	155	140	157	176	158.60	13.28	36

(a) Calculate the values of $\bar{\bar{x}}$, $\bar{s}$, and $\bar{R}$.
(b) Construct an $\bar{x}$-chart using the value of A_3 and $\bar{s}$.
(c) Construct an s-chart.
(d) Construct an $\bar{x}$-chart using A_2 and $\bar{R}$.
(e) Construct an R-chart.
(f) Do the charts indicate that viscosity is in statistical control?

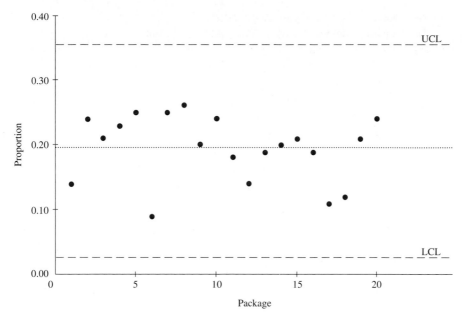

FIGURE 5.8-6: Proportions of Yellow Candies

5.8-2 It is necessary to control the percentage of solids in a product, so samples are taken randomly throughout the day and the percentage of solids is measured. Use the following 20 sets of 5 observations for this exercise.

					$\bar{x}$	s	R
69.8	71.3	65.6	66.3	70.1	68.62	2.51	5.7
71.9	69.6	71.9	71.1	71.7	71.24	0.97	2.3
71.9	69.8	66.8	68.3	64.4	68.24	2.86	7.5
64.2	65.1	63.7	66.2	61.9	64.22	1.61	4.3
66.1	62.9	66.9	67.3	63.3	65.30	2.06	4.4
63.4	67.2	67.4	65.5	66.2	65.94	1.61	4.0
67.5	67.3	66.9	66.5	65.5	66.74	0.79	2.0
63.9	64.6	62.3	66.2	67.2	64.84	1.92	4.9
66.0	69.8	69.7	71.0	69.8	69.26	1.90	5.0
66.0	70.3	65.5	67.0	66.8	67.12	1.88	4.8
67.6	68.6	66.5	66.2	70.4	67.86	1.71	4.2
68.1	64.3	65.2	68.0	65.1	66.14	1.78	3.8
64.5	66.6	65.2	69.3	62.0	65.52	2.69	7.3
67.1	68.3	64.0	64.9	68.2	66.50	1.96	4.3
67.1	63.8	71.4	67.5	63.7	66.70	3.17	7.7
60.7	63.5	62.9	67.0	69.6	64.74	3.53	8.9
71.0	68.6	68.1	67.4	71.7	69.36	1.88	4.3
69.5	61.5	63.7	66.3	68.6	65.92	3.34	8.0
66.7	75.2	79.0	75.3	79.2	75.08	5.07	12.5
77.3	67.2	69.3	67.9	65.6	69.46	4.58	11.7

(a) Calculate the values of $\bar{\bar{x}}$, $\bar{s}$, and $\bar{R}$.
(b) Construct an $\bar{x}$-chart using the value of A_3 and $\bar{s}$.
(c) Construct an s-chart.
(d) Construct an $\bar{x}$-chart using A_2 and $\bar{R}$.
(e) Construct an R-chart.
(f) Do the charts indicate that the percentage of solids in this product is in statistical control?

5.8-3 It is important to control the net weight of a packaged item; thus items are selected randomly throughout the day from the production line and their weights are recorded. Use the following 20 sets of 5 weights (in grams) for this exercise. (Note that a weight recorded here is the actual weight minus 330.)

					$\bar{x}$	s	R
7.97	8.10	7.73	8.26	7.30	7.866	0.3723	0.96
8.11	7.26	7.99	7.88	8.88	8.024	0.5800	1.62
7.60	8.23	8.07	8.51	8.05	8.092	0.3309	0.91
8.44	4.35	4.33	4.48	3.89	5.098	1.8815	4.55
5.11	4.05	5.62	4.13	5.01	4.784	0.6750	1.57
4.79	5.25	5.19	5.23	3.97	4.886	0.5458	1.28
4.47	4.58	5.35	5.86	5.61	5.174	0.6205	1.39
5.82	4.51	5.38	5.01	5.54	5.252	0.5077	1.31
5.06	4.98	4.13	4.58	4.35	4.620	0.3993	0.93
4.74	3.77	5.05	4.03	4.29	4.376	0.5199	1.28
4.05	3.71	4.73	3.51	4.76	4.152	0.5748	1.25
3.94	5.72	5.07	5.09	4.61	4.886	0.6599	1.78
4.63	3.79	4.69	5.13	4.66	4.580	0.4867	1.34
4.30	4.07	4.39	4.63	4.47	4.372	0.2079	0.56
4.05	4.14	4.01	3.95	4.05	4.040	0.0693	0.19
4.20	4.50	5.32	4.42	5.24	4.736	0.5094	1.12
4.54	5.23	4.32	4.66	3.86	4.522	0.4999	1.37
5.02	4.10	5.08	4.94	5.18	4.864	0.4360	1.08
4.80	4.73	4.82	4.69	4.27	4.662	0.2253	0.55
4.55	4.76	4.45	4.85	4.02	4.526	0.3249	0.83

(a) Calculate the values of $\bar{\bar{x}}$, $\bar{s}$, and $\bar{R}$.
(b) Construct an $\bar{x}$-chart using the value of A_3 and $\bar{s}$.
(c) Construct an s-chart.
(d) Construct an $\bar{x}$-chart using A_2 and $\bar{R}$.
(e) Construct an R-chart.
(f) Do the charts indicate that these fill weights are in statistical control?

5.8-4 To give some indication of how the values in Table 5.8-1 are calculated, values of A_3 are found in this exercise. Let $X_1, X_2, \ldots, X_n$ be a random sample of size n from the normal distribution $N(\mu, \sigma^2)$. Let S^2 equal the sample variance of this random sample.

(a) Show that $E[S^2] = \sigma^2$, using the fact that the distribution of $Y = (n-1)S^2/\sigma^2$ has a distribution that is $\chi^2(n-1)$.

(b) Using the $\chi^2(n-1)$ p.d.f., find the value of $E(\sqrt{Y})$.

(c) Show that

$$E\left[\frac{\sqrt{n-1}\,\Gamma\left(\dfrac{n-1}{2}\right)}{\sqrt{2}\,\Gamma\left(\dfrac{n}{2}\right)}\,S\right] = \sigma.$$

(d) Verify that

$$\frac{3}{\sqrt{n}}\left[\frac{\sqrt{n-1}\,\Gamma\left(\dfrac{n-1}{2}\right)}{\sqrt{2}\,\Gamma\left(\dfrac{n}{2}\right)}\right] = A_3,$$

found in Table 5.8-1 for $n = 5$ and $n = 6$. Thus $A_3\bar{s}$ approximates $3\sigma/\sqrt{n}$.

5.8-5 A company has been producing bolts that are about $\bar{p} = 0.02$ defective, and this is satisfactory. To monitor the quality of this process, 100 bolts are selected at random each hour and the number of defective bolts counted. With $\bar{p} = 0.02$, compute the UCL and LCL of the $\bar{p}$-chart. Then suppose that, over the next 24 hours, the following numbers of defective bolts are observed:

$$4\ 1\ 1\ 0\ 5\ 2\ 1\ 3\ 4\ 3\ 1\ 0\ 0\ 4\ 1\ 1\ 6\ 2\ 0\ 0\ 2\ 8\ 7\ 5$$

Would any action have been required during this time?

5.8-6 In the past, $n = 50$ fuses are tested each hour and $\bar{p} = 0.03$ have been defective. Calculate the UCL and LCL. After an intervention, say the true p shifts to $p = 0.05$.

(a) What is the probability that the next observation exceeds the UCL?

(b) What is the probability that at least one of the next five observations exceeds the UCL?

HINT: Assume independence and compute the probability that none of the next five exceeds the UCL.

5.8-7 In a woolens mill, 100-yard pieces are inspected. In the last 20 observations the following numbers of flaws were found:

$$2\ 4\ 0\ 1\ 0\ 3\ 4\ 1\ 1\ 2\ 4\ 0\ 0\ 1\ 0\ 3\ 2\ 3\ 5\ 0$$

(a) Compute the control limits of the c-chart.

(b) Is the process in statistical control?

5.8-8 In 50-foot tin strips, we find that the number of blemishes average about $\bar{c} = 1.4$. Calculate the control limits. Say the process has gone out of control and this average has increased to 3.

(a) What is the probability that the next observation will exceed the UCL?

(b) What is the probability that at least 1 of the next 10 observations will exceed the UCL?

★5.9   Transformations of Random Variables

In this section we consider another important method of constructing models. As a matter of fact, to go very far in theory or application of statistics, one must know something about transformations of random variables. We saw how important this was in dealing with the normal distribution when we noted that if X is $N(\mu, \sigma^2)$, then $Z = u(X) = (X - \mu)/\sigma$ is $N(0, 1)$; this simple transformation allows us to use one table for probabilities associated with all normal distributions. In the proof of this, we found that the distribution function of Z was given by the integral

$$G(z) = P\left(\frac{X - \mu}{\sigma} \le z\right) = P(X \le z\sigma + \mu)$$

$$= \int_{-\infty}^{z\sigma + \mu} \frac{1}{\sigma\sqrt{2\pi}} \exp\left[-\frac{(x - \mu)^2}{2\sigma^2}\right] dx.$$

In Section 4.4 we changed variables, $x = w\sigma + \mu$, in the integral to determine the p.d.f. of Z; but let us now simply differentiate the integral with respect to z. If we recall from calculus that the derivative

$$D_z\left[\int_a^{v(z)} f(t)\, dt\right] = f[v(z)]v'(z),$$

then, since certain assumptions are satisfied in our case, we have

$$g(z) = G'(z) = \frac{1}{\sigma\sqrt{2\pi}} \exp\left[-\frac{(z\sigma + \mu - \mu)^2}{2\sigma^2}\right] \frac{d(z\sigma + \mu)}{dz}$$

$$= \frac{1}{\sqrt{2\pi}} \exp\left(-\frac{z^2}{2}\right), \qquad -\infty < z < \infty.$$

That is, Z has the p.d.f. of a standard normal distribution.

Motivated by the preceding argument we note that, in general, if X is a continuous-type random variable with p.d.f. $f(x)$ on support $a < x < b$ and if $Y = u(X)$ and its inverse $X = v(Y)$ are increasing continuous functions, then the p.d.f. of Y is

$$g(y) = f[v(y)]v'(y)$$

on the support given by $a < v(y) < b$ or, equivalently, $u(a) < y < u(b)$. Moreover, if u and v are decreasing functions, then the p.d.f. is

$$g(y) = f[v(y)][-v'(y)], \qquad u(b) < y < u(a).$$

Hence, to cover both cases, we can simply write

$$g(y) = |v'(y)|\, f[v(y)], \qquad c < y < d,$$

where the support $c < y < d$ corresponds to $a < x < b$ through the transformation $x = v(y)$.

Example 5.9-1 Let the positive random variable X have the p.d.f. $f(x) = e^{-x}$, $0 < x < \infty$, which is skewed to the right. To find a distribution that is more symmetric than that of X, statisticians frequently use the square root transformation, namely $Y = \sqrt{X}$. Here $y = \sqrt{x}$ corresponds to $x = y^2$, which has derivative $2y$. Thus the p.d.f. of Y is

$$g(y) = 2ye^{-y^2}, \qquad 0 < y < \infty,$$

which is of the Weibull type (see Section 4.5). The graphs of $f(x)$ and $g(y)$ should convince the reader that the latter is more symmetric than the former (see Figure 5.9-1).

Example 5.9-2 Let X be binomial with parameters n and p. Since X has a discrete distribution, $Y = u(X)$ will also have a discrete distribution with the same probabilities as those in the support of X. For illustration, with $n = 3$, $p = 1/4$, and $Y = X^2$, we have

$$g(y) = \binom{3}{\sqrt{y}} \left(\frac{1}{4}\right)^{\sqrt{y}} \left(\frac{3}{4}\right)^{3-\sqrt{y}}, \qquad y = 0, 1, 4, 9.$$

For a more interesting problem with the binomial random variable X, suppose we were to search for a transformation $u(X/n)$ of the relative frequency X/n that would have a variance very little dependent on p itself when n is large. That is, we want the variance of $u(X/n)$ to be essentially a constant.

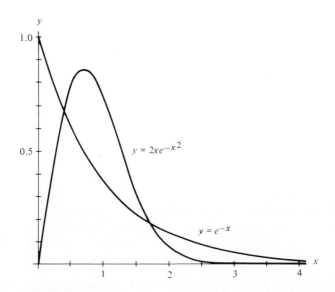

$$y = 2xe^{-x^2}$$

$$y = e^{-x}$$

FIGURE 5.9-1

Consider the function $u(X/n)$ and find, using two terms of Taylor's expansion about p, that

$$u\left(\frac{X}{n}\right) \approx u(p) + u'(p)\left(\frac{X}{n} - p\right).$$

Here terms of higher powers can be disregarded if n is large enough so that X/n is close enough to p. Thus

$$\operatorname{Var}\left[u\left(\frac{X}{n}\right)\right] \approx [u'(p)]^2 \operatorname{Var}\left(\frac{X}{n} - p\right) = [u'(p)]^2 \frac{p(1-p)}{n}. \qquad (5.9\text{-}1)$$

However, if $\operatorname{Var}[u(X/n)]$ is to be constant with respect to p, then

$$[u'(p)]^2 p(1-p) = k \qquad \text{or} \qquad u'(p) = \frac{c}{\sqrt{p(1-p)}},$$

where k and c are constants. We see that

$$u(p) = 2c \operatorname{Arcsin} \sqrt{p}$$

is a solution to this differential equation. Thus with $c = 1/2$, we frequently see, in the literature, use of the arcsine transformation, namely

$$Y = \operatorname{Arcsin} \sqrt{\frac{X}{n}},$$

which, with large n, has an approximate normal distribution with mean

$$\mu = \operatorname{Arcsin} \sqrt{p}$$

and variance [using formula (5.9-1)]

$$\sigma^2 = [D_p(\operatorname{Arcsin} \sqrt{p})]^2 \frac{p(1-p)}{n} = \frac{1}{4n}.$$

There should be one note of warning here: If the function $Y = u(X)$ does not have a single-valued inverse, the determination of the distribution of Y will not be as simple. We did consider one such example in Section 4.4 by finding the distribution of Z^2, where Z is $N(0, 1)$. In this case, there were two inverse functions and special care was exercised. In our examples, we will *not* consider problems with many inverses; however, we thought that such a warning should be issued here.

When two or more random variables are involved, many interesting problems can result. In the case of a single-valued inverse, the rule is about the same as that in the one-variable case, with the derivative being replaced by the Jacobian. Namely, if X_1 and X_2 are two continuous-type random variables with joint p.d.f. $f(x_1, x_2)$ and if $Y_1 = u_1(X_1, X_2)$, $Y_2 = u_2(X_1, X_2)$ has the single-valued inverse $X_1 = v_1(Y_1, Y_2)$, $X_2 = v_2(Y_1, Y_2)$, then the joint p.d.f. of Y_1 and Y_2 is

$$g(y_1, y_2) = |J| \, f[v_1(y_1, y_2), v_2(y_1, y_2)],$$

where the Jacobian J is the determinant

$$J = \begin{vmatrix} \dfrac{\partial x_1}{\partial y_1} & \dfrac{\partial x_1}{\partial y_2} \\[2ex] \dfrac{\partial x_2}{\partial y_1} & \dfrac{\partial x_2}{\partial y_2} \end{vmatrix}.$$

Of course, we find the support of Y_1, Y_2 by considering the mapping of the support of X_1, X_2 under the transformation $y_1 = u_1(x_1, x_2)$, $y_2 = u_2(x_1, x_2)$.

Example 5.9-3 Let X_1 and X_2 have independent gamma distributions with parameters α, θ and β, θ, respectively. That is, the joint p.d.f. of X_1 and X_2 is

$$f(x_1, x_2) = \frac{1}{\Gamma(\alpha)\Gamma(\beta)\theta^{\alpha+\beta}} x_1^{\alpha-1} x_2^{\beta-1} \exp\left(-\frac{x_1 + x_2}{\theta}\right),$$

$$0 < x_1 < \infty, \quad 0 < x_2 < \infty.$$

Consider

$$Y_1 = \frac{X_1}{X_1 + X_2}, \qquad Y_2 = X_1 + X_2$$

or, equivalently,

$$X_1 = Y_1 Y_2, \qquad X_2 = Y_2 - Y_1 Y_2.$$

The Jacobian is

$$J = \begin{vmatrix} y_2 & y_1 \\ -y_2 & 1 - y_1 \end{vmatrix} = y_2(1 - y_1) + y_1 y_2 = y_2.$$

Thus the joint p.d.f. $g(y_1, y_2)$ of Y_1 and Y_2 is

$$g(y_1, y_2) = |y_2| \frac{1}{\Gamma(\alpha)\Gamma(\beta)\theta^{\alpha+\beta}} (y_1 y_2)^{\alpha-1}(y_2 - y_1 y_2)^{\beta-1} e^{-y_2/\theta},$$

where the support is $0 < y_1 < 1$, $0 < y_2 < \infty$. The marginal p.d.f. of Y_1 is

$$g_1(y_1) = \frac{y_1^{\alpha-1}(1 - y_1)^{\beta-1}}{\Gamma(\alpha)\Gamma(\beta)} \int_0^\infty \frac{y_2^{\alpha+\beta-1}}{\theta^{\alpha+\beta}} e^{-y_2/\theta} \, dy_2.$$

But the integral in this expression is that of a gamma p.d.f. with parameters $\alpha + \beta$ and θ, except for $\Gamma(\alpha + \beta)$ in the denominator; hence the integral equals $\Gamma(\alpha + \beta)$ and

$$g_1(y_1) = \frac{\Gamma(\alpha + \beta)}{\Gamma(\alpha)\Gamma(\beta)} y_1^{\alpha-1}(1 - y_1)^{\beta-1}, \qquad 0 < y_1 < 1.$$

We say that Y_1 has a beta p.d.f. with parameters α and β (see Figure 5.9-2).

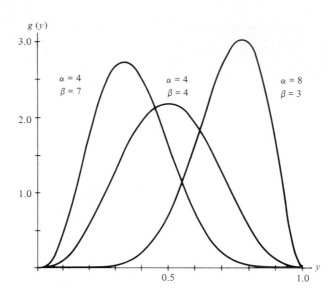

FIGURE 5.9-2

Example 5.9-4   (Box-Muller Transformation)   Let the distribution of X be $N(\mu, \sigma^2)$. It is not easy to simulate an observation of X using $Y = F(X)$, where Y is uniform $U(0, 1)$ and F is the distribution function of the desired normal distribution, because we cannot express the normal distribution function $F(x)$ in closed form. Consider the following transformation, however, where X_1 and X_2 are observations of a random sample from $U(0, 1)$:

$$Z_1 = \sqrt{-2 \ln X_1} \, \cos(2\pi X_2), \qquad Z_2 = \sqrt{-2 \ln X_1} \, \sin(2\pi X_2)$$

or, equivalently,

$$X_1 = \exp\left(-\frac{Z_1^2 + Z_2^2}{2}\right) = e^{-q/2}, \qquad X_2 = \frac{1}{2\pi} \operatorname{Arctan}\left(\frac{Z_2}{Z_1}\right),$$

which has Jacobian

$$J = \begin{vmatrix} -z_1 e^{-q/2} & -z_2 e^{-q/2} \\ \dfrac{-z_2}{2\pi(z_1^2 + z_2^2)} & \dfrac{z_1}{2\pi(z_1^2 + z_2^2)} \end{vmatrix} = \frac{-1}{2\pi} e^{-q/2}.$$

Since the joint p.d.f. of X_1 and X_2 is

$$f(x_1, x_2) = 1, \qquad 0 < x_1 < 1, \quad 0 < x_2 < 1,$$

we have that the joint p.d.f. of Z_1 and Z_2 is

$$g(z_1, z_2) = \left| \frac{-1}{2\pi} e^{-q/2} \right| \quad (1)$$

$$= \frac{1}{2\pi} \exp\left(-\frac{z_1^2 + z_2^2}{2}\right), \qquad -\infty < z_1 < \infty, \quad -\infty < z_2 < \infty.$$

(The student should note that there is some difficulty with the definition of the transformation, particularly when $Z_1 = 0$. However, these difficulties occur at events with probability zero and hence cause no problems; see Exercise 5.9-12.) To summarize, from two independent $U(0, 1)$ random variables, we have generated two independent $N(0, 1)$ random variables through this **Box-Muller transformation.**

The techniques described for two random variables can be extended to three or more random variables. We do not give any details here but mention, for illustration, that with three random variables X_1, X_2, X_3 of the continuous type, we need three "new" random variables Z_1, Z_2, Z_3 so that the corresponding Jacobian of the single-valued inverse transformation is the nonzero determinant

$$J = \begin{vmatrix} \dfrac{\partial x_1}{\partial z_1} & \dfrac{\partial x_1}{\partial z_2} & \dfrac{\partial x_1}{\partial z_3} \\[2mm] \dfrac{\partial x_2}{\partial z_1} & \dfrac{\partial x_2}{\partial z_2} & \dfrac{\partial x_2}{\partial z_3} \\[2mm] \dfrac{\partial x_3}{\partial z_1} & \dfrac{\partial x_3}{\partial z_2} & \dfrac{\partial 3_3}{\partial 3_3} \end{vmatrix}.$$

Exercises

5.9-1 Let the p.d.f. of X be defined by $f(x) = x^3/4$, $0 < x < 2$. Find the p.d.f. of $Y = X^2$.

5.9-2 Let the p.d.f. of X be defined $f(x) = 1/\pi$, $-\pi/2 < x < \pi/2$. Find the p.d.f. of $Y = \tan X$. We say that Y has a Cauchy distribution.

5.9-3 Let the p.d.f. of X be defined by $f(x) = (3/2)x^2$, $-1 < x < 1$. Find the p.d.f. of $Y = (X^3 + 1)/2$.

5.9-4 If Y has a uniform distribution on the interval $(0, 1)$, find the p.d.f. of $X = (2Y - 1)^{1/3}$.

5.9-5 Let X_1, X_2 denote a random sample from a distribution $\chi^2(2)$. Find the joint p.d.f. of $Y_1 = X_1$ and $Y_2 = X_2 + X_1$. Here note that the support of Y_1, Y_2 is $0 < y_1 < y_2 < \infty$. Also find the marginal p.d.f. of each of Y_1 and Y_2. Are Y_1 and Y_2 independent?

5.9-6 Let Z_1, Z_2 be a random sample from the standard normal distribution $N(0, 1)$. Use the transformation defined by polar coordinates $Z_1 = X_1 \cos X_2$, $Z_2 = X_1 \sin X_2$.
(a) Show that the Jacobian equals x_1. (This explains the factor r of $r\,dr\,d\theta$ in the usual polar coordinate notation.)

(b) Find the joint p.d.f. of X_1 and X_2.

(c) Are X_1 and X_2 independent?

5.9-7 Let the independent random variables X_1 and X_2 be $N(0, 1)$ and $\chi^2(r)$, respectively. Let $Y_1 = X_1/\sqrt{X_2/r}$ and $Y_2 = X_2$.

(a) Find the joint p.d.f. of Y_1 and Y_2.

(b) Determine the marginal p.d.f. of Y_1 and show that Y_1 has a Student's t distribution.

5.9-8 Let X_1 and X_2 be independent chi-square random variables with r_1 and r_2 degrees of freedom, respectively. Let $Y_1 = (X_1/r_1)/(X_2/r_2)$ and $Y_2 = X_2$.

(a) Find the joint p.d.f. of Y_1 and Y_2.

(b) Determine the marginal p.d.f. of Y_1 and show that Y_1 has an F distribution.

5.9-9 Let X have a Poisson distribution with mean λ. Find a transformation $u(X)$ so that Var $[u(X)]$ is about free of λ, for large values of λ.

HINT: $u(X) \approx u(\lambda) + [u'(\lambda)](X - \lambda)$, provided $[u''(\lambda)](X - \lambda)^2/2$ and higher terms can be neglected when λ is large. A solution is $u(X) = \sqrt{X}$, and the latter restriction should be checked.

5.9-10 Generalize Exercise 5.9-9 by assuming that the variance of a distribution is equal to $c\mu^p$, where c is a constant (note in Exercise 5.9-9 that this is the case with $p = 1$). In particular, the transformation $Y = u(X) = X^{1-p/2}$, when $p \neq 2$, or $Y = u(X) = \ln X$, when $p = 2$, seems to produce a random variable Y whose variance is about free of μ. This is the reason transformations like X, $\ln X$, and more generally X^b are so popular in applications.

5.9-11 Let Z_1 and Z_2 be independent standard normal random variables $N(0, 1)$. Show that $Y_1 = Z_1/Z_2$ has a Cauchy distribution. Here, let $Y_2 = Z_2$.

5.9-12 In Example 5.9-4 verify that the given transformation maps $\{(x_1, x_2):0 < x_1 < 1, 0 < x_2 < 1\}$ onto $\{(z_1, z_2): -\infty < z_1 < \infty, -\infty < z_2 < \infty\}$ except for a set of points that has probability 0.

HINT: What is the image of vertical line segments? What is the image of horizontal line segments?

5.9-13 Let X_1 and X_2 be independent chi-square random variables with r_1 and r_2 degrees of freedom, respectively. Show that

(a) $U = X_1/(X_1 + X_2)$ has a beta distribution with $\alpha = r_1/2$ and $\beta = r_2/2$.

(b) $V = X_2/(X_1 + X_2)$ has a beta distribution with $\alpha = r_2/2$ and $\beta = r_1/2$.

5.9-14 (a) Let X have a beta distribution with parameters α and β (see Example 5.9-3). Show that the mean and variance of X are

$$\mu = \frac{\alpha}{\alpha + \beta} \quad \text{and} \quad \sigma^2 = \frac{\alpha\beta}{(\alpha + \beta + 1)(\alpha + \beta)^2}.$$

(b) Show that the beta p.d.f. has a mode at $x = (\alpha - 1)/(\alpha + \beta - 2)$.

5.9-15 Determine the constant c such that $f(x) = cx^3(1 - x)^6$, $0 < x < 1$, is a p.d.f.

5.9-16 When α and β are integers and $0 < p < 1$, then

$$\int_0^p \frac{\Gamma(\alpha + \beta)}{\Gamma(\alpha)\Gamma(\beta)}\, y^{\alpha-1}(1 - y)^{\beta-1}\, dy = \sum_{y=\alpha}^n \binom{n}{y} p^y(1 - p)^{n-y},$$

where $n = \alpha + \beta - 1$. Verify this formula when $\alpha = 4$ and $\beta = 3$.

5.9-17 Evaluate

$$\int_0^{0.4} \frac{\Gamma(7)}{\Gamma(4)\Gamma(3)}\, y^3(1 - y)^2\, dy$$

(a) Using integration.

(b) Using the result of Exercise 5.9-16.

6

Estimation

6.1 Properties of Estimators

In earlier chapters we have alluded to estimating characteristics of the distribution from the corresponding ones of the sample, hoping that the latter would be reasonably close to the former. For example, the sample mean $\bar{x}$ can be thought of as an estimate of the distribution mean μ, and the sample variance s^2 can be used as an estimate of the distribution variance σ^2. Even the relative frequency histogram associated with a sample can be taken as an estimate of the p.d.f. of the underlying distribution. But how good are these estimates? What makes an estimate good? Can we say anything about the closeness of an estimate to an unknown parameter?

In this chapter we consider random variables for which the functional form of the p.d.f. is known, but the distribution depends on an unknown parameter, say θ, that may have any value in a set, say Ω, which is called the **parameter space**. For example, perhaps it is known that $f(x; \theta) = (1/\theta)e^{-x/\theta}$, $0 < x < \infty$ and that $\theta \in \Omega = \{\theta: 0 < \theta < \infty\}$. In certain instances, it might be necessary for the experimenter to select precisely one member of the family $\{f(x, \theta), \theta \in \Omega\}$ as the most likely p.d.f. of the random variable. That is, the experimenter needs a point estimate of the parameter θ, namely the value of the parameter that corresponds to the selected p.d.f.

In estimation we take a random sample from the distribution to elicit some information about the unknown parameter θ. That is, we repeat the experiment n independent times, observe the sample, $X_1, X_2, \ldots, X_n$, and try to guess the value of θ using the observations $x_1, x_2, \ldots, x_n$. The function of $X_1, X_2, \ldots, X_n$ used to

334

guess θ, say the statistic $u(X_1, X_2, \ldots, X_n)$, is called an **estimator** of θ. We want it to be such that the computed **estimate** $u(x_1, x_2, \ldots, x_n)$ is usually close to θ.

Let $Y = u(X_1, X_2, \ldots, X_n)$ be an estimator of θ. What properties should Y enjoy to be a good estimator of θ? Note that Y is a random variable and thus has a probability distribution. If Y is to be a good estimator of θ, a very desirable property is that its mean be equal to θ, namely, $E(Y) = \theta$.

DEFINITION 6.1-1 *If $E[u(X_1, X_2, \ldots, X_n)] = \theta$, the statistic $u(X_1, X_2, \ldots, X_n)$ is called an **unbiased estimator** of θ. Otherwise, it is said to be **biased**.*

Example 6.1-1 Let $X_1, X_2, \ldots, X_n$ be a random sample of size n from a normal distribution $N(\mu, \sigma^2)$. It seems reasonable to use the sample mean $\overline{X}$ to estimate the distribution mean μ. Recalling that the distribution of $\overline{X}$ is $N(\mu, \sigma^2/n)$, we see that $E(\overline{X}) = \mu$ and thus $\overline{X}$ is an unbiased estimator of μ.

In Theorem 5.3-4 we showed that the distribution of $(n - 1)S^2/\sigma^2$ is $\chi^2(n-1)$. Thus

$$E(S^2) = E\left[\frac{\sigma^2}{n - 1} \frac{(n - 1)S^2}{\sigma^2} \right] = \frac{\sigma^2}{n - 1}(n - 1) = \sigma^2.$$

That is, the sample variance

$$S^2 = \frac{1}{n - 1} \sum_{i=1}^{n} (X_i - \overline{X})^2$$

is an unbiased estimator of σ^2.

In Example 6.1-1 we showed that $\overline{X}$ and S^2 are unbiased estimators of μ and σ^2, respectively, when sampling from a normal distribution. This is also true when sampling from any population with finite variance σ^2. That is, $E(\overline{X}) = \mu$ and $E(S^2) = \sigma^2$, provided that the sample arises from a distribution with variance $\sigma^2 < \infty$ (see Exercise 6.1-4). Although S^2 is an unbiased estimator of σ^2, S is a biased estimator of σ. In Exercise 5.8-4, you were asked to show that, when sampling from a normal distribution, cS is an unbiased estimator for σ where $c = \sqrt{n - 1}\Gamma[(n - 1)/2]/\sqrt{2}\Gamma(n/2)$. This would be an appropriate time to work that exercise if you didn't do it earlier.

Not only do we want an estimator whose expectation is about equal to θ, but we would also like the variance of the estimator to be as small as possible. If we have, for example, two unbiased estimators of θ, we would probably choose the one with the smaller variance. In general, with a random sample $X_1, X_2, \ldots, X_n$ of a fixed sample size n, a statistician might like to find that estimator $Y = u(X_1, X_2, \ldots, X_n)$ of an unknown parameter θ which minimizes the mean (expected) value of the square of the error (difference) $Y - \theta$; that is, minimizes

$$E[(Y - \theta)^2] = E\{[u(X_1, X_2, \ldots, X_n) - \theta]^2\}.$$

DEFINITION 6.1-2 *The statistic Y that minimizes $E[(Y - \theta)^2]$ is the one with* **minimum mean square error**. *If we restrict our attention to unbiased estimators only, then $\text{Var}(Y) = E[(Y - \theta)^2]$, and the unbiased statistic Y that minimizes this expression is said to be the* **unbiased minimum variance estimator** *of θ.*

Two illustrations will help reinforce these new ideas.

Example 6.1-2 Say that our observations are restricted to a random sample X_1, X_2, X_3 of size $n = 3$ from a distribution with unknown mean μ, $-\infty < \mu < \infty$, where the variance σ^2 is a known positive number. Of course, $\overline{X}$ is unbiased and has variance $\sigma^2/3$. Let us, however, consider another possible estimator, namely

$$Y = \frac{X_1 + 2X_2 + 3X_3}{6},$$

which is the weighted average of the three sample items. It is possible that some statistician might like a statistic such as Y because it places the most weight on the last observation, namely X_3. Note that

$$E(Y) = \frac{1}{6}E(X_1 + 2X_2 + 3X_3) = \frac{1}{6}(\mu + 2\mu + 3\mu) = \mu$$

and

$$\text{Var}(Y) = \left(\frac{1}{6}\right)^2\sigma^2 + \left(\frac{2}{6}\right)^2\sigma^2 + \left(\frac{3}{6}\right)^2\sigma^2 = \frac{7\sigma^2}{18}.$$

Thus we see that Y is also unbiased but that $\overline{X}$ is a better unbiased estimator of μ, since $\text{Var}(\overline{X}) < \text{Var}(Y)$.

Example 6.1-3 We have seen in Example 6.1-1 that the sample variance S^2 of the sample that arises from the normal distribution $N(\mu, \sigma^2)$, $-\infty < \mu < \infty$, $0 < \sigma^2 < \infty$, is an unbiased estimator of σ^2. The variance of the empirical distribution,

$$V = \frac{1}{n}\sum_{i=1}^{n}(X_i - \overline{X})^2,$$

is a biased estimator of σ^2 because

$$E(V) = E\left(\frac{n-1}{n}S^2\right) = \frac{n-1}{n}\sigma^2.$$

Suppose now that we wish to find a constant c such that $E[(cS^2 - \sigma^2)^2]$ is minimized. That is, we want to find the minimum mean square error estimator of σ^2, which is of the form cS^2. If we recall that $E(W^2) = \mu_W^2 + \sigma_W^2$ for every random variable (provided that these expectations exist), we have, with $W = cS^2 - \sigma^2$, that

$$E[(cS^2 - \sigma^2)^2] = [E(cS^2 - \sigma^2)]^2 + \text{Var}(cS^2 - \sigma^2).$$

Since $E(S^2) = \sigma^2$ and $\text{Var}(S^2) = 2\sigma^4/(n-1)$, we have that

$$E[(cS^2 - \sigma^2)^2] = (c\sigma^2 - \sigma^2)^2 + \frac{2c^2\sigma^4}{(n-1)}.$$

The derivative of this expression with respect to c equated to zero provides the equation

$$2(c\sigma^2 - \sigma^2)\sigma^2 + \frac{4\sigma^4 c}{(n-1)} = 0,$$

the solution of which is $c = (n-1)/(n+1)$. It is an easy exercise to verify that this value of c actually minimizes $E[(cS^2 - \sigma^2)^2]$. That is, the minimum mean square error estimator of σ^2 that is of the form cS^2 is

$$\frac{n-1}{n+1}S^2 = \frac{1}{n+1}\sum_{i=1}^{n}(X_i - \bar{X})^2.$$

We have given two properties that a good estimator should have. However, we have not yet indicated how to go about finding possible estimators for parameters. One of the oldest procedures for estimating parameters is the method of moments. We introduce this method now with an example. In Section 6.6 we present another method for finding an estimator of an unknown parameter, namely the method of maximum likelihood.

Example 6.1-4 Let $X_1, X_2, \ldots, X_n$ be a random sample of size n from the distribution with p.d.f. $f(x; \theta) = \theta x^{\theta-1}$, $0 < x < 1$, $0 < \theta < \infty$. Sketch the graphs of this p.d.f. for $\theta = 1/4$, 1, and 4. Note that sets of observations for these three values of θ would look very different. However, how do we estimate the value of θ? The mean of this distribution is given by

$$E(X) = \int_0^1 x\theta x^{\theta-1}\,dx = \frac{\theta}{\theta+1}.$$

We shall set the population mean equal to the sample mean and solve for θ. We have

$$\bar{x} = \frac{\theta}{\theta+1}$$

or, solving for θ, we obtain the method of moments estimator, say

$$\tilde{\theta} = \frac{\bar{X}}{1-\bar{X}}.$$

Thus an estimate of θ by the method of moments is $\bar{x}/(1-\bar{x})$.

In general, in the **method of moments**, if there are k parameters that have to be estimated, the first k sample moments are set equal to the first k population moments that are given in terms of the unknown parameters. These k equations are then solved simultaneously for the unknown parameters.

Example 6.1-5   Let the distribution of X be $N(\mu, \sigma^2)$. Then

$$E(X) = \mu \quad \text{and} \quad E(X^2) = \sigma^2 + \mu^2.$$

Given a random sample of size n, the first two moments are given by

$$m_1 = \frac{1}{n} \sum_{i=1}^{n} x_i \quad \text{and} \quad m_2 = \frac{1}{n} \sum_{i=1}^{n} x_i^2.$$

We set $m_1 = E(X)$ and $m_2 = E(X^2)$ and solve for μ and σ^2. That is,

$$\frac{1}{n} \sum_{i=1}^{n} x_i = \mu \quad \text{and} \quad \frac{1}{n} \sum_{i=1}^{n} x_i^2 = \sigma^2 + \mu^2.$$

The first equation yields $\bar{x}$ as the estimate of μ. Replacing μ^2 with $\bar{x}^2$ in the second equation and solving for σ^2, we obtain

$$\frac{1}{n} \sum_{i=1}^{n} x_i^2 - \bar{x}^2 = v$$

for the solution of σ^2. Thus the method of moment estimators for μ and σ^2 are $\tilde{\mu} = \bar{X}$ and $\tilde{\sigma}^2 = V$. Of course, $\tilde{\mu} = \bar{X}$ is unbiased whereas $\tilde{\sigma}^2 = V$ is biased.

With two different estimators, say $\hat{\theta}$ and $\tilde{\theta}$, for a parameter θ, which one do we use? Most statisticians select the one that has smallest mean square error; that is, if, for example,

$$E[(\hat{\theta} - \theta)^2] < E[(\tilde{\theta} - \theta)^2],$$

then $\hat{\theta}$ seems to be preferred. This means that if $E(\hat{\theta}) = E(\tilde{\theta}) = \theta$, then we would select the one with the smallest variance.

We have here given some properties of a good estimator, and there are more listed in Sections 6.7 and 9.7. Moreover, we have also given one method for finding an estimator, namely the method of moments. But given an estimate for a parameter, how accurate is the estimate? That is, how confident are we about the closeness of the estimate to the unknown parameter? We now begin to answer these questions.

Given a random sample $X_1, X_2, \ldots, X_n$ from a normal distribution $N(\mu, \sigma^2)$, we shall now consider the closeness of $\bar{X}$, the unbiased estimator of μ, to the unknown μ. To do this, we use the error structure (distribution) of $\bar{X}$, namely that $\bar{X}$ is $N(\mu, \sigma^2/n)$, to construct what is called a confidence interval for the unknown parameter μ, when the variance σ^2 is known. For the probability $1 - \alpha$, we can find a number $z_{\alpha/2}$ from Table V in the Appendix such that

$$P\left(-z_{\alpha/2} \leq \frac{\overline{X} - \mu}{\sigma/\sqrt{n}} \leq z_{\alpha/2}\right) = 1 - \alpha.$$

For example, if $1 - \alpha = 0.95$, then $z_{\alpha/2} = z_{0.025} = 1.96$ and if $1 - \alpha = 0.90$, then $z_{\alpha/2} = z_{0.05} = 1.645$. Now recalling that $\sigma > 0$, we see that the following inequalities are equivalent:

$$-z_{\alpha/2} \leq \frac{\overline{X} - \mu}{\sigma/\sqrt{n}} \leq z_{\alpha/2},$$

$$-z_{\alpha/2}\left(\frac{\sigma}{\sqrt{n}}\right) \leq \overline{X} - \mu \leq z_{\alpha/2}\left(\frac{\sigma}{\sqrt{n}}\right),$$

$$-\overline{X} - z_{\alpha/2}\left(\frac{\sigma}{\sqrt{n}}\right) \leq -\mu \leq -\overline{X} + z_{\alpha/2}\left(\frac{\sigma}{\sqrt{n}}\right),$$

and

$$\overline{X} + z_{\alpha/2}\left(\frac{\sigma}{\sqrt{n}}\right) \geq \mu \geq \overline{X} - z_{\alpha/2}\left(\frac{\sigma}{\sqrt{n}}\right).$$

Thus, since the probability of the first of these is $1 - \alpha$, the probability of the last must also be $1 - \alpha$ because the latter is true if and only if the former is true. That is, we have

$$P\left[\overline{X} - z_{\alpha/2}\left(\frac{\sigma}{\sqrt{n}}\right) \leq \mu \leq \overline{X} + z_{\alpha/2}\left(\frac{\sigma}{\sqrt{n}}\right)\right] = 1 - \alpha.$$

So, the probability that the random interval

$$\left[\overline{X} - z_{\alpha/2}\left(\frac{\sigma}{\sqrt{n}}\right), \overline{X} + z_{\alpha/2}\left(\frac{\sigma}{\sqrt{n}}\right)\right]$$

includes the unknown mean μ is $1 - \alpha$.

Once the sample is observed and the sample mean computed to equal $\overline{x}$, the interval $[\overline{x} - z_{\alpha/2}(\sigma/\sqrt{n}), \overline{x} + z_{\alpha/2}(\sigma/\sqrt{n})]$ is a known interval. Since the probability that the *random* interval covers μ before the sample is drawn is equal to $1 - \alpha$, we now call the computed interval, $\overline{x} \pm z_{\alpha/2}(\sigma/\sqrt{n})$ (for brevity), a $100(1 - \alpha)\%$ **confidence interval** for the unknown mean μ. For illustration, $\overline{x} \pm 1.96(\sigma/\sqrt{n})$ is a 95% confidence interval for μ. The number $100(1 - \alpha)\%$, or equivalently, $1 - \alpha$, is called the **confidence coefficient**.

We see that the confidence interval for μ is centered at the point estimate $\overline{x}$ and is completed by subtracting and adding the quantity $z_{\alpha/2}\sigma/\sqrt{n}$.

Note that as n increases, $z_{\alpha/2}\sigma/\sqrt{n}$ decreases, resulting in a shorter confidence interval with the same confidence coefficient $1 - \alpha$. A shorter confidence interval indicates that we have more reliance in $\bar{x}$ as an estimate of μ. Statisticians who are not restricted by time, money, effort, or availability of observations can obviously make the confidence interval as short as they like by increasing the sample size n. For a fixed sample size n, the length of the confidence interval can also be shortened by decreasing the confidence coefficient $1 - \alpha$. But if this is done, we achieve a shorter confidence interval by losing some confidence.

Example 6.1-6 Let X equal the length of life of a 60-watt light bulb marketed by a certain manufacturer of light bulbs. Assume that the distribution of X is $N(\mu, 1296)$. If a random sample of $n = 27$ bulbs were tested until they burned out, yielding a sample mean of $\bar{x} = 1478$ hours, then a 95% confidence interval for μ is

$$\left[\bar{x} - z_{0.025}\left(\frac{\sigma}{\sqrt{n}}\right), \bar{x} + z_{0.025}\left(\frac{\sigma}{\sqrt{n}}\right)\right]$$

$$= \left[1478 - 1.96\left(\frac{36}{\sqrt{27}}\right), 1478 + 1.96\left(\frac{36}{\sqrt{27}}\right)\right]$$

$$= [1478 - 13.58, 1478 + 13.58]$$

$$= [1464.42, 1491.58].$$

The next example will help to give a better intuitive feeling for the interpretation of a confidence interval.

Example 6.1-7 Let $\bar{x}$ be the observed sample mean of 16 items of a random sample from the normal distribution $N(\mu, 23.04)$. A 90% confidence interval for the unknown mean μ is

$$\left[\bar{x} - 1.645\sqrt{\frac{23.04}{16}}, \bar{x} + 1.645\sqrt{\frac{23.04}{16}}\right].$$

For a particular sample this interval either does or does not contain the mean μ. However, if many such intervals were calculated, it should be true that about 90% of them contain the mean μ. Fifteen random samples of size 16 from the normal distribution $N(5, 23.04)$ were simulated on a computer. A 90% confidence interval was calculated for each random sample, as if the mean were unknown. Figure 6.1-1 lists these 15 intervals and depicts each of them as a line segment. For the 15 intervals, 13 (or 86.7%) of them contain the mean.

If we cannot assume that the distribution from which the sample arose is normal, we can still obtain an approximate confidence interval for μ. By the Central Limit Theorem the ratio $(\bar{X} - \mu)/(\sigma/\sqrt{n})$ has, provided that n is large enough, the approximate normal distribution $N(0, 1)$ when the underlying distribution is not normal. In this case

$$P\left(-z_{\alpha/2} \le \frac{\overline{X} - \mu}{\sigma/\sqrt{n}} \le z_{\alpha/2}\right) \approx 1 - \alpha,$$

and

$$\left[\overline{x} - z_{\alpha/2}\left(\frac{\sigma}{\sqrt{n}}\right), \overline{x} + z_{\alpha/2}\left(\frac{\sigma}{\sqrt{n}}\right)\right]$$

is an approximate $100(1 - \alpha)\%$ confidence interval for μ.

The closeness of the approximate probability $1 - \alpha$ to the exact probability depends on both the underlying distribution and the sample size. When the underlying distribution is unimodal (has only one mode) and continuous, the approximation is usually quite good for even small n, such as $n = 5$. As the underlying distribution becomes "less normal" (i.e., badly skewed or discrete), a larger sample size might be required to keep a reasonably accurate approximation. But, in all cases, an n of at least 30 is usually quite adequate.

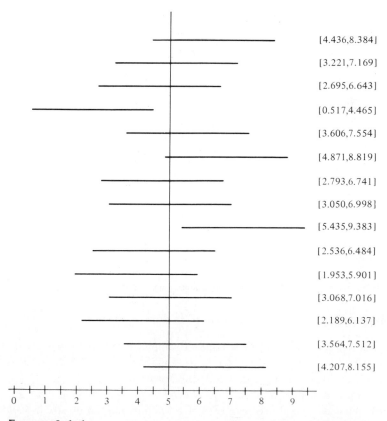

FIGURE 6.1-1

Example 6.1-8 Let X equal the amount of orange juice (in grams per day) consumed by an American. Suppose it is known that the standard deviation of X is $\sigma = 96$. To estimate the mean μ of X, an orange growers' association took a random sample of $n = 576$ Americans and found that they consumed, on the average, $\bar{x} = 133$ grams of orange juice per day. Thus an approximate 90% confidence interval for μ is

$$133 \pm 1.645\left(\frac{96}{\sqrt{576}}\right) \quad \text{or} \quad [133 - 6.58, \ 133 + 6.58] = [126.42, \ 139.58].$$

If σ^2 is unknown and the sample size n is 30 or greater, we shall use the fact that the ratio $(\bar{X} - \mu)/(S/\sqrt{n})$ has an approximate normal distribution $N(0, 1)$. This statement is true whether or not the underlying distribution is normal. However, if the underlying distribution is badly skewed or contaminated with occasional outliers, most statisticians would prefer to have a larger sample size, say 50 or more, and even that might not produce good results. In the next section we consider what to do when n is small.

Example 6.1-9 Lake Macatawa, an inlet lake on the east side of Lake Michigan, is divided into an east basin and a west basin. To measure the effect on the lake of salting city streets in the winter, students took 32 samples of water from the west basin and measured the amount of sodium in parts per million in order to make a statistical inference about the unknown mean μ. They obtained the following data:

13.0	18.5	16.4	14.8	19.4
17.3	23.2	24.9	20.8	19.3
18.8	23.1	15.2	19.9	19.1
18.1	25.1	16.8	20.4	17.4
25.2	23.1	15.3	19.4	16.0
21.7	15.2	21.3	21.5	16.8
15.6	17.6			

For these data $\bar{x} = 19.07$ and $s^2 = 10.60$. Thus an approximate 95% confidence interval for μ is

$$\bar{x} \pm 1.96\left(\frac{s}{\sqrt{n}}\right) \quad \text{or} \quad 19.07 \pm 1.96\sqrt{\frac{10.60}{32}} \quad \text{or} \quad [17.94, \ 20.20].$$

Exercises

6.1-1 Let $X_1, X_2, \ldots, X_n$ be a random sample of size n from the exponential distribution whose p.d.f. is $f(x; \theta) = (1/\theta)e^{-x/\theta}$, $0 < x < \infty$, $0 < \theta < \infty$.
(a) Show that $\bar{X}$ is an unbiased estimator of θ.
(b) Show that the variance of $\bar{X}$ is θ^2/n.
(c) What is a good estimate of θ if a random sample of size 5 yielded the sample values 3.5, 8.1, 0.9, 4.4, and 0.5?

6.1-2 Let $X_1, X_2, \ldots, X_n$ be a random sample of size n from a geometric distribution for which p is the probability of success.
(a) Use the method of moments to find a point estimate for p.
(b) Explain intuitively why your estimate makes good sense.
(c) Use the following data to give a point estimate of p:

$$
\begin{array}{cccccccccc}
3 & 34 & 7 & 4 & 19 & 2 & 1 & 19 & 43 & 2 \\
22 & 4 & 19 & 11 & 7 & 1 & 2 & 21 & 15 & 16
\end{array}
$$

6.1-3 Let $X_1, X_2, \ldots, X_n$ denote a random sample from $b(1, p)$. Thus

$$
Y = \sum_{i=1}^{n} X_i \quad \text{is} \quad b(n, p).
$$

(a) Show that $\overline{X} = Y/n$ is an unbiased estimator of p.
(b) Show that $\mathrm{Var}(\overline{X}) = p(1 - p)/n$.
(c) Show that $E[\overline{X}(1 - \overline{X})/n] = (n - 1)[p(1 - p)/n^2]$.
(d) Find the value of c so that $c\overline{X}(1 - \overline{X})$ is an unbiased estimator of
$p(1 - p)/n = \mathrm{Var}(\overline{X})$.

6.1-4 Let $X_1, X_2, \ldots, X_n$ be a random sample from a distribution having finite
variance σ^2. Show that

$$
S^2 = \sum_{i=1}^{n} \frac{(X_i - \overline{X})^2}{n - 1}
$$

is an unbiased estimator of σ^2.

HINT: Write

$$
S^2 = \frac{\displaystyle\sum_{i=1}^{n} X_i^2 - n\overline{X}^2}{n - 1}
$$

and compute $E(S^2)$.

6.1-5 Given a random sample of size n from a normal distribution, an unbiased
estimator for σ is cS, where $c = \sqrt{n - 1}\,\Gamma[(n - 1)/2]/\sqrt{2}\,\Gamma(n/2)$. Find the
value of c when
(a) $n = 5$.
(b) $n = 6$.

6.1-6 Let $X_1, X_2, \ldots, X_n$ be a random sample from a uniform distribution on the
interval $(\theta - 1, \theta + 1)$.
(a) Find the method of moments estimator for θ.
(b) Is your estimator in part (a) an unbiased estimator for θ?
(c) Given the following $n = 5$ observations of X, give a point estimate of θ:
6.61, 7.70, 6.98, 8.36, 7.26.

(d) The method of moments estimator actually has greater variance than the estimator $[\min(X_i) + \max(X_i)]/2$. Compute the value of this estimator for the $n = 5$ observations in (c).

6.1-7 Let $X_1, X_2, \ldots, X_n$ be a random sample of size n from a gamma distribution with mean $\mu = \alpha\theta$ and variance $\sigma^2 = \alpha\theta^2$. Use the method of moments to find estimates of α and θ.

6.1-8 Let $X_1, X_2, \ldots, X_n$ be a random sample of size n from $N(\mu, \sigma^2 = \theta)$, $0 < \theta < \infty$, where μ is known. Show that $Y = (1/n) \sum_{i=1}^{n} (X_i - \mu)^2$ is an unbiased estimator of θ.

6.1-9 Let $f(x; \theta) = \theta x^{\theta-1}$, $0 < x < 1$, $\theta \in \Omega = \{\theta: 0 < \theta < \infty\}$.
(a) For each of the following three sets of 10 observations from this distribution, calculate the value of the method of moments estimate for θ.
(b) For each set of data, sketch the empirical and theoretical distribution functions (using your estimate as the value of θ) on the same graph.

(i) 0.0256	0.3051	0.0278	0.8971	0.0739
0.3191	0.7379	0.3671	0.9763	0.0102
(ii) 0.9960	0.3125	0.4374	0.7464	0.8278
0.9518	0.9924	0.7112	0.2228	0.8609
(iii) 0.4698	0.3675	0.5991	0.9513	0.6049
0.9917	0.1551	0.0710	0.2110	0.2154

6.1-10 During each lecture in a statistics class, let X equal the number of times that Professor Tanis collides with a computer table at the front of the classroom. Assume that the distribution of X is Poisson with mean λ.
(a) Given n observations of X, find the method of moments estimate of λ.
(b) Give a point estimate of λ using the following 11 observations of X that were collected by Chris:

$$1 \quad 0 \quad 1 \quad 3 \quad 3 \quad 0 \quad 2 \quad 2 \quad 4 \quad 1 \quad 1$$

(c) Compare the values of $\bar{x}$ and s^2. Does this information support the assumption that X has a Poisson distribution?

6.1-11 A Geiger counter was set up in the physics laboratory to record the number of alpha particle emissions of carbon-14 in half a second. Ten observations yielded the following data:

$$4 \quad 6 \quad 9 \quad 6 \quad 10 \quad 11 \quad 6 \quad 3 \quad 7 \quad 10$$

Give the value of an unbiased estimate for λ, the mean number of counts per half second, assuming that these are observations of a Poisson random variable.

6.1-12 An urn contains 64 balls of which n_1 are orange and n_2 are blue. A random sample of $r = 8$ balls is selected from the urn without replacement and X is equal to the number of orange balls in the sample. This experiment was repeated 30 times (the 8 balls being returned to the urn before each repetition) yielding the following data:

3	0	0	1	1	1	1	3	1	1	2	0	1	3	1
0	1	0	2	1	1	2	3	2	2	4	3	1	1	2

Using these data, guess the value of n_1 and give a reason for your guess.

6.1-13 A random sample of size 16 from $N(\mu, 25)$ yielded $\bar{x} = 73.8$. Find a 95% confidence interval for μ.

6.1-14 A random sample of size 8 from $N(\mu, 72)$ yielded $\bar{x} = 85$. Find the following confidence intervals for μ.
  (a) 99%.   (b) 95%.   (c) 90%.   (d) 80%.

6.1-15 A pet store sells gerbil food in "2-pound" bags that are weighed on the platform of an old 25-pound scale. Suppose it is known that the standard deviation of weights is $\sigma = 0.12$ pound. If a sample of $n = 16$ bags of gerbil food were weighed carefully in a laboratory and the average weight was $\bar{x} = 2.09$ pounds, find an approximate 95% confidence interval for μ, the mean weight of gerbil food in the "2-pound" bags sold by the pet store.

6.1-16 Let X equal the weight in grams of a "52-gram" snack pack of candies. Assume that the distribution of X is $N(\mu, 4)$. A random sample of $n = 10$ observations of X yielded the following data:

$$55.95 \quad 56.54 \quad 57.58 \quad 55.13 \quad 57.48$$
$$56.06 \quad 59.93 \quad 58.30 \quad 52.57 \quad 58.46$$

(a) Give a point estimate for μ.
(b) Find the endpoints for a 95% confidence interval for μ.
(c) Based on these very limited data, what is the probability that an individual snack pack selected at random is filled with less than 52 grams of candy?

6.1-17 As a clue to the amount of organic waste in Lake Macatawa (see Example 6.1-9), a count was made of the number of bacteria colonies in 100 milliliters of water. The number of colonies, in hundreds, for $n = 30$ samples of water from the east basin yielded

93	140	8	120	3	120
33	70	91	61	7	100
19	98	110	23	14	94
57	9	66	53	28	76
58	9	73	49	37	92

Find an approximate 90% confidence interval for the mean number of colonies in 100 milliliters of water in the east basin, μ_E.

6.1-18 To determine whether bacteria count was lower in the west basin of Lake Macatawa (Exercise 6.1-17) than in the east basin, $n = 37$ samples of water were taken from the west basin, and the number of bacteria colonies in 100 milliliters of water was counted. The sample characteristics were $\bar{x} = 11.95$ and $s = 11.80$, measured in hundreds of colonies. Find the approximate 95% confidence interval for the mean number of colonies in 100 milliliters of water in the west basin, μ_W.

6.2   Confidence Intervals for Means

In the last section we found a confidence interval for the mean μ of a normal distribution, assuming that the value of the standard deviation σ is known. However, in most applications, we do not know the value of the standard deviation, although in some cases we might have a very good idea about its value. For illustration, a manufacturer of light bulbs probably has a good notion from past experience of the value of the standard deviation of the length of life of different types of light bulbs. But certainly, most of the time, the investigator will not have any more idea about the standard deviation than about the mean—and frequently less. Let us consider how to proceed under these circumstances.

Suppose that the underlying distribution is normal and that σ^2 is unknown. We showed in Section 5.7 that given a random sample $X_1, X_2, \ldots, X_n$ from a normal distribution, the statistic

$$T = \frac{\bar{X} - \mu}{S/\sqrt{n}}$$

has a t distribution with $r = n - 1$ degrees of freedom, where S^2 is the usual unbiased estimator of σ^2. Select $t_{\alpha/2}(n-1)$ so that $P[T \geq t_{\alpha/2}(n-1)] = \alpha/2$. Then

$$1 - \alpha = P\left[-t_{\alpha/2}(n-1) \leq \frac{\bar{X} - \mu}{S/\sqrt{n}} \leq t_{\alpha/2}(n-1) \right]$$

$$= P\left[-t_{\alpha/2}(n-1)\frac{S}{\sqrt{n}} \leq \bar{X} - \mu \leq t_{\alpha/2}(n-1)\frac{S}{\sqrt{n}} \right]$$

$$= P\left[-\bar{X} - t_{\alpha/2}(n-1)\frac{S}{\sqrt{n}} \leq -\mu \leq -\bar{X} + t_{\alpha/2}(n-1)\frac{S}{\sqrt{n}} \right]$$

$$= P\left[\bar{X} - t_{\alpha/2}(n-1)\frac{S}{\sqrt{n}} \leq \mu \leq \bar{X} + t_{\alpha/2}(n-1)\frac{S}{\sqrt{n}} \right].$$

Thus the observations of a random sample provide $\bar{x}$ and s^2 and

$$\left[\bar{x} - t_{\alpha/2}(n-1)\frac{s}{\sqrt{n}}, \bar{x} + t_{\alpha/2}(n-1)\frac{s}{\sqrt{n}} \right]$$

is a $100(1 - \alpha)\%$ confidence interval for μ.

Example 6.2-1   Let X equal the amount of butterfat in pounds produced by a typical cow during a 305-day milk production period between her first and second calves. Assume that the distribution of X is $N(\mu, \sigma^2)$. To estimate μ

a farmer measured the butterfat production for $n = 20$ cows yielding the following data:

$$
\begin{array}{ccccccc}
481 & 537 & 513 & 583 & 453 & 510 & 570 \\
500 & 457 & 555 & 618 & 327 & 350 & 643 \\
499 & 421 & 505 & 637 & 599 & 392 &
\end{array}
$$

For these data, $\bar{x} = 507.50$ and $s = 89.75$. Thus a point estimate of μ is $\bar{x} = 507.50$. Since $t_{0.05}(19) = 1.729$, a 90% confidence interval for μ is

$$
507.50 \pm 1.729 \left(\frac{89.75}{\sqrt{20}} \right),
$$

507.50 ± 34.70, or equivalently, $[472.80, 542.20]$.

Let T have a t distribution with $n - 1$ degrees of freedom. Then $t_{\alpha/2}(n-1) > z_{\alpha/2}$. Consequently, we would expect the interval $\bar{x} \pm z_{\alpha/2}\sigma/\sqrt{n}$ to be shorter than the interval $\bar{x} \pm t_{\alpha/2}(n-1)s/\sqrt{n}$. After all, we have more information, namely the value of σ, in constructing the first interval. However, the length of the second interval is very much dependent on the value of s. If the observed s is smaller than σ, a shorter confidence interval could result by the second scheme. But on the average, $\bar{x} \pm z_{\alpha/2}\sigma/\sqrt{n}$ is the shorter of the two confidence intervals (Exercise 6.2-9).

If we are not able to assume that the underlying distribution is normal but μ and σ are both unknown, approximate confidence intervals for μ can still be constructed using

$$
T = \frac{\bar{X} - \mu}{S/\sqrt{n}},
$$

which now only has an approximate t distribution. Generally, this approximation is quite good for many nonnormal distributions (i.e., it is robust), in particular, if the underlying distribution is symmetric, unimodal, and of the continuous type. However, if the distribution is highly skewed, there is great danger using this approximation. In such a situation, it would be safer to use certain nonparametric methods for finding a confidence interval for the median of the distribution, one of which is given in Section 10.2.

Suppose that we are interested in comparing the means of two independent normal distributions. Let $X_1, X_2, \ldots, X_n$ and $Y_1, Y_2, \ldots, Y_m$ be, respectively, two random samples of sizes n and m from the two independent normal distributions $N(\mu_X, \sigma_X^2)$ and $N(\mu_Y, \sigma_Y^2)$. Suppose, for now, that σ_X^2 and σ_Y^2 are known. If $\bar{X}$ and $\bar{Y}$ denote the respective sample means, we know that the distribution of $W = \bar{X} - \bar{Y}$ is $N(\mu_X - \mu_Y, \sigma_X^2/n + \sigma_Y^2/m)$. Thus

$$
P\left(-z_{\alpha/2} \leq \frac{(\bar{X} - \bar{Y}) - (\mu_X - \mu_Y)}{\sqrt{\sigma_X^2/n + \sigma_Y^2/m}} \leq z_{\alpha/2} \right) = 1 - \alpha,
$$

which can be rewritten as

$$
P[(\bar{X} - \bar{Y}) - z_{\alpha/2}\sigma_W \leq \mu_X - \mu_Y \leq (\bar{X} - \bar{Y}) + z_{\alpha/2}\sigma_W] = 1 - \alpha,
$$

where $\sigma_W = \sqrt{\sigma_X^2/n + \sigma_Y^2/m}$ is the standard deviation of $\overline{X} - \overline{Y}$. Once the experiments have been performed and the means $\overline{x}$ and $\overline{y}$ computed, then

$$[\overline{x} - \overline{y} - z_{\alpha/2}\sigma_W, \overline{x} - \overline{y} + z_{\alpha/2}\sigma_W]$$

or, equivalently, $\overline{x} - \overline{y} \pm z_{\alpha/2}\sigma_W$ provides a $100(1 - \alpha)\%$ confidence interval for $\mu_X - \mu_Y$. Note that this interval is centered at the point estimate $\overline{x} - \overline{y}$ of $\mu_X - \mu_Y$ and completed by subtracting and adding the product of $z_{\alpha/2}$ and the standard deviation of the point estimator.

> **Example 6.2-2** In the preceding discussion, let $n = 15$, $m = 8$, $\overline{x} = 70.1$, $\overline{y} = 75.3$, $\sigma_X^2 = 60$, $\sigma_Y^2 = 40$, and $1 - \alpha = 0.90$. Thus $1 - \alpha/2 = 0.95 = \Phi(1.645)$. Hence
>
> $$1.645\sigma_W = 1.645\sqrt{\frac{60}{15} + \frac{40}{8}} = 4.935,$$
>
> and, since $\overline{x} - \overline{y} = -5.2$, we have that
>
> $$[-5.2 - 4.935, -5.2 + 4.935] = [-10.135, -0.265]$$
>
> is a 90% confidence interval for $\mu_X - \mu_Y$. Since the confidence interval does not include zero, we suspect that μ_Y is greater than μ_X.

If the sample sizes are large, and σ_X^2 and σ_Y^2 are unknown, we can replace σ_X^2 and σ_Y^2 with s_x^2 and s_y^2, where s_x^2 and s_y^2 are the values of the respective unbiased estimates of the variances. This means that

$$\overline{x} - \overline{y} \pm z_{\alpha/2}\sqrt{\frac{s_x^2}{n} + \frac{s_y^2}{m}}$$

serves as an approximate $100(1 - \alpha)\%$ confidence interval for $\mu_X - \mu_Y$.

Now consider the problem of constructing confidence intervals for the difference of the means of two independent normal distributions when the variances are unknown but the sample sizes are small. Let $X_1, X_2, \ldots, X_n$ and $Y_1, Y_2, \ldots, Y_m$ be random samples from independent distributions $N(\mu_X, \sigma_X^2)$ and $N(\mu_Y, \sigma_Y^2)$, respectively. If the sample sizes are not large (say considerably smaller than 30), this problem can be a difficult one. However, even in these cases, if we can assume common, but unknown, variances, say $\sigma_X^2 = \sigma_Y^2 = \sigma^2$, there is a way out of our difficulty. In Section 5.6 we showed that

$$T = \frac{\overline{X} - \overline{Y} - (\mu_X - \mu_Y)}{\sqrt{\{[(n - 1)S_X^2 + (m - 1)S_Y^2]/(n + m - 2)\}[(1/n) + (1/m)]}}$$

has a t distribution with $r = n + m - 2$ degrees of freedom. Thus, with $t_0 = t_{\alpha/2}(n+m-2)$,

$$P(-t_0 \leq T \leq t_0) = 1 - \alpha.$$

Solving the inequality for $\mu_X - \mu_Y$ yields

$$P\left(\overline{X} - \overline{Y} - t_0 S_P\sqrt{\frac{1}{n} + \frac{1}{m}} \le \mu_X - \mu_Y \le \overline{X} - \overline{Y} + t_0 S_P\sqrt{\frac{1}{n} + \frac{1}{m}}\right) = 1 - \alpha$$

where the pooled estimator of the common standard deviation is

$$S_P = \sqrt{\frac{(n-1)S_X^2 + (m-1)S_Y^2}{n + m - 2}}.$$

If $\overline{x}$, $\overline{y}$, and s_p are the observed values of $\overline{X}$, $\overline{Y}$, and S_P, then

$$\left[\overline{x} - \overline{y} - t_0 s_p\sqrt{\frac{1}{n} + \frac{1}{m}}, \ \overline{x} - \overline{y} + t_0 s_p\sqrt{\frac{1}{n} + \frac{1}{m}}\right]$$

is a $100(1 - \alpha)\%$ confidence interval for $\mu_X - \mu_Y$.

> **Example 6.2-3** Suppose that scores on a standardized test in mathematics taken by students from large and small high schools are $N(\mu_X, \sigma^2)$ and $N(\mu_Y, \sigma^2)$, respectively, where σ^2 is unknown. If a random sample of $n = 9$ students from large high schools yielded $\overline{x} = 81.31$, $s_x^2 = 60.76$ and a random sample of $m = 15$ students from small high schools yielded $\overline{y} = 78.61$, $s_y^2 = 48.24$, the endpoints for a 95% confidence interval for $\mu_X - \mu_Y$ are given by
>
> $$81.31 - 78.61 \pm 2.074\sqrt{\frac{8(60.76) + 14(48.24)}{22}}\sqrt{\frac{1}{9} + \frac{1}{15}},$$
>
> because $t_{0.025}(22) = 2.074$. The 95% confidence interval is $[-3.65, 9.05]$.

The assumption of equal variances, namely $\sigma_X^2 = \sigma_Y^2$, can be modified somewhat so that we are still able to find a confidence interval for $\mu_X - \mu_Y$. That is, if we know the ratio σ_X^2/σ_Y^2 of the variances, we can still make this type of statistical inference using a random variable with a t distribution (see Exercise 6.2-17). However, if we do not know the ratio of the variances and yet suspect that the unknown σ_X^2 and σ_Y^2 differ by a great deal, what do we do? It is safest to return to

$$\frac{\overline{X} - \overline{Y} - (\mu_X - \mu_Y)}{\sqrt{\sigma_X^2/n + \sigma_Y^2/m}}$$

for the inference about $\mu_X - \mu_Y$, but replacing σ_X^2 and σ_Y^2 by their respective estimators S_X^2 and S_Y^2. That is, consider

$$W = \frac{\overline{X} - \overline{Y} - (\mu_X - \mu_Y)}{\sqrt{S_X^2/n + S_Y^2/m}}.$$

What is the distribution of W? As before, we note that if n and m are large enough and the underlying distributions are close to normal (or at least not badly skewed),

W has an approximate normal distribution and a confidence interval for $\mu_X - \mu_Y$ can be found by considering

$$P(-z_{\alpha/2} \leq W \leq z_{\alpha/2}) \approx 1 - \alpha.$$

However, if n and m are smaller, then Welch has proposed a Student-t distribution as the approximating one for W. Welch's proposal was later modified by Aspin. [See A. A. Aspin, "Tables for Use in Comparisons Whose Accuracy Involves Two Variances, Separately Estimated," *Biometrika*, **36** (1949), pp. 290–296, with an appendix by B. L. Welch in which he makes the suggestion used here.] The approximating Student-t distribution has r degrees of freedom where

$$\frac{1}{r} = \frac{c^2}{n-1} + \frac{(1-c)^2}{m-1} \quad \text{and} \quad c = \frac{s_x^2/n}{s_x^2/n + s_y^2/m}.$$

An equivalent formula for r is

$$r = \frac{\left(\dfrac{s_x^2}{n} + \dfrac{s_y^2}{m} \right)^2}{\dfrac{1}{n-1}\left(\dfrac{s_x^2}{n} \right)^2 + \dfrac{1}{m-1}\left(\dfrac{s_y^2}{m} \right)^2}.$$

In particular, the assignment of r by this rule provides protection in the case in which the smaller sample size is associated with the larger variance by greatly reducing the number of degrees of freedom from the usual $n + m - 2$. Of course, this increases the value of $t_{\alpha/2}$. If r is not an integer, then use the greatest integer in r; that is, $[r]$ is the number of degrees of freedom associated with the approximating Student-t distribution. An approximate $100(1 - \alpha)$ percent confidence interval for $\mu_X - \mu_Y$ is given by

$$\bar{x} - \bar{y} \pm t_{\alpha/2}(r)\sqrt{\frac{s_x^2}{n} + \frac{s_y^2}{m}}.$$

In some applications, two measurements, say X and Y, are taken on the same subject. In these cases, X and Y are dependent random variables. Many times these are "before" and "after" measurements, such as weight before and after participating in a diet and exercise program. To compare the means of X and Y, it is not permissible to use the t statistics and confidence intervals that we just developed, because there X and Y are independent. Instead we proceed as follows.

Let (X_1, Y_1), (X_2, Y_2), . . . , (X_n, Y_n) be n pairs of dependent measurements. Let $D_i = X_i - Y_i$, $i = 1, 2, \ldots, n$. Suppose that $D_1, D_2, \ldots, D_n$ could be thought of as a random sample from $N(\mu_D, \sigma_D^2)$, where μ_D and σ_D are the mean and standard deviation of each difference. To form a confidence interval for $\mu_D = \mu_X - \mu_Y$, use

$$T = \frac{\bar{D} - \mu_D}{S_D/\sqrt{n}},$$

where $\overline{D}$ and S_D are the sample mean and sample standard deviation of the n differences. Thus T is a t statistic with $n - 1$ degrees of freedom. The endpoints for a $100(1 - \alpha)\%$ confidence interval for $\mu_D = \mu_X - \mu_Y$ are then

$$\overline{d} \pm t_{\alpha/2}(n-1)\frac{s_d}{\sqrt{n}},$$

where $\overline{d}$ and s_d are the observed mean and standard deviation of the sample. Of course, this is the same as the first confidence interval for a single mean in this section.

Example 6.2-4 An experiment was conducted to compare people's reaction times to a red light versus a green light. When signaled with either the red or the green light, the subject was asked to hit a switch to turn off the light. When the switch was hit, a clock was turned off and the reaction time in seconds was recorded. The following results give the reaction times for eight subjects.

Subject	Red (X)	Green (Y)	$D = X - Y$
1	0.30	0.43	−0.13
2	0.23	0.32	−0.09
3	0.41	0.58	−0.17
4	0.53	0.46	0.07
5	0.24	0.27	−0.03
6	0.36	0.41	−0.05
7	0.38	0.38	0.00
8	0.51	0.61	−0.10

For these data, $\overline{d} = -0.0625$ and $s_d = 0.0765$. To form a 95% confidence interval for $\mu_D = \mu_X - \mu_Y$, we find from Appendix Table VI that $t_{0.025}(7) = 2.365$. Thus the endpoints for the confidence interval are

$$-0.0625 \pm 2.365\frac{0.0765}{\sqrt{8}} \quad \text{or} \quad [-0.1265, 0.0015].$$

In this very limited data set, zero is included in the confidence interval but is close to the endpoint 0.0015. We suspect that if more data were taken, zero might not be included in the confidence interval. Accordingly, if this actually happens, it would seem that people react faster to a red light.

Exercises

6.2-1 Let X equal the thickness of peppermint gum that is manufactured for vending machines. Assume that the distribution of X is $N(\mu, \sigma^2)$. The target thickness is 7.5 hundredths of an inch. The following $n = 10$ thicknesses, in hundredths of an inch, were made on pieces of gum that were selected randomly from the production line:

7.50   7.55   7.55   7.40   7.45   7.35   7.45   7.45   7.45   7.50.

(a) Give point estimates of μ and σ.
(b) Find a 95% confidence interval for μ.

6.2-2   Thirteen tons of cheese is stored in some old gypsum mines, including "22-pound" wheels (label weight). A random sample of $n = 9$ of these wheels yielded the following weights in pounds:

$$21.50 \quad 18.95 \quad 18.55 \quad 19.40 \quad 19.15$$
$$22.35 \quad 22.90 \quad 22.20 \quad 23.10$$

Assuming that the distribution of the weights of the wheels of cheese is $N(\mu, \sigma^2)$, find a 95% confidence interval for μ.

6.2-3   Assume that the yield per acre for a particular variety of soybeans is $N(\mu, \sigma^2)$. For a random sample of $n = 5$ plots, the yields in bushels per acre were 37.4, 48.8, 46.9, 55.0, and 44.0.
(a) Give a point estimate for μ.
(b) Find a 90% confidence interval for μ.

6.2-4   In a study of maximal aerobic capacity [*Journal of Applied Physiology 65, 6 (December 1988), p. 2696*], 12 women were used as subjects, and one measurement that was made was blood plasma volume. The following data give their blood plasma volumes in liters:

$$3.15 \quad 2.99 \quad 2.77 \quad 3.12 \quad 2.45 \quad 3.85$$
$$2.99 \quad 3.87 \quad 4.06 \quad 2.94 \quad 3.53 \quad 3.20$$

Assume that these are observations of a normally distributed random variable X that has mean μ and standard deviation σ.
(a) Give the value of a point estimate of μ.
(b) Determine a point estimate of σ.
(c) Find a 90% confidence interval for μ.

6.2-5   One measurement of the effectiveness of a baseball catcher is his release time, namely, the time from the impact of the ball into the catcher's glove to the impact of the ball into the infielder's glove when an opponent is attempting to steal second base. Four measurements (in seconds) for a particular major league catcher (Parrish) were 1.81, 1.89, 1.84, and 1.76. Assuming that these four measurements are a random sample of a normally distributed random variable that has a mean of μ, find a 95% confidence interval for μ.

6.2-6   During the Friday night shift, $n = 28$ mints were selected at random from a production line and weighed. They had an average weight of $\bar{x} = 21.45$ grams and $s = 0.31$ grams. Give the endpoints for a 90% confidence interval for μ, the mean weight of all the mints.

6.2-7   A farm grows grapes for jelly. The following data are measurements of sugar in the grapes of a sample taken from each of 30 truckloads:

16.0	15.2	12.0	16.9	14.4	16.3	15.6	12.9	15.3	15.1
15.8	15.5	12.5	14.5	14.9	15.1	16.0	12.5	14.3	15.4
15.4	13.0	12.6	14.9	15.1	15.3	12.4	17.2	14.7	14.8

Assume that these are observations of a random variable X that has mean μ and standard deviation σ.
(a) Give point estimates of μ and σ.
(b) Find an approximate 90% confidence interval for μ.

6.2-8   A cherry farmer in Suttons Bay, Mich., transports tart cherries in large tanks. Let X equal the weight in pounds of cherries in a tank. The following data give 30 observations of X.

1023.02	981.27	949.95	931.92	940.46	931.92
992.65	1015.43	947.10	1136.90	1132.16	1122.67
1030.61	963.24	922.43	971.78	1062.88	943.31
967.98	918.63	1003.09	917.68	939.51	967.98
1051.49	1064.78	985.06	1016.38	985.06	958.49

(a) Give point estimates of μ and σ, the mean and the variance of X.
(b) Find an approximate 95% confidence interval for μ.

6.2-9   Let $X_1, X_2, \ldots, X_n$ be a random sample of size n from the normal distribution $N(\mu, \sigma^2)$. Calculate the expected length of a 95% confidence interval for μ assuming that $n = 5$ and the variance is
(a) known.
(b) unknown.

HINT: To find $E(S)$, determine $E[\sqrt{(n-1)S^2/\sigma^2}]$, recalling $(n-1)S^2/\sigma^2$ is $\chi^2(n-1)$. Also see Exercise 5.8-4.

6.2-10   Students took $n = 35$ samples of water from the east basin of Lake Macatawa (see Example 6.1-9) and measured the amount of sodium in parts per million. For their data they calculated $\bar{x} = 24.11$ and $s^2 = 24.44$. Find an approximate 90% confidence interval for μ, the mean of the amount of sodium in parts per million.

6.2-11   A manufacturer of soap powder packages the soap in "6-pound" boxes. To check the filling machine, they took a sample of $n = 1219$ boxes and weighed them. Given that $\bar{x} = 6.05$ pounds and $s = 0.02$ pounds, give the endpoints for a 99% confidence interval for μ, the mean weight of the boxes of soap filled by this machine.

6.2-12   The length of life of brand X light bulbs is assumed to be $N(\mu_X, 784)$. The length of life of brand Y light bulbs is assumed to be $N(\mu_Y, 627)$ and independent of that of X. If a random sample of $n = 56$ brand X light bulbs yielded a mean of $\bar{x} = 937.4$ hours and a random sample of size $m = 57$ brand Y light bulbs yielded a mean of $\bar{y} = 988.9$ hours, find a 90% confidence interval for $\mu_X - \mu_Y$.

6.2-13 Let $X_1, X_2, \ldots, X_5$ be a random sample of SAT mathematics scores, assumed to be $N(\mu_X, \sigma^2)$, and let $Y_1, Y_2, \ldots, Y_8$ be an independent random sample of SAT verbal scores, assumed to be $N(\mu_Y, \sigma^2)$. If the following data are observed, find a 90% confidence interval for $\mu_X - \mu_Y$:

$$x_1 = 644 \quad x_2 = 493 \quad x_3 = 532 \quad x_4 = 462 \quad x_5 = 565$$
$$y_1 = 623 \quad y_2 = 472 \quad y_3 = 492 \quad y_4 = 661 \quad y_5 = 540$$
$$y_6 = 502 \quad y_7 = 549 \quad y_8 = 518$$

6.2-14 Independent random samples of the heights of adult males living in two countries yielded the following results: $n = 12$, $\bar{x} = 65.7$ inches, $s_x = 4$ inches and $m = 15$, $\bar{y} = 68.2$ inches, $s_y = 3$ inches. Find an approximate 98% confidence interval for the difference $\mu_X - \mu_Y$ of the means of the populations of heights. Assume that $\sigma_X^2 = \sigma_Y^2$.

6.2-15 [*Medicine and Science in Sports and Exercise* (January 1990.)] Let X and Y equal, respectively, the blood volumes in milliliters for a male who is a paraplegic and participates in vigorous physical activities and a male who is able-bodied and participates in normal activities. Assume that X is $N(\mu_X, \sigma_X^2)$ and Y is $N(\mu_Y, \sigma_Y^2)$, respectively. Using the following $n = 7$ observations of X:

$$1612 \quad 1352 \quad 1456 \quad 1222 \quad 1560 \quad 1456 \quad 1924$$

and $m = 10$ observations of Y:

$$1082 \quad 1300 \quad 1092 \quad 1040 \quad 910$$
$$1248 \quad 1092 \quad 1040 \quad 1092 \quad 1288$$

(a) Give a point estimate for $\mu_X - \mu_Y$.
(b) Find a 95% confidence interval for $\mu_X - \mu_Y$. Since the variances σ_X^2 and σ_Y^2 might not be equal, use Welch's T.

6.2-16 A biologist who studies spiders was interested in comparing the lengths of female and male green lynx spiders. Assume that the length X of the male spider is approximately $N(\mu_X, \sigma_X^2)$ and the length Y of the female spider is approximately $N(\mu_Y, \sigma_Y^2)$. Find an approximate 95% confidence interval for $\mu_X - \mu_Y$ using $n = 30$ observations of X:

$$
\begin{array}{cccccc}
5.20 & 4.70 & 5.75 & 7.50 & 6.45 & 6.55 \\
4.70 & 4.80 & 5.95 & 5.20 & 6.35 & 6.95 \\
5.70 & 6.20 & 5.40 & 6.20 & 5.85 & 6.80 \\
5.65 & 5.50 & 5.65 & 5.85 & 5.75 & 6.35 \\
5.75 & 5.95 & 5.90 & 7.00 & 6.10 & 5.80 \\
\end{array}
$$

and $m = 30$ observations of Y:

8.25	9.95	5.90	7.05	8.45	7.55
9.80	10.80	6.60	7.55	8.10	9.10
6.10	9.30	8.75	7.00	7.80	8.00
9.00	6.30	8.35	8.70	8.00	7.50
9.50	8.30	7.05	8.30	7.95	9.60

where the measurements are in millimeters.

6.2-17 Let $\overline{X}$, $\overline{Y}$, S_X^2, and S_Y^2 be the respective sample means and unbiased estimates of the variances using samples of sizes n and m from the independent normal distributions $N(\mu_X, \sigma_X^2)$ and $N(\mu_Y, \sigma_Y^2)$, where μ_X, μ_Y, σ_X^2, and σ_Y^2 are unknown. If, however, $\sigma_X^2/\sigma_Y^2 = d$, a known constant, argue that

(a) $\dfrac{(\overline{X} - \overline{Y}) - (\mu_X - \mu_Y)}{\sqrt{d\sigma_Y^2/n + \sigma_Y^2/m}}$ is $N(0, 1)$.

(b) $\dfrac{(n - 1)S_X^2}{d\sigma_Y^2} + \dfrac{(m - 1)S_Y^2}{\sigma_Y^2}$ is $\chi^2(n+m-2)$.

(c) The two random variables in (a) and (b) are independent.

(d) With these results, construct a random variable (not depending upon σ_Y^2) that has a t distribution and can be used to construct a confidence interval for $\mu_X - \mu_Y$.

6.2-18 Students are weighed (in kilograms) at the beginning and the end of a semester-long health-fitness program. Let the random variable D equal the weight change for a student, postweight minus preweight. Assume that the distribution of D is $N(\mu_D, \sigma_D^2)$. A random sample of $n = 12$ female students yielded the following observations of D

2.0	−0.5	1.4	−2.2	0.3	−0.8
3.7	−0.1	0.6	0.2	0.9	−0.1

(a) Give a point estimate of μ_D.

(b) Find a 95% confidence interval for μ_D.

6.2-19 Twenty-four 9th- and 10th-grade high school girls were put on an ultraheavy rope-jumping program. The following data give the time difference for each girl—"before program time" minus "after program time"— for the 40-yard dash.

0.28	0.01	0.13	0.33	−0.03	0.07	−0.18	−0.14
−0.33	0.01	0.22	0.29	−0.08	0.23	0.08	0.04
−0.30	−0.08	0.09	0.70	0.33	−0.34	0.50	0.06

(a) Give a point estimate of μ_D, the mean of the difference in race times.

(b) Find a 95% confidence interval for μ_D.

(c) Does it look like the rope-jumping program was effective?

6.3   Confidence Intervals for Variances

In this section we find confidence intervals for the variance of a normal distribution and for the ratio of the variances of two independent normal distributions. The confidence interval for the variance σ^2 is based on the sample variance

$$S^2 = \frac{1}{n-1} \sum_{i=1}^{n} (X_i - \bar{X})^2.$$

We use the fact that the distribution of $(n-1)S^2/\sigma^2$ is $\chi^2(n-1)$ to find a confidence interval for σ^2. Select constants a and b from Table IV in the Appendix with $n-1$ degrees of freedom such that

$$P\left(a \leq \frac{(n-1)S^2}{\sigma^2} \leq b \right) = 1 - \alpha.$$

One way to do this is by selecting a and b so that $a = \chi^2_{1-\alpha/2}(n-1)$ and $b = \chi^2_{\alpha/2}(n-1)$. That is, select a and b so that the probabilities in the two tails are equal. Then, solving the inequalities, we have

$$1 - \alpha = P\left(\frac{a}{(n-1)S^2} \leq \frac{1}{\sigma^2} \leq \frac{b}{(n-1)S^2} \right)$$

$$= P\left(\frac{(n-1)S^2}{b} \leq \sigma^2 \leq \frac{(n-1)S^2}{a} \right).$$

Thus the probability that the random interval $[(n-1)S^2/b, (n-1)S^2/a]$ contains the unknown σ^2 is $1 - \alpha$. Once the values of $X_1, X_2, \ldots, X_n$ are observed to be $x_1, x_2, \ldots, x_n$ and s^2 computed, then the interval $[(n-1)s^2/b, (n-1)s^2/a]$ is a $100(1 - \alpha)\%$ confidence interval for σ^2. It follows that $[\sqrt{(n-1)/bs}, \sqrt{(n-1)/as}]$ is a $100(1 - \alpha)\%$ confidence interval for σ, the standard deviation.

> **Example 6.3-1**   Assume that the time in days required for maturation of seeds of a species of *Guardiola,* a flowering plant found in Mexico, is $N(\mu, \sigma^2)$. A random sample of $n = 13$ seeds, both parents having narrow leaves, yielded $\bar{x} = 18.97$ days and
>
> $$12s^2 = \sum_{i=1}^{13} (x_i - \bar{x})^2 = 128.41.$$
>
> A 90% confidence interval for σ^2 is
>
> $$\left[\frac{128.41}{21.03}, \frac{128.41}{5.226} \right] = [6.11, 24.57]$$
>
> because $5.226 = \chi^2_{0.95}(12)$ and $21.03 = \chi^2_{0.05}(12)$ from Table IV in the Appendix. The corresponding 90% confidence interval for σ is
>
> $$[\sqrt{6.11}, \sqrt{24.57}] = [2.47, 4.96].$$

Although a and b are generally selected so that the probabilities in the two tails are equal, the resulting $100(1 - \alpha)\%$ confidence interval is not the shortest that can be formed using the available data. Table X in the Appendix gives solutions for a and b that yield confidence intervals of minimum length for the standard deviation. (See Exercise 6.3-13.)

Example 6.3-2 Using the data in Example 6.3-1, a 90% confidence interval for σ of minimum length is (since $a = 5.940$ and $b = 24.202$ from Table X)

$$\left[\sqrt{\frac{128.41}{24.202}}, \ \sqrt{\frac{128.41}{5.940}} \right] = [2.30, 4.65].$$

The length of this interval is 2.35, whereas the length of the interval given in Example 6.3-1 is 2.49. To see why this new interval is shorter, carefully sketch a graph of the $\chi^2(12)$ p.d.f. and compare $\chi^2_{0.95}(12)$ to a and $\chi^2_{0.05}(12)$ to b on the graph.

There are occasions when it is of interest to compare the variances of two independent normal distributions. We do this by finding a confidence interval for σ_X^2/σ_Y^2 using the ratio of $(n - 1)S_X^2/\sigma_X^2$ and $(m - 1)S_Y^2/\sigma_Y^2$, where S_X^2 and S_Y^2 are the two sample variances based on two samples of respective sizes n and m. Since $(n - 1)S_X^2/\sigma_X^2$ and $(m - 1)S_Y^2/\sigma_Y^2$ are independent chi-square variables, we know from Section 5.7 that

$$F = \frac{(m - 1)S_Y^2/\sigma_Y^2(m - 1)}{(n - 1)S_X^2/\sigma_X^2(n - 1)} = \frac{S_Y^2/\sigma_Y^2}{S_X^2/\sigma_X^2}$$

has an F distribution with $r_1 = m - 1$ and $r_2 = n - 1$ degrees of freedom. To form the confidence interval, select constants c and d from Table VII in the Appendix so that

$$1 - \alpha = P\left(c \leq \frac{S_Y^2/\sigma_Y^2}{S_X^2/\sigma_X^2} \leq d \right)$$

$$= P\left(c\frac{S_X^2}{S_Y^2} \leq \frac{\sigma_X^2}{\sigma_Y^2} \leq d\frac{S_X^2}{S_Y^2} \right).$$

Because of the limitations of Table VII, we generally let $c = F_{1-\alpha/2}(m-1, n-1) = 1/F_{\alpha/2}(n-1, m-1)$ and $d = F_{\alpha/2}(m-1, n-1)$. If s_x^2 and s_y^2 are the observed values of S_X^2 and S_Y^2, respectively, then

$$\left[\frac{1}{F_{\alpha/2}(n-1, m-1)}\frac{s_x^2}{s_y^2}, \ F_{\alpha/2}(m-1, n-1)\frac{s_x^2}{s_y^2} \right]$$

is a $100(1 - \alpha)\%$ confidence interval for σ_X^2/σ_Y^2. By taking square roots of both endpoints, we would obtain a $100(1 - \alpha)\%$ confidence interval for σ_X/σ_Y.

Example 6.3-3 In Example 6.3-1, denote σ^2 by σ_X^2. There $(n-1)s_x^2 = 12s^2 = 128.41$. Assume that the time in days required for maturation of seeds of a species of *Guardiola*, both parents having broad leaves, is $N(\mu_Y, \sigma_Y^2)$. A random sample of size $m = 9$ seeds yielded $\bar{y} = 23.20$ and

$$8s_y^2 = \sum_{i=1}^{9} (y_i - \bar{y})^2 = 36.72.$$

A 98% confidence interval for σ_X^2/σ_Y^2 is given by

$$\left[\left(\frac{1}{5.67}\right)\frac{(128.41)/12}{(36.72)/8}, \ (4.50)\frac{(128.41)/12}{(36.72)/8}\right] = [0.41, 10.49]$$

because $F_{0.01}(12, 8) = 5.67$ and $F_{0.01}(8, 12) = 4.50$. It follows that a 98% confidence interval for σ_X/σ_Y is

$$[\sqrt{0.41}, \sqrt{10.49}] = [0.64, 3.24].$$

Although we are able to formally find a confidence interval for the ratio of two distribution variances and/or standard deviations, we should point out that these intervals are generally not too good because they are often very wide.

Exercises

6.3-1 Let X equal the length (in centimeters) of a certain species of fish when caught in the spring. A random sample of $n = 13$ observations of X are

13.1	5.1	18.0	8.7	16.5	9.8	6.8
12.0	17.8	25.4	19.2	15.8	23.0	

(a) Give a point estimate of the standard deviation σ of this species of fish.
(b) Find a 95% confidence interval for σ.

6.3-2 A random sample of $n = 9$ wheels of cheese yielded the following weights in pounds, assumed to be $N(\mu, \sigma^2)$:

21.50	18.95	18.55	19.40	19.15
22.35	22.90	22.20	23.10	

(a) Give a point estimate of σ.
(b) Find a 95% confidence interval for σ.
(c) Find a 90% confidence interval for σ.

6.3-3 A student who works in a blood lab tested 25 men for cholesterol levels and found the following values:

164	272	261	248	235	192	203	278	268
230	242	305	286	310	345	289	326	
335	297	328	400	228	194	338	252	

Assume that these values represent observations of a random sample taken from $N(\mu, \sigma^2)$.

(a) Calculate the sample mean and sample variance for these data.

(b) Find a 90% confidence interval for σ^2.

(c) Find a 90% confidence interval for σ.

(d) Find a 90% confidence interval for σ that has minimum length.

(e) Does the assumption of normality seem to be valid?

6.3-4 For the data in Exercise 6.2-4, find a 95% confidence interval for

(a) σ^2.

(b) σ.

(c) σ having minimum length.

6.3-5 Let $X_1, X_2, X_3, \ldots, X_n$ be a random sample from $N(\mu, \sigma^2)$, with known mean μ. Describe how you would construct a confidence interval for the unknown variance σ^2.

HINT: Use the fact that $\Sigma_{i=1}^{n} (X_i - \mu)^2/\sigma^2$ is $\chi^2(n)$.

6.3-6 Let $X_1, X_2, \ldots, X_n$ be a random sample of size n from an exponential distribution with unknown mean of $\mu = \theta$.

(a) Show that the distribution of $W = (2/\theta) \Sigma_{i=1}^{n} X_i$ is $\chi^2(2n)$.

HINT: Find the moment-generating function of W.

(b) Use W to construct a $100(1 - \alpha)\%$ confidence interval for θ.

(c) If $n = 7$ and $\bar{x} = 93.6$, give the endpoints for a 90% confidence interval for θ.

6.3-7 Let X and Y equal the weights of a phosphorus-free laundry detergent in a "6-pound" box and a "12-pound" box, respectively. Assume that the distributions of X and Y are $N(\mu_X, \sigma_X^2)$ and $N(\mu_Y, \sigma_Y^2)$, respectively. A random sample of $n = 10$ observations of X yielded a sample mean of $\bar{x} = 6.10$ pounds with a sample variance of $s_x^2 = 0.0040$, while a random sample of $m = 9$ observations of Y yielded a sample mean of $\bar{y} = 12.10$ pounds with a sample variance of $s_y^2 = 0.0076$.

(a) Give a point estimate of σ_X^2/σ_Y^2.

(b) Find a 95% confidence interval for σ_X^2/σ_Y^2.

6.3-8 Let X and Y equal the number of milligrams of tar in filtered and nonfiltered cigarettes, respectively. Assume that the distributions of X and Y are $N(\mu_X, \sigma_X^2)$ and $N(\mu_Y, \sigma_Y^2)$, respectively. A random sample of $n = 9$ observations of X were

0.9	1.1	0.1	0.7	0.3	0.9	0.8	1.0	0.4

and a random sample of $m = 11$ observations of Y were

1.5	0.9	1.6	0.5	1.4	1.9	1.0	1.2	1.3	1.6	2.1

(a) Give a point estimate of σ_X^2/σ_Y^2.

(b) Find a 98% confidence interval for σ_X^2/σ_Y^2.

6.3-9   A candy maker produces mints that have a label weight of 20.4 grams. For quality assurance, $n = 16$ mints were selected at random from the Wednesday morning shift, resulting in the statistics $\bar{x} = 21.95$ grams and $s_x = 0.197$. On Wednesday afternoon $m = 13$ mints were selected at random, giving $\bar{y} = 21.88$ grams and $s_y = 0.318$. Find a 90% confidence interval for σ_X/σ_Y, the ratio of the standard deviations of the mints produced by the morning and by the afternoon shifts, respectively.

6.3-10   Let X and Y equal the concentration in parts per billion of chromium in the blood for healthy persons and for persons with a suspected disease, respectively. Assume that the distributions of X and Y are $N(\mu_X, \sigma_X^2)$ and $N(\mu_Y, \sigma_Y^2)$, respectively. Using $n = 8$ observations of X:

$$15 \quad 23 \quad 12 \quad 18 \quad 9 \quad 28 \quad 11 \quad 10$$

and $m = 10$ observations of Y:

$$25 \quad 20 \quad 35 \quad 15 \quad 40 \quad 16 \quad 10 \quad 22 \quad 18 \quad 32$$

(a) Give a point estimate of σ_X^2/σ_Y^2.
(b) Find a 95% confidence interval for σ_X^2/σ_Y^2.

6.3-11   Some nurses were interested in the effect of prenatal care on the birthweight of babies. Mothers were divided into two groups, and their babies' weights were compared. The birthweight in ounces of babies of mothers who had received 5 or fewer prenatal visits were

$$
\begin{array}{ccccccc}
49 & 108 & 110 & 82 & 93 & 114 & 134 \\
114 & 96 & 52 & 101 & 114 & 120 & 116
\end{array}
$$

and the birthweights of babies of mothers who had received 6 or more prenatal visits were

$$
\begin{array}{ccccccc}
133 & 108 & 93 & 119 & 119 & 98 & 106 \\
87 & 153 & 116 & 129 & 97 & 110 & 131
\end{array}
$$

Assuming that these are respectively independent observations of X and Y, which are $N(\mu_X, \sigma_X^2)$ and $N(\mu_Y, \sigma_Y^2)$, find a 95% confidence interval for
(a) σ_X^2/σ_Y^2.
(b) σ_X/σ_Y.

6.3-12   Using a random sample of size n from the normal distribution $N(\mu, \sigma^2)$, let $\bar{X}$ and S^2 be the unbiased estimators of μ and σ^2, respectively. Find the $100(1 - \alpha)\%$ confidence interval for μ of minimum length based on the statistic $T = (\bar{X} - \mu)/S/\sqrt{n}$.

HINT: (i) Show that $1 - \alpha = P(a \le T \le b) = P(\bar{X} - bS/\sqrt{n} \le \mu \le \bar{X} - aS/\sqrt{n})$, (ii) the length of the resulting confidence interval is $L = (s/\sqrt{n})(b - a)$. (iii) Minimize L subject to the condition $\int_a^b g(t)\, dt = 1 - \alpha$, where $g(t)$ is the p.d.f. of a t random variable with $r = n - 1$ degrees of freedom.

6.3-13 Let $X_1, X_2, \ldots, X_n$ be a random sample of size n from a normal distribution, $N(\mu, \sigma^2)$. Select a and b so that

$$P\left(a \le \frac{(n-1)S^2}{\sigma^2} \le b\right) = 1 - \alpha.$$

So a $100(1 - \alpha)\%$ confidence interval for σ is $[\sqrt{(n-1)/bs}, \sqrt{(n-1)/as}]$. Find values of a and b that minimize the length of this confidence interval. That is, minimize

$$k = s\sqrt{n-1}\left(\frac{1}{\sqrt{a}} - \frac{1}{\sqrt{b}}\right)$$

under the restriction

$$G(b) - G(a) = \int_a^b g(u)du = 1 - \alpha,$$

where $G(u)$ and $g(u)$ are the distribution function and p.d.f. of a $\chi^2(n-1)$ distribution, respectively.

HINT: Due to the restriction, b is a function of a. In particular, by taking derivatives of the restricting equation with respect to a, show that $\dfrac{db}{da} = \dfrac{g(a)}{g(b)}$. Determine $\dfrac{dk}{da}$. By setting $\dfrac{dk}{da} = 0$, show that a and b must satisfy

$$a^{n/2}e^{-a/2} - b^{n/2}e^{-b/2} = 0.$$

This condition, along with the restriction, was used to calculate the values in Appendix Table X.

6.3-14 Let X and Y equal the playing times for a $33\frac{1}{3}$ rpm long-playing record album and a compact disc, respectively. Observe that $n = 9$ observations of X were

| 40.83 | 43.18 | 35.72 | 38.68 | 37.17 | 39.75 | 24.76 | 34.58 | 33.98 |

and $m = 13$ observations of Y were

| 42.82 | 64.42 | 56.92 | 39.92 | 72.38 | 47.26 | 64.58 |
| 38.20 | 72.75 | 39.09 | 39.07 | 33.70 | 62.02 | |

(a) Determine a 95% confidence interval for σ_X^2/σ_Y^2, assuming that the independent distributions of X and Y are $N(\mu_X, \sigma_X^2)$ and $N(\mu_Y, \sigma_Y^2)$, respectively.

(b) Does the assumption of normality seem to be valid for these data? Why?

6.4   Confidence Intervals for Proportions

We have suggested that the histogram is a good description of how the items of a random sample are distributed. We might naturally inquire about the accuracy of those relative frequencies (or percentages) associated with the various classes.

For illustration, in Example 1.2-1 concerning $n = 200$ weights, we found that the relative frequency of the class interval $(172.5, 184.5)$ was 0.18, that is, 18%. If we think of this collection of 200 weights as a random sample observed from a larger population of weights, how close is 18% to the true percentage (or 0.18 to the true proportion) of weights in that class interval throughout the entire population?

In considering this problem, we generalize it somewhat by treating the class interval $(172.5, 184.5)$ as "success." That is, there is some true probability of success p, namely the proportion of the population in that interval. Let Y equal the frequency of measurements in the interval out of the n observations so that, under the assumptions of independence and constant probability p, Y has the binomial distribution $b(n, p)$. Thus the problem is to determine the accuracy of the relative frequency Y/n as an estimator of p. We solve this by finding for the unknown p a confidence interval based on Y/n.

In general, when observing n Bernoulli trials with probability p of success on each trial, we shall find a confidence interval for p based on Y/n, where Y is the number of successes and Y/n is an unbiased point estimator for p.

In Section 5.5 we noted that

$$\frac{Y - np}{\sqrt{np(1 - p)}} = \frac{(Y/n) - p}{\sqrt{p(1 - p)/n}}$$

has an approximate normal distribution $N(0, 1)$, provided that n is large enough. This means that, for a given probability $1 - \alpha$, we can find a $z_{\alpha/2}$ in Table V in the Appendix such that

$$P\left[-z_{\alpha/2} \leq \frac{(Y/n) - p}{\sqrt{p(1 - p)/n}} \leq z_{\alpha/2}\right] \approx 1 - \alpha. \tag{6.4-1}$$

If we proceed as we did when we found a confidence interval for μ in Section 6.2, we would obtain

$$P\left[\frac{Y}{n} - z_{\alpha/2}\sqrt{\frac{p(1 - p)}{n}} \leq p \leq \frac{Y}{n} + z_{\alpha/2}\sqrt{\frac{p(1 - p)}{n}}\right] \approx 1 - \alpha.$$

Unfortunately, the unknown parameter p appears in the endpoints of this inequality. There are two ways out of this dilemma. We could make an additional approximation, namely replacing p with Y/n in $p(1 - p)/n$ in the endpoints. That is, it is still true, if n is large enough, that

$$P\left[\frac{Y}{n} - z_{\alpha/2}\sqrt{\frac{(Y/n)(1 - Y/n)}{n}} \leq p \leq \frac{Y}{n} + z_{\alpha/2}\sqrt{\frac{(Y/n)(1 - Y/n)}{n}}\right] \approx 1 - \alpha.$$

Thus, for large n, if the observed Y equals y, the interval

$$\left[\frac{y}{n} - z_{\alpha/2}\sqrt{\frac{(y/n)(1 - y/n)}{n}}, \frac{y}{n} + z_{\alpha/2}\sqrt{\frac{(y/n)(1 - y/n)}{n}}\right]$$

serves as an approximate $100(1 - \alpha)\%$ confidence interval for p. Frequently, this is written as

$$\frac{y}{n} \pm z_{\alpha/2}\sqrt{\frac{(y/n)(1 - y/n)}{n}}, \tag{6.4-2}$$

for brevity. This clearly notes, as does $\bar{x} \pm z_{\alpha/2}\sigma/\sqrt{n}$ in Section 6.1, the reliability of the estimate y/n; namely, we are $100(1 - \alpha)\%$ confident that p is within $z_{\alpha/2}\sqrt{(y/n)(1 - y/n)/n}$ of y/n.

A second way to solve for p, in the inequality in equation (6.4-1), is to note that

$$\frac{|Y/n - p|}{\sqrt{p(1 - p)/n}} \leq z_{\alpha/2}$$

is equivalent to

$$H(p) = \left(\frac{Y}{n} - p\right)^2 - \frac{z_{\alpha/2}^2 p(1 - p)}{n} \leq 0. \tag{6.4-3}$$

But $H(p)$ is a quadratic expression in p, and we can find those values of p for which $H(p) \leq 0$ by finding the two zeros of $H(p)$. Letting $\hat{p} = Y/n$ and $z_0 = z_{\alpha/2}$ in equation (6.4-3), we have

$$H(p) = \left(1 + \frac{z_0^2}{n}\right)p^2 - \left(2\hat{p} + \frac{z_0^2}{n}\right)p + \hat{p}^2.$$

By the quadratic formula, the zeros of $H(p)$ are, after simplifying,

$$\frac{\hat{p} + z_0^2/2n \pm z_0\sqrt{\hat{p}(1 - \hat{p})/n + z_0^2/4n^2}}{1 + z_0^2/n}, \tag{6.4-4}$$

and these zeros give the endpoints for an approximate $100(1 - \alpha)\%$ confidence interval for p. If n is large, $z_0^2/2n$, $z_0^2/4n^2$, and z_0^2/n are small. Thus the confidence intervals given by equations (6.4-2) and (6.4-4) are approximately equal when n is large.

Example 6.4-1   Let us return to the example of the histogram with $n = 200$ and $y/n = 0.18$. If $1 - \alpha = 0.95$ so that $z_{\alpha/2} = 1.96$, then, using equation (6.4-2),

$$0.18 \pm 1.96\sqrt{\frac{(0.18)(0.82)}{200}}$$

serves as an approximate 95% confidence interval for the true fraction p. That is, $[0.127, 0.233]$, which is the same as $[12.7\%, 23.3\%]$, is an approximate 95% confidence interval for the percentage of weights of the entire population in the interval $(172.5, 184.5)$. If we had used the endpoints given by equation (6.4-4), the confidence interval would be $[0.128, 0.234]$, which differs very little from the first interval.

Example 6.4-2   In a certain political campaign, one candidate has a poll taken at random among the voting population. The results are $y = 185$ out of $n = 351$ voters favor this candidate. Even though $y/n = 185/351 = 0.527$, should the candidate feel very confident of winning? An approximate 95% confidence interval for the fraction p of the voting population who favor this candidate is, using equation 6.4-2,

$$0.527 \pm 1.96\sqrt{\frac{(0.527)(0.473)}{351}}$$

or, equivalently, $[0.475, 0.579]$. Thus there is a good possibility that p is less than 50%, and the candidate should certainly take this into account in campaigning.

Frequently, there are two (or more) possible independent ways of performing an experiment; suppose these have probabilities of success p_1 and p_2, respectively. Let n_1 and n_2 be the number of independent trials associated with these two methods, and let us say they result in Y_1 and Y_2 successes, respectively. In order to make a statistical inference about the difference $p_1 - p_2$, we proceed as follows.

Since the independent random variables Y_1/n_1 and Y_2/n_2 have respective means p_1 and p_2 and variances $p_1(1 - p_1)/n_1$ and $p_2(1 - p_2)/n_2$, we know from Section 5.2 that the difference $Y_1/n_1 - Y_2/n_2$ must have mean $p_1 - p_2$ and variance

$$\frac{p_1(1 - p_1)}{n_1} + \frac{p_2(1 - p_2)}{n_2}.$$

(Recall that the variances are added to get the variance of a difference of two independent random variables.) Moreover, the fact that Y_1/n_1 and Y_2/n_2 have approximate normal distributions would suggest that the difference

$$\frac{Y_1}{n_1} - \frac{Y_2}{n_2}$$

would have an approximate normal distribution with the above mean and variance (see Theorem 5.3-5). That is,

$$\frac{(Y_1/n_1) - (Y_2/n_2) - (p_1 - p_2)}{\sqrt{p_1(1 - p_1)/n_1 + p_2(1 - p_2)/n_2}}$$

has an approximate normal distribution $N(0, 1)$. If we now replace p_1 and p_2 in the denominator of this ratio by Y_1/n_1 and Y_2/n_2, respectively, it is still true for large enough n_1 and n_2, that the new ratio will be approximately $N(0, 1)$. Thus, for a given $1 - \alpha$, we can find $z_{\alpha/2}$ from Table V in the Appendix so that

$$P\left[-z_{\alpha/2} \leq \frac{(Y_1/n_1) - (Y_2/n_2) - (p_1 - p_2)}{\sqrt{(Y_1/n_1)(1 - Y_1/n_1)/n_1 + (Y_2/n_2)(1 - Y_2/n_2)/n_2}} \leq z_{\alpha/2}\right] \approx 1 - \alpha.$$

This can be solved to obtain, once Y_1 and Y_2 are observed to be y_1 and y_2, an approximate $100(1 - \alpha)\%$ confidence interval

$$\frac{y_1}{n_1} - \frac{y_2}{n_2} \pm z_{\alpha/2} \sqrt{\frac{(y_1/n_1)(1 - y_1/n_1)}{n_1} + \frac{(y_2/n_2)(1 - y_2/n_2)}{n_2}}$$

for the unknown difference $p_1 - p_2$. Note again how this form indicates the reliability of the estimate $y_1/n_1 - y_2/n_2$ of the difference $p_1 - p_2$.

> **Example 6.4-3** Two detergents were tested for their ability to remove stains of a certain type. An inspector judged the first one to be successful on 63 out of 91 independent trials and the second one to be successful on 42 out of 79 independent trials. The respective relative frequencies of success are 0.692 and 0.532. An approximate 90% confidence interval for the difference $p_1 - p_2$ of the two detergents is
>
> $$0.692 - 0.532 \pm 1.645 \sqrt{\frac{(0.692)(0.308)}{91} + \frac{(0.532)(0.468)}{79}},$$
>
> or, equivalently, $[0.038, 0.282]$. Accordingly, it seems that the first detergent is definitely better than the second one for removing these stains.

Exercises

6.4-1 A machine shop manufactures toggle levers. A lever is flawed if a standard nut cannot be screwed onto the threads. Let p equal the proportion of flawed toggle levers that they manufacture. If there were 24 flawed levers out of a sample of 642 that were selected randomly from the production line,
(a) Give a point estimate of p.
(b) Find an approximate 95% confidence interval for p using equation 6.4-2.
(c) Find an approximate 95% confidence interval for p using equation 6.4-4.

6.4-2 Let p equal the proportion of letters mailed in the Netherlands that are delivered the next day. If $y = 142$ out of a random sample of $n = 200$ letters were delivered the day after they were mailed, find an approximate 90% confidence interval for p.

6.4-3 Let p equal the proportion of adult Americans who favor a law requiring a teenager to have her parents' consent before having an abortion. In a survey of 1000 adult Americans (conducted by *Time*/CNN and reported in *Time* on July 9, 1990), 690 said they favored such a law.
(a) Give a point estimate of p.
(b) Find an approximate 95% confidence interval for p.

6.4-4 Let p equal the proportion of Americans who favor the death penalty. If a random sample of $n = 1234$ Americans yielded $y = 864$ who favored the death penalty, find an approximate 95% confidence interval for p.

6.4-5   Let p equal the proportion of triathletes who suffered a training-related overuse injury during the past year. Out of 330 triathletes who responded to a survey, 167 indicated that they had suffered a training-related injury during the past year. Using these data,
(a) Give a point estimate of p.
(b) Find an approximate 90% confidence interval for p.
(c) Do you think that the 330 triathletes who responded to the survey may be considered as a random sample from the population of triathletes?

6.4-6   Let p equal the proportion of Americans who select jogging as one of their recreational activities. If 1497 out of a random sample of 5757 selected jogging, find an approximate 98% confidence interval for p.

6.4-7   In order to estimate the percentage of a large class of college freshmen that had high school GPAs from 3.2 to 3.6 inclusive, a sample of $n = 50$ students was taken, and $y = 9$ students fell in this class. Give a 95% confidence interval for the percentage of this freshman class having a high school GPA of 3.2 to 3.6.

6.4-8   A proportion, p, that many public opinion polls estimate is the number of Americans who would say yes to the question, ''If something were to happen to the President of the United States, do you think that the Vice President would be qualified to take over as President?'' In one such random sample of 1022 adults, 388 said *yes*.
(a) Based on the given data, find a point estimate of p.
(b) Find an approximate 90% confidence interval for p.
(c) Give updated answers to this question if new poll results are available.

6.4-9   To obtain an estimate of the proportion, p, of New York City residents who feel that the quality of life in New York City has become worse in the past few years, a telephone poll by *Time*/CNN on August 2–5, 1990, revealed that 686 out of 1009 residents said that life has become worse.
(a) Give a point estimate of p.
(b) Find an approximate 98% confidence interval for p.

6.4-10   In developing countries in Africa and the Americas, respectively, let p_1 and p_2 be the proportions of women with nutritional anemia. Find an approximate 90% confidence interval for $p_1 - p_2$ given that a random sample of $n_1 = 2100$ African women yielded $y_1 = 840$ with nutritional anemia and a random sample of $n_2 = 1900$ women from the Americas yielded $y_2 = 323$ women with nutritional anemia.

6.4-11   A candy manufacturer selects mints at random from the production line and weighs them. For one week, the day shift weighed $n_1 = 194$ mints, and the night shift weighed $n_2 = 162$ mints. The numbers of these mints that weighed at most 21 grams was $y_1 = 28$ for the day shift and $y_2 = 11$ for the night shift. Let p_1 and p_2 denote the proportions of mints that weigh at most 21 grams for the day and night shifts, respectively.
(a) Give a point estimate of p_1.

(b) Give the endpoints for a 95% confidence interval for p_1.

(c) Give a point estimate of $p_1 - p_2$.

(d) Give the endpoints for a 95% confidence interval for $p_1 - p_2$.

6.4-12 Consider the following two groups of women: Group 1—Women who spend less than \$500 annually on clothes; Group 2—Women who spend over \$1000 annually on clothes. Let p_1 and p_2 equal the proportions of women in these two groups, respectively, who believe that clothes are too expensive. If 1009 out of a random sample of 1230 women from group 1 and 207 out of a random sample 340 from group 2 believe that clothes are too expensive,

(a) Give a point estimate of $p_1 - p_2$.

(b) Find an approximate 95% confidence interval for $p_1 - p_2$.

6.4-13 For developing countries in Asia (excluding China) and Africa, respectively, let p_1 and p_2 be the proportions of preschool children with chronic malnutrition (stunting). If respective random samples of $n_1 = 1300$ and $n_2 = 1100$ yielded $y_1 = 520$ and $y_2 = 385$ children with chronic malnutrition, find an approximate 95% confidence interval for $p_1 - p_2$.

6.4-14 The following question was asked in a *Newsweek* poll: "Would you prefer to live in a neighborhood with mostly whites, with mostly blacks, or in a neighborhood mixed half and half?" Let p_1 and p_2 equal the proportion of black and white adult respondents, respectively, who prefer "half and half." If 207 out of 305 black adults and 291 out of 632 white adults prefer "half and half,"

(a) Give a point estimate of $p_1 - p_2$.

(b) Find an approximate 90% confidence interval for $p_1 - p_2$.

6.4-15 An environmental survey contained a question asking what the respondent thought was the major cause of air pollution in this country, giving the choices "automobiles," "factories," and "incinerators." Two versions of the test, A and B, were used. Let p_A and p_B be the proportions of people using forms A and B who select "factories." If 170 out of 460 people who used version A chose "factories" and 141 out of 440 people who used version B chose "factories,"

(a) Find a 95% confidence interval for $p_A - p_B$.

(b) Do the forms seem to be consistent concerning this answer? Why?

6.5 Sample Size

In statistical consulting, the first question frequently asked is, "How large should the sample size be to estimate a mean?" In order to convince the inquirer that the answer will depend on the variation associated with the random variable under observation, the statistician could correctly respond, "Only one item is needed, provided that the standard deviation of the distribution is zero." That is, if σ equals zero, then the value of that one item would necessarily equal the unknown mean of the distribution. This, of course, is an extreme case and one that is not met in practice; however, it should help convince persons that the smaller the variance, the smaller the sample size needed to achieve a given degree of accuracy. This will become clearer as we consider several examples. Let us begin with a problem that involves a statistical inference about the unknown mean of a distribution.

Example 6.5-1   A mathematics department wishes to evaluate a new method of teaching calculus that does mathematics using a computer. At the end of the course, the evaluation will be made on the basis of scores of the participating students on a standard test. There is particular interest in estimating the mean score, μ, for students taking calculus using the computer. Thus there is a desire to determine the number of students, n, who are to be selected at random from a larger group of students to take the course taught using the computer. Since new computing equipment must be purchased, they cannot afford to let all of the students take calculus the new way. In addition, some of the staff question the value of this approach and hence do not want to expose every student to this new procedure. So, let us find the sample size n such that we are fairly confident that $\bar{x} \pm 1$ contains the unknown test mean μ. From past experience it is believed that the standard deviation associated with this type of test is about 15. (The mean is also known when students take the standard calculus course.) Accordingly, using the fact that the sample mean of the test scores, $\bar{X}$, is approximately $N(\mu, \sigma^2/n)$, we see that the interval given by $\bar{x} \pm 1.96(15/\sqrt{n})$ will serve as an approximate 95% confidence interval for μ. That is, we want

$$1.96\left(\frac{15}{\sqrt{n}}\right) = 1$$

or, equivalently,

$$\sqrt{n} = 29.4 \qquad \text{and thus} \qquad n \approx 864.36$$

or $n = 865$ because n must be an integer.

It is quite likely that it had not been anticipated that as many as 865 students would be needed in this study. If that is the case, the statistician must discuss with those involved in the experiment whether or not the accuracy and the confidence level could be relaxed some. For illustration, rather than requiring $\bar{x} \pm 1$ to be a 95% confidence interval for μ, possibly $\bar{x} \pm 2$ would be a satisfactory 80% one. If this modification is acceptable, we now have

$$1.282\left(\frac{15}{\sqrt{n}}\right) = 2$$

or, equivalently,

$$\sqrt{n} = 9.615 \qquad \text{and} \qquad n \approx 92.4.$$

Since n must be an integer, we would probably use 93 in practice. Most likely, the persons involved in this project would find this a more reasonable sample size. Of course, any sample size greater than 93 could be used. Then either the length of the confidence interval could be decreased from that of $\bar{x} \pm 2$ or the confidence coefficient could be increased from 80% or a combination of both. Also, since there might be some question of whether the standard deviation σ actually equals 15, the sample standard deviation s would no doubt be used in the construction of the interval. For example, suppose that the sample characteristics observed are

$$n = 145, \qquad \bar{x} = 77.2, \qquad s = 13.2;$$

then

$$\bar{x} \pm \frac{1.282s}{\sqrt{n}} \qquad \text{or} \qquad 77.2 \pm 1.41$$

provides an approximate 80% confidence interval for μ.

In general, if we want the $100(1 - \alpha)\%$ confidence interval for μ, $\bar{x} \pm z_{\alpha/2}(\sigma/\sqrt{n})$, to be no longer than that given by $\bar{x} \pm \varepsilon$, the sample size n is the solution of

$$\varepsilon = \frac{z_{\alpha/2}\sigma}{\sqrt{n}}, \qquad \text{where} \qquad \Phi(z_{\alpha/2}) = 1 - \frac{\alpha}{2}.$$

That is,

$$n = \frac{z_{\alpha/2}^2\sigma^2}{\varepsilon^2}, \tag{6.5-1}$$

where it is assumed that σ^2 is known. We sometimes call $\varepsilon = z_{\alpha/2}\sigma/\sqrt{n}$ the **maximum error of the estimate**. If the experimenter has no idea about the value of σ^2, it may be necessary to first take a preliminary sample to estimate σ^2.

The type of statistic we see most often in newspapers and magazines is an estimate of a proportion p. We might, for example, want to know the percentage of the labor force that is unemployed or the percentage of voters favoring a certain candidate. Sometimes extremely important decisions are made on the basis of these estimates. If this is the case, we would most certainly desire short confidence intervals for p with large confidence coefficients. We recognize that these conditions will require a large sample size. On the other hand, if the fraction p being estimated is not too important, an estimate associated with a longer confidence interval with a smaller confidence coefficient is satisfactory; and thus a smaller sample size can be used.

Example 6.5-2   Suppose we know that the unemployment rate has been about 8% (0.08). However, we wish to update our estimate in order to make an important decision about the national economic policy. Accordingly, let us say we wish to be 99% confident that the new estimate of p is within 0.001 of the true p. If we assume Bernoulli trials (an assumption that might be questioned), the relative frequency y/n, based upon a large sample size n, provides the approximate 99% confidence interval

$$\frac{y}{n} \pm 2.576\sqrt{\frac{(y/n)(1 - y/n)}{n}}.$$

Although we do not know y/n exactly before sampling, we do know, since y/n will be near 0.08, that

$$2.576\sqrt{\frac{(y/n)(1 - y/n)}{n}} \approx 2.576\sqrt{\frac{(0.08)(0.92)}{n}},$$

and we want this number to equal 0.001. That is,

$$2.576\sqrt{\frac{(0.08)(0.92)}{n}} = 0.001$$

or, equivalently,

$$\sqrt{n} = 2576\sqrt{0.0736} \quad \text{and} \quad n \approx 488,394.$$

That is, under our assumptions, such a sample size is needed in order to achieve the reliability and the accuracy desired.

From the preceding example we hope that the student will recognize how important it is to know the sample size (or length of the confidence interval and confidence coefficient) before he or she can place much weight on a statement such as 51% of the voters seem to favor candidate A, 46% favor candidate B, and 3% are undecided. Is this statement based on a sample of 100 or 2000 or 10,000 voters? If we assume Bernoulli trials, the approximate 95% confidence intervals for the fraction of voters favoring candidate A in these cases are, respectively, [0.41, 0.61], [0.49, 0.53], and [0.50, 0.52]. Quite obviously, the first interval, with $n = 100$, does not assure candidate A of the support of at least half the voters, whereas the interval with $n = 10,000$ is more convincing.

In general, to find the required sample size to estimate p, recall that the point estimate of p is $\hat{p} = y/n$ and an approximate $1 - \alpha$ confidence interval for p is

$$\hat{p} \pm z_{\alpha/2}\sqrt{\frac{\hat{p}(1 - \hat{p})}{n}}.$$

Suppose we want an estimate of p that is within ε of the unknown p with $100(1 - \alpha)\%$ confidence where $\varepsilon = z_{\alpha/2}\sqrt{\hat{p}(1 - \hat{p})}/n$ is the **maximum error of the point estimate** $\hat{p} = y/n$. Since $\hat{p}$ is unknown before the experiment is run, we cannot use the value of $\hat{p}$ in our determination of n. However, if it is known that p is about equal to p^*, the necessary sample size n is the solution of

$$\varepsilon = \frac{z_{\alpha/2}\sqrt{p^*(1 - p^*)}}{\sqrt{n}}.$$

That is,

$$n = \frac{z_{\alpha/2}^2 p^*(1 - p^*)}{\varepsilon^2}. \tag{6.5-2}$$

It is often true, however, that we do not have a strong prior idea about p, as we did in Example 6.5-2 about the rate of unemployment. It is interesting to observe that no matter what value p takes between zero and one, it is always true that $p^*(1 - p^*) \leq 1/4$. Hence,

$$n = \frac{z_{\alpha/2}^2 p^*(1 - p^*)}{\varepsilon^2} \leq \frac{z_{\alpha/2}^2}{4\varepsilon^2}.$$

Thus, if we want the $100(1 - \alpha)\%$ confidence interval for p to be no longer than that given by $y/n \pm \varepsilon$, a solution for n that provides this protection is

$$n = \frac{z_{\alpha/2}^2}{4\varepsilon^2}. \qquad (6.5\text{-}3)$$

Example 6.5-3 A possible gubernatorial candidate wants to assess initial support among the voters before making a candidacy announcement. If the fraction p of voters who are favorable, without any advance publicity, is around 0.15, the candidate will enter the race. From a poll of n voters selected at random, the candidate would like the estimate y/n to be within 0.03 of p. That is, the decision will be based on a 95% confidence interval of the form $y/n \pm 0.03$. Since the candidate has no idea about the magnitude of p, a consulting statistician formulates the equation

$$n = \frac{(1.96)^2}{4(0.03)^2} = 1067.11.$$

Thus the sample size should be around 1068 to achieve the desired reliability and accuracy. Suppose that 1068 voters around the state were selected at random and interviewed and $y = 214$ express support for this candidate. Then $\hat{p} = 214/1068 = 0.20$ is a point estimate of p and an approximate 95% confidence interval for p is

$$0.20 \pm 1.96\sqrt{\frac{(0.20)(0.80)}{1068}} \quad \text{or} \quad 0.20 \pm 0.024.$$

That is, we are 95% confident that p belongs to the interval $[0.176, 0.224]$. On the basis of this sample, the candidate decided to run for office. Note that for a confidence coefficient of 95%, we found a sample size so that the maximum error of the estimate would be 0.03. From the data that were collected, the maximum error of the estimate is only 0.024. We ended up with a smaller error because we found the sample size assuming that $p = 0.50$ while, in fact, p is closer to 0.20.

Suppose that you want to estimate the proportion p of a student body that favors a new policy. How large should the sample be? If p is close to 1/2 and you want to be 95% confident that the maximum error of the estimate is $\varepsilon = 0.02$, then

$$n = \frac{(1.96)^2}{4(0.02)^2} = 2401.$$

Such a sample size makes sense at a large university. However, if you are a student at a small college, the entire enrollment could be less than 2401. Thus we now give a procedure that can be used to determine the sample size when the population is small relative to the desired sample size.

Let N equal the size of a population and assume that n_1 of them have a certain characteristic C (e.g., favor a new policy). Let $p = n_1/N$. Then $1 - p = 1 - n_1/N$. If we take a sample of size n without replacement, then X, the number of observations with the characteristic C, has a hypergeometric distribution. The mean and variance of X are (with $N = n$ and $n = r$ in the formulas that were found in Section 3.4)

$$\mu = n\left(\frac{n_1}{N}\right) = np$$

and

$$\sigma^2 = n\left(\frac{n_1}{N}\right)\left(1 - \frac{n_1}{N}\right)\left(\frac{N - n}{N - 1}\right) = np(1 - p)\left(\frac{N - n}{N - 1}\right).$$

The mean and variance of X/n are

$$E\left(\frac{X}{n}\right) = \frac{\mu}{n} = p$$

and

$$\mathrm{Var}\left(\frac{X}{n}\right) = \frac{\sigma^2}{n^2} = \frac{p(1 - p)}{n}\left(\frac{N - n}{N - 1}\right).$$

To find an approximate confidence interval for p, we can use the normal approximation:

$$P\left[-z_{\alpha/2} \le \frac{(X/n) - p}{\sqrt{\dfrac{p(1 - p)}{n}\left(\dfrac{N - n}{N - 1}\right)}} \le z_{\alpha/2}\right] \approx 1 - \alpha.$$

Thus $1 - \alpha \approx$

$$P\left[\frac{X}{n} - z_{\alpha/2}\sqrt{\frac{p(1 - p)}{n}\left(\frac{N - n}{N - 1}\right)} \le p \le \frac{X}{n} + z_{\alpha/2}\sqrt{\frac{p(1 - p)}{n}\left(\frac{N - n}{N - 1}\right)}\right].$$

An approximate $1 - \alpha$ confidence interval for p is, replacing p under the radical with $\hat{p} = x/n$,

$$\hat{p} \pm z_{\alpha/2}\sqrt{\frac{\hat{p}(1 - \hat{p})}{n}\left(\frac{N - n}{N - 1}\right)}.$$

This is similar to the confidence interval for p when the distribution of X is $b(n, p)$. If N is large relative to n, then

$$\frac{N - n}{N - 1} = \frac{1 - n/N}{1 - 1/N} \approx 1,$$

so in this case the two intervals are essentially equal.

Suppose now that we are interested in determining the sample size n that is required to have $1 - \alpha$ confidence that the maximum error of the estimate of p is ε. We let

$$\varepsilon = z_{\alpha/2} \sqrt{\frac{p(1-p)}{n}\left(\frac{N-n}{N-1}\right)}$$

and solve for n. After some simplification we obtain

$$n = \frac{Nz_{\alpha/2}^2 p(1-p)}{(N-1)\varepsilon^2 + z_{\alpha/2}^2 p(1-p)}$$

$$= \frac{z_{\alpha/2}^2 p(1-p)/\varepsilon^2}{\left(\dfrac{N-1}{N}\right) + \dfrac{z_{\alpha/2}^2 p(1-p)/\varepsilon^2}{N}}.$$

If we let

$$m = \frac{z_{\alpha/2}^2 p^*(1-p^*)}{\varepsilon^2}$$

as given by equation (6.5-2), we then choose for our sample size n

$$n = \frac{m}{1 + \dfrac{m-1}{N}}.$$

If we know nothing about p, set $p^* = 1/2$ to determine m. For example, if the size of the student body is $N = 4000$ and, $1 - \alpha = 0.95$, $\varepsilon = 0.02$, and we let $p^* = 1/2$, then $m = 2401$ and

$$n = \frac{2401}{1 + 2400/4000} = 1501$$

rounded up to the nearest integer. Thus we would sample approximately 37.5% of the student body.

Example 6.5-4 Suppose that a college of $N = 3000$ students is interested in assessing student support for a new form for teacher evaluation. To estimate the proportion p in favor of the new form, how large a sample is required so that with 95% confidence the maximum error of the estimate of p is $\varepsilon = 0.03$? If we assume that p is completely unknown, we use $p^* = 1/2$ to obtain

$$m = \frac{(1.96)^2}{4(0.03)^2} = 1068,$$

rounding up to the nearest integer. Thus the desired sample size is

$$n = \frac{1068}{1 + 1067/3000} = 788,$$

rounding up to the nearest integer.

Exercises

6.5-1 Let X equal the tarsus length for a male grackle. Assume that the distribution of X is $N(\mu, 4.84)$. Find the sample size n that is needed so that we are 95% confident that the maximum error of the estimate of μ is 0.4.

6.5-2 Let X equal the excess weight of soap in a "1000-gram" bottle. Assume that the distribution of X is $N(\mu, 169)$. What sample size is required so that we have 95% confidence that the maximum error of the estimate of μ is 1.5?

6.5-3 A company packages powdered soap in "6-pound" boxes. The sample mean and standard deviation of the soap in these boxes are currently 6.09 and 0.02 pounds. If the mean fill can be lowered by 0.01 pounds, \$14,000 would be saved per year. Adjustments were made in the filling equipment.
  (a) How large a sample is needed so that the maximum error of the estimate of the new μ is $\varepsilon = 0.001$ with 90% confidence?
  (b) A random sample of size $n = 1219$ yielded $\bar{x} = 6.048$ and $s = 0.022$. Calculate a 90% confidence interval for μ.
  (c) Estimate the savings per year with these new adjustments.
  (d) Estimate the proportion of boxes that will now weigh less than 6 pounds.

6.5-4 The length in centimeters of $n = 29$ fish (species nezumia) yielded an average length of $\bar{x} = 16.82$ and $s^2 = 34.9$. Determine the size of a new sample so that $\bar{x} \pm 0.5$ is an approximate 95% confidence interval for μ.

6.5-5 A quality engineer wanted to be 98% confident that the maximum error of the estimate of the mean strength, μ, of the left hinge on a vanity cover molded by a machine is 0.25. A preliminary sample of size $n = 32$ parts yielded a sample mean of $\bar{x} = 35.68$ and a standard deviation of $s = 1.723$.
  (a) How large a sample is required?
  (b) Does this seem to be a reasonable sample size, noting that destructive testing is needed to obtain the data?

6.5-6 A light bulb manufacturer sells a light bulb that has a mean life of 1450 hours with a standard deviation of 33.7 hours. A new manufacturing process is being tested and there is interest in knowing the mean life μ of the new bulbs. How large a sample is required so that $\bar{x} \pm 5$ is a 95% confidence interval for μ? You may assume that the change in the standard deviation is minimal.

6.5-7 For a public opinion poll for a close presidential election, let p denote the proportion of voters who favor candidate A. How large a sample should be taken if we want the maximum error of the estimate of p to be equal to
  (a) 0.03 with 95% confidence?

(b) 0.02 with 95% confidence?

(c) 0.03 with 90% confidence?

6.5-8 Let p equal the proportion of all college and university students who would say *yes* to the question, "Would you drink from the same glass as your friend if you suspected that this friend were an AIDS virus carrier?" Find the sample size required to be 95% confident that the maximum error of the estimate of p is 0.025. (For your information, it was reported in *Sociology and Social Research* **72**, 2 [January 1988] that 30 out of 375 San Diego State University students answered *yes* to this question.)

6.5-9 A die has been loaded to change the probability of rolling a 6. In order to estimate p, the new probability of rolling a 6, how many times must the die be rolled so that we are 99% confident that the maximum error of the estimate of p is $\varepsilon = 0.02$?

6.5-10 Some college professors and students examined 137 Canadian Geese for patent schistosome in the year they hatched. Of these 137 birds, 54 were infected. They were interested in estimating p, the proportion of infected birds of this type. For future studies determine the sample size n so that the estimate of p is within $\varepsilon = 0.04$ of the unknown p with 90% confidence.

6.5-11 According to the Department of Health Care of the Elderly, Sherwood Hospital, Nottingham, England, only 25% of the patients who were using canes had canes of the correct length. Suppose that you were interested in estimating the proportion p of Americans who used canes of the correct length. How large a sample is required so that, with 95% confidence, the maximum error of the estimate of p is 0.04?

6.5-12 A seed distributor claims that 80% of its beet seeds will germinate. How many seeds must be tested for germination in order to estimate p, the true proportion that will germinate, so that the maximum error of the estimate is $\varepsilon = 0.03$ with 90% confidence?

6.5-13 Some dentists were interested in studying the fusion of embryonic rat palates by using a standard transplantation technique. When no treatment is used, the probability of fusion approximately equals 0.89. They would like to estimate p, the probability of fusion, when vitamin A is lacking.

(a) How large a sample n of rat embryos is needed for $y/n \pm 0.10$ to be a 95% confidence interval for p?

(b) If $y = 44$ out of $n = 60$ palates showed fusion, give a 95% confidence interval for p.

6.5-14 Let p equal the proportion of New York City residents who feel that the quality of life in New York City has become worse in the past few years. To update the estimate given in Exercise 6.4-9, how large a sample is required to be 98% confident that the maximum error of the estimate of p is 0.025?

6.5-15 Let p equal the proportion of triathletes who suffered a training-related overuse injury during the past year. (See Exercise 6.4-5.) How large a sample would be required to estimate p so that with 95% confidence the maximum error of the estimate of p is 0.04?

6.5-16 Let p equal the proportion of college students who favor a new policy for alcohol consumption on campus. How large a sample is required to estimate p so that with 95% confidence the maximum error of the estimate of p is 0.04 when the size of the student body is
(a) $N = 1500$?
(b) $N = 15,000$?
(c) $N = 25,000$?

6.5-17 Out of 1000 welds that have been made on a tower, it is suspected that 15% of the welds are defective. To estimate p, the proportion of defective welds, how many welds must be inspected to have 95% confidence, approximately, that the maximum error of the estimate of p is 0.04?

6.5-18 If Y_1/n and Y_2/n are the respective relative frequencies of successes associated with the two independent binomial distributions $b(n, p_1)$ and $b(n, p_2)$, compute n such that the approximate probability that the random interval $Y_1/n - Y_2/n \pm 0.05$ covers $p_1 - p_2$ is at least 0.80.

6.6   Maximum Likelihood Estimation

We have used the method of equating empirical and theoretical moments to find estimates for parameters and have also discussed some desirable properties for an estimator. We now discuss the method of maximum likelihood estimation, which is an important method for finding estimates of parameters. This method and some new terminology are introduced using a distribution that is both important and familiar to you.

Let p equal the probability of success in a sequence of Bernoulli trials or the proportion of a large population with a certain characteristic. The method of moments estimate for p is the relative frequency of success (having that characteristic). We now show that the maximum likelihood estimate for p is also the relative frequency of success.

Suppose that X is $b(1, p)$ so that the p.d.f. of X is

$$f(x; p) = p^x(1 - p)^{1-x}, \qquad x = 0, 1, 0 \le p \le 1.$$

Sometimes we write $p \in \Omega = \{p: 0 \le p \le 1\}$ where we use Ω to represent the parameter space, that is, the space of all possible values of the parameter. A random sample $X_1, X_2, \ldots, X_n$ is taken, and the problem is to find an estimator $u(X_1, X_2, \ldots, X_n)$ such that $u(x_1, x_2, \ldots, x_n)$ is a good point estimate of p, where $x_1, x_2, \ldots, x_n$ are the observed values of the random sample. Now the probability that $X_1, X_2, \ldots, X_n$ takes these particular values is

$$P(X_1 = x_1, \ldots, X_n = x_n) = \prod_{i=1}^{n} p^{x_i}(1 - p)^{1-x_i} = p^{\Sigma x_i}(1 - p)^{n - \Sigma x_i},$$

which is the joint p.d.f. of $X_1, X_2, \ldots, X_n$ evaluated at the observed values. One reasonable way to proceed toward finding a good estimate of p is to regard this probability (or joint p.d.f.) as a function of p and find the value of p that maximizes it. That is, find the p value most likely to have produced these sample values. The joint p.d.f., when regarded as a function of p, is frequently called the **likelihood function**. Thus here the likelihood function is

$$
\begin{aligned}
L(p) &= L(p; x_1, x_2, \ldots, x_n) \\
&= f(x_1; p)f(x_2; p) \cdots f(x_n; p) \\
&= p^{\Sigma x_i} (1 - p)^{n - \Sigma x_i}, \qquad 0 \le p \le 1.
\end{aligned}
$$

To find the value of p that maximizes $L(p)$ we first take its derivative for $0 < p < 1$:

$$
\frac{dL(p)}{dp} = \left(\sum x_i \right) p^{\Sigma x_i - 1} (1 - p)^{n - \Sigma x_i} - \left(n - \sum x_i \right) p^{\Sigma x_i} (1 - p)^{n - \Sigma x_i - 1}.
$$

Setting this first derivative equal to zero gives us

$$
p^{\Sigma x_i} (1 - p)^{n - \Sigma x_i} \left[\frac{\Sigma x_i}{p} - \frac{n - \Sigma x_i}{1 - p} \right] = 0.
$$

Since $0 < p < 1$, this equals zero when

$$
\frac{\Sigma x_i}{p} - \frac{n - \Sigma x_i}{1 - p} = 0. \tag{6.6-1}
$$

Multiplying each member of this equation by $p(1 - p)$ and simplifying, we obtain

$$
\sum x_i - np = 0
$$

or, equivalently,

$$
p = \frac{\Sigma x_i}{n} = \bar{x}.
$$

The corresponding statistic, namely $\Sigma X_i/n = \overline{X}$, is called the **maximum likelihood estimator** and is denoted by $\hat{p}$; that is,

$$
\hat{p} = \frac{1}{n} \sum_{i=1}^{n} X_i = \overline{X}.
$$

When finding a maximum likelihood estimator, it is often easier to find the value of the parameter that maximizes the natural logarithm of the likelihood function rather than the value of the parameter that maximizes the likelihood function itself. Because the natural logarithm function is an increasing function, the solutions will be the same. To see this, the example we have been considering gives us, for $0 < p < 1$,

$$
\ln L(p) = \left(\sum_{i=1}^{n} x_i \right) \ln p + \left(n - \sum_{i=1}^{n} x_i \right) \ln (1 - p).
$$

To find the maximum, we set the first derivative equal to zero to obtain

$$\frac{d[\ln L(p)]}{dp} = \left(\sum_{i=1}^{n} x_i\right)\left(\frac{1}{p}\right) + \left(n - \sum_{i=1}^{n} x_i\right)\left(\frac{-1}{1-p}\right) = 0,$$

which is the same as equation (6.6-1). Thus the solution is $p = \bar{x}$ and the maximum likelihood estimator for p is $\hat{p} = \bar{X}$.

Motivated by the preceding illustration, we present the formal definition of maximum likelihood estimators. This definition is used in both the discrete and continuous cases.

Let $X_1, X_2, \ldots, X_n$ be a random sample from a distribution with p.d.f. $f(x; \theta_1, \theta_2, \ldots, \theta_m)$, which depends on one or more unknown parameters θ_1, $\theta_2, \ldots, \theta_m$. Suppose $(\theta_1, \theta_2, \ldots, \theta_m)$ is restricted to a given parameter space Ω. Then the joint p.d.f. of $X_1, X_2, \ldots, X_n$, namely

$$L(\theta_1, \theta_2, \ldots, \theta_m) = f(x_1; \theta_1, \ldots, \theta_m)f(x_2; \theta_1, \ldots, \theta_m)$$
$$\cdots f(x_n; \theta_1, \ldots, \theta_m), \qquad (\theta_1, \theta_2, \ldots, \theta_m) \in \Omega,$$

when regarded as a function of $\theta_1, \theta_2, \ldots, \theta_m$, is called the **likelihood function**. Say

$$[u_1(x_1, \ldots, x_n), u_2(x_1, \ldots, x_n), \ldots, u_m(x_1, \ldots, x_n)]$$

is that m-tuple in Ω that maximizes $L(\theta_1, \theta_2, \ldots, \theta_m)$. Then

$$\hat{\theta}_1 = u_1(X_1, \ldots, X_n),$$
$$\hat{\theta}_2 = u_2(X_1, \ldots, X_n),$$
$$\vdots$$
$$\hat{\theta}_m = u_m(X_1, \ldots, X_n)$$

are maximum likelihood estimators of $\theta_1, \theta_2, \ldots, \theta_m$, respectively; and the corresponding observed values of these statistics, namely

$$u_1(x_1, \ldots, x_n), u_2(x_1, \ldots, x_n), \ldots, u_m(x_1, \ldots, x_n),$$

are called **maximum likelihood estimates**. In many practical cases, these estimators (and estimates) are unique.

For many applications there is just one unknown parameter. In these cases the likelihood function is given by

$$L(\theta) = \prod_{i=1}^{n} f(x_i; \theta).$$

Some additional examples will help clarify these definitions.

Example 6.6-1 Let $X_1, X_2, \ldots, X_n$ be a random sample from the exponential distribution with p.d.f.

$$f(x; \theta) = \frac{1}{\theta} e^{-x/\theta}, \qquad 0 < x < \infty, \quad \theta \in \Omega = \{\theta: 0 < \theta < \infty\}.$$

The likelihood function is given by

$$L(\theta) = L(\theta; x_1, x_2, \ldots, x_n)$$

$$= \left(\frac{1}{\theta}e^{-x_1/\theta}\right)\left(\frac{1}{\theta}e^{-x_2/\theta}\right)\cdots\left(\frac{1}{\theta}e^{-x_n/\theta}\right)$$

$$= \frac{1}{\theta^n}\exp\left(\frac{-\sum_{i=1}^{n}x_i}{\theta}\right), \qquad 0 < \theta < \infty.$$

The natural logarithm of $L(\theta)$ is

$$\ln L(\theta) = -(n)\ln(\theta) - \frac{1}{\theta}\sum_{1}^{n}x_i, \qquad 0 < \theta < \infty.$$

Thus

$$\frac{d[\ln L(\theta)]}{d\theta} = \frac{-n}{\theta} + \frac{\sum_{1}^{n}x_i}{\theta^2} = 0.$$

The solution of this equation for θ is

$$\theta = \frac{1}{n}\sum_{1}^{n}x_i = \bar{x}.$$

Note that

$$\frac{d[\ln L(\theta)]}{d\theta} = \frac{1}{\theta}\left(-n + \frac{n\bar{x}}{\theta}\right) \begin{array}{ll} >0, & \theta < \bar{x}, \\ =0, & \theta = \bar{x}, \\ <0, & \theta > \bar{x}. \end{array}$$

Hence $\ln L(\theta)$ does have a maximum at $\bar{x}$, and thus the maximum likelihood estimator for θ is

$$\hat{\theta} = \bar{X} = \frac{1}{n}\sum_{i=1}^{n}X_i.$$

This is both an unbiased estimator and the method of moments estimator for θ.

Example 6.6-2   Let $X_1, X_2, \ldots, X_n$ be a random sample from the geometric distribution with p.d.f. $f(x; p) = (1 - p)^{x-1}p$, $x = 1, 2, 3, \ldots$. The likelihood function is given by

$$L(p) = (1 - p)^{x_1-1}p(1 - p)^{x_2-1}p \cdots (1 - p)^{x_n-1}p$$
$$= p^n(1 - p)^{\sum x_i - n}, \qquad 0 \le p \le 1.$$

The natural logarithm of $L(p)$ is

$$\ln L(p) = n \ln p + \left(\sum_{i=1}^{n}x_i - n\right)\ln(1 - p), \qquad 0 < p < 1.$$

Thus, restricting p to $0 < p < 1$ so as to be able to take the derivative, we have

$$\frac{d \ln L(p)}{dp} = \frac{n}{p} - \frac{\sum_{i=1}^{n} x_i - n}{1 - p} = 0.$$

Solving for p, we obtain

$$p = \frac{n}{\sum_{i=1}^{n} x_i} = \frac{1}{\bar{x}}.$$

So the maximum likelihood estimator of p is

$$\widehat{p} = \frac{n}{\sum_{i=1}^{n} X_i} = \frac{1}{\bar{X}}.$$

Again this estimator is the method of moments estimator, and it agrees with our intuition because, in n observations of a geometric random variable, there are n successes in the $\sum_{i=1}^{n} x_i$ trials. Thus the estimate of p is the number of successes divided by the total number of trials.

In the following important example we find the maximum likelihood estimators of the parameters associated with the normal distribution.

Example 6.6-3 Let $X_1, X_2, \ldots, X_n$ be a random sample from $N(\theta_1, \theta_2)$, where

$$\Omega = \{(\theta_1, \theta_2): -\infty < \theta_1 < \infty, \quad 0 < \theta_2 < \infty\}.$$

That is, here we let $\theta_1 = \mu$ and $\theta_2 = \sigma^2$. Then

$$L(\theta_1, \theta_2) = \prod_{i=1}^{n} \frac{1}{\sqrt{2\pi\theta_2}} \exp\left[-\frac{(x_i - \theta_1)^2}{2\theta_2} \right]$$

or, equivalently,

$$L(\theta_1, \theta_2) = \left(\frac{1}{\sqrt{2\pi\theta_2}} \right)^n \exp\left[\frac{-\sum_{1}^{n} (x_i - \theta_1)^2}{2\theta_2} \right], \qquad (\theta_1, \theta_2) \in \Omega.$$

The natural logarithm of the likelihood function is

$$\ln L(\theta_1, \theta_2) = -\frac{n}{2} \ln (2\pi\theta_2) - \frac{\sum_{1}^{n} (x_i - \theta_1)^2}{2\theta_2}.$$

The partial derivatives with respect to θ_1 and θ_2 are

$$\frac{\partial(\ln L)}{\partial\theta_1} = \frac{1}{\theta_2} \sum_1^n (x_i - \theta_1)$$

and

$$\frac{\partial(\ln L)}{\partial\theta_2} = \frac{-n}{2\theta_2} + \frac{1}{2\theta_2^2} \sum_1^n (x_i - \theta_1)^2.$$

The equation $\partial(\ln L)/\partial\theta_1 = 0$ has the solution $\theta_1 = \bar{x}$. Setting $\partial(\ln L)/\partial\theta_2 = 0$ and replacing θ_1 by $\bar{x}$ yields

$$\theta_2 = \frac{1}{n} \sum_1^n (x_i - \bar{x})^2.$$

By considering the usual condition on the second partial derivatives, these solutions do provide a maximum. Thus the maximum likelihood estimators of $\mu = \theta_1$ and $\sigma^2 = \theta_2$ are

$$\widehat{\theta}_1 = \bar{X} \qquad \text{and} \qquad \widehat{\theta}_2 = \frac{1}{n} \sum_1^n (X_i - \bar{X})^2 = V.$$

From Examples 6.1-5 and 6.6-3, we see that the method of moments estimators and the maximum likelihood estimators for μ and σ^2 are the same. But this is not always the case, as illustrated by Example 6.1-4 and Exercise 6.6-6. If they are not the same, which is better? In Section 6.7, we discuss the fact that the maximum likelihood estimator $\widehat{\theta}$ of θ has an approximate normal distribution with mean θ and a variance that is equal to a certain lower bound. Thus, at least approximately, $\widehat{\theta}$ is that unbiased minimum variance estimator. Accordingly, most statisticians prefer the maximum likelihood estimator $\widehat{\theta}$ to $\overline{\theta}$, the estimator found using the method of moments.

Exercises

6.6-1   Let $X_1, X_2, \ldots, X_n$ be a random sample from $N(\mu, \sigma^2)$ where the mean $\theta = \mu$ is such that $-\infty < \theta < \infty$ and σ^2 is a known positive number. Show that the maximum likelihood estimator for θ is $\widehat{\theta} = \bar{X}$.

6.6-2   A random sample $X_1, X_2, \ldots, X_n$ of size n is taken from $N(\mu, \sigma^2)$, where the variance $\theta = \sigma^2$ is such that $0 < \theta < \infty$ and μ is a known real number. Show that the maximum likelihood estimator for θ is $\widehat{\theta} = (1/n) \sum (X_i - \mu)^2$.

6.6-3   A random sample $X_1, X_2, \ldots, X_n$ of size n is taken from a Poisson distribution with a mean of λ, $0 < \lambda < \infty$.
(a) Show that the maximum likelihood estimator for λ is $\widehat{\lambda} = \bar{X}$.

(b) Let X equal the number of flaws per 100 feet of a used computer tape. Assume that X has a Poisson distribution with a mean of λ. If 40 observations of X yielded 5 zeros, 7 ones, 12 twos, 9 threes, 5 fours, 1 five, and 1 six, find the maximum likelihood estimate of λ.

6.6-4 Let $X_1, X_2, \ldots, X_n$ be a random sample from distributions with the following probability density functions. In each case find the maximum likelihood estimator $\widehat{\theta}$.

(a) $f(x; \theta) = (1/\theta^2)xe^{-x/\theta}$, $0 < x < \infty$, $0 < \theta < \infty$.

(b) $f(x; \theta) = (1/2\theta^3)x^2e^{-x/\theta}$, $0 < x < \infty$, $0 < \theta < \infty$.

(c) $f(x; \theta) = (1/2)e^{-|x-\theta|}$, $-\infty < x < \infty$, $-\infty < \theta < \infty$.

6.6-5 Find the maximum likelihood estimates for $\theta_1 = \mu$ and $\theta_2 = \sigma^2$ if a random sample of size 15 from $N(\mu, \sigma^2)$ yielded the following values:

$$
\begin{array}{ccccc}
31.5 & 36.9 & 33.8 & 30.1 & 33.9 \\
35.2 & 29.6 & 34.4 & 30.5 & 34.2 \\
31.6 & 36.7 & 35.8 & 34.5 & 32.7
\end{array}
$$

6.6-6 Let $f(x; \theta) = \theta x^{\theta-1}$, $0 < x < 1$, $\theta \in \Omega = \{\theta: 0 < \theta < \infty\}$. Let X_1, $X_2, \ldots, X_n$ denote a random sample of size n from this distribution.

(a) Sketch the p.d.f. of X for (i) $\theta = 1/2$, (ii) $\theta = 1$, and (iii) $\theta = 2$.

(b) Show that $\widehat{\theta} = -n/\ln \prod_{i=1}^{n} X_i$ is the maximum likelihood estimator of θ.

(c) For each of the following three sets of 10 observations, calculate the maximum likelihood estimate (note that in Exercise 6.1-9 you were asked to find the method of moments estimates for θ):

$$
\begin{array}{llllll}
\text{(i)} & 0.0256 & 0.3051 & 0.0278 & 0.8971 & 0.0739 \\
& 0.3191 & 0.7379 & 0.3671 & 0.9763 & 0.0102 \\
\text{(ii)} & 0.9960 & 0.3125 & 0.4374 & 0.7464 & 0.8278 \\
& 0.9518 & 0.9924 & 0.7112 & 0.2228 & 0.8609 \\
\text{(iii)} & 0.4698 & 0.3675 & 0.5991 & 0.9513 & 0.6049 \\
& 0.9917 & 0.1551 & 0.0710 & 0.2110 & 0.2154
\end{array}
$$

(d) Sketch the empirical and theoretical distribution functions (using $\widehat{\theta}$ as the value of the parameter) on the same graph for each set of data. Comment on the fit.

6.6-7 Out of 50,000,000 instant winner lottery tickets, the proportion of winning tickets is p. Each day, for 20 consecutive days, a bettor purchased tickets, one at a time, until a winning ticket was purchased. The numbers of tickets that were purchased each day to obtain the winning ticket were

$$
\begin{array}{cccccccccc}
1 & 26 & 19 & 6 & 6 & 1 & 2 & 3 & 1 & 23 \\
19 & 3 & 6 & 8 & 4 & 1 & 18 & 34 & 1 & 8
\end{array}
$$

By making reasonable assumptions, find the maximum likelihood estimate of p based on these data.

6.6-8 Let $f(x; \theta) = (1/\theta)x^{(1-\theta)/\theta}$, $0 < x < 1$, $0 < \theta < \infty$.
  (a) Show that the maximum likelihood estimator of θ is $\widehat{\theta} = -(1/n) \sum_{i=1}^{n} \ln X_i$.
  (b) Show that $E(\widehat{\theta}) = \theta$ and thus $\widehat{\theta}$ is an unbiased estimator of θ.

★6.7 Asymptotic Distributions of Maximum Likelihood Estimators

Let us consider a distribution with p.d.f. $f(x; \theta)$ such that the parameter θ is not involved in the support of the distribution. Moreover, we want $f(x; \theta)$ to enjoy a number of mathematical properties that we do not list here. However, in particular, we want to be able to find the maximum likelihood estimator $\widehat{\theta}$ by solving

$$\frac{\partial[\ln L(\theta)]}{\partial \theta} = 0,$$

where here we use a partial derivative sign because $L(\theta)$ involves $x_1, x_2, \ldots, x_n$ too.

That is,

$$\frac{\partial[\ln L(\widehat{\theta})]}{\partial \theta} = 0,$$

where now, with $\widehat{\theta}$ in this expression, $L(\widehat{\theta}) = f(X_1; \widehat{\theta})f(X_2; \widehat{\theta}) \cdots f(X_n; \widehat{\theta})$. We can approximate the left-hand member of this latter equation by a linear function found from the first two terms of a Taylor's series expanded about θ, namely

$$\frac{\partial[\ln L(\theta)]}{\partial \theta} + (\widehat{\theta} - \theta)\frac{\partial^2[\ln L(\theta)]}{\partial \theta^2} \approx 0,$$

when $L(\theta) = f(X_1; \theta)f(X_2; \theta) \cdots f(X_n; \theta)$.

Obviously, this approximation is good enough only if $\widehat{\theta}$ is close to θ, and an adequate mathematical proof involves those conditions that we have not given here. But a heuristic argument can be made by solving for $\widehat{\theta} - \theta$ to obtain

$$\widehat{\theta} - \theta = \frac{\dfrac{\partial[\ln L(\theta)]}{\partial \theta}}{-\dfrac{\partial^2[\ln L(\theta)]}{\partial \theta^2}}. \tag{6.7-1}$$

Recall that

$$\ln L(\theta) = \ln f(X_1; \theta) + \ln f(X_2; \theta) + \cdots + \ln f(X_n; \theta)$$

and

$$\frac{\partial \ln L(\theta)}{\partial \theta} = \sum_{i=1}^{n} \frac{\partial[\ln f(X_i; \theta)]}{\partial \theta}, \tag{6.7-2}$$

which is the numerator of (6.7-1). However, this expression (6.7-2) is the sum of the n independent and identically distributed random variables

$$Y_i = \frac{\partial[\ln f(X_i; \theta)]}{\partial \theta}, \quad i = 1, 2, \ldots, n;$$

and thus, by the Central Limit Theorem, has an approximate normal distribution with mean (in the continuous case) equal to

$$\int_{-\infty}^{\infty} \frac{\partial[\ln f(x; \theta)]}{\partial \theta} f(x; \theta) \, dx = \int_{-\infty}^{\infty} \frac{\partial[f(x; \theta)]}{\partial \theta} \frac{f(x; \theta)}{f(x; \theta)} \, dx$$

$$= \int_{-\infty}^{\infty} \frac{\partial[f(x; \theta)]}{\partial \theta} \, dx$$

$$= \frac{\partial}{\partial \theta} \left[\int_{-\infty}^{\infty} f(x; \theta) \, dx \right]$$

$$= \frac{\partial}{\partial \theta}[1]$$

$$= 0.$$

Clearly, we need the mathematical condition that it is permissible to interchange the operations of integration and differentiation in those last steps. Of course, the integral of $f(x; \theta)$ is equal to one because it is a p.d.f.

Since we now know that the mean of each Y is

$$\int_{-\infty}^{\infty} \frac{\partial[\ln f(x; \theta)]}{\partial \theta} f(x; \theta) dx = 0,$$

let us take derivatives of each member of this equation with respect to θ, obtaining

$$\int_{-\infty}^{\infty} \left\{ \frac{\partial^2[\ln f(x; \theta)]}{\partial \theta^2} f(x; \theta) + \frac{\partial[\ln f(x; \theta)]}{\partial \theta} \frac{\partial[f(x; \theta)]}{\partial \theta} \right\} dx = 0.$$

However,

$$\frac{\partial[f(x; \theta)]}{\partial \theta} = \frac{\partial[\ln f(x; \theta)]}{\partial \theta} f(x; \theta);$$

so

$$\int_{-\infty}^{\infty} \left\{ \frac{\partial[\ln f(x; \theta)]}{\partial \theta} \right\}^2 f(x; \theta) \, dx = -\int_{-\infty}^{\infty} \frac{\partial^2[\ln f(x; \theta)]}{\partial \theta^2} f(x; \theta) \, dx.$$

Since $E(Y) = 0$, this last expression provides the variance of $Y = \partial[\ln f(X; \theta)]/\partial \theta$. Then the variance of expression (6.7-2) is n times this value, namely

$$-nE\left\{ \frac{\partial^2[\ln f(X; \theta)]}{\partial \theta^2} \right\}.$$

Let us rewrite (6.7-1) as

$$\frac{\sqrt{n}(\hat{\theta} - \theta)}{1/\sqrt{-E\{\partial^2[\ln f(X;\ \theta)]/\partial \theta^2\}}} = \frac{\dfrac{\partial[\ln L(\theta)]/\partial \theta}{\sqrt{-nE\{\partial^2[\ln f(X;\ \theta)]/\partial \theta^2\}}}}{\dfrac{-\dfrac{1}{n}\dfrac{\partial^2[\ln L(\theta)]}{\partial \theta^2}}{E\{-\partial^2[\ln f(X;\ \theta)]/\partial \theta^2\}}}. \qquad (6.7\text{-}3)$$

The numerator of (6.7-3) has an approximate $N(0, 1)$ distribution; and those un-stated mathematical conditions require, in some sense, for

$$-\frac{1}{n}\frac{\partial^2[\ln L(\theta)]}{\partial \theta^2} \qquad \text{to converge to} \qquad E\{-\partial^2[\ln f(X;\ \theta)]/\partial \theta^2\}.$$

Accordingly, the ratios given in equation (6.7-3) must be approximately $N(0, 1)$. That is, $\hat{\theta}$ has an approximate normal distribution with mean θ and standard deviation

$$\frac{1}{\sqrt{-nE\{\partial^2[\ln f(X;\ \theta)]/\partial \theta^2\}}}.$$

Example 6.7-1 (continuation of Example 6.6-1). With the underlying exponential p.d.f.

$$f(x;\ \theta) = \frac{1}{\theta}e^{-x/\theta},\ 0 < x < \infty,\ \theta \in \Omega = \{\theta:\ 0 < \theta < \infty\},$$

$\bar{X}$ is the maximum likelihood estimator. Since

$$\ln f(x;\ \theta) = -\ln \theta - \frac{x}{\theta}$$

and

$$\frac{\partial[\ln f(x;\ \theta)]}{\partial \theta} = -\frac{1}{\theta} + \frac{x}{\theta^2} \qquad \text{and} \qquad \frac{\partial^2[\ln f(x;\ \theta)]}{\partial \theta^2} = \frac{1}{\theta^2} - \frac{2x}{\theta^3},$$

we have

$$-E\left[\frac{1}{\theta^2} - \frac{2X}{\theta^3}\right] = -\frac{1}{\theta^2} + \frac{2\theta}{\theta^3} = \frac{1}{\theta^2}$$

because $E(X) = \theta$. That is, $\bar{X}$ has an approximate normal distribution with mean θ and standard deviation $\theta/\sqrt{n}$. Thus the random interval $\bar{X} \pm 1.96\ (\theta/\sqrt{n})$ has an approximate probability of 0.95 for covering θ. Substituting the observed $\bar{x}$ for θ, as well as for $\bar{X}$, we say that $\bar{x} \pm 1.96\bar{x}/\sqrt{n}$ is an approximate 95% confidence interval for θ.

Example 6.7-2 (continuation of Exercise 6.6-3). The maximum likelihood estimator for λ in

$$f(x; \lambda) = \frac{\lambda^x e^{-\lambda}}{x!}, \quad x = 0, 1, 2, \ldots; \; \theta \in \Omega = \{\theta: 0 < \theta < \infty\},$$

is $\widehat{\lambda} = \overline{X}$. Now

$$\ln f(x; \lambda) = x \ln \lambda - \lambda - \ln x!$$

and

$$\frac{\partial[\ln f(x; \lambda)]}{\partial \lambda} = \frac{x}{\lambda} - 1 \quad \text{and} \quad \frac{\partial^2[\ln f(x; \lambda)]}{\partial \lambda^2} = -\frac{x}{\lambda^2}.$$

Thus

$$-E\left(-\frac{X}{\lambda^2}\right) = \frac{\lambda}{\lambda^2} = \frac{1}{\lambda}$$

and $\widehat{\lambda} = \overline{X}$ has an approximate normal distribution with mean λ and standard deviation $\sqrt{\lambda/n}$. Finally $\overline{x} \pm 1.645\sqrt{\overline{x}/n}$ serves as an approximate 90% confidence interval for λ. With the data in Exercise 6.6-3, $\overline{x} = 2.225$ and hence this interval is from 1.887 to 2.563.

It is interesting that there is another theorem which is somewhat related to the preceding result in that the variance of $\widehat{\theta}$ serves as a lower bound for the variance of every unbiased estimator of θ. Thus we know that if a certain unbiased estimator has a variance equal to that lower bound, we cannot find a better one and hence it is the best in the sense of being the unbiased minimum variance estimator. So, in the limit, the maximum likelihood estimator is this type of best estimator.

We describe this **Rao–Cramér Inequality** here without proof. Let $X_1, X_2, \ldots,$ X_n be a random sample from a distribution with p.d.f. $f(x; \theta)$, $\theta \in \Omega = \{\theta: c < \theta < d\}$, where the support of X does not depend upon θ so that we can differentiate, with respect to θ, under integral signs like that in the following integral:

$$\int_{-\infty}^{\infty} f(x; \theta) \, dx = 1.$$

If $Y = u(X_1, X_2, \ldots, X_n)$ is an unbiased estimator of θ, then

$$\mathrm{Var}(Y) \geq \frac{1}{n \int_{-\infty}^{\infty} \{[\partial \ln f(x; \theta)/\partial \theta]\}^2 f(x; \theta) \, dx}$$

$$= \frac{-1}{n \int_{-\infty}^{\infty} [\partial^2 \ln f(x; \theta)/\partial \theta^2] f(x; \theta) \, dx}.$$

Note that the two integrals in the respective denominators are the expectations

$$E\left\{\left[\frac{\partial \ln f(X; \theta)}{\partial \theta}\right]^2\right\} \quad \text{and} \quad E\left[\frac{\partial^2 \ln f(X; \theta)}{\partial \theta^2}\right];$$

sometimes one is easier to compute than the other.

Note that we have computed this lower bound for each of two distributions: exponential and Poisson. Those respective lower bounds were θ^2/n and λ/n; see Examples 6.7-1 and 6.7-2. Since, in each case, the variance of $\overline{X}$ equals the lower bound, then $\overline{X}$ is the unbiased minimum variance estimator.

Let us consider another example.

> **Example 6.7-3**   (continuation of Exercise 6.6-6). Since the sample arises from a distribution with p.d.f.
>
> $$f(x; \theta) = \theta x^{\theta-1}, \ 0 < x < 1, \ \theta \in \Omega = \{\theta: 0 < \theta < \infty\},$$
>
> we have
>
> $$\ln f(x; \theta) = \ln \theta + (\theta - 1)\ln x,$$
>
> $$\frac{\partial \ln f(x; \theta)}{\partial \theta} = \frac{1}{\theta} + \ln x,$$
>
> and
>
> $$\frac{\partial^2 \ln f(x; \theta)}{\partial \theta^2} = -\frac{1}{\theta^2}$$
>
> Since $E(-1/\theta^2) = -1/\theta^2$, the lower bound of the variance of every unbiased estimator of θ is θ^2/n. Moreover, the maximum likelihood estimator $\widehat{\theta} = -n/\ln \prod_{i=1}^{n} X_i$ has an approximate normal distribution with mean θ and variance θ^2/n. Thus, in a limiting sense, $\widehat{\theta}$ is the unbiased minimum variance estimator of θ.

To measure the value of estimators, their variances are compared to the Rao–Cramér lower bound. The ratio of the Rao–Cramér lower bound to the actual variance of any unbiased estimator is called the **efficiency** of that estimator. An estimator with efficiency of 50%, say, requires that $1/0.5 = 2$ times as many sample observations are needed to do as well in estimation as can be done with the unbiased minimum variance estimator (the 100% efficient estimator).

Another property of a good estimator, that of sufficiency, is considered in Section 9.7.

Exercises

6.7-1   Let $X_1, X_2, \ldots, X_n$ be a random sample of size n from the exponential distribution whose p.d.f. is $f(x; \theta) = (1/\theta)e^{-x/\theta}, 0 < x < \infty, 0 < \theta < \infty$. We know that $\overline{X}$ is an unbiased estimator for θ and the variance of $\overline{X}$ is θ^2/n. (See Exercise 6.1-1.) Show that the Rao–Cramér lower bound is also equal to θ^2/n so that $\overline{X}$ is the best unbiased estimator for θ.

6.7-2   Let $X_1, X_2, \ldots, X_n$ be a random sample from $N(\theta, \sigma^2)$, where σ^2 is known.
(a) Show that $Y = (X_1 + X_2)/2$ is an unbiased estimator of θ.

    (b) Find the Rao–Cramér lower bound for the variance of an unbiased estimator of θ.

    (c) If $n = 10$, what is the efficiency of Y in part (a)?

6.7-3 Let $X_1, X_2, \ldots, X_n$ denote a random sample from $b(1, p)$. We know that $\overline{X}$ is an unbiased estimator of p and that $\text{Var}(\overline{X}) = p(1 - p)/n$. (See Exercise 6.1-3.)

    (a) Find the Rao–Cramér lower bound for $\overline{X}$.

    (b) What is the efficiency of $\overline{X}$ as an estimator of p?

6.7-4 (Continuation of Exercise 6.6-2.) When sampling from a normal distribution with known mean μ, $\widehat{\theta} = \Sigma_{i=1}^{n} (X_i - \mu)^2/n$ is the maximum likelihood estimator of $\theta = \sigma^2$.

    (a) Determine the Rao–Cramér lower bound.

    (b) What is the approximate distribution of $\widehat{\theta}$?

    (c) What is the exact distribution of $n\widehat{\theta}/\theta$, where $\theta = \sigma^2$?

6.7-5 Find the Rao–Cramér lower bound and thus the asymptotic variance of the maximum likelihood estimator $\widehat{\theta}$ if the random sample $X_1, X_2, \ldots, X_n$ is taken from each of the distributions having the following p.d.f.'s:

    (a) $f(x; \theta) = (1/\theta^2)xe^{-x/\theta}, \ 0 < x < \infty, \ 0 < \theta < \infty$.

    (b) $f(x; \theta) = (1/2\theta^3)x^2e^{-x/\theta}, \ 0 < x < \infty, \ 0 < \theta < \infty$.

    (c) $f(x; \theta) = (1/\theta)x^{(1-\theta)/\theta}, \ 0 < x < 1, \ 0 < \theta < \infty$.

★6.8 Chebyshev's Inequality

    In this section we use Chebyshev's inequality to show, in another sense, that the sample mean, $\bar{x}$, is a good statistic to use to estimate a population mean μ; the relative frequency of success in n Bernoulli trials, y/n, is a good statistic for estimating p; and the empirical distribution function, $F_n(x)$, can be used to estimate the theoretical distribution function $F(x)$. The effect of the sample size n on these estimates is discussed.

    We begin by showing that Chebyshev's inequality gives added significance to the standard deviation in terms of bounding certain probabilities. The inequality is valid for all distributions for which the standard deviation exists. The proof is given for the discrete case, but it holds for the continuous case with integrals replacing summations.

THEOREM 6.8-1 **(Chebyshev's Inequality)** *If the random variable X has a mean μ and variance σ^2, then for every $k \geq 1$,*

$$P(|X - \mu| \geq k\sigma) \leq \frac{1}{k^2}.$$

PROOF: Let $f(x)$ denote the p.d.f. of X. Then

$$\sigma^2 = E[(X - \mu)^2] = \sum_{x \in R} (x - \mu)^2 f(x)$$

$$= \sum_{x \in A} (x - \mu)^2 f(x) + \sum_{x \in A'} (x - \mu)^2 f(x), \qquad (6.8\text{-}1)$$

where

$$A = \{x \colon |x - \mu| \ge k\sigma\}.$$

The second term in the righthand member of equation (6.8-1) is the sum of nonnegative numbers and thus is greater than or equal to zero. Hence

$$\sigma^2 \ge \sum_{x \in A} (x - \mu)^2 f(x).$$

However, in A, $|x - \mu| \ge k\sigma$; so

$$\sigma^2 \ge \sum_{x \in A} (k\sigma)^2 f(x) = k^2 \sigma^2 \sum_{x \in A} f(x).$$

But the latter summation equals $P(X \in A)$, and thus

$$\sigma^2 \ge k^2 \sigma^2 P(X \in A) = k^2 \sigma^2 P(|X - \mu| \ge k\sigma).$$

That is,

$$P(|X - \mu| \ge k\sigma) \le \frac{1}{k^2}. \qquad \square$$

COROLLARY 6.8-1   *If $\varepsilon = k\sigma$, then*

$$P(|X - \mu| \ge \varepsilon) \le \frac{\sigma^2}{\varepsilon^2}.$$

     In words, Chebyshev's inequality states that the probability that X differs from its mean by at least k standard deviations is less than or equal to $1/k^2$. It follows that the probability that X differs from its mean by less than k standard deviations is at least $1 - 1/k^2$. That is,

$$P(|X - \mu| < k\sigma) \ge 1 - \frac{1}{k^2}.$$

From the corollary, it also follows that

$$P(|X - \mu| < \varepsilon) \ge 1 - \frac{\sigma^2}{\varepsilon^2}.$$

Thus Chebyshev's inequality can be used as a bound for certain probabilities. However, in many instances, the bound is not very close to the true probability.

> Example 6.8-1   If it is known that X has a mean of 25 and a variance of 16, then, since $\sigma = 4$, a lower bound for $P(17 < X < 33)$ is given by
>
> $$P(17 < X < 33) = P(|X - 25| < 8)$$
>
> $$= P(|X - \mu| < 2\sigma) \geq 1 - \frac{1}{4} = 0.75$$
>
> and an upper bound for $P(|X - 25| \geq 12)$ is found to be
>
> $$P(|X - 25| \geq 12) = P(|X - \mu| \geq 3\sigma) \leq \frac{1}{9}.$$

Note that the results of the last example hold for any distribution with mean 25 and standard deviation 4. But, even stronger, the probability that any random variable X differs from its mean by 3 or more standard deviations is at most 1/9 by letting $k = 3$ in the theorem. Also the probability that any random variable X differs from its mean by less than 2 standard deviations is at least 3/4 by letting $k = 2$.

The following consideration partially indicates the value of Chebyshev's inequality in theoretical discussions. If Y is the number of successes in n Bernoulli trials with probability p of success on each trial, then Y is $b(n, p)$. Furthermore, Y/n gives the relative frequency of success, and, when p is unknown, Y/n can be used as an estimate of p. To gain some insight into the closeness of Y/n to p, we shall use Chebyshev's inequality. With $\varepsilon > 0$, we note that

$$P\left(\left|\frac{Y}{n} - p\right| \geq \varepsilon\right) = P(|Y - np| \geq n\varepsilon)$$

$$= P\left(|Y - np| \geq \frac{\sqrt{n\varepsilon}}{\sqrt{pq}}\sqrt{npq}\right).$$

However, $\mu = np$ and $\sigma = \sqrt{npq}$ are the mean and the standard deviation of Y so that, with $k = \sqrt{n}\varepsilon/\sqrt{pq}$, we have

$$P\left(\left|\frac{Y}{n} - p\right| \geq \varepsilon\right) = P(|Y - \mu| \geq k\sigma) \leq \frac{1}{k^2} = \frac{pq}{n\varepsilon^2}$$

or, equivalently,

$$P\left(\left|\frac{Y}{n} - p\right| < \varepsilon\right) \geq 1 - \frac{pq}{n\varepsilon^2}. \qquad (6.8\text{-}2)$$

When p is completely unknown, we can use the fact that $pq = p(1 - p)$ is a maximum when $p = 1/2$ in order to find a lower bound for the probability in equation (6.8-2). That is,

$$1 - \frac{pq}{n\varepsilon^2} \geq 1 - \frac{(1/2)(1/2)}{n\varepsilon^2}.$$

For illustration, if $\varepsilon = 0.05$ and $n = 400$,

$$P\left(\left|\frac{Y}{400} - p\right| < 0.05\right) \geq 1 - \frac{(1/2)(1/2)}{400(0.0025)} = 0.75.$$

If, on the other hand, it is known that p is close to $1/10$, we would have

$$P\left(\left|\frac{Y}{400} - p\right| < 0.05\right) \geq 1 - \frac{(0.1)(0.9)}{400(0.0025)} = 0.91.$$

Note that Chebyshev's inequality is applicable to all distributions with a finite variance, and thus the bound is not always a tight one; that is, the bound is not necessarily close to the true probability.

In general, however, it should be noted that with fixed $\varepsilon > 0$ and $0 < p < 1$, we have that

$$\lim_{n\to\infty} P\left(\left|\frac{Y}{n} - p\right| < \varepsilon\right) \geq \lim_{n\to\infty}\left(1 - \frac{pq}{n\varepsilon^2}\right) = 1.$$

But since the probability of every event is less than or equal to 1, it must be that

$$\lim_{n\to\infty} P\left(\left|\frac{Y}{n} - p\right| < \varepsilon\right) = 1.$$

That is, the probability that the relative frequency Y/n is within ε of p is close to 1 when n is large enough. This is one form of the **law of large numbers**, which we alluded to in our description of the relative frequency interpretation of probability in Section 1.1.

This theoretical result has something to contribute to our understanding of the properties of the empirical distribution function, $F_n(x)$. For a fixed x, $F_n(x)$ is the proportion of sample observations that are less than or equal to x. That is, if "success" is an observed X being less than or equal to the fixed x, then $F_n(x)$ is the relative frequency of success. The law of large numbers states that this converges, in that probabilistic sense, to the true probability $P(X \leq x)$ of success, namely $p = F(x)$. Thus $F_n(x)$ does approach $F(x)$ in this sense and hence provides an estimate of the distribution function. This property is used in Section 10.7.

A more general form of the law of large numbers is found by considering the mean $\overline{X}$ of a random sample from a distribution with mean μ and variance σ^2. This is more general because the relative frequency Y/n can be thought of as $\overline{X}$ when the sample arises from a Bernoulli distribution. We know that

$$E(\overline{X}) = \mu \qquad \text{and} \qquad \text{Var}(\overline{X}) = \frac{\sigma^2}{n}.$$

Thus, for every $\varepsilon > 0$,

$$P[|\overline{X} - \mu| \geq \varepsilon] = P\left[|\overline{X} - \mu| \geq \left(\frac{\varepsilon\sqrt{n}}{\sigma}\right)\left(\frac{\sigma}{\sqrt{n}}\right)\right].$$

But the standard deviation of $\overline{X}$ is $\sigma/\sqrt{n}$. Hence, using Chebyshev's inequality with $k = \varepsilon\sqrt{n}/\sigma$, we have

$$P[|\overline{X} - \mu| \geq \varepsilon] \leq \frac{\sigma^2}{\varepsilon^2 n}.$$

Since probability is nonnegative, it follows that

$$0 \leq \lim_{n \to \infty} P(|\overline{X} - \mu| \geq \varepsilon) \leq \lim_{n \to \infty} \frac{\sigma^2}{\varepsilon^2 n} = 0.$$

This implies that

$$\lim_{n \to \infty} P(|\overline{X} - \mu| \geq \varepsilon) = 0,$$

or, equivalently,

$$\lim_{n \to \infty} P(|\overline{X} - \mu| < \varepsilon) = 1.$$

The preceding discussion shows that the probability associated with the distribution of $\overline{X}$ becomes concentrated in an arbitrarily small interval centered at μ as n increases. This is a more general form of the law of large numbers.

Although Chebyshev's inequality is quite useful in theoretical discussion, it also shows that in a collection of numbers, say $x_1, x_2, \ldots, x_n$, a certain proportion of them must be within $k\sqrt{v}$ of $\overline{x}$, where $\overline{x}$ and v are the respective mean and variance of the empirical distribution defined by the numbers. That is,

$$\overline{x} = \frac{1}{n} \sum_{i=1}^{n} x_i \qquad \text{and} \qquad v = \frac{1}{n} \sum_{i=1}^{n} (x_i - \overline{x})^2.$$

To show this, think of the empirical probability (relative frequency) as defining the probability distribution; thus $\overline{x}$ and v are the mean and the variance of this distribution. Hence, for every $k \geq 1$, we have

$$\frac{\#\{x_i: |x_i - \overline{x}| \geq k\sqrt{v}\}}{n} \leq \frac{1}{k^2}$$

or, equivalently,

$$\frac{\#\{x_i: |x_i - \overline{x}| < k\sqrt{v}\}}{n} \geq 1 - \frac{1}{k^2}.$$

In words, the second inequality says: "The proportion of the numbers $x_1, x_2, \ldots, x_n$ that lie within $k\sqrt{v}$ of the mean $\overline{x}$ is at least $1 - 1/k^2$."

Exercises

6.8-1 If X is a random variable with mean 33 and variance 16, use Chebyshev's inequality to find
(a) A lower bound for $P(23 < X < 43)$.
(b) An upper bound for $P(|X - 33| \geq 14)$.

6.8-2   If $E(X) = 17$ and $E(X^2) = 298$, use Chebyshev's inequality to determine
   (a) A lower bound for $P(10 < X < 24)$.
   (b) An upper bound for $P(|X - 17| \geq 16)$.

6.8-3   Let X denote the outcome when rolling a fair die. Then $\mu = 7/2$ and $\sigma^2 = 35/12$. Note that the maximum deviation of X from μ equals $5/2$. Express this deviation in terms of number of standard deviations; that is, find k where $k\sigma = 5/2$. Determine a lower bound for $P(|X - 3.5| < 2.5)$.

6.8-4   If Y is $b(n, 0.5)$, give a lower bound for $P(|Y/n - 0.5| < 0.08)$ when
   (a) $n = 100$.
   (b) $n = 500$.
   (c) $n = 1000$.

6.8-5   If Y is $b(n, 0.25)$, give a lower bound for $P(|Y/n - 0.25| < 0.05)$ when
   (a) $n = 100$.
   (b) $n = 500$.
   (c) $n = 1000$.

6.8-6   Let $\overline{X}$ be the mean of a random sample of size $n = 15$ from a distribution with mean $\mu = 80$ and variance $\sigma^2 = 60$. Use Chebyshev's inequality to find a lower bound for $P(75 < \overline{X} < 85)$.

6.8-7   The characteristics of the empirical distribution of test scores of 900 students are $\overline{x} = 83$ and $v = 36$, respectively. At least how many students received test scores between 71 and 95?

7

Tests of Statistical Hypotheses

7.1 Tests About Proportions

A first major area of statistical inference, namely estimation of parameters, was introduced in Chapter 6. In this chapter we consider a second major one, **tests of statistical hypotheses**. This very important topic is introduced through an illustration.

Suppose a manufacturer of a certain printed circuit observes that about $p = 0.05$ of the circuits fail. An engineer and statistician working together suggest some changes that might improve the design of the product. To test this new procedure, it was agreed that $n = 200$ circuits would be produced using the proposed method and then checked. Let Y equal the number of these 200 circuits that fail. Clearly, if the number of failures, Y, is such that $Y/200$ is about equal to 0.05, then it seems that the new procedure has not resulted in an improvement. On the other hand, if Y is small so that $Y/200$ is about 0.01 or 0.02, we might believe that the new method is better than the old. On the other hand, if $Y/200$ is 0.08 or 0.09, the proposed method has perhaps caused a greater proportion of failures.

What we need to establish is a formal rule that tells us when to accept the new procedure as an improvement. In addition, we must know the consequences of this rule. For an example of a rule, we could accept the new procedure as an improvement if $Y \leq 5$ or $Y/n \leq 0.025$. We do note, however, that the probability of failure could still be about $p = 0.05$ even with the new procedure, and yet we could observe 5 or fewer failures in $n = 200$ trials. That is, we would accept the new method as being an improvement when, in fact, it was not. This decision is a mistake which we call a **Type I error**. On the other hand, the new procedure might actually improve the product so that p is much smaller, say $p = 0.02$, and yet we could observe $y = 7$ failures so that $y/200 = 0.035$. Thus we would not accept the new method as resulting in an improvement when in fact it had. This decision would also be a mistake which we call a **Type II error**. We must study the probabilities of these two types of errors to understand fully the consequences of our rule.

Let us begin by modeling the situation. If we believe these trials, using the new procedure, are independent and have about the same probability of failure on each trial, then Y is binomial $b(200, p)$. We wish to make a statistical inference about p using the unbiased estimator $\hat{p} = Y/200$. Of course, we could construct a confidence interval, say one that has 95% confidence, obtaining

$$\hat{p} \pm 1.96 \sqrt{\frac{\hat{p}(1 - \hat{p})}{200}}.$$

This inference is very appropriate and many statisticians simply do this. If the limits of this confidence interval contain 0.05, they would not say the new procedure is necessarily better, at least until more data are taken. If, on the other hand, the upper limit of this confidence interval is less than 0.05, then they feel 95% confident that the true p is now less than 0.05. Hence they would support the fact that the new procedure has improved the manufacturing of these printed circuits.

While this use of confidence intervals is highly appropriate and later we indicate the relationship of confidence intervals and tests of hypotheses, every student of statistics should also have some understanding of the basic concepts in the latter area. Here, in our illustration, we are testing whether or not the probability of failure has or has not decreased from 0.05 when the new manufacturing procedure is used. The *no change* hypothesis, $H_0: p = 0.05$, is called the **null hypothesis**. Since $H_0: p = 0.05$ completely specifies the distribution it is called a **simple hypothesis**; thus $H_0: p = 0.05$ is a **simple null hypothesis**. The *research worker's* (here the engineer and/or the statistician) hypothesis $H_1: p < 0.05$ is called the **alternative hypothesis**. Since $H_1: p < 0.05$ does not completely specify the distribution, it is a composite hypothesis because it is composed of many simple hypotheses. Our rule of rejecting H_0 and accepting H_1 if $Y \leq 5$, and otherwise accepting H_0 is called a **test of a statistical hypothesis**. We now see that the two types of errors can be recorded as follows.

Type I error: Rejecting H_0 and accepting H_1 when H_0 is true;
Type II error: Accepting H_0 when H_1 is true, that is, when H_0 is false.

Since, in our illustration, we make a Type I error if $Y \leq 5$ when in fact $p = 0.05$, we can calculate the probability of this error, which we denote by α and call the **significance level of the test**. Under our assumptions, it is

$$\alpha = P(Y \leq 5; p = 0.05) = \sum_{y=0}^{5} \binom{200}{y}(0.05)^y(0.95)^{200-y}.$$

Since n is rather large and p is small, these binomial probabilities can be approximated extremely well by Poisson probabilities with $\lambda = 200(0.05) = 10$. That is, from the Poisson table, the probability of the Type I error is

$$\alpha \approx \sum_{y=0}^{5} \frac{10^y e^{-10}}{y!} = 0.067;$$

thus, the approximate significance level of this test is $\alpha = 0.067$.

This value of α is reasonably small. However, what about the probability of Type II error in case p has been improved to 0.02, say? This error occurs if $Y > 5$ when, in fact, $p = 0.02$; hence its probability, denoted by β, is

$$\beta = P(Y > 5; p = 0.02) = \sum_{y=6}^{200} \binom{200}{y}(0.02)^y(0.98)^{200-y}.$$

Again we use the Poisson approximation, here with $\lambda = 200(0.02) = 4$, to obtain

$$\beta \approx 1 - \sum_{y=0}^{5} \frac{4^y e^{-4}}{y!} = 1 - 0.785 = 0.215.$$

The engineer and the statistician who created this new procedure probably are not too pleased with this answer. That is, they note that if their new procedure of manufacturing circuits has actually decreased the probability of failure to 0.02 from 0.05 (*a big improvement*), there is still a good chance, 0.215, that H_0: $p = 0.05$ is accepted and their improvement rejected. Thus, in their eyes, this test of H_0: $p = 0.05$ against H_1: $p = 0.02$ is unsatisfactory. More will be said about modifying tests in the next section so that satisfactory values of the probabilities of the two types of errors, namely α and β, can be obtained.

Without worrying more about the probability of the Type II error here, we present a frequently used procedure for testing H_0: $p = p_0$, where p_0 is some specified probability of success. This test is based upon the fact that the number of successes, Y, in n independent Bernoulli trials is such that Y/n has an approximate normal distribution, $N[p_0, p_0(1 - p_0)/n]$, provided H_0: $p = p_0$ is true and n is large. Suppose the alternative hypothesis is H_0: $p > p_0$; that is, it has been hypothesized by a research worker that something has been done to increase the probability of success. Consider the test of H_0: $p = p_0$ against H_1: $p > p_0$ that rejects H_0 and accepts H_1 if and only if

$$Z = \frac{Y/n - p_0}{\sqrt{p_0(1 - p_0)/n}} \geq z_\alpha.$$

That is, if Y/n exceeds p_0 by z_α standard deviations of Y/n, we reject H_0 and accept the hypothesis H_1: $p > p_0$. Since, under H_0, Z is approximately $N(0, 1)$, the approximate probability of this occurring when H_0: $p = p_0$ is true is α. That is, the significance level of this test is approximately α.

If the alternative is H_1: $p < p_0$ instead of H_1: $p > p_0$, then the appropriate α-level test is given by $Z \leq -z_\alpha$. That is, if Y/n is smaller than p_0 by z_α standard deviations of Y/n, we accept H_1: $p < p_0$.

Example 7.1-1 It was claimed that many commercially manufactured dice are not fair because the "spots" are really indentations so that, for example, the 6-side is lighter than the 1-side. Let p equal the probability of rolling a 6 with one of these dice. To test H_0: $p = 1/6$ against the alternative hypothesis H_1: $p > 1/6$, several of these dice will be rolled to yield a total of $n = 8000$ observations. The test statistic is

$$Z = \frac{Y/n - 1/6}{\sqrt{(1/6)(5/6)/n}} = \frac{Y/8000 - 1/6}{\sqrt{(1/6)(5/6)/8000}}.$$

If we use a significance level of $\alpha = 0.05$, the critical region is

$$z \geq z_{0.05} = 1.645.$$

The results of the experiment yielded $y = 1389$, so that the calculated value of the test statistic is

$$z = \frac{1389/8000 - 1/6}{\sqrt{(1/6)(5/6)/8000}} = 1.670.$$

Since

$$z = 1.670 > 1.645,$$

the null hypothesis is rejected and these experimental results indicate that these dice favor a 6 more than a fair die would. (You could perform your own experiment to check out other dice. Also see Exercise 7.1-6.)

Tests of H_0: $p = p_0$ against H_1: $p < p_0$ or H_0: $p = p_0$ against H_1: $p > p_0$ are called *one-sided* tests because the alternative hypotheses are *one-sided*. There are times when *two-sided* alternatives and tests are appropriate as in H_1: $p \neq p_0$. For illustration, suppose that the pass rate in the usual beginning statistics course is p_0. There has been an intervention (say some new teaching method) and it is not known whether the pass rate will increase, decrease, or stay about the same. Thus we test the null (no change) hypothesis H_0: $p = p_0$ against the two-sided alternative H_1: $p \neq p_0$. A test of the approximate significance level α for doing this is to reject H_0: $p = p_0$ if

$$|Z| = \frac{|Y/n - p_0|}{\sqrt{p_0(1 - p_0)/n}} \geq z_{\alpha/2},$$

since, under H_0, $P(|Z| \geq z_{\alpha/2}) \approx \alpha$. These tests of approximate significance level α are summarized in Table 7.1-1. The rejection region is often called the **critical region** of the test, and we use that terminology in the table.

Often there is interest in tests about p_1 and p_2, the probabilities of success for two different distributions or the proportions of two different populations having a certain characteristic. For example, if p_1 and p_2 denote the respective proportions of homeowners and renters who vote in favor of a proposal to reduce property tax, a politician might be interested in testing H_0: $p_1 = p_2$ against the one-sided alternative hypothesis H_1: $p_1 > p_2$.

TABLE 7.1-1

H_0	H_1	Critical Region
$p = p_0$	$p > p_0$	$z = \dfrac{y/n - p_0}{\sqrt{p_0(1 - p_0)/n}} \geq z_\alpha$
$p = p_0$	$p < p_0$	$z = \dfrac{y/n - p_0}{\sqrt{p_0(1 - p_0)/n}} \leq -z_\alpha$
$p = p_0$	$p \neq p_0$	$|z| = \dfrac{|y/n - p_0|}{\sqrt{p_0(1 - p_0)/n}} \geq z_{\alpha/2}$

Let Y_1 and Y_2 represent, respectively, the numbers of observed successes in n_1 and n_2 independent trials with probabilities of success p_1 and p_2. Recall that the distribution of $\hat{p}_1 = Y_1/n_1$ is approximately $N[p_1, p_1(1 - p_1)/n_1]$ and the distribution of $\hat{p}_2 = Y_2/n_2$ is approximately $N[p_2, p_2(1 - p_2)/n_2]$. Thus the distribution of $\hat{p}_1 - \hat{p}_2 = Y_1/n_1 - Y_2/n_2$ is approximately $N[p_1 - p_2, p_1(1 - p_1)/n_1 + p_2(1 - p_2)/n_2]$. It follows that the distribution of

$$Z = \frac{Y_1/n_1 - Y_2/n_2 - (p_1 - p_2)}{\sqrt{p_1(1 - p_1)/n_1 + p_2(1 - p_2)/n_2}} \tag{7.1-1}$$

is approximately $N(0, 1)$. To test H_0: $p_1 - p_2 = 0$ or, equivalently, H_0: $p_1 = p_2$, let $p = p_1 = p_2$ be the common value under H_0 and estimate p with $\hat{p} = (Y_1 + Y_2)/(n_1 + n_2)$. Replacing p_1 and p_2 in the denominator of expression (7.1-1) with this estimate, we obtain the test statistic

$$Z = \frac{\hat{p}_1 - \hat{p}_2 - 0}{\sqrt{\hat{p}(1 - \hat{p})(1/n_1 + 1/n_2)}},$$

which has an approximate $N(0, 1)$ distribution when the null hypothesis is true.

The three possible alternative hypotheses and their critical regions are summarized in Table 7.1-2.

TABLE 7.1-2

H_0	H_1	Critical Region
$p_1 = p_2$	$p_1 > p_2$	$z = \dfrac{\hat{p}_1 - \hat{p}_2}{\sqrt{\hat{p}(1 - \hat{p})(1/n_1 + 1/n_2)}} \geq z_\alpha$
$p_1 = p_2$	$p_1 < p_2$	$z = \dfrac{\hat{p}_1 - \hat{p}_2}{\sqrt{\hat{p}(1 - \hat{p})(1/n_1 + 1/n_2)}} \leq -z_\alpha$
$p_1 = p_2$	$p_1 \neq p_2$	$|z| = \dfrac{|\hat{p}_1 - \hat{p}_2|}{\sqrt{\hat{p}(1 - \hat{p})(1/n_1 + 1/n_2)}} \geq z_{\alpha/2}$

REMARK In testing both H_0: $p = p_0$ and H_0: $p_1 = p_2$, statisticians often use different denominators for z. In the first, $\sqrt{p_0(1 - p_0)/n}$ is replaced by $\sqrt{(y/n)(1 - y/n)/n}$; and, in the second, the following denominator is used:

$$\sqrt{\frac{\hat{p}_1(1 - \hat{p}_1)}{n_1} + \frac{\hat{p}_2(1 - \hat{p}_2)}{n_2}}.$$

We do not have strong preference one way or the other since the two methods provide about the same numerical result. The substitutions do provide better estimates of the standard deviations of the numerators when the null hypotheses are clearly false. There is some advantage to this if the null hypothesis is likely to be false. In addition, the substitutions also tie together the use of confidence intervals and two-sided tests because, for example,

$$\frac{|y/n - p_0|}{\sqrt{\dfrac{(y/n)(1 - y/n)}{n}}} < z_{\alpha/2} \quad \text{and} \quad p_0 \in \left(\frac{y}{n} \pm z_{\alpha/2} \sqrt{\frac{(y/n)(1 - y/n)}{n}} \right),$$

where the latter is an open confidence interval, are equivalent. However, using the forms given in Tables 7.1-1 and 7.1-2, we do get better approximations to α-level significance tests. Thus there are trade-offs and it is difficult to say one is better than the other. Fortunately, the numerical answers are about the same.

Exercises

7.1-1 Bowl A contains 100 red balls and 200 white balls; bowl B contains 200 red balls and 100 white balls. Let p denote the probability of drawing a red ball from a bowl, but say p is unknown, since it is unknown whether bowl A or bowl B is being used. We shall test the simple null hypothesis H_0: $p = 1/3$ against the simple alternative hypothesis H_1: $p = 2/3$. Draw three balls at random, one at a time and with replacement from the selected bowl. Let X equal the number of red balls drawn. Then let the critical region be $C = \{x: x = 2, 3\}$. What are the values of α and β, the probabilities of Type I and Type II errors, respectively?

7.1-2 A bowl contains two red balls, two white balls, and fifth ball that is either red or white. Let p denote the probability of drawing a red ball from the bowl. We shall test the simple null hypothesis H_0: $p = 3/5$ against the simple alternative hypothesis H_1: $p = 2/5$. Draw four balls at random from the bowl, one at a time and with replacement. Let X equal the number of red balls drawn.
(a) Define a critical region C for this test in terms of X.
(b) For the critical region C defined in part (a), find the values of α and β.

7.1-3 Let Y be $b(100, p)$. To test H_0: $p = 0.08$ against H_1: $p < 0.08$, we reject H_0 and accept H_1 if and only if $Y \leq 6$.
(a) Determine the significance level α of the test.
(b) Find the probability of the Type II error if in fact $p = 0.04$.

7.1-4 Let p denote the probability that, for a particular tennis player, the first serve is good. Since $p = 0.40$, this player decided to take lessons in order to increase p. When the lessons are completed, the hypothesis H_0: $p = 0.40$ will be tested against H_1: $p > 0.40$ based on $n = 25$ trials. Let y equal the number of first serves that are good, and let the critical region be defined by $C = \{y: y \geq 13\}$.
   (a) Determine $\alpha = P(Y \geq 13; p = 0.40)$. Use Table II in the Appendix.
   (b) Find $\beta = P(Y < 13)$ when $p = 0.60$; that is, $\beta = P(Y \leq 12; p = 0.60)$. Use Table II.

7.1-5 Let Y be $b(192, p)$. We reject H_0: $p = 0.75$ and accept H_1: $p > 0.75$ if and only if $Y \geq 152$. Use the normal approximation to determine
   (a) $\alpha = P(Y \geq 152; p = 0.75)$.
   (b) $\beta = P(Y < 152)$ when $p = 0.80$.

7.1-6 To determine whether the 1-side on a commercially manufactured die is heavy and the 6-side is light (see Example 7.1-1), some students kept track of the number of observed 1s when observing $n = 8000$ rolls of the dice. Let p equal the probability of rolling a 1 with such a die. We shall test the null hypothesis H_0: $p = 1/6$ against the alternative hypothesis H_1: $p < 1/6$.
   (a) Define the test statistic and an $\alpha = 0.05$ critical region.
   (b) If $y = 1265$ 1s were observed in 8000 rolls, calculate the value of the test statistic and state your conclusion.

7.1-7 If a newborn baby has a birth weight that is less than 2500 grams (5.5 pounds), we say that the baby has a low birth weight. The proportion of babies with a low birth weight is an indicator of nutrition (or lack of nutrition) for the mothers. For the United States, approximately 7% of babies have a low birth weight. Let p equal the proportion of babies born in the Sudan who weigh less than 2500 grams. We shall test the null hypothesis H_0: $p = 0.07$ against the alternative hypothesis H_1: $p > 0.07$. If $y = 23$ babies out of a random sample of $n = 209$ babies weighed less than 2500 grams, what is your conclusion at a significance level of
   (a) $\alpha = 0.05$?
   (b) $\alpha = 0.01$?

7.1-8 It was claimed that 75% of all dentists recommend a certain brand of gum for their gum-chewing patients. A consumer group doubted this claim and decided to test H_0: $p = 0.75$ against the alternative hypothesis H_1: $p < 0.75$, where p is the proportion of dentists who recommend this brand of gum. A survey of 390 dentists found that 273 recommended this brand of gum. Which hypothesis would you accept if the significance level is
   (a) $\alpha = 0.05$?
   (b) $\alpha = 0.01$?

7.1-9 It was claimed that the proportion of Americans who select jogging as one of their recreational activities is $p = 0.25$. A shoe manufacturer thought that p was larger than 0.25. They decided to test the null hypothesis H_0: $p = 0.25$

against the alternative hypothesis $H_1: p > 0.25$. If $y = 1497$ out of a random sample of $n = 5757$ selected jogging, what is your conclusion at a significance level of

(a) $\alpha = 0.05$?

(b) $\alpha = 0.025$?

7.1-10 Let p equal the proportion of drivers who use a seat belt in a state that does not have a mandatory seat belt law. It was claimed that $p = 0.14$. An advertising campaign was conducted to increase this proportion. Two months after the campaign, $y = 104$ out of a random sample of $n = 590$ drivers were wearing their seat belts. Was the campaign successful?

(a) Define the null and alternative hypotheses.

(b) Define a critical region with an $\alpha = 0.01$ significance level.

(c) What is your conclusion?

7.1-11 The management of the Tiger baseball team decided to sell only low-alcohol beer in their ballpark to help combat rowdy fan conduct. They claimed that more than 40% of the fans would approve of this decision. Let p equal the proportion of Tiger fans on opening day who approved of this decision. We shall test the null hypothesis $H_0: p = 0.40$ against the alternative hypothesis $H_1: p > 0.40$.

(a) Define a critical region that has an $\alpha = 0.05$ significance level.

(b) If out of a random sample of $n = 1278$ fans, $y = 550$ said that they approved of this new policy, what is your conclusion?

7.1-12 An official with the NCAA claimed that only $p = 0.01 = 1\%$ of college football players used anabolic steroids. Others believed that $p > 0.01$.

(a) Define the test statistic and critical region that has an approximate significance level of $\alpha = 0.005$.

(b) If $y = 17$ out of a random sample of $n = 546$ football players tested positive in an offseason testing program, what is your conclusion?

7.1-13 Let p equal the proportion of women who agree that ''men are basically selfish and self-centered.'' Suppose that in the past, it was believed that $p = 0.40$. It is now claimed that p has increased.

(a) Define the null and alternative hypotheses, a test statistic and critical region that has an approximate $\alpha = 0.01$ significance level. Sketch a standard normal p.d.f. and illustrate this critical region.

(b) The *Detroit Free Press* (April 26, 1990) reported that $y = 1260$ out of a random sample of $n = 3000$ women agree with the statement. What is the conclusion of your test? Locate the calculated value of your test statistic on the p.d.f. in part (a).

7.1-14 According to a population census in 1986, the percentage of males who are 18 or 19 years old that are married was 3.7%. We shall test whether this percentage increased from 1986 to 1988.

(a) Define the null and alternative hypotheses.

(b) Define a critical region that has an approximate significance level of $\alpha = 0.01$. Sketch a standard normal p.d.f. to illustrate this critical region.

(c) If $y = 20$ out of a random sample of $n = 300$ males, each 18 or 19 years old, were married (*U.S. Bureau of the Census, Statistical Abstract of the United States: 1988*), what is your conclusion? Show the calculated value of the test statistic on your figure in part (c).

7.1-15 Let p equal the proportion of yellow candies in a package of mixed colors. It is claimed that $p = 0.20$.

(a) Define the test statistic and critical region with a significance level of $\alpha = 0.05$ for testing H_0: $p = 0.20$ against a two-sided alternative hypothesis.

(b) To perform the test, each of 20 students counted the number of yellow candies, y, and the total number, n, in a 48.1-gram package, yielding the following ratios, y/n: 8/56, 13/55, 12/58, 13/56, 14/57, 5/54, 14/56, 15/57, 11/54, 13/55, 10/57, 8/59, 10/54, 11/55, 12/56, 11/57, 6/54, 7/58, 12/58, 14/58. If each individual makes a test of H_0: $p = 0.20$, what proportion of the students rejected the null hypothesis?

(c) If we may assume that the null hypothesis is true, what proportion of the students would you have expected to reject the null hypothesis?

(d) For each of the 20 ratios in part (b), a 95% confidence interval for p can be calculated. What proportion of these 95% confidence intervals contain $p = 0.20$?

(e) If the 20 results are pooled so that $\Sigma_{i=1}^{20} y_i$ equals the number of yellow candies and $\Sigma_{i=1}^{20} n_i$ equals the total sample size, do we reject H_0: $p = 0.20$?

7.1-16 A machine shop that manufactures toggle levers has both a day and a night shift. A toggle lever is defective if a standard nut cannot be screwed onto the threads. Let p_1 and p_2 be the proportion of defective levers among those manufactured by the day and night shifts, respectively. We shall test the null hypothesis, H_0: $p_1 = p_2$, against a two-sided alternative hypothesis based on two random samples, each of 1000 levers taken from the production of the respective shifts.

(a) Define the test statistic and a critical region that has an $\alpha = 0.05$ significance level. Sketch a standard normal p.d.f. illustrating this critical region.

(b) If $y_1 = 37$ and $y_2 = 53$ defectives were observed for the day and night shifts, respectively, calculate the value of the test statistic. Locate the calculated test statistic on your figure in part (a) and state your conclusion.

7.1-17 Let p_1 and p_2 equal the respective proportions of male and female college students who make their beds. We shall test the null hypothesis H_0: $p_1 = p_2$ against the alternative hypothesis H_1: $p_1 < p_2$.

(a) Define a critical region that has a significance level of $\alpha = 0.05$.

(b) If 27 out of 90 men and 51 out of 98 women made their beds on a particular Monday morning, calculate the value of the test statistic and give your conclusion.

7.1-18 For developing countries in Africa and the Americas, let p_1 and p_2 be the respective proportions of babies with a low birth weight (below 2500 grams). We shall test H_0: $p_1 = p_2$ against the alternative hypothesis H_1: $p_1 > p_2$.
(a) Define a critical region that has an $\alpha = 0.05$ significance level.
(b) If random samples of sizes $n_1 = 900$ and $n_2 = 700$ yielded $y_1 = 135$ and $y_2 = 77$ babies with a low birth weight, what is your conclusion?
(c) What would your decision be with a significance level of $\alpha = 0.01$?

7.2   Power and Sample Size

In Section 7.1 we introduced some terminology connected with tests of statistical hypotheses. With two more examples we continue to examine a number of concepts associated with tests of statistical hypotheses.

Example 7.2-1 Let X equal the breaking strength of a steel bar. If the steel bar is manufactured by process I, X is $N(50, 36)$. It is hoped that if process II (a new process) is used, X will be $N(55, 36)$. Given a large number of steel bars manufactured by process II, how could we test whether the increase in the mean breaking strength was realized?

That is, we are assuming X is $N(\mu, 36)$ and μ is equal to 50 or 55. We want to test the null hypothesis H_0: $\mu = 50$ against alternative hypothesis H_1: $\mu = 55$. Note that each of these hypotheses completely specifies the distribution of X. That is, H_0 states that X is $N(50, 36)$, and H_1 states that X is $N(55, 36)$. An hypothesis that completely specifies the distribution of X is called a simple hypothesis; otherwise it is called a **composite hypothesis** (composed of at least two simple hypotheses). For example, H_1: $\mu > 50$ would be a composite hypothesis because it is composed of all normal distributions with $\sigma^2 = 36$ and means greater than 50. In order to test which of the two hypotheses, H_0 or H_1, is true, we shall set up a rule based on the breaking strengths $x_1, x_2, \ldots, x_n$ of n bars (the observed values of a random sample of size n from this new normal distribution). The rule leads to a decision to accept or reject H_0; so it is necessary to partition the sample space into two parts, say C and C', so that if $(x_1, x_2, \ldots x_n) \in C$, H_0 is rejected, and if $(x_1, x_2, \ldots, x_n) \in C'$, H_0 is accepted (not rejected). The rejection region C for H_0 is called the critical region for the test. Often the partitioning of the sample space is specified in terms of the values of a statistic called the **test statistic**. In this illustration we could let $\overline{X}$ be the test statistic and, for example, take $C = \{(x_1, x_2, \ldots, x_n): \overline{x} \geq 53\}$. We could then define the critical region as those values of the test statistic for which H_0 is rejected. That is, the given critical region is equivalent to defining $C = \{\overline{x}: \overline{x} \geq 53\}$ in the $\overline{x}$ space. If $(x_1, x_2, \ldots, x_n) \in C$ when H_0 is true, H_0 would be rejected when it is true.

Such an error is called a Type I error. If $(x_1, x_2, \ldots, x_n) \in C'$ when H_1 is true, H_0 would be accepted when in fact H_1 is true. Such an error is called a Type II error. The probability of a Type I error is called the significance level of the test and is denoted by α. That is,

$$\alpha = P[(X_1, X_2, \ldots, X_n) \in C; H_0]$$

is the probability that $(X_1, X_2, \ldots, X_n)$ falls in C when H_0 is true. The probability of a Type II error is denoted by β; that is,

$$\beta = P[(X_1, X_2, \ldots X_n) \in C'; H_1]$$

is the probability of accepting (failing to reject) H_0 when it is false. For illustration, suppose $n = 16$ bars were tested and $C = \{\bar{x}: \bar{x} \geq 53\}$. Then $\bar{X}$ is $N(50, 36/16)$ when H_0 is true and is $N(55, 36/16)$ when H_1 is true. Thus

$$\alpha = P(\bar{X} \geq 53; H_0) = P\left(\frac{\bar{X} - 50}{6/4} \geq \frac{53 - 50}{6/4}; H_0\right)$$

$$= 1 - \Phi(2) = 0.0228$$

and

$$\beta = P(\bar{X} < 53; H_1) = P\left(\frac{\bar{X} - 55}{6/4} < \frac{53 - 55}{6/4}; H_1\right)$$

$$= \Phi\left(-\frac{4}{3}\right) = 1 - 0.9087 = 0.0913.$$

See Figure 7.2-1 for the graphs of the probability density functions of $\bar{X}$ when H_0 and H_1 are true, respectively. Note that a decrease in the size of α leads to an increase in the size of β, and vice versa. Both α and β can be decreased if the sample size n is increased.

The next example introduces a new concept using a test about p, the probability of success. The sample size is kept small so that Appendix Table II can be used for finding probabilities. The application is one that you can actually perform.

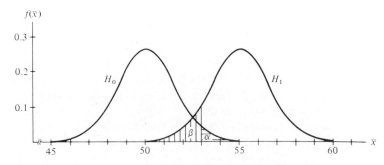

FIGURE 7.2-1

Example 7.2-2   Assume that a person, when given a name tag, puts it on either the right or left side. Let p equal the probability that the name tag is placed on the right side. We shall test the null hypothesis H_0: $p = 1/2$ against the composite alternative hypothesis H_1: $p < 1/2$. (Included with the null hypothesis are those values of p that are greater than 1/2. That is, we could think of H_0 as H_0: $p \geq 1/2$.) We shall give name tags to a random sample of $n = 20$ people, denoting the placements of their name tags with Bernoulli random variables, $X_1, X_2, \ldots, X_{20}$, where $X_i = 1$ if a person places the name tag on the right and $X_i = 0$ if a person places the name tag on the left. We can then use for our test statistic $Y = \Sigma_{i=1}^{20} X_i$ which has a binomial distribution, $b(20, p)$. Say the critical region is defined by $C = \{y: y \leq 6\}$ or, equivalently, by $\{(x_1, x_2, \ldots, x_{20}): \Sigma_{i=1}^{20} x_i \leq 6\}$. Since Y is $b(20, 1/2)$ if $p = 1/2$, the significance level of the corresponding test is

$$\alpha = P\left(Y \leq 6; p = \frac{1}{2}\right) = \sum_{y=0}^{6} \binom{20}{y}\left(\frac{1}{2}\right)^{20} = 0.0577,$$

using Table II in the Appendix. Of course the probability β of the Type II error has different values with different values of p selected from the composite alternative hypothesis H_1: $p < 1/2$. For illustration, with $p = 1/4$,

$$\beta = P\left(7 \leq Y \leq 20; p = \frac{1}{4}\right) = \sum_{y=7}^{20} \binom{20}{y}\left(\frac{1}{4}\right)^{y}\left(\frac{3}{4}\right)^{20-y} = 0.2142,$$

whereas with $p = 1/10$.

$$\beta = P\left(7 \leq Y \leq 20; p = \frac{1}{10}\right) = \sum_{y=7}^{20} \binom{20}{y}\left(\frac{1}{10}\right)^{y}\left(\frac{9}{10}\right)^{20-y} = 0.0024.$$

Instead of considering the probability β of accepting H_0 when H_1 is true, we could compute the probability K of rejecting H_0 when H_1 is true. After all, β and $K = 1 - \beta$ provide the same information. Since K is a function of p, we denote this explicitly by writing $K(p)$. The probability

$$K(p) = \sum_{y=0}^{6} \binom{20}{y} p^y (1 - p)^{20-y}, \qquad 0 < p \leq \frac{1}{2},$$

is called the **power function of the test**. Of course, $\alpha = K(1/2) = 0.0577$, $1 - K(1/4) = 0.2142$ and $1 - K(1/10) = 0.0024$. The value of the power function at a specified p is called the **power** of the test at that point. For illustration, $K(1/4) = 0.7858$ and $K(1/10) = 0.9976$ are the powers at $p = 1/4$ and $p = 1/10$, respectively. An acceptable power function is one that assumes small values when H_0 is true and larger values when p differs much from $p = 1/2$.

In Example 7.2-2 we introduced the new concept of the power function of a test. We now show how the sample size can be selected so as to create a test with appropriate power.

Let $X_1, X_2, \ldots, X_n$ be a random sample of size n from the normal distribution $N(\mu, 100)$, which we can suppose is a possible distribution of scores of students in a statistics course that uses a new method of teaching (e.g., computer-related materials). We wish to decide between H_0: $\mu = 60$ (*no change* hypothesis because, let us say, this was the mean by the previous method of teaching) and the research worker's hypothesis H_1: $\mu > 60$. Let us consider a sample of size $n = 25$. Of course, the sample mean $\overline{X}$ is the maximum likelihood estimator of μ, and thus it seems reasonable to base our decision on this statistic. Initially, we use the rule to reject H_0 and accept H_1 if and only if $\bar{x} \geq 62$. What are the consequences of this test? These are summarized in the power function of the test.

We first find the probability of rejecting H_0: $\mu = 60$ for various values of $\mu \geq 60$. The probability of rejecting H_0 is given by

$$K(\mu) = P(\overline{X} \geq 62; \mu)$$

because this test calls for the rejection of H_0: $\mu = 60$ when $\bar{x} \geq 62$. When the new process has the general mean μ, we know that $\overline{X}$ has the normal distribution $N(\mu, 100/25 = 4)$. Accordingly,

$$K(\mu) = P\left(\frac{\overline{X} - \mu}{2} \geq \frac{62 - \mu}{2}; \mu\right)$$

$$= 1 - \Phi\left(\frac{62 - \mu}{2}\right), \qquad 60 \leq \mu,$$

is the probability of rejecting H_0: $\mu = 60$ using this particular test. Several values of $K(\mu)$ are given in Table 7.2-1. Figure 7.2-2 depicts the graph of the function $K(\mu)$.

The probability $K(\mu)$ of rejecting H_0: $\mu = 60$ is called the power function of the test. At the value μ_1 of the parameter, $K(\mu_1)$ is the power at μ_1. The power at $\mu = 60$ is $K(60) = 0.1587$, and this is the probability of rejecting H_0: $\mu = 60$ when H_0 is true. That is, $K(60) = 0.1587 = \alpha$ is the probability of the Type I error and is called the significance level of the test.

TABLE 7.2-1

μ	$K(\mu)$
60	0.1587
61	0.3085
62	0.5000
63	0.6915
64	0.8413
65	0.9332
66	0.9772

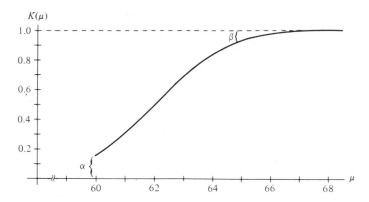

FIGURE 7.2-2

The power at $\mu = 65$ is $K(65) = 0.9332$, and this is the probability of making the correct decision (namely, rejecting H_0: $\mu = 60$ when $\mu = 65$). Hence we are pleased that here it is large. When $\mu = 65$, $1 - K(65) = 0.0668$ is the probability of not rejecting H_0: $\mu = 60$ when $\mu = 65$; that is, it is the probability of a Type II error and is denoted by $\beta = 0.0668$. These α and β values are displayed on Figure 7.2-2. Clearly, the probability $\beta = 1 - K(\mu_1)$ of the Type II error depends on which value, say μ_1, is taken in the alternative hypothesis H_1: $\mu > 60$. Thus while $\beta = 0.0668$ when $\mu = 65$, β is equal to $1 - K(63) = 0.3085$ when $\mu = 63$.

Frequently, statisticians like to have the significance level α smaller than 0.1587, say around 0.05 or less, because it is a probability of an error, namely the Type I error. Hence if we would like $\alpha = 0.05$, then we can no longer use, with $n = 25$, the critical region $\bar{x} \geq 62$, but $\bar{x} \geq c$, where c is selected so that

$$K(60) = P(\bar{X} \geq c; \mu = 60) = 0.05.$$

However, when $\mu = 60$, $\bar{X}$ is $N(60, 4)$ and

$$K(60) = P\left(\frac{\bar{X} - 60}{2} \geq \frac{c - 60}{2}; \mu = 60\right)$$

$$= 1 - \Phi\left(\frac{c - 60}{2}\right) = 0.05.$$

From Table Va in the Appendix we have that

$$\frac{c - 60}{2} = 1.645 = z_{0.05} \qquad \text{and} \qquad c = 60 + 3.29 = 63.29.$$

Although this change reduces α from 0.1587 to 0.05, it increases β from 0.0668 at $\mu = 65$ to

$$\beta = 1 - P(\bar{X} \geq 63.29; \mu = 65)$$

$$= 1 - P\left(\frac{\bar{X} - 65}{2} \geq \frac{63.29 - 65}{2}; \mu = 65\right)$$

$$= \Phi(-0.855) = 0.1963.$$

In general, without changing the sample size or the type of test of the hypothesis, a decrease in α causes an increase in β, and a decrease in β causes an increase in α. Both probabilities α and β of the two types of errors can be decreased only by increasing the sample size or, in some way, constructing a better test of the hypothesis.

For example, if $n = 100$ and we desire a test with significance level $\alpha = 0.05$, then

$$\alpha = P(\overline{X} \geq c; \mu = 60) = 0.05$$

means, since $\overline{X}$ is $N(\mu, 100/100 = 1)$,

$$P\left(\frac{\overline{X} - 60}{1} \geq \frac{c - 60}{1}; \mu = 60\right) = 0.05$$

and $c - 60 = 1.645$. Thus $c = 61.645$. The power function is

$$K(\mu) = P(\overline{X} \geq 61.645; \mu)$$
$$= P\left(\frac{\overline{X} - \mu}{1} \geq \frac{61.645 - \mu}{1}; \mu\right) = 1 - \Phi(61.645 - \mu).$$

In particular, this means that β at $\mu = 65$ is

$$\beta = 1 - K(\mu) = \Phi(61.645 - 65) = \Phi(-3.355) \approx 0;$$

so, with $n = 100$, both α and β have decreased from their respective original values of 0.1587 and 0.0668 when $n = 25$.

Rather than guess at the value of n, an ideal power function determines the sample size. Let us use a critical region of the form $\bar{x} \geq c$. Further, suppose that we want $\alpha = 0.025$ and, when $\mu = 65$, $\beta = 0.05$. Thus, since $\overline{X}$ is $N(\mu, 100/n)$,

$$0.025 = P(\overline{X} \geq c; \mu = 60) = 1 - \Phi\left(\frac{c - 60}{10/\sqrt{n}}\right)$$

and

$$0.05 = 1 - P(\overline{X} \geq c; \mu = 65) = \Phi\left(\frac{c - 65}{10/\sqrt{n}}\right).$$

That is,

$$\frac{c - 60}{10/\sqrt{n}} = 1.96 \quad \text{and} \quad \frac{c - 65}{10/\sqrt{n}} = -1.645.$$

Solving these equations simultaneously for c and $10/\sqrt{n}$, we obtain

$$c = 60 + 1.96\frac{5}{3.605} = 62.718;$$

$$\frac{10}{\sqrt{n}} = \frac{5}{3.605}.$$

Thus

$$\sqrt{n} = 7.21 \qquad \text{and} \qquad n = 51.98.$$

Since n must be an integer, we would use $n = 52$ and obtain $\alpha = 0.025$ and $\beta = 0.05$, approximately.

For a number of years there has been another value associated with a statistical test, and most statistical computer programs automatically print this out; it is called the **probability value** or, for brevity, p-value. The **p-value** associated with a test is the probability that we obtain the observed value of the test statistic or a value that is more extreme in the direction of the alternative hypothesis, calculated when H_0 is true. Rather than select the critical region ahead of time, the p-value of a test can be reported and the reader then makes a decision.

Say we are testing H_0: $\mu = 60$ against H_1: $\mu > 60$ with a sample mean $\overline{X}$ based on $n = 52$ observations. Suppose that we obtain the observed sample mean of $\overline{x} = 62.75$. If we compute the probability of obtaining an $\overline{x}$ of that value of 62.75 or greater when $\mu = 60$, then we obtain the p-value associated with $\overline{x} = 62.75$. That is,

$$p\text{-value} = P(\overline{X} \geq 62.75; \mu = 60)$$

$$= P\left(\frac{\overline{X} - 60}{10/\sqrt{52}} \geq \frac{62.75 - 60}{10/\sqrt{52}}; \mu = 60\right)$$

$$= 1 - \Phi\left(\frac{62.75 - 60}{10/\sqrt{52}}\right) = 1 - \Phi(1.983) = 0.0237.$$

If this p-value is small, we tend to reject the hypothesis H_0: $\mu = 60$. For example, rejection of H_0: $\mu = 60$ if the p-value is less than or equal to 0.025 is exactly the same as rejection if $\overline{x} \geq 62.718$. That is, $\overline{x} = 62.718$ has a p-value of 0.025. To help the reader keep the definition of p-value in mind, we note that it can be thought of as that **tail-end probability**, under H_0, of the distribution of the statistic, here $\overline{X}$, beyond the observed value of the statistic. See Figure 7.2-3 for the p-value associated with $\overline{x} = 62.75$.

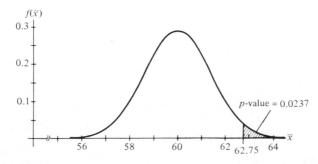

FIGURE 7.2-3

Example 7.2-3   Suppose that in the past, a golfer's scores have been (approximately) normally distributed with mean $\mu = 90$ and $\sigma^2 = 9$. After taking some lessons, the golfer has reason to believe that the mean μ has decreased. (We assume that σ^2 is still about 9.) To test the null hypothesis H_0: $\mu = 90$ against the alternative hypothesis H_1: $\mu < 90$, the golfer plays 16 games, computing the sample mean $\bar{x}$. If $\bar{x}$ is small, say $\bar{x} \leq c$, then H_0 is rejected and H_1 accepted; that is, it seems as if the mean μ has actually decreased after the lessons. If $c = 88.5$, then the power function of the test is

$$K(\mu) = P(\overline{X} \leq 88.5; \mu)$$

$$= P\left(\frac{\overline{X} - \mu}{3/4} \leq \frac{88.5 - \mu}{3/4}; \mu\right) = \Phi\left(\frac{88.5 - \mu}{3/4}\right)$$

because 9/16 is the variance of $\overline{X}$. In particular,

$$\alpha = K(90) = \Phi(-2) = 1 - 0.9772 = 0.0228.$$

If, in fact, the true mean is equal to $\mu = 88$ after the lessons, the power is $K(88) = \Phi(2/3) = 0.7475$. If $\mu = 87$, then $K(87) = \Phi(2) = 0.9772$. An observed sample mean of $\bar{x} = 88.25$ has a

$$p\text{-value} = P(\overline{X} \leq 88.25; \mu = 90)$$

$$= \Phi\left(\frac{88.25 - 90}{3/4}\right) = \Phi\left(-\frac{7}{3}\right) = 0.0098,$$

and this would lead to rejection at $\alpha = 0.0228$ (or even $\alpha = 0.01$).

The next example is an extension of Example 7.2-2.

Example 7.2-4   To test H_0: $p = 1/2$ against H_1: $p < 1/2$, we take a random sample of Bernoulli trials, $X_1, X_2, \ldots, X_n$, and use for our test statistic $Y = \sum_{i=1}^{n} X_i$ which has a binomial distribution $b(n, p)$. Let the critical region be defined by $C = \{y: y \leq c\}$. The power function for this test is defined by $K(p) = P(Y \leq c; p)$. We shall find the values of n and c so that $K(1/2) = 0.05$ and $K(1/4) = 0.90$, approximately. That is, we would like the significance level to be $\alpha = K(1/2) = 0.05$ and the power at $p = 1/4$ to equal 0.90. We proceed as follows:

$$0.05 = P\left(Y \leq c; p = \frac{1}{2}\right) = P\left(\frac{Y - n/2}{\sqrt{n(1/2)(1/2)}} \leq \frac{c - n/2}{\sqrt{n(1/2)(1/2)}}\right)$$

implies that

$$(c - n/2)/\sqrt{n/4} \approx -1.645;$$

and

$$0.90 = P\left(Y \leq c; p = \frac{1}{4}\right) = P\left(\frac{Y - n/4}{\sqrt{n(1/4)(3/4)}} \leq \frac{c - n/4}{\sqrt{n(1/4)(3/4)}}\right)$$

implies that

$$(c - n/4)/\sqrt{3n/16} \approx 1.282.$$

Therefore,

$$\frac{n}{4} \approx 1.645\sqrt{\frac{n}{4}} + 1.282\sqrt{\frac{3n}{16}} \quad \text{and} \quad \sqrt{n} \approx 4(1.378) = 5.512.$$

Thus n is about equal to 31, and from either of the first two approximate equalities we find that c is about equal to 10.9. Using $n = 31$ and $c = 10.9$ means that $K(1/2) = 0.05$ and $K(1/4) = 0.90$ are only approximate. In fact, since Y must be an integer, we could let $c = 10.5$. Then, with $n = 31$,

$$\alpha = K\!\left(\frac{1}{2}\right) = P\!\left(Y \le 10.5; p = \frac{1}{2}\right) \approx 0.0362;$$

$$K\!\left(\frac{1}{4}\right) = P\!\left(Y \le 10.5; p = \frac{1}{4}\right) \approx 0.8729.$$

Or, we could let $c = 11.5$ and $n = 31$, in which case

$$\alpha = K\!\left(\frac{1}{2}\right) = P\!\left(Y \le 11.5; p = \frac{1}{2}\right) \approx 0.0558;$$

$$K\!\left(\frac{1}{4}\right) = P\!\left(Y \le 11.5; p = \frac{1}{4}\right) \approx 0.9235.$$

Exercises

7.2-1　A certain size bag is designed to hold 25 pounds of potatoes. A farmer fills such bags in the field. Assume that the weight X of potatoes in a bag is $N(\mu, 9)$. We shall test the null hypothesis H_0: $\mu = 25$ against the alternative hypothesis H_1: $\mu < 25$. Let X_1, X_2, X_3, X_4 be a random sample of size 4 from this distribution, and let the critical region C for this test be defined by $\bar{x} \le 22.5$, where $\bar{x}$ is the observed value of $\bar{X}$.
   (a) What is the power function $K(\mu)$ of this test? In particular, what is the significance level $\alpha = K(25)$ for your test?
   (b) If the random sample of four bags of potatoes yielded the values $x_1 = 21.24$, $x_2 = 24.81$, $x_3 = 23.62$, $x_4 = 26.82$, would you accept or reject H_0 using your test?
   (c) What is the p-value associated with the $\bar{x}$ in part (b)?

7.2-2　Let X equal the number of milliliters in a bottle that has a label volume of 350 milliliters. Assume that the distribution of X is $N(\mu, 4)$. To test the null hypothesis H_0: $\mu = 355$ against the alternative hypothesis H_1: $\mu < 355$, let the critical region be defined by $C = \{\bar{x}: \bar{x} \le 354.05\}$, where $\bar{x}$ is the sample mean of the contents of a random sample of $n = 12$ bottles.
   (a) Define the power function $K(\mu)$ for this test.
   (b) What is the (approximate) significance level of this test?

(c) Find the values of $K(354.05)$ and $K(353.1)$ and sketch the graph of the power function.

(d) Use the following 12 observations to state your conclusion of this test.

$$
\begin{array}{cccccc}
350 & 353 & 354 & 356 & 353 & 352 \\
354 & 355 & 357 & 353 & 354 & 355
\end{array}
$$

(e) What is the approximate p-value of this test?

7.2-3 Assume that SAT mathematics scores of students who attend small liberal arts colleges are $N(\mu, 8100)$. We shall test H_0: $\mu = 530$ against the alternative hypothesis H_1: $\mu < 530$. Given a random sample of size $n = 36$ SAT mathematics scores, let the critical region be defined by $C = \{\bar{x}: \bar{x} \le 510.77\}$, where $\bar{x}$ is the observed mean of the sample.
(a) Define the power function, $K(\mu)$, for this test.
(b) What is the value of the significance level of this test?
(c) What is the value of $K(510.77)$?
(d) Sketch the graph of the power function.
(e) What is the p-value associated with (i) $\bar{x} = 507.35$; (ii) $\bar{x} = 497.45$?

7.2-4 Let X be $N(\mu, 100)$. To test H_0: $\mu = 80$ against H_1: $\mu > 80$, let the critical region be defined by $C = \{(x_1, x_2, \ldots, x_{25}): \bar{x} \ge 83\}$, where $\bar{x}$ is the sample mean of a random sample of size $n = 25$ from this distribution.
(a) How is the power function $K(\mu)$ defined for this test?
(b) What is the significance level of this test?
(c) What are the values of $K(80)$, $K(83)$, and $K(86)$?
(d) Sketch the graph of the power function.
(e) What is the p-value corresponding to $\bar{x} = 83.41$?

7.2-5 Let X equal the yield of alfalfa in tons per acre per year. Assume that X is $N(1.5, 0.09)$. It is hoped that new fertilizer will increase the average yield. We shall test the null hypothesis H_0: $\mu = 1.5$ against the alternative hypothesis H_1: $\mu > 1.5$. Assume that the variance continues to equal $\sigma^2 = 0.09$ with the new fertilizer. Using $\bar{X}$, the mean of a random sample of size n, as the test statistic, reject H_0 if $\bar{x} \ge c$. Find n and c so that the power function $K(\mu) = P(\bar{X} \ge c: \mu)$ is such that $\alpha = K(1.5) = 0.05$ and $K(1.7) = 0.95$.

7.2-6 Let X equal the number of pounds of butterfat produced by a Holstein cow during the 305-day milking period following the birth of a calf. Assume that the distribution of X is $N(\mu, 140^2)$. To test the null hypothesis H_0: $\mu = 715$ against the alternative hypothesis H_1: $\mu < 715$, let the critical region be defined by $C = \{\bar{x}: \bar{x} \le 668.94\}$, where $\bar{x}$ is the sample mean of $n = 25$ butterfat weights from 25 cows selected at random.
(a) Define the power function $K(\mu)$ for this test.
(b) What is the significance level of this test?
(c) What are the values of $K(668.94)$ and $K(622.88)$?
(d) Sketch a graph of the power function.
(e) What is your conclusion using the following 25 observations of X?

425	710	661	664	732	714	934	761	744
653	725	657	421	573	535	602	537	405
874	791	721	849	567	468	975		

(f) What is the approximate p-value of this test?

7.2-7 In Exercise 7.2-6, let $C = \{\bar{x}: \bar{x} \leq c\}$ be the critical region. Find values for n and c so that the significance level of this test is $\alpha = 0.05$ and the power at $\mu = 650$ is 0.90.

7.2-8 Let X have a Bernoulli distribution with p.d.f.

$$f(x; p) = p^x(1 - p)^{1-x}, \qquad x = 0, 1, \qquad 0 \leq p \leq 1.$$

We would like to test the null hypothesis $H_0: p \leq 0.4$ against the alternative hypothesis $H_1: p > 0.4$. For the test statistic use $Y = \sum_{i=1}^{n} X_i$, where $X_1, X_2, \ldots, X_n$ is a random sample of size n from this Bernoulli distribution. Let the critical region be of the form $C = \{y: y \geq c\}$.
(a) Let $n = 100$. On the same set of axes, sketch the graphs of the power functions corresponding to the three critical regions, $C_1 = \{y: y \geq 40\}$, $C_2 = \{y: y \geq 50\}$, and $C_3 = \{y: y \geq 60\}$. Use the normal approximation to compute the probabilities.
(b) Let $C = \{y: y \geq 0.45n\}$. On the same set of axes, sketch the graphs of the power functions corresponding to the three samples of sizes 10, 100, and 1000.

7.2-9 Let p denote the probability that, for a particular tennis player, the first serve is good. Since $p = 0.40$, this player decided to take lessons in order to increase p. When the lessons are completed, the hypothesis $H_0: p = 0.40$ will be tested against $H_1: p > 0.40$ based on $n = 25$ trials. Let y equal the number of first serves that are good, and let the critical region be defined by $C = \{y: y \geq 14\}$.
(a) Define the power function $K(p)$ for this test.
(b) What is the value of the significance level, $\alpha = K(0.40)$? Use Appendix Table II.
(c) Evaluate $K(p)$ at $p = 0.45, 0.50, 0.60, 0.70, 0.80$, and 0.90. Use Table II.
(d) Sketch the graph of the power function.
(e) If $y = 15$ first serves were good following the lessons, would H_0 be rejected?
(f) What is the p-value associated with $y = 15$?

7.2-10 Let $X_1, X_2, \ldots, X_8$ be a random sample of size $n = 8$ from a Poisson distribution with mean λ. Reject the simple null hypothesis $H_0: \lambda = 0.5$ and accept $H_1: \lambda > 0.5$ if the observed sum $\sum_{i=1}^{8} x_i \geq 8$.
(a) Compute the significance level α of the test.
(b) Find the power function $K(\lambda)$ of the test as a sum of Poisson probabilities.
(c) Using Table III in the Appendix, determine $K(0.75), K(1)$, and $K(1.25)$.

7.2-11 Let p equal the fraction defective of a certain manufactured item. To test $H_0: p = 1/26$ against $H_1: p > 1/26$, we inspect n items selected at random and let Y be the number of defective items in this sample. We reject H_0 if the observed $y \geq c$. Find n and c so that $\alpha = K(1/26) = 0.05$ and $K(1/10) = 0.90$ approximately, where $K(p) = P(Y \geq c; p)$.

HINT: Use either the normal or Poisson approximation to help solve this exercise.

7.2-12. Let X_1, X_2, X_3 be a random sample of size $n = 3$ from an exponential distribution with mean $\theta > 0$. Reject the simple null hypothesis $H_0: \theta = 2$ and accept the composite alternative hypothesis $H_1: \theta < 2$ if the observed sum $\sum_{i=1}^{3} x_i \leq 2$.
(a) What is the power function $K(\theta)$ written as an integral?
(b) Using integration by parts, define the power function as a summation.
(c) With the help of Table III in the Appendix, determine $\alpha = K(2)$, $K(1)$, $K(1/2)$, and $K(1/4)$.

7.3 Tests About One Mean and One Variance

In several of the examples and exercises in Section 7.2 we assumed that we were sampling from a normal distribution and that the variance was known. The null hypothesis was generally of the form $H_0: \mu = \mu_0$. There are essentially three possibilities for the alternative hypothesis, namely that μ has increased, (i) $H_1:$ $\mu > \mu_0$; μ has decreased, (ii) $H_1: \mu < \mu_0$; or μ has changed, but it is not known if it has increased or decreased, which leads to a two-sided alternative hypothesis (iii) $H_1: \mu \neq \mu_0$.

To test $H_0: \mu = \mu_0$ against one of these three alternative hypotheses, a random sample is taken from the distribution, and an observed sample mean, $\bar{x}$, that is close to μ_0 supports H_0. The closeness of $\bar{x}$ to μ_0 is measured in terms of standard deviations of $\bar{X}$, $\sigma/\sqrt{n}$, which is sometimes called the **standard error of the mean**. Thus the test statistic could be defined by

$$Z = \frac{\bar{X} - \mu_0}{\sqrt{\sigma^2/n}} = \frac{\bar{X} - \mu_0}{\sigma/\sqrt{n}}, \tag{7.3-1}$$

and the critical regions, at a significance level α, for the three respective alternative hypotheses would be (i) $z \geq z_\alpha$, (ii) $z \leq -z_\alpha$, and (iii) $|z| \geq z_{\alpha/2}$. In terms of $\bar{x}$ these three critical regions become (i) $\bar{x} \geq \mu_0 + z_\alpha \sigma/\sqrt{n}$, (ii) $\bar{x} \leq \mu_0 - z_\alpha \sigma/\sqrt{n}$, and (iii) $|\bar{x} - \mu_0| \geq z_{\alpha/2}\sigma/\sqrt{n}$.

These tests and critical regions are summarized in Table 7.3-1. The underlying assumption is that the distribution is $N(\mu, \sigma^2)$ and σ^2 is known.

Thus far we have assumed that the variance σ^2 was known. We now take a more realistic position and assume that the variance is unknown. Suppose our null hypothesis is $H_0: \mu = \mu_0$ and the two-sided alternative hypothesis is $H_1: \mu \neq \mu_0$. If a random sample $X_1, X_2, \ldots, X_n$ is taken from a normal distribution $N(\mu, \sigma^2)$, we recall from Section 6.2 that a confidence interval for μ was

TABLE 7.3-1

H_0	H_1	Critical Region
$\mu = \mu_0$	$\mu > \mu_0$	$z \geq z_\alpha$ or $\bar{x} \geq \mu_0 + z_\alpha \sigma/\sqrt{n}$
$\mu = \mu_0$	$\mu < \mu_0$	$z \leq -z_\alpha$ or $\bar{x} \leq \mu_0 - z_\alpha \sigma/\sqrt{n}$
$\mu = \mu_0$	$\mu \neq \mu_0$	$|z| \geq z_{\alpha/2}$ or $|\bar{x} - \mu_0| \geq z_{\alpha/2} \sigma/\sqrt{n}$

based on

$$T = \frac{\bar{X} - \mu}{\sqrt{S^2/n}} = \frac{\bar{X} - \mu}{S/\sqrt{n}}.$$

This suggests that T might be a good statistic to use for the test of H_0: $\mu = \mu_0$ with μ replaced by μ_0. In addition, it is the natural statistic to use if we replace σ^2/n by its unbiased estimator S^2/n in $(\bar{X} - \mu_0)/\sqrt{\sigma^2/n}$ in equation (7.3-1). If $\mu = \mu_0$, we know that T has a t distribution with $n - 1$ degrees of freedom. Thus, with $\mu = \mu_0$,

$$P[|T| \geq t_{\alpha/2}(n-1)] = P\left[\frac{|\bar{X} - \mu_0|}{S/\sqrt{n}} \geq t_{\alpha/2}(n-1)\right] = \alpha.$$

Accordingly, if $\bar{x}$ and s are the sample mean and sample standard deviation, the rule that rejects H_0: $\mu = \mu_0$ if and only if

$$|t| = \frac{|\bar{x} - \mu_0|}{s/\sqrt{n}} \geq t_{\alpha/2}(n-1)$$

provides a test of this hypothesis with significance level α. It should be noted that this rule is equivalent to rejecting H_0: $\mu = \mu_0$ if μ_0 is not in the open $100(1 - \alpha)\%$ confidence interval

$$(\bar{x} - t_{\alpha/2}(n-1)s/\sqrt{n}, \ \bar{x} + t_{\alpha/2}(n-1)s/\sqrt{n}).$$

Table 7.3-2 summarizes tests of hypotheses for a single mean, along with the three possible alternative hypotheses, when the underlying distribution is $N(\mu, \sigma^2)$, σ^2 is unknown, $t = (\bar{x} - \mu_0)/(s/\sqrt{n})$, and $n \leq 31$. If $n > 31$, use Table 7.3-1 for approximate tests with σ replaced by s.

TABLE 7.3-2

H_0	H_1	Critical Region
$\mu = \mu_0$	$\mu > \mu_0$	$t \geq t_\alpha(n-1)$ or $\bar{x} \geq \mu_0 + t_\alpha(n-1)s/\sqrt{n}$
$\mu = \mu_0$	$\mu < \mu_0$	$t \leq -t_\alpha(n-1)$ or $\bar{x} \leq \mu_0 - t_\alpha(n-1)s/\sqrt{n}$
$\mu = \mu_0$	$\mu \neq \mu_0$	$|t| \geq t_{\alpha/2}(n-1)$ or $|\bar{x} - \mu_0| \geq t_{\alpha/2}(n-1)s/\sqrt{n}$

Example 7.3-1   Let X (in millimeters) equal the growth in 15 days of a tumor induced in a mouse. Assume that the distribution of X is $N(\mu, \sigma^2)$. We shall test the null hypothesis H_0: $\mu = \mu_0 = 4.0$ millimeters against the two-sided alternative hypothesis H_1: $\mu \neq 4.0$. If we use $n = 9$ observations and a significance level of $\alpha = 0.10$, the critical region is

$$|t| = \frac{|\bar{x} - 4.0|}{s/\sqrt{9}} \geq t_{\alpha/2}(8) = t_{0.05}(8) = 1.860.$$

If we are given that $n = 9$, $\bar{x} = 4.3$, and $s = 1.2$, we see that

$$t = \frac{4.3 - 4.0}{1.2/\sqrt{9}} = \frac{0.3}{0.4} = 0.75.$$

Thus

$$|t| = |0.75| < 1.860$$

and we accept (do not reject) H_0: $\mu = 4.0$ at the $\alpha = 10\%$ significance level. See Figure 7.3-1.

REMARK   In discussing the test of a statistical hypothesis, the word *accept* might better be replaced by *do not reject*. That is, if, in Example 7.3-1, $\bar{x}$ is close enough to 4.0 so that we accept $\mu = 4.0$, we do not want that acceptance to imply that μ is actually equal to 4.0. We want to say that the data do not deviate enough from $\mu = 4.0$ for us to reject that hypothesis; that is, we do not reject $\mu = 4.0$ with these observed data. With this understanding, we sometimes use *accept*, and sometimes *fail to reject* or *do not reject,* the null hypothesis.

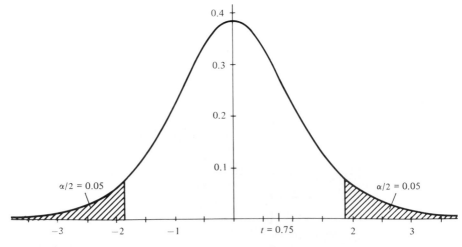

FIGURE 7.3-1

The next example illustrates the use of the t-statistic with a one-sided alternative hypothesis.

Example 7.3-2　　In attempting to control the strength of the wastes discharged into a nearby river, a paper firm has taken a number of measures. Members of the firm believe that they have reduced the oxygen-consuming power of their wastes from a previous mean μ of 500 (measured in permanganate in parts per million). They plan to test H_0: $\mu = 500$ against H_1: $\mu < 500$, using readings taken on $n = 25$ consecutive days. If these 25 values can be treated as a random sample, then the critical region, for a significance level of $\alpha = 0.01$, is

$$ t = \frac{\bar{x} - 500}{s/\sqrt{25}} \le -t_{0.01}(24) = -2.492. $$

The observed values of the sample mean and sample standard deviation were $\bar{x} = 308.8$ and $s = 115.15$. Since

$$ t = \frac{308.8 - 500}{115.15/\sqrt{25}} = -8.30 < -2.492, $$

we clearly reject the null hypothesis and accept H_1: $\mu < 500$. It should be noted, however, that although an improvement has been made, there still might exist the question of whether the improvement is adequate. The 95% confidence interval $308.8 \pm 2.064(115.15/5)$ or $[261.27, 356.33]$ for μ might help the company answer that question.

Many times there is interest in comparing the means of two different distributions or populations. We must consider two situations, that in which X and Y are dependent, and that in which X and Y are independent. We consider the independent case in the next section.

If X and Y are dependent, let $W = X - Y$, and the hypothesis that $\mu_X = \mu_Y$ would be replaced with the hypothesis H_0: $\mu_W = 0$. For example, suppose that X and Y equal the resting pulse rate for a person before and after taking an 8-week program in aerobic dance. We would be interested in testing H_0: $\mu_W = 0$ (no change) against H_1: $\mu_W > 0$ (the aerobic dance program decreased resting pulse rate). Because X and Y are measurements on the same person, X and Y are clearly dependent. If we can assume that the distribution of W is (approximately) $N(\mu_W, \sigma^2)$, then the appropriate t-test for a single mean could be used, selecting from Table 7.3-2.

Example 7.3-3　　Twenty-four girls in the 9th and 10th grades were put on an ultraheavy rope-jumping program. Someone thought that such a program would increase their speed when running the 40-yard dash. Let W equal the difference in time to run the 40-yard dash—the "before program time" minus the "after program time." Assume that the distribution of W is (approximately) $N(\mu_W, \sigma_W^2)$. We shall test the null hypothesis H_0: $\mu_W = 0$ against the

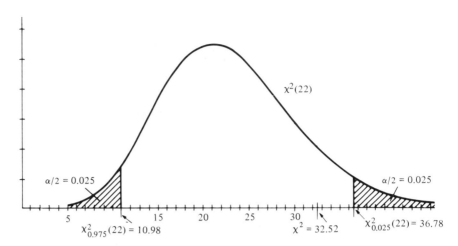

FIGURE 7.3-2

If H_1: $\sigma^2 > 100$ had been the alternative hypothesis, H_0: $\sigma^2 = 100$ would have been rejected if

$$\chi^2 = \frac{22s^2}{100} \geq \chi_\alpha^2(22) = 33.92$$

or, equivalently,

$$s^2 \geq \frac{100\chi_\alpha^2(22)}{22} = \frac{(100)(33.92)}{22} = 154.18.$$

Because

$$\chi^2 = 32.52 < 33.92 \qquad \text{and} \qquad s^2 = 147.82 < 154.18,$$

H_0 would not be rejected in favor of this one-sided alternative hypothesis for $\alpha = 0.05$, although we suspect a slightly larger data set might lead to rejection.

Table 7.3-3 summarizes tests of hypotheses for a single variance. The critical region is given in terms of the sample variance. The critical region is also given in terms of the chi-square test statistic

$$\chi^2 = \frac{(n-1)s^2}{\sigma_0^2}.$$

There are two ways of viewing a statistical test. One of these is through the p-value of the test; this is becoming more popular and is included in most computer printouts. (This was discussed in Section 7.2.) After observing the test statistic, the p-value is the probability, under H_0, of all values of the test statistic that are as

TABLE 7.3-3

H_0	H_1	Critical Region
$\sigma^2 = \sigma_0^2$	$\sigma^2 > \sigma_0^2$	$s^2 \geq \dfrac{\sigma_0^2 \chi_\alpha^2(n-1)}{n-1}$ or $\chi^2 \geq \chi_\alpha^2(n-1)$
$\sigma^2 = \sigma_0^2$	$\sigma^2 < \sigma_0^2$	$s^2 \leq \dfrac{\sigma_0^2 \chi_{1-\alpha}^2(n-1)}{n-1}$ or $\chi^2 \leq \chi_{1-\alpha}^2(n-1)$
$\sigma^2 = \sigma_0^2$	$\sigma^2 \neq \sigma_0^2$	$s^2 \leq \dfrac{\sigma_0^2 \chi_{1-\alpha/2}^2(n-1)}{n-1}$ or $s^2 \geq \dfrac{\sigma_0^2 \chi_{\alpha/2}^2(n-1)}{n-1}$
		or $\chi^2 \leq \chi_{1-\alpha/2}^2(n-1)$ or $\chi^2 \geq \chi_{\alpha/2}^2(n-1)$

extreme (in the direction of rejection of H_0) as the observed one. For illustration, in Example 7.2-3, if the golfer averaged $\bar{x} = 87.94$, then

$$p\text{-value} = P(\overline{X} \leq 87.94) = P\left(\frac{\overline{X} - 90}{3/4} \leq \frac{87.94 - 90}{3/4}\right) = 0.0027.$$

The fact that the p-value < 0.05 is equivalent to the fact that $\bar{x} < 88.77$ because $P(\overline{X} \leq 88.77; \mu = 90) = 0.05$. Since $\bar{x} = 87.94$ is an observed value of a random variable, namely $\overline{X}$, then the p-value, a function of $\bar{x}$, is also an observed value of a random variable. That is, before the random experiment is performed, the probability that the p-value is less than α is equal to α when the null hypothesis is true. Many statisticians believe that the observed p-value provides an understandable measure of the truth of H_0: The smaller the p-value, the less they believe in H_0.

Three additional examples of the p-value are given by referring to Examples 7.3-1, 7.3-2, and 7.3-4. In two-sided tests for means and proportions, the p-value is the probability of the extreme values in both directions. With the mouse data (Example 7.3-1), the p-value is

$$p\text{-value} = P(|T| \geq 0.75).$$

In Table VI we see that if T has a t-distribution with 8 degrees of freedom, then $P(T \geq 0.706) = 0.25$. Thus $P(|T| \geq 0.706) = 0.50$ and the p-value will be a little smaller than 0.50. In fact, $P(|T| \geq 0.75) = 0.47$ (a probability that can be found with certain computer programs), which is not less than $\alpha = 0.10$, and thus we do not reject H_0 at that significance level. In the example concerned with waste (Example 7.3-2), the p-value is essentially zero, since $P(Z \leq -8.30) \approx 0$, where Z is a standardized normal variable, and we reject H_0. In the example about the variance of IQ scores (Example 7.3-4), with the one-sided alternative, H_1: $\sigma^2 > 100$, the p-value is

$$p\text{-value} = P(W \geq 32.52),$$

where $W = 22S^2/100$ has a chi-square distribution with 22 degrees of freedom when H_0 is true. From Table IV we see that

$$P(W \geq 30.81) = 0.10 \quad \text{and} \quad P(W \geq 33.92) = 0.05.$$

Thus

$$0.05 < p\text{-value} = P(W \geq 32.52) < 0.10$$

and the null hypothesis would not be rejected at an $\alpha = 0.05$ significance level. To find the p-value for a two-sided alternative for tests about the variance, double the tail probability beyond the chi-square test statistic if the sample variance is larger than σ_0^2, or double the tail probability below the chi-square test statistic if the sample variance is below σ_0^2. So in Example 7.3-4, since $s^2 = 147.82 > 100$, the p-value equals $2[P(W \geq 32.52)]$ and thus

$$0.10 < p\text{-value} < 0.20.$$

The other way of looking at tests of hypotheses is through the consideration of confidence intervals, particularly for two-sided alternatives and the corresponding tests. For example, with the mouse data (Example 7.3-1), a 90% confidence interval for the unknown mean is

$$4.3 \pm (1.86)(1.2)/\sqrt{9} \quad \text{or} \quad [3.56, 5.04],$$

since $t_{0.05}(8) = 1.86$. Note that this confidence interval covers the hypothesized value $\mu = 4.0$ and we do not reject H_0: $\mu = 4.0$. If the confidence interval did not cover $\mu = 4.0$, then we would have rejected H_0: $\mu = 4.0$. Many statisticians believe that estimation is much more important than tests of hypotheses and accordingly approach statistical tests through confidence intervals. For one-sided tests, we could use one-sided confidence intervals, but we do not trouble the reader with this additional concept in this book.

Exercises

7.3-1   Assume that IQ scores for a certain population are approximately $N(\mu, 100)$. To test H_0: $\mu = 110$ against the one-sided alternative hypothesis H_1: $\mu > 110$, we take a random sample of size $n = 16$ from this population and observe $\bar{x} = 113.5$. Do we accept or reject H_0 at the
(a) 5% significance level?
(b) 10% significance level?
(c) What is the p-value of this test?

7.3-2 Assume that the weight of cereal in a "10-ounce box" is $N(\mu, \sigma^2)$. To test H_0: $\mu = 10.1$ against H_1: $\mu > 10.1$, we take a random sample of size $n = 16$ and observe that $\bar{x} = 10.4$ and $s = 0.4$.
(a) Do we accept or reject H_0 at the 5% significance level?
(b) What is the approximate p-value of this test?

7.3-3 Let X equal the brinell hardness measurement of ductile iron subcritically annealed. Assume that the distribution of X is $N(\mu, 100)$. We shall test the null hypothesis H_0: $\mu = 170$ against the alternative hypothesis H_1: $\mu > 170$ using $n = 25$ observations of X.

(a) Define the test statistic and a critical region that has a significance level of $\alpha = 0.05$. Sketch a figure showing this critical region.

(b) A random sample of $n = 25$ observations of X yielded the following measurements:

170	167	174	179	179	156	163	156	187
156	183	179	174	179	170	156	187	
179	183	174	187	167	159	170	179	

Calculate the value of the test statistic and clearly give your conclusion.

(c) Give the approximate p-value of this test.

7.3-4 Let X equal the thickness of spearmint gum manufactured for vending machines. Assume that the distribution of X is $N(\mu, \sigma^2)$. The target thickness is 7.5 hundredths of an inch. We shall test the null hypothesis H_0: $\mu = 7.5$ against a two-sided alternative hypothesis using 10 observations.

(a) Define the test statistic and critical region for an $\alpha = 0.05$ significance level. Sketch a figure illustrating this critical region.

(b) Calculate the value of the test statistic and clearly give your decision using the following $n = 10$ thicknesses in hundredths of an inch for pieces of gum that were selected randomly from the production line:

7.65	7.60	7.65	7.70	7.55
7.55	7.40	7.40	7.50	7.50

(c) Is $\mu = 7.50$ contained in a 95% confidence interval for μ?

7.3-5 The mean birth weight in the United States is $\mu = 3315$ grams with a standard deviation of $\sigma = 575$. Let X equal the birth weight in grams in Jerusalem. Assume that the distribution of X is $N(\mu, \sigma^2)$. We shall test the null hypothesis H_0: $\mu = 3315$ against the alternative hypothesis H_1: $\mu < 3315$ using a random sample of size $n = 30$.

(a) Define a critical region that has a significance level of $\alpha = 0.05$.

(b) If the random sample of $n = 30$ yielded $\bar{x} = 3189$ and $s = 488$, what is your conclusion?

(c) What is the approximate p-value of your test?

7.3-6 Let X equal the forced vital capacity (FVC) in liters for a female college student. (This is the amount of air that a student can force out of her lungs.) Assume that the distribution of X is (approximately) $N(\mu, \sigma^2)$. Suppose it is known that $\mu = 3.4$ liters. A volleyball coach claims that the FVC of volleyball players is greater than 3.4. She plans to test her claim using a random sample of size $n = 9$.

(a) Define the null hypothesis.

(b) Define the alternative (coach's) hypothesis.

(c) Define the test statistic.

(d) Define a critical region for which $\alpha = 0.05$. Draw a figure illustrating your critical region.

(e) Calculate the value of the test statistic given that the random sample yielded the following forced vital capacities: 3.4, 3.6, 3.8, 3.3, 3.4, 3.5, 3.7, 3.6, 3.7.

(f) What is your conclusion?

(g) What is the approximate p-value of this test?

7.3-7 Vitamin B_6 is one of the vitamins in a multiple vitamin pill manufactured by a pharmaceutical company. The pills are produced with a mean of 50 milligrams of vitamin B_6 per pill. The company believes that there is a deterioration of 1 milligram per month, so that after 3 months they expect that $\mu = 47$. A consumer group suspects that $\mu < 47$ after 3 months.

(a) Define a critical region to test H_0: $\mu = 47$ against H_1: $\mu < 47$ at an $\alpha = 0.05$ significance level based on a random sample of size $n = 20$.

(b) If the 20 pills yielded a mean of $\bar{x} = 46.94$ with a standard deviation of $s = 0.15$, what is your conclusion?

(c) What is the approximate p-value of this test?

7.3-8 Assume that the birthweight in grams of a baby born in the United States is $N(3315, 525^2)$, boys and girls combined. Let X equal the weight of a baby girl who is born at home in Ottawa County and assume that the distribution of X is $N(\mu_X, \sigma_X^2)$.

(a) Using 11 observations of X, define the test statistic and critical region, $\alpha = 0.01$, for testing H_0: $\mu_X = 3315$ against the alternative hypothesis H_1: $\mu_X > 3315$ (home-born babies are heavier).

(b) Calculate the value of the test statistic and give your conclusion using the following weights:

3119	2657	3459	3629	3345	3629
3515	3856	3629	3345	3062	

(c) What is the approximate p-value of the test?

(d) Define the test statistic and critical region, $\alpha = 0.05$, for testing H_0: $\sigma_X^2 = 525^2$ against the alternative hypothesis H_1: $\sigma_X^2 < 525^2$ (less variation of weights of home-born babies).

(e) Calculate the value of your test statistic and state your conclusion.

(f) What is the approximate p-value of this second test?

7.3-9 Let Y equal the weight in grams of a baby boy who is born at home in Ottawa County and assume that the distribution of Y is $N(\mu_Y, \sigma_Y^2)$. Using the following weights:

4082	3686	4111	3686	3175	4139
3686	3430	3289	3657	4082	

answer the questions in Exercise 7.3-8. What other hypothesis is suggested by these two exercises?

7.3-10 A company that manufactures brackets for an auto maker regularly selects brackets from the production line and performs a torque test. The goal is for mean torque to equal 125. Let X equal the torque and assume that X is

$N(\mu, \sigma^2)$. We shall use a sample of size $n = 15$ to test H_0: $\mu = 125$ against a two-sided alternative hypothesis.

(a) Define the test statistic and a critical region with significance level $\alpha = 0.05$. Sketch a figure illustrating the critical region.

(b) Use the following observations to calculate the value of the test statistic and state your conclusion:

128	149	136	114	126	142	124	136
122	118	122	129	118	122	129	

7.3-11 Let X equal the number of pounds of butterfat produced by a Holstein cow during the 305-day milking period following the birth of a calf. In Exercise 7.2-6 we assumed that the distribution of X is $N(\mu, 140^2)$. We shall now test the null hypothesis H_0: $\sigma^2 = 140^2$ against the alternative hypothesis H_1: $\sigma^2 > 140^2$.

(a) Define the test statistic and a critical region that has a significance level of $\alpha = 0.05$, assuming that there are $n = 25$ observations.

(b) Using the data in Exercise 7.2-6, calculate the value of the test statistic and provide your conclusion.

7.3-12 In May the fill weights of 6-pound boxes of laundry soap had a mean of 6.13 pounds with a standard deviation of 0.095. The goal was to decrease the standard deviation. The company decided to adjust the filling machines and then test H_0: $\sigma = 0.095$ against H_1: $\sigma < 0.095$. In June a random sample of size $n = 20$ yielded $\bar{x} = 6.10$ and $s = 0.065$.

(a) At an $\alpha = 0.05$ significance level, was the company successful?

(b) What is the approximate p-value of your test?

7.3-13 The mean birth weight in the United States is $\mu = 3315$ grams with a standard deviation of $\sigma = 575$. Let X equal the birth weight in Rwanda. Assume that the distribution of X is $N(\mu, \sigma^2)$. We shall test the hypothesis H_0: $\sigma = 575$ against the alternative hypothesis H_1: $\sigma < 575$ at an $\alpha = 0.10$ significance level.

(a) What is your decision if a random sample of size $n = 81$ yielded $\bar{x} = 2819$ and $s = 496$?

(b) What is the approximate p-value of this test?

7.3-14 Let $X_1, X_2, \ldots, X_{23}$ be a random sample from a normal distribution that has variance $\sigma^2 = 100$. Let $S^2 = (1/22)\Sigma_{i=1}^{23}(X_i - \bar{X})^2$ be the sample variance. Show that $\text{Var}(S^2) = 10,000/11$. Thus the standard deviation of S^2 is 30.15, and this helps explain the critical region $s^2 \leq 49.91$ or $s^2 \geq 167.18$ in Example 7.3-4.

7.3-15 Let X be $N(\mu, \sigma^2)$.

(a) Is the hypothesis H_0: $\sigma^2 = 0.04$ rejected in favor of the two-sided alternative hypothesis H_1: $\sigma^2 \neq 0.04$ if a random sample of size $n = 13$ yielded $s^2 = 0.058$? Let $\alpha = 0.05$.

(b) Is $\sigma^2 = 0.04$ contained in a 95% confidence interval for σ^2?

7.3-16 Let $X_1, X_2, \ldots, X_{19}$ be a random sample of size $n = 19$ from the normal distribution $N(\mu, \sigma^2)$.

(a) Define a critical region, C, of size $\alpha = 0.05$ for testing H_0: $\sigma^2 = 30$ against H_1: $\sigma^2 = 80$.

(b) Find the approximate value of β, the probability of Type II error, for the critical region C of part (a).

7.3-17 Each of 51 golfers hit three golf balls of brand X and three golf balls of brand Y in a random order. Let X_i and Y_i equal the averages of the distances traveled by the brand X and brand Y golf balls hit by the ith golfer, $i = 1, 2,$ $\ldots, 51$. Let $W_i = X_i - Y_i$, $i = 1, 2, \ldots, 51$. Test H_0: $\mu_W = 0$ against H_1: $\mu_W > 0$, where μ_W is the mean of the differences. If $\overline{w} = 2.07$ and $s_w^2 = 84.63$, would H_0 be accepted or rejected at an $\alpha = 0.05$ significance level?

7.3-18 To test whether a golf ball of brand A can be hit a greater distance off the tee than a golf ball of brand B, each of 17 golfers hit a ball of each brand, eight hitting ball A before ball B and nine hitting ball B before ball A. Assume that the differences of the paired A distance and B distance are approximately normally distributed and test the null hypothesis H_0: $\mu_D = 0$ against the alternative hypothesis H_1: $\mu_D > 0$ using a t-test with the 17 differences. Let $\alpha = 0.05$.

Golfer	Distance for Ball A	Distance for Ball B
1	265	252
2	272	276
3	246	243
4	260	246
5	274	275
6	263	246
7	255	244
8	258	245
9	276	259
10	274	260
11	274	267
12	269	267
13	244	251
14	212	222
15	235	235
16	254	255
17	224	231

7.3-19 A vendor of milk products produces and sells low-fat dry milk to a company that uses it to produce baby formula. In order to determine the fat content of the milk, both the company and the vendor take a sample from each lot and test it for fat content in percent. Ten sets of paired test results are

Lot Number	Company Test Results (X)	Vendor Test Results (Y)
1	0.50	0.79
2	0.58	0.71
3	0.90	0.82
4	1.17	0.82
5	1.14	0.73
6	1.25	0.77
7	0.75	0.72
8	1.22	0.79
9	0.74	0.72
10	0.80	0.91

Let μ_D denote the mean of the differences. Test H_0: $\mu_D = 0$ against H_1: $\mu_D > 0$ using a t-test for paired differences. Let $\alpha = 0.05$.

7.3-20 A company that manufactures motors receives reels of 10,000 terminals per reel. Before using a reel of terminals, 20 terminals are randomly selected to be tested. The test is the amount of pressure needed to pull the terminal apart from its mate. This amount of pressure should continue to increase from test to test as the terminal is "roughed up." (Since this is destructive testing, a terminal that is tested cannot be used in a motor.) Let W equal the difference of the pressures: "test No. 1 pressure" minus "test No. 2 pressure." Assume that the distribution of W is $N(\mu_W, \sigma_W^2)$. We shall test the null hypothesis H_0: $\mu_W = 0$ against the alternative hypothesis H_1: $\mu_W < 0$ using 20 pairs of observations.

(a) Define the test statistic and a critical region that has a significance level of $\alpha = 0.05$. Sketch a figure illustrating this critical region.
(b) Use the following data to calculate the value of the test statistic and clearly state your conclusion.

Terminal	Test No. 1	Test No. 2	Terminal	Test No. 1	Test No. 2
1	2.5	3.8	11	7.3	8.2
2	4.0	3.9	12	7.2	6.6
3	5.2	4.7	13	5.9	6.8
4	4.9	6.0	14	7.5	6.6
5	5.2	5.7	15	7.1	7.5
6	6.0	5.7	16	7.2	7.5
7	5.2	5.0	17	6.1	7.3
8	6.6	6.2	18	6.3	7.1
9	6.7	7.3	19	6.5	7.2
10	6.6	6.5	20	6.5	6.7

(c) What would the conclusion be if $\alpha = 0.01$?
(d) What is the approximate p-value of this test?

7.4   Tests of the Equality of Two Independent Normal Distributions

Let X and Y have independent normal distributions $N(\mu_X, \sigma_X^2)$ and $N(\mu_Y, \sigma_Y^2)$, respectively. There are times when we are interested in testing whether the distributions of X and Y are the same. So if the assumption of normality is valid, we would be interested in testing whether the two variances are equal and whether the two means are equal.

We first consider a test of the equality of the two means. When X and Y are independent and normally distributed, we can test hypotheses about their means using the same t-statistic that was used for constructing a confidence interval for $\mu_X - \mu_Y$ in Section 6.2. Recall that the t-statistic used for constructing the confidence interval assumed that the variances of X and Y are equal. That is why we shall later consider a test for the equality of two variances.

We begin with an example and then give a table that lists some hypotheses and critical regions. A botanist is interested in comparing the growth response of dwarf pea stems to two different levels of the hormone indoleacetic acid (IAA). Using 16-day-old pea plants, the botanist obtains 5-millimeter sections and floats these sections on solutions with different hormone concentrations to observe the effect of the hormone on the growth of the pea stem. Let X and Y denote, respectively, the independent growths that can be attributed to the hormone during the first 26 hours after sectioning for $(0.5)(10)^{-4}$ and 10^{-4} levels of concentration of IAA. The botanist would like to test the null hypothesis $H_0: \mu_X - \mu_Y = 0$ against the alternative hypothesis $H_1: \mu_X - \mu_Y < 0$. If we can assume X and Y are independent and normally distributed with common variance, respective random samples of sizes n and m give a test based on the statistic

$$T = \frac{\bar{X} - \bar{Y}}{\sqrt{\{[(n-1)S_X^2 + (m-1)S_Y^2]/(n+m-2)\}(1/n + 1/m)}} \qquad (7.4\text{-}1)$$

$$= \frac{\bar{X} - \bar{Y}}{S_P\sqrt{1/n + 1/m}}$$

where

$$S_P = \sqrt{\frac{(n-1)S_X^2 + (m-1)S_Y^2}{n+m-2}}. \qquad (7.4\text{-}2)$$

T has a t distribution with $r = n + m - 2$ degrees of freedom when H_0 is true and the variances are (approximately) equal. The hypothesis H_0 will be rejected in favor of H_1 if the observed value of T is less than $-t_\alpha(n+m-2)$.

> **Example 7.4-1** In the preceding discussion, the botanist measured the growths of pea stem segments, in millimeters, for $n = 11$ observations of X:
>
> 0.8   1.8   1.0   0.1   0.9   1.7   1.0   1.4   0.9   1.2   0.5

and $m = 13$ observations of Y:

$$
\begin{array}{ccccccc}
1.0 & 0.8 & 1.6 & 2.6 & 1.3 & 1.1 & 2.4 \\
1.8 & 2.5 & 1.4 & 1.9 & 2.0 & 1.2 &
\end{array}
$$

For these data, $\bar{x} = 1.03$, $s_x^2 = 0.24$, $\bar{y} = 1.66$, and $s_y^2 = 0.35$. The critical region for testing H_0: $\mu_X - \mu_Y = 0$ against H_1: $\mu_X - \mu_Y < 0$ is $t \leq -t_{0.05}(22) = -1.717$, where t is given in equation (7.4-1). Since

$$
t = \frac{1.03 - 1.66}{\sqrt{\{[10(0.24) + 12(0.35)]/(11 + 13 - 2)\}(1/11 + 1/13)}}
$$

$$
= -2.81 < -1.717,
$$

H_0 is clearly rejected at an $\alpha = 0.05$ significance level. Notice that the approximate p-value of this test is 0.005 because $-t_{0.005}(22) = -2.819$. See Figure 7.4-1. Also, the sample variances do not differ too much; thus most statisticians would use this two-sample t-test.

It is also instructive to construct box-and-whisker diagrams to gain a visual comparison of the two samples. For these two sets of data, the five-number summaries (minimum, three quartiles, maximum) are

$$
\begin{array}{ccccc}
0.1 & 0.8 & 1.0 & 1.4 & 1.8
\end{array}
$$

for the X sample and

$$
\begin{array}{ccccc}
0.8 & 1.15 & 1.6 & 2.2 & 2.6
\end{array}
$$

for the Y sample. The two box plots are shown in Figure 7.4-2.

Based on random samples of sizes n and m, let $\bar{x}$, $\bar{y}$, and s_p^2 represent the observed unbiased estimates of the respective parameters μ_X, μ_Y, and $\sigma_X^2 = \sigma_Y^2$ of two inde-

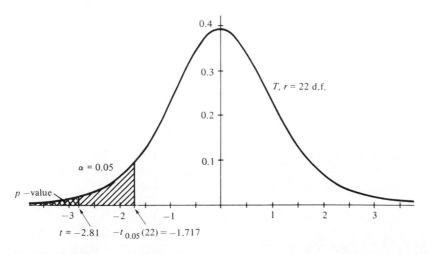

FIGURE 7.4-1

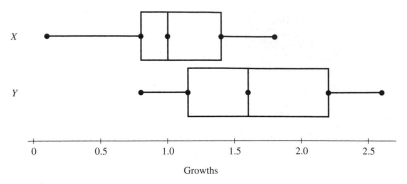

FIGURE 7.4-2

pendent normal distributions with common variance. Then α-level tests of certain hypotheses are given in Table 7.4-1, when $\sigma_X^2 = \sigma_Y^2$. If the common variance assumption is violated, but not too badly, the test is satisfactory but the significance levels are only approximate. The t-statistic along with s_p are given in equations (7.4-1) and (7.4-2), respectively.

Example 7.4-2   A product is packaged using a machine with 24 filler heads numbered 1 to 24, with the odd numbered heads on one side of the machine and the even on the other side. Let X and Y equal the fill weights in grams when a package is filled by an odd-numbered head and an even-numbered head, respectively. Assume that the distributions of X and Y are $N(\mu_X, \sigma^2)$ and $N(\mu_Y, \sigma^2)$, respectively, and that X and Y are independent. We would like to test the null hypothesis H_0: $\mu_X - \mu_Y = 0$ against the alternative hypothesis H_1: $\mu_X - \mu_Y \neq 0$. To perform the test, after the machine has been set up and is running, we shall select one package at random from each filler head and weigh it; we do this 12 times. The test statistic is that given by equation (7.4-1) with $n = m = 12$. At an $\alpha = 0.10$ significance level, the critical region is $|t| \geq t_{0.05}(22) = 1.717$.

TABLE 7.4-1

H_0	H_1	Critical Region
$\mu_X = \mu_Y$	$\mu_X > \mu_Y$	$t \geq t_\alpha(n+m-2)$ or $\bar{x} - \bar{y} \geq t_\alpha(n+m-2)s_p\sqrt{1/n + 1/m}$
$\mu_X = \mu_Y$	$\mu_X < \mu_Y$	$t \leq -t_\alpha(n+m-2)$ or $\bar{x} - \bar{y} \leq -t_\alpha(n+m-2)s_p\sqrt{1/n + 1/m}$
$\mu_X = \mu_Y$	$\mu_X \neq \mu_Y$	$|t| \geq t_{\alpha/2}(n+m-2)$ or $|\bar{x} - \bar{y}| \geq t_{\alpha/2}(n+m-2)s_p\sqrt{1/n + 1/m}$

For the $n = 12$ observations of X,

1071	1076	1070	1083	1082	1067
1078	1080	1075	1084	1075	1080

$\bar{x} = 1076.75$ and $s_x^2 = 29.30$. For the $m = 12$ observations of Y,

1074	1069	1075	1067	1068	1079
1082	1064	1070	1073	1072	1075

$\bar{y} = 1072.33$ and $s_y^2 = 26.24$. The calculated value of the test statistic is

$$t = \frac{1076.75 - 1072.33}{\sqrt{\dfrac{11(29.30) + 11(26.24)}{22}\left(\dfrac{1}{12} + \dfrac{1}{12}\right)}} = 2.05.$$

Since

$$|t| = |2.05| = 2.05 > 1.717,$$

the null hypothesis is rejected at an $\alpha = 0.10$ significance level. Note, however, that

$$|t| = 2.05 < 2.074 = t_{0.025}(22)$$

so that the null hypothesis would not be rejected at an $\alpha = 0.05$ significance level. That is, the p-value is between 0.05 and 0.10.

Again it is instructive to construct box plots on the same graph for these two sets of data. The box plots in Figure 7.4-3 were constructed using the five-number summary for the observations of X—1067, 1072, 1077, 1081.5, 1084—and the five-number summary for the observations of Y—1064, 1068.25, 1072.5, 1075, 1082. It looks like additional sampling would be advisable to test that the filler heads on the two sides of the machine are filling in a similar manner. If not, some corrective action needs to be taken.

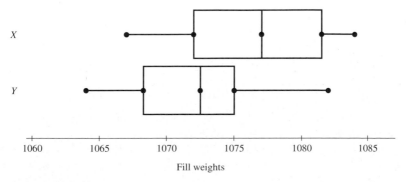

FIGURE 7.4-3

We would like to give two modifications of tests about two means. If we are able to assume that we know the variances of X and Y, then the appropriate test statistic to use for testing H_0: $\mu_X = \mu_Y$ is

$$Z = \frac{\overline{X} - \overline{Y}}{\sqrt{\dfrac{\sigma_X^2}{n} + \dfrac{\sigma_Y^2}{m}}}, \tag{7.4-3}$$

which has a standard normal distribution when the null hypothesis is true and, of course, the populations are normally distributed. If the variances are unknown and the sample sizes are large, replace σ_X^2 with S_X^2 and σ_Y^2 with S_Y^2 in equation (7.4-3). The resulting statistic will have an approximate $N(0, 1)$ distribution.

Example 7.4-3   The target thickness for Fruit Flavored Gum and for Fruit Flavored Bubble Gum is 6.7 hundredths of an inch. Let the independent random variables X and Y equal the respective thicknesses of these gums in hundredths of an inch and assume that their distributions are $N(\mu_X, \sigma_X^2)$ and $N(\mu_Y, \sigma_Y^2)$, respectively. Because bubble gum has more elasticity than regular gum, it seems as if it would be harder to roll it out to the correct thickness. Thus we shall test the null hypothesis H_0: $\mu_X = \mu_Y$ against the alternative hypothesis H_1: $\mu_X < \mu_Y$ using samples of sizes $n = 50$ and $m = 40$.

Because the variances are unknown and the sample sizes are large, the test statistic that is used is

$$Z = \frac{\overline{X} - \overline{Y}}{\sqrt{\dfrac{S_X^2}{50} + \dfrac{S_Y^2}{40}}}.$$

At an approximate significance level of $\alpha = 0.01$, the critical region is

$$z \leq -z_{0.01} = -2.326.$$

The observed values of X were

6.85	6.60	6.70	6.75	6.75	6.90	6.85	6.90	6.70	6.85
6.60	6.70	6.75	6.70	6.70	6.70	6.55	6.60	6.95	6.95
6.80	6.80	6.70	6.75	6.60	6.70	6.65	6.55	6.55	6.60
6.60	6.70	6.80	6.75	6.60	6.75	6.50	6.75	6.70	6.65
6.70	6.70	6.55	6.65	6.60	6.65	6.60	6.65	6.80	6.60

for which $\overline{x} = 6.701$ and $s_x = 0.108$. The observed values of Y were

7.10	7.05	6.70	6.75	6.90	6.90	6.65	6.60	6.55	6.55
6.85	6.90	6.60	6.85	6.95	7.10	6.95	6.90	7.15	7.05
6.70	6.90	6.85	6.95	7.05	6.75	6.90	6.80	6.70	6.75
6.90	6.90	6.70	6.70	6.90	6.90	6.70	6.70	6.90	6.95

for which $\overline{y} = 6.841$ and $s_y = 0.155$. Since the calculated value of the test statistic is

$$z = \frac{6.701 - 6.841}{\sqrt{0.108^2/50 + 0.155^2/40}} = -4.848 < -2.326,$$

the null hypothesis is clearly rejected.

The box-and-whisker diagrams in Figure 7.4-4 were constructed using the five-number summary of the observations of X—6.50, 6.60, 6.70, 6.75, 6.95—and the five-number summary of the observations of Y—6.55, 6.70, 6.90, 6.94, 7.15. This graphical display also confirms our conclusion.

REMARKS To have satisfactory tests, our assumptions must be satisfied reasonably well. As long as the underlying distributions are not highly skewed, the normal assumptions are not too critical as $\overline{X}$ and $\overline{Y}$ have approximate normal distributions by the Central Limit Theorem. As distributions become nonnormal and highly skewed, the sample mean and sample variance become more dependent and that causes problems using the Student-t as an approximating distribution for T. In these cases, nonparametric methods described in Chapter 10 could be used.

When the distributions are close to normal but the variances seem to differ by a great deal, the t-statistic should again be avoided, particularly if the sample sizes are also different. In that case use Z or the modification by substituting the sample variances for the distribution variances. In the latter situation, if n and m are large enough, there is no problem. With small n and m, most statisticians would use Welch's suggestion (or other modifications of it); that is, use an approximating Student-t distribution with $[r]$ degrees of freedom, where

$$\frac{1}{r} = \frac{c^2}{n-1} + \frac{(1-c)^2}{m-1} \quad \text{and} \quad c = \frac{s_x^2}{s_x^2 + s_y^2}.$$

An equivalent formula for r is

$$r = \frac{\left(\dfrac{s_x^2}{n} + \dfrac{s_y^2}{m}\right)^2}{\dfrac{1}{n-1}\left(\dfrac{s_x^2}{n}\right)^2 + \dfrac{1}{m-1}\left(\dfrac{s_y^2}{m}\right)^2}.$$

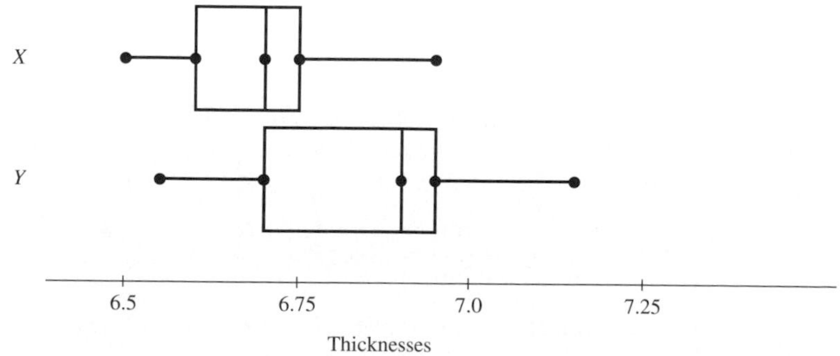

FIGURE 7.4-4

We actually give a test for the equality of variances. This could be used to decide whether to use T or a modification of Z. However, most statisticians do not place much confidence in this test of $\sigma_X^2 = \sigma_Y^2$ and would use a modification of Z (possibly Welch's) if they suspected that the variances differ greatly.

We now give a test for the equality of two variances when sampling from normal populations. Let the independent random variables X and Y have respective distributions that are $N(\mu_X, \sigma_X^2)$ and $N(\mu_Y, \sigma_Y^2)$. To test the null hypothesis H_0: $\sigma_X^2/\sigma_Y^2 = 1$ (or, equivalently, $\sigma_X^2 = \sigma_Y^2$), take random samples of n observations of X and m observations of Y. Recall that $(n-1)S_X^2/\sigma_X^2$ and $(m-1)S_Y^2/\sigma_Y^2$ have independent chi-square distributions $\chi^2(n-1)$ and $\chi^2(m-1)$, respectively. Thus, when H_0 is true,

$$F = \frac{(n-1)S_X^2/[\sigma_X^2(n-1)]}{(m-1)S_Y^2/[\sigma_Y^2(m-1)]} = \frac{S_X^2}{S_Y^2}$$

has an F distribution with $r_1 = n - 1$ and $r_2 = m - 1$ degrees of freedom. This F statistic is our test statistic. When H_0 is true, we would expect the observed value of F to be close to 1.

Three possible alternative hypotheses along with critical regions of size α are summarized in Table 7.4-2. Recall that $1/F$, the reciprocal of F, has an F distribution with $m - 1$ and $n - 1$ degrees of freedom so all critical regions may be written in terms of right-tail rejection regions so that the critical values can be selected easily from Appendix Table VII.

Example 7.4-4   A biologist who studies spiders believes that not only do female green lynx spiders tend to be longer than their male counterparts but also that the lengths of the female spiders seem to vary more than those of the male spiders. We shall test whether this latter belief is true. Suppose that the distribution of the length X of male spiders is $N(\mu_X, \sigma_X^2)$ and the length Y of female spiders is $N(\mu_Y, \sigma_Y^2)$, and X and Y are independent. We shall test H_0: $\sigma_X^2/\sigma_Y^2 = 1$ (i.e., $\sigma_X^2 = \sigma_Y^2$) against the alternative hypothesis H_1: $\sigma_X^2/\sigma_Y^2 < 1$ (i.e., $\sigma_X^2 < \sigma_Y^2$.) If we use $n = 30$ and $m = 30$ observations of X and Y, respectively, a critical region that has a significance level of $\alpha = 0.01$ is

$$\frac{s_y^2}{s_x^2} \geq F_{0.01}(29, 29) = 2.42,$$

approximately, using interpolation in Appendix Table VII. In Exercise 6.2-16, $n = 30$ observations of X yielded $\bar{x} = 5.917$ and $s_x^2 = 0.4399$, while $m = 30$ observations of Y yielded $\bar{y} = 8.153$ and $s_y^2 = 1.4100$. Since

$$\frac{s_y^2}{s_x^2} = \frac{1.4100}{0.4399} = 3.2053 > 2.42,$$

the null hypothesis is rejected in favor of the biologist's belief. (See Exercise 7.4-11.)

TABLE 7.4-2

H_0	H_1	Critical Region
$\sigma_X^2 = \sigma_Y^2$	$\sigma_X^2 > \sigma_Y^2$	$\dfrac{s_x^2}{s_y^2} \geq F_\alpha(n-1, m-1)$
$\sigma_X^2 = \sigma_Y^2$	$\sigma_X^2 < \sigma_Y^2$	$\dfrac{s_y^2}{s_x^2} \geq F_\alpha(m-1, n-1)$
$\sigma_X^2 = \sigma_Y^2$	$\sigma_X^2 \neq \sigma_Y^2$	$\dfrac{s_x^2}{s_y^2} \geq F_{\alpha/2}(n-1, m-1)$ or $\dfrac{s_y^2}{s_x^2} \geq F_{\alpha/2}(m-1, n-1)$

In Examples 7.4-1 and 7.4-2, we used a t-statistic for testing the equality of means that assumed the variances were equal. In the next two examples we shall test whether that assumption is valid.

Example 7.4-5 For Example 7.4-1, given $n = 11$ observations of X and $m = 13$ observations of Y, where X is $N(\mu_X, \sigma_X^2)$ and Y is $N(\mu_Y, \sigma_Y^2)$, we shall test the null hypothesis $H_0: \sigma_X^2/\sigma_Y^2 = 1$ against a two-sided alternative hypothesis. At an $\alpha = 0.05$ significance level, H_0 is rejected if

$$s_x^2/s_y^2 \geq F_{0.025}(10, 12) = 3.37$$

or

$$s_y^2/s_x^2 \geq F_{0.025}(12, 10) = 3.62.$$

Using the data in Example 7.4-1, we obtain

$$s_x^2/s_y^2 = 0.24/0.35 = 0.686 \qquad \text{and} \qquad s_y^2/s_x^2 = 1.458,$$

so we do not reject H_0. Thus the assumption of equal variances for the t-statistic that was used in Example 7.4-1 seems to be valid. See Figure 7.4-5 noting that $F_{0.975}(10, 12) = 1/F_{0.025}(12, 10) = 1/3.62 = 0.276$.

Example 7.4-6 For Example 7.4-2, we shall test the null hypothesis H_0: $\sigma_X^2/\sigma_Y^2 = 1$ against a two-sided alternative hypothesis. Since

$$\frac{s_x^2}{s_y^2} = \frac{29.30}{26.24} = 1.12 < F_{0.025}(11, 11)$$

because $F_{0.025}(11, 11)$ must be between $F_{0.025}(12, 12) = 3.28$ and $F_{0.025}(10, 10) = 3.72$ from Appendix Table VII, the assumption of equal variances is confirmed.

Exercises

7.4-1 The botanist in Example 7.4-1 is really interested in testing for synergistic interaction. That is, given two hormones, gibberellin (GA_3) and indoleacetic

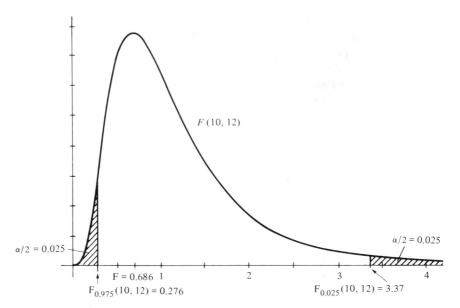

$F\,(10,\,12)$

$\alpha/2 = 0.025$

$\alpha/2 = 0.025$

F = 0.686 1   2   3   4

$F_{0.975}(10,\,12) = 0.276$ $F_{0.025}(10,\,12) = 3.37$

FIGURE 7.4-5

acid (IAA), let X_1 and X_2 equal the growth responses (in millimeters) of dwarf pea stem segments to GA_3 and IAA, respectively and separately. Let $X = X_1 + X_2$. Let Y equal the growth response when both hormones are present. Assuming that X is $N(\mu_X,\ \sigma^2)$ and Y is $N(\mu_Y,\ \sigma^2)$, the botanist is interested in testing the hypothesis H_0: $\mu_X = \mu_Y$ against the alternative hypothesis of synergistic interaction H_1: $\mu_X < \mu_Y$.

(a) Using $n = m = 10$ observations of X and Y, define the test statistic and critical region. Sketch a figure of the t p.d.f. and show the critical region on your figure. Let $\alpha = 0.05$.

(b) Given $n = 10$ observations of X:

2.1	2.6	2.6	3.4	2.1	1.7	2.6	2.6	2.2	1.2

and $m = 10$ observations of Y:

3.5	3.9	3.0	2.3	2.1	3.1	3.6	1.8	2.9	3.3

calculate the value of the test statistic and state your conclusion. Locate the test statistic on your figure.

(c) Construct two box plots on the same figure. Does this confirm your conclusion?

7.4-2   Let X and Y denote the weights in grams of male and female gallinules, respectively. Assume that X is $N(\mu_X,\ \sigma_X^2)$ and Y is $N(\mu_Y,\ \sigma_Y^2)$.

(a) Given $n = 16$ observations of X and $m = 13$ observations of Y, define a test statistic and critical region for testing the null hypothesis H_0: $\mu_X = \mu_Y$

against the one-sided alternative hypothesis $H_1: \mu_X > \mu_Y$. Let $\alpha = 0.01$. (Assume the variances are equal.)

(b) Given that $\bar{x} = 415.16$, $s_x^2 = 1356.75$, $\bar{y} = 347.40$, and $s_y^2 = 692.21$, calculate the value of the test statistic and state your conclusion.

(c) Test whether the assumption of equal variances is valid. Let $\alpha = 0.05$.

(d) Despite the fact that $\sigma_X^2 = \sigma_Y^2$ is accepted in part (c), let us say we suspect the equality is not valid. Thus use the test proposed by Welch.

7.4-3 Among the data collected for the World Health Organization air quality monitoring project is a measure of suspended particles in $\mu g/m^3$. Let X and Y equal the concentration of suspended particles in $\mu g/m^3$ in the city center (commercial district), for Melbourne and Houston, respectively. Using $n = 13$ observations of X and $m = 16$ observations of Y, we shall test $H_0: \mu_X = \mu_Y$ against $H_1: \mu_X < \mu_Y$.

(a) Define the test statistic and critical region, assuming that the variances are equal. Let $\alpha = 0.05$.

(b) If $\bar{x} = 72.9$, $s_x = 25.6$, $\bar{y} = 81.7$, and $s_y = 28.3$, calculate the value of the test statistic and state your conclusion.

(c) Give limits for the p-value of this test.

(d) Test whether the assumption of equal variances is valid. Let $\alpha = 0.05$.

7.4-4 Some nurses in County Public Health conducted a survey of women who had received inadequate prenatal care. They used information from birth certificates to select mothers for the survey. The mothers that were selected were divided into two groups: 14 mothers who said they had 5 or fewer prenatal visits and 14 mothers who said they had 6 or more prenatal visits. Let X and Y equal the respective birthweights of the babies from these two sets of mothers and assume that the distribution of X is $N(\mu_X, \sigma^2)$ and the distribution of Y is $N(\mu_Y, \sigma^2)$.

(a) Define the test statistic and critical region for testing $H_0: \mu_X - \mu_Y = 0$ against $H_1: \mu_X - \mu_Y < 0$. Let $\alpha = 0.05$.

(b) Given that the observations of X were

49	108	110	82	93	114	134
114	96	52	101	114	120	116

and the observations of Y were

133	108	93	119	119	98	106
131	87	153	116	129	97	110

calculate the value of the test statistic and state your conclusion.

(c) Approximate the p-value.

(d) Construct box plots on the same figure for these two sets of data. Do the box plots support your conclusion?

(e) Test whether the assumption of equal variances is valid. Let $\alpha = 0.05$.

7.4-5 Let X and Y equal the forces required to pull stud No. 3 and stud No. 4 out of a window that has been manufactured for an automobile. Assume that the distributions of X and Y are $N(\mu_X, \sigma_X^2)$ and $N(\mu_Y, \sigma_Y^2)$, respectively.

(a) If $m = n = 10$ observations are selected randomly, define a test statistic and critical region for testing H_0: $\mu_X - \mu_Y = 0$ against a two-sided alternative hypothesis. Let $\alpha = 0.05$. Assume that the variances are equal.

(b) Given $n = 10$ observations of X:

| 111 | 120 | 139 | 136 | 138 | 149 | 143 | 145 | 111 | 123 |

and $m = 10$ observations of Y:

| 152 | 155 | 133 | 134 | 119 | 155 | 142 | 146 | 157 | 149 |

calculate the value of the test statistic and clearly state your conclusion.

(c) What is the approximate p-value of this test?

(d) Construct box plots on the same figure for these two sets of data. Do the box plots confirm your decision in part (b)?

(e) Test whether the assumption of equal variances is valid.

7.4-6 Let X and Y equal the number of milligrams of tar in filtered and nonfiltered cigarettes, respectively. Assume that the distributions of X and Y are $N(\mu_X, \sigma_X^2)$ and $N(\mu_Y, \sigma_Y^2)$, respectively. We shall test the null hypothesis H_0: $\mu_X - \mu_Y = 0$ against the alternative hypothesis H_1: $\mu_X - \mu_Y < 0$ using random samples of sizes $n = 9$ and $m = 11$ observations of X and Y, respectively.

(a) Define the test statistic and a critical region that has an $\alpha = 0.01$ significance level. Sketch a figure illustrating this critical region.

(b) Given the $n = 9$ observations of X:

| 0.9 | 1.1 | 0.1 | 0.7 | 0.4 | 0.9 | 0.8 | 1.0 | 0.4 |

and the $m = 11$ observations of Y:

| 1.5 | 0.9 | 1.6 | 0.5 | 1.4 | 1.9 | 1.0 | 1.2 | 1.3 | 1.6 | 2.1 |

calculate the value of the test statistic and clearly state your conclusion. Locate the value of the test statistic on your figure.

7.4-7 In Exercise 6.2-15, we let X and Y equal the blood volumes in milliliters for males who are paraplegics participating in vigorous physical activities and males who are able bodied participating in normal activities. We assumed that X was $N(\mu_X, \sigma_X^2)$ and Y was $N(\mu_Y, \sigma_Y^2)$. Using $n = 7$ observations of X

| 1612 | 1352 | 1456 | 1222 | 1560 | 1456 | 1924 |

and $m = 10$ observations of Y

| 1082 | 1300 | 1092 | 1040 | 910 |
| 1248 | 1092 | 1040 | 1092 | 1288 |

you were asked to find a 95% confidence interval for $\mu_X - \mu_Y$.

(a) To construct the confidence interval, you would like to assume that the variances are equal. Is this a valid assumption? Test the hypothesis that the variances are equal against a two-sided alternative hypothesis. Let $\alpha = 0.05$.

(b) Test the hypothesis that the means are equal against a two-sided alternative hypothesis. Let $\alpha = 0.05$.

(c) Did 0 belong to the 95% confidence interval in Exercise 6.2-15? Did you reject the null hypothesis in this exercise? Are your answers to these questions compatible?

7.4-8   An office furniture manufacturer installed a new adhesive application process. To compare the new process to the old process, random samples were selected from the two processes, and "pull tests" were performed to determine the number of pounds of pressure that were required to pull apart the glued parts (destructive testing). Let X and Y denote the pounds of pressure needed for the new and old processes, respectively.

(a) Based on $n = m = 24$ observations, define the test statistic and critical region for testing $H_0: \mu_X - \mu_Y = 0$ against $H_1: \mu_X - \mu_Y > 0$. Let $\alpha = 0.05$. State assumptions.

(b) Use the following $n = 24$ observations of X:

1250	1210	990	1310	1320	1200	1290	1360
1120	1360	1310	1110	1320	980	950	1430
960	1050	1310	1240	1420	1170	1470	1060

and the following $m = 24$ observations of Y:

1180	1360	1310	1190	920	1060	1440	1010
1310	980	1310	1030	960	800	1280	1080
930	1050	1010	1310	940	860	1450	1070

to calculate the value of the test statistic and clearly state your conclusion.

(c) What is the approximate p-value of this test?

(d) Construct two box plots on the same graph. Do the box plots support your conclusion?

7.4-9   Let X and Y denote the tarsus lengths of male and female grackles, respectively. Assume that X is $N(\mu_X, \sigma_X^2)$ and Y is $N(\mu_Y, \sigma_Y^2)$. Given that $n = 25$, $\bar{x} = 33.80$, $s_x^2 = 4.88$, $m = 29$, $\bar{y} = 31.66$, and $s_y^2 = 5.81$, test

(a) $H_0: \sigma_X^2/\sigma_Y^2 = 1$ against a two-sided alternative with $\alpha = 0.02$.

(b) $H_0: \mu_X = \mu_Y$ against $H_1: \mu_X > \mu_Y$ with $\alpha = 0.01$.

7.4-10   Let X equal the fill weight in April and Y the fill weight in June for an 8-pound box of bleach. We shall test the null hypothesis $H_0: \mu_X - \mu_Y = 0$ against the alternative hypothesis $H_1: \mu_X - \mu_Y > 0$ given that $n = 90$ observations of X yielded $\bar{x} = 8.10$, $s_x = 0.117$ and $m = 110$ observations of Y yielded $\bar{y} = 8.07$ and $s_y = 0.054$.

(a) What is your conclusion if $\alpha = 0.05$?

HINT: Do the variances seem to be equal?

(b) What is the approximate p-value of this test?

7.4-11   Let X and Y denote the lengths of male and female green lynx spiders, respectively. Assume that the distributions of X and Y are $N(\mu_X, \sigma_X^2)$ and

$N(\mu_Y, \sigma_Y^2)$. In Example 7.4-4, we concluded that $\sigma_Y^2 > \sigma_X^2$. Thus, use the modification of Z to test the hypothesis $H_0: \mu_X - \mu_Y = 0$ against the alternative hypothesis $H_1: \mu_X - \mu_Y < 0$.

(a) Define the test statistic and a critical region that has a significance level of $\alpha = 0.025$.

(b) Using the data in Exercise 6.2-16, calculate the value of the test statistic and state your conclusion.

(c) Draw two box-and-whisker diagrams on the same figure. Does this confirm both the conclusion of this exercise and that of Example 7.4-4?

7.4-12 To measure air pollution in a home, let X and Y equal the amount of suspended particulate matter (in $\mu g/m^3$) measured during a 24-hour period in a home in which there is no smoker and a home in which there is a smoker, respectively. (Assume that the distributions of X and Y are $N(\mu_X, \sigma_X^2)$ and $N(\mu_Y, \sigma_Y^2)$, respectively.) We shall test the null hypothesis $H_0: \sigma_X^2/\sigma_Y^2 = 1$ against the one-sided alternative hypothesis $H_1: \sigma_X^2/\sigma_Y^2 > 1$. If a random sample of size $n = 9$ yielded $\bar{x} = 93$ and $s_x = 12.9$ while a random sample of size $m = 11$ yielded $\bar{y} = 132$ and $s_y = 7.1$, define a critical region and give your conclusion if $\alpha = 0.05$. Now test $H_0: \mu_X = \mu_Y$ against $H_1: \mu_x < \mu_Y$ if $\alpha = 0.05$.

7.4-13 Let X and Y equal the times in days required for maturation of Guardiola seeds from narrow-leaved and broad-leaved parents, respectively. Assume that X is $N(\mu_X, \sigma_X^2)$ and Y is $N(\mu_Y, \sigma_Y^2)$. Test the hypothesis $H_0: \sigma_X^2/\sigma_Y^2 = 1$ against the alternative hypothesis $H_1: \sigma_X^2/\sigma_Y^2 > 1$ if a sample size $n = 13$ yielded $\bar{x} = 18.97$, $s_x^2 = 9.88$ and a sample of size $m = 9$ yielded $\bar{y} = 23.20$, $s_y^2 = 4.08$. Let $\alpha = 0.05$.

7.4-14 A random sample of $n = 7$ brand X light bulbs yielded $\bar{x} = 891$ hours and $s_x^2 = 9201$. A random sample of $m = 10$ brand Y light bulbs yielded $\bar{y} = 592$ hours and $s_y^2 = 4856$. Use these data to test $H_0: \sigma_X^2/\sigma_Y^2 = 1$ against $H_1: \sigma_X^2/\sigma_Y^2 > 1$. Let $\alpha = 0.05$ and state assumptions.

★7.5   Best Critical Regions

In this section we consider the properties a satisfactory test (or critical region) should possess. To introduce our investigation, we begin with a nonstatistical example.

Example 7.5-1 Say that you have α dollars with which to buy books. Suppose further that you are not interested in the books themselves but only in filling as much of your bookshelves as possible. How do you decide which books to buy? Does the following approach seem reasonable? First of all, take all the available free books. Then start choosing those books for which the cost of filling an inch of bookshelf is smallest. That is, choose those books for which the ratio c/w is a minimum, where w is the width of the book in inches and c is the cost of the book. Continue choosing books this way until you have spent the α dollars.

To see how Example 7.5-1 provides the background for selecting a good critical region of size α, let us consider a test of the simple hypothesis H_0: $\theta = \theta_0$ against a simple alternative hypothesis H_1: $\theta = \theta_1$. In this discussion we assume that the random variables $X_1, X_2, \ldots, X_n$ under consideration have a joint p.d.f. of the discrete type, which we here denote by $L(\theta; x_1, x_2, \ldots, x_n)$. That is,

$$P(X_1 = x_1, X_2 = x_2, \ldots, X_n = x_n) = L(\theta; x_1, x_2, \ldots, x_n).$$

A critical region C of size α is a set of points $(x_1, x_2, \ldots, x_n)$ with the probability of α when $\theta = \theta_0$. For a good test, this set C of points should have a large probability when $\theta = \theta_1$ because, under H_1: $\theta = \theta_1$, we wish to reject H_0: $\theta = \theta_0$. Accordingly, the first point we would place in the critical region C is the one with the smallest ratio

$$\frac{L(\theta_0; x_1, x_2, \ldots, x_n)}{L(\theta_1; x_1, x_2, \ldots, x_n)}.$$

That is, the "cost" in terms of probability under H_0: $\theta = \theta_0$ is small compared to the probability that we can "buy" if $\theta = \theta_1$. The next point to add to C would be the one with the next smallest ratio. We would continue to add points to C in this manner until the probability of C, under H_0: $\theta = \theta_0$, equals α. In this way, we have achieved, for the given significance level α, the region C with the largest probability when H_1: $\theta = \theta_1$ is true. We now formalize this discussion by defining a best critical region and proving the well-known Neyman-Pearson lemma.

DEFINITION 7.5-1 *Consider the test of the simple null hypothesis H_0: $\theta = \theta_0$ against the simple alternative hypothesis H_1: $\theta = \theta_1$. Let C be a critical region of size α; that is, $\alpha = P(C; \theta_0)$. Then C is a **best critical region of size** α if, for every other critical region D of size $\alpha = P(D; \theta_0)$, we have that*

$$P(C; \theta_1) \geq P(D; \theta_1).$$

That is, when H_1: $\theta = \theta_1$ is true, the probability of rejecting H_0: $\theta = \theta_0$ using the critical region C is at least as great as the corresponding probability using any other critical region D of size α.

Thus a best critical region of size α is the critical region that has the greatest power among all critical regions of size α. The Neyman-Pearson lemma gives sufficient conditions for a best critical region of size α.

THEOREM 7.5-1 **(Neyman–Pearson Lemma)** *Let $X_1, X_2, \ldots, X_n$ be a random sample of size n from a distribution with p.d.f. $f(x; \theta)$, where θ_0 and θ_1 are two possible values of θ. Denote the joint p.d.f. of $X_1, X_2, \ldots, X_n$ by the likelihood function*

$$L(\theta) = L(\theta; x_1, x_2, \ldots, x_n) = f(x_1; \theta) f(x_2; \theta) \cdots f(x_n; \theta).$$

If there exist a positive constant k and a subset C of the sample space such that

(a) $P[(X_1, X_2, \ldots, X_n) \in C; \theta_0] = \alpha$,

(b) $\dfrac{L(\theta_0)}{L(\theta_1)} \leq k$ *for* $(x_1, x_2, \ldots, x_n) \in C$,

(c) $\dfrac{L(\theta_0)}{L(\theta_1)} \geq k$ *for* $(x_1, x_2, \ldots, x_n) \in C'$,

then C is a best critical region of size α for testing the simple null hypothesis H_0: $\theta = \theta_0$ against the simple alternative hypothesis H_1: $\theta = \theta_1$.

PROOF: We prove the theorem when the random variables are of the continuous type; for discrete-type random variables, replace the integral signs by summation signs. To simplify the exposition, we shall use the following notation:

$$\int_B L(\theta) = \int \cdots \int_B L(\theta; x_1, x_2, \ldots, x_n) \, dx_1 \, dx_2 \cdots dx_n.$$

Assume that there exists another critical region of size α, say D, such that, in this new notation,

$$\alpha = \int_C L(\theta_0) = \int_D L(\theta_0).$$

So we have

$$0 = \int_C L(\theta_0) - \int_D L(\theta_0)$$

$$= \int_{C \cap D'} L(\theta_0) + \int_{C \cap D} L(\theta_0) - \int_{C \cap D} L(\theta_0) - \int_{C' \cap D} L(\theta_0)$$

and hence

$$0 = \int_{C \cap D'} L(\theta_0) - \int_{C' \cap D} L(\theta_0).$$

By hypothesis (b), $kL(\theta_1) \geq L(\theta_0)$ at each point in C and therefore in $C \cap D'$; thus

$$k \int_{C \cap D'} L(\theta_1) \geq \int_{C \cap D'} L(\theta_0).$$

By hypothesis (c), $kL(\theta_1) \leq L(\theta_0)$ at each point in C', and therefore in $C' \cap D$; thus we obtain

$$k \int_{C' \cap D} L(\theta_1) \leq \int_{C' \cap D} L(\theta_0).$$

Therefore,

$$0 = \int_{C \cap D'} L(\theta_0) - \int_{C' \cap D} L(\theta_0) \le (k) \left\{ \int_{C \cap D'} L(\theta_1) - \int_{C' \cap D} L(\theta_1) \right\}.$$

That is,

$$0 \le (k) \left\{ \int_{C \cap D'} L(\theta_1) + \int_{C \cap D} L(\theta_1) - \int_{C \cap D} L(\theta_1) - \int_{C' \cap D} L(\theta_1) \right\}$$

or, equivalently,

$$0 \le (k) \left\{ \int_C L(\theta_1) - \int_D L(\theta_1) \right\}.$$

Thus

$$\int_C L(\theta_1) \ge \int_D L(\theta_1);$$

that is, $P(C; \theta_1) \ge P(D; \theta_1)$. Since that is true for every critical region D of size α, C is a best critical region of size α. $\qquad\square$

For a realistic application of the Neyman–Pearson lemma, consider the following, in which the test is based on a random sample from a normal distribution.

Example 7.5-2 Let $X_1, X_2, \ldots, X_n$ be a random sample from the normal distribution $N(\mu, 36)$. We shall find the best critical region for testing the simple hypothesis H_0: $\mu = 50$ against the simple alternative hypothesis H_1: $\mu = 55$. Using the ratio of the likelihood functions, namely $L(50)/L(55)$, we shall find those points in the sample space for which this ratio is less than or equal to some constant k. That is, we shall solve the following inequality:

$$\frac{L(50)}{L(55)} = \frac{(72\pi)^{-n/2} \exp\left[-\left(\frac{1}{72}\right) \sum_1^n (x_i - 50)^2 \right]}{(72\pi)^{-n/2} \exp\left[-\left(\frac{1}{72}\right) \sum_1^n (x_i - 55)^2 \right]}$$

$$= \exp\left[-\left(\frac{1}{72}\right) \left(10 \sum_1^n x_i + n50^2 - n55^2 \right) \right] \le k.$$

If we take the natural logarithm of each member of the inequality, we find that

$$-10 \sum_1^n x_i - n50^2 + n55^2 \le (72) \ln k.$$

Thus

$$\frac{1}{n} \sum_{1}^{n} x_i \geq -\frac{1}{10n}[n50^2 - n55^2 + (72) \ln k]$$

or equivalently,

$$\bar{x} \geq c,$$

where $c = -(1/10n)[n50^2 - n55^2 + (72) \ln k]$. Thus $L(50)/L(55) \leq k$ is equivalent to $\bar{x} \geq c$. A best critical region is, according to the Neyman–Pearson lemma,

$$C = \{(x_1, x_2, \ldots, x_n): \bar{x} \geq c\},$$

where c is selected so that the size of the critical region is α. Say $n = 16$ and $c = 53$. Since $\bar{X}$ is $N(50, 36/16)$ under H_0, we have

$$\alpha = P(\bar{X} \geq 53; \mu = 50)$$

$$= P\left(\frac{\bar{X} - 50}{6/4} \geq \frac{3}{6/4}; \mu = 50\right) = 1 - \Phi(2) = 0.0228.$$

This last example illustrates what is often true, namely, that the inequality

$$L(\theta_0)/L(\theta_1) \leq k$$

can be expressed in terms of a function $u(x_1, x_2, \ldots, x_n)$, say

$$u(x_1, \ldots, x_n) \leq c_1$$

or

$$u(x_1, \ldots, x_n) \geq c_2,$$

where c_1 or c_2 is selected so that the size of the critical region is α. Thus the test can be based on the statistic $u(X_1, \ldots, X_n)$. Also, for illustration, if we want α to be a given value, say 0.05, we could then choose our c_1 or c_2. In Example 7.5-2, with $\alpha = 0.05$, we want

$$0.05 = P(\bar{X} \geq c; \mu = 50)$$

$$= P\left(\frac{\bar{X} - 50}{6/4} \geq \frac{c - 50}{6/4}; \mu = 50\right) = 1 - \Phi\left(\frac{c - 50}{6/4}\right).$$

Hence it must be true that $(c - 50)/(3/2) = 1.645$, or equivalently,

$$c = 50 + \frac{3}{2}(1.645) \approx 52.47.$$

Example 7.5-3 Let $X_1, X_2, \ldots, X_n$ denote a random sample of size n from a Poisson distribution with mean λ. A best critical region for testing H_0: $\lambda = 2$ against H_1: $\lambda = 5$ is given by

$$\frac{L(2)}{L(5)} = \frac{2^{\Sigma \, x_i} e^{-2n}}{x_1! x_2! \cdots x_n!} \frac{x_1! x_2! \cdots x_n!}{5^{\Sigma \, x_i} e^{-5n}} \le k.$$

This inequality is equivalent to

$$\left(\frac{2}{5}\right)^{\Sigma \, x_i} e^{3n} \le k \qquad \text{and} \qquad \left(\sum x_i\right) \ln \left(\frac{2}{5}\right) + 3n \le \ln k.$$

Since $\ln (2/5) < 0$, this is the same as

$$\sum_{i=1}^{n} x_i \ge \frac{\ln k - 3n}{\ln (2/5)} = c.$$

If $n = 4$ and $c = 13$, then

$$\alpha = P\left(\sum_{i=1}^{4} X_i \ge 13; \lambda = 2\right) = 1 - 0.936 = 0.064,$$

from Table III in the Appendix, since $\Sigma_{i=1}^{4} X_i$ has a Poisson distribution with mean 8 when $\lambda = 2$.

When H_0: $\theta = \theta_0$ and H_1: $\theta = \theta_1$ are both simple hypotheses, a critical region of size α is a best critical region if the probability of rejecting H_0 when H_1 is true is a maximum when compared with all other critical regions of size α. The test using the best critical region is called a **most powerful test** because it has the greatest value of the power function at $\theta = \theta_1$ when compared with that of other tests of significance level α. If H_1 is a composite hypothesis, the power of a test depends on each simple alternative in H_1.

DEFINITION 7.5-2 *A test, defined by a critical region C of size α, is a* **uniformly most powerful test** *if it is a most powerful test against each simple alternative in H_1. The critical region C is called a* **uniformly most powerful critical region of size α.**

Let us reconsider Example 7.5-2 when the alternative hypothesis is composite.

Example 7.5-4 Let $X_1, \ldots, X_n$ be a random sample from $N(\mu, 36)$. We have seen that when testing H_0: $\mu = 50$ against H_1: $\mu = 55$, a best critical region C is defined by $C = \{(x_1, x_2, \ldots, x_n): \bar{x} \ge c\}$, where c is selected so that the significance level is α. Now consider testing H_0: $\mu = 50$ against the one-sided composite alternative hypothesis H_1: $\mu > 50$. For each simple hypothesis in H_1, say $\mu = \mu_1$, the quotient of the likelihood functions is

$$\frac{L(50)}{L(\mu_1)} = \frac{(72\pi)^{-n/2} \exp\left[-\left(\frac{1}{72}\right)\sum_1^n (x_i - 50)^2\right]}{(72\pi)^{-n/2} \exp\left[-\left(\frac{1}{72}\right)\sum_1^n (x_i - \mu_1)^2\right]}$$

$$= \exp\left[-\frac{1}{72}\left\{2(\mu_1 - 50)\sum_1^n x_i + n(50^2 - \mu_1^2)\right\}\right].$$

Now $L(50)/L(\mu_1) \le k$ if and only if

$$\bar{x} \ge \frac{(-72)\ln(k)}{2n(\mu_1 - 50)} + \frac{50 + \mu_1}{2} = c.$$

Thus the best critical region of size α for testing H_0: $\mu = 50$ against H_1: $\mu = \mu_1$, where $\mu_1 > 50$, is given by $C = \{(x_1, x_2, \ldots, x_n): \bar{x} \ge c\}$, where c is selected such that $P(\bar{X} \ge c; H_0: \mu = 50) = \alpha$. Note that the same value of c can be used for each $\mu_1 > 50$, but (of course) k does not remain the same. Since the critical region C defines a test that is most powerful against each simple alternative $\mu_1 > 50$, this is a uniformly most powerful test, and C is a uniformly most powerful critical region of size α. Again if $\alpha = 0.05$, then $c \approx 52.47$.

Example 7.5-5   Let Y have the binomial distribution $b(n, p)$. To find a uniformly most powerful test of the simple null hypothesis H_0: $p = p_0$ against the one-sided alternative hypothesis H_1: $p > p_0$, consider, with $p_1 > p_0$,

$$\frac{L(p_0)}{L(p_1)} = \frac{\binom{n}{y} p_0^y (1 - p_0)^{n-y}}{\binom{n}{y} p_1^y (1 - p_1)^{n-y}} \le k.$$

This is equivalent to

$$\left[\frac{p_0(1 - p_1)}{p_1(1 - p_0)}\right]^y \left[\frac{1 - p_0}{1 - p_1}\right]^n \le k$$

and

$$y \ln\left[\frac{p_0(1 - p_1)}{p_1(1 - p_0)}\right] \le \ln k - n \ln\left[\frac{1 - p_0}{1 - p_1}\right].$$

Since $p_0 < p_1$, $p_0(1 - p_1) < p_1(1 - p_0)$ and it follows that $\ln [p_0(1 - p_1)/p_1(1 - p_0)] < 0$. Thus we have

$$\frac{y}{n} \ge \frac{\ln k - n \ln [(1 - p_0)/(1 - p_1)]}{n \ln [p_0(1 - p_1)/p_1(1 - p_0)]} = c,$$

for each $p_1 > p_0$.

It is interesting to note that if the alternative hypothesis is the one-sided H_1: $p < p_0$, then a uniformly most powerful test is of the form $(y/n) \leq c$. Thus we see that the tests of H_0: $p = p_0$ against the one-sided alternatives given in Table 7.1-1 are uniformly most powerful.

Exercise 7.5-5 will demonstrate that uniformly most powerful tests do not always exist; in particular, they do not usually exist when the composite alternative hypothesis is two-sided.

Exercises

7.5-1 Let X_1, X_2, . . . , X_n be a random sample from a normal distribution $N(\mu, 64)$.
(a) Show that $C = \{(x_1, x_2, . . . , x_n): \bar{x} \leq c\}$ is a best critical region for testing H_0: $\mu = 80$ against H_1: $\mu = 76$.
(b) Find n and c so that $\alpha = 0.05$ and $\beta = 0.05$, approximately.

7.5-2 Let X_1, X_2, . . . , X_n be a random sample from $N(0, \sigma^2)$.
(a) Show that $C = \{(x_1, x_2, . . . , x_n): \Sigma_1^n x_i^2 \geq c\}$ is a best critical region for testing H_0: $\sigma^2 = 4$ against H_1: $\sigma^2 = 16$.
(b) If $n = 15$, find the value of c so that $\alpha = 0.05$.

HINT: Recall that $\Sigma_1^n X_i^2 / \sigma^2$ is $\chi^2(n)$.

(c) If $n = 15$ and c is the value found in part (b), find the approximate value of $\beta = P(\Sigma_1^n X_i^2 < c; \sigma^2 = 16)$.

7.5-3 Let X have an exponential distribution with a mean of θ; that is, the p.d.f. of X is $f(x; \theta) = (1/\theta)e^{-x/\theta}$, $0 < x < \infty$.
(a) Show that a best critical region for testing H_0: $\theta = 3$ against H_1: $\theta = 5$ can be based on the statistic $\Sigma_1^n X_i$.
(b) If $n = 12$, use the fact that $(2/\theta) \Sigma_1^{12} X_i$ is $\chi^2(24)$ to find a best critical region of size $\alpha = 0.10$.
(c) If $n = 12$, find a best critical region of size $\alpha = 0.10$ for testing H_0: $\theta = 3$ against H_1: $\theta = 7$.
(d) If H_1: $\theta > 3$, is the common region found in parts (b) and (c) a uniformly most powerful critical region of size $\alpha = 0.10$?

7.5-4 Let X_1, X_2, . . . , X_n be a random sample of Bernoulli trials $b(1, p)$.
(a) Show that a best critical region for testing H_0: $p = 0.9$ against H_1: $p = 0.8$ can be based on the statistic $Y = \Sigma_1^n X_i$, which is $b(n, p)$.
(b) If $C = \{(x_1, x_2, . . . , x_n): \Sigma_{i=1}^n x_i \leq n(0.85)\}$ and $Y = \Sigma_1^n X_i$, find the value of n such that $\alpha = 0.10 = P[Y \leq n(0.85); p = 0.9]$, approximately.

HINT: Use the normal approximation for the binomial distribution.

(c) What is the approximate value of $\beta = P[Y > n(0.85); p = 0.8]$ for the test given in part (b)?
(d) Is the test of part (b) a uniformly most powerful test when the alternative hypothesis is H_1: $p < 0.9$?

7.5-5   Let $X_1, X_2, \ldots, X_n$ be a random sample from the normal distribution $N(\mu, 36)$.
   (a) Show that a uniformly most powerful critical region for testing H_0: $\mu = 50$ against H_1: $\mu < 50$ is given by $C_2 = \{\bar{x}: \bar{x} \leq c\}$.
   (b) With this result and that of Example 7.5-4, argue that a uniformly most powerful test for testing H_0: $\mu = 50$ against H_1: $\mu \neq 50$ does not exist.

7.5-6   Let $X_1, X_2, \ldots, X_n$ be a random sample from the normal distribution $N(\mu, 9)$. To test the hypothesis H_0: $\mu = 80$ against H_1: $\mu \neq 80$, consider the following three critical regions: $C_1 = \{\bar{x}: \bar{x} \geq c_1\}$, $C_2 = \{\bar{x}: \bar{x} \leq c_2\}$, and $C_3 = \{\bar{x}: |\bar{x} - 80| \geq c_3\}$.
   (a) If $n = 16$, find the values of c_1, c_2, c_3 such that the size of each critical region is 0.05. That is, find c_1, c_2, c_3 such that

$$0.05 = P(\bar{X} \in C_1; \mu = 80) = P(\bar{X} \in C_2; \mu = 80)$$
$$= P(\bar{X} \in C_3; \mu = 80).$$

   (b) Sketch, on the same graph paper, the power functions for these three critical regions.

7.5-7   Let $X_1, X_2, \ldots, X_{10}$ be a random sample of size 10 from a Poisson distribution with mean μ.
   (a) Show that a uniformly most powerful critical region for testing H_0: $\mu = 0.5$ against H_1: $\mu > 0.5$ can be defined using the statistic $\Sigma_1^{10} X_i$.
   (b) What is a uniformly most powerful critical region of size $\alpha = 0.068$? Recall that $\Sigma_1^{10} X_i$ has a Poisson distribution with mean 10μ.
   (c) Sketch the power function of this test.

★7.6   Likelihood Ratio Tests

In this section we consider a general test-construction method that is applicable when both the null and alternative hypotheses, say H_0 and H_1, are composite. We continue to assume that the functional form of the p.d.f. is known but that it depends on an unknown parameter or unknown parameters. That is, we assume that the p.d.f. of X is $f(x; \theta)$, where θ represents one or more unknown parameters. We let Ω denote the total parameter space, that is, the set of all possible values of the parameter θ given by either H_0 or H_1. These hypotheses will be stated as follows:

$$H_0: \theta \in \omega, \qquad H_1: \theta \in \omega',$$

where ω is a subset of Ω and ω' is the complement of ω with respect to Ω. The test will be constructed by using a ratio of likelihood functions that have been maximized in ω and Ω, respectively. In a sense, this is a natural generalization of the ratio appearing in the Neyman–Pearson lemma when the two hypotheses were simple.

DEFINITION 7.6-1   *The **likelihood ratio** is the quotient*

$$\lambda = \frac{L(\hat{\omega})}{L(\hat{\Omega})},$$

where $L(\hat{\omega})$ is the maximum of the likelihood function with respect to θ when $\theta \in \omega$ and $L(\hat{\Omega})$ is the maximum of the likelihood function with respect to θ when $\theta \in \Omega$.

Because λ is the quotient of nonnegative functions, $\lambda \geq 0$. In addition, since $\omega \subset \Omega$, we have that $L(\hat{\omega}) \leq L(\hat{\Omega})$ and hence $\lambda \leq 1$. Thus $0 \leq \lambda \leq 1$. If the maximum of L in ω is much smaller than that in Ω, it would seem that the data $x_1, x_2, \ldots, x_n$ do not support the hypothesis H_0: $\theta \in \omega$. That is, a small value of the ratio $\lambda = L(\hat{\omega})/L(\hat{\Omega})$ would lead to the rejection of H_0. On the other hand, a value of the ratio λ that is close to one would support the null hypothesis H_0. This leads us to the following definition.

DEFINITION 7.6-2 *To test H_0: $\theta \in \omega$ against H_1: $\theta \in \omega'$, the **critical region for the likelihood ratio test** is the set of points in the sample space for which*

$$\lambda = \frac{L(\hat{\omega})}{L(\hat{\Omega})} \leq k,$$

where $0 < k < 1$ and k is selected so that the test has a desired significance level α.

The following example illustrates these definitions.

Example 7.6-1 Assume that the weight X in ounces of a "10-pound" bag of sugar is $N(\mu, 5)$. We shall test the hypothesis H_0: $\mu = 162$ against the alternative hypothesis H_1: $\mu \neq 162$. Thus $\Omega = \{\mu: -\infty < \mu < \infty\}$ and $\omega = \{162\}$. To find the likelihood ratio, we need $L(\hat{\omega})$ and $L(\hat{\Omega})$. When H_0 is true, μ can take on only one value, namely $\mu = 162$. Thus $L(\hat{\omega}) = L(162)$. To find $L(\hat{\Omega})$, we must find the value of μ that maximizes $L(\mu)$. We recall that $\hat{\mu} = \bar{x}$ is the maximum likelihood estimate of μ. Thus $L(\hat{\Omega}) = L(\bar{x})$ and the likelihood ratio $\lambda = L(\hat{\omega})/L(\hat{\Omega})$ is given by

$$\lambda = \frac{(10\pi)^{-n/2} \exp\left[-\left(\dfrac{1}{10}\right) \sum_1^n (x_i - 162)^2\right]}{(10\pi)^{-n/2} \exp\left[-\left(\dfrac{1}{10}\right) \sum_1^n (x_i - \bar{x})^2\right]}$$

$$= \frac{\exp\left[-\left(\dfrac{1}{10}\right) \sum_1^n (x_i - \bar{x})^2 - \left(\dfrac{n}{10}\right)(\bar{x} - 162)^2\right]}{\exp\left[-\left(\dfrac{1}{10}\right) \sum_1^n (x_i - \bar{x})^2\right]}$$

$$= \exp\left[-\frac{n}{10}(\bar{x} - 162)^2\right].$$

A value of $\bar{x}$ close to 162 would tend to support H_0 and in that case λ is close to 1. On the other hand, an $\bar{x}$ that differs from 162 by too much would tend to support H_1. See Figure 7.6-1 for the graph of this likelihood ratio when $n = 5$.

A likelihood ratio critical region is given by $\lambda \leq k$, where k is selected so that the significance level of the test is α. Using this criterion and simplifying the inequality as we do using the Neyman–Pearson lemma, we have that $\lambda \leq k$ is equivalent to each of the following inequalities:

$$-\left(\frac{n}{10}\right)(\bar{x} - 162)^2 \leq \ln k,$$

$$(\bar{x} - 162)^2 \geq -\left(\frac{10}{n}\right) \ln k,$$

$$\frac{|\bar{x} - 162|}{\sigma/\sqrt{n}} \geq \frac{\sqrt{-(10/n) \ln k}}{\sigma/\sqrt{n}} = c.$$

Since $Z = (\bar{X} - 162)/(\sigma/\sqrt{n})$ is $N(0, 1)$ when $H_0: \mu = 162$ is true, let $c = z_{\alpha/2}$. Thus the critical region is

$$C = \left\{\bar{x}: \frac{|\bar{x} - 162|}{\sigma/\sqrt{n}} \geq z_{\alpha/2}\right\}$$

and, for illustration, if $\alpha = 0.05$, we have that $z_{0.025} = 1.96$.

As illustrated in Example 7.6-1, the inequality $\lambda \leq k$ can often be expressed in terms of a statistic whose distribution is known. Also, note that although the likelihood ratio test is an intuitive test, it leads to the same critical region as that given by the Neyman–Pearson lemma when H_0 and H_1 are both simple hypotheses.

Suppose now that the random sample $X_1, X_2, \ldots, X_n$ arises from the normal population $N(\mu, \sigma^2)$, where both μ and σ^2 are unknown. Let us consider the

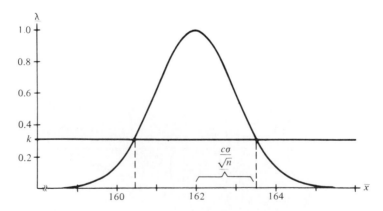

FIGURE 7.6-1

likelihood ratio test of the null hypothesis H_0: $\mu = \mu_0$ against the two-sided alternative hypothesis H_1: $\mu \neq \mu_0$. For this test

$$\omega = \{(\mu, \sigma^2): \mu = \mu_0, \ 0 < \sigma^2 < \infty\}$$

and

$$\Omega = \{(\mu, \sigma^2): -\infty < \mu < \infty, \ 0 < \sigma^2 < \infty\}$$

If $(\mu, \sigma^2) \in \Omega$, the observed maximum likelihood estimates are $\hat{\mu} = \bar{x}$ and $\widehat{\sigma^2} = (1/n) \sum_1^n (x_i - \bar{x})^2$. Thus

$$L(\hat{\Omega}) = \left[\frac{1}{2\pi\left(\dfrac{1}{n}\right)\sum_1^n (x_i - \bar{x})^2} \right]^{n/2} \exp\left[-\frac{\sum_1^n (x_i - \bar{x})^2}{\left(\dfrac{2}{n}\right)\sum_1^n (x_i - \bar{x})^2} \right]$$

$$= \left[\frac{ne^{-1}}{2\pi \sum_1^n (x_i - \bar{x})^2} \right]^{n/2}.$$

Similarly, if $(\mu, \sigma^2) \in \omega$, the observed maximum likelihood estimates are $\hat{\mu} = \mu_0$ and $\widehat{\sigma^2} = (1/n) \sum_1^n (x_i - \mu_0)^2$. Thus

$$L(\hat{\omega}) = \left[\frac{1}{2\pi\left(\dfrac{1}{n}\right)\sum_1^n (x_i - \mu_0)^2} \right]^{n/2} \exp\left[-\frac{\sum_1^n (x_i - \mu_0)^2}{\left(\dfrac{2}{n}\right)\sum_1^n (x_i - \mu_0)^2} \right]$$

$$= \left[\frac{ne^{-1}}{2\pi \sum_1^n (x_i - \mu_0)^2} \right]^{n/2}.$$

The likelihood ratio $\lambda = L(\hat{\omega})/L(\hat{\Omega})$[mv,-1.5) for this test is

$$\lambda = \frac{\left[ne^{-1}/2\pi \sum_1^n (x_i - \mu_0)^2 \right]^{n/2}}{\left[ne^{-1}/2\pi \sum_1^n (x_i - \bar{x})^2 \right]^{n/2}} = \left[\frac{\sum_1^n (x_i - \bar{x})^2}{\sum_1^n (x_i - \mu_0)^2} \right]^{n/2}.$$

However, note that

$$\sum_1^n (x_i - \mu_0)^2 = \sum_1^n (x_i - \bar{x} + \bar{x} - \mu_0)^2 = \sum_1^n (x_i - \bar{x})^2 + n(\bar{x} - \mu_0)^2.$$

If this substitution is made in the denominator of λ, we have

$$
\lambda = \left[\frac{\sum_{1}^{n} (x_i - \bar{x})^2}{\sum_{1}^{n} (x_i - \bar{x})^2 + n(\bar{x} - \mu_0)^2} \right]^{n/2}
$$

$$
= \left[\frac{1}{1 + n(\bar{x} - \mu_0)^2 \Big/ \sum_{1}^{n} (x_i - \bar{x})^2} \right]^{n/2} .
$$

Note that λ is close to one when $\bar{x}$ is close to μ_0 and it is small when $\bar{x}$ and μ_0 differ by a great deal. The likelihood ratio test, given by the inequality $\lambda \leq k$, is the same as

$$
\frac{1}{1 + n(\bar{x} - \mu_0)^2 \Big/ \sum_{1}^{n} (x_i - \bar{x})^2} \leq k^{2/n}
$$

or, equivalently,

$$
\frac{n(\bar{x} - \mu_0)^2}{\sum_{1}^{n} (x_i - \bar{x})^2/(n - 1)} \geq (n - 1)(k^{-2/n} - 1).
$$

When H_0 is true, $\sqrt{n}(\bar{X} - \mu_0)/\sigma$ is $N(0, 1)$, and $\sum_{1}^{n} (X_i - \bar{X})^2/\sigma^2$ has an independent chi-square distribution $\chi^2(n-1)$. Hence, under H_0,

$$
T = \frac{\sqrt{n}(\bar{X} - \mu_0)/\sigma}{\sqrt{\sum_{1}^{n} (X_i - \bar{X})^2/[\sigma^2(n - 1)]}} = \frac{\sqrt{n}(\bar{X} - \mu_0)}{\sqrt{\sum_{1}^{n} (X_i - \bar{X})^2/(n - 1)}} = \frac{\bar{X} - \mu_0}{S/\sqrt{n}}
$$

has a t distribution with $r = n - 1$ degrees of freedom. In accordance with the likelihood ratio test criterion, H_0 is rejected if the observed

$$
T^2 \geq (n - 1)(k^{-2/n} - 1).
$$

That is, we reject H_0: $\mu = \mu_0$ and accept H_1: $\mu \neq \mu_0$ if the observed $|T| \geq t_{\alpha/2}(n-1)$.

The reader should observe that this test is exactly the same as that listed in Table 7.3-2 for testing H_0: $\mu = \mu_0$ against H_1: $\mu \neq \mu_0$. That is, the test listed there is a likelihood ratio test. As a matter of fact, all six of the tests given in Tables 7.3-2 and 7.4-1 are likelihood ratio tests. Thus the examples and exercises associated with those tables are illustrations of the use of likelihood ratio tests.

The final development of this section concerns a test about the variance of a normal population. Let $X_1, X_2, \ldots, X_n$ be a random sample from $N(\mu, \sigma^2)$, where μ and σ^2 are unknown. We wish to test $H_0: \sigma^2 = \sigma_0^2$ against $H_1: \sigma^2 \neq \sigma_0^2$. For this, we have

$$\omega = \{(\mu, \sigma^2): -\infty < \mu < \infty, \sigma^2 = \sigma_0^2\}.$$

and

$$\Omega = \{(\mu, \sigma^2): -\infty < \mu < \infty, 0 < \sigma^2 < \infty\}.$$

As in the test concerning the mean, we obtain

$$L(\hat{\Omega}) = \left[\frac{ne^{-1}}{2\pi \sum\limits_1^n (x_i - \bar{x})^2} \right]^{n/2}.$$

If $(\mu, \sigma^2) \in \omega$, then $\hat{\mu} = \bar{x}$ and $\widehat{\sigma^2} = \sigma_0^2$; thus

$$L(\hat{\omega}) = \left(\frac{1}{2\pi\sigma_0^2} \right)^{n/2} \exp \left[-\frac{\sum\limits_1^n (x_i - \bar{x})^2}{2\sigma_0^2} \right].$$

Accordingly, the likelihood ratio test $\lambda = L(\hat{\omega})/L(\hat{\Omega})$ is

$$\lambda = \left(\frac{w}{n} \right)^{n/2} \exp\left(-\frac{w}{2} + \frac{n}{2} \right) \leq k,$$

where $w = \sum_1^n (x_i - \bar{x})^2/\sigma_0^2$. Solving this inequality for w, we obtain a solution of the form $w \leq c_1$ or $w \geq c_2$, where the constants c_1 and c_2 are appropriate functions of the constants k and n so as to achieve the desired significance level α. Since $W = \sum_1^n (X_i - \bar{X})^2/\sigma_0^2$ is $\chi^2(n-1)$ if $H_0: \sigma^2 = \sigma_0^2$ is true, most statisticians modify this test slightly by taking $c_1 = \chi_{1-\alpha/2}^2(n-1)$ and $c_2 = \chi_{\alpha/2}^2(n-1)$. That is, this test and the other tests listed in Table 7.3-3 are either likelihood ratio tests or slight modifications of likelihood ratio tests. As a matter of fact, most tests involving normal assumptions are likelihood ratio tests or modifications of them. Some of these tests, like those involving regression and analysis of variance, are considered later in this book.

Exercises

7.6-1 In Example 7.6-1, if $n = 20$ and $\bar{x} = 161.1$, is H_0 accepted at a significance level of size α
(a) If $\alpha = 0.10$?
(b) If $\alpha = 0.05$?
(c) What is the p-value of this test?

7.6-2 Assume that the weight X in ounces of a "10-ounce" box of corn flakes is $N(\mu, 0.03)$.

(a) To test the hypothesis H_0: $\mu \geq 10.35$ against the alternative hypothesis H_1: $\mu < 10.35$, what is the critical region of size $\alpha = 0.05$ specified by the likelihood ratio test criterion?

HINT: If $\mu \geq 10.35$ and $\bar{x} < 10.35$, note that $\hat{\mu} = 10.35$.

(b) If a random sample of $n = 50$ boxes yielded a sample mean of $\bar{x} = 10.31$, is H_0 rejected?

HINT: Find the critical value z_α when H_0 is true by taking $\mu = 10.35$, which is the extreme value in $\mu \geq 10.35$.

(c) What is the p-value of this test?

7.6-3 Let X be $N(\mu, 100)$.

(a) To test H_0: $\mu = 230$ against H_1: $\mu > 230$, what is the critical region specified by the likelihood ratio test criterion?

(b) Is this test uniformly most powerful?

(c) If a random sample of $n = 16$ yielded $\bar{x} = 232.6$, is H_0 accepted at a significance level of $\alpha = 0.10$?

(d) What is the p-value of this test?

7.6-4 Let X be $N(\mu, 225)$.

(a) To test H_0: $\mu = 59$ against H_1: $\mu \neq 59$, what is the critical region of size $\alpha = 0.05$ specified by the likelihood ratio test criterion?

(b) If a sample of size $n = 100$ yielded $\bar{x} = 56.13$, is H_0 accepted?

(c) What is the p-value of this test?

7.6-5 It is desired to test the hypothesis H_0: $\mu = 30$ against the alternative hypothesis H_1: $\mu \neq 30$, where μ is the mean of a normal distribution and σ^2 is unknown. If a random sample of size $n = 9$ has $\bar{x} = 32.8$ and $s = 4$, is H_0 accepted at an $\alpha = 0.05$ significance level? What is the approximate p-value of this test?

7.6-6 To test H_0: $\mu = 335$ against H_1: $\mu < 335$, under normal assumptions, a random sample of size 17 yielded $\bar{x} = 324.8$ and $s = 40$. Is H_0 accepted at an $\alpha = 0.10$ significance level?

7.6-7 Let X have a normal distribution, in which μ and σ^2 are both unknown. It is desired to test H_0: $\mu = 1.80$ against H_1: $\mu > 1.80$ at an $\alpha = 0.10$ significance level. If a random sample of size $n = 121$ yielded $\bar{x} = 1.84$ and $s = 0.20$, is H_0 accepted or rejected? What is the p-value of this test?

7.6-8 Let $X_1, X_2, \ldots, X_n$ be a random sample from an exponential distribution with mean θ. Show that the likelihood ratio test of H_0: $\theta = \theta_0$ against H_1: $\theta \neq \theta_0$ has a critical region of the form $\sum_1^n x_i \leq c_1$ or $\sum_1^n x_i \geq c_2$. How would you modify this so that chi-square tables can be used easily?

7.6-9 Let random samples of sizes n and m be taken respectively from two independent normal distributions with unknown means μ_X and μ_Y and unknown variances σ_X^2 and σ_Y^2.

(a) Show that the likelihood ratio for testing H_0: $\mu_X = \mu_Y$ against H_1: $\mu_X \neq \mu_Y$ when $\sigma_X^2 = \sigma_Y^2$ is a function of the usual two-sample t-statistic.

(b) Show that the likelihood ratio for testing H_0: $\sigma_X^2 = \sigma_Y^2$ against H_1: $\sigma_X^2 \neq \sigma_Y^2$ is a function of the usual two-sample F-statistic.

The tests described in parts (a) and (b) are listed (the second is modified slightly) in Tables 7.4-1 and 7.4-2, respectively.

8

Linear Models

8.1 Tests of the Equality of Several Means

Frequently, experimenters want to compare more than two treatments: yields of several different corn hybrids, results due to three or more teaching techniques, or miles per gallon obtained from many different types of compact cars. Sometimes the different treatment distributions of the resulting observations are due to changing the level of a certain factor, (e.g., different doses of a given drug). Thus the consideration of the equality of the different means of the various distributions comes under the analysis of a **one-factor experiment**.

In Section 7.4 we discussed how to compare two independent normal distributions. More generally, let us consider m independent normal distributions with unknown means $\mu_1, \mu_2, \ldots, \mu_m$, respectively, and an unknown but common variance σ^2. One inference that we wish to consider is a test of the equality of the m means, namely $H_0: \mu_1 = \mu_2 = \cdots = \mu_m = \mu$, μ unspecified, against all possible alternative hypotheses H_1. In order to test this hypothesis we shall take a random sample from each distribution. Let $X_{i1}, X_{i2}, \ldots, X_{in_i}$ represent a random sample of size n_i from the normal distribution $N(\mu_i, \sigma^2)$, $i = 1, 2, \ldots, m$. In Table 8.1-1 we have indicated these random samples along with the row means (sample means) where, with $n = n_1 + n_2 + \cdots + n_m$,

$$\overline{X}_{..} = \frac{1}{n} \sum_{i=1}^{m} \sum_{j=1}^{n_i} X_{ij} \quad \text{and} \quad \overline{X}_{i.} = \frac{1}{n_i} \sum_{j=1}^{n_i} X_{ij}, \quad i = 1, 2, \ldots, m.$$

TABLE 8.1-1

					Means
X_1:	X_{11}	X_{12}	$\cdots$	X_{1n_1}	$\overline{X}_1.$
X_2:	X_{21}	X_{22}	$\cdots$	X_{2n_2}	$\overline{X}_2.$
$\vdots$	$\vdots$	$\vdots$	$\vdots$	$\vdots$	$\vdots$
X_m:	X_{m1}	X_{m2}	$\cdots$	X_{mn_m}	$\overline{X}_m.$
Grand mean:					$\overline{X}..$

The dot in the notation for the means, $\overline{X}..$ and $\overline{X}_i.$, indicates the index over which the average is taken. Here $\overline{X}..$ is an average taken over both indices while $\overline{X}_i.$ is just taken over the index j.

To determine a critical region for a test of H_0, we shall first partition the sum of squares associated with the variance of the combined samples into two parts. This sum of squares is given by

$$\text{SS(TO)} = \sum_{i=1}^{m} \sum_{j=1}^{n_i} (X_{ij} - \overline{X}..)^2$$

$$= \sum_{i=1}^{m} \sum_{j=1}^{n_i} (X_{ij} - \overline{X}_i. + \overline{X}_i. - \overline{X}..)^2$$

$$= \sum_{i=1}^{m} \sum_{j=1}^{n_i} (X_{ij} - \overline{X}_i.)^2 + \sum_{i=1}^{m} \sum_{j=1}^{n_i} (\overline{X}_i. - \overline{X}..)^2$$

$$+ 2 \sum_{i=1}^{m} \sum_{j=1}^{n_i} (X_{ij} - \overline{X}_i.)(\overline{X}_i. - \overline{X}..).$$

The last term of the righthand member of this identity may be written as

$$2 \sum_{i=1}^{m} \left[(\overline{X}_i. - \overline{X}..) \sum_{j=1}^{n_i} (X_{ij} - \overline{X}_i.) \right] = 2 \sum_{i=1}^{m} (\overline{X}_i. - \overline{X}..)(n_i \overline{X}_i. - n_i \overline{X}_i.) = 0,$$

and the preceding term may be written as

$$\sum_{i=1}^{m} \sum_{j=1}^{n_i} (\overline{X}_i. - \overline{X}..)^2 = \sum_{i=1}^{m} n_i (\overline{X}_i. - \overline{X}..)^2.$$

Thus

$$\text{SS(TO)} = \sum_{i=1}^{m} \sum_{j=1}^{n_i} (X_{ij} - \overline{X}_i.)^2 + \sum_{i=1}^{m} n_i (\overline{X}_i. - \overline{X}..)^2.$$

For notation let

$$SS(TO) = \sum_{i=1}^{m} \sum_{j=1}^{n_i} (X_{ij} - \bar{X}..)^2, \qquad \text{the total sum of squares.}$$

$$SS(E) = \sum_{i=1}^{m} \sum_{j=1}^{n_i} (X_{ij} - \bar{X}_{i.})^2, \qquad \text{the sum of squares within treatments,}$$
groups, or classes, often called the error
sum of squares.

$$SS(T) = \sum_{i=1}^{m} n_i(\bar{X}_{i.} - \bar{X}..)^2, \qquad \text{the sum of squares among the different}$$
treatments, groups, or classes, often called
the between treatment sum of squares.

Thus

$$SS(TO) = SS(E) + SS(T).$$

When H_0 is true, we may regard X_{ij}, $i = 1, 2, \ldots, m$, $j = 1, 2, \ldots, n_i$, as a random sample of size $n = n_1 + n_2 + \cdots + n_m$ from $N(\mu, \sigma^2)$. Then $SS(TO)/(n - 1)$ is an unbiased estimator of σ^2 because $SS(TO)/\sigma^2$ is $\chi^2(n-1)$ so that $E[SS(TO)/\sigma^2] = n - 1$ and $E[SS(TO)/(n - 1)] = \sigma^2$. An unbiased estimator of σ^2 based only on the sample from the ith distribution is

$$W_i = \frac{\sum_{j=1}^{n_i} (X_{ij} - \bar{X}_{i.})^2}{n_i - 1} \qquad \text{for } i = 1, 2, \ldots, m,$$

because $(n_i - 1)W_i/\sigma^2$ is $\chi^2(n_i-1)$. Thus

$$E\left[\frac{(n_i - 1)W_i}{\sigma^2}\right] = n_i - 1,$$

and so

$$E(W_i) = \sigma^2, \qquad i = 1, 2, \ldots, m.$$

It follows that the sum of m of these independent chi-square random variables, namely

$$\sum_{i=1}^{m} \frac{(n_i - 1)W_i}{\sigma^2} = \frac{SS(E)}{\sigma^2},$$

is also chi-square with $(n_1 - 1) + (n_2 - 1) + \cdots + (n_m - 1) = n - m$ degrees of freedom. Hence $SS(E)/(n - m)$ is an unbiased estimator of σ^2. We now have that

$$\frac{SS(TO)}{\sigma^2} = \frac{SS(E)}{\sigma^2} + \frac{SS(T)}{\sigma^2},$$

where

$$\frac{\text{SS(TO)}}{\sigma^2} \text{ is } \chi^2(n-1) \quad \text{and} \quad \frac{\text{SS(E)}}{\sigma^2} \text{ is } \chi^2(n-m).$$

Because $\text{SS(T)} \geq 0$, there is a theorem (see following remark) that states that SS(E) and SS(T) are independent and the distribution of $\text{SS(T)}/\sigma^2$ is $\chi^2(m-1)$.

REMARK The sums of squares SS(T), SS(E), and SS(TO) are examples of **quadratic forms** in the variables X_{ij}, $i = 1, 2, \ldots, m, j = 1, 2, \ldots, n_i$. That is, each term in these sums of squares is of second degree in X_{ij}. Furthermore the coefficients of the variables are real numbers, so these sums of squares are called **real quadratic forms**. The following theorem, stated without proof, is used in this chapter. [For a proof, see R. V. Hogg and A. T. Craig, *Introduction to Mathematical Statistics,* 4th ed. (New York: Macmillan, 1978).]

THEOREM 8.1-1 *Let $Q = Q_1 + Q_2 + \cdots + Q_k$, where $Q, Q_1, \ldots, Q_k$ are $k + 1$ real quadratic forms in n mutually independent random variables normally distributed with the same variance σ^2. Let $Q/\sigma^2, Q_1/\sigma^2, \ldots, Q_{k-1}/\sigma^2$ have chi-square distributions with $r, r_1, \ldots, r_{k-1}$ degrees of freedom, respectively. If Q_k is nonnegative, then*
(a) $Q_1, \ldots, Q_k$ are mutually independent, and hence,
(b) Q_k/σ^2 has a chi-square distribution with $r - (r_1 + \cdots + r_{k-1}) = r_k$ degrees of freedom.

Since under H_0, $\text{SS(T)}/\sigma^2$ is $\chi^2(m-1)$, we have $E[\text{SS(T)}/\sigma^2] = m - 1$ and hence $E[\text{SS(T)}/(m - 1)] = \sigma^2$. Now the estimator of σ^2, which is based on SS(E), namely $\text{SS(E)}/(n - m)$, is always unbiased whether H_0 is true or false. However, if the means $\mu_1, \mu_2, \ldots, \mu_m$ are not equal, the expected value of the estimator based on SS(T) will be greater than σ^2. To make this last statement clear, we have

$$E[\text{SS(T)}] = E\left[\sum_{i=1}^{m} n_i(\overline{X}_{i.} - \overline{X}_{..})^2\right] = E\left[\sum_{i=1}^{m} n_i\overline{X}_{i.}^2 - n\overline{X}_{..}^2\right]$$

$$= \sum_{i=1}^{m} n_i\{\text{Var}(\overline{X}_{i.}) + [E(\overline{X}_{i.})]^2\} - n\{\text{Var}(\overline{X}_{..}) + [E(\overline{X}_{..})]^2\}$$

$$= \sum_{i=1}^{m} n_i\left\{\frac{\sigma^2}{n_i} + \mu_i^2\right\} - n\left\{\frac{\sigma^2}{n} + \overline{\mu}^2\right\}$$

$$= (m - 1)\sigma^2 + \sum_{i=1}^{m} n_i(\mu_i - \overline{\mu})^2,$$

where $\overline{\mu} = (1/n) \sum_{i=1}^{m} n_i\mu_i$. If $\mu_1 = \mu_2 = \cdots = \mu_m = \mu$,

$$E\left(\frac{\text{SS(T)}}{m - 1}\right) = \sigma^2.$$

If the means are not all equal,

$$E\left[\frac{SS(T)}{m-1}\right] = \sigma^2 + \sum_{i=1}^{m} n_i \frac{(\mu_i - \overline{\mu})^2}{m-1} > \sigma^2.$$

Exercise 8.1-4 also illustrates the fact that the estimator using $SS(T)$ is usually greater than that using $SS(E)$ when H_0 is false.

We can base our test of H_0 on the ratio of $SS(T)/(m-1)$ and $SS(E)/(n-m)$, both of which are unbiased estimators of σ^2, provided that H_0: $\mu_1 = \mu_2 = \cdots = \mu_m$ is true so that, under H_0, the ratio would assume values near one. However, as the means $\mu_1, \mu_2, \ldots, \mu_m$ begin to differ, this ratio tends to become large, since $E[SS(T)/(m-1)]$ gets larger. Under H_0, the ratio

$$\frac{SS(T)/(m-1)}{SS(E)/(n-m)} = \frac{[SS(T)/\sigma^2]/(m-1)}{[SS(E)/\sigma^2]/(n-m)} = F$$

has an F distribution with $m-1$ and $n-m$ degrees of freedom because $SS(T)/\sigma^2$ and $SS(E)/\sigma^2$ are independent chi-square variables. We would reject H_0 if the observed value of F is too large because this would indicate that we have a relatively large $SS(T)$, which suggests that the means are unequal. Thus the critical region is of the form $F \geq F_\alpha(m-1, n-m)$.

The information for tests of the equality of several means is often summarized in an **analysis-of-variance** table or **ANOVA** table like that given in Table 8.1-2, where MS is SS divided by the degrees of freedom.

> **Example 8.1-1**   Let X_1, X_2, X_3, X_4 be independent random variables that are $N(\mu_i, \sigma^2)$, $i = 1, 2, 3, 4$. We shall test
>
> $$H_0:\ \mu_1 = \mu_2 = \mu_3 = \mu_4 = \mu$$
>
> against all alternatives based on a random sample of size $n_i = 3$ from each of the four distributions. A critical region of size $\alpha = 0.05$ is given by
>
> $$F = \frac{SS(T)/(4-1)}{SS(E)/(12-4)} \geq 4.07 = F_{0.05}(3, 8).$$

TABLE 8.1-2

Source	Sum of Squares (SS)	Degrees of Freedom	Mean Square (MS)	F-Ratio
Treatment	SS(T)	$m-1$	$MS(T) = \dfrac{SS(T)}{m-1}$	$\dfrac{MS(T)}{MS(E)}$
Error	SS(E)	$n-m$	$MS(E) = \dfrac{SS(E)}{n-m}$	
Total	SS(TO)	$n-1$		

TABLE 8.1-3

				$\overline{X}_{i.}$
X_1	13	8	9	10
X_2	15	11	13	13
X_3	8	12	7	9
X_4	11	15	10	12
$\overline{X}_{..}$				11

The observed data are given in Table 8.1-3. (Clearly, these data are not observations from normal distributions. They were selected to illustrate the calculations.)

For these data, the calculated SS(TO), SS(E), and SS(T) are

$$SS(TO) = (13 - 11)^2 + (8 - 11)^2 + \cdots + (15 - 11)^2 + (10 - 11)^2 = 80;$$
$$SS(E) = (13 - 10)^2 + (8 - 10)^2 + \cdots + (15 - 12)^2 + (10 - 12)^2 = 50;$$
$$SS(T) = 3[(10 - 11)^2 + (13 - 11)^2 + (9 - 11)^2 + (12 - 11)^2] = 30.$$

Note that since SS(TO) = SS(E) + SS(T), only two of the three values need to be calculated from the data directly. Here the computed value of F is

$$\frac{30/3}{50/8} = 1.6 < 4.07,$$

and H_0 is not rejected. The p-value is the probability, under H_0, of observing an F that is at least as large as this observed F. It is often given by computer programs.

The information for this example is summarized in the ANOVA table, Table 8.1-4. Again we note that (here and elsewhere) the F statistic is the ratio of two appropriate mean squares.

TABLE 8.1-4

Source	Sum of Squares (SS)	Degrees of Freedom	Mean Square (MS)	F	p-value
Treatment	30	3	30/3	1.6	0.264
Error	50	8	50/8		
Total	80	11			

Formulas that sometimes simplify the calculations of SS(TO), SS(T), and SS(E) are

$$\text{SS(TO)} = \sum_{i=1}^{m} \sum_{j=1}^{n_i} X_{ij}^2 - \frac{1}{n} \left[\sum_{i=1}^{m} \sum_{j=1}^{n_i} X_{ij} \right]^2,$$

$$\text{SS(T)} = \sum_{i=1}^{m} \frac{1}{n_i} \left[\sum_{j=1}^{n_i} X_{ij} \right]^2 - \frac{1}{n} \left[\sum_{i=1}^{m} \sum_{j=1}^{n_i} X_{ij} \right]^2,$$

and

$$\text{SS(E)} = \text{SS(TO)} - \text{SS(T)}.$$

It is interesting to note that in these formulas each square is divided by the number of items in the sum being squared: X_{ij}^2 by one, $(\sum_{j=1}^{n_i} X_{ij})^2$ by n_i, and $(\sum_{i=1}^{m} \sum_{j=1}^{n_i} X_{ij})^2$ by n. These formulas are used in Example 8.1-2.

If the sample sizes are all at least equal to 7, insight can be gained by plotting on the same figure box-and-whisker diagrams for each of the samples. This is also illustrated in Example 8.1-2.

Example 8.1-2 A window that is manufactured for an automobile has five studs for attaching it. A company that manufactures these windows performs "pull out tests" to determine the force needed to pull a stud out of the window. Let X_i, $i = 1, 2, 3, 4, 5$, equal the force required at position i and then assume that the distribution of X_i is $N(\mu_i, \sigma^2)$. We shall test the null hypothesis H_0: $\mu_1 = \mu_2 = \mu_3 = \mu_4 = \mu_5$ using 7 observations at each position. At an $\alpha = 0.01$ significance level, H_0 is rejected if the computed

$$F = \frac{\text{SS(T)}/(5-1)}{\text{SS(E)}/(35-5)} \geq 4.02 = F_{0.01}(4, 30).$$

The observed data along with certain sums are given in Table 8.1-5.

TABLE 8.1-5

								$\sum_{j=1}^{7} x_{ij}$	$\sum_{j=1}^{7} x_{ij}^2$
X_1:	92	90	87	105	86	83	102	645	59,847
X_2:	100	108	98	110	114	97	94	721	74,609
X_3:	143	149	138	136	139	120	145	970	134,936
X_4:	147	144	160	149	152	131	134	1017	148,367
X_5:	142	155	119	134	133	146	152	981	138,174
Totals								4334	556,174

TABLE 8.1-6

Source	Sum of Squares (SS)	Degrees of Freedom	Mean Square (MS)	F
Treatment	16,672.11	4	4,168.03	44.20
Error	2,828.86	30	94.30	
Total	19,500.97	34		

For the data in Table 8.1-5,

$$SS(TO) = 556,174 - \tfrac{1}{35}(4334)^2 = 19,500.97$$
$$SS(T) = \tfrac{1}{7}[645^2 + 721^2 + 970^2 + 1017^2 + 981^2]$$
$$- \tfrac{1}{35}(4334)^2 = 16,672.11$$
$$SS(E) = 19,500.97 - 16,672.11 = 2828.86.$$

Since the computed F is

$$F = \frac{16,672.11/4}{2,828.86/30} = 44.20,$$

the null hypothesis is clearly rejected. This information is summarized in Table 8.1-6.

But why is H_0 rejected? The box-and-whisker diagrams shown in Figure 8.1-1 help to answer this question. It looks like the forces required to pull out studs in positions 1 and 2 are similar and those in positions 3, 4, and 5 are quite similar but different from positions 1 and 2. (See Exercise 8.1-12.) An examination of the window would confirm that this is the case.

REMARK   As with the two sample t-test, the F-test works quite well even if the underlying distributions are nonnormal, unless they are highly skewed or the variances are quite different. In these latter cases, we might need to transform the observations to make the data more symmetric with about the same variances or to use certain nonparametric methods that are beyond the scope of this course.

Exercises

8.1-1   Let μ_1, μ_2, μ_3 be, respectively, the means of three independent normal distributions with a common but unknown variance σ^2. In order to test, at the $\alpha = 5\%$ significance level, the hypothesis H_0: $\mu_1 = \mu_2 = \mu_3$ against all possible alternative hypotheses, we take a random sample of size 4 from each of

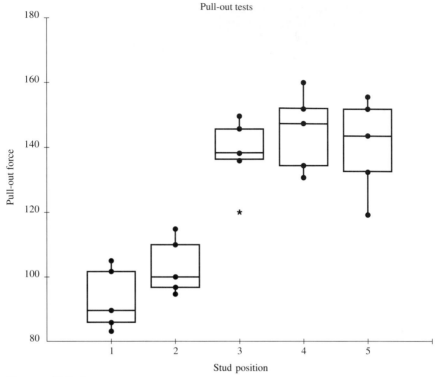

Pull-out tests

FIGURE 8.1-1

these distributions. Determine whether we accept or reject H_0 if the observed values from these three distributions are, respectively,

$$
\begin{array}{lcccc}
X_1: & 5 & 9 & 6 & 8 \\
X_2: & 11 & 13 & 10 & 12 \\
X_3: & 10 & 6 & 9 & 9
\end{array}
$$

8.1-2   Let μ_i be the average yield in bushels per acre of variety i of corn, $i = 1, 2,$ 3, 4. In order to test, at the 5% significance level, the hypothesis H_0: $\mu_1 = \mu_2 = \mu_3 = \mu_4$, four test plots for each of the four varieties of corn are planted. Determine whether we accept or reject H_0 if the yield in bushels per acre of the four varieties of corn are, respectively,

$$
\begin{array}{lcccc}
X_1: & 68.82 & 76.99 & 74.30 & 78.73 \\
X_2: & 86.84 & 75.69 & 77.87 & 76.18 \\
X_3: & 90.16 & 78.84 & 80.65 & 83.58 \\
X_4: & 61.58 & 73.51 & 74.57 & 70.75
\end{array}
$$

8.1-3   Four groups of three pigs each were fed individually four different feeds for a specified length of time to test the hypothesis H_0: $\mu_1 = \mu_2 = \mu_3 = \mu_4,$

where μ_i is the mean weight gain for each of the feeds, $i = 1, 2, 3, 4$. Determine whether the null hypothesis is accepted or rejected at a 5% significance level if the observed weight gains are, respectively,

$$X_1: \quad 194.11 \quad 182.80 \quad 187.43$$
$$X_2: \quad 216.06 \quad 203.50 \quad 216.88$$
$$X_3: \quad 178.10 \quad 189.20 \quad 181.33$$
$$X_4: \quad 197.11 \quad 202.68 \quad 209.18$$

8.1-4 For the following set of data show that the computed $SS(E)/(n - m) = 1$ and $SS(T)/(m - 1) = 75$. This suggests that the unbiased estimate of σ^2 based on $SS(T)$ is usually greater than σ^2 when the true means are unequal.

$$X_1: \quad 4 \quad 5 \quad 6$$
$$X_2: \quad 9 \quad 10 \quad 11$$
$$X_3: \quad 14 \quad 15 \quad 16$$

8.1-5 In a tulip display area, several varieties of tulips have long stems. The lengths of the flower stems were measured for four varieties of tulips, the colors being white, pink, red trimmed with yellow, and dark red. We shall call them varieties 1, 2, 3, and 4. Let μ_i, $i = 1, 2, 3, 4$, denote the mean of the lengths of the stems for variety i. Test the hypothesis $H_0: \mu_1 = \mu_2 = \mu_3 = \mu_4$ against all alternatives, making reasonable assumptions. Use a 5% significance level. Clearly specify the critical region for this test.

$$X_1: \quad 13.75 \quad 13.00 \quad 14.25 \quad 12.00 \quad 12.25$$
$$X_2: \quad 13.75 \quad 14.50 \quad 12.75 \quad 14.50 \quad 13.25$$
$$X_3: \quad 12.75 \quad 12.50 \quad 12.00 \quad 14.00 \quad 13.00$$
$$X_4: \quad 16.75 \quad 14.25 \quad 14.50 \quad 15.50 \quad 13.75$$

8.1-6 Use the data of Example 1.4-3 concerning the strengths of three different types of beams to make a formal F-test of the equality of the three means. Use $\alpha = 0.05$.

8.1-7 Montgomery considers the strengths of a synthetic fiber that is possibly affected by the percentage of cotton in the fiber. Five levels of this percentage are considered with five observations taken at each level.

Percentage of Cotton	Tensile Strength (pounds per square inch)				
15	7	7	15	11	9
20	12	17	12	18	18
25	14	18	18	19	19
30	19	25	22	19	23
35	7	10	11	15	11

Use the F-test, with $\alpha = 0.05$, to see if there are differences in the breaking strengths due to the percentages of cotton used. [D. C. Montgomery, *Design and Analysis of Experiments,* 2nd ed. (New York: Wiley, 1984), p. 51.]

8.1-8   Let X_1, X_2, X_3, X_4 equal the cholesterol level of a woman under the age of 50, a man under 50, a woman 50 or older, and a man 50 or older, respectively. Assume that the distribution of X_i is $N(\mu_i, \sigma^2)$, $i = 1, 2, 3, 4$. We shall test the null hypothesis H_0: $\mu_1 = \mu_2 = \mu_3 = \mu_4$ using 7 observations of each X_i.
(a) Define a critical region for an $\alpha = 0.05$ significance level.
(b) Construct an ANOVA summary table and state your conclusion using the following data:

X_1:	221	213	202	183	185	197	162
X_2:	271	192	189	209	227	236	142
X_3:	262	193	224	201	161	178	265
X_4:	192	253	248	278	232	267	289

(c) Give bounds on the p-value for this test.
(d) Construct box-and-whisker diagrams for each set of data on the same figure and give an interpretation.

8.1-9   Let X_i, $i = 1, 2, 3, 4$ equal the distance that a golf ball travels when hit from a tee, where i denotes the index of the ith manufacturer. Assume that the distribution of X_i is $N(\mu_i, \sigma^2)$, $i = 1, 2, 3, 4$ when hit by a certain golfer. We shall test the null hypothesis H_0: $\mu_1 = \mu_2 = \mu_3 = \mu_4$ using 3 observations of each random variable.
(a) Define a critical region for an $\alpha = 0.05$ significance level.
(b) Construct an ANOVA summary table and state your conclusion using the following data:

X_1:	240	221	265
X_2:	286	256	272
X_3:	259	245	232
X_4:	239	215	223

(c) What would your conclusion be if $\alpha = 0.025$?
(d) What is the approximate p-value of this test?

8.1-10   Different sizes of nails are packaged in "1-pound" boxes. Let X_i equal the weight of a box with nail size $4i$C, $i = 1, 2, 3, 4, 5$, where 4C, 8C, 12C, 16C, and 20C are the sizes of the sinkers from smallest to largest. Assume that the distribution of X_i is $N(\mu_i, \sigma^2)$. To test the null hypothesis that the mean weights of "1-pound" boxes are all equal for different sizes of nails, we shall use random samples of size 7, weighing the nails to the nearest hundredth of a pound.
(a) Define a critical region for an $\alpha = 0.05$ significance level.

(b) Construct an ANOVA summary table and state your conclusion using the following data:

X_1:	1.03	1.04	1.07	1.03	1.08	1.06	1.07
X_2:	1.03	1.10	1.08	1.05	1.06	1.06	1.05
X_3:	1.03	1.08	1.06	1.02	1.04	1.04	1.07
X_4:	1.10	1.10	1.09	1.09	1.06	1.05	1.08
X_5:	1.04	1.06	1.07	1.06	1.05	1.07	1.05

(c) Construct box-and-whisker diagrams for each set of data on the same figure and give an interpretation.

8.1-11 The driver of a diesel-powered automobile decided to test the quality of three types of diesel fuel sold in the area based on mpg. Test the null hypothesis that the three means are equal using the following data. Make the usual assumptions and take $\alpha = 0.05$.

Brand A:	38.7	39.2	40.1	38.9	
Brand B:	41.9	42.3	41.3		
Brand C:	40.8	41.2	39.5	38.9	40.3

8.1-12 Based on the box-and-whisker diagrams in Figure 8.1-1, it looks like the means of X_1 and X_2 could be equal and also that the means of X_3, X_4, and X_5 could be equal but different from the first two.
(a) Using the data in Example 8.1-2, test H_0: $\mu_1 = \mu_2$ against a two-sided alternative hypothesis using a t-test and an F-test. Let $\alpha = 0.05$. Do F and t tests give the same result?
(b) Using the data in Example 8.1-2, test H_0: $\mu_3 = \mu_4 = \mu_5$. Let $\alpha = 0.05$.

8.1-13 For an aerosol product, there are three weights: the tare weight (container weight), the concentrate weight, and the propellant weight. Let X_1, X_2, X_3 denote the propellant weights on three different days. Assume that each of these independent random variables has a normal distribution with common variance and respective means, μ_1, μ_2, μ_3. We shall test the null hypothesis H_0: $\mu_1 = \mu_2 = \mu_3$ using 9 observations of each of the random variables.
(a) Define a critical region for an $\alpha = 0.01$ significance level.
(b) Construct an ANOVA summary table and state your conclusion using the following data:

X_1:	43.06	43.32	42.63	42.86	43.05
	42.87	42.94	42.80	42.36	
X_2:	42.33	42.81	42.13	42.41	42.39
	42.10	42.42	41.42	42.52	
X_3:	42.83	42.57	42.96	43.16	42.25
	42.24	42.20	41.97	42.61	

(c) Construct box-and-whisker diagrams for each set of data on the same figure and give an interpretation.

8.2 Two-Factor Analysis of Variance

The test of the equality of several means, considered in Section 8.1 is an example of a statistical inference method called the analysis of variance (ANOVA). This method derives its name from the fact that the quadratic form SS(TO) = $(n - 1)S^2$, the total sum of squares about the combined sample mean, is decomposed into component parts and analyzed. In this section, other problems in the analysis of variance will be investigated; here we restrict our considerations to the two-factor case, but the reader can see how it can be extended to three-factor and other cases.

Consider a situation in which it is desirable to investigate the effects of two factors that influence an outcome of an experiment. For example, a teaching method (lecture, discussion, computer-assisted, television, etc.) and the size of a class might influence a student's score on a standard test; or the type of car and the grade of gasoline used might change the number of miles per gallon. In this latter example, if the number of miles per gallon is not affected by the grade of gasoline, we would no doubt use the least expensive grade.

The first analysis of variance model that we discuss is referred to as a **two-way classification with one observation per cell**. In particular, assume that there are two factors (attributes), one of which has a levels and the other b levels. There are thus $n = ab$ possible combinations, each of which determines a cell. Let us think of these cells as being arranged in a rows and b columns. Here we take one observation per cell, and we denote the observation in the ith row and jth column by X_{ij}. Further assume that X_{ij}, $i = 1, 2, \ldots, a$, and $j = 1, 2, \ldots, b$, are n random variables that are mutually independent and are $N(\mu_{ij}, \sigma^2)$, respectively. [The assumptions of normality and homogeneous (same) variances can be *somewhat* relaxed in applications with little change in the significance levels of the resulting tests.] We shall assume that the means μ_{ij} are composed of a row effect, a column effect, and an overall effect in some additive way, namely $\mu_{ij} = \mu + \alpha_i + \beta_j$, where $\Sigma_{i=1}^a \alpha_i = 0$ and $\Sigma_{j=1}^b \beta_j = 0$. The parameter α_i represents the ith row effect, and the parameter β_j represents the jth column effect.

REMARK   There is no loss in generality in assuming that

$$\sum_1^a \alpha_i = \sum_1^b \beta_j = 0.$$

To see this, let $\mu_{ij} = \mu' + \alpha_i' + \beta_j'$. Write

$$\overline{\alpha}' = \left(\frac{1}{a}\right) \sum_1^a \alpha_i' \quad \text{and} \quad \overline{\beta}' = \left(\frac{1}{b}\right) \sum_1^b \beta_j'.$$

We have

$$\mu_{ij} = (\mu' + \overline{\alpha}' + \overline{\beta}') + (\alpha_i' - \overline{\alpha}') + (\beta_j' - \overline{\beta}') = \mu + \alpha_i + \beta_j,$$

where $\Sigma_1^a \alpha_i = 0$ and $\Sigma_1^b \beta_j = 0$. The reader is asked to find μ, α_i, and β_j for one display of μ_{ij} in Exercise 8.2-2.

To test the hypothesis that there is no row effect, we would test H_A: $\alpha_1 = \alpha_2 = \cdots = \alpha_a = 0$, since $\Sigma_1^a \alpha_i = 0$. Similarly, to test that there is no column effect, we would test H_B: $\beta_1 = \beta_2 = \cdots = \beta_b = 0$, since $\Sigma_1^b \beta_j = 0$. To test these hypotheses, we shall again partition the total sum of squares into several component parts. Letting

$$\bar{X}_{i.} = \frac{1}{b} \sum_{j=1}^{b} X_{ij}, \qquad \bar{X}_{.j} = \frac{1}{a} \sum_{i=1}^{a} X_{ij}, \qquad \bar{X}_{..} = \frac{1}{ab} \sum_{j=1}^{b} \sum_{i=1}^{a} X_{ij},$$

we have

$$SS(TO) = \sum_{j=1}^{b} \sum_{i=1}^{a} (X_{ij} - \bar{X}_{..})^2$$

$$= \sum_{j=1}^{b} \sum_{i=1}^{a} [(\bar{X}_{i.} - \bar{X}_{..}) + (\bar{X}_{.j} - \bar{X}_{..}) + (X_{ij} - \bar{X}_{i.} - \bar{X}_{.j} + \bar{X}_{..})]^2$$

$$= b \sum_{i=1}^{a} (\bar{X}_{i.} - \bar{X}_{..})^2 + a \sum_{j=1}^{b} (\bar{X}_{.j} - \bar{X}_{..})^2$$

$$+ \sum_{j=1}^{b} \sum_{i=1}^{a} (X_{ij} - \bar{X}_{i.} - \bar{X}_{.j} + \bar{X}_{..})^2$$

$$= SS(A) + SS(B) + SS(E),$$

where SS(A) is the sum of squares among levels of factor A or among rows, SS(B) is the sum of squares among levels of factor B or among columns, and SS(E) is the error or residual sum of squares. In Exercise 8.2-4 the reader is asked to show that the three "cross-product" terms in the square of the trinomial sum to zero. The distribution of the error sum of squares does not depend on the mean μ_{ij}, provided that the additive model is correct, and hence its distribution is the same whether H_A or H_B is true or not; thus SS(E) acts as a "measuring stick" as did SS(E) in Section 8.1. This can be seen more clearly by writing

$$SS(E) = \sum_{j=1}^{b} \sum_{i=1}^{a} (X_{ij} - \bar{X}_{i.} - \bar{X}_{.j} + \bar{X}_{..})^2$$

$$= \sum_{j=1}^{b} \sum_{i=1}^{a} [X_{ij} - (\bar{X}_{i.} - \bar{X}_{..}) - (\bar{X}_{.j} - \bar{X}_{..}) - \bar{X}_{..}]^2$$

and noting the similarity of the summand in the righthand member with

$$X_{ij} - \mu_{ij} = X_{ij} - \alpha_i - \beta_j - \mu.$$

We now show that SS(A)/σ^2, SS(B)/σ^2, and SS(E)/σ^2 are independent chi-square variables, provided that both H_A and H_B are true, that is, when all the means

μ_{ij} have a common value μ. To do this, we first note that $SS(TO)/\sigma^2$ is $\chi^2(ab-1)$. In addition, from Section 8.1, we see that expressions such as $SS(A)/\sigma^2$ and $SS(B)/\sigma^2$ are chi-square variables, namely $\chi^2(a-1)$ and $\chi^2(b-1)$, by replacing the n_i of Section 8.1 by a and b, respectively. Obviously, $SS(E) \geq 0$, and hence by Theorem 8.1-1 we have that $SS(A)/\sigma^2$, $SS(B)/\sigma^2$, and $SS(E)/\sigma^2$ are independent chi-square variables with $a - 1$, $b - 1$, and $ab - 1 - (a - 1) - (b - 1) = (a - 1)(b - 1)$ degrees of freedom, respectively.

To test the hypothesis $H_A: \alpha_1 = \alpha_2 = \cdots = \alpha_a = 0$, we shall use the row sum of squares $SS(A)$ and the residual sum of squares $SS(E)$. When H_A is true, $SS(A)/\sigma^2$ and $SS(E)/\sigma^2$ are independent chi-square variables with $a - 1$ and $(a - 1)(b - 1)$ degrees of freedom, respectively. Thus $SS(A)/(a - 1)$ and $SS(E)/[(a - 1)(b - 1)]$ are both unbiased estimators of σ^2 when H_A is true. However, $E[SS(A)/(a - 1)] > \sigma^2$ when H_A is not true, and hence we would reject H_A when

$$F_A = \frac{SS(A)/[\sigma^2(a - 1)]}{SS(E)/[\sigma^2(a - 1)(b - 1)]} = \frac{SS(A)/(a - 1)}{SS(E)/[(a - 1)(b - 1)]}$$

is "too large." Since F_A has an F distribution with $a - 1$ and $(a - 1)(b - 1)$ degrees of freedom when H_A is true, H_A is rejected if the observed value of $F_A \geq F_\alpha[a-1, (a-1)(b-1)]$.

Similarly, the test of the hypothesis $H_B: \beta_1 = \beta_2 = \cdots = \beta_b = 0$ against all alternatives can be based on

$$F_B = \frac{SS(B)/[\sigma^2(b - 1)]}{SS(E)/[\sigma^2(a - 1)(b - 1)]} = \frac{SS(B)/(b - 1)}{SS(E)/[(a - 1)(b - 1)]},$$

which has an F distribution with $b - 1$ and $(a - 1)(b - 1)$ degrees of freedom, provided that H_B is true.

In Table 8.2-1 we give the ANOVA table that summarizes the information needed for these tests of hypotheses. The formulas for F_A and F_B show that each of them is a ratio of two mean squares.

TABLE 8.2-1

Source	Sum of Squares (SS)	Degrees of Freedom	Mean Square (MS)	F
Factor A row	SS(A)	$a - 1$	$MS(A) = \dfrac{SS(A)}{a - 1}$	$\dfrac{MS(A)}{MS(E)}$
Factor B column	SS(B)	$b - 1$	$MS(B) = \dfrac{SS(B)}{b - 1}$	$\dfrac{MS(B)}{MS(E)}$
Error	SS(E)	$(a - 1)(b - 1)$	$MS(E) = \dfrac{SS(E)}{(a - 1)(b - 1)}$	
Total	SS(TO)	$ab - 1$		

Example 8.2-1   Each of three cars is driven with each of four different brands of gasoline. The number of miles per gallon driven for each of the $ab = (3)(4) = 12$ different combinations is recorded in Table 8.2-2.

We would like to test whether we can expect the same mileage for each of these four brands of gasoline. In our notation, we test the hypothesis

$$H_B: \beta_1 = \beta_2 = \beta_3 = \beta_4 = 0$$

against all alternatives. At a 1% significance level we shall reject H_B if the computed F, namely

$$\frac{SS(B)/(4-1)}{SS(E)/[(3-1)(4-1)]} \geq 9.78 = F_{0.01}(3, 6).$$

We have

$$SS(B) = 3[(15-17)^2 + (16-17)^2 + (19-17)^2 + (18-17)^2] = 30;$$
$$SS(E) = (16-19-15+17)^2 + (14-16-15+17)^2 + \cdots$$
$$+ (16-16-18+17)^2 = 4.$$

Hence the computed F is

$$\frac{30/3}{4/6} = 15 > 9.78,$$

and the hypothesis H_B is rejected. That is, the gasolines seem to give different performances (at least with these three cars).

The information for this example is summarized in the ANOVA table, Table 8.2-3.

In a two-way classification problem, particular combinations of the two factors might interact differently from that expected from the additive model. For example, in Example 8.2-1 gasoline 3 seemed to be the best gasoline and car 1 the best car; however, it sometimes happens that the two best do not "mix" well and the joint performance is poor. That is, there might be a strange interaction between this combination of car and gasoline and accordingly the joint performance is not as

TABLE 8.2-2

	Gasoline				
Car	1	2	3	4	$\overline{X}_{i.}$
1	16	18	21	21	19
2	14	15	18	17	16
3	15	15	18	16	16
$\overline{X}_{.j}$	15	16	19	18	17

TABLE 8.2-3

Source	Sum of Squares (SS)	Degrees of Freedom	Mean Square (MS)	F	p-value
Row (A)	24	2	12	18	0.003
Column (B)	30	3	10	15	0.003
Error	4	6	2/3		
Total	58	11			

good as expected. Sometimes it can happen that we get good results from a combination of some of the poorer levels of each factor. This is called interaction, and it frequently occurs in practice (e.g., in chemistry). In order to test for possible interaction, we shall consider a two-way classification problem in which $c > 1$ independent observations are taken per cell.

Assume that X_{ijk}, $i = 1, 2, \ldots, a$; $j = 1, 2, \ldots, b$; and $k = 1, 2, \ldots, c$, are $n = abc$ random variables that are mutually independent and have normal distributions with a common, but unknown, variance σ^2. The mean of each X_{ijk}, $k = 1, 2, \ldots, c$, is $\mu_{ij} = \mu + \alpha_i + \beta_j + \gamma_{ij}$, where $\Sigma_1^a \alpha_i = 0$, $\Sigma_1^b \beta_j = 0$, $\Sigma_{i=1}^a \gamma_{ij} = 0$, and $\Sigma_{j=1}^b \gamma_{ij} = 0$. The parameter γ_{ij} is called the **interaction** associated with cell (i, j). That is, the interaction between the ith level of one classification and the jth level of the other classification is γ_{ij}. The reader is asked to determine μ, α_i, β_j, and γ_{ij} for some given μ_{ij} in Exercise 8.2-6.

To test the hypotheses that (a) the row effects are equal to zero, (b) the column effects are equal to zero, and (c) there is no interaction, we shall again partition the total sum of squares into several component parts. Letting

$$\overline{X}_{ij\cdot} = \frac{1}{c} \sum_{k=1}^c X_{ijk},$$

$$\overline{X}_{i\cdot\cdot} = \frac{1}{bc} \sum_{j=1}^b \sum_{k=1}^c X_{ijk},$$

$$\overline{X}_{\cdot j\cdot} = \frac{1}{ac} \sum_{i=1}^a \sum_{k=1}^c X_{ijk},$$

$$\overline{X}_{\cdots} = \frac{1}{abc} \sum_{i=1}^a \sum_{j=1}^b \sum_{k=1}^c X_{ijk},$$

we have

$$SS(TO) = \sum_{i=1}^a \sum_{j=1}^b \sum_{k=1}^c (X_{ijk} - \overline{X}_{\cdots})^2$$

$$= bc \sum_{i=1}^{a} (\overline{X}_{i..} - \overline{X}_{...})^2 + ac \sum_{j=1}^{b} (\overline{X}_{.j.} - \overline{X}_{...})^2$$

$$+ c \sum_{i=1}^{a} \sum_{j=1}^{b} (\overline{X}_{ij.} - \overline{X}_{i..} - \overline{X}_{.j.} + \overline{X}_{...})^2 + \sum_{i=1}^{a} \sum_{j=1}^{b} \sum_{k=1}^{c} (X_{ijk} - \overline{X}_{ij.})^2$$

$$= SS(A) + SS(B) + SS(AB) + SS(E),$$

where SS(A) is the row sum of squares or the sum of squares among levels of factor A, SS(B) is the column sum of squares or the sum of squares among levels of factor B, SS(AB) is the interaction sum of squares, and SS(E) is the error sum of squares. Again we can show that the cross-product terms sum to zero.

To consider the joint distribution of SS(A), SS(B), SS(AB), and SS(E), let us assume that all the means equal the same value μ. Of course, we know that $SS(TO)/\sigma^2$ is $\chi^2(abc-1)$. Also, by letting the n_i of Section 8.1 equal bc and ac, respectively, we know that $SS(A)/\sigma^2$ and $SS(B)/\sigma^2$ are $\chi^2(a-1)$ and $\chi^2(b-1)$. Moreover,

$$\frac{\sum_{k=1}^{c} (X_{ijk} - \overline{X}_{ij.})^2}{\sigma^2}$$

is $\chi^2(c-1)$; hence $SS(E)/\sigma^2$ is the sum of ab independent chi-square variables such as this and thus is $\chi^2[ab(c-1)]$. Of course $SS(AB) \geq 0$; so, according to Theorem 8.1-1, we have that $SS(A)/\sigma^2$, $SS(B)/\sigma^2$, $SS(AB)/\sigma^2$, and $SS(E)/\sigma^2$ are mutually independent chi-square variables with $a - 1$, $b - 1$, $(a - 1)(b - 1)$, and $ab(c - 1)$ degrees of freedom, respectively.

To test the hypotheses concerning row, column, and interaction effects, we form F statistics in which the numerators are affected by deviations from the respective hypotheses whereas the denominator is a function of SS(E), whose distribution depends only on the value of σ^2 and not on the values of the cell means. Hence SS(E) acts as our measuring stick here.

The statistic for testing the hypothesis

$$H_{AB}: \gamma_{ij} = 0, \qquad i = 1, 2, \ldots, a, \quad j = 1, 2, \ldots, b,$$

against all alternatives is

$$F_{AB} = \frac{c \sum_{i=1}^{a} \sum_{j=1}^{b} (\overline{X}_{ij.} - \overline{X}_{i..} - \overline{X}_{.j.} + \overline{X}_{...})^2 / [\sigma^2(a - 1)(b - 1)]}{\sum_{i=1}^{a} \sum_{j=1}^{b} \sum_{k=1}^{c} (X_{ijk} - \overline{X}_{ij.})^2 / [\sigma^2 ab(c - 1)]}$$

$$= \frac{SS(AB)/[(a - 1)(b - 1)]}{SS(E)/[ab(c - 1)]},$$

which has an F distribution with $(a - 1)(b - 1)$ and $ab(c - 1)$ degrees of freedom when H_{AB} is true. If the computed $F_{AB} \geq F_{\alpha}[(a-1)(b-1), ab(c-1)]$, we reject H_{AB}

and say there is a difference among the means, since there seems to be interaction. Most statisticians do *not* proceed to test row and column effects if H_{AB} is rejected.

The statistic for testing the hypothesis

$$H_A: \alpha_1 = \alpha_2 = \cdots = \alpha_a = 0$$

against all alternatives is

$$F_A = \frac{bc \sum\limits_{i=1}^{a} (\overline{X}_{i..} - \overline{X}_{...})^2/[\sigma^2(a - 1)]}{\sum\limits_{i=1}^{a} \sum\limits_{j=1}^{b} \sum\limits_{k=1}^{c} (X_{ijk} - \overline{X}_{ij.})^2/[\sigma^2 ab(c - 1)]} = \frac{SS(A)/(a - 1)}{SS(E)/[ab(c - 1)]},$$

which has an F distribution with $a - 1$ and $ab(c - 1)$ degrees of freedom when H_A is true. The statistic for testing the hypothesis

$$H_B: \beta_1 = \beta_2 = \cdots = \beta_b = 0$$

against all alternatives is

$$F_B = \frac{ac \sum\limits_{j=1}^{b} (\overline{X}_{.j.} - \overline{X}_{...})^2/[\sigma^2(b - 1)]}{\sum\limits_{i=1}^{a} \sum\limits_{j=1}^{b} \sum\limits_{k=1}^{c} (X_{ijk} - \overline{X}_{ij.})^2/[\sigma^2 ab(c - 1)]} = \frac{SS(B)/(b - 1)}{SS(E)/[ab(c - 1)]},$$

which has an F distribution with $b - 1$ and $ab(c - 1)$ degrees of freedom when H_B is true. Each of these hypotheses is rejected if the observed value of F is greater than a given constant that is selected to yield the desired significance level.

In Table 8.2-4 we give the ANOVA table that summarizes the information needed for these tests of hypotheses.

REMARK It should be noted that when the hypotheses are not true, each of F_{AB}, F_A, and F_B has a noncentral F distribution. Although they are not too common, tables of probabilities for the *non*central F distribution are available in extensive statistical libraries.

> **Example 8.2-2** Consider the following experiment. One hundred eight people were randomly divided into six groups with 18 people in each group. Each person was given sets of three numbers to add. The three numbers were either in a "down array" or an "across array," representing the two levels of factor A. The levels of factor B are determined by the number of digits in the numbers to be added: one-digit, two-digit, or three-digit numbers. Table 8.2-5 illustrates this with a sample problem for each cell; note, however, that an individual person only works problems of one of these types. Each person was placed in one of the six groups and was told to work as many problems as possible in 90 seconds. The measurement that was recorded was the average number of problems worked correctly in two trials.

TABLE 8.2-4

Source	Sum of Squares (SS)	Degrees of Freedom	Mean Square (MS)	F
Factor A row	SS(A)	$a - 1$	$\text{MS(A)} = \dfrac{\text{SS(A)}}{a - 1}$	$\dfrac{\text{MS(A)}}{\text{MS(E)}}$
Factor B column	SS(B)	$b - 1$	$\text{MS(B)} = \dfrac{\text{SS(B)}}{b - 1}$	$\dfrac{\text{MS(B)}}{\text{MS(E)}}$
Factor AB interaction	SS(AB)	$(a - 1)(b - 1)$	$\text{MS(AB)} = \dfrac{\text{SS(AB)}}{(a - 1)(b - 1)}$	$\dfrac{\text{MS(AB)}}{\text{MS(E)}}$
Error	SS(E)	$ab(c - 1)$	$\text{MS(E)} = \dfrac{\text{SS(E)}}{ab(c - 1)}$	
Total	SS(TO)	$abc - 1$		

Whenever this many subjects are used, a computer becomes an invaluable tool. A computer program provided the summary in Table 8.2-6 of the sample means of the rows, the columns, and the six cells. Each cell mean is the average for 18 people.

Simply considering these means, we can see clearly that there is a column effect. It is not surprising that it is easier to add one-digit than three-digit numbers.

The most interesting feature of these results is that they show the possibility of interaction. The largest cell mean occurs when adding one-digit numbers in an across array. Note, however, that for two- and three-digit numbers the down arrays have larger means than the across arrays.

The computer program provided the ANOVA table given in Table 8.2-7. The number of degrees of freedom for SS(E) is not in our F table. However,

TABLE 8.2-5

Type of Array	Number of Digits		
	1	2	3
Down	5	25	259
	3	69	567
	8	37	130
Across	$5 + 3 + 8 =$	$25 + 69 + 37 =$	$259 + 567 + 130 =$

TABLE 8.2-6

	Number of Digits			
Type of Array	1	2	3	Row Means
Down	23.806	10.694	6.278	13.593
Across	26.056	6.750	3.944	12.250
Column means	24.931	8.722	5.111	

TABLE 8.2-7

Source	Sum of Squares	Degrees of Freedom	Mean Square	F	p-value
Factor A array	48.678	1	48.68	2.885	0.089
Factor B number of digits	8022.73	2	4011.36	237.78	<0.001
Interaction	185.92	2	92.96	5.51	0.006
Error	1720.76	102	16.87		
Total	9978.08	107			

the right column, obtained from the computer printout, provides the p-value of each test, namely the probability of obtaining an F as large as or larger than the calculated F-ratio. Note, for example, that to test for interaction $F = 5.51$ and the p-value is 0.006. Thus the hypothesis of no interaction would be rejected at the $\alpha = 0.05$ or $\alpha = 0.01$ significance level but not with $\alpha = 0.001$.

Exercises

8.2-1   For the data given in Example 8.2-1, test the hypothesis H_A: $\alpha_1 = \alpha_2 = \alpha_3 = 0$ against all alternatives at the 5% significance level.

8.2-2   With $a = 3$ and $b = 4$, find μ, α_i, and β_j if μ_{ij}, $i = 1, 2, 3$ and $j = 1, 2, 3, 4$, are given by

$$\begin{array}{cccc} 6 & 3 & 7 & 8 \\ 10 & 7 & 11 & 12 \\ 8 & 5 & 9 & 10 \end{array}$$

Note that in an "additive" model such as this one, one row (column) can be determined by adding a constant value to each of the elements of another row (column).

8.2-3   We wish to compare compressive strengths of concrete corresponding to $a = 3$ different drying methods (treatments). Concrete is mixed in batches that are just large enough to produce three cylinders. Although care is taken to achieve uniformity, we expect some variability among the $b = 5$ batches used to obtain the following compressive strengths. (There is little reason to suspect interaction and hence only one observation is taken in each cell.)

			Batch		
Treatment	B_1	B_2	B_3	B_4	B_5
A_1	52	47	44	51	42
A_2	60	55	49	52	43
A_3	56	48	45	44	38

(a) Use the 5% significance level and test H_A: $\alpha_1 = \alpha_2 = \alpha_3 = 0$ against all alternatives.

(b) Use the 5% significance level and test H_B: $\beta_1 = \beta_2 = \beta_3 = \beta_4 = \beta_5 = 0$ against all alternatives. [See R. V. Hogg and J. Ledolter, *Applied Statistics for Engineers and Physical Scientists*, 2nd ed., (New York: Macmillan, 1992).]

8.2-4   Show that the cross-product terms formed from $\overline{X}_{i.} - \overline{X}_{..}$, $\overline{X}_{.j} - \overline{X}_{..}$, and $X_{ij} - \overline{X}_{i.} - \overline{X}_{.j} + \overline{X}_{..}$ sum to zero, $i = 1, 2, \ldots, a$ and $j = 1, 2, \ldots, b$. HINT: For example, write

$$\sum_{i=1}^{a} \sum_{j=1}^{b} (\overline{X}_{.j} - \overline{X}_{..})(X_{ij} - \overline{X}_{i.} - \overline{X}_{.j} + \overline{X}_{..})$$

$$= \sum_{j=1}^{b} (\overline{X}_{.j} - \overline{X}_{..}) \sum_{i=1}^{a} [(X_{ij} - \overline{X}_{.j}) - (\overline{X}_{i.} - \overline{X}_{..})]$$

and sum each term in the inner summation as grouped here to get zero.

8.2-5   A psychology student was interested in testing how food consumption by rats would be affected by a particular drug. She used two levels of one attribute, namely drug and placebo, and four levels of a second attribute, namely male (M), castrated (C), female (F), and ovariectomized (O). For each cell she observed five rats. The amount of food consumed in grams per 24 hours is listed in Table 8.2-8. Test the hypotheses using a 5% significance level for each.

TABLE 8.2-8

	M	C	F	O
Drug	22.56	16.54	18.58	18.20
	25.02	24.64	15.44	14.56
	23.66	24.62	16.12	15.54
	17.22	19.06	16.88	16.82
	22.58	20.12	17.58	14.56
Placebo	25.64	22.50	17.82	19.74
	28.84	24.48	15.76	17.48
	26.00	25.52	12.96	16.46
	26.02	24.76	15.00	16.44
	23.24	20.62	19.54	15.70

(a) H_{AB}: $\gamma_{ij} = 0$, $\quad i = 1, 2,$ $\quad j = 1, 2, 3, 4$.

(b) H_A: $\alpha_1 = \alpha_2 = 0$.

(c) H_B: $\beta_1 = \beta_2 = \beta_3 = \beta_4 = 0$.

(d) How could you modify this model so that there are three attributes of classification, each with two levels?

8.2-6 With $a = 3$ and $b = 4$, find μ, α_i, β_j, and γ_{ij}, if μ_{ij}, $i = 1, 2, 3$ and $j = 1, 2, 3, 4$, are given by

$$
\begin{array}{cccc}
6 & 7 & 7 & 12 \\
10 & 3 & 11 & 8 \\
8 & 5 & 9 & 10
\end{array}
$$

Note the difference between the layout here and that in Exercise 8.2-2. Does the interaction help explain these differences?

8.2-7 In order to test whether four brands of gasoline give equal performance in terms of mileage, each of three cars was driven with each of the four brands of gasoline. Then each of the $(3)(4) = 12$ possible combinations was repeated four times. The number of miles per gallon for each of the four repetitions in each cell is recorded in Table 8.2-9. Test each of the hypotheses H_{AB}, H_A, H_B at the 5% significance level.

8.2-8 The data in Table 8.2-10 could represent the outcomes of some experiment, but in actuality they were generated on a computer. Recall that $\mu_{ij} = \mu + \alpha_i + \beta_j + \gamma_{ij}$ is the mean of cell (i, j) and in the simulation, we took $\mu = 15$, $\alpha_1 = -3$, $\alpha_2 = 2$, $\alpha_3 = 1$, $\beta_1 = -1.5$, $\beta_2 = 1.5$, $\gamma_{11} = 4$, $\gamma_{12} = -4$, $\gamma_{21} = -2$, $\gamma_{22} = 2$, $\gamma_{31} = -2$, $\gamma_{32} = 2$, and $\sigma^2 = 9$. Use the data and the two-factor analysis of variance model with four repetitions per cell to test the hypotheses H_{AB}: no interaction, H_A: no row effect, and H_B: no column effect, each at the $\alpha = 0.05$ significance level.

TABLE 8.2-9

Car		1		2		3		4
	Brand of Gasoline							
1	21.0	14.9	16.3	20.0	15.8	19.4	17.8	17.3
	16.2	18.8	15.2	21.6	14.5	14.8	18.2	20.4
2	20.6	19.5	15.5	16.8	16.6	13.7	18.1	17.1
	20.8	18.9	17.4	19.4	18.2	16.1	21.5	19.1
3	14.2	13.1	17.4	18.1	15.2	16.7	16.3	16.4
	16.8	17.4	16.4	16.9	17.7	18.1	17.9	18.8

TABLE 8.2-10

14.552	13.980
17.024	5.777
10.132	7.638
16.979	12.089
11.378	20.825
12.578	17.333
17.709	20.329
11.037	25.277
12.345	22.223
11.044	21.844
9.458	16.176
16.156	17.946

8.2-9 Hogg and Ledolter report on an engineer in a textile mill who studies the effects of temperature and time in a process involving dye on the brightness of a synthetic fabric (brightness is measured on a 50-point scale). Three observations were taken at each combination of temperature and time.

	Temperature		
Time (cycles)	350°F	375°F	400°F
40	38, 32, 30	37, 35, 40	36, 39, 43
50	40, 45, 36	39, 42, 46	39, 48, 47

Construct the ANOVA table and conduct tests for the interaction first and then, if appropriate, the mean effects. Use $\alpha = 0.05$ for each test. [R. V.

Hogg and J. Ledolter, *Applied Statistics for Engineers and Physical Scientists,* 2nd ed. (New York: Macmillan, 1992).]

8.2-10 There is another way of looking at Exercise 8.1-8, namely as a two-factor analysis of variance problem with the levels of sex being female and male, the levels of age being less than 50 and at least 50, and the measurement for each subject being their cholesterol level. The data would then be set up as follows:

	Age	
Sex	<50	≥50
Female	221	262
	213	193
	202	224
	183	201
	185	161
	197	178
	162	265
Male	271	192
	192	253
	189	248
	209	278
	227	232
	236	267
	142	289

(a) Test H_{AB}: $\gamma_{ij} = 0$, $i = 1, 2$; $j = 1, 2$ (no interaction).
(b) Test H_A: $\alpha_1 = \alpha_2 = 0$ (no row effect).
(c) Test H_B: $\beta_1 = \beta_2 = 0$ (no column effect).
Use a 5% significance level for each test.

★8.3 General Factorial and 2^k Factorial Designs

In Section 8.2 we studied two-factor experiments in which the A factor is performed at a levels and the B factor has b levels. Without replications, we need ab level combinations and, with c replications with each of these combinations, we need a total of abc experiments.

Let us consider a situation with three factors, say A, B, and C with a, b, and c levels, respectively. Here there are a total of abc level combinations; and if at each of these combinations we have d replications, there is need for $abcd$ experiments. Once these experiments are run, in some random order, and the data collected, there are computer programs available to calculate the entries in the ANOVA table, as in Table 8.3-1.

The main effects (A, B, and C) and the two-factor interactions (AB, AC, and BC) have the same interpretations as in the two-factor ANOVA. The three-factor interaction represents that part of the model for the means μ_{ijh}; $i = 1, 2, \ldots, a$; $j = 1, 2, \ldots, b$; $h = 1, 2, \ldots, c$ that cannot be explained by a model including only the main effects and two-factor interactions. In particular, if for each fixed h, the "plane" created by μ_{ijh} is "parallel" to the "plane" created by every other fixed h, then the three-factor interaction is equal to zero. Usually, higher-order interactions tend to be small.

In the testing sequence, we test the three-factor interaction first by checking to see whether or not

$$MS(ABC)/MS(E) \geq F_\alpha[(a-1)(b-1)(c-1), \, abc(d-1)].$$

If this inequality holds, the ABC interaction is significant at the α level. We would then not continue testing the two-factor interactions and the main effects with those F values, but analyze the data otherwise. For example, we could, for each fixed h, look at a two-factor ANOVA for factors A and B. Of course, if the inequality does not hold, we next check the two-factor interactions with the appropriate F values. If these are not significant, we check the main effects, A, B, and C.

Factorial experiments with three or more factors require many experiments, particularly if each factor has several levels. Often, in the health, social, and physical sciences, experimenters want to consider several factors (maybe as many as 10 or 20 or hundreds), and they cannot afford to run that many experiments. This is particularly true with preliminary or screening investigations in which they want to detect the factors that seem most important. In these cases they often consider factorial experiments such that each of k factors is run only at two levels, frequently without replication. We consider only this situation, although the reader should recognize that there are many variations of this. In particular, there are methods for investigating only *fractions of these 2^k designs*. The reader interested in more infor-

TABLE 8.3-1

Source	SS	d.f.	MS	F
A	SS(A)	$a - 1$	MS(A)	MS(A)/MS(E)
B	SS(B)	$b - 1$	MS(B)	MS(B)/MS(E)
C	SS(C)	$c - 1$	MS(C)	MS(C)/MS(E)
AB	SS(AB)	$(a - 1)(b - 1)$	MS(AB)	MS(AB)/MS(E)
AC	SS(AC)	$(a - 1)(c - 1)$	MS(AC)	MS(AC)/MS(E)
BC	SS(BC)	$(b - 1)(c - 1)$	MS(BC)	MS(BC)/MS(E)
ABC	SS(ABC)	$(a - 1)(b - 1)(c - 1)$	MS(ABC)	MS(ABC)/MS(E)
Error	SS(E)	$abc(d - 1)$	MS(E)	
Total	SS(TO)	$abcd - 1$		

mation should refer to a good book on design of experiments, such as that by Box, Hunter, and Hunter. Many statisticians in industry believe these statistical methods are the most useful in improving product and process designs. Hence it is clearly an extremely important topic as many industries are greatly concerned about the quality of their products.

In factorial experiments in which each of the k factors is considered at only two levels, those levels are selected at some reasonable low and high values. That is, with the help of someone in the field, the typical range of each factor is considered. For illustration, if we are considering baking temperatures in the range of 300° to 375°, a representative low is selected, say 320°, and a representative high is selected, say 355°. There is no formula for these selections, and someone familiar with the experiment would help make these selections. Often it happens that only two different types of material are considered, say fabric, and one is called low and the other high.

Thus we select a low and high for each factor; these will be coded as -1 and $+1$ or, more simply, as $-$ and $+$, respectively. We give three 2^k designs, for $k = 2, 3,$ and 4, in standard order in Tables 8.3-2, 8.3-3, and 8.3-4, respectively. From these

TABLE 8.3-2

2^2 Design

Run	A	B	Observation
1	$-$	$-$	X_1
2	$+$	$-$	X_2
3	$-$	$+$	X_3
4	$+$	$+$	X_4

TABLE 8.3-3

2^3 Design

Run	A	B	C	Observation
1	$-$	$-$	$-$	X_1
2	$+$	$-$	$-$	X_2
3	$-$	$+$	$-$	X_3
4	$+$	$+$	$-$	X_4
5	$-$	$-$	$+$	X_5
6	$+$	$-$	$+$	X_6
7	$-$	$+$	$+$	X_7
8	$+$	$+$	$+$	X_8

TABLE 8.3-4

	2^4 Design				
Run	A	B	C	D	Observation
1	−	−	−	−	X_1
2	+	−	−	−	X_2
3	−	+	−	−	X_3
4	+	+	−	−	X_4
5	−	−	+	−	X_5
6	+	−	+	−	X_6
7	−	+	+	−	X_7
8	+	+	+	−	X_8
9	−	−	−	+	X_9
10	+	−	−	+	X_{10}
11	−	+	−	+	X_{11}
12	+	+	−	+	X_{12}
13	−	−	+	+	X_{13}
14	+	−	+	+	X_{14}
15	−	+	+	+	X_{15}
16	+	+	+	+	X_{16}

three tables, we can easily note what is meant by standard order. The A column starts with a minus sign and then the sign alternates. The B column begins with two minus signs, then the signs alternate in blocks of two. The C column has 4 minus signs and then 4 plus signs and so on. The D column starts with 8 minus signs and then 8 plus signs. It is easy to extend this idea to 2^k designs, where $k \geq 5$. For illustration, under the E column in a 2^5 design, we have 16 minus signs followed by 16 plus signs, which account for the 32 experiments.

To be absolutely certain what these runs mean, consider run number 12 in Table 8.3-4: A is set at its high level, B at its high, C at its low, and D at its high level. The value X_{12} is the random observation resulting from this one combination of these four settings. It must be emphasized, however, that the runs are not necessarily performed in the order 1, 2, 3, . . . , 2^k; but they should be performed in a random order, if at all possible. That is, in a 2^3 design, we might perform the experiment in the order: 3, 2, 8, 6, 5, 1, 4, 7 if this, in fact, was a random selection of a permutation of the first eight positive integers.

Once all 2^k experiments have been run, it is possible to consider the total sum of squares

$$\sum_{i=1}^{2^k} (X_i - \overline{X})^2$$

Table 8.3-5

| | 2³ Design | | | | | | | |
Run	A	B	C	AB	AC	BC	ABC	Observations
1	−	−	−	+	+	+	−	X_1
2	+	−	−	−	−	+	+	X_2
3	−	+	−	−	+	−	+	X_3
4	+	+	−	+	−	−	−	X_4
5	−	−	+	+	−	−	+	X_5
6	+	−	+	−	+	−	−	X_6
7	−	+	+	−	−	+	−	X_7
8	+	+	+	+	+	+	+	X_8

and decompose it very easily into $2^k - 1$ parts, which represent the respective measurements (estimators) of the k main effects, $\binom{k}{2}$ two-factor interactions, $\binom{k}{3}$ three-factor interactions, and so on until we have the one k-factor interaction. We illustrate this decomposition with the 2^3 design in Table 8.3-5. Note the column AB is found by formally multiplying the elements of column A by the corresponding ones in B. Likewise, AC is found by multiplying column A and column C, and so on until column ABC is the product of the corresponding elements of columns A, B, C. We then construct seven linear forms using these seven columns of signs with the corresponding observations. The resulting measures (estimates) of the main effects (A, B, C), the two-factor interactions (AB, AC, BC), and the three-factor interaction (ABC) are then found by dividing these linear forms by $2^k = 2^3 = 8$ (some statisticians divide by $2^{k-1} = 2^{3-1} = 4$). These are denoted by

$$[A] = (-X_1 + X_2 - X_3 + X_4 - X_5 + X_6 - X_7 + X_8)/8,$$
$$[B] = (-X_1 - X_2 + X_3 + X_4 - X_5 - X_6 + X_7 + X_8)/8,$$
$$[C] = (-X_1 - X_2 - X_3 - X_4 + X_5 + X_6 + X_7 + X_8)/8,$$
$$[AB] = (+X_1 - X_2 - X_3 + X_4 + X_5 - X_6 - X_7 + X_8)/8,$$
$$[AC] = (+X_1 - X_2 + X_3 - X_4 - X_5 + X_6 - X_7 + X_8)/8,$$
$$[BC] = (+X_1 + X_2 - X_3 - X_4 - X_5 - X_6 + X_7 + X_8)/8,$$
$$[ABC] = (-X_1 + X_2 + X_3 - X_4 + X_5 - X_6 - X_7 + X_8)/8.$$

With assumptions of normality, mutual independence, and common variance σ^2, under the overall null hypothesis of the equality of all the means, each of these measures has a normal distribution with mean zero and variance $\sigma^2/8$ (in general $\sigma^2/2^k$). This implies that the square of each measure divided by $\sigma^2/8$ is $\chi^2(1)$. Moreover, it can be shown (see Exercise 8.3-2) that

$$\sum_{i=1}^{8} (X_i - \overline{X})^2$$

$$= 8([A]^2 + [B]^2 + [C]^2 + [AB]^2 + [AC]^2 + [BC]^2 + [ABC]^2).$$

So by Theorem 8.1-1, the terms on the righthand side, divided by σ^2, are mutually independent random variables, each being $\chi^2(1)$. While it requires a little more theory, it follows that the linear forms [A], [B], [C], [AB], [AC], [BC], and [ABC] are mutually independent $N(0, \sigma^2/8)$ random variables.

Since we have assumed that we have not run any replications, how can we obtain an estimate of σ^2 to see if any of the main effects or interactions are significant? To help us, we fall back on the use of a q–q plot because, under the overall null hypothesis, those seven measures are mutually independently normally distributed with the same mean and variance. Thus a q–q plot of the normal percentiles against the corresponding ordered values of the measures should be about on a straight line if, in fact, the null hypothesis is true. If one of these points is "out of line," we might believe that the overall null hypothesis is not true and that the effect associated with it is significant. It is possible that two or three points might be out of line, then all corresponding effects (main or interaction) should be investigated. Clearly, this is not a formal test, but it has been extremely successful in practice.

Example 8.3-1 As an illustration, we use the data from an experiment designed to evaluate the effects of laundering on a certain fire-retardant treatment for fabrics. These data, somewhat modified were taken from *Experimental Statistics, National Bureau of Standards Handbook 91* by Mary G. Natrella (Washington, D.C.: U.S. Government Printing Office, 1963). Factor A is the type of fabric (sateen or monk's cloth), factor B corresponds to two different fire-retardant treatments, and factor C describes the laundering conditions (no laundering, after one laundering). The observation are inches burned, measured on a standard size fabric after a flame test. They are, in standard order,

$$x_1 = 41, \quad x_2 = 30.5, \quad x_3 = 47.5, \quad x_4 = 27,$$
$$x_5 = 39.5, \quad x_6 = 26.5, \quad x_7 = 48, \quad x_8 = 27.5.$$

Thus the measures of the effects are

$$[A] = (-41 + 30.5 - 47.5 + 27 - 39.5 + 26.5 - 48 + 27.5)/8 = -8.06,$$
$$[B] = (-41 - 30.5 + 47.5 + 27 - 39.5 - 26.5 + 48 + 27.5)/8 = 1.56,$$
$$[C] = (-41 - 30.5 - 47.5 - 27 + 39.5 + 26.5 + 48 + 27.5)/8 = 0.56,$$
$$[AB] = (+41 - 30.5 - 47.5 + 27 + 39.5 - 26.5 - 48 + 27.5)/8 = -2.19,$$
$$[AC] = (+41 - 30.5 + 47.5 - 27 - 39.5 + 26.5 - 48 + 27.5)/8 = -0.31,$$
$$[BC] = (+41 + 30.5 - 47.5 - 27 - 39.5 - 26.5 + 48 + 27.5)/8 = 0.81,$$
$$[ABC] = (-41 + 30.5 + 47.5 - 27 + 39.5 - 26.5 - 48 + 27.5)/8 = 0.31.$$

TABLE 8.3-6

Identity of Effect	Ordered Effect	Percentile	Percentile from $N(0, 1)$
[A]	−8.06	12.5	−1.15
[AB]	−2.19	25.0	−0.67
[AC]	−0.31	37.5	−0.32
[ABC]	0.31	50.0	0.00
[C]	0.56	62.5	0.32
[BC]	0.81	75.0	0.67
[B]	1.56	87.5	1.15

In Table 8.3-6 we order these seven measures, determine their percentiles, and find the corresponding percentiles of the standard normal distribution.

The q–q plot is given in Figure 8.3-1. Each point has been identified with its effect. A straight line fits six of those points reasonably well, but the point associated with [A] = −8.06 is far from this straight line. Hence the main effect of factor A (the type of fabric) seems to be significant. It is interesting to note that the laundering factor, C, does not seem to be a significant factor.

Exercises

8.3-1 Write out a 2^2 design, displaying the A, B, and AB columns for the four runs.

(a) If X_1, X_2, X_3, X_4 are the four observations for the respective runs in standard order, write out the three linear forms, [A], [B], and [AB], that measure the two main effects and the interaction. These linear forms should include the divisor $2^2 = 4$.

(b) Show that $\sum_{i=1}^{4} (X_i - \overline{X})^2 = 4([A]^2 + [B]^2 + [AB]^2)$.

(c) Under the null hypothesis that all the means are equal and with the usual assumptions (normality, mutually independent, and common variance), what can you say about the distributions of the expressions in (b) after each is divided by σ^2?

8.3-2 Show that in a 2^3 design, it is true that

$$\sum_{i=1}^{8} (X_i - \overline{X})^2 = 8([A]^2 + [B]^2 + [C]^2 + [AB]^2$$
$$+ [AC]^2 + [BC]^2 + [ABC]^2).$$

HINT: Since both the right and the left members of this equation are symmetric in the variables X_1, X_2, . . . , X_8, it is only necessary to show that the corresponding coefficients of $X_1 X_i$, $i = 1, 2, \ldots , 8$, are the same in each member of the equation. Of course, recall $\overline{X} = (X_1 + X_2 + \cdots + X_8)/8$.

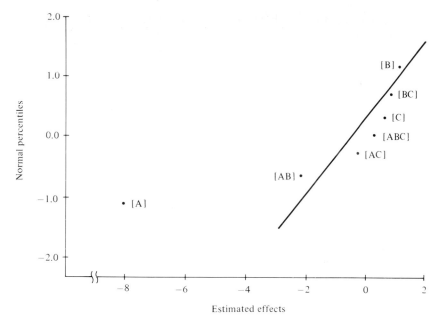

FIGURE 8.3-1

8.3-3 Show that the unbiased estimator of the variance σ^2 from a sample of size $n = 2$ is one-half of the square of the difference of the two observations. Thus, if a 2^k design is replicated, say with X_{i1} and X_{i2}, $i = 1, 2, \ldots, 2^k$, show that the estimate of the common σ^2 is

$$\frac{1}{2^{k+1}} \sum_{i=1}^{2^k} (X_{i1} - X_{i2})^2 = MS(E).$$

Under the usual assumptions, this implies that each of $2^k[A]^2/MS(E)$, $2^k[B]^2/MS(E)$, $2^k[AB]^2/MS(E)$, and so on has an $F(1, 2^k)$ distribution under the null hypothesis. This, of course, would provide tests for the significance of the various effects, including interactions.

8.3-4 Hogg and Ledolter note that percent yields from a certain chemical reaction for changing temperature (factor A), reaction time (factor B), and concentration (factor C) are $x_1 = 79.7$, $x_2 = 74.3$, $x_3 = 76.7$, $x_4 = 70.0$, $x_5 = 84.0$, $x_6 = 81.3$, $x_7 = 87.3$, $x_8 = 73.7$ in standard order with a 2^3 design.

(a) Estimate the main effects, the three two-factor interactions, and the three-factor interaction.

(b) Construct an appropriate q–q plot to see if any of these effects seem to be significantly larger than the others. [R. V. Hogg and J. Ledolter, *Applied Statistics for Engineers and Physical Scientists*, 2nd ed. (New York: Macmillan, 1992).]

8.3-5   Box, Hunter, and Hunter studied the effects of catalyst charge (10 pounds $=$ -1, 20 pounds $= +1$), temperature ($220°C = -1$, $240°C = +1$), pressure (50 psi $= -1$, 80 psi $= +1$), and concentration ($10\% = -1$, $12\% = +1$) on percent conversion (X) of a certain chemical. The results of a 2^4 design, in standard order, are

$$x_1 = 71, \; x_2 = 61, \; x_3 = 90, \; x_4 = 82, \; x_5 = 68, \; x_6 = 61,$$
$$x_7 = 87, \; x_8 = 80, \; x_9 = 61, \; x_{10} = 50, \; x_{11} = 89,$$
$$x_{12} = 83, \; x_{13} = 59, \; x_{14} = 51, \; x_{15} = 85, \; x_{16} = 78.$$

(a) Estimate the main effects and the two-, three-, and four-factor interactions.

(b) Construct an appropriate q–q plot and assess the significance of the various effects.

8.4   A Simple Regression Problem

There is often interest in the relation between two variables, for example, a student's scholastic aptitude test score in mathematics and this same student's grade in calculus. Frequently, one of these variables, say x, is known in advance of the other, and hence there is interest in predicting a future random variable Y. Since Y is a random variable, we cannot predict its future observed value $Y = y$ with certainty. Thus let us first concentrate on the problem of estimating the mean of Y, that is, $E(Y)$. Now $E(Y)$ is usually a function of x; for example, in our illustration with the calculus grade, say Y, we would expect $E(Y)$ to increase with increasing mathematics aptitude score x. Sometimes $E(Y) = \mu(x)$ is assumed to be of a given form, such as linear or quadratic or exponential; that is, $\mu(x)$ could be assumed to be equal to $\alpha + \beta x$ or $\alpha + \beta x + \gamma x^2$ or $\alpha e^{\beta x}$. To estimate $E(Y) = \mu(x)$, or equivalently the parameters α, β, and γ, we observe the random variable Y for each of n different values of x, say $x_1, x_2, \ldots, x_n$. Once the n independent experiments have been performed, we have n pairs of known numbers $(x_1, y_1), (x_2, y_2), \ldots, (x_n, y_n)$. These pairs are then used to estimate the mean $E(Y)$. Problems like this are often classified under **regression** because $E(Y) = \mu(x)$ is frequently called a regression curve.

REMARK   A model for the mean like $\alpha + \beta x + \gamma x^2$, is called a **linear model** because it is linear in the parameters, α, β, and γ. Thus $\alpha e^{\beta x}$ is not a linear model because it is not linear in α and β. Note, in Sections 8.1, 8.2, and 8.3, all the means were linear in the parameters and hence linear models. That is the reason for the title of this chapter.

Let us begin with the case in which $E(Y) = \mu(x)$ is a linear function. The data points are $(x_1, y_1), (x_2, y_2), \ldots, (x_n, y_n)$; so the first problem is that of fitting a straight line to the set of data (see Figure 8.4-1). Recall that we considered a similar problem in Section 1.6. There we used the method of least squares to find the

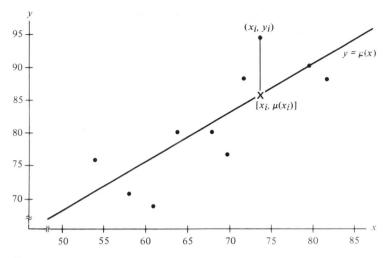

FIGURE 8.4-1

best-fitting line. In this chapter we will not only find the line of best fit, but will also make some statistical inferences about the unknown parameters.

In addition to assuming that the mean of Y is a linear function, we assume that for a particular value of x, the value of Y will differ from its mean by a random amount ε. We further assume that the distribution of ε is $N(0, \sigma^2)$. So we have for our linear model

$$Y_i = \alpha_1 + \beta x_i + \varepsilon_i,$$

where ε_i, for $i = 1, 2, \ldots, n$, are independent and $N(0, \sigma^2)$.

We shall now find point estimates for α_1, β, and σ^2. For convenience we let $\alpha_1 = \alpha - \beta \bar{x}$, so that

$$Y_i = \alpha + \beta(x_i - \bar{x}) + \varepsilon_i, \qquad \text{where } \bar{x} = \frac{1}{n} \sum_{i=1}^{n} x_i.$$

Then Y_i is equal to a constant, $\alpha + \beta(x_i - \bar{x})$, plus a normal random variable ε_i. Hence $Y_1, Y_2, \ldots, Y_n$ are mutually independent normal variables with respective means $\alpha + \beta(x_i - \bar{x})$, $i = 1, 2, \ldots, n$, and unknown variance σ^2. Their joint p.d.f. is therefore the product of the individual probability density functions; that is, the likelihood function equals

$$L(\alpha, \beta, \sigma^2) = \prod_{i=1}^{n} \frac{1}{\sqrt{2\pi\sigma^2}} \exp\left\{ -\frac{[y_i - \alpha - \beta(x_i - \bar{x})]^2}{2\sigma^2} \right\}$$

$$= \left(\frac{1}{2\pi\sigma^2} \right)^{n/2} \exp\left\{ -\frac{\sum_{i=1}^{n} [y_i - \alpha - \beta(x_i - \bar{x})]^2}{2\sigma^2} \right\}.$$

To maximize $L(\alpha, \beta, \sigma^2)$, or, equivalently, to minimize

$$-\ln L(\alpha, \beta, \sigma^2) = \frac{n}{2} \ln (2\pi\sigma^2) + \frac{\sum_{i=1}^{n} [y_i - \alpha - \beta(x_i - \bar{x})]^2}{2\sigma^2},$$

we must select α and β to minimize

$$H(\alpha, \beta) = \sum_{i=1}^{n} [y_i - \alpha - \beta(x_i - \bar{x})]^2.$$

Since $|y_i - \alpha - \beta(x_i - \bar{x})| = |y_i - \mu(x_i)|$ is the vertical distance from the point (x_i, y_i) to the line $y = \mu(x)$, we note that $H(\alpha, \beta)$ represents the sum of the squares of those distances. Thus selecting α and β so that the sum of the squares is minimized means that we are fitting the straight line to the data by the **method of least squares**.

To minimize $H(\alpha, \beta)$, we find the two first partial derivatives

$$\frac{\partial H(\alpha, \beta)}{\partial \alpha} = 2 \sum_{i=1}^{n} [y_i - \alpha - \beta(x_i - \bar{x})](-1)$$

and

$$\frac{\partial H(\alpha, \beta)}{\partial \beta} = 2 \sum_{i=1}^{n} [y_i - \alpha - \beta(x_i - \bar{x})][-(x_i - \bar{x})].$$

Setting $\partial H(\alpha, \beta)/\partial\alpha = 0$, we obtain

$$\sum_{i=1}^{n} y_i - n\alpha - \beta \sum_{i=1}^{n} (x_i - \bar{x}) = 0.$$

Since

$$\sum_{i=1}^{n} (x_i - \bar{x}) = 0,$$

we have that

$$\sum_{i=1}^{n} y_i - n\alpha = 0$$

and thus

$$\hat{\alpha} = \bar{Y}.$$

The equation $\partial H(\alpha, \beta)/\partial\beta = 0$ yields, with α replaced by $\bar{y}$,

$$\sum_{i=1}^{n} (y_i - \bar{y})(x_i - \bar{x}) - \beta \sum_{i=1}^{n} (x_i - \bar{x})^2 = 0$$

or, equivalently,

$$\widehat{\beta} = \frac{\sum_{i=1}^{n} (Y_i - \bar{Y})(x_i - \bar{x})}{\sum_{i=1}^{n} (x_i - \bar{x})^2} = \frac{\sum_{i=1}^{n} Y_i(x_i - \bar{x})}{\sum_{i=1}^{n} (x_i - \bar{x})^2}.$$

Thus, to find the mean line of best fit, $\mu(x) = \alpha + \beta(x - \bar{x})$, we use

$$\widehat{\alpha} = \bar{y} \tag{8.4-1}$$

and

$$\widehat{\beta} = \frac{\sum_{i=1}^{n} y_i(x_i - \bar{x})}{\sum_{i=1}^{n} (x_i - \bar{x})^2} = \frac{\sum_{i=1}^{n} x_i y_i - \left(\sum_{i=1}^{n} x_i\right)\left(\sum_{i=1}^{n} y_i\right)\Big/ n}{\sum_{i=1}^{n} x_i^2 - \left(\sum_{i=1}^{n} x_i\right)^2\Big/ n}. \tag{8.4-2}$$

It is also true that $\widehat{\beta} = r s_y/s_x$ (see Sections 1.6 and 9.4).

To find the maximum likelihood estimator of σ^2, consider the partial derivative

$$\frac{\partial[-\ln L(\alpha, \beta, \sigma^2)]}{\partial(\sigma^2)} = \frac{n}{2\sigma^2} - \frac{\sum_{i=1}^{n} [y_i - \alpha - \beta(x_i - \bar{x})]^2}{2(\sigma^2)^2}.$$

Setting this equal to zero and replacing α and β by their solutions $\widehat{\alpha}$ and $\widehat{\beta}$, we obtain

$$\widehat{\sigma^2} = \frac{1}{n} \sum_{i=1}^{n} [Y_i - \widehat{\alpha} - \widehat{\beta}(x_i - \bar{x})]^2. \tag{8.4-3}$$

A formula for $n\widehat{\sigma^2}$ useful in its calculation is

$$n\widehat{\sigma^2} = \sum_{i=1}^{n} y_i^2 - \left(\sum_{i=1}^{n} y_i\right)^2\Big/ n - \widehat{\beta}\left(\sum_{i=1}^{n} x_i y_i\right) + \widehat{\beta}\left(\sum_{i=1}^{n} x_i\right)\left(\sum_{i=1}^{n} y_i\right)\Big/ n. \tag{8.4-4}$$

Note that the summand in the expression (8.4-3) for $\widehat{\sigma^2}$ is the difference between the value of Y_i and the predicted mean of Y_i. Let $\widehat{Y}_i = \widehat{\alpha} + \widehat{\beta}(x_i - \bar{x})$, the predicted mean value of Y_i. The difference

$$Y_i - \widehat{Y}_i = Y_i - \widehat{\alpha} - \widehat{\beta}(x_i - \bar{x})$$

is called the ith **residual**, $i = 1, 2, \ldots, n$. The maximum likelihood estimate of σ^2 is then the sum of the squares of the residuals divided by n. It should always be true that the sum of the residuals is equal to zero. However, in practice, due to round off, the sum of the observed residuals, $y_i - \widehat{y}_i$, sometimes differs slightly from zero.

Example 8.4-1   The data plotted in Figure 8.4-1 are 10 pairs of test scores
of 10 students in a psychology class, x being the score on a preliminary test
and y the score on the final examination. The values of x and y are shown in
Table 8.4-1. The sums that are needed to calculate estimates of the parameters
are also given. Of course, the estimates of α and β have to be found before the
residuals can be calculated. (The sum of the squares of the y values is included
for use in a later section.)

Thus $\hat{\alpha} = 813/10 = 81.3$,

$$\hat{\beta} = \frac{56{,}089 - (683)(813)/10}{47{,}405 - (683)(683)/10} = \frac{561.1}{756.1} = 0.742.$$

Hence the least squares regression line is

$$\hat{y} = 81.3 + (0.742)(x - 68.3).$$

The maximum likelihood estimate of σ^2 is

$$\widehat{\sigma^2} = \frac{217.709038}{10} = 21.7709.$$

We shall now consider the problem of finding the distributions of $\hat{\alpha}$, $\hat{\beta}$, and $\widehat{\sigma^2}$ (or
distributions of functions of these estimators). We would like to be able to say
something about the error of the estimates, find confidence intervals for the parameters, and in the next section, test some hypotheses and predict future values of Y.

During the preceding discussion we have treated $x_1, x_2, \ldots, x_n$ as constants. Of
course, many times they can be set by the experimenter; for example, a chemist in
experimentation might produce a compound at many different temperatures. But it
is also true that these numbers might be observations on an earlier random variable,

TABLE 8.4-1

x	y	x^2	xy	y^2	$\hat{y}$	$y - \hat{y}$	$(y - \hat{y})^2$
70	77	4,900	5,390	5,929	82.561566	−5.561566	30.931016
74	94	5,476	6,956	8,836	85.529956	8.470044	71.741645
72	88	5,184	6,336	7,744	84.045761	3.954239	15.636006
68	80	4,624	5,440	6,400	81.077371	−1.077371	1.160728
58	71	3,364	4,118	5,041	73.656395	−2.656395	7.056434
54	76	2,916	4,104	5,776	70.688004	5.311996	28.217302
82	88	6,724	7,216	7,744	91.466737	−3.466737	12.018265
64	80	4,096	5,120	6,400	78.108980	1.891020	3.575957
80	90	6,400	7,200	8,100	89.982542	0.017458	0.000305
61	69	3,721	4,209	4,761	75.882687	−6.882687	47.371380
683	813	47,405	56,089	66,731		0.000001	217.709038

such as an SAT score or preliminary test grade (as in Example 8.4-1), but we consider the problem on the *condition* that these x values are given in either case. Thus, in finding the distributions of $\hat{\alpha}$, $\hat{\beta}$, and $\widehat{\sigma^2}$, the only random variables are Y_1, Y_2, . . . , Y_n. More will be said about the other situation in Section 9.4.

Since $\hat{\alpha}$ is a linear function of independent and normally distributed random variables, $\hat{\alpha}$ has a normal distribution with mean

$$E(\hat{\alpha}) = E\left(\frac{1}{n} \sum_{i=1}^{n} Y_i\right) = \frac{1}{n} \sum_{i=1}^{n} E(Y_i)$$

$$= \frac{1}{n} \sum_{i=1}^{n} [\alpha + \beta(x_i - \bar{x})] = \alpha,$$

and variance

$$\text{Var}(\hat{\alpha}) = \sum_{i=1}^{n} \left(\frac{1}{n}\right)^2 \text{Var}(Y_i) = \frac{\sigma^2}{n}.$$

The estimator $\hat{\beta}$ is also a linear function of $Y_1, Y_2, \ldots, Y_n$ and hence has a normal distribution with mean

$$E(\hat{\beta}) = \frac{\displaystyle\sum_{i=1}^{n} (x_i - \bar{x})E(Y_i)}{\displaystyle\sum_{i=1}^{n} (x_i - \bar{x})^2}$$

$$= \frac{\displaystyle\sum_{i=1}^{n} (x_i - \bar{x})[\alpha + \beta(x_i - \bar{x})]}{\displaystyle\sum_{i=1}^{n} (x_i - \bar{x})^2}$$

$$= \frac{\alpha \displaystyle\sum_{i=1}^{n} (x_i - \bar{x}) + \beta \displaystyle\sum_{i=1}^{n} (x_i - \bar{x})^2}{\displaystyle\sum_{i=1}^{n} (x_i - \bar{x})^2} = \beta$$

and variance

$$\text{Var}(\hat{\beta}) = \sum_{i=1}^{n} \left[\frac{x_i - \bar{x}}{\displaystyle\sum_{i=1}^{n} (x_i - \bar{x})^2}\right]^2 \text{Var}(Y_i)$$

$$= \frac{\displaystyle\sum_{i=1}^{n} (x_i - \bar{x})^2}{\left[\displaystyle\sum_{i=1}^{n} (x_i - \bar{x})^2\right]^2} \sigma^2 = \frac{\sigma^2}{\displaystyle\sum_{i=1}^{n} (x_i - \bar{x})^2}.$$

It can be shown (Exercise 8.4-4) that

$$\sum_{i=1}^{n} [Y_i - \alpha - \beta(x_i - \bar{x})]^2 = \sum_{i=1}^{n} \{(\hat{\alpha} - \alpha) + (\hat{\beta} - \beta)(x_i - \bar{x})$$
$$+ [Y_i - \hat{\alpha} - \hat{\beta}(x_i - \bar{x})]\}^2$$
$$= n(\hat{\alpha} - \alpha)^2 + (\hat{\beta} - \beta)^2 \sum_{i=1}^{n} (x_i - \bar{x})^2$$
$$+ \sum_{i=1}^{n} [Y_i - \hat{\alpha} - \hat{\beta}(x_i - \bar{x})]^2. \qquad (8.4-5)$$

From the fact that Y_i, $\hat{\alpha}$, and $\hat{\beta}$ have normal distributions, we know that each of

$$\frac{[Y_i - \alpha - \beta(x_i - \bar{x})]^2}{\sigma^2}, \qquad \frac{(\hat{\alpha} - \alpha)^2}{\sigma^2/n}, \qquad \frac{(\hat{\beta} - \beta)^2}{\sigma^2 \Big/ \sum_{i=1}^{n} (x_i - \bar{x})^2}$$

has a chi-square distribution with one degree of freedom. Since $Y_1, Y_2, \ldots, Y_n$ are mutually independent, then

$$\frac{\sum_{i=1}^{n} [Y_i - \alpha - \beta(x_i - \bar{x})]^2}{\sigma^2}$$

is $\chi^2(n)$. That is, the left-hand member of equation (8.4-5) divided by σ^2 is $\chi^2(n)$ and is equal to the sum of two $\chi^2(1)$ variables and

$$\frac{\sum_{i=1}^{n} [Y_i - \hat{\alpha} - \hat{\beta}(x_i - \bar{x})]^2}{\sigma^2} = \frac{n\widehat{\sigma^2}}{\sigma^2} \geq 0.$$

Thus we might then guess that $n\widehat{\sigma^2}/\sigma^2$ is $\chi^2(n-2)$. This is true, and, moreover, $\hat{\alpha}$, $\hat{\beta}$, and $\widehat{\sigma^2}$ are mutually independent (see Theorem 8.1-1).

Suppose now that we are interested in forming a confidence interval for β, the slope of the line. We can use the fact that

$$T_1 = \frac{\sqrt{\sum_{i=1}^{n} (x_i - \bar{x})^2}(\hat{\beta} - \beta)/\sigma}{\sqrt{n\widehat{\sigma^2}/[\sigma^2(n-2)]}} = \frac{\hat{\beta} - \beta}{\sqrt{n\widehat{\sigma^2} \Big/ \left[(n-2) \sum_{i=1}^{n} (x_i - \bar{x})^2\right]}}$$

is a Student's t with $n - 2$ degrees of freedom. Thus

$$P\left[-t_{\alpha/2}(n-2) \leq \frac{\hat{\beta} - \beta}{\sqrt{n\widehat{\sigma^2}/(n-2) \sum_{i=1}^{n} (x_i - \bar{x})^2}} \leq t_{\alpha/2}(n-2)\right] = 1 - \alpha.$$

It follows that

$$\left[\hat{\beta} - t_{\alpha/2}(n-2)\sqrt{n\widehat{\sigma^2}/(n-2) \sum_{i=1}^{n} (x_i - \bar{x})^2},\right.$$

$$\left.\hat{\beta} + t_{\alpha/2}(n-2)\sqrt{n\widehat{\sigma^2}/(n-2) \sum_{i=1}^{n} (x_i - \bar{x})^2}\right]$$

is a $100(1 - \alpha)\%$ confidence interval for β.
  Similarly,

$$T_2 = \frac{\sqrt{n}(\hat{\alpha} - \alpha)/\sigma}{\sqrt{n\widehat{\sigma^2}/[\sigma^2(n-2)]}}$$

$$= \frac{\hat{\alpha} - \alpha}{\sqrt{\widehat{\sigma^2}/(n-2)}}$$

has a t distribution with $n - 2$ degrees of freedom. Thus T_2 can be used to make inferences about α (see Exercise 8.4-5). The fact that $n\widehat{\sigma^2}/\sigma^2$ has a chi-square distribution with $n - 2$ degrees of freedom can be used to make inferences about the variance σ^2 (see Exercise 8.4-6).

Exercises

8.4-1 The midterm and final exam scores of 10 students in a statistics course are tabulated below.
  (a) Calculate the least squares regression line for these data.
  (b) Plot the points and the least squares regression line on the same graph.
  (c) Find the value of $\widehat{\sigma^2}$.

Midterm	Final	Midterm	Final
70	87	67	73
74	79	70	83
80	88	64	79
84	98	74	91
80	96	82	94

8.4-2 The final course grade in calculus was predicted on the basis of the student's high school grade point average in mathematics, Scholastic Aptitude Test (SAT) score in mathematics, and score on a mathematics entrance examination. The predicted grades X and the earned grades Y for 10 students are given (2.0 represents a C, 2.3 a C+, 2.7 a B−, etc.).
(a) Calculate the least squares regression line for these data.
(b) Plot the points and the least squares regression line on the same graph.
(c) Find the value of $\widehat{\sigma^2}$.

x	y	x	y
2.0	1.3	2.7	3.0
3.3	3.3	4.0	4.0
3.7	3.3	3.7	3.0
2.0	2.0	3.0	2.7
2.3	1.7	2.3	3.0

8.4-3 Students' scores on the mathematics portion of the ACT examination, X, and on the final examination in first semester calculus (200 points possible), Y, are given.
(a) Calculate the least squares regression line for these data.
(b) Plot the points and the least squares regression line on the same graph.

x	y	x	y
25	138	20	100
20	84	25	143
26	104	26	141
26	112	28	161
28	88	25	124
28	132	31	118
29	90	30	168
32	183		

8.4-4 Show that

$$\sum_{i=1}^{n} [Y_i - \alpha - \beta(x_i - \bar{x})]^2 = n(\widehat{\alpha} - \alpha)^2 + (\widehat{\beta} - \beta)^2 \sum_{i=1}^{n} (x_i - \bar{x})^2$$

$$+ \sum_{i=1}^{n} [Y_i - \widehat{\alpha} - \widehat{\beta}(x_i - \bar{x})]^2.$$

REMARK In Exercises 8.4-5 to 8.4-12, to make the required statistical inferences, we need the distributional assumptions made in Section 8.4.

8.4-5 Show that the endpoints for a $100(1 - \alpha)\%$ confidence interval for α are

$$\hat{\alpha} \pm t_{\alpha/2}(n-2)\sqrt{\widehat{\sigma^2}/(n - 2)}.$$

8.4-6 Show that a $100(1 - \alpha)\%$ confidence interval for σ^2 is

$$\left[\frac{n\widehat{\sigma^2}}{\chi^2_{\alpha/2}(n-2)}, \frac{n\widehat{\sigma^2}}{\chi^2_{1-\alpha/2}(n-2)}\right].$$

8.4-7 Find 95% confidence intervals for α, β, and σ^2 for the data in Exercise 8.4-1.

8.4-8 Find 95% confidence intervals for α, β, and σ^2 for the data in Exercise 8.4-2.

8.4-9 In a "48.1-gram" package of candies, let X equal the number of pieces of candy and let Y equal the total weight of the candies. For each of 5 different values of X, 4 observations of Y are given.

x	y	x	y
54	48.8	56	50.8
54	49.4	56	50.9
54	49.2	57	50.7
54	50.4	57	51.6
55	49.5	57	51.3
55	49.0	57	50.8
55	50.2	58	52.1
55	48.9	58	51.3
56	49.9	58	51.4
56	50.1	58	52.0

(a) Calculate the least squares regression line for these data.
(b) Plot the points and the least squares regression line on the same graph.
(c) Find the value of $\widehat{\sigma^2}$.
(d) Find a 95% confidence interval for β under the usual assumptions.

8.4-10 The Federal Trade Commission measured the number of milligrams of tar and carbon monoxide (CO) per cigarette for all domestic cigarettes. Let x and y equal the measurements of tar and CO, respectively, for 100-millimeter filtered and mentholated cigarettes. A sample of 12 brands yielded the following data:

Brand	x	y
Capri	9	6
Carlton	4	6
Kent	14	14
Kool Milds	12	12
Marlboro Lights	10	12
Merit Ultras	5	7
Now	3	4
Salem	17	18
Triumph	6	8
True	7	8
Vantage	8	13
Virginia Slims	15	13

(a) Calculate the least squares regression line for these data.
(b) Plot the points and the least squares regression line on the same graph.
(c) Find point estimates for α, β, and σ^2.
(d) Find 95% confidence intervals for α, β, and σ^2 under the usual assumptions.

8.4-11   Let X and Y equal the lengths in inches of a foot and a hand, respectively. The following measurements were made on 15 women.

x	y	x	y	x	y
9.00	6.50	10.00	7.00	9.25	7.00
8.50	6.25	9.50	6.50	10.00	7.50
9.25	7.25	9.00	7.00	10.00	7.25
9.75	7.00	9.25	7.00	9.75	7.25
9.00	6.75	9.50	7.00	9.50	7.25

(a) Calculate the least squares regression line for these data.
(b) Plot the points and the least squares regression line on the same graph.
(c) Find point estimates for α, β, and σ^2.
(d) Find 95% confidence intervals for α, β, and σ^2 under the usual assumptions.

8.4-12   Let x and y equal the ACT scores in social science and natural science for a student who is applying for admission to a small liberal arts college. A sample of $n = 15$ such students yielded the following data.

x	y	x	y	x	y
32	28	30	27	26	32
23	25	17	23	16	22
23	24	20	30	21	28
23	32	17	18	24	31
26	31	18	18	30	26

(a) Calculate the least squares regression line for these data.
(b) Plot the points and the least squares regression line on the same graph.
(c) Find point estimates for α, β, and σ^2.
(d) Find 95% confidence intervals for α, β, and σ^2 under the usual assumptions.

8.5   More Regression

In Section 8.4 we considered the estimation of the parameters of a very simple regression curve, namely a straight line. In addition to the point and interval estimates, we can perform tests of hypotheses about these parameters. For illustration, with the same model as that in Section 8.4, we could test the hypothesis $H_0: \beta = \beta_0$ by using a Student's t random variable, namely

$$T_1 = \frac{\hat{\beta} - \beta_0}{\sqrt{n\widehat{\sigma^2} \Big/ \left[(n-2) \sum_{i=1}^{n} (x_i - \bar{x})^2 \right]}}.$$

The null hypothesis along with three possible alternative hypotheses are given in Table 8.5-1.

Often we let $\beta_0 = 0$ and test the hypothesis $H_0: \beta = 0$. That is, we test the null hypothesis that there is no linear relationship.

Example 8.5-1   Let x equal a student's preliminary test score in a psychology course and y the same student's score on the final examination. With

TABLE 8.5-1

H_0	H_1	Critical Region
$\beta = \beta_0$	$\beta > \beta_0$	$t_1 \geq t_\alpha(n-2)$
$\beta = \beta_0$	$\beta < \beta_0$	$t_1 \leq -t_\alpha(n-2)$
$\beta = \beta_0$	$\beta \neq \beta_0$	$|t_1| \geq t_{\alpha/2}(n-2)$

$n = 10$ students, we shall test H_0: $\beta = 0$ against H_1: $\beta \neq 0$. At an $\alpha = 0.01$ significance level, the critical region is $|t_1| \geq t_{0.005}(8) = 3.355$. Using the data in Example 8.4-1, the observed value of T_1 is

$$t_1 = \frac{0.742 - 0}{\sqrt{10(21.7709)/8(756.1)}} = \frac{0.742}{0.1897} = 3.911.$$

Thus we reject H_0.

There is another way of looking at tests of H_0: $\beta = 0$ that gives additional insight into determining how much of the variation in the data is explained by the linear model. This approach is similar to that used in the study of analysis of variance.

We take the total sum of squares and partition it into the variation due to the linear model and an error sum of squares. Within the total sum of squares, SS(TO), we add and subtract the same quantity, recalling that $\hat{\alpha} = \bar{Y}$. We have

$$\text{SS(TO)} = \sum_{i=1}^{n} (Y_i - \bar{Y})^2$$

$$= \sum_{i=1}^{n} [\hat{\beta}(x_i - \bar{x}) + Y_i - \hat{\alpha} - \hat{\beta}(x_i - \bar{x})]^2$$

$$= \sum_{i=1}^{n} \hat{\beta}^2 (x_i - \bar{x})^2 + \sum_{i=1}^{n} [Y_i - \hat{\alpha} - \hat{\beta}(x_i - \bar{x})]^2$$

$$= \text{SS(R)} + \text{SS(E)}. \tag{8.5-1}$$

The cross product term is equal to zero. (See Exercise 8.5-6.) When $\beta = 0$, $Y_i = \alpha + \varepsilon_i$; SS(TO)/$(n - 1)$ is an unbiased estimator of σ^2; and SS(TO)/σ^2 is $\chi^2(n-1)$. Also,

$$\frac{\text{SS(R)}}{\sigma^2} = \frac{\hat{\beta}^2 \sum_{i=1}^{n} (x_i - \bar{x})^2}{\sigma^2}$$

$$= \left[\frac{\hat{\beta} - 0}{\sigma \Big/ \sqrt{\sum_{i=1}^{n} (x_i - \bar{x})^2}} \right]^2.$$

We have noted near the end of Section 8.4 that the distribution of $\hat{\beta}$ is $N[\beta, \sigma^2/\sum_{i=1}^{n} (x_i - \bar{x})^2]$. Since $\beta = 0$, we see that SS(R)/σ^2 is the square of a standard normal random variable and thus is $\chi^2(1)$. Furthermore, SS(E) ≥ 0, so by Theorem 8.1-1 we see that the distribution of SS(E)/σ^2 is $\chi^2(n-2)$ and that SS(R)/σ^2 and SS(E)/σ^2 are independent random variables. Also, if $\beta = 0$,

$$E[MS(R)] = E[SS(R)/1]$$
$$= E[\{\sigma^2\}SS(R)/\sigma^2] = \sigma^2;$$
$$E[MS(E)] = E[SS(E)/(n-2)]$$
$$= E[\{\sigma^2\}SS(E)/\sigma^2(n-2)]$$
$$= \sigma^2.$$

Thus both MS(R) and MS(E) provide unbiased estimates of σ^2 when $\beta = 0$. MS(R) and MS(E) are called the **mean square due to regression** and the **mean square due to error**.

Because MS(R) and MS(E) are independent chi-square random variables,

$$F = \frac{[SS(R)/\sigma^2]/1}{[SS(E)/\sigma^2]/(n-2)} = \frac{MS(R)}{MS(E)}$$

has an F distribution with 1 and $n-2$ degrees of freedom under H_0: $\beta = 0$.

If $\beta \neq 0$, it is still true that $MS(E) = SS(E)/(n-2)$ is an unbiased estimator of σ^2. However, MS(R) will tend to overestimate σ^2. Thus to test H_0: $\beta = 0$ against H_1: $\beta \neq 0$, we can define the critical region by

$$F = \frac{MS(R)}{MS(E)} \geq F_\alpha(1, n-2).$$

(Note that $T_1^2 = F$ with $\beta_0 = 0$ and thus T_1^2 has an F distribution with 1 and $n-2$ degrees of freedom under H_0. See Exercise 5.7-5.)

We can summarize this test using an ANOVA table like that given in Table 8.5-2.

Example 8.5-2 We repeat Example 8.4-1 by constructing an ANOVA table. From that example,

$$SS(R) = \hat\beta^2 \sum_{i=1}^{n} (x_i - \bar{x})^2 = \left(\frac{561.1}{756.1}\right)^2 (756.1) = 416.391,$$

$$SS(E) = \sum_{i=1}^{n} (y_i - \hat{y}_i)^2 = 217.709,$$

$$SS(TO) = \sum_{i=1}^{n} y_i^2 - \left(\sum_{i=1}^{n} y_i\right)^2 \Big/ 10 = 66731 - 66096.9 = 634.100.$$

TABLE 8.5-2

Source	SS	d.f.	MS	F
Regression	SS(R)	1	MS(R) = SS(R)/1	$\dfrac{MS(R)}{MS(E)}$
Error	SS(E)	$n-2$	MS(E) = SS(E)/(n-2)	
Total	SS(TO)	$n-1$		

These are summarized in an ANOVA table:

Source	SS	d.f.	MS	F
Regression	416.391	1	416.3910	15.3444
Error	217.709	8	27.13635	
Total	634.100	9		

Since $F = 15.3444 > 11.26 = F_{0.01}(1, 8)$, the null hypothesis H_0: $\beta = 0$ is rejected at the $\alpha = 0.01$ significance level. Note that $t_1^2 = (3.9111)^2 = 15.2967 \approx 15.3444 = F$, the difference being due to round-off error.

We have noted that $\hat{Y} = \hat{\alpha} + \hat{\beta}(x - \bar{x})$ is a point estimate for the mean of Y for some given x, or we could think of this as a prediction of the value of Y for this given x. But how close is $\hat{Y}$ to the mean of Y or to Y itself? We shall now find a confidence interval for $E(Y|x) = \mu_{Y|x}$ and a prediction interval for Y, given a particular value of x.

To find a confidence interval for

$$\mu(x) = \mu_{Y|x} = \alpha + \beta(x - \bar{x}),$$

let

$$\hat{Y} = \hat{\alpha} + \hat{\beta}(x - \bar{x}).$$

Recall that $\hat{Y}$ is a linear combination of normally distributed random variables so that $\hat{Y}$ has a normal distribution. Furthermore

$$E(\hat{Y}) = E[\hat{\alpha} + \hat{\beta}(x - \bar{x})]$$
$$= \alpha + \beta(x - \bar{x})$$

and

$$\text{Var}(\hat{Y}) = \text{Var}[\hat{\alpha} + \hat{\beta}(x - \bar{x})]$$

$$= \frac{\sigma^2}{n} + \frac{\sigma^2}{\sum_{i=1}^{n} (x_i - \bar{x})^2} (x - \bar{x})^2$$

$$= \sigma^2 \left[\frac{1}{n} + \frac{(x - \bar{x})^2}{\sum_{i=1}^{n} (x_i - \bar{x})^2} \right].$$

Recall that the distribution of $n\widehat{\sigma^2}/\sigma^2$ is $\chi^2(n-2)$. Since $\hat{\alpha}$ and $\hat{\beta}$ are independent of $\widehat{\sigma^2}$, we can form the t statistic

$$T = \cfrac{\cfrac{\hat{\alpha} + \hat{\beta}(x - \bar{x}) - [\alpha + \beta(x - \bar{x})]}{\sigma\sqrt{1/n + (x - \bar{x})^2 \Big/ \sum_{i=1}^{n}(x_i - \bar{x})^2}}}{\sqrt{\cfrac{\widehat{n\sigma^2}}{(n - 2)\sigma^2}}}$$

which has a t distribution with $r = n - 2$ degrees of freedom. Select $t_{\alpha/2}(n-2)$ from Table VI in the Appendix so that

$$P[-t_{\alpha/2}(n-2) \leq T \leq t_{\alpha/2}(n-2)] = 1 - \alpha.$$

This becomes

$$P[\hat{\alpha} + \hat{\beta}(x - \bar{x}) - ct_{\alpha/2}(n-2) \leq \alpha + \beta(x - \bar{x}) \leq$$
$$\hat{\alpha} + \hat{\beta}(x - \bar{x}) + ct_{\alpha/2}(n-2)] = 1 - \alpha,$$

where

$$c = \sqrt{\frac{\widehat{n\sigma^2}}{n - 2}}\sqrt{\frac{1}{n} + \frac{(x - \bar{x})^2}{\sum_{i=1}^{n}(x_i - \bar{x})^2}}.$$

Thus the endpoints for a $100(1 - \alpha)\%$ confidence interval for $\mu(x) = \alpha + \beta(x - \bar{x})$ are

$$\hat{\alpha} + \hat{\beta}(x - \bar{x}) \pm ct_{\alpha/2}(n-2).$$

Note that the width of this interval depends on the particular value of x because c depends on x (see Example 8.5-3).

We have used $(x_1, y_1), (x_2, y_2), \ldots, (x_n, y_n)$ to estimate α and β. Suppose that we are given a value of x, say x_{n+1}. A point estimate of the corresponding value of Y is

$$\hat{y}_{n+1} = \hat{\alpha} + \hat{\beta}(x_{n+1} - \bar{x}).$$

However, $\hat{y}_{n+1}$ is just one possible value of the random variable

$$Y_{n+1} = \alpha + \beta(x_{n+1} - \bar{x}) + \varepsilon_{n+1}.$$

What can we say about possible values for Y_{n+1}? We shall now obtain a prediction interval for Y_{n+1} when $x = x_{n+1}$ that is similar to the confidence interval for the mean of Y when $x = x_{n+1}$.

We have that

$$Y_{n+1} = \alpha + \beta(x_{n+1} - \bar{x}) + \varepsilon_{n+1},$$

where ε_{n+1} is $N(0, \sigma^2)$ and $\bar{x} = (1/n) \sum_{i=1}^{n} x_i$. Now

$$W = Y_{n+1} - \hat{\alpha} - \hat{\beta}(x_{n+1} - \bar{x})$$

is a linear combination of normally distributed random variables, so W has a normal distribution. The mean of W is

$$E(W) = E[Y_{n+1} - \hat{\alpha} - \hat{\beta}(x_{n+1} - \bar{x})]$$
$$= \alpha + \beta(x_{n+1} - \bar{x}) - \alpha - \beta(x_{n+1} - \bar{x}) = 0;$$

and since Y_{n+1}, $\hat{\alpha}$ and $\hat{\beta}$ are independent, the variance of W is

$$\text{Var}(W) = \sigma^2 + \frac{\sigma^2}{n} + \frac{\sigma^2}{\sum_{i=1}^{n} (x_i - \bar{x})^2}(x_{n+1} - \bar{x})^2$$

$$= \sigma^2\left[1 + \frac{1}{n} + \frac{(x_{n+1} - \bar{x})^2}{\sum_{i=1}^{n} (x_i - \bar{x})^2}\right].$$

Recall that $n\widehat{\sigma^2}/[(n-2)\sigma^2]$ is $\chi^2(n-2)$. Since Y_{n+1}, $\hat{\alpha}$, and $\hat{\beta}$ are independent of $\widehat{\sigma^2}$, we can form the t-statistic

$$T = \frac{\dfrac{Y_{n+1} - \hat{\alpha} - \hat{\beta}(x_{n+1} - \bar{x})}{\sigma\sqrt{1 + 1/n + (x_{n+1} - \bar{x})^2 \Big/ \sum_{i=1}^{n} (x_i - \bar{x})^2}}}{\sqrt{\dfrac{n\widehat{\sigma^2}}{(n-2)\sigma^2}}}$$

which has a t-distribution with $r = n - 2$ degrees of freedom. Select a constant $t_{\alpha/2}(n-2)$ from Table VI in the Appendix so that

$$P[-t_{\alpha/2}(n - 2) \leq T \leq t_{\alpha/2}(n - 2)] = 1 - \alpha.$$

Solving this inequality for Y_{n+1}, we have

$$P[\hat{\alpha} + \hat{\beta}(x_{n+1} - \bar{x}) - dt_{\alpha/2}(n-2) \leq Y_{n+1} \leq$$
$$\hat{\alpha} + \hat{\beta}(x_{n+1} - \bar{x}) + dt_{\alpha/2}(n-2)] = 1 - \alpha,$$

where

$$d = \sqrt{\frac{n\widehat{\sigma^2}}{n - 2}}\sqrt{1 + \frac{1}{n} + \frac{(x_{n+1} - \bar{x})^2}{\sum_{i=1}^{n} (x_i - \bar{x})^2}}.$$

Thus the endpoints for a $100(1 - \alpha)\%$ prediction interval for Y_{n+1} are

$$\hat{\alpha} + \hat{\beta}(x_{n+1} - \bar{x}) \pm dt_{\alpha/2}(n-2).$$

We shall now illustrate a 95% confidence interval for $\mu_{Y|x}$ and a 95% prediction interval for Y for a given value of x using the data in Example 8.4-1. To find such intervals, the calculating formulas that are given in equations (8.4-1), (8.4-2), and (8.4-4) can be used.

Example 8.5-3   To find a 95% confidence interval for $\mu_{Y|x}$ using the data in Example 8.4-1, note that we have already found that $\bar{x} = 68.3$, $\hat{\alpha} = 81.3$, $\hat{\beta} = 561.1/756.1 = 0.7421$, and $\hat{\sigma}^2 = 21.7709$. We also need

$$\sum_{i=1}^{n} (x_i - \bar{x})^2 = \sum_{i=1}^{n} x_i^2 - \left(\sum_{i=1}^{n} x_i \right)^2 \Big/ n$$

$$= 47,405 - \frac{683^2}{10} = 756.1.$$

For 95% confidence, $t_{0.025}(8) = 2.306$. When $x = 60$, the endpoints for a 95% confidence interval for $\mu_{Y|60}$ are

$81.3 + 0.7421(60 - 68.3)$

$$\pm \left[\sqrt{\frac{10(21.7709)}{8}} \sqrt{\frac{1}{10} + \frac{(60 - 68.3)^2}{756.1}} \right](2.306)$$

or

$$75.1406 \pm 5.2589.$$

Similarly, when $x = 70$ the endpoints for a 95% confidence interval for $\mu_{Y|70}$ are

$$82.5616 \pm 3.8761.$$

Note that the lengths of these intervals depend on the particular value of x. A 95% confidence band for $\mu_{Y|x}$ is graphed in Figure 8.5-1 along with the scatter diagram and $\hat{y} = \hat{\alpha} + \hat{\beta}(x - \bar{x})$.

The endpoints for a 95% prediction interval for Y when $x = 60$ are

$81.3 + 0.7421(60 - 68.3)$

$$\pm \left[\sqrt{\frac{10(21.7709)}{8}} \sqrt{1.1 + \frac{(60 - 68.3)^2}{756.1}} \right](2.306)$$

or

$$75.1406 \pm 13.1289.$$

Note that this interval is much wider than that for $\mu_{Y|x}$. In Figure 8.5-2 the 95% prediction band for Y is graphed along with the scatter diagram and the least squares regression line.

We now generalize the simple regression model to the **multiple regression** case. Say we observe several x-values, say $x_1, x_2, \ldots, x_k$, along with the y-value. For example, suppose that x_1 equals the student's ACT composite score, x_2 equals the

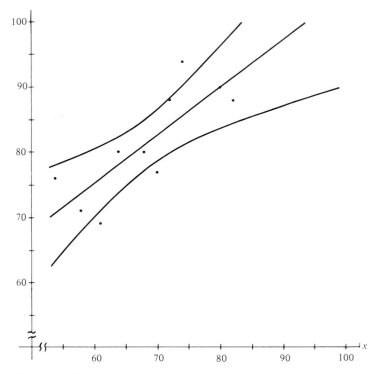

FIGURE 8.5-1

student's high school class rank, and y equals the student's first-year GPA in college. We want to estimate a regression function $E(Y) = \mu(x_1, x_2, \ldots, x_k)$ from some observed data. If

$$\mu(x_1, x_2, \ldots, x_k) = \beta_1 x_1 + \beta_2 x_2 + \cdots + \beta_k x_k,$$

then we say that we have a **linear model** because this expression is linear in the coefficients $\beta_1, \beta_2, \ldots, \beta_k$.

Note, for illustration, that the model in Section 8.4 is linear in $\alpha = \beta_1$ and $\beta = \beta_2$, with $x_1 = 1$ and $x_2 = x$, giving the mean $\alpha + \beta x$. (For convenience, there the mean of the x-values was subtracted from x.) Suppose, however, that we had wished to use the cubic function $\beta_1 + \beta_2 x + \beta_3 x^2 + \beta_4 x^3$ as the mean. This cubic expression still provides a linear model (i.e., linear in the β-values), and we would take $x_1 = 1$, $x_2 = x$, $x_3 = x^2$, and $x_4 = x^3$.

Say our n observation points are

$$(x_{1j}, x_{2j}, \ldots, x_{kj}, y_j), \qquad j = 1, 2, \ldots, n.$$

To fit the linear model $\beta_1 x_1 + \beta_2 x_2 + \cdots + \beta_k x_k$ by the method of least squares, we minimize

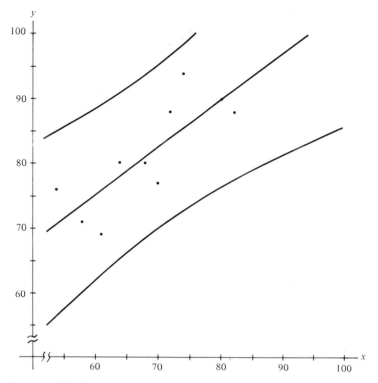

FIGURE 8.5-2

$$G = \sum_{j=1}^{n} (y_j - \beta_1 x_{1j} - \beta_2 x_{2j} - \cdots - \beta_k x_{kj})^2.$$

If we equate to zero the k first partial derivatives

$$\frac{\partial G}{\partial \beta_i} = \sum_{j=1}^{n} (-2)(y_j - \beta_1 x_{1j} - \cdots - \beta_k x_{kj})(x_{ij}), \qquad i = 1, 2, \ldots, k,$$

we obtain the k **normal equations**

$$\beta_1 \sum_{j=1}^{n} x_{1j}^2 + \beta_2 \sum_{j=1}^{n} x_{1j} x_{2j} + \cdots + \beta_k \sum_{j=1}^{n} x_{1j} x_{kj} = \sum_{j=1}^{n} x_{1j} y_j,$$

$$\beta_1 \sum_{j=1}^{n} x_{2j} x_{1j} + \beta_2 \sum_{j=1}^{n} x_{2j}^2 + \cdots + \beta_k \sum_{j=1}^{n} x_{2j} x_{kj} = \sum_{j=1}^{n} x_{2j} y_j,$$

$$\vdots \qquad\qquad\qquad\qquad \vdots$$

$$\beta_1 \sum_{j=1}^{n} x_{kj} x_{1j} + \beta_2 \sum_{j=1}^{n} x_{kj} x_{2j} + \cdots + \beta_k \sum_{j=1}^{n} x_{kj}^2 = \sum_{j=1}^{n} x_{kj} y_j.$$

The solution of these k equations provides the least squares estimates of β_1, β_2, . . . , β_k. These estimates are also maximum likelihood estimates of β_1, β_2, . . . , β_k, provided the random variables Y_1, Y_2, . . . , Y_n are mutually independent and Y_j is $N(\beta_1 x_{1j} + \cdots + \beta_k x_{kj}, \sigma^2)$, $j = 1, 2, \ldots, k$.

Example 8.5-4   By the method of least squares, fit $y = \beta_1 x_1 + \beta_2 x_2 + \beta_3 x_3$ to the five observed points (x_1, x_2, x_3, y):

$$(1, 1, 0, 4), (1, 0, 1, 3), (1, 2, 3, 2), (1, 3, 0, 6), (1, 0, 0, 1).$$

Note that $x_1 = 1$ in each point; so we are really fitting $y = \beta_1 + \beta_2 x_2 + \beta_3 x_3$. Since

$$\sum_{j=1}^{5} x_{1j}^2 = 5, \qquad \sum_{j=1}^{5} x_{1j} x_{2j} = 6, \qquad \sum_{j=1}^{5} x_{1j} x_{3j} = 4, \qquad \sum_{j=1}^{5} x_{1j} y_j = 16,$$

$$\sum_{j=1}^{5} x_{2j} x_{1j} = 6, \qquad \sum_{j=1}^{5} x_{2j}^2 = 14, \qquad \sum_{j=1}^{5} x_{2j} x_{3j} = 6, \qquad \sum_{j=1}^{5} x_{2j} y_j = 26,$$

$$\sum_{j=1}^{5} x_{3j} x_{1j} = 4, \qquad \sum_{j=1}^{5} x_{3j} x_{2j} = 6, \qquad \sum_{j=1}^{5} x_{3j}^2 = 10, \qquad \sum_{j=1}^{5} x_{3j} y_j = 9,$$

the normal equations are

$$5\beta_1 + 6\beta_2 + 4\beta_3 = 16,$$
$$6\beta_1 + 14\beta_2 + 6\beta_3 = 26,$$
$$4\beta_1 + 6\beta_2 + 10\beta_3 = 9.$$

Solving these three linear equations in three unknowns, we obtain

$$\hat{\beta}_1 = \frac{274}{112}, \qquad \hat{\beta}_2 = \frac{127}{112}, \qquad \hat{\beta}_3 = -\frac{85}{112}.$$

Thus the least squares fit is

$$y = \frac{274 x_1 + 127 x_2 - 85 x_3}{112}.$$

If x_1 always equals 1, then the equation reads

$$y = \frac{274 + 127 x_2 - 85 x_3}{112}.$$

It is interesting to observe that the usual two-sample problem is actually a linear model. Let $\beta_1 = \mu_1$ and $\beta_2 = \mu_2$ and consider n pairs of (x_1, x_2) that equal $(1, 0)$ and m pairs that equal $(0, 1)$. This would require each of the first n Y variables, namely Y_1, Y_2, . . . , Y_n, to have the mean

$$\beta_1 \cdot 1 + \beta_2 \cdot 0 = \beta_1 = \mu_1$$

and the next m Y variables, namely $Y_{n+1}, Y_{n+2}, \ldots, Y_{n+m}$, to have the mean

$$\beta_1 \cdot 0 + \beta_2 \cdot 1 = \beta_2 = \mu_2.$$

This is the background of the two-sample problem with the usual X_1, $X_2, \ldots, X_n$ and $Y_1, Y_2, \ldots, Y_m$ replaced by $Y_1, Y_2, \ldots, Y_n$ and Y_{n+1}, $Y_{n+2}, \ldots, Y_{n+m}$.

Clearly, this two-sample problem can be extended to three or more samples. For illustration, let $\beta_1 = \mu_1$, $\beta_2 = \mu_2$, and $\beta_3 = \mu_3$ and have n_1 triples of (x_1, x_2, x_3) equal to $(1, 0, 0)$, n_2 triples equal to $(0, 1, 0)$, and n_3 triples equal to $(0, 0, 1)$.

Exercises

8.5-1 For the data given in Exercise 8.4-1,
   (a) Test H_0: $\beta = 0$ against H_1: $\beta > 0$ at the $\alpha = 0.025$ significance level using a t-test.
   (b) Test H_0: $\beta = 0$ against H_1: $\beta \neq 0$ at the $\alpha = 0.05$ significance level by setting up the ANOVA table and using an F statistic.
   (c) Find a 95% confidence interval for $\mu_{Y|x}$ when $x = 68$, 75, and 82.
   (d) Find a 95% prediction interval for Y when $x = 68$, 75, and 82.

8.5-2 For the data given in Exercise 8.4-2,
   (a) Test H_0: $\beta = 0$ against H_1: $\beta > 0$ at the $\alpha = 0.025$ significance level using a t-test.
   (b) Test H_0: $\beta = 0$ against H_1: $\beta \neq 0$ at the $\alpha = 0.05$ significance level by setting up the ANOVA table and using an F statistic.
   (c) Find a 95% confidence interval for $\mu_{Y|x}$ when $x = 2$, 3, and 4.
   (d) Find a 95% prediction interval for Y when $x = 2$, 3, and 4.

8.5-3 For the data given in Exercise 8.4-3,
   (a) Test H_0: $\beta = 0$ against H_1: $\beta > 0$ at the $\alpha = 0.025$ significance level using a t-test.
   (b) Test H_0: $\beta = 0$ against H_1: $\beta \neq 0$ at the $\alpha = 0.05$ significance level by setting up the ANOVA table and using an F statistic.
   (c) Find a 95% confidence interval for $\mu_{Y|x}$ when $x = 20$, 25, and 30.
   (d) Find a 95% prediction interval for Y when $x = 20$, 25, and 30.

8.5-4 For the candy data given in Exercise 8.4-9, with the usual assumptions,
   (a) Test H_0: $\beta = 0$ against H_1: $\beta > 0$ at an $\alpha = 0.01$ significance level.
   (b) Find a 90% confidence interval for $\mu_{Y|x}$ when $x = 54$, 56, and 58.
   (c) Determine a 90% prediction interval for Y when $x = 54$, 56, and 58.

8.5-5 For the cigarette data in Exercise 8.4-10, with the usual assumptions,
   (a) Find a 95% confidence interval for $\mu_{Y|x}$ when $x = 5$, 10, and 15.
   (b) Determine a 95% prediction interval for Y when $x = 5$, 10, and 15.

8.5-6 Show that the cross product term in equation (8.5-1) is equal to zero, i.e.,

$$\sum_{i=1}^{n} [\hat{\beta}(x_i - \bar{x})][Y_i - \hat{\alpha} - \hat{\beta}(x_i - \bar{x})] = 0.$$

8.5-7 For the ACT scores in Exercise 8.4-12, with the usual assumptions,
(a) Test H_0: $\beta = 0$ against H_1: $\beta > 0$ at an $\alpha = 0.05$ significance level.
(b) Find a 95% confidence interval for $\mu_{Y|x}$ when $x = 17, 20, 23, 26,$ and 29.
(c) Determine a 90% prediction interval for Y when $x = 17, 20, 23, 26,$ and 29.

8.5-8 A computer center recorded the number of programs it maintained during each of 10 consecutive years.
(a) Calculate the least squares regression line for these data.
(b) Plot the points and the line on the same graph.
(c) Find a 95% prediction interval for the number of programs in year 11 under the usual assumptions.

Year	Number of Programs
1	430
2	480
3	565
4	790
5	885
6	960
7	1200
8	1380
9	1530
10	1591

8.5-9 By the method of least squares, fit the cubic $y = \beta_1 + \beta_2 x + \beta_3 x^2 + \beta_4 x^3$ to the 10 observed data points (x, y): $(0, 1)$, $(-1, -3)$, $(0, 3)$, $(1, 3)$, $(-1, -1)$, $(2, 10)$, $(0, 0)$, $(-2, -9)$, $(-1, -2)$, $(2, 8)$.

8.5-10 By the method of least squares, fit the regression plane $y = \beta_1 x_1 + \beta_2 x_2$ to the 12 observations of (x_1, x_2, y): $(1, 1, 6)$, $(0, 2, 3)$, $(3, 0, 10)$, $(-2, 0, -4)$, $(-1, 2, 0)$, $(0, 0, 1)$, $(2, 1, 8)$, $(-1, -1, -2)$, $(0, -3, -3)$, $(2, 1, 5)$, $(1, 1, 1)$, $(-1, 0, -2)$.

8.5-11 Explain why the model $\mu(x) = \beta_1 e^{\beta_2 x}$ is not a linear model. Would the logarithm, $\ln \mu(x)$, be a linear model?

9

Multivariate Distributions

9.1   The Correlation Coefficient

Section 5.1 introduced some properties of multivariate distributions and gave the definitions of joint probability density functions, marginal probability density functions, dependent and independent random variables. We now extend that discussion.

Recall that if the joint p.d.f. of $X_1, X_2, \ldots, X_n$ is $f(x_1, x_2, \ldots, x_n)$, then, in the discrete case, the marginal p.d.f. of X_k is

$$f_k(x_k) = \sum_{x_1} \cdots \sum_{x_{k-1}} \sum_{x_{k+1}} \cdots \sum_{x_n} f(x_1, x_2, \ldots, x_n), \quad x_k \in R_k.$$

This can be extended to the joint marginal distribution of 2 or more of n random variables. A joint marginal p.d.f. of X_j and X_k is found by summing $f(x_1, x_2, \ldots, x_n)$ over all x values except x_j and x_k; that is, summing over the support with x_j and x_k fixed, obtaining

$$f_{j,k}(x_j, x_k) = \sum_{x_1} \cdots \sum_{x_{j-1}} \sum_{x_{j+1}} \cdots \sum_{x_{k-1}} \sum_{x_{k+1}} \cdots \sum_{x_n} f(x_1, x_2, \ldots, x_n), \quad x_j \in R_j, x_k \in R_k.$$

Extensions of these marginal probability density functions to more than two random variables are made in an obvious way. For distributions of the continuous type, integrals would replace summations.

These ideas of marginal distributions are illustrated using extensions of the hypergeometric and the binomial distributions.

Example 9.1-1   Consider a population of 200 students who have just finished a first course in calculus. Of these 200, 40 have earned A's, 60 B's, 70 C's, 20 D's, and 10 F's. A sample of size 25 is taken at random and without replacement from this population so that each possible sample has probability

$$\frac{1}{\binom{200}{25}}$$

of being selected. Within the sample of 25, let X_1 be the number of A students, X_2 the number of B students, X_3 the number of C students, X_4 the number of D students, and $25 - X_1 - X_2 - X_3 - X_4$ the number of F students. The space R of (X_1, X_2, X_3, X_4) is defined by the collection of ordered 4-tuplets of nonnegative integers (x_1, x_2, x_3, x_4) such that $x_1 + x_2 + x_3 + x_4 \leq 25$. The joint p.d.f. of $X_1, X_2, X_3,$ and X_4 is

$$f(x_1, x_2, x_3, x_4) = \frac{\binom{40}{x_1}\binom{60}{x_2}\binom{70}{x_3}\binom{20}{x_4}\binom{10}{25 - x_1 - x_2 - x_3 - x_4}}{\binom{200}{25}},$$

for $(x_1, x_2, x_3, x_4) \in R$, where it is understood that $\binom{k}{j} = 0$ if $j > k$. Without actually summing, we know that the marginal p.d.f. of X_3 is

$$f_3(x_3) = \frac{\binom{70}{x_3}\binom{130}{25 - x_3}}{\binom{200}{25}}, \qquad x_3 = 0, 1, 2, \ldots, 25,$$

since X_3 alone has a hypergeometric distribution. Similarly, the joint marginal p.d.f. of X_1 and X_2 is

$$f_{12}(x_1, x_2) = \frac{\binom{40}{x_1}\binom{60}{x_2}\binom{100}{25 - x_1 - x_2}}{\binom{200}{25}}, \quad 0 \leq x_1, 0 \leq x_2, x_1 + x_2 \leq 25.$$

Of course, $f_3(x_3)$ is a hypergeometric p.d.f. and $f_{12}(x_1, x_2)$ and $f(x_1, x_2, x_3, x_4)$ are extensions of that type of p.d.f. Since

$$f(x_1, x_2, x_3, x_4) \neq f_1(x_1)f_2(x_2)f_3(x_3)f_4(x_4),$$

$X_1, X_2, X_3,$ and X_4 are dependent. Note also that the space R is not "rectangular," which would also imply that the random variables are dependent.

The distribution in Example 9.1-1 illustrates an extension of the hypergeometric distribution. In general, instead of two classes, suppose that each of n objects can be placed into one of s disjoint classes so that n_1 objects are in the first class, n_2 in the second, and so on until we find n_s in the sth class. Thus, $n = n_1 + n_2 + \cdots + n_s$. A collection of r objects is selected from these n at random and without replacement. Let the random variable X_i denote the number of observed objects in the sample belonging to the ith class. Find the probability that exactly x_i objects belong to the ith class, $i = 1, 2, \ldots, s$. Here

$$0 \le x_i \le n_i \qquad \text{and} \qquad x_1 + x_2 + \cdots + x_s = r.$$

We can select x_i objects from the ith class in any one of $\binom{n_i}{x_i}$ ways, $i = 1, 2, \ldots, s$. By the multiplication principle, the product

$$\binom{n_1}{x_1}\binom{n_2}{x_2}\cdots\binom{n_s}{x_s}$$

equals the number of ways the joint operation can be performed. If we assume that each of the $\binom{n}{r}$ ways of selecting r objects from $n = n_1 + n_2 + \cdots + n_s$ objects has the same probability, we have that the probability of selecting exactly x_i objects from the ith class, $i = 1, 2, \ldots, s$, is

$$P(X_1 = x_1, X_2 = x_2, \ldots, X_s = x_s) = \frac{\binom{n_1}{x_1}\binom{n_2}{x_2}\cdots\binom{n_s}{x_s}}{\binom{n}{r}},$$

$$0 \le x_i \le n_i \qquad \text{and} \qquad x_1 + x_2 + \cdots + x_s = r.$$

Example 9.1-2 The probability that a 13-card bridge hand (selected at random and without replacement) contains two clubs, four diamonds, three hearts, and four spades is

$$\frac{\binom{13}{2}\binom{13}{4}\binom{13}{3}\binom{13}{4}}{\binom{52}{13}} = \frac{11,404,407,300}{635,013,559,600} = 0.018.$$

We now consider an extension of the binomial distribution, namely the multinomial distribution. Consider a sequence of repetitions of an experiment for which the following conditions are satisfied:

(a) The experiment has k possible outcomes that are mutually exclusive and exhaustive, say $A_1, A_2, \ldots, A_k$.
(b) n independent trials of this experiment are observed.
(c) $P(A_i) = p_i$, $i = 1, 2, \ldots, k$, on each trial with $\sum_{i=1}^k p_i = 1$.
(d) The random variable X_i is equal to the number of times A_i occurs in the n trials, $i = 1, 2, \ldots, k$.

If $x_1, x_2, \ldots, x_k$ are nonnegative integers such that their sum equals n, then, for such a sequence, the probability that A_i occurs x_i times, $i = 1, 2, \ldots, k$, is given by

$$P(X_1 = x_1, X_2 = x_2, \ldots, X_k = x_k) = \frac{n!}{x_1!x_2! \cdots x_k!} \, p_1^{x_1}p_2^{x_2} \cdots p_k^{x_k}.$$

(9.1-1)

To see that this is correct, note that the number of distinguishable arrangements of $x_1 \, A_1$'s, $x_2 \, A_2$'s, $\ldots, x_k \, A_k$'s is

$$\binom{n}{x_1, x_2, \ldots, x_k} = \frac{n!}{x_1!x_2! \cdots x_k!}$$

and that the probability of each of these distinguishable arrangements is

$$p_1^{x_1}p_2^{x_2} \cdots p_k^{x_k}.$$

Hence the product for these two latter expressions gives the correct probability, which is in agreement with equation (9.1-1).

We say that $X_1, X_2, \ldots, X_k$ have a **multinomial distribution**. The reason is that

$$\sum \frac{n!}{x_1!x_2! \cdots x_k!} \, p_1^{x_1}p_2^{x_2} \cdots p_k^{x_k} = (p_1 + p_2 + \cdots + p_k)^n = 1,$$

where the summation is over the set of all nonnegative integers $x_1, x_2, \ldots, x_k$ whose sum is n. That is, $P(X_1 = x_1, X_2 = x_2, \ldots, X_k = x_k)$ is a typical term in the expansion of the nth power of the multinomial $(p_1 + p_2 + \cdots + p_k)$.

When $k = 3$, we often let $X = X_1$ and $Y = X_2$; then $n - X - Y = X_3$. We say that X and Y have a **trinomial distribution**. The joint p.d.f. of X and Y is

$$f(x, y) = \frac{n!}{x!y!(n - x - y)!} \, p_1^x p_2^y (1 - p_1 - p_2)^{n-x-y},$$

where x and y are nonnegative integers such that $x + y \leq n$. Since the marginal distributions of X and Y are, respectively, $b(n, p_1)$ and $b(n, p_2)$, the product of their probability density functions does not equal $f(x, y)$, and hence they are dependent random variables. Also note that the support of X and Y is triangular, so the random variables must be dependent.

Example 9.1-3   In manufacturing a certain item, it is found that in normal production about 95% of the items are good ones, 4% are "seconds," and 1% are defective. This particular company has a program of quality control by statistical method; and each hour an on-line inspector observes 20 items selected at random, counting the number X of seconds, and the number Y of defectives. If, in fact, the production is normal, we shall find the probability that in this sample of size $n = 20$, at least two seconds or at least two defective items are found. If we let $A = \{(x, y): x \geq 2 \text{ or } y \geq 2\}$, then

$$P(A) = 1 - P(A')$$
$$= 1 - P(X = 0 \text{ or } 1 \text{ and } Y = 0 \text{ or } 1)$$
$$= 1 - \frac{20!}{0!0!20!} (0.04)^0 (0.01)^0 (0.95)^{20} - \frac{20!}{1!0!19!} (0.04)^1 (0.01)^0 (0.95)^{19}$$
$$- \frac{20!}{0!1!19!} (0.04)^0 (0.01)^1 (0.95)^{19} - \frac{20!}{1!1!18!} (0.04)^1 (0.01)^1 (0.95)^{18}$$
$$= 0.204.$$

We now consider the mathematical expectation of functions of n random variables, $u(X_1, X_2, \ldots, X_n)$. These expectations are introduced with random variables of the discrete type. For random variables of the continuous type, integrals replace summations. Let the joint p.d.f. be $f(x_1, x_2, \ldots, x_n)$ with space R. Then

$$E[u(X_1, X_2, \ldots, X_n)] = \sum_{(x_1, \ldots, x_n)} \cdots \sum u(x_1, x_2, \ldots, x_n) f(x_1, x_2, \ldots, x_n),$$

if it exists, is called the **mathematical expectation** (or **expected value**) of

$$u(X_1, X_2, \ldots, X_n).$$

Example 9.1-4   There are eight similar chips in a bowl: one marked $(0, 0)$, two marked $(1, 0)$, two marked $(0, 1)$, and three marked $(1, 1)$. A player selects a chip at random and is given the sum of the two coordinates in dollars. If X_1 and X_2 represent those two coordinates, respectively, their joint p.d.f. is

$$f(x_1, x_2) = \frac{1 + x_1 + x_2}{8}, \qquad x_1 = 0, 1 \text{ and } x_2 = 0, 1.$$

Thus

$$E(X_1 + X_2) = \sum_{x_2=0}^{1} \sum_{x_1=0}^{1} (x_1 + x_2) \frac{1 + x_1 + x_2}{8}$$
$$= (0)\left(\frac{1}{8}\right) + (1)\left(\frac{2}{8}\right) + (1)\left(\frac{2}{8}\right) + (2)\left(\frac{3}{8}\right) = \frac{5}{4}.$$

That is, the expected payoff is $1.25.

For both discrete and continuous random variables, the following mathematical expectations, subject to their existence, have special names:

(a) If $u_1(X_1, X_2, \ldots, X_n) = X_i$, then

$$E[u_1(X_1, X_2, \ldots, X_n)] = E(X_i) = \mu_i$$

is called the **mean** of X_i, $i = 1, 2, \ldots, n$.

(b) If $u_2(X_1, X_2, \ldots, X_n) = (X_i - \mu_i)^2$, then

$$E[u_2(X_1, X_2, \ldots, X_n)] = E[(X_i - \mu_i)^2] = \sigma_i^2 = \text{Var}(X_i)$$

is called the **variance** of X_i, $i = 1, 2, \ldots, n$.

(c) If $u_3(X_1, X_2, \ldots, X_n) = (X_i - \mu_i)(X_j - \mu_j)$, $i \neq j$, then

$$E[u_3(X_1, X_2, \ldots, X_n)] = E[(X_i - \mu_i)(X_j - \mu_j)] = \sigma_{ij} = \text{Cov}(X_i, X_j)$$

is called the **covariance** of X_i and X_j.

(d) If the standard deviations σ_i and σ_j are positive, then

$$\rho_{ij} = \frac{\text{Cov}(X_i, X_j)}{\sigma_i \sigma_j} = \frac{\sigma_{ij}}{\sigma_i \sigma_j}$$

is called the **correlation coefficient** of X_i and X_j.

It is convenient that the mean and the variance of X_i can be computed from either the joint p.d.f. or the marginal p.d.f. of X_i. For example, if $n = 2$,

$$\mu_1 = E(X_1) = \sum_{x_1} \sum_{x_2} x_1 f(x_1, x_2)$$

$$= \sum_{x_1} x_1 \left[\sum_{x_2} f(x_1, x_2) \right] = \sum_{x_1} x_1 f_1(x_1).$$

Before considering the meaning of the covariance and the correlation coefficient, let us note a few simple facts. With $i \neq j$,

$$E[(X_i - \mu_i)(X_j - \mu_j)] = E(X_i X_j - \mu_i X_j - \mu_j X_i + \mu_i \mu_j)$$
$$= E(X_i X_j) - \mu_i E(X_j) - \mu_j E(X_i) + \mu_i \mu_j$$

because it is true that even in the multivariate situation, E is still a linear or distributive operator (see Exercise 9.1-10). Thus

$$\text{Cov}(X_i, X_j) = E(X_i X_j) - \mu_i \mu_j - \mu_j \mu_i + \mu_i \mu_j = E(X_i X_j) - \mu_i \mu_j.$$

Since $\rho_{ij} = \text{Cov}(X_i, X_j)/\sigma_i\sigma_j$, we also have

$$E(X_iX_j) = \mu_i\mu_j + \rho_{ij}\sigma_i\sigma_j.$$

That is, the expected value of the product of two random variables is equal to the product $\mu_i\mu_j$ of their expectations plus their covariance $\rho_{ij}\sigma_i\sigma_j$.

A simple example at this point would be helpful.

Example 9.1-5 Let X_1 and X_2 have the joint p.d.f.

$$f(x_1, x_2) = \frac{x_1 + 2x_2}{18}, \qquad x_1 = 1, 2, \quad x_2 = 1, 2.$$

The marginal probability density functions are, respectively,

$$f_1(x_1) = \sum_{x_2=1}^{2} \frac{x_1 + 2x_2}{18} = \frac{2x_1 + 6}{18}, \qquad x_1 = 1, 2,$$

and

$$f_2(x_2) = \sum_{x_1=1}^{2} \frac{x_1 + 2x_2}{18} = \frac{3 + 4x_2}{18}, \qquad x_2 = 1, 2.$$

Since $f(x_1, x_2) \neq f_1(x_1)f_2(x_2)$, X_1 and X_2 are dependent. The mean and the variance of X_1 are

$$\mu_1 = \sum_{x_1=1}^{2} x_1\frac{2x_1 + 6}{18} = (1)\left(\frac{8}{18}\right) + (2)\left(\frac{10}{18}\right) = \frac{14}{9},$$

and

$$\sigma_1^2 = \sum_{x_1=1}^{2} x_1^2\frac{2x_1 + 6}{18} - \left(\frac{14}{9}\right)^2 = \frac{24}{9} - \frac{196}{81} = \frac{20}{81}.$$

The mean and the variance of X_2 are

$$\mu_2 = \sum_{x_2=1}^{2} x_2\frac{3 + 4x_2}{18} = (1)\left(\frac{7}{18}\right) + (2)\left(\frac{11}{18}\right) = \frac{29}{18},$$

and

$$\sigma_2^2 = \sum_{x_2=1}^{2} x_2^2\frac{3 + 4x_2}{18} - \left(\frac{29}{18}\right)^2 = \frac{51}{18} - \frac{841}{324} = \frac{77}{324}.$$

The covariance of X_1 and X_2 is

$$\text{Cov}(X_1, X_2) = \sum_{x_2=1}^{2} \sum_{x_1=1}^{2} x_1 x_2 \frac{x_1 + 2x_2}{18} - \left(\frac{14}{9}\right)\left(\frac{29}{18}\right)$$

$$= (1)(1)\left(\frac{3}{18}\right) + (2)(1)\left(\frac{4}{18}\right) + (1)(2)\left(\frac{5}{18}\right)$$

$$+ (2)(2)\left(\frac{6}{18}\right) - \left(\frac{14}{9}\right)\left(\frac{29}{18}\right)$$

$$= \frac{45}{18} - \frac{406}{162} = -\frac{1}{162}.$$

Hence the correlation coefficient is

$$\rho = \frac{-1/162}{\sqrt{(20/81)(77/324)}} = \frac{-1}{\sqrt{1540}} = -0.025.$$

Insight into the correlation coefficient ρ of two discrete random variables X and Y may be gained by thoughtfully examining its definition

$$\rho = \frac{\sum_{R} (x - \mu_X)(y - \mu_Y) f(x, y)}{\sigma_X \sigma_Y},$$

where μ_X, μ_Y, σ_X, and σ_Y denote the respective means and standard deviations. If positive probability is assigned to pairs (x, y) in which both x and y are either simultaneously above or simultaneously below their respective means, the corresponding terms in the summation that defines ρ are positive because both factors $(x - \mu_X)$ and $(y - \mu_Y)$ will be positive or both will be negative. If pairs (x, y), which yield large positive products $(x - \mu_X)(y - \mu_Y)$, contain most of the probability of the distribution, the correlation coefficient will tend to be positive. If, on the other hand, the points (x, y), in which one component is below its mean and the other above its mean, have most of the probability, then the coefficient of correlation will tend to be negative because the products $(x - \mu_X)(y - \mu_Y)$ are negative (see Example 9.1-6). This interpretation of the sign of the correlation coefficient will play an important role in subsequent work.

To gain additional insight into the meaning of the correlation coefficient ρ, consider the following problem. Think of the points (x, y) in the space R and their corresponding probabilities. Let us consider all possible lines in two-dimensional space, each with finite slope, that pass through the point associated with the means, namely (μ_X, μ_Y). These lines are of the form $y - \mu_Y = b(x - \mu_X)$ or, equivalently, $y = \mu_Y + b(x - \mu_X)$. For each point in R, say (x_0, y_0) so that $f(x_0, y_0) > 0$, consider the vertical distance from that point to one of these lines. Since y_0 is the height of the point above the x axis and $\mu_Y + b(x_0 - \mu_X)$ is the height of the point on the line that is directly above or below the point (x_0, y_0), then the absolute value of the difference of these two heights is the vertical distance from point (x_0, y_0) to the line $y = \mu_Y + b(x - \mu_X)$. That is, the required distance is $|y_0 - \mu_Y - b(x_0 - \mu_X)|$. Let us now square this distance and take the weighted average of all such squares; that is, let us consider the mathematical expectation

$$E\{[(Y - \mu_Y) - b(X - \mu_X)]^2\} = K(b).$$

The problem is to find that line (or that b) which minimizes this expectation of the square $\{Y - \mu_Y - b(X - \mu_X)\}^2$. This is an application of the principle of least squares, and the line is sometimes called the least squares regression line.

The solution of the problem is very easy, since

$$K(b) = E\{(Y - \mu_Y)^2 - 2b(X - \mu_X)(Y - \mu_Y) + b^2(X - \mu_X)^2\}$$
$$= \sigma_Y^2 - 2b\rho\sigma_X\sigma_Y + b^2\sigma_X^2,$$

because E is a linear operator and $E[(X - \mu_X)(Y - \mu_Y)] = \rho\sigma_X\sigma_Y$. Accordingly, the derivative

$$K'(b) = -2\rho\sigma_X\sigma_Y + 2b\sigma_X^2$$

equals zero at $b = \rho\sigma_Y/\sigma_X$; and we see that $K(b)$ obtains its minimum for that b since $K''(b) = 2\sigma_X^2 > 0$. Consequently the **least squares regression line** (the line of the given form that is the best fit in the foregoing sense) is

$$y = \mu_Y + \rho\frac{\sigma_Y}{\sigma_X}(x - \mu_X).$$

Of course, if $\rho > 0$, the slope of the line is positive; but if $\rho < 0$, the slope is negative.

It is also instructive to note the value of the minimum of

$$K(b) = E\{[(Y - \mu_Y) - b(X - \mu_X)]^2\} = \sigma_Y^2 - 2b\rho\sigma_X\sigma_Y + b^2\sigma_X^2.$$

It is

$$K\left(\rho\frac{\sigma_Y}{\sigma_X}\right) = \sigma_Y^2 - 2\rho\frac{\sigma_Y}{\sigma_X}\rho\sigma_X\sigma_Y + \left(\rho\frac{\sigma_Y}{\sigma_X}\right)^2\sigma_X^2$$
$$= \sigma_Y^2 - 2\rho^2\sigma_Y^2 + \rho^2\sigma_Y^2 = \sigma_Y^2(1 - \rho^2).$$

Since $K(b)$ is the expected value of a square, it must be nonnegative for all b, and we see that $\sigma_Y^2(1 - \rho^2) \geq 0$; that is $\rho^2 \leq 1$, and hence $-1 \leq \rho \leq 1$, which is an important property of the correlation coefficient ρ. If $\rho = 0$, then $K(\rho\sigma_Y/\sigma_X) = \sigma_Y^2$; on the other hand, $K(\rho\sigma_Y/\sigma_X)$ is relatively small if ρ is close to 1 or negative 1. That is, the vertical deviations of the points with positive density from the line $y = \mu_Y + \rho(\sigma_Y/\sigma_X)(x - \mu_X)$ are small if ρ is close to 1 or negative 1 because $K(\rho\sigma_Y/\sigma_X)$ is the expectation of the square of those deviations. Thus ρ measures, in this sense, the amount of *linearity* in the probability distribution. As a matter of fact, in the discrete case, all the points of positive density lie on this straight line if and only if ρ is equal to 1 or negative 1.

REMARK   More generally, we could have fitted the line $y = a + bx$ by the same application of the principle of least squares. We would then have proved that the "best" line actually passes through the point (μ_X, μ_Y). Recall that in the discussion above we assumed our line to be of that form. Students will find this derivation to be an interesting exercise using partial derivatives (see Exercise 9.1-11).

The following example illustrates a joint discrete distribution for which ρ is negative. In Figure 9.1-1 the line of best fit or the least squares regression line is also drawn.

Example 9.1-6   Let X equal the number of ones and Y the number of twos and threes when a pair of fair four-sided dice are rolled. Then X and Y have a trinomial distribution with p.d.f.

$$f(x, y) = \frac{2!}{x!y!(2 - x - y)!} \left(\frac{1}{4}\right)^x \left(\frac{2}{4}\right)^y \left(\frac{1}{4}\right)^{2-x-y}, \qquad 0 \le x + y \le 2,$$

where x and y are nonnegative integers. Since the marginal p.d.f. of X is $b(2, 1/4)$ and the marginal p.d.f. of Y is $b(2, 1/2)$, we know that $\mu_X = 1/2$, $\mathrm{Var}(X) = 6/16$, $\mu_Y = 1$, and $\mathrm{Var}(Y) = 1/2$. Since $E(XY) = (1)(1)(4/16) = 4/16$, $\mathrm{Cov}(X, Y) = 4/16 - (1/2)(1) = -4/16$; therefore, the correlation coefficient is $\rho = -1/\sqrt{3}$. Using these values for the parameters, we obtain the line of best fit, namely

$$y = 1 + \left(-\frac{1}{\sqrt{3}}\right) \sqrt{\frac{1/2}{3/8}} \left(x - \frac{1}{2}\right) = -\frac{2}{3}x + \frac{4}{3}.$$

The joint p.d.f. is displayed in Figure 9.1-1. On this figure we have drawn horizontal and vertical lines through (μ_X, μ_Y) and also the line of best fit.

Suppose that X and Y are independent so that $f(x, y) \equiv f_1(x)f_2(y)$ and we want to find the expected value of the product $(X - \mu_X)(Y - \mu_Y)$. Subject to the existence of the expectations, we know that

$$E[u(X)v(Y)] = \sum_R \sum u(x)v(y)f(x, y)$$

$$= \sum_{R_1} \sum_{R_2} u(x)v(y)f_1(x)f_2(y)$$

$$= \sum_{R_1} u(x)f_1(x) \sum_{R_2} v(y)f_2(y)$$

$$= E[u(X)]E[v(Y)].$$

This can be used to show that the correlation coefficient of two independent variables is zero. For, in a standard notation, we have

$$\mathrm{Cov}(X, Y) = E[(X - \mu_X)(Y - \mu_Y)]$$
$$= E(X - \mu_X)E(Y - \mu_Y) = 0.$$

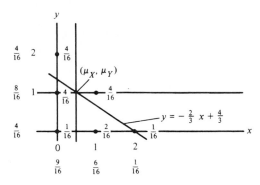

FIGURE 9.1-1

The converse of this fact is not necessarily true, however; zero correlation does not in general imply independence. It is most important to keep this straight: independence implies zero correlation, but zero correlation does not necessarily imply independence. The latter is now illustrated.

Example 9.1-7 Let X and Y have the joint p.d.f.

$$f(x, y) = \frac{1}{3}, \qquad (x, y) = (0, 1), (1, 0), (2, 1).$$

Since the support is not "rectangular," X and Y must be dependent. The means of X and Y are $\mu_X = 1$ and $\mu_Y = 2/3$, respectively. Hence

$$\text{Cov}(X, Y) = E(XY) - \mu_X\mu_Y$$

$$= (0)(1)\left(\frac{1}{3}\right) + (1)(0)\left(\frac{1}{3}\right) + (2)(1)\left(\frac{1}{3}\right) - (1)\left(\frac{2}{3}\right) = 0.$$

That is $\rho = 0$, but X and Y are dependent.

Exercises

9.1-1 A football team consisted of 84 players. Of these athletes, there were 30 freshmen, 23 sophomores, 22 juniors, and 9 seniors. If Coach Smith plans to select four players randomly to run a blocking drill, find the probability that
(a) He will select a player from each class.
(b) He will select four sophomores.

9.1-2 A box contains 100 Christmas tree light bulbs of which 30 are red, 35 are blue, 15 are white, and 20 are green. Fifteen bulbs are to be drawn at random from the box to fill a string with 15 sockets. Let X_1 denote the number of red, X_2 the number of blue, and X_3 the number of white bulbs drawn.
(a) Give $f(x_1, x_2, x_3)$, the joint p.d.f. of X_1, X_2, and X_3.
(b) Describe the set of points for which $f(x_1, x_2, x_3) > 0$.

(c) Determine $f_1(x_1)$, the marginal p.d.f. of X_1, and find $P(X_1 = 10)$.

(d) Find $f_{12}(x_1, x_2)$, the joint marginal p.d.f. of X_1 and X_2.

9.1-3 Draw 13 cards at random and without replacement from an ordinary deck of playing cards. Among these 13 cards, let X_1 be the number of spades, X_2 the number of hearts, X_3 the number of diamonds, and $13 - X_1 - X_2 - X_3$ the number of clubs.

(a) Give the joint p.d.f. of X_1, X_2, X_3 and describe the support in three-space.

(b) Determine the p.d.f. of X_1 and X_2 and describe the support in two-space.

(c) Find the p.d.f. of X_3, including its support.

9.1-4 In a biology laboratory, the mating of fruit flies is used to illustrate the Mendelian theory of inheritance. It is claimed that when two particular red-eyed fruit flies mate, the eyes of their offspring will be red-eyed, brown-eyed, scarlet-eyed, and white-eyed in the ratio $9:3:3:1$. Out of 254 offspring, let X_1, X_2, X_3, and $X_4 = 254 - X_1 - X_2 - X_3$ denote the numbers in the four categories, respectively, and assume that the theory is true.

(a) Give the joint p.d.f. of X_1, X_2, and X_3 and describe the support in three-space.

(b) Give the marginal p.d.f. of X_3.

(c) Give the joint marginal p.d.f. of X_1 and X_2 and describe the support in two-space.

(d) Find the mean and variance of X_i, $i = 1, 2, 3, 4$.

9.1-5 A manufactured item is classified as good, a ''second,'' or defective, with probabilities 6/10, 3/10, and 1/10, respectively. Fifteen such items are selected at random from the production line. Let X denote the number of good items, Y the number of seconds, and $15 - X - Y$ the number of defective items.

(a) Give the joint p.d.f. of X and Y, $f(x, y)$.

(b) Sketch the set of points for which $f(x, y) > 0$. From the shape of this region, can X and Y be independent? Why?

(c) Find $P(X = 10, Y = 4)$.

(d) Give the marginal p.d.f. of X.

(e) Find $P(X \le 11)$.

9.1-6 In a smoking survey among boys between the ages of 12 and 17, 78% prefer to date nonsmokers, 1% prefer to date smokers, and 21% don't care. If seven 12–17-year-old boys are selected randomly and X equals the number who prefer to date nonsmokers and Y equals the number who prefer to date smokers,

(a) Determine the joint p.d.f. of X and Y. Be sure to include the domain of the p.d.f.

(b) Find the marginal p.d.f. of X. Again include the domain.

9.1-7 Let the random variables X and Y have the joint p.d.f.

$$f(x, y) = \frac{x + y}{32}, \qquad x = 1, 2, \quad y = 1, 2, 3, 4.$$

Find the means μ_X and μ_Y, the variances σ_X^2 and σ_Y^2, and the correlation coefficient ρ. Are X and Y independent or dependent?

9.1-8 Let X and Y have the joint p.d.f. defined by $f(0, 0) = f(1, 2) = 0.2$, $f(0, 1) = f(1, 1) = 0.3$.
(a) Depict the points and corresponding probabilities on a graph.
(b) Give the marginal p.d.f.'s in the "margins."
(c) Compute μ_X, μ_Y, σ_X^2, σ_Y^2, $\text{Cov}(X, Y)$, and ρ.
(d) Find the equation of the least squares regression line and draw it on your graph. Does the line make sense to you intuitively?

9.1-9 Roll a fair four-sided die twice. Let X equal the outcome on the first roll and let Y equal the sum of the two rolls.
(a) Display the joint p.d.f. on a graph along with the marginal probabilities.
(b) Determine μ_X, μ_Y, σ_X^2, σ_Y^2, $\text{Cov}(X, Y)$, and ρ.
(c) Find the equation of the least squares regression line and draw it on your graph. Does the line make sense to you intuitively?

9.1-10 In the multivariate situation, show that E is a linear or distributive operator. For convenience, let $n = 2$ and show that

$$E[a_1u_1(X_1, X_2) + a_2u_2(X_1, X_2)] = a_1E[u_1(X_1, X_2)] + a_2E[u_2(X_1, X_2)].$$

9.1-11 Let X and Y be random variables with respective means μ_X and μ_Y, respective variances σ_X^2 and σ_Y^2, and correlation coefficient ρ. Fit the line $y = a + bx$ by the method of least squares to the probability distribution by minimizing the expectation

$$K(a, b) = E[(Y - a - bX)^2]$$

with respect to a and b.

HINT: Consider $\partial K/\partial a = 0$ and $\partial K/\partial b = 0$ and solve simultaneously.

9.1-12 Let X and Y have a trinomial distribution with parameters $n = 3$, $p_1 = 1/6$, and $p_2 = 1/2$. Find
(a) $E(X)$.
(b) $E(Y)$.
(c) $\text{Var}(X)$.
(d) $\text{Var}(Y)$.
(e) $\text{Cov}(X, Y)$.
(f) ρ.
Note that $\rho = -\sqrt{p_1p_2/(1 - p_1)(1 - p_2)}$.

9.1-13 Let the joint p.d.f. of X and Y be $f(x, y) = 1/4$, $(x, y) \in R = \{(0, 0), (1, 1), (1, -1), (2, 0)\}$.
(a) Are X and Y independent?
(b) Calculate $\text{Cov}(X, Y)$ and ρ.
This also illustrates the fact that dependent random variables can have a correlation coefficient of zero.

9.1-14 The joint p.d.f. of X and Y is $f(x, y) = 1/6, 0 \leq x + y \leq 2$, where x and y are integers.
   (a) Sketch the support of X and Y.
   (b) Record the marginal p.d.f.'s $f_1(x)$ and $f_2(y)$ in the "margins."
   (c) Compute $\text{Cov}(X, Y)$.
   (d) Determine ρ, the correlation coefficient.
   (e) Find the best-fitting line and draw it on your figure.

9.1-15 Let X_1, X_2 be a random sample of size $n = 2$ from the distribution with the binomial p.d.f.

$$f(x) = \binom{2}{x}\left(\frac{1}{3}\right)^x \left(\frac{2}{3}\right)^{2-x}, \qquad x = 0, 1, 2.$$

Find the joint p.d.f. of $Y = X_1$ and $W = X_1 + X_2$, determine the marginal p.d.f. of W and compute the correlation coefficient of Y and W.

HINT: Map the nine points (x_1, x_2) in the space of X_1, X_2 into the nine points (y, w) in the space of Y, W along with the corresponding probabilities and proceed as in earlier exercises.

9.1-16 Let X and Y be random variables of the continuous type having the joint p.d.f. $f(x, y) = 2, 0 \leq y \leq x \leq 1$. Draw a graph that illustrates the domain of this p.d.f.
   (a) Find the marginal p.d.f.'s of X and Y.
   (b) Compute $\mu_X, \mu_Y, \sigma_X^2, \sigma_Y^2, \text{Cov}(X, Y)$, and ρ.
   (c) Determine the equation of the least squares regression line and draw it on your graph. Does the line make sense to you intuitively?

9.2 Conditional Distributions

Let X and Y have a joint discrete distribution with p.d.f. $f(x, y)$ on space R. Say the marginal probability density functions are $f_1(x)$ and $f_2(y)$ with spaces R_1 and R_2, respectively. Let event $A = \{X = x\}$ and event $B = \{Y = y\}, (x, y) \in R$. Thus $A \cap B = \{X = x, Y = y\}$. Because

$$P(A \cap B) = P(X = x, Y = y) = f(x, y)$$

and

$$P(B) = P(Y = y) = f_2(y) > 0 \qquad \text{(since } y \in R_2\text{)},$$

we see that the conditional probability of event A given event B is

$$P(A \mid B) = \frac{P(A \cap B)}{P(B)} = \frac{f(x, y)}{f_2(y)}.$$

This leads to the following definition.

DEFINITION 9.2-1 *The* **conditional probability density function of** *X, given that* $Y = y$, *is defined by*

$$g(x|y) = \frac{f(x, y)}{f_2(y)}, \qquad provided\ that\ f_2(y) > 0.$$

Similarly, the **conditional probability density function of** *Y, given that* $X = x$, *is defined by*

$$h(y|x) = \frac{f(x, y)}{f_1(x)}, \qquad provided\ that\ f_1(x) > 0.$$

Example 9.2-1 Let X and Y have the joint p.d.f.

$$f(x, y) = \frac{x + y}{21}, \qquad x = 1, 2, 3, \quad y = 1, 2.$$

In Example 5.1-2 we showed that

$$f_1(x) = \frac{2x + 3}{21}, \qquad x = 1, 2, 3,$$

and

$$f_2(y) = \frac{3y + 6}{21}, \qquad y = 1, 2.$$

Thus the conditional p.d.f. of X, given $Y = y$, is equal to

$$g(x|y) = \frac{(x + y)/21}{(3y + 6)/21} = \frac{x + y}{3y + 6}, \qquad x = 1, 2, 3, \text{ when } y = 1 \text{ or } 2.$$

For example,

$$P(X = 2 | Y = 2) = g(2|2) = \frac{4}{12} = \frac{1}{3}.$$

Similarly, the conditional p.d.f. of Y, given $X = x$, is equal to

$$h(y|x) = \frac{x + y}{2x + 3}, \qquad y = 1, 2, \text{ when } x = 1, 2, \text{ or } 3.$$

The joint p.d.f. $f(x, y)$ is depicted in Figure 9.2-1 along with the marginal p.d.f.'s. Conditionally, if $y = 2$, we would expect the outcomes of x, namely 1, 2, and 3, to occur in the ratios $3:4:5$. This is precisely what $g(x|y)$ does:

$$g(1|2) = \frac{1 + 2}{12}, \qquad g(2|2) = \frac{2 + 2}{12}, \qquad g(3|2) = \frac{3 + 2}{12}.$$

Figure 9.2-2 displays $g(x|1)$ and $g(x|2)$, while Figure 9.2-3 gives $h(y|1)$, $h(y|2)$, and $h(y|3)$. Compare the probabilities in Figure 9.2-3, with those in

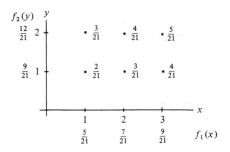

FIGURE 9.2-1

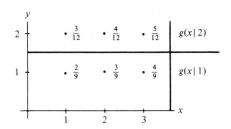

FIGURE 9.2-2

Figure 9.2-1. They should agree with your intuition as well as with the formula for $h(y|x)$.

Note that $0 \leq h(y|x)$. If we sum $h(y|x)$ over y for that fixed x, we obtain

$$\sum_y h(y|x) = \sum_y \frac{f(x, y)}{f_1(x)} = \frac{f_1(x)}{f_1(x)} = 1.$$

Thus $h(y|x)$ satisfies the conditions of a probability density function, and so we can compute conditional probabilities such as

$$P(a < Y < b | X = x) = \sum_{\{y:a<y<b\}} h(y|x)$$

and conditional expectations such as

$$E[u(Y)|X = x] = \sum_y u(y)h(y|x)$$

in a manner similar to those associated with probabilities and expectations.

Two special conditional expectations are the **conditional mean** of Y, given $X = x$, defined by

$$\mu_{Y|x} = E(Y|x) = \sum_y yh(y|x),$$

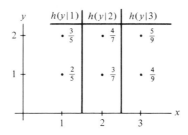

FIGURE 9.2-3

and the **conditional variance** of Y, given $X = x$, defined by

$$\sigma^2_{Y|x} = E\{[Y - E(Y|x)]^2 | x\} = \sum_y [y - E(Y|x)]^2 h(y|x),$$

which can be computed using

$$\sigma^2_{Y|x} = E(Y^2|x) - [E(Y|x)]^2.$$

The conditional mean $\mu_{X|y}$ and the conditional variance $\sigma^2_{X|y}$ are given by similar expressions.

Example 9.2-2   We use the background of Example 9.2-1 and compute $\mu_{Y|x}$ and $\sigma^2_{Y|x}$, when $x = 3$:

$$\mu_{Y|3} = E(Y|X = 3) = \sum_{y=1}^{2} yh(y|3)$$

$$= \sum_{y=1}^{2} y\left(\frac{3+y}{9}\right) = 1\left(\frac{4}{9}\right) + 2\left(\frac{5}{9}\right) = \frac{14}{9},$$

and

$$\sigma^2_{Y|3} = E\left[\left(Y - \frac{14}{9}\right)^2 \middle| X = 3\right] = \sum_{y=1}^{2}\left(y - \frac{14}{9}\right)^2\left(\frac{3+y}{9}\right)$$

$$= \frac{25}{81}\left(\frac{4}{9}\right) + \frac{16}{81}\left(\frac{5}{9}\right) = \frac{20}{81}.$$

The conditional mean of X, given $Y = y$, is a function of y alone; the conditional mean of Y, given $X = x$, is a function of x alone. Suppose that the latter conditional mean is a linear function of x; that is, $E(Y|x) = a + bx$. Let us find the constants a and b in terms of characteristics μ_X, μ_Y, σ^2_X, σ^2_Y, and ρ. This development will shed additional light on the correlation coefficient ρ; accordingly we assume that the respective standard deviations σ_X and σ_Y are both positive so that the correlation coefficient will exist.

It is given that

$$\sum_{y} y\, h(y|x) = \sum_{y} y\frac{f(x,\, y)}{f_1(x)} = a + bx, \quad \text{for } x \in R_1,$$

where R_1 is the space of X and R_2 is the space of Y. Hence

$$\sum_{y} y\, f(x,\, y) = (a + bx)f_1(x), \quad \text{for } x \in R_1, \tag{9.2-1}$$

and

$$\sum_{x \in R_1} \sum_{y} y\, f(x,\, y) = \sum_{x \in R_1} (a + bx)f_1(x);$$

that is, with μ_X and μ_Y representing the respective means, we have

$$\mu_Y = a + b\mu_X. \tag{9.2-2}$$

In addition, if we multiply both members of equation (9.2-1) by x and sum, we obtain

$$\sum_{x \in R_1} \sum_{y} xy f(x,\, y) = \sum_{x \in R_1} (ax + bx^2)f_1(x);$$

that is,

$$E(XY) = aE(X) + bE(X^2)$$

or, equivalently,

$$\mu_X\mu_Y + \rho\sigma_X\sigma_Y = a\mu_X + b(\mu_X^2 + \sigma_X^2). \tag{9.2-3}$$

The solution of equations (9.2-2) and (9.2-3) is

$$a = \mu_Y - \rho\frac{\sigma_Y}{\sigma_X}\mu_X \quad \text{and} \quad b = \rho\frac{\sigma_Y}{\sigma_X},$$

which implies that if $E(Y|x)$ is linear, it is given by

$$E(Y|x) = \mu_Y + \rho\frac{\sigma_Y}{\sigma_X}(x - \mu_X).$$

That is, if the conditional mean of Y, given $X = x$, is linear, it is exactly the same as the best-fitting line (least squares regression line) considered in Section 9.1.

By symmetry, if the conditional mean of X, given $Y = y$, is linear, it is given by

$$E(X|y) = \mu_X + \rho\frac{\sigma_X}{\sigma_Y}(y - \mu_Y).$$

We see that the point $[x = \mu_X, E(Y|x) = \mu_Y]$ satisfies the expression for $E(Y|x)$; and $[E(X|y) = \mu_X, y = \mu_Y]$ satisfies the expression for $E(X|y)$. That is, the point (μ_X, μ_Y) is on each of the two lines. In addition, we note that the product of the coefficient of x in $E(Y|x)$ and the coefficient of y in $E(X|y)$ equals ρ^2 and the ratio of these two coefficients equals σ_Y^2/σ_X^2. These observations sometimes prove useful in particular problems.

Example 9.2-3 Let X and Y have the trinomial p.d.f. with parameters n, p_1, p_2, and $1 - p_1 - p_2 = p_3$. That is,

$$f(x, y) = \frac{n!}{x!y!(n - x - y)!} \, p_1^x p_2^y p_3^{n-x-y},$$

where x and y are nonnegative integers such that $x + y \leq n$. From the development of the trinomial distribution, we note that X and Y have marginal binomial distributions $b(n, p_1)$ and $b(n, p_2)$, respectively. Thus

$$h(y|x) = \frac{f(x, y)}{f_1(x)} = \frac{(n - x)!}{y!(n - x - y)!} \left(\frac{p_2}{1 - p_1}\right)^y \left(\frac{p_3}{1 - p_1}\right)^{n-x-y},$$

$$y = 0, 1, 2, \ldots, n - x.$$

That is, the conditional p.d.f. of Y, given $X = x$, is binomial

$$b\left[n - x, \frac{p_2}{1 - p_1}\right]$$

and thus has conditional mean

$$E(Y|x) = (n - x) \frac{p_2}{1 - p_1}.$$

In a similar manner, we obtain

$$E(X|y) = (n - y) \frac{p_1}{1 - p_2}.$$

Since each of the conditional means is linear, the product of the respective coefficients of x and y is

$$\rho^2 = \left(\frac{-p_2}{1 - p_1}\right)\left(\frac{-p_1}{1 - p_2}\right) = \frac{p_1 p_2}{(1 - p_1)(1 - p_2)}.$$

However, ρ must be negative because the coefficients of x and y are negative, thus

$$\rho = -\sqrt{\frac{p_1 p_2}{(1 - p_1)(1 - p_2)}}.$$

Although we have used random variables of the discrete type to introduce the new definitions, they also hold for random variables of the continuous type. Let X and Y have a distribution of the continuous type with joint p.d.f. $f(x, y)$ and marginal p.d.f.'s $f_1(x)$ and $f_2(y)$, respectively. We have that the conditional p.d.f., mean, and variance of Y, given $X = x$, are, respectively,

$$h(y\,|\,x) = \frac{f(x,\,y)}{f_1(x)}, \qquad \text{provided that } f_1(x) > 0,$$

$$E(Y\,|\,x) = \int_{-\infty}^{\infty} yh(y\,|\,x)\,dy,$$

and

$$\begin{aligned} \text{Var}(Y\,|\,x) &= E\{[Y - E(Y\,|\,x)]^2\,|\,x\} \\ &= \int_{-\infty}^{\infty} [y - E(Y\,|\,x)]^2 h(y\,|\,x)\,dy \\ &= E[Y^2\,|\,x] - [E(Y\,|\,x)]^2. \end{aligned}$$

Similar expressions are associated with the conditional distribution of X, given $Y = y$.

Example 9.2-4   Let X and Y be the random variables of Example 5.1-8. Thus

$$\begin{aligned} f(x,\,y) &= 2, & 0 \le x \le y \le 1, \\ f_1(x) &= 2(1 - x), & 0 \le x \le 1, \end{aligned}$$

and

$$f_2(y) = 2y, \qquad 0 \le y \le 1.$$

Before we actually find the conditional p.d.f. of Y, given $X = x$, we shall give an intuitive argument. The joint p.d.f. is constant over the triangular region shown in Figure 5.1-3. If the value of X is known, say $X = x$, then the possible values of Y are between x and 1. Furthermore, we would expect Y to be uniformly distributed on the interval $[x, 1]$. That is, we would anticipate that $h(y\,|\,x) = 1/(1 - x)$, $x \le y \le 1$. More formally now, we have by definition that

$$h(y\,|\,x) = \frac{f(x,\,y)}{f_1(x)} = \frac{2}{2(1 - x)} = \frac{1}{1 - x}, \qquad x \le y \le 1, \quad 0 \le x \le 1.$$

The conditional mean of Y, given $X = x$, is

$$E(Y\,|\,x) = \int_x^1 y\,\frac{1}{1 - x}\,dy = \left[\frac{y^2}{2(1 - x)}\right]_x^1 = \frac{1 + x}{2}, \qquad 0 \le x \le 1.$$

Similarly, it could be shown that

$$E(X\,|\,y) = \frac{y}{2}, \qquad 0 \le y \le 1.$$

The conditional variance of Y, given $X = x$, is

$$E\{[Y - E(Y|x)]^2 | x\} = \int_x^1 \left(y - \frac{1+x}{2}\right)^2 \frac{1}{1-x}\, dy$$

$$= \left[\frac{1}{3(1-x)}\left(y - \frac{1+x}{2}\right)^3\right]_x^1$$

$$= \frac{(1-x)^2}{12}.$$

Recall that if a random variable W is $U(a, b)$, then $E(W) = (a + b)/2$, and $\text{Var}(W) = (b - a)^2/12$. Since the conditional distribution of Y, given $X = x$, is $U(x, 1)$, we could have written down immediately that $E(Y|x) = (x + 1)/2$ and $\text{Var}(Y|x) = (1 - x)^2/12$.

An illustration of a computation of a conditional probability is

$$P\left(\frac{3}{4} \le Y \le \frac{7}{8} \,\middle|\, X = \frac{1}{4}\right) = \int_{3/4}^{7/8} h\left(y \,\middle|\, \frac{1}{4}\right) dy$$

$$= \int_{3/4}^{7/8} \frac{1}{3/4}\, dy = \frac{1}{6}.$$

In general, if $E(Y|x)$ is linear, it is equal to

$$E(Y|x) = \mu_Y + \rho\left(\frac{\sigma_Y}{\sigma_X}\right)(x - \mu_X).$$

If $E(X|y)$ is linear, then

$$E(X|y) = \mu_X + \rho\left(\frac{\sigma_X}{\sigma_Y}\right)(y - \mu_Y).$$

Thus, in Example 9.2-4, we see that the product of the coefficients of x in $E(Y|x)$ and y in $E(X|y)$ is $\rho^2 = 1/4$. Thus $\rho = 1/2$ since each coefficient is positive. Since the ratio of those coefficients is equal to $\sigma_Y^2/\sigma_X^2 = 1$, we have that $\sigma_X^2 = \sigma_Y^2$.

Exercises

9.2-1 Let X and Y have the joint p.d.f.

$$f(x, y) = \frac{x + y}{32}, \qquad x = 1, 2, \quad y = 1, 2, 3, 4.$$

(a) Display the joint p.d.f. and the marginal p.d.f.'s on a graph like Figure 9.2-1.
(b) Find $g(x|y)$ and draw a figure like Figure 9.2-2, depicting the conditional p.d.f.'s for $y = 1, 2, 3,$ and 4.
(c) Find $h(y|x)$ and draw a figure like Figure 9.2-3, depicting the conditional p.d.f.'s for $x = 1$ and 2.

(d) Find $P(1 \leq Y \leq 3 | X = 1)$, $P(Y \leq 2 | X = 2)$, and $P(X = 2 | Y = 3)$.

(e) Find $E(Y | X = 1)$ and $\text{Var}(Y | X = 1)$.

9.2-2 Let the joint p.d.f. $f(x, y)$ of X and Y be given by the following:

(x, y)	$f(x, y)$
$(1, 1)$	$3/8$
$(2, 1)$	$1/8$
$(1, 2)$	$1/8$
$(2, 2)$	$3/8$

Find the two conditional probability density functions and the corresponding means and variances.

9.2-3 Let W equal the weight of laundry soap in a 1-kilogram box that is distributed in Southeast Asia. Suppose that $P(W < 1) = 0.02$ and $P(W > 1.072) = 0.08$. Call a box of soap light, good, or heavy depending on whether $W < 1$, $1 \leq W \leq 1.072$, or $W > 1.072$, respectively. In a random sample of $n = 50$ boxes, let X equal the number of light boxes and Y the number of good boxes.

(a) What is the joint p.d.f. of X and Y?

(b) Give the name of the distribution of Y along with the values of the parameters of this distribution.

(c) Given that $X = 3$, how is Y distributed conditionally?

(d) Determine $E(Y | X = 3)$.

(e) Find ρ, the correlation coefficient of X and Y.

9.2-4 The genes for eye color for a certain male fruit fly are (R, W). The genes for eye color for the mating female fruit fly are (R, W). Their offspring receive one gene for eye color from each parent. If an offspring ends up with either (R, R), (R, W), or (W, R), its eyes will look red. Let X equal the number of offspring having red eyes. Let Y equal the number of red-eyed offspring having (R, W) or (W, R) genes.

(a) If the total number of offspring is $n = 400$, how is X distributed?

(b) Give the values of $E(X)$ and $\text{Var}(X)$.

(c) Given that $X = 300$, how is Y distributed?

(d) Give the values of $E(Y)$ and $\text{Var}(Y)$.

9.2-5 Let X and Y have a trinomial distribution with $n = 2$, $p_1 = 1/4$, and $p_2 = 1/2$.

(a) Give $E(Y | x)$.

(b) Compare your answer in part (a) with the equation of the line of best fit in Example 9.1-6. Are they the same? Why?

9.2-6 Roll a pair of four-sided dice for which the outcome is 1, 2, 3, or 4 on each die. Let X denote the smaller and Y the larger outcome on the dice. Then the joint p.d.f. of X and Y is

$$f(x, y) = \begin{cases} \dfrac{1}{16}, & 1 \le x = y \le 4, \\[2mm] \dfrac{2}{16}, & 1 \le x < y \le 4. \end{cases}$$

(a) Find $E(Y|x)$ for $x = 1, 2, 3, 4$.
(b) Show that the least squares regression line is $y = 0.45x + 2.27$.
(c) Do the points $[x, E(Y|x)]$, $x = 1, 2, 3, 4$, lie on the least squares regression line?

REMARK Recall that we discovered that if the conditional mean of Y, given $X = x$, is linear, then it is exactly the same as the least squares regression line.

9.2-7 Using the joint p.d.f. given in Exercise 9.1-9, find the value of $E(Y|x)$ for $x = 1, 2, 3, 4$. Is this linear? Do these points lie on the best-fitting line?

9.2-8 An unbiased six-sided die is cast 30 independent times. Let X be the number of one's and Y the number of two's.
(a) What is the joint p.d.f. of X and Y?
(b) Find the conditional p.d.f. of X, given $Y = y$.
(c) Compute $E(X^2 - 4XY + 3Y^2)$.

9.2-9 Let X and Y have a uniform distribution on the set of points with integer coordinates in $R = \{(x, y): 0 \le x \le 7, x \le y \le x + 2\}$. That is, $f(x, y) = 1/24$, $(x, y) \in R$, and both x and y are integers. Find
(a) $f_1(x)$.
(b) $h(y|x)$.
(c) $E(Y|x)$.
(d) $\sigma^2_{Y|x}$.
(e) $f_2(y)$.

9.2-10 Let $f_1(x) = 1/10$, $x = 0, 1, 2, \ldots, 9$, and $h(y|x) = 1/(10 - x)$, $y = x$, $x + 1, \ldots, 9$. Find
(a) $f(x, y)$.
(b) $f_2(y)$.
(c) $E(Y|x)$.

9.2-11 The joint p.d.f. of X and Y is $f(x, y) = \frac{1}{6}$, $0 \le x + y \le 2$, where x and y are nonnegative integers.
(a) Sketch the domain of $f(x, y)$.
(b) Determine the marginal p.d.f.'s of X and Y.
(c) Find $E(Y|x)$ and plot $[x, E(Y|x)]$ on the sketch in part (a).
(d) Find $E(X|y)$ and plot $[E(X|y), y]$ on the sketch in part (a).
(e) Find ρ, the correlation coefficient, using your answers to parts (c) and (d).

9.2-12 For the random variables defined in Example 9.2-4, calculate the correlation coefficient directly from the definition

$$\rho = \frac{\text{Cov}(X, Y)}{\sigma_X \sigma_Y}.$$

9.2-13 Let $f(x, y) = 1/40$, $0 \le x \le 10$, $10 - x \le y \le 14 - x$, be the joint p.d.f. of X and Y.
(a) Sketch the region for which $f(x, y) > 0$.
(b) Find $f_1(x)$, the marginal p.d.f. of X.
(c) Determine $h(y|x)$, the conditional p.d.f. of Y, given $X = x$.
(d) Calculate $E(Y|x)$, the conditional mean of Y, given $X = x$.

9.2-14 Let $f(x, y) = 1/8$, $0 \le y \le 4$, $y \le x \le y + 2$, be the joint p.d.f. of X and Y.
(a) Sketch the region for which $f(x, y) > 0$.
(b) Find $f_1(x)$, the marginal p.d.f. of X.
(c) Determine $h(y|x)$, the conditional p.d.f. of Y, given $X = x$.
(d) Compute $E(Y|x)$, the conditional mean of Y, given $X = x$.
(e) Graph $y = E(Y|x)$ on your sketch in part (a). Is $y = E(Y|x)$ linear?

9.2-15 Let X have a uniform distribution $U(0, 2)$, and let the conditional distribution of Y, given $X = x$, be $U(0, x^2)$.
(a) Determine the joint p.d.f. of X and Y, $f(x, y)$.
(b) Calculate $f_2(y)$, the marginal p.d.f. of Y.
(c) Compute $E(X|y)$, the conditional mean of X, given $Y = y$.
(d) Find $E(Y|x)$, the conditional mean of Y, given $X = x$.

9.2-16 Let X have a uniform distribution on the interval $(0, 1)$. Given $X = x$, let Y have a uniform distribution on the interval $(0, x)$.
(a) Define the conditional p.d.f. of Y, given that $X = x$. Be sure to include the domain.
(b) Find $E(Y|x)$.
(c) Determine the joint p.d.f. of X and Y.
(d) Find the marginal p.d.f. of Y.

9.2-17 The marginal distribution of X is $U(0, 1)$. The conditional distribution of Y, given $X = x$, is $U(0, e^x)$.
(a) Determine $h(y|x)$, the conditional p.d.f. of Y, given $X = x$.
(b) Find $E(Y|x)$.
(c) Display the joint p.d.f. of X and Y. Sketch the region where $f(x, y) > 0$.
(d) Find $f_2(y)$, the marginal p.d.f. of Y.

9.3 The Bivariate Normal Distribution

Let X and Y be random variables with joint p.d.f. $f(x, y)$ of the continuous type. Many applications are concerned with the conditional distribution of one of the random variables, say Y, given that $X = x$. For example, X and Y might be a student's grade point averages from high school and from the first year in college, respectively. Persons in the field of educational testing and measurement are extremely interested in the conditional distribution of Y, given $X = x$, in such situations.

Suppose that we have an application in which we can make the following three assumptions about the conditional distribution of Y, given $X = x$:

(a) It is normal for each real x.
(b) Its mean $E(Y|x)$ is a linear function of x.
(c) Its variance is constant; that is, it does not depend upon the given value of x.

Of course, assumption (b), along with a result given in Section 9.2 implies that

$$E(Y|x) = \mu_Y + \rho \frac{\sigma_Y}{\sigma_X} (x - \mu_X).$$

Let us consider the implication of assumption (c). The conditional variance is given by

$$\sigma_{Y|x}^2 = \int_{-\infty}^{\infty} \left[y - \mu_Y - \rho \frac{\sigma_Y}{\sigma_X} (x - \mu_X) \right]^2 h(y|x) \, dy,$$

where $h(y|x)$ is the conditional p.d.f. of Y given $X = x$. Multiply each member of this equation of $f_1(x)$ and integrate on x. Since $\sigma_{Y|x}^2$ is a constant, the lefthand member is equal to $\sigma_{Y|x}^2$. Thus we have

$$\sigma_{Y|x}^2 = \int_{-\infty}^{\infty} \int_{-\infty}^{\infty} \left[(y - \mu_Y) - \rho \frac{\sigma_Y}{\sigma_X} (x - \mu_X) \right]^2 h(y|x) f_1(x) \, dy \, dx.$$

However, $h(y|x)f_1(x) = f(x, y)$; hence the righthand member is just an expectation and the equation can be written as

$$\sigma_{Y|x}^2 = E\left\{ (Y - \mu_Y)^2 - 2\rho \frac{\sigma_Y}{\sigma_X} (X - \mu_X)(Y - \mu_Y) + \rho^2 \frac{\sigma_Y^2}{\sigma_X^2} (X - \mu_X)^2 \right\}.$$

But using the fact that the expectation E is a linear operator, we have, recalling $E[(X - \mu_X)(Y - \mu_Y)] = \rho \sigma_X \sigma_Y$, that

$$\sigma_{Y|x}^2 = \sigma_Y^2 - 2\rho \frac{\sigma_Y}{\sigma_X} \rho \sigma_X \sigma_Y + \rho^2 \frac{\sigma_Y^2}{\sigma_X^2} \sigma_X^2$$

$$= \sigma_Y^2 - 2\rho^2 \sigma_Y^2 + \rho^2 \sigma_Y^2 = \sigma_Y^2(1 - \rho^2).$$

That is, the conditional variance of Y, for each given x, is $\sigma_Y^2(1 - \rho^2)$. These facts about the conditional mean and variance, along with assumption (a), require that the conditional p.d.f. of Y, given $X = x$, be

$$h(y|x) = \frac{1}{\sigma_Y \sqrt{2\pi} \sqrt{1 - \rho^2}} \exp\left[-\frac{[y - \mu_Y - \rho(\sigma_Y/\sigma_X)(x - \mu_X)]^2}{2\sigma_Y^2(1 - \rho^2)} \right],$$

$$-\infty < y < \infty, \text{ for every real } x.$$

Before we make any assumptions about the distribution of X, we give an example and figure to illustrate the implications of our current assumptions.

Example 9.3-1   Let $\mu_X = 10$, $\sigma_X^2 = 9$, $\mu_Y = 15$, $\sigma_Y^2 = 16$, and $\rho = 0.8$. We have seen that assumptions (a), (b), and (c) imply that the conditional distribution of Y, given $X = x$, is

$$N\left[15 + (0.8)\left(\frac{4}{3}\right)(x - 10),\ 16(1 - 0.8^2)\right].$$

In Figure 9.3-1 the conditional mean line

$$E(Y|x) = 15 + (0.8)\left(\frac{4}{3}\right)(x - 10) = \left(\frac{3.2}{3}\right)x + \left(\frac{13}{3}\right)$$

has been graphed. For each of $x = 5$, 10, and 15, the p.d.f. of Y, given $X = x$, is given.

Up to this point, nothing has been said about the distribution of X other than that it has mean μ_X and positive variance σ_X^2. Suppose, in addition, we assume that this distribution is also normal; that is, the marginal p.d.f. of X is

$$f_1(x) = \frac{1}{\sigma_X\sqrt{2\pi}}\ \exp\left[-\frac{(x - \mu_X)^2}{2\sigma_X^2}\right],\qquad -\infty < x < \infty.$$

Hence the joint p.d.f. of X and Y is given by the product

$$f(x, y) = h(y|x)f_1(x) = \frac{1}{2\pi\sigma_X\sigma_Y\sqrt{1 - \rho^2}}\ \exp\left[-\frac{q(x, y)}{2}\right],\qquad (9.3\text{-}1)$$

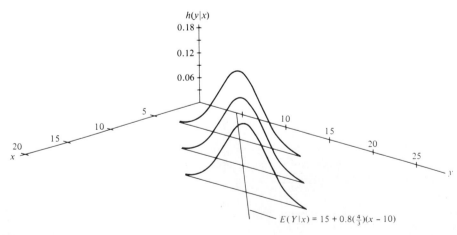

Figure 9.3-1

where it can be shown (see Exercise 9.3-2) that

$$q(x, y) = \frac{1}{1 - \rho^2}\left[\left(\frac{x - \mu_X}{\sigma_X}\right)^2 - 2\rho\left(\frac{x - \mu_X}{\sigma_X}\right)\left(\frac{y - \mu_Y}{\sigma_Y}\right) + \left(\frac{y - \mu_Y}{\sigma_Y}\right)^2\right].$$

A joint p.d.f. of this form is called a **bivariate normal p.d.f.**

> **Example 9.3-2** Let us assume that in a certain population of college students, the respective grade point averages, say X and Y, in high school and the first year in college have an approximate bivariate normal distribution with parameters $\mu_X = 2.9$, $\mu_Y = 2.4$, $\sigma_X = 0.4$, $\sigma_Y = 0.5$, and $\rho = 0.8$.
> Then, for illustration,
>
> $$P(2.1 < Y < 3.3) = P\left(\frac{2.1 - 2.4}{0.5} < \frac{Y - 2.4}{0.5} < \frac{3.3 - 2.4}{0.5}\right)$$
> $$= \Phi(1.8) - \Phi(-0.6) = 0.6898.$$
>
> Since the conditional p.d.f. of Y, given $X = 3.2$, is normal with mean
>
> $$2.4 + (0.08)\left(\frac{0.5}{0.4}\right)(3.2 - 2.9) = 2.7$$
>
> and standard deviation $(0.5)\sqrt{1 - 0.64} = 0.3$, we have that
>
> $$P(2.1 < Y < 3.3 \mid X = 3.2)$$
> $$= P\left(\frac{2.1 - 2.7}{0.3} < \frac{Y - 2.7}{0.3} < \frac{3.3 - 2.7}{0.3} \;\middle|\; X = 3.2\right)$$
> $$= \Phi(2) - \Phi(-2) = 0.9544.$$
>
> From a practical point of view, however, the reader should be warned that the correlation coefficient of these grade point averages is, in many instances, much smaller than 0.8.

Since x and y enter the bivariate normal p.d.f. in a similar manner, the roles of X and Y could have been interchanged. That is, Y could have been assigned the marginal normal p.d.f. $N(\mu_Y, \sigma_Y^2)$, and the conditional p.d.f. of X, given $Y = y$, would have then been normal, with mean $\mu_X + \rho(\sigma_X/\sigma_Y)(y - \mu_Y)$ and variance $\sigma_X^2(1 - \rho^2)$. Although this is fairly obvious, we do want to make special note of it.

In order to have a better understanding of the geometry of the bivariate normal distribution, consider the graph of $z = f(x, y)$, where $f(x, y)$ is given by equation (9.3-1). If we intersect this surface with planes parallel to the yz plane, that is, with $x = x_0$, we have

$$f(x_0, y) = f_1(x_0)h(y \mid x_0).$$

In this equation $f_1(x_0)$ is a constant, and $h(y \mid x_0)$ is a normal p.d.f. Thus $z = f(x_0, y)$ is bell-shaped, that is, has the shape of a normal p.d.f. However, note that it is not necessarily a p.d.f. because of the factor $f_1(x_0)$. Similarly, intersections of the surface $z = f(x, y)$ with planes $y = y_0$, parallel to the xz plane will be bell-shaped. If

$$0 < z_0 < \frac{1}{2\pi\sigma_X\sigma_Y\sqrt{1 - \rho^2}},$$

then

$$0 < z_0 2\pi\sigma_X\sigma_Y\sqrt{1 - \rho^2} < 1.$$

If we intersect $z = f(x, y)$ with the plane $z = z_0$, which is parallel to the xy plane, we have

$$z_0 2\pi\sigma_X\sigma_Y\sqrt{1 - \rho^2} = \exp\left[\frac{-q(x, y)}{2}\right].$$

Taking the natural logarithm of each side, we obtain

$$\left(\frac{x - \mu_X}{\sigma_X}\right)^2 - 2\rho\left(\frac{x - \mu_X}{\sigma_X}\right)\left(\frac{y - \mu_Y}{\sigma_Y}\right) + \left(\frac{y - \mu_Y}{\sigma_Y}\right)^2$$
$$= -2(1 - \rho^2) \ln (z_0 2\pi\sigma_X\sigma_Y\sqrt{1 - \rho^2}).$$

Thus we see that these intersections are ellipses.

Example 9.3-3 With $\mu_X = 10$, $\sigma_X^2 = 9$, $\mu_Y = 15$, $\sigma_Y^2 = 16$, and $\rho = 0.8$, the bivariate normal p.d.f. has been graphed in Figure 9.3-2. For $\rho = 0.8$, the level curves for $z_0 = 0.001, 0.006, 0.011, 0.016$, and 0.021 are given in Figure 9.3-3. The conditional mean line,

$$E(Y|x) = 15 + (0.8)\left(\frac{4}{3}\right)(x - 10) = \frac{3.2}{3}x + \frac{13}{3},$$

is also drawn on Figure 9.3-3. Note that this line intersects the level curves at points through which vertical tangents can be drawn to the ellipses.

We close this section by observing another important property of the correlation coefficient ρ if X and Y have a bivariate normal distribution. In equation (9.3-1) of the product $h(y|x)f_1(x)$, let us consider the factor $h(y|x)$ if $\rho = 0$. We see that this product, which is the joint p.d.f. of X and Y, equals $f_1(x)f_2(y)$ because $h(y|x)$ is, when $\rho = 0$, a normal p.d.f. with mean μ_Y and variance σ_Y^2. That is, if $\rho = 0$, the joint p.d.f. factors into the product of the two marginal probability density functions, and, hence, X and Y are independent random variables. Of course, if X and Y are any independent random variables (not necessarily normal), we know that ρ, if it exists, is always equal to zero. Thus we have proved the following.

THEOREM 9.3-1 *If X and Y have a bivariate normal distribution with correlation coefficient ρ, then X and Y are independent if and only if $\rho = 0$.*

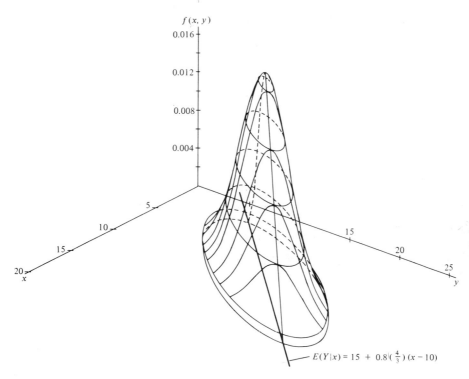

FIGURE **9.3-2**

Thus, in the bivariate normal case, $\rho = 0$ does imply independence of X and Y.

It should be mentioned here than these characteristics of the bivariate normal distribution can be extended to the trivariate normal distribution or, more generally, the multivariate normal distribution. This is done in more advanced texts assuming some knowledge of matrices; for illustration, see Chapter 12 of Hogg and Craig (1978).

Exercises

9.3-1 Let X and Y have a bivariate normal distribution with parameters $\mu_X = -3$, $\mu_Y = 10$, $\sigma_X^2 = 25$, $\sigma_Y^2 = 9$, and $\rho = 3/5$. Compute
(a) $P(-5 < X < 5)$.   (b) $P(-5 < X < 5 | Y = 13)$.
(c) $P(7 < Y < 16)$.   (d) $P(7 < Y < 16 | X = 2)$.

9.3-2 Show that the expression in the exponent of equation (9.3-1) is equal to the function $q(x, y)$ given in the text.

9.3-3 Let X and Y have a bivariate normal distribution with parameters $\mu_X = 2.8$, $\mu_Y = 110$, $\sigma_X^2 = 0.16$, $\sigma_Y^2 = 100$, and $\rho = 0.6$. Compute
(a) $P(106 < Y < 124)$.   (b) $P(106 < Y < 124 | X = 3.2)$.

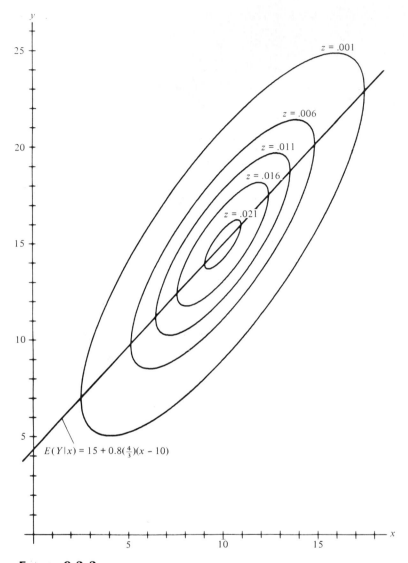

$$E(Y|x) = 15 + 0.8(\tfrac{4}{3})(x - 10)$$

FIGURE 9.3-3

9.3-4 Let X and Y have a bivariate normal distribution with $\mu_X = 70$, $\sigma_X^2 = 100$, $\mu_Y = 80$, $\sigma_Y^2 = 169$, and $\rho = 5/13$. Find

(a) $E(Y|X = 72)$.

(b) $\mathrm{Var}(Y|X = 72)$.

(c) $P(Y \le 84 | X = 72)$.

9.3-5 Let X denote the height in centimeters and Y the weight in kilograms of male college students. Assume that X and Y have a bivariate normal distribution with parameters $\mu_X = 185$, $\sigma_X^2 = 100$, $\mu_Y = 84$, $\sigma_Y^2 = 64$, and $\rho = 3/5$.

(a) Determine the conditional distribution of Y, given that $X = 190$.

(b) Find $P(86.4 < Y < 95.36 | X = 190)$.

9.3-6 For a freshman taking introductory statistics and majoring in psychology, let X equal the student's ACT mathematics score and Y the student's ACT verbal score. Assume that X and Y have a bivariate normal distribution with $\mu_X = 22.7$, $\sigma_X^2 = 17.64$, $\mu_Y = 22.7$, $\sigma_Y^2 = 12.25$, and $\rho = 0.78$. Find

(a) $P(18.5 < Y < 25.5)$.

(b) $E(Y|x)$.

(c) $\text{Var}(Y|x)$.

(d) $P(18.5 < Y < 25.5 | X = 23)$.

(e) $P(18.5 < Y < 25.5 | X = 25)$.

(f) For $x = 21$, 23, and 25, draw a graph of $z = h(y|x)$ similar to Figure 9.3-1.

9.3-7 For a pair of gallinules, let X equal the weight in grams of the male and Y the weight in grams of the female. Assume that X and Y have a bivariate normal distribution with $\mu_X = 415$, $\sigma_X^2 = 611$, $\mu_Y = 347$, $\sigma_Y^2 = 689$, and $\rho = -0.25$. Find

(a) $P(309.2 < Y < 380.6)$.

(b) $E(Y|x)$.

(c) $\text{Var}(Y|x)$.

(d) $P(309.2 < Y < 380.6 | X = 385.1)$.

9.3-8 Let X and Y have a bivariate normal distribution with parameters $\mu_X = 10$, $\sigma_X^2 = 9$, $\mu_Y = 15$, $\sigma_Y^2 = 16$, and $\rho = 0$. Find

(a) $P(13.6 < Y < 17.2)$.

(b) $E(Y|x)$.

(c) $\text{Var}(Y|x)$.

(d) $P(13.6 < Y < 17.2 | X = 9.1)$.

9.3-9 Let X and Y have a bivariate normal distribution. Find two different lines, $a(x)$ and $b(x)$, parallel to and equidistant from $E(Y|x)$, such that

$$P[a(x) < Y < b(x) | X = x] = 0.9544$$

for all real x. Plot $a(x)$, $b(x)$, and $E(Y|x)$ when $\mu_X = 2$, $\mu_Y = -1$, $\sigma_X = 3$, $\sigma_Y = 5$, and $\rho = 3/5$.

9.3-10 In a college health fitness program, let X denote the weight in kilograms of a male freshman at the beginning of the program and let Y denote his weight change during a semester. Assume that X and Y have a bivariate normal distribution with $\mu_X = 72.30$, $\sigma_X^2 = 110.25$, $\mu_Y = 2.80$, $\sigma_Y^2 = 2.89$, and $\rho = -0.57$. (The lighter students tend to gain weight, while the heavier students tend to lose weight.) Find

(a) $P(2.80 \le Y \le 5.35)$.

(b) $P(2.76 \le Y \le 5.34 | X = 82.3)$.

9.3-11 For a female freshman in a health fitness program, let X equal her percentage of body fat at the beginning of the program and let Y equal the change in her percentage of body fat measured at the end of the program. Assume that X and Y have a bivariate normal distribution with $\mu_X = 24.5$, $\sigma_X^2 = 4.8^2 = 23.04$, $\mu_Y = -0.2$, $\sigma_Y^2 = 3.0^2 = 9.0$, and $\rho = -0.32$. Find

(a) $P(1.3 \le Y \le 5.8)$.

(b) $\mu_{Y|x}$, the conditional mean of Y, given $X = x$.

(c) $\sigma_{Y|x}^2$, the conditional variance of Y, given $X = x$.

(d) $P(1.3 \le Y \le 5.8 | X = 18)$.

9.3-12   For a male freshman in a health fitness program, let X equal his percentage of body fat at the beginning of the program and let Y equal the change in his percentage of body fat measured at the end of the program. Assume that X and Y have a bivariate normal distribution with $\mu_X = 15.00$, $\sigma_X^2 = 4.5^2$, $\mu_Y = -1.55$, $\sigma_Y^2 = 1.5^2$, and $\rho = -0.60$. Find

(a) $P(0.205 \le Y \le 0.805)$.

(b) $P(0.21 \le Y \le 0.81 | X = 20)$.

9.4   Correlation Analysis

We consider tests about the correlation coefficient ρ of a bivariate normal distribution. Let X and Y have a bivariate normal distribution. We know that, if the correlation coefficient ρ is zero, then X and Y are independent random variables. Furthermore, the value of ρ gives a measure of the linear relationship between X and Y. We now give methods for using the sample correlation coefficient to test the hypothesis H_0: $\rho = 0$ and also to form a confidence interval for ρ.

Let (X_1, Y_1), (X_2, Y_2), . . . , (X_n, Y_n) denote a random sample from a bivariate normal distribution with parameters μ_X, μ_Y, σ_X^2, σ_Y^2, and ρ. Recall that the sample correlation coefficient is

$$R = \frac{[1/(n-1)] \sum_{i=1}^{n} (X_i - \bar{X})(Y_i - \bar{Y})}{\sqrt{[1/(n-1)] \sum_{i=1}^{n} (X_i - \bar{X})^2} \sqrt{[1/(n-1)] \sum_{i=1}^{n} (Y_i - \bar{Y})^2}} = \frac{S_{XY}}{S_X S_Y}.$$

We note that

$$R \frac{S_Y}{S_X} = \frac{S_{XY}}{S_X^2} = \frac{[1/(n-1)] \sum_{i=1}^{n} (X_i - \bar{X})(Y_i - \bar{Y})}{[1/(n-1)] \sum_{i=1}^{n} (X_i - \bar{X})^2}$$

is exactly the solution that we obtained for $\hat{\beta}$ in Section 8.4 when the X values were fixed at $X_1 = x_1$, $X_2 = x_2$, . . . , $X_n = x_n$. Let us consider these values fixed temporarily so that we are considering conditional distributions, given $X_1 = x_1$, . . . , $X_n = x_n$. Moreover, if H_0: $\rho = 0$ is true, then the distributions of Y_1, Y_2, . . . , Y_n are independent of x_1, x_2, . . . , x_n and thus $\beta = \rho \sigma_Y / \sigma_X = 0$. Under these conditions, the conditional distribution of

$$\hat{\beta} = \frac{\displaystyle\sum_{i=1}^{n} (x_i - \bar{x})(Y_i - \bar{Y})}{\displaystyle\sum_{i=1}^{n} (x_i - \bar{x})^2}$$

is $N[0, \sigma_Y^2/(n - 1)s_x^2]$ when $s_x^2 > 0$. Moreover, recall from Section 8.4 that the conditional distribution, given $X_1 = x_1, \ldots, X_n = x_n$, of (see Exercise 9.4-7)

$$\frac{\displaystyle\sum_{i=1}^{n} [Y_i - \bar{Y} - (S_{xY}/s_x^2)(x_i - \bar{x})]^2}{\sigma_Y^2} = \frac{(n - 1)S_Y^2(1 - R^2)}{\sigma_Y^2}$$

is $\chi^2(n-2)$ and is independent of $\hat{\beta}$. Thus, when $\rho = 0$, the conditional distribution of

$$T = \frac{(RS_Y/s_x)/(\sigma_Y/\sqrt{n - 1}s_x)}{\sqrt{[(n - 1)S_Y^2(1 - R^2)/\sigma_Y^2][1/(n - 2)]}} = \frac{R\sqrt{n - 2}}{\sqrt{1 - R^2}}$$

is Student's t with $n - 2$ degrees of freedom. However, since the conditional distribution of T, given $X_1 = x_1, \ldots, X_n = x_n$, does not depend on $x_1, x_2, \ldots, x_n$, the unconditional distribution of T must be Student's t with $n - 2$ degrees of freedom and T and $(X_1, X_2, \ldots, X_n)$ are independent when $\rho = 0$.

REMARK It is interesting to note that in the discussion about the distribution of T, nothing was said about the distribution of $X_1, X_2, \ldots, X_n$. This means that if X and Y are independent and Y has a normal distribution, then T has a Student's t distribution whatever the distribution of X. Obviously, the roles of X and Y can be reversed in all of this development. In particular, if X and Y are independent, then T and $Y_1, Y_2, \ldots, Y_n$ are also independent.

Now T can be used to test $H_0: \rho = 0$. If the alternative hypothesis is $H_1: \rho > 0$, we would use the critical region defined by observed $T \geq t_\alpha(n-2)$, since large T implies large R. Obvious modifications would be made for the alternative hypotheses $H_1: \rho < 0$ and $H_1: \rho \neq 0$, the latter leading to a two-sided test.

Using the p.d.f. $h(t)$ of T, we can find the distribution function and p.d.f. of R when $-1 < r < 1$, provided that $\rho = 0$:

$$G(r) = P(R \leq r) = P\left(T \leq \frac{r\sqrt{n - 2}}{\sqrt{1 - r^2}}\right)$$

$$= \int_{-\infty}^{r\sqrt{n-2}/\sqrt{1-r^2}} h(t) \, dt$$

$$= \int_{-\infty}^{r\sqrt{n-2}/\sqrt{1-r^2}} \frac{\Gamma[(n - 1)/2]}{\Gamma(1/2)\Gamma[(n - 2)/2]} \frac{1}{\sqrt{n - 2}} \left(1 + \frac{t^2}{n - 2}\right)^{-(n-1)/2} dt.$$

The derivative of $G(r)$, with respect to r, is (see Appendix A.4)

$$g(r) = h\left(\frac{r\sqrt{n-2}}{\sqrt{1-r^2}}\right) \frac{d(r\sqrt{n-2}/\sqrt{1-r^2})}{dr},$$

which equals

$$g(r) = \frac{\Gamma[(n-1)/2]}{\Gamma(1/2)\Gamma[(n-2)/2]} (1-r^2)^{(n-4)/2}, \qquad -1 < r < 1.$$

Thus, for example, to test the hypothesis H_0: $\rho = 0$ against the alternative hypothesis H_1: $\rho \neq 0$ at a significance level α, select either a constant $r_{\alpha/2}(n-2)$ or a constant $t_{\alpha/2}(n-2)$ so that

$$\alpha = P(|R| \geq r_{\alpha/2}(n-2); H_0) = P(|T| \geq t_{\alpha/2}(n-2); H_0),$$

depending on the availability of R or T tables.

It is interesting to graph the p.d.f of R. Note in particular that if $n = 4$, $g(r) = 1/2$, $-1 < r < 1$, and if $n = 6$, $g(r) = (3/4)(1-r^2)$, $-1 < r < 1$. The graphs of the p.d.f. of R when $n = 8$ and when $n = 14$ are given in Figure 9.4-1. Recall that this is the p.d.f. of R when $\rho = 0$. As n increases R is more likely to equal values close to 0.

Table XI in the Appendix lists selected values of the distribution function of R when $\rho = 0$. For example, if $n = 8$, the number of degrees of freedom is 6 and $P(R \leq 0.7887) = 0.99$. Also, if $\alpha = 0.10$, then $r_{\alpha/2}(6) = r_{0.05}(6) = 0.6215$. See Figure 9.4-1(a).

It is also possible to obtain an approximate test of size α by using the fact that

$$W = \frac{1}{2} \ln \frac{1+R}{1-R}$$

has an approximate normal distribution with mean $(1/2) \ln [(1+\rho)/(1-\rho)]$ and variance $1/(n-3)$. We accept this statement without proof (see Exercise 9.4-9). Thus a test of H_0: $\rho = \rho_0$ can be based on the statistic

$$Z = \frac{(1/2) \ln [(1+R)/(1-R)] - (1/2) \ln [(1+\rho_0)/(1-\rho_0)]}{\sqrt{1/(n-3)}},$$

which has a distribution that is approximately $N(0, 1)$.

Example 9.4-1 We would like to test the hypothesis H_0: $\rho = 0$ against H_1: $\rho \neq 0$ at an $\alpha = 0.05$ significance level. A random sample of size 18 from a bivariate normal distribution yielded a sample correlation coefficient of $r = 0.35$. Using Table XI in the Appendix, since $0.35 < 0.4683$, H_0 is accepted (not rejected) at an $\alpha = 0.05$ significance level. Using the t distribution, we would reject H_0 if $|t| \geq 2.120 = t_{0.025}(16)$. Since

$$t = \frac{0.35\sqrt{16}}{\sqrt{1-(0.35)^2}} = 1.495,$$

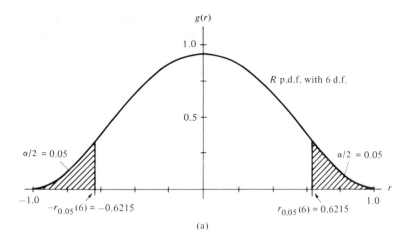

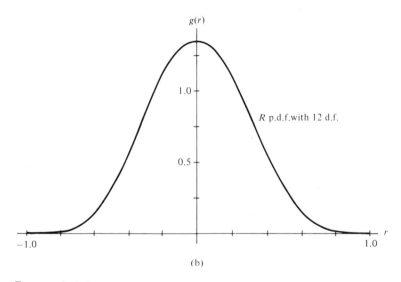

FIGURE 9.4-1

H_0 is not rejected. If we had used the normal approximation for Z, H_0 would be rejected if $|z| \geq 1.96$. Since

$$z = \frac{(1/2) \ln \left[(1 + 0.35)/(1 - 0.35)\right] - 0}{\sqrt{1/(18 - 3)}} = 1.415,$$

H_0 is not rejected.

To develop an approximate $100(1 - \alpha)\%$ confidence interval for ρ, we use the normal approximation for the distribution of Z. Thus we select a constant $c = z_{\alpha/2}$ from Table V in the Appendix so that

$$P\left(-c \le \frac{(1/2) \ln [(1 + R)/(1 - R)] - (1/2) \ln [(1 + \rho)/(1 - \rho)]}{\sqrt{1/(n - 3)}} \le c \right) \approx 1 - \alpha.$$

After several steps, this becomes

$$P\left(\frac{1 + R - (1 - R) \exp(2c/\sqrt{n - 3})}{1 + R + (1 - R) \exp(2c/\sqrt{n - 3})} \le \rho \le \right.$$
$$\left. \frac{1 + R - (1 - R) \exp(-2c/\sqrt{n - 3})}{1 + R + (1 - R) \exp(-2c/\sqrt{n - 3})} \right) \approx 1 - \alpha.$$

Example 9.4-2 Suppose that a random sample of size 12 from a bivariate normal distribution yielded a correlation coefficient of $r = 0.6$. An approximate 95% confidence interval for ρ would be

$$\left[\frac{1 + 0.6 - (1 - 0.6) \exp[2(1.96)/3]}{1 + 0.6 + (1 - 0.6) \exp[2(1.96)/3]}, \right.$$
$$\left. \frac{1 + 0.6 - (1 - 0.6) \exp[-2(1.96)/3]}{1 + 0.6 + (1 - 0.6) \exp[-2(1.96)/3]} \right] = [0.040, 0.873].$$

If the sample size had been $n = 39$ and $r = 0.6$, the approximate 95% confidence interval would have been $[0.351, 0.770]$.

Exercises

9.4-1 A random sample of size $n = 27$ from a bivariate normal distribution yielded a sample correlation coefficient of $r = -0.45$. Would the hypothesis H_0: $\rho = 0$ be rejected in favor of H_1: $\rho \ne 0$ at an $\alpha = 0.05$ significance level?

9.4-2 When bowling it is often possible to score well on the first game and then bowl poorly on the second game, or vice versa. The following six pairs of numbers give the scores of the first and second games bowled by the same person on six consecutive Tuesday evenings. Assume a bivariate normal distribution and use these scores to test the hypothesis H_0: $\rho = 0$ against H_1: $\rho \ne 0$ at $\alpha = 0.10$.

Game 1:	170	190	200	183	187	178
Game 2:	197	178	150	176	205	153

9.4-3 A random sample of size 28 from a bivariate normal distribution yielded a sample correlation coefficient of $r = 0.65$. Find an approximate 90% confidence interval for ρ.

9.4-4 For each part of Example 1.6-2, find an approximate 95% confidence interval for ρ.

9.4-5 The Board of Trustees of a small liberal arts college was interested in learning whether tuition, X, and enrollment, Y, are independent. Let ρ denote the correlation coefficient of X and Y. We shall test H_0: $\rho = 0$ againt H_1: $\rho \neq 0$. If $n = 62$ and $r = 0.283$, what is the conclusion at an $\alpha = 0.05$ significance level?

9.4-6 To help determine whether gallinules selected their mate on the basis of weight, 14 pairs of gallinules were captured and weighed. Test the null hypothesis H_0: $\rho = 0$ against a two-sided alternative at an $\alpha = 0.01$ significance level. Given that the male and female weights for the $n = 14$ pairs of birds yielded a sample correlation coefficient of $r = -0.252$, would H_0 be rejected?

9.4-7 By squaring the binomial expression $[(Y_i - \bar{Y}) - (S_{xY}/s_x^2)(x_i - \bar{x})]$, show that

$$\sum_{i=1}^{n} [(Y_i - \bar{Y}) - (S_{xY}/s_x^2)(x_i - \bar{x})]^2$$

$$= \sum_{i=1}^{n} (Y_i - \bar{Y})^2 - 2\left(\frac{S_{xY}}{s_x^2}\right) \sum_{i=1}^{n} (x_i - \bar{x})(Y_i - \bar{Y}) + \frac{S_{xY}^2}{s_x^4} \sum_{i=1}^{n} (x_i - \bar{x})^2$$

equals $(n - 1)S_Y^2(1 - R^2)$, where $X_1 = x_1$, $X_2 = x_2$, . . . , $X_n = x_n$.

HINT: Replace $S_{xY} = \sum_{i=1}^{n} (x_i - \bar{x})(Y_i - \bar{Y})/(n - 1)$ by $Rs_x S_Y$.

9.4-8 Show that the points of inflection for the graph of the p.d.f. of R when $\rho = 0$ are at $r = \pm 1/\sqrt{n - 5}$ for $n > 5$.

9.4-9 It is true, when sampling from a bivariate normal distribution, that the correlation coefficient R has an approximate normal distribution $N[\rho, (1 - \rho^2)^2/n]$ if the sample size n is large. Since, for large n, R is close to ρ, use two terms of the Taylor's expansion of $u(R)$ about ρ and determine that function $u(R)$ such that it has a variance (essentially) free of ρ. (The solution of this exercise explains why the transformation $(1/2) \ln [(1 + R)/(1 - R)]$ was suggested.)

9.4-10 Let X and Y have a bivariate normal distribution with correlation coefficient ρ. To test H_0: $\rho = 0$ against H_1: $\rho \neq 0$, a random sample of n pairs of observations is selected. Suppose that the sample correlation coefficient is $r = 0.68$. Using a significance level of $\alpha = 0.05$, find the smallest value of the sample size n so that H_0 is rejected.

9.4-11 In a college health fitness program, let X equal the weight in kilograms of a female freshman at the beginning of the program and let Y equal her weight change during the semester. We shall use the following data for $n = 16$ observations of (x, y) to test the null hypothesis H_0: $\rho = 0$ against a two-sided alternative hypothesis.

(61.4, −3.2)	(62.9, 1.4)	(58.7, 1.3)	(49.3, 0.6)
(71.3, 0.2)	(81.5, −2.2)	(60.8, 0.9)	(50.2, 0.2)
(60.3, 2.0)	(54.6, 0.3)	(51.1, 3.7)	(53.3, 0.2)
(81.0, −0.5)	(67.6, −0.8)	(71.4, −0.1)	(72.1, −0.1)

(a) What is the conclusion if $\alpha = 0.10$?

(b) What is the conclusion if $\alpha = 0.05$?

9.4-12 For the data in Exercise 8.4-11, test $H_0: \rho = 0$ against $H_1: \rho > 0$. Use $\alpha = 0.01$.

9.4-13 For the data in Exercise 8.4-12, test $H_0: \rho = 0$ against $H_1: \rho > 0$. In particular, approximate the p-value for this test.

9.5   Testing Probabilistic Models

We now consider applications of the very important chi-square statistic first proposed by Karl Pearson in 1900. As the reader will see, it is a very adaptable test statistic and can be used for many different types of tests. In particular, one application of it allows us to test the appropriateness of different probabilistic models and, in this sense, is a competitor of the Kolmogorov–Smirnov test (see Section 10.7).

We begin our study by considering the basic chi-square statistic, which, interestingly enough, has only an approximate chi-square distribution. It is based on some random variables $Y_1, Y_2, \ldots, Y_k$ that have a multinomial distribution with parameters n and $p_1, p_2, \ldots, p_k$. So that the reader can get some idea why Pearson first proposed it, we start with the binomial case.

Let Y_1 be $b(n, p_1)$, where $0 < p_1 < 1$. According to the Central Limit Theorem,

$$Z = \frac{Y_1 - np_1}{\sqrt{np_1(1 - p_1)}}$$

has a distribution that is approximately $N(0, 1)$ for large n, particularly when $np_1 \geq 5$ and $n(1 - p_1) \geq 5$. Thus it is not surprising that $Q_1 = Z^2$ is approximately $\chi^2(1)$. If we let $Y_2 = n - Y_1$ and $p_2 = 1 - p_1$, we see that Q_1 may be written as

$$Q_1 = \frac{(Y_1 - np_1)^2}{np_1(1 - p_1)} = \frac{(Y_1 - np_1)^2}{np_1} + \frac{(Y_1 - np_1)^2}{n(1 - p_1)}.$$

Since

$$(Y_1 - np_1)^2 = (n - Y_2 - n + np_2)^2 = (Y_2 - np_2)^2,$$

we have

$$Q_1 = \frac{(Y_1 - np_1)^2}{np_1} + \frac{(Y_2 - np_2)^2}{np_2}.$$

Let us now carefully consider each term in this last expression for Q_1. Of course, Y_1 is the number of "successes," and np_1 is the expected number of "successes";

that is, $E(Y_1) = np_1$. Likewise, Y_2 and np_2 are, respectively, the number and the expected number of "failures." So each numerator consists of the square of a difference of the observed number and expected number. Note that Q_1 can be written as

$$Q_1 = \sum_{i=1}^{2} \frac{(Y_i - np_i)^2}{np_i},$$

and we have seen intuitively that it has an approximate chi-square distribution with one degree of freedom. In a sense, Q_1 measure the "closeness" of the observed numbers to the corresponding expected numbers. For example, if the observed values of Y_1 and Y_2 equal their expected values, then the computed Q_1 is equal to $q_1 = 0$; but if they differ much from them, then the computed $Q_1 = q_1$ is relatively large.

To generalize, we let an experiment have k (instead of only two) mutually exclusive and exhaustive outcomes, say $A_1, A_2, \ldots, A_k$. Let $p_i = P(A_i)$ and thus $\Sigma_{i=1}^{k} p_i = 1$. The experiment is repeated n independent times, and we let Y_i represent the number of times the experiment results in A_i, $i = 1, 2, \ldots, k$. Then $Y_1, Y_2, \ldots, Y_k$ have a multinomial distribution with parameters n, p_1, $p_2, \ldots, p_k$. Say Y_i is observed to be equal to y_i. A statistic that measures the "closeness" of the observed number y_i of occurrences A_i to the expected number $e_i = np_i$, $i = 1, 2, \ldots, k$, is

$$q_{k-1} = \sum_{i=1}^{k} \frac{(y_i - np_i)^2}{np_i} = \sum_{i=1}^{k} \frac{(y_i - e_i)^2}{e_i}.$$

The corresponding random variable, namely

$$Q_{k-1} = \sum_{i=1}^{k} \frac{(Y_i - np_i)^2}{np_i},$$

can be shown to have, for large n, an approximate chi-square distribution with $k - 1$ degrees of freedom. The proof of this is beyond the level of this text; however, it is based upon the fact that $Y_1, Y_2, \ldots, Y_{k-1}$ have an approximate multivariate normal distribution. It is similar to the argument given in the case $k = 2$ in which Y_1 has an approximate normal distribution with mean np_1 and variance $np_1(1 - p_1)$.

Some writers suggest that n should be large enough so that $np_i \geq 5$, $i = 1, 2, \ldots, k$, to be certain that the approximating distribution is adequate. This is probably good advice for the beginner to follow, although we have seen the approximation work very well when $np_i \geq 1$, $i = 1, 2, \ldots, k$. The important thing to guard against is allowing some particular np_i to become so small that the corresponding term in q_{k-1}, namely $(y_i - np_i)^2/np_i$, tends to dominate the others because of its small denominator. In any case it is important to realize that Q_{k-1} has only an approximate chi-square distribution.

We shall now show how we can use the fact that Q_{k-1} is approximately $\chi^2(k-1)$ to test hypotheses about probabilities of various outcomes. Let an experiment have k mutually exclusive and exhaustive outcomes, $A_1, A_2, \ldots, A_k$. We would like to test whether $p_i = P(A_i)$ is equal to a known number p_{i0}, $i = 1, 2, \ldots, k$. That is, we shall test the hypothesis

$$H_0: p_i = p_{i0}, \ i = 1, 2, \ldots, k.$$

In order to test such a hypothesis, we shall take a sample of size n, that is repeat the experiment n independent times. We tend to favor H_0 if the observed number of times that A_i occurred, say y_i, and the number of times A_i was expected to occur if H_0 were true, namely np_{i0}, are approximately equal. That is, if

$$q_{k-1} = \sum_{i=1}^{k} \frac{(y_i - np_{i0})^2}{np_{i0}}$$

is "small," we tend to favor H_0. Since the distribution of Q_{k-1} is approximately $\chi^2(k-1)$, we shall reject H_0 if $q_{k-1} \geq \chi_\alpha^2(k-1)$, where α is the desired significance level of the test.

> **Example 9.5-1** If persons are asked to record a string of random digits, such as
>
> $$3 \quad 7 \quad 2 \quad 4 \quad 1 \quad 9 \quad 7 \quad 2 \quad 1 \quad 5 \quad 0 \quad 8 \quad \ldots,$$
>
> we usually find that they are reluctant to record the same or even the two closest numbers in adjacent positions. And yet, in true random digit generation, the probability of the next digit being the same as the preceding one is $p_{10} = 1/10$, the probability of the next being only one away from the preceding (assuming that 0 is one away from 9) is $p_{20} = 2/10$, and the probability of all other possibilities is $p_{30} = 7/10$. We shall test one person's concept of a random sequence by asking him to record a string of 51 digits that seems to represent a random generation. Thus we shall test
>
> $$H_0: p_1 = p_{10} = \frac{1}{10}, \ p_2 = p_{20} = \frac{2}{10}, \ p_3 = p_{30} = \frac{7}{10}.$$
>
> The critical region for an $\alpha = 0.05$ significance level is $q_2 \geq \chi_{0.05}^2(2) = 5.991$. The sequence of digits were as follows:
>
> | 5 | 8 | 3 | 1 | 9 | 4 | 6 | 7 | 9 | 2 | 6 | 3 |
> |---|---|---|---|---|---|---|---|---|---|---|---|
> | 0 | 8 | 7 | 5 | 1 | 3 | 6 | 2 | 1 | 9 | 5 | 4 |
> | 8 | 0 | 3 | 7 | 1 | 4 | 6 | 0 | 4 | 3 | 8 | 2 |
> | 7 | 3 | 9 | 8 | 5 | 6 | 1 | 8 | 7 | 0 | 3 | 5 |
> | 2 | 5 | 2 | | | | | | | | | |

We went through this listing and observed how many times the next digit was the same as or was one away from the preceding one.

	Frequency	Expected Number
Same	0	$50(1/10) = 5$
One away	8	$50(2/10) = 10$
Other	42	$50(7/10) = 35$
Total	50	50

The computed chi-square statistic is

$$\frac{(0-5)^2}{5} + \frac{(8-10)^2}{10} + \frac{(42-35)^2}{35} = 6.8 > 5.991 = \chi^2_{0.05}(2).$$

Thus we would say that this string of 51 digits does not seem to be random.

One major disadvantage in the use of the chi-square test is that it is a many-sided test. That is, the alternative hypothesis is very general, and it would be difficult to restrict alternatives to situations such as, with $k = 3$, H_1: $p_1 > p_{10}$, $p_2 > p_{20}$, $p_3 < p_{30}$. As a matter of fact, some statisticians would probably test H_0 against this particular alternative H_1 by using a linear function of Y_1, Y_2, and Y_3. However, this sort of discussion is beyond the scope of this book because it involves knowing more about the distributions of linear functions of the dependent random variables Y_1, Y_2, and Y_3. In any case, the student who truly recognizes that this chi-square statistic tests H_0: $p_i = p_{i0}$, $i = 1, 2, \ldots, k$ against *all* alternatives can usually appreciate the fact that it is more difficult to reject H_0 at a given significance level α using the chi-square statistic than it would be if some appropriate "one-sided" test statistic were available.

Many experiments yield a set of data, say $x_1, x_2, \ldots, x_n$, and the experimenter is often interested in determining whether these data can be treated as the observed values of a random sample $X_1, X_2, \ldots, X_n$ from a given distribution. That is, would this proposed distribution be a reasonable probabilistic model for these sample items? To see how the chi-square test can help us answer questions of this sort, consider a very simple example.

Example 9.5-2 Let X denote the number of heads that occur when four coins are tossed at random. Under the assumptions that the four coins are independent and the probability of heads on each coin is $1/2$, X is $b(4, 1/2)$. One hundred repetitions of this experiment resulted in 0, 1, 2, 3, and 4 heads being observed on 7, 18, 40, 31, and 4 trials, respectively. Do these results support the assumptions? That is, is $b(4, 1/2)$ a reasonable model for the distribution of X? To answer this, we begin by letting $A_1 = \{0\}$, $A_2 = \{1\}$, $A_3 = \{2\}$, $A_4 = \{3\}$, $A_5 = \{4\}$. If $p_{i0} = P(X \in A_i)$ when X is $b(4, 1/2)$, then

$$p_{10} = p_{50} = \binom{4}{0}\left(\frac{1}{2}\right)^4 = \frac{1}{16} = 0.0625,$$

$$p_{20} = p_{40} = \binom{4}{1}\left(\frac{1}{2}\right)^4 = \frac{4}{16} = 0.25,$$

$$p_{30} = \binom{4}{2}\left(\frac{1}{2}\right)^4 = \frac{6}{16} = 0.375.$$

At an approximate $\alpha = 0.05$ significance level, the null hypothesis

$$H_0: p_i = p_{i0}, \quad i = 1, 2, \ldots, 5,$$

is rejected if the observed value of Q_4 is greater than $\chi^2_{0.05}(4) = 9.488$. If we use the 100 repetitions of this experiment that resulted in the observed values of $Y_1, Y_2, \ldots, Y_5$ of $y_1 = 7$, $y_2 = 18$, $y_3 = 40$, $y_4 = 31$, and $y_5 = 4$, the computed value of Q_4 is

$$q_4 = \frac{(7 - 6.25)^2}{6.25} + \frac{(18 - 25)^2}{25} + \frac{(40 - 37.5)^2}{37.5} + \frac{(31 - 25)^2}{25}$$

$$+ \frac{(4 - 6.25)^2}{6.25}$$

$$= 4.47.$$

Since $4.47 < 9.488$, the hypothesis is not rejected. That is, the data support the hypothesis that $b(4, 1/2)$ is a reasonable probabilistic model for X.

Thus far all the hypotheses H_0 tested with the chi-square statistic Q_{k-1} have been simple ones (i.e., completely specified, namely, in H_0: $p_i = p_{i0}$, $i = 1, 2, \ldots, k$, each p_{i0} has been known). This is not always the case, and it frequently happens that $p_{10}, p_{20}, \ldots, p_{k0}$ are functions of one or more unknown parameters. For example, suppose that the hypothesized model for X in Example 9.5-2 was H_0: X is $b(4, p)$, $0 < p < 1$. Then

$$p_{i0} = P(X \in A_i) = \frac{4!}{(i-1)!(5-i)!} p^{i-1}(1-p)^{5-i}, \quad i = 1, 2, \ldots, 5,$$

which is a function of the unknown parameter p. Of course, if H_0: $p_i = p_{i0}$, $i = 1, 2, \ldots, 5$, is true, for large n,

$$Q_4 = \sum_{i=1}^{5} \frac{(Y_i - np_{i0})^2}{np_{i0}}$$

still has an approximate chi-square distribution with four degrees of freedom. The difficulty is that when $Y_1, Y_2, \ldots, Y_5$ are observed to be equal to $y_1, y_2, \ldots, y_5$,

Q_4 cannot be computed, since $p_{10}, p_{20}, \ldots, p_{50}$ (and hence Q_4) are functions of the unknown parameter p.

One way out of the difficulty would be to estimate p from the data and then carry out the computations using this estimate. It is interesting to note the following. Say the estimation of p is carried out by minimizing Q_4 with respect to p yielding $\tilde{p}$. This $\tilde{p}$ is sometimes called a **minimum chi-square estimator** of p. If then this $\tilde{p}$ is used in Q_4, the statistic Q_4 still has an approximate chi-square distribution but with only $4 - 1 = 3$ degrees of freedom. That is, the number of degrees of freedom of the approximating chi-square distribution is reduced by one for each parameter estimated by the minimum chi-square technique. We accept this result without proof (as it is a rather difficult one). Although we have considered this when p_{i0}, $i = 1, 2, \ldots, k$, is a function of only one parameter, it holds when there is more than one unknown parameter, say r. Hence, in a more general situation, the test would be completed by computing Q_{k-1} using y_i and the estimated p_{i0}, $i = 1, 2, \ldots, k$, to obtain q_{k-1} (i.e., q_{k-1} is the minimized chi-square). This value q_{k-1} would then be compared to a critical value $\chi_\alpha^2(k-1-r)$. In our special case, the computed (minimized) chi-square q_4 would be compared to $\chi_\alpha^2(3)$.

There is still one trouble with all of this: It is usually very difficult to find minimum chi-square estimators. Hence most statisticians usually use some reasonable method (maximum likelihood is satisfactory) of estimating the parameters. They then compute q_{k-1}, recognizing that it is somewhat larger than the minimized chi-square, and compare it with $\chi_\alpha^2(k-1-r)$. Note that this provides a slightly larger probability of rejecting H_0 than would the scheme in which the minimized chi-square were used because this computed q_{k-1} is larger than the minimum q_{k-1}.

Example 9.5-3 Let X denote the number of alpha particles emitted by barium-133 in $1/10$ of a second. The following 50 observations of X were taken with a Geiger counter in a fixed position:

7	4	3	6	4	4	5	3	5	3
5	5	3	2	5	4	3	3	7	6
6	4	3	11	9	6	7	4	5	4
7	3	2	8	6	7	4	1	9	8
4	8	9	3	9	7	7	9	3	10

The experimenter is interested in determining whether X has a Poisson distribution. To test H_0: X is Poisson, we first estimate the mean of X, say λ, with the sample mean, $\bar{x} = 5.4$, of these 50 observations. we then partition the set of outcomes for this experiment into the sets $A_1 = \{0, 1, 2, 3\}$, $A_2 = \{4\}$, $A_3 = \{5\}$, $A_4 = \{6\}$, $A_5 = \{7\}$, and $A_6 = \{8, 9, 10, \ldots\}$. We combine $0, 1, 2, 3$ into one set A_1 and $8, 9, 10, \ldots$ into another A_6 so that the expected number of outcomes for each set is at least five when H_0 is true. In Table 9.5-1 the data are grouped, and the estimated probabilities specified by the hypothesis that X has a Poisson distribution with an estimated $\lambda = 5.4$ are

TABLE 9.5-1

	Outcome					
	A_1	A_2	A_3	A_4	A_5	A_6
Frequency	13	9	6	5	7	10
Probability	0.213	0.160	0.173	0.156	0.120	0.178

given. Since one parameter was estimated, Q_{6-1} has an approximate chi-square distribution with $r = 5 - 1 = 4$ degrees of freedom. Since

$$q_5 = \frac{[13 - 50(0.213)]^2}{50(0.213)} + \cdots + \frac{[10 - 50(0.178)]^2}{50(0.178)}$$

$$= 2.763 < 9.488 = \chi^2_{0.05}(4),$$

H_0 is not rejected at the 5% significance level. That is, with only these data, we are quite willing to accept the model that X has a Poisson distribution.

Let us now consider the problem of testing a model for the distribution of a random variable W of the continuous type. That is, if $F(w)$ is the distribution function of W, we wish to test

$$H_0: F(w) = F_0(w),$$

where $F_0(w)$ is some known distribution function of the continuous type. In Section 10.7 we will use a Kolmogorov–Smirnov test for H_0. Also recall that we have considered problems of this type using q–q plots. In order to use the chi-square statistic, we must partition the set of possible values of W into k sets. One way this can be done is as follows. Partition the interval $[0, 1]$ into k sets with the points $b_0, b_1, b_2, \ldots, b_k$, where

$$0 = b_0 < b_1 < b_2 < \cdots < b_k = 1.$$

Let $a_i = F_0^{-1}(b_i)$, $i = 1, 2, \ldots, k - 1$; $A_1 = (-\infty, a_1]$, $A_i = (a_{i-1}, a_i]$ for $i = 2, 3, \ldots, k - 1$, and $A_k = (a_{k-1}, \infty)$; $p_i = P(W \in A_i)$, $i = 1, 2, \ldots, k$. Let Y_i denote the number of times the observed value of W belongs to A_i, $i = 1, 2, \ldots, k$, in n independent repetitions of the experiment. Then $Y_1, Y_2, \ldots, Y_k$ have a multinomial distribution with parameters $n, p_1, p_2, \ldots, p_k$. Also let $p_{i0} = P(W \in A_i)$ when the distribution function of W is $F_0(w)$. The hypothesis that we actually test is a modification of H_0, namely

$$H_0': p_i = p_{i0}, \qquad i = 1, 2, \ldots, k.$$

This hypothesis is rejected if the observed value of the chi-square statistic

$$Q_{k-1} = \sum_{i=1}^{k} \frac{(Y_i - np_{i0})^2}{np_{i0}}$$

is at least as great as $\chi_\alpha^2(k-1)$. If the hypothesis H_0': $p_i = p_{i0}$, $i = 1, 2, \ldots, k$, is not rejected, we do not reject the hypothesis H_0: $F(w) = F_0(w)$.

Example 9.5-4   Let W denote the outcome of a random experiment. Let $F(w)$ denote the distribution function of W, and let

$$
F_0(w) = \begin{cases} 0, & w < -1, \\ \dfrac{1}{2}(w^3 + 1), & -1 \le w < 1, \\ 1, & 1 \le w. \end{cases}
$$

Note that $f_0(w) = F_0'(w) = (3/2)w^2$, $-1 < w < 1$, is the hypothesized p.d.f. of W. We shall test the hypothesis H_0: $F(w) = F_0(w)$ against all alternatives. The interval $[0, 1]$ can be partitioned into 10 sets of equal probability with the points $b_i = i/10$, $i = 0, 1, \ldots, 10$. If $a_i = F_0^{-1}(b_i) = (2b_i - 1)^{1/3}$, $i = 1, 2, \ldots, 9$, then the sets $A_1 = [-1, a_1]$, $A_2 = (a_1, a_2], \ldots, A_{10} = (a_9, 1]$ will each have probability $1/10$ when H_0 is true. If a random sample of size $n = 50$ is observed, then $50(1/10) = 5$ is the expected number of outcomes for each set A_i, $i = 1, 2, \ldots, 10$. Letting Y_i denote the number of outcomes in A_i, we see that the hypothesis is rejected if the calculated value of

$$
Q_9 = \sum_{i=1}^{10} \frac{(Y_i - 5)^2}{5}
$$

is greater than 16.92 at an approximate 5% significance level. The outcomes of 50 independent repetitions of this experiment simulated on the computer were

-0.75	-0.28	-0.59	0.26	-0.89	-0.60	0.63
-0.68	-0.62	0.75	-0.81	0.99	0.76	-0.88
0.35	-0.87	0.68	0.81	-0.82	-0.90	0.65
0.43	0.56	0.62	0.96	0.72	-0.67	-0.40
-0.97	-0.96	0.83	0.90	0.83	-0.55	0.69
0.75	0.91	0.88	-0.70	0.96	-0.98	-0.93
-0.77	-0.96	-0.29	0.83	-0.93	0.96	0.79
-0.76						

A summary of these data is given in Table 9.5-2. The values of $a_0, a_1, \ldots, a_{10}$ are -1.00, -0.928, -0.843, -0.737, -0.585, 0.00, 0.585, 0.737, 0.843, 0.928, and 1.00, respectively. The calculated value of Q_9 is

$$
q_9 = \frac{(6 - 5)^2}{5} + \frac{(4 - 5)^2}{5} + \cdots + \frac{(4 - 5)^2}{5} = 4.0 \le 16.92 = \chi_{0.05}^2(9),
$$

and thus the hypothesis is not rejected.

TABLE 9.5-2

Outcome	Frequency	Outcome	Frequency
A_1	6	A_6	4
A_2	4	A_7	6
A_3	5	A_8	8
A_4	6	A_9	3
A_5	4	A_{10}	4

Let $X_1, X_2, \ldots, X_n$ denote a random sample of size n from a distribution with mean μ and positive variance σ^2. Let $W = (\overline{X} - \mu)/(\sigma/\sqrt{n})$, where $\overline{X}$ is the sample mean. From the Central Limit Theorem, we claim that W has a distribution that is approximately $N(0, 1)$ when n is sufficiently large. For what values of n can we say that W is approximately $N(0, 1)$? The answer will, of course, depend on the distribution from which the sample is taken. For example, if the underlying distribution is normal, $n = 1$ is sufficient. In the next example, we shall give a partial answer for the U-shaped distribution used in Example 9.5-4.

Example 9.5-5 Let $X_1, X_2, \ldots, X_n$ denote a random sample of size n from the distribution with p.d.f. $f(x) = (3/2)x^2$, $-1 < x < 1$. For this distribution $\mu = 0$ and $\sigma^2 = 3/5$. We shall test the hypothesis that $W = \overline{X}/\sqrt{3/(5n)}$ is approximately $N(0, 1)$ when $n = 7$. One thousand observations of W were generated by the computer. Table 9.5-3 gives 19 sets, their frequencies, and their expected values when sampling from $N(0, 1)$. For these data,

$$q_{18} = \frac{(3 - 5.4)^2}{5.4} + \frac{(4 - 6.8)^2}{6.8} + \cdots + \frac{(5 - 5.4)^2}{5.4} = 39.988.$$

TABLE 9.5-3

A_i	Y_i	$1000p_{i0}$	A_i	Y_i	$1000p_{i0}$
$(-\infty, -2.55]$	3	5.4	$(0.15, 0.45]$	115	114.0
$(-2.55, -2.25]$	4	6.8	$(0.45, 0.75]$	94	99.8
$(-2.25, -1.95]$	21	13.4	$(0.75, 1.05]$	83	79.7
$(-1.95, -1.65]$	29	23.9	$(1.05, 1.35]$	77	58.4
$(-1.65, -1.35]$	43	39.0	$(1.35, 1.65]$	29	39.0
$(-1.35, -1.05]$	79	58.4	$(1.65, 1.95]$	18	23.9
$(-1.05, -0.75]$	90	79.7	$(1.95, 2.25]$	9	13.4
$(-0.75, -0.45]$	85	99.8	$(2.25, 2.55]$	7	6.8
$(-0.45, -0.15]$	122	114.0	$(2.55, +\infty)$	5	5.4
$(-0.15, 0.15]$	87	119.2			

Since $39.988 > 28.87 = \chi^2_{0.05}(18)$, the hypothesis is rejected when $n = 7$. In Exercise 9.5-19 you will be asked to test this hypothesis for other values of n. It would be instructive for you to draw a relative frequency histogram for the data in Table 9.5-3 with the $N(0, 1)$ p.d.f. superimposed on the same graph (see Figure 5.4-4).

It is also true, in dealing with models of random variables of the continuous type, that we must frequently estimate unknown parameters. For example, let H_0 be that W is $N(\mu, \sigma^2)$, where μ and σ^2 are unknown. With a random sample W_1, $W_2, \ldots, W_n$, we first can estimate μ and σ^2, possibly with $\overline{w}$ and s_w^2. We partition the space $\{w: -\infty < w < \infty\}$ into k mutually disjoint sets $A_1, A_2, \ldots, A_k$. We then use the estimated μ and σ^2 to estimate

$$p_{i0} = \int_{A_i} \frac{1}{\sigma\sqrt{2\pi}} \exp\left[-\frac{(w - \mu)^2}{2\sigma^2} \right] dw,$$

$i = 1, 2, \ldots, k$. Using the observed frequencies $y_1, y_2, \ldots, y_k$ of A_1, A_2, $\ldots, A_k$, respectively, from the observed random sample $w_1, w_2, \ldots, w_n$, and $\widehat{p}_{10}, \widehat{p}_{20}, \ldots, \widehat{p}_{k0}$ estimated with $\overline{w}$ and s_w^2, we compare the computed

$$q_{k-1} = \sum_{i=1}^{k} \frac{(y_i - n\widehat{p}_{i0})^2}{n\widehat{p}_{i0}}$$

to $\chi^2_\alpha(k-1-2)$. This value q_{k-1} will again be somewhat larger than that which would be found using minimum chi-square estimation, and certain caution should be observed. Several exercises illustrate the procedure in which one or more parameters must be estimated.

Finally, it should be noted that the methods given in this section frequently are classified under the more general title of goodness of fit tests. In particular, then, the tests in this section would be **chi-square goodness of fit tests.**

Exercises

9.5-1   A 1-pound bag of candy-coated chocolate-covered peanuts contained 224 pieces of candy colored brown, orange, green, and yellow. Test the null hypothesis that the four colors of candy are equally likely; that is, test

$$H_0: p_B = p_O = p_G = p_Y = \frac{1}{4}.$$

The observed values were 42 brown, 64 orange, 53 green, and 65 yellow. You may select the significance level.

9.5-2   A particular brand of candy-coated chocolate comes in five different colors that we shall denote as $A_1 = \{\text{brown}\}$, $A_2 = \{\text{yellow}\}$, $A_3 = \{\text{orange}\}$, $A_4 = \{\text{green}\}$, and $A_5 = \{\text{coffee}\}$. Let p_i equal the probability that the color of a piece of candy selected at random belongs to A_i, $i = 1, 2, \ldots, 5$. Test the null hypothesis

$$H_0: p_1 = 0.4, \, p_2 = 0.2, \, p_3 = 0.2, \, p_4 = 0.1, \, p_5 = 0.1$$

using a random sample of $n = 580$ pieces of candy whose colors yielded the respective frequencies 224, 119, 130, 48, and 59. You may select the significance level.

9.5-3   In the Michigan Daily Lottery, each weekday a three-digit integer is generated one digit at a time. Let p_i denote the probability of generating digit i, $i = 0, 1, \ldots, 9$. Use the following 50 digits to test $H_0: p_0 = p_1 = \cdots = p_9 = 1/10$. Let $\alpha = 0.05$.

1	6	9	9	3	8	5	0	6	7
4	7	5	9	4	6	5	6	4	4
4	8	0	9	3	2	1	5	4	5
7	3	2	1	4	6	7	1	3	4
4	8	8	6	1	6	1	2	8	8

9.5-4   In a large bin of crocus bulbs it is claimed that 1/4 will produce yellow crocuses, 1/4 will produce white crocuses, and 1/2 will produce purple crocuses. If 40 bulbs produced 6 yellow, 7 white, and 27 purple crocuses, would the claim be rejected?

9.5-5   In a biology laboratory the mating of two red-eyed fruit flies yielded $n = 432$ offspring, among which 254 were red-eyed, 69 were brown-eyed, 87 were scarlet-eyed, and 22 were white-eyed. Use these data to test, with $\alpha = 0.05$, the hypothesis that the ratio among the offspring would be $9:3:3:1$, respectively.

9.5-6   In a biology laboratory students test the Mendelian theory of inheritance using corn. The Mendelian theory of inheritance claims that frequencies of the four categories smooth and yellow, wrinkled and yellow, smooth and purple, and wrinkled and purple will occur in the ratio $9:3:3:1$. If a student counted 124, 30, 43, and 11, respectively, for these four categories, would these data support the Mendelian theory? Let $\alpha = 0.05$.

9.5-7   Let X equal the number of female children in a 3-child family. We shall use a chi-square goodness of fit statistic to test the null hypothesis that the distribution of X is $b(3, 0.5)$.
   (a) Define the test statistic and critical region using an $\alpha = 0.05$ significance level.
   (b) Among students who were taking statistics, 52 came from families with 3 children. For these families, $x = 0, 1, 2$, and 3 for 5, 17, 24, and 6 families, respectively. Calculate the value of the test statistic and state your conclusion.

9.5-8   For the data in Exercise 3.5-15, it is claimed that the distribution of X is $b(9, 0.8)$. Test whether this is true using a chi-square goodness of fit test with $\alpha = 0.01$.

9.5-9 In Example 3.7-3, it is claimed that X has a Poisson distribution and 100 observations of X are given in Table 3.7-1. Use a chi-square goodness of fit statistic to test whether this is true. Since $\bar{x} = 5.59$, let the estimate of λ be 5.6 so that you can use the Poisson probability table in the Appendix. Let $A_1 = \{x \leq 1\}$ and $A_{11} = \{x \geq 11\}$. Use an $\alpha = 0.05$ significance level.

9.5-10 There is a rare type of heredity change in *E. coli* which causes this bacterium to become resistant to the drug streptomycin. This type of change, called mutation, can be detected by plating many bacteria on petri dishes containing antibiotic medium. Any colonies that grow on this medium result from a single mutant cell. A sample of $n = 150$ petri plates of streptomycin agar were each plated with 10^6 bacteria and the numbers of colonies were counted on each petri dish. The observed results were that 92 plates had 0 colonies, 46 had 1, 8 had 2, 3 had 3 and 1 plate had 4 colonies. Let X equal the number of colonies per plate. Test the hypothesis that X has a Poisson distribution. Use $\bar{x} = 0.5$ as an estimate of λ. Let $\alpha = 0.01$.

9.5-11 In Exercise 3.7-13, we are asked whether the data look like observations of a Poisson random variable with mean $\lambda = 3$. Use a chi-square goodness of fit test to answer this question. Use $\alpha = 0.05$.

9.5-12 While testing a used tape for bad records, a computer operator counted the number of flaws per 100 feet of tape. Let X equal this random variable. Test the null hypothesis that X has a Poisson distribution with a mean of $\lambda = 2.4$ given that 40 observations of X yielded 5 zeros, 7 ones, 12 two, 9 threes, 5 fours, 1 five, and 1 six. Let $\alpha = 0.05$.

HINT: Combine five and six into one set; that is, the last set would be all x values ≥ 5.

9.5-13 Let X equal the distance between bad records on a used computer tape. Test the hypothesis that the distribution of X is exponential using the following 90 observations of X and 10 classes of equal probability. Use $\bar{x} = 42.4$ as an estimate of θ. Let $\alpha = 0.05$.

30	79	38	47	22	52	36	36	7	57
3	22	30	14	8	32	15	21	12	12
6	67	6	7	35	78	28	74	5	9
37	1	3	3	44	160	50	27	61	15
39	44	130	18	6	1	32	116	23	12
58	101	68	53	58	21	21	7	79	41
80	33	71	81	17	10	13	49	21	56
107	21	17	64	14	36	26	1	54	207
64	238	25	51	82	8	2	3	43	87

9.5-14 In Exercise 4.2-16, we are asked whether the times that bees spend in low-density flower patches are observations of an exponential random variable. Test whether this is true using $k = 10$ classes of equal probability. Note that θ must be estimated using $\bar{x}$. Let $\alpha = 0.05$.

9.5-15 In Exercise 4.4-12, we are asked whether the force required to pull a stud out of a window is an observation of a normally distributed random variable. Test whether this is true using $k = 10$ classes. Note that both μ and σ must be estimated using $\bar{x}$ and s, respectively. Let $\alpha = 0.05$.

9.5-16 Let X equal the amount of butterfat (in pounds) produced by 90 cows during a 305-day milk production period following their first calf. Test the hypothesis that the distribution of X is $N(\mu, \sigma^2)$. You may use $\bar{x} = 511.633$ and $s_x = 87.576$ as estimates of μ and σ, respectively.

486	537	513	583	453	510	570	500	458	555
618	327	350	643	500	497	421	505	637	599
392	574	492	635	460	696	593	422	499	524
539	339	472	427	532	470	417	437	388	481
537	489	418	434	466	464	544	475	608	444
573	611	586	613	645	540	494	532	691	478
513	583	457	612	628	516	452	501	453	643
541	439	627	619	617	394	607	502	395	470
531	526	496	561	491	380	345	274	672	509

9.5-17 A sample of 100 2.2K ohms resistors was tested, yielding the following measurements:

2.17	2.18	2.17	2.19	2.23	2.18	2.16	2.23	2.29	2.20
2.24	2.13	2.18	2.22	2.21	2.22	2.23	2.18	2.23	2.25
2.24	2.20	2.21	2.20	2.25	2.15	2.20	2.25	2.16	2.19
2.20	2.18	2.18	2.19	2.22	2.19	2.20	2.19	2.22	2.21
2.22	2.18	2.18	2.28	2.19	2.23	2.21	2.19	2.21	2.21
2.19	2.18	2.21	2.17	2.20	2.18	2.18	2.21	2.22	2.18
2.22	2.23	2.19	2.23	2.18	2.19	2.18	2.21	2.18	2.15
2.20	2.23	2.20	2.20	2.23	2.20	2.19	2.22	2.17	2.20
2.20	2.17	2.19	2.19	2.25	2.19	2.19	2.20	2.20	2.19
2.21	2.14	2.24	2.21	2.19	2.23	2.18	2.22	2.23	2.19

(a) Group these data into 9 classes with the class boundaries for the first class being 2.125–2.145.

(b) Use a chi-square goodness of fit statistic to test whether these are observations of a normally distributed random variable, after grouping the first two classes and the last two classes to avoid small expected numbers. Estimate μ and σ with the mean and standard deviation of the grouped data. Let $\alpha = 0.05$.

9.5-18 Let $X_1, X_2, \ldots, X_n$ denote a random sample of size n from the distribution with p.d.f. $f(x) = (x + 1)/2, -1 < x < 1$. For this distribution $\mu = 1/3$, and $\sigma^2 = 2/9$. Let $W_n = (\bar{X} - 1/3)/\sqrt{2/9n}$. One thousand observations of W_n were simulated by the computer when $n = 2, 7$, and 12. These data were grouped and are listed in Table 9.5-4. Test the hypothesis H_0: W_n is $N(0, 1)$, at $\alpha = 0.05$, for

TABLE 9.5-4

A_i	$Y_i(n = 2)$	$Y_i(n = 7)$	$Y_i(n = 12)$
$(-\infty, -2.55]$	10	6	7
$(-2.55, -2.25]$	9	10	11
$(-2.25, -1.95]$	18	18	16
$(-1.95, -1.65]$	24	29	24
$(-1.65, -1.35]$	35	42	34
$(-1.35, -1.05]$	47	60	60
$(-1.05, -0.75]$	71	80	88
$(-0.75, -0.45]$	82	87	98
$(-0.45, -0.15]$	93	112	100
$(-0.15, 0.15]$	102	108	121
$(0.15, 0.45]$	125	120	108
$(0.45, 0.75]$	124	101	92
$(0.75, 1.05]$	93	86	81
$(1.05, 1.35]$	89	64	74
$(1.35, 1.65]$	51	36	42
$(1.65, 1.95]$	27	22	26
$(1.95, 2.25]$	0	15	10
$(2.25, 2.55]$	0	3	7
$(2.55, +\infty)$	0	1	1

(a) $n = 2$.
(b) $n = 7$.
(c) $n = 12$.
(d) Based on these limited data, for what values of n would you accept H_0 when the underlying distribution is this skewed one?

REMARK Figure 5.4-3 uses the data listed for $n = 2$ and $n = 12$.

9.5-19 Let $X_1, X_2, \ldots , X_n$ denote a random sample of size n from the distribution with p.d.f. $f(x) = (3/2)x^2$, $-1 < x < 1$. Let $W_n = (\overline{X} - 0)/\sqrt{3/5n}$. One thousand observations of W_n were simulated using a computer when $n = 2, 8$, and 12. These data were grouped and are listed in Table 9.5-5. Test the hypothesis H_0: W_n is $N(0, 1)$, at $\alpha = 0.05$, for
(a) $n = 2$.
(b) $n = 8$.
(c) $n = 12$.
(d) Draw the relative frequency histogram for each value of n.
(e) Based on these limited data, for what values of n would you accept the hypothesis that W_n is about $N(0, 1)$ when the underlying distribution has this U-shape?

REMARK Figure 5.4-4 uses the data listed for $n = 2$ and $n = 12$.

TABLE 9.5-5

A_i	$Y_i(n = 2)$	$Y_i(n = 8)$	$Y_i(n = 12)$
$(-\infty, -2.55]$	0	5	3
$(-2.55, -2.25]$	0	6	5
$(-2.25, -1.95]$	0	14	11
$(-1.95, -1.65]$	28	11	27
$(-1.65, -1.35]$	111	51	51
$(-1.35, -1.05]$	70	41	62
$(-1.05, -0.75]$	28	89	86
$(-0.75, -0.45]$	20	93	99
$(-0.45, -0.15]$	126	97	113
$(-0.15, 0.15]$	213	138	120
$(0.15, 0.45]$	100	117	102
$(0.45, 0.75]$	23	108	94
$(0.75, 1.05]$	34	78	81
$(1.05, 1.35]$	87	49	59
$(1.35, 1.65]$	126	51	39
$(1.65, 1.95]$	34	25	27
$(1.95, 2.25]$	0	16	10
$(2.25, 2.55]$	0	9	9
$(2.55, +\infty)$	0	2	2

9.6 Contingency Tables

In this section we demonstrate the flexibility of the chi-square test. We first look at a method for testing whether two or more multinomial distributions are equal, sometimes called a test for homogeneity. Then we consider a test for independence of attributes of classification. Both of these lead to a similar test statistic.

Suppose that each of two independent experiments can end in one of the k mutually exclusive and exhaustive events $A_1, A_2, \ldots, A_k$. Let

$$p_{ij} = P(A_i), \qquad i = 1, 2, \ldots, k, \qquad j = 1, 2.$$

That is, $p_{11}, p_{21}, \ldots, p_{k1}$ are the probabilities of the events in the first experiment and $p_{12}, p_{22}, \ldots, p_{k2}$ are those associated with the second experiment. Let the experiments be repeated n_1 and n_2 independent times, respectively. Also let $Y_{11}, Y_{21}, \ldots, Y_{k1}$ be the frequencies of $A_1, A_2, \ldots, A_k$ associated with the n_1 independent trials of the first experiment. Similarly, let $Y_{12}, Y_{22}, \ldots, Y_{k2}$ be the respective frequencies associated with the n_2 trials of the second experiment. Of course, $\Sigma_{i=1}^{k} Y_{ij} = n_j, j = 1, 2$. From the sampling distribution theory corresponding to the basic chi-square test, we know that each of

$$\sum_{i=1}^{k} \frac{(Y_{ij} - n_j p_{ij})^2}{n_j p_{ij}}, \qquad j = 1, 2,$$

has an approximate chi-square distribution with $k - 1$ degrees of freedom. Since the two experiments are independent (and thus the two chi-square statistics are independent), the sum

$$\sum_{j=1}^{2} \sum_{i=1}^{k} \frac{(Y_{ij} - n_j p_{ij})^2}{n_j p_{ij}}$$

is approximately chi-square with $k - 1 + k - 1 = 2k - 2$ degrees of freedom.

Usually, p_{ij}, $i = 1, 2, \ldots, k$, $j = 1, 2$, are unknown, but frequently we wish to test the hypothesis

$$H_0: p_{11} = p_{12}, \; p_{21} = p_{22}, \; \ldots, \; p_{k1} = p_{k2};$$

that is, the hypothesis that the corresponding probabilities associated with the two independent experiments are equal. Under H_0, we can estimate the unknown

$$p_{i1} = p_{i2}, \qquad i = 1, 2, \ldots, k,$$

by using the relative frequency $(Y_{i1} + Y_{i2})/(n_1 + n_2)$, $i = 1, 2, \ldots, k$. That is, if H_0 is true, we can say that the two experiments are actually parts of a large one in which $Y_{i1} + Y_{i2}$ is the frequency of the event A_i, $i = 1, 2, \ldots, k$. Note that we only have to estimate the $k - 1$ probabilities $p_{i1} = p_{i2}$ using

$$\frac{Y_{i1} + Y_{i2}}{n_1 + n_2}, \qquad i = 1, 2, \ldots, k - 1,$$

since the sum of the k probabilities must equal one. That is, the estimator of $p_{k1} = p_{k2}$ is

$$1 - \frac{Y_{11} + Y_{12}}{n_1 + n_2} - \cdots - \frac{Y_{k-1,1} + Y_{k-1,2}}{n_1 + n_2} = \frac{Y_{k1} + Y_{k2}}{n_1 + n_2}.$$

Substituting these estimators, we have that

$$Q = \sum_{j=1}^{2} \sum_{i=1}^{k} \frac{[Y_{ij} - n_j(Y_{i1} + Y_{i2})/(n_1 + n_2)]^2}{n_j(Y_{i1} + Y_{i2})/(n_1 + n_2)}$$

has an approximate chi-square distribution with $2k - 2 - (k - 1) = k - 1$ degrees of freedom. Here $k - 1$ is subtracted from $2k - 2$ because that is the number of estimated parameters. The critical region for testing H_0 is of the form

$$q \geq \chi_\alpha^2(k-1).$$

Example 9.6-1 To test two methods of instruction, 50 students are selected at random for each of two groups. At the end of the instruction period each student is assigned a grade (A, B, C, D, or F) by an evaluating team. The data are recorded as follows:

	A	B	C	D	F	Totals
Group I	8	13	16	10	3	50
Group II	4	9	14	16	7	50

(Grade columns: A B C D F)

Accordingly, if the hypothesis H_0 that the corresponding probabilities are equal is true, the respective estimates of the probabilities are

$$\frac{8 + 4}{100} = 0.12, \ 0.22, \ 0.30, \ 0.26, \ \frac{3 + 7}{100} = 0.10.$$

Thus the estimates of $n_1 p_{i1} = n_2 p_{i2}$ are 6, 11, 15, 13, 5, respectively. Hence the computed value of Q is

$$q = \frac{(8 - 6)^2}{6} + \frac{(13 - 11)^2}{11} + \frac{(16 - 15)^2}{15} + \frac{(10 - 13)^2}{13} + \frac{(3 - 5)^2}{5}$$

$$+ \frac{(4 - 6)^2}{6} + \frac{(9 - 11)^2}{11} + \frac{(14 - 15)^2}{15} + \frac{(16 - 13)^2}{13} + \frac{(7 - 5)^2}{5}$$

$$= \frac{4}{6} + \frac{4}{11} + \frac{1}{15} + \frac{9}{13} + \frac{4}{5} + \frac{4}{6} + \frac{4}{11} + \frac{1}{15} + \frac{9}{13} + \frac{4}{5} = 5.18.$$

Now, under H_0, Q has an approximate chi-square distribution with $k - 1 = 4$ degrees of freedom so the $\alpha = 0.05$ critical region is $q \geq 9.488 = \chi^2_{0.05}(4)$. Here $q = 5.18 < 9.488$, and hence H_0 is not rejected at the 5% significance level. Furthermore the p-value for $q = 5.18$ is 0.268, which is greater than most significance levels. Thus with these data, we cannot say there is a difference between the two methods of instruction.

It is fairly obvious how this procedure can be extended to testing the equality of h independent multinomial distributions. That is, let

$$p_{ij} = P(A_i), \qquad i = 1, 2, \ldots, k, \ j = 1, 2, \ldots, h,$$

and test

$$H_0: p_{i1} = p_{i2} = \cdots = p_{ih} = p_i, \qquad i = 1, 2, \ldots, k.$$

Repeat the jth experiment n_j independent times and let $Y_{1j}, Y_{2j}, \ldots, Y_{kj}$ denote the frequencies of the respective events $A_1, A_2, \ldots, A_k$. Now

$$Q = \sum_{j=1}^{h} \sum_{i=1}^{k} \frac{(Y_{ij} - n_j p_{ij})^2}{n_j p_{ij}}$$

has an approximate chi-square distribution with $h(k - 1)$ degrees of freedom. Under H_0, we must estimate $k - 1$ probabilities using

$$\widehat{p}_i = \frac{\displaystyle\sum_{j=1}^{h} Y_{ij}}{\displaystyle\sum_{j=1}^{h} n_j}, \qquad i = 1, 2, \ldots, k - 1,$$

because the estimate of p_k follows from $\widehat{p}_k = 1 - \widehat{p}_1 - \widehat{p}_2 - \cdots - \widehat{p}_{k-1}$. We use these estimates to obtain

$$Q = \sum_{j=1}^{h} \sum_{i=1}^{k} \frac{(Y_{ij} - n_j \widehat{p}_i)^2}{n_j \widehat{p}_i},$$

which has an approximate chi-square distribution with $h(k - 1) - (k - 1) = (h - 1)(k - 1)$ degrees of freedom.

Let us see how we can use the above procedures to test the equality of two or more independent distributions that are not necessarily multinomial. Suppose first that we are given random variables U and V with distribution functions $F(u)$ and $G(v)$, respectively. It is sometimes of interest to test the hypothesis H_0: $F(x) = G(x)$ for all x. We have previously considered tests such as $\mu_U = \mu_V$, $\sigma_U^2 = \sigma_V^2$, and have used q–q plots for graphical comparisons of two random samples. Now we shall only assume that the distributions are independent and of the continuous types.

We are interested in testing the hypothesis H_0: $F(x) = G(x)$ for all x. This hypothesis will be replaced by another one. Partition the real line into k mutually disjoint sets $A_1, A_2, \ldots, A_k$. Let

$$p_{i1} = P(U \in A_i), \qquad i = 1, 2, \ldots, k$$

and

$$p_{i2} = P(V \in A_i), \qquad i = 1, 2, \ldots, k.$$

We observe that if $F(x) = G(x)$ for all x, then $p_{i1} = p_{i2}$, $i = 1, 2, \ldots, k$. We replace the hypothesis H_0: $F(x) = G(x)$ with the less restrictive hypothesis H_0': $p_{i1} = p_{i2}$, $i = 1, 2, \ldots, k$. That is, we are now essentially interested in testing the equality of two multinomial distributions.

Let n_1 and n_2 denote the number of independent observations of U and V, respectively. For $i = 1, 2, \ldots, k$, let Y_{ij} denote the number of these observations of U and V, $j = 1, 2$, respectively, that fall into a set A_i. At this point, we proceed to make the test of H_0' as described earlier. Of course, if H_0' is rejected at the (approximate) significance level α, then H_0 is rejected with the same probability. However, if H_0' is true, H_0 is not necessarily true. Thus if H_0' is accepted, it is probably better to say that we do not reject H_0 than to say H_0 is accepted.

In applications, the question of how to select $A_1, A_2, \ldots, A_k$ is frequently raised. Obviously, there is not a single choice for k nor the dividing marks of the partition. But it is interesting to observe that the combined sample can be used in this selection without upsetting the approximate distribution of Q. For example, suppose that $n_1 = n_2 = 20$. We could easily select the dividing marks of the partition so that $k = 4$ and one fourth of the combined sample falls in each of the four sets.

Example 9.6-2   Select, at random, 20 cars of each of two comparable major-brand models. All 40 cars are submitted to accelerated life testing; that is, they are driven many miles over very poor roads in a short time and their failure times are recorded (in weeks):

Brand U: 25   31   20   42   39   19   35   36   44   26
          38   31   29   41   43   36   28   31   25   38

Brand V: 28   17   33   25   31   21   16   19   31   27
          23   19   25   22   29   32   24   20   34   26

If we use 23.5, 28.5, and 34.5 as dividing marks, we note that exactly one fourth of the 40 cars fall in each of the resulting four sets. Thus the data can be summarized as follows:

	A_1	A_2	A_3	A_4	Totals
Brand U	2	4	4	10	20
Brand V	8	6	6	0	20

The estimate of each p_i is $10/40 = 1/4$, which multiplied by $n_j = 20$ gives 5. Hence the computed Q is

$$q = \frac{(2-5)^2}{5} + \frac{(4-5)^2}{5} + \frac{(4-5)^2}{5} + \frac{(10-5)^2}{5} + \frac{(8-5)^2}{5}$$
$$+ \frac{(6-5)^2}{5} + \frac{(6-5)^2}{5} + \frac{(0-5)^2}{5}$$
$$= \frac{72}{5} = 14.4 > 7.815 = \chi^2_{0.05}(3).$$

Also the p-value is 0.0028. Hence it seems that the two brands of cars have different distributions for the length of life under accelerated life testing.

Again it should be clear how this can be extended to more than two distributions, and this extension will be illustrated in the exercises.

Now let us suppose that a random experiment results in an outcome that can be classified by two different attributes, such as height and weight. Assume that the first attribute is assigned to one and only one of k mutually exclusive and exhaustive events, say $A_1, A_2, \ldots, A_k$, and the second attribute falls in one and only one of

h mutually exclusive and exhaustive events, say $B_1, B_2, \ldots, B_h$. Let the probability of $A_i \cap B_j$ be defined by

$$p_{ij} = P(A_i \cap B_j), \qquad i = 1, 2, \ldots, k, \qquad j = 1, 2, \ldots, h.$$

The random experiment is to be repeated n independent times, and Y_{ij} will denote the frequency of the event $A_i \cap B_j$. Since there are kh such events as $A_i \cap B_j$, the random variable

$$Q_{kh-1} = \sum_{j=1}^{h} \sum_{i=1}^{k} \frac{(Y_{ij} - np_{ij})^2}{np_{ij}}$$

has an approximate chi-square distribution with $kh - 1$ degrees of freedom, provided n is large.

Suppose that we wish to test the hypothesis of the independence of the A and B attributes, namely,

$$H_0: P(A_i \cap B_j) = P(A_i)P(B_j), \qquad i = 1, 2, \ldots, k, \qquad j = 1, 2, \ldots, h.$$

Let us denote $P(A_i)$ by $p_{i\cdot}$ and $P(B_j)$ by $p_{\cdot j}$; that is,

$$p_{i\cdot} = \sum_{j=1}^{h} p_{ij} = P(A_i), \qquad \text{and} \qquad p_{\cdot j} = \sum_{i=1}^{k} p_{ij} = P(B_j),$$

and, of course,

$$1 = \sum_{j=1}^{h} \sum_{i=1}^{k} p_{ij} = \sum_{j=1}^{h} p_{\cdot j} = \sum_{i=1}^{k} p_{i\cdot}.$$

Then the hypothesis can be formulated as

$$H_0: p_{ij} = p_{i\cdot}p_{\cdot j}, \qquad i = 1, 2, \ldots, k, \qquad j = 1, 2, \ldots, h.$$

To test H_0, we can use Q_{kh-1} with p_{ij} replaced by $p_{i\cdot}p_{\cdot j}$. But if $p_{i\cdot}$, $i = 1, 2, \ldots, k$, and $p_{\cdot j}$, $j = 1, 2, \ldots, h$, are unknown, as they usually are in the applications, we cannot compute Q_{kh-1} once the frequencies are observed. In such a case we estimate these unknown parameters by

$$\widehat{p}_{i\cdot} = \frac{y_{i\cdot}}{n}, \qquad \text{where } y_{i\cdot} = \sum_{j=1}^{h} y_{ij}$$

is the observed frequency of A_i, $i = 1, 2, \ldots, k$; and

$$\widehat{p}_{\cdot j} = \frac{y_{\cdot j}}{n}, \qquad \text{where } y_{\cdot j} = \sum_{i=1}^{k} y_{ij}$$

is the observed frequency of B_j, $j = 1, 2, \ldots, h$. Since $\Sigma_{i=1}^{k} p_{i\cdot} = \Sigma_{j=1}^{h} p_{\cdot j} = 1$, we actually estimate only $k - 1 + h - 1 = k + h - 2$ parameters. So if these estimates are used in Q_{kh-1}, with $p_{ij} = p_{i\cdot}p_{\cdot j}$, then, according to the rule stated earlier, the random variable

$$Q = \sum_{j=1}^{h} \sum_{i=1}^{k} \frac{[Y_{ij} - n(Y_{i\cdot}/n)(Y_{\cdot j}/n)]^2}{n(Y_{i\cdot}/n)(Y_{\cdot j}/n)}$$

has an approximate chi-square distribution with $kh - 1 - (k + h - 2) = (k - 1)(h - 1)$ degrees of freedom, provided that H_0 is true. The hypothesis H_0 is rejected if the computed value of this statistic exceeds $\chi^2_\alpha[(k-1)(h-1)]$.

Example 9.6-3 A random sample of 400 undergraduate students at the University of Iowa was taken. The students in the sample were classified according to the college in which they were enrolled and according to their sex. These results are recorded in Table 9.6-1; this table is called a $k \times h$ **contingency table,** where here $k = 2$ and $h = 5$. (Do not be concerned about the numbers in parentheses at this point.) Incidentally, these numbers do actually reflect the composition of the undergraduate colleges at Iowa, but they were modified a little to make the computations easier in this first example.

We desire to test the null hypothesis H_0: $p_{ij} = p_{i\cdot}p_{\cdot j}$, $1 = 1, 2$ and $j = 1, 2, 3, 4, 5$, that the college in which a student enrolls is independent of the sex of that student. Under H_0, estimates of the probabilities are

$$\hat{p}_{1\cdot} = \frac{190}{400} = 0.475 \qquad \text{and} \qquad \hat{p}_{2\cdot} = \frac{210}{400} = 0.525$$

and

$$\hat{p}_{\cdot 1} = \frac{35}{400} = 0.0875, \ \hat{p}_{\cdot 2} = 0.05, \ \hat{p}_{\cdot 3} = 0.8, \ \hat{p}_{\cdot 4} = 0.0375, \ \hat{p}_{\cdot 5} = 0.025.$$

The expected numbers $n(y_{i\cdot}/n)(y_{\cdot j}/n)$ are computed as follows:

$$400(0.475)(0.0875) = 16.625,$$
$$400(0.525)(0.0875) = 18.375,$$
$$400(0.475)(0.05) = 9.5,$$

and so on. These are the values recorded in the parentheses in Table 9.6-1. The computed chi-square statistic is

$$q = \frac{(21 - 16.625)^2}{16.625} + \frac{(14 - 18.375)^2}{18.375} + \cdots + \frac{(4 - 5.25)^2}{5.25}$$
$$= 1.15 + 1.04 + 4.45 + 4.02 + 0.32 + 0.29 + 3.69$$
$$+ 3.34 + 0.33 + 0.30 = 18.93.$$

Since the number of degrees of freedom equals $(k - 1)(h - 1) = 4$, this $q = 18.93 > 13.28 = \chi^2_{0.01}(4)$, and we reject H_0 at the $\alpha = 0.01$ significance level. Moreover, since the first two terms of q come from the business col-

TABLE 9.6-1

			College			
Sex	Business	Engineering	Liberal Arts	Nursing	Pharmacy	Totals
Male	21	16	145	2	6	190
	(16.625)	(9.5)	(152)	(7.125)	(4.75)	
Female	14	4	175	13	4	210
	(18.375)	(10.5)	(168)	(7.875)	(5.25)	
Totals	35	20	320	15	10	400

lege, the next two from engineering, and so on, it is clear that the enrollments in engineering and nursing are more highly dependent on sex than in the other colleges because they have contributed the most to the value of the chi-square statistic. It is also interesting to note that one expected number is less than five, namely 4.75. However, as the associated term in q does not contribute an unusual amount to the chi-square value, it does not concern us.

It is fairly obvious how to extend the testing procedure above to more than two attributes. For example, if the third attribute falls in one and only one of m mutually exclusive and exhaustive events, say $C_1, C_2, \ldots, C_m$, then we test the independence of the three attributes by using

$$Q = \sum_{r=1}^{m} \sum_{j=1}^{h} \sum_{i=1}^{k} \frac{[Y_{ijr} - n(Y_{i..}/n)(Y_{.j.}/n)(Y_{..r}/n)]^2}{n(Y_{i..}/n)(Y_{.j.}/n)(Y_{..r}/n)},$$

where Y_{ijr}, $Y_{i..}$, $Y_{.j.}$, and $Y_{..r}$ are the respective observed frequencies of the events $A_i \cap B_j \cap C_r$, A_i, B_j, and C_r in n independent trials of the experiment. If n is large and if the three attributes are independent, then Q has an approximate chi-square distribution with $khm - 1 - (k - 1) - (h - 1) - (m - 1) = khm - k - h - m + 2$ degrees of freedom.

Rather than explore this extension further, it is more instructive to note some interesting uses of contingency tables.

Example 9.6-4 Say we observed 30 values $x_1, x_2, \ldots, x_{30}$ said to represent the items of a random sample. That is, the corresponding random variables $X_1, X_2, \ldots, X_{30}$ were supposed to be mutually independent and to have the same distribution. Say, however, by looking at the 30 values we detect an upward trend that indicates there might have been some dependence and/or the items did not actually have the same distribution. One simple way to test if they could be thought of as being observed items of a random sample is the following. Mark each x high (H) or low (L) depending on whether it is above or below the sample median. Then divide the x values into three groups: $x_1, \ldots, x_{10}$; $x_{11}, \ldots, x_{20}$; $x_{21}, \ldots, x_{30}$. Certainly if the items are those of a random sample we would expect five H's and five L's in each group. That is, the attribute classified as H or L should be independent of the group number. The summary of these data provides a 3×2 contingency table. For example, say the 30 values are

5.6	8.2	7.8	4.8	5.5	8.1	6.7	7.7	9.3	6.9
8.2	10.1	7.5	6.9	11.1	9.2	8.7	10.3	10.7	10.0
9.2	11.6	10.3	11.7	9.9	10.6	10.0	11.4	10.9	11.1

The median can be taken to be the average of the two middle items in magnitude, namely, 9.2 and 9.3. Marking each item H or L after comparing it with this median, we obtain the following 3×2 contingency table.

Group	L	H	Totals
1	9	1	10
2	5	5	10
3	1	9	10
Totals	15	15	30

Here each $n(y_{i.}/n)(y_{.j}/n) = 30(10/30)(15/30) = 5$ so that the computed value of Q is

$$q = \frac{(9-5)^2}{5} + \frac{(1-5)^2}{5} + \frac{(5-5)^2}{5} + \frac{(5-5)^2}{5} + \frac{(1-5)^2}{5} + \frac{(9-5)^2}{5}$$
$$= 12.8 > 5.991 = \chi^2_{0.05}(2),$$

since here $(k-1)(h-1) = 2$ degrees of freedom. (The p-value is 0.0017.) Hence we reject the conjecture that these 30 values could be the items of a random sample. Obviously, modifications could be made to this scheme: dividing the sample into more (or less) than three groups and rating items differently, such as low (L), middle (M), and high (H).

It cannot be emphasized enough that the chi-square statistic can be used fairly effectively in almost any situation in which there should be independence. For illustration, suppose that we have a group of workers who have essentially *the same qualifications* (training, experience, etc.). Many believe that salary and sex of the workers should be independent attributes; yet there have been several claims in special cases that there is a dependence—or discrimination—in calculations associated with such a problem.

Example 9.6-5   Two groups of workers have the same qualifications for a particular type of work. Their experience in salaries is summarized by the following 2×5 contingency table, in which the upper bound of each salary range is not included in that listing.

Group	Salary (Thousands of Dollars)					Totals
	8–10	10–12	12–14	14–16	16–	
1	6	11	16	14	13	60
2	5	9	8	6	2	30
Totals	11	20	24	20	15	90

To test if the group assignment and the salaries seem to be independent with these data at the $\alpha = 0.05$ significance level, we compute

$$q = \frac{[6 - 90(60/90)(11/90)]^2}{90(60/90)(11/90)} + \cdots + \frac{[2 - 90(30/90)(15/90)]^2}{90(30/90)(15/90)}$$
$$= 4.752 < 9.488 = \chi^2_{0.05}(4).$$

Also, the p-value is 0.313. Hence with these limited data, group assignment and salaries seem to be independent.

Before turning to the exercises, note that we could have thought of the last two examples in this section as testing the equality of two or more multinomial distributions. In Example 9.6-4 the three groups define three binomial distributions; and in Example 9.6-5 the two groups define two multinomial distributions. What would have happened if we had used the computations outlined earlier in this section? It is interesting to note that we obtain exactly the same value of chi-square and in each case the number of degrees of freedom is equal to $(k - 1)(h - 1)$. Hence it makes no difference whether we think of it as a test of independence or a test of the equality of several multinomial distributions. Our advice is to use the terminology that seems most natural for the particular situations.

Exercises

9.6-1   We wish to test to see if two groups of nurses distribute their time in six different categories about the same way. That is, the hypothesis under consideration is H_0: $p_{i1} = p_{i2}$, $i = 1, 2, \ldots, 6$. To test this, nurses are observed at random throughout several days, each observation resulting in a mark in one of the six categories. The summary is given by the following frequency table:

		Category					
	1	2	3	4	5	6	Totals
Group I	95	36	71	21	45	32	300
Group II	53	26	43	18	32	28	200

Use a chi-square test with $\alpha = 0.05$.

9.6-2 Suppose that a third group of nurses was observed along with groups I and II of Exercise 9.6-1, resulting in the respective frequencies 130, 75, 136, 33, 61, and 65. Test H_0: $p_{i1} = p_{i2} = p_{i3}$, $i = 1, 2, \ldots , 6$, at the $\alpha = 0.025$ significance level.

9.6-3 Each of two comparable classes of 15 students responded to two different methods of instructions with the following scores on a standardized test:

Class U:	91	42	39	62	55	82	67	44
	51	77	61	52	76	41	59	

Class V:	80	71	55	67	61	93	49	78
	57	88	79	81	63	51	75	

Use a chi-square test with $\alpha = 0.05$ to test the equality of the distributions of test scores by dividing the combined sample into three equal parts (low, middle, high).

9.6-4 Suppose that a third class (W) of 15 students was observed along with classes U and V of Exercise 9.6-3, resulting in scores of

91	73	67	83	59	98	87	69
78	80	65	94	82	74	85	

Again use a chi-square test with $\alpha = 0.05$ to test the equality of the three distributions by dividing the combined sample into three equal parts.

9.6-5 Test the hypothesis that the distribution of letters is the same
(a) In a psychology textbook and a natural science textbook.
(b) On the editorial page and sports page of a newspaper.
(c) In two types of written material of your choice.
    Since some letters do not occur very frequently, you might want to group $\{j, k\}$, $\{p, q\}$, and $\{x, y, z\}$. Take a sample of 600 to 1000 letters and let $\alpha = 0.05$.

9.6-6 In a contingency table, 1015 individuals are classified by sex and by whether they favor, oppose, or have no opinion on a complete ban on smoking in public places. Test the null hypothesis that sex and opinion on smoking in public places are independent. Give the approximate p-value of this test.

	Smoking in Public Places			
Sex	Favor	Oppose	No Opinion	Totals
Male	262	231	10	503
Female	302	205	5	512
Totals	564	436	15	1015

9.6-7 A random survey of 100 students asked each student to select the most preferred form of recreational activity from 5 choices. Test whether the choice is independent of the sex of the respondent. Approximate the p-value of the test. Would we reject at $\alpha = 0.05$?

	Recreational Choice					
Sex	Basketball	Baseball Softball	Swimming	Jogging Running	Tennis	Totals
Male	21	5	9	12	13	60
Female	9	3	1	15	12	40
Totals	30	8	10	27	25	100

9.6-8 A random sample of 100 students were classified by sex and by the "instrument" that they "played." Test whether the selection of instrument is independent of the sex of the respondent. Approximate the p-value of this test.

	Instrument					
Sex	Piano	Woodwind	Brass	String	Vocal	Totals
Male	4	11	15	6	9	45
Female	7	18	6	6	18	55
Totals	11	29	21	12	27	100

9.6-9 A student who uses the college recreational facilities was interested in whether there is a difference between the facilities used by men and women. Test the null hypothesis that facility and sex are independent attributes using the following data. Use $\alpha = 0.05$.

	Facility		
Sex	Racquetball Court	Track	Totals
Male	51	30	81
Female	43	48	91
Totals	94	78	172

9.6-10 A survey of high school girls classified them by two attributes of classification: whether or not they participated in sports and whether or not they had (an) older brother(s). Test the null hypothesis that these two attributes of classification are independent.

	Participated in Sports		
Older Brother(s)	Yes	No	Totals
Yes	12	8	20
No	13	27	40
Totals	25	35	60

Approximate the *p*-value of this test. Do we reject the null hypothesis if $\alpha = 0.05$?

9.6-11 A random sample of 50 women who were tested for cholesterol were classified according to age and cholesterol level and grouped in the following contingency table.

	Cholesterol Level			
Age	<180	180–210	>210	Totals
<50	5	11	9	25
≥50	4	3	18	25
Totals	9	14	27	50

Test the null hypothesis H_0: age and cholesterol level are independent attributes of classification. What is your conclusion if $\alpha = 0.01$?

9.6-12 Although high school grades and testing scores, such as SAT or ACT, can be used to predict first-year college grade-point average (GPA), many educators claim that a more important factor influencing that GPA is the living conditions of students. In particular, it is claimed that the roommate of the student will have a great influence on his or her grades. To test this, suppose we selected at random 200 students and classified each according to the following two attributes:

(i) Ranking of the student's roommate from 1 to 5, from a person who was difficult to live with and discouraged scholarship to one who was congenial but encouraged scholarship.

(ii) The student's first-year GPA.

Say this gives the following 5 × 4 contingency table.

	Grade-Point Average				
Rank of Roommate	Under 2.00	2.00–2.69	2.70–3.19	3.20–4.00	Totals
1	8	9	10	4	31
2	5	11	15	11	42
3	6	7	20	14	47
4	3	5	22	23	53
5	1	3	11	12	27
Totals	23	35	78	64	200

Compute the chi-square statistic used to test the independence of the two attributes and compare it to the critical value associated with $\alpha = 0.05$.

9.6-13 A random sample of 1004 people of various ages were asked whether they approved or disapproved of the way Ronald Reagan was handling his job as president (September 1987 Gallup Report No. 264). Test the null hypothesis H_0: opinion of performance and age are independent attributes. Approximate the p-value of this test.

	Opinion of Performance			
Age Group	Approve	Disapprove	No Opinion	Totals
18–29	128	95	28	251
30–49	207	164	28	399
≥50	152	166	36	354
Totals	487	425	92	1004

9.6-14   A study was conducted to determine the media credibility for reporting news. Those surveyed were asked to give their age, sex, education, and the most credible medium. Test whether

(a) media credibility and age are independent.

(b) media credibility and sex are independent.

(c) media credibility and education are independent.

(d) Give the approximate p-value for each test.

	Most Credible Medium			
Age	Newspaper	Television	Radio	Totals
Under 35	30	68	10	108
35–54	61	79	20	160
Over 54	98	43	21	162
Totals	189	190	51	430

	Most Credible Medium			
Sex	Newspaper	Television	Radio	Totals
Male	92	108	19	219
Female	97	81	32	210
Totals	189	189	51	429

	Most Credible Medium			
Education	Newspaper	Television	Radio	Totals
Grade School	45	22	6	73
High School	94	115	30	239
College	49	52	13	114
Totals	188	189	49	426

9.6-15   In a psychology experiment 140 students were divided into majors emphasizing left-hemisphere brain skills (e.g., philosophy, physics, and mathematics), and into majors emphasizing right-hemisphere skills (e.g., art, music, theatre, and dance). They were also classified into one of three groups on the basis of hand posture (right noninverted, left inverted, left noninverted).

	LH	RH
RN	89	29
LI	5	4
LN	5	8

Do these data show sufficient evidence to reject the claim that the choice of college major is independent of hand posture? Let $\alpha = 0.025$.

9.6-16   A random sample of $n = 1362$ persons were classified according to the respondent's education level and whether the respondent was Protestant, Catholic, or Jewish. Use these data to test at an $\alpha = 0.05$ significance level the hypothesis that these attributes of classification are independent.

Education Level	Protestant	Catholic	Jewish
Less than high school	359	140	5
High school or junior college	462	200	17
Bachelor's degree	88	39	2
Graduate degree	37	10	3

9.6-17   In the Michigan Daily Lottery one three-digit integer is generated 6 days a week. Twenty weeks of lottery numbers have been classified by the day of the week and the magnitude of the number. At the 10% significance level, test whether these attributes of classification are independent.

	Magnitude of Number		
Days	000–499	500–999	Totals
Monday and Tuesday	22	18	40
Wednesday and Thursday	19	21	40
Friday and Saturday	13	27	40
Totals	54	66	120

★9.7   Sufficient Statistics

This and the next section would have been introduced earlier. However, the notion of conditional distributions is the key concept associated with both sufficient statistics and Bayesian methods. Therefore we delayed these discussions until that concept was introduced.

We first define a sufficient statistic $Y = u(X_1, X_2, \ldots, X_n)$ for a parameter θ using a statement that, in most books, is given as a necessary and sufficient condition for sufficiency, namely the well-known Fisher–Neyman Factorization Theorem. Using this as a definition, we shall note, by examples, the implications of this definition, one of which is sometimes used as the definition.

> **DEFINITION 9.7-1**   (**Factorization Theorem**)   *Let $X_1, X_2, \ldots,$ X_n denote random variables with joint p.d.f. $f(x_1, \ldots, x_n; \theta)$, which depends on the parameter θ. The statistic $Y = u(X_1, X_2, \ldots, X_n)$ is* **sufficient** *for θ if and only if*
>
> $$f(x_1, x_2, \ldots, x_n; \theta) = \phi[u(x_1, \ldots, x_n); \theta]h(x_1, \ldots, x_n),$$
>
> *where ϕ depends on $x_1, \ldots, x_n$ only through $u(x_1, \ldots, x_n)$ and $h(x_1, \ldots, x_n)$ does not depend on θ.*

Let us consider several important examples and consequences of this definition. We first note, however, that in all instances in this book the random variables $X_1, X_2, \ldots, X_n$ will be of a random sample and hence their joint p.d.f. will be of the form

$$f(x_1; \theta)f(x_2; \theta) \cdots f(x_n; \theta).$$

Example 9.7-1   Let $X_1, X_2, \ldots, X_n$ denote a random sample from a Poisson distribution with parameter $\lambda > 0$. Then

$$f(x_1; \lambda)f(x_2; \lambda) \cdots f(x_n; \lambda) = \frac{\lambda^{\Sigma x_i}e^{-n\lambda}}{x_1!x_2! \cdots x_n!} = (\lambda^{n\bar{x}}e^{-n\lambda})\left(\frac{1}{x_1!x_2! \cdots x_n!}\right),$$

where $\bar{x} = (1/n) \Sigma x_i$. Thus, from the Factorization Theorem (definition), it is clear that the sample mean $\bar{X}$ is a sufficient statistic for λ. It can easily be shown that the maximum likelihood estimator for λ is also $\bar{X}$. (See Exercise 6.6-3.)

In Example 9.7-1, if we replace $n\bar{x}$ by Σx_i, it is quite obvious that the sum ΣX_i is also a sufficient statistic for λ. This certainly agrees with our intuition because if we know one of the statistics, $\bar{X}$ or ΣX_i, we can easily find the other. If we generalize this, we see that if Y is sufficient for a parameter θ, then every single-valued function of Y, not involving θ but with a single-valued inverse, is also a sufficient statistic for θ. Again the reason is that knowing either Y or that function of Y, we know the other. More formally, if $W = v(Y) = v[u(X_1, \ldots, X_n)]$ is that function and $Y = v^{-1}(W)$ is the single-valued inverse, then the display of the Factorization Theorem can be written as

$$f(x_1, \ldots, x_n; \theta) = \phi[v^{-1}\{v[u(x_1, \ldots, x_n)]\}; \theta]h(x_1, \ldots, x_n).$$

The first factor of the righthand member of this equation depends on $x_1, x_2, \ldots,$ x_n through $v[u(x_1, \ldots, x_n)]$, so $W = v[u(X_1, \ldots, X_n)]$ is a sufficient statistic for θ. We illustrate this fact and the Factorization Theorem with an underlying distribution of the continuous type.

Example 9.7-2   Let $X_1, X_2, \ldots, X_n$ be a random sample from $N(\mu, 1)$, $-\infty < \mu < \infty$. The joint p.d.f. of these random variables is

$$\frac{1}{(2\pi)^{n/2}} \exp\left[-\frac{1}{2} \sum_{i=1}^{n} (x_i - \mu)^2 \right]$$

$$= \frac{1}{(2\pi)^{n/2}} \exp\left[-\frac{1}{2} \sum_{i=1}^{n} [(x_i - \bar{x}) + (\bar{x} - \mu)]^2 \right]$$

$$= \left\{ \exp\left[-\frac{n}{2} (\bar{x} - \mu)^2 \right] \right\} \left\{ \frac{1}{(2\pi)^{n/2}} \exp\left[-\frac{1}{2} \sum_{i=1}^{n} (x_i - \bar{x})^2 \right] \right\}.$$

From the Factorization Theorem we see that $\bar{X}$ is sufficient for μ. Now $\bar{X}^3$ is also sufficient for θ because knowing $\bar{X}^3$ is equivalent to having knowledge of the value of $\bar{X}$. However, $\bar{X}^2$ does not have this property, and it is not sufficient for θ.

One consequence of the sufficiency of a statistic Y is that the conditional probability of any given event A in the support of $X_1, X_2, \ldots, X_n$, given $Y = y$, does not depend on θ. This is sometimes used as the definition of sufficiency and is illustrated by the following example.

Example 9.7-3   Let $X_1, X_2, \ldots, X_n$ be a random sample from a distribution with p.d.f.

$$f(x; p) = p^x (1 - p)^{1-x}, \qquad x = 0, 1,$$

where the parameter p is between zero and one. We know that

$$Y = X_1 + X_2 + \cdots + X_n$$

is $b(n, p)$ and Y is sufficient for p because the joint p.d.f. of $X_1, X_2, \ldots, X_n$ is

$$p^{x_1}(1 - p)^{1-x_1} \cdots p^{x_n}(1 - p)^{1-x_n} = [p^{\sum x_i}(1 - p)^{n - \sum x_i}](1),$$

where $\phi(y; p) = p^y(1 - p)^{n-y}$ and $h(x_1, x_2, \ldots, x_n) = 1$. What then is the conditional probability $P(X_1 = x_1, \ldots, X_n = x_n | Y = y)$, where $y = 0, 1,$ $\ldots, n - 1$, or n? Unless the sum of the nonnegative integers $x_1, x_2, \ldots,$ x_n equals y, this conditional probability is obviously equal to zero, which does not depend on p. Hence it is only interesting to consider the solution when $y = x_1 + \cdots + x_n$. From the definition of conditional probability we have

$$P(X_1 = x_1, \ldots, X_n = x_n | Y = y) = \frac{P(X_1 = x_1, \ldots, X_n = x_n)}{P(Y = y)}$$

$$= \frac{p^{x_1}(1 - p)^{1-x_1} \cdots p^{x_n}(1 - p)^{1-x_n}}{\binom{n}{y} p^y (1 - p)^{n-y}}$$

$$= \frac{1}{\binom{n}{y}},$$

where $y = x_1 + \cdots + x_n$. Since y equals the number of ones in the collection $x_1, x_2, \ldots, x_n$, this answer is only the probability of selecting a particular arrangement, namely $x_1, x_2, \ldots, x_n$ of y ones and $n - y$ zeros, and does not depend on the parameter p. That is, given that the sufficient statistic $Y = y$, the conditional probability of $X_1 = x_1, X_2 = x_2, \ldots, X_n = x_n$ does not depend on the parameter p.

It is interesting to observe that the underlying p.d.f. in Examples 9.7-1, 9.7-2, and 9.7-3 can be written in the exponential form

$$f(x; \theta) = \exp[K(x)p(\theta) + S(x) + q(\theta)],$$

where the support is free of θ. That is, we have, respectively,

$$\frac{e^{-\lambda}\lambda^x}{x!} = \exp\{x \ln \lambda - \ln x! - \lambda\}, \qquad x = 0, 1, 2, \ldots,$$

$$\frac{1}{\sqrt{2\pi}} e^{-(x-\mu)^2/2} = \exp\left\{x\mu - \frac{x^2}{2} - \frac{\mu^2}{2} - \frac{1}{2} \ln (2\pi)\right\}, \qquad -\infty < x < \infty,$$

and

$$p^x(1 - p)^{1-x} = \exp\left\{x \ln \left(\frac{p}{1 - p}\right) + \ln (1 - p)\right\}, \qquad x = 0, 1.$$

In each of these examples, the sum ΣX_i of the items of the random sample was the sufficient statistic for the parameter. This is generalized by Theorem 9.7-1.

THEOREM 9.7-1   *Let $X_1, X_2, \ldots, X_n$ be a random sample from a distribution with a p.d.f. of the exponential form*

$$f(x; \theta) = \exp[K(x)p(\theta) + S(x) + q(\theta)]$$

on a support free of θ. The statistic $\Sigma_{i=1}^n K(X_i)$ is sufficient for θ.

PROOF: The joint p.d.f. of $X_1, X_2, \ldots, X_n$ is

$$\exp\left[p(\theta) \sum_1^n K(x_i) + \sum_1^n S(x_i) + nq(\theta)\right]$$

$$= \left\{\exp\left[p(\theta) \sum_1^n K(x_i) + nq(\theta)\right]\right\}\left\{\exp\left[\sum_1^n S(x_i)\right]\right\}.$$

In accordance with the Factorization Theorem, the $\Sigma K(X_i)$ is sufficient for θ.

$\square$

In many cases Theorem 9.7-1 permits the student to find the sufficient statistic for the parameter with very little effort, as shown by the following example.

Example 9.7-4   Let $X_1, X_2, \ldots, X_n$ be a random sample from an exponential distribution with p.d.f.

$$f(x; \theta) = \frac{1}{\theta} e^{-x/\theta} = \exp\left[x\left(-\frac{1}{\theta}\right) - \ln \theta\right], \qquad 0 < x < \infty,$$

provided $0 < \theta < \infty$. Here $K(x) = x$. Thus $\Sigma_1^n X_i$ is sufficient for θ; of course, $\overline{X} = \Sigma_1^n X_i/n$ is also sufficient.

We should also note that if there is a sufficient statistic for the parameter under consideration and if the maximum likelihood estimator of this parameter is unique, then the maximum likelihood estimator is a function of the sufficient statistic. To see this heuristically, consider the following. If a sufficient statistic exists, then the likelihood function is

$$L(\theta) = f(x_1, x_2, \ldots, x_n; \theta) = \phi[u(x_1, \ldots, x_n); \theta]h(x_1, \ldots, x_n).$$

Since $h(x_1, \ldots, x_n)$ does not depend on θ, we maximize $L(\theta)$ by maximizing $\phi[u(x_1, \ldots, x_n); \theta]$. But ϕ is a function of $x_1, x_2, \ldots, x_n$ only through $u(x_1, \ldots, x_n)$. Thus, if there is a unique value of θ that maximizes ϕ, then it must be a function of $u(x_1, \ldots, x_n)$. That is, $\hat{\theta}$ is a function of the sufficient statistic $u(X_1, X_2, \ldots, X_n)$. This fact was alluded to in Example 9.7-1, but it could be checked using other examples and exercises.

Exercises

9.7-1   Let $X_1, X_2, \ldots, X_n$ be a random sample from the distribution with p.d.f. $f(x; p) = p(1 - p)^{x-1}$, $x = 1, 2, 3, \ldots$ where $0 < p < 1$.
(a) Show that $Y = \Sigma_{i=1}^n X_i$ is a sufficient statistic for p.
(b) Find a function of $Y = \Sigma_{i=1}^n X_i$ that is an unbiased estimator of $\theta = 1/p$.

9.7-2   Let $X_1, X_2, \ldots, X_n$ be a random sample from a Poisson distribution with mean $\lambda > 0$. Find the conditional probability $P(X_1 = x_1, \ldots, X_n = x_n | Y = y)$, where $Y = X_1 + \cdots + X_n$ and the nonnegative integers $x_1, x_2, \ldots, x_n$ sum to y, showing that this probability does not depend on λ.

9.7-3   Let $X_1, X_2, \ldots, X_n$ be a random sample from $N(0, \sigma^2)$.
(a) find the sufficient statistic Y for σ^2.
(b) Show that the maximum likelihood estimator for σ^2 is a function of Y.
(c) Is the maximum likelihood estimator for σ^2 unbiased?

9.7-4   Let $X_1, X_2, \ldots, X_n$ be a random sample from a distribution with p.d.f. $f(x; \theta) = \theta x^{\theta-1}$, $0 < x < 1$, where $0 < \theta$.

(a) Find the sufficient statistic Y for θ.

(b) Show that the maximum likelihood estimator $\widehat{\theta}$ is a function of Y.

(c) Argue that $\widehat{\theta}$ is also sufficient for θ.

★9.8 Bayesian Estimation

We now describe another approach to estimation that is used by a group of statisticians who call themselves Bayesians. To understand fully their approach would require more text than we can allocate to this topic, but let us begin this brief introduction by considering a simple application of the theorem of the Reverend Thomas Bayes (see Section 2.5).

Example 9.8-1 Suppose we know that we are going to select an observation from a Poisson distribution with mean λ equal to 2 or 4. Moreover, we believe prior to performing the experiment that $\lambda = 2$ has about four times as much chance of being the parameter as does $\lambda = 4$; that is, the prior probabilities are $P(\lambda = 2) = 0.8$ and $P(\lambda = 4) = 0.2$. The experiment is now performed and we observe $x = 6$. At this point, our intuition now tells us that $\lambda = 2$ seems less likely than before as the observation $x = 6$ is much more probable with $\lambda = 4$ than with $\lambda = 2$, namely, in an obvious notation,

$$P(X = 6 | \lambda = 2) = 0.995 - 0.983 = 0.012$$

and

$$P(X = 6 | \lambda = 4) = 0.889 - 0.785 = 0.104,$$

from Table III in the Appendix. Our intuition can be supported by computing the conditional probability of $\lambda = 2$, given $X = 6$;

$$P(\lambda = 2 | X = 6) = \frac{P(\lambda = 2, X = 6)}{P(X = 6)}$$

$$= \frac{P(\lambda = 2)P(X = 6 | \lambda = 2)}{P(\lambda = 2)P(X = 6 | \lambda = 2) + P(\lambda = 4)P(X = 6 | \lambda = 4)}$$

$$= \frac{(0.8)(0.012)}{(0.8)(0.012) + (0.2)(0.104)} = 0.316.$$

This conditional probability is called the posterior probability of $\lambda = 2$, given the data (here $x = 6$). In a similar fashion, the posterior probability of $\lambda = 6$ is found to be 0.684. Thus we see that the probability of $\lambda = 2$ has decreased from 0.8 (the prior probability) to 0.316 (the posterior probability) with the observation of $x = 6$.

In a more practical application the parameter, say θ, can possibly take many more than two values as in Example 9.8-1. Somehow Bayesians must assign prior probabilities to this total parameter space through a prior p.d.f. $h(\theta)$. They have developed procedures for assessing these prior probabilities, and we clearly cannot do justice to these methods here. Somehow $h(\theta)$ reflects the prior weights that the Bayesian wants to assign to the various possible values of θ. In some instances, if $h(\theta)$ is a constant and thus θ has the uniform prior distribution, we say that the Bayesian has a *noninformative* prior. If, in fact, some knowledge of θ exists in advance of experimentation, noninformative priors should be avoided, if at all possible.

Also, in more practical examples, we usually take several observations, not just one. That is, we take a random sample and there is frequently a good statistic, say Y, for the parameter θ. The p.d.f. of Y, say $g(y; \theta)$, can be thought of as the conditional p.d.f. of Y, given θ; and henceforth in this section we write $g(y; \theta) = g(y \mid \theta)$. Thus we can treat

$$g(y \mid \theta)h(\theta) = k(y, \theta)$$

as the joint p.d.f. of the statistic Y and the parameter. Of course, the marginal p.d.f. of Y is

$$k_1(y) = \int_{-\infty}^{\infty} h(\theta)g(y \mid \theta) \, d\theta.$$

Thus

$$\frac{k(y, \theta)}{k_1(y)} = \frac{g(y \mid \theta)h(\theta)}{k_1(y)} = k(\theta \mid y)$$

would serve as the conditional p.d.f. of the parameter, given $Y = y$. This is essentially Bayes' Theorem, and $k(\theta \mid y)$ is called the *posterior p.d.f. of θ*, given $Y = y$.

Bayesians believe that everything that needs to be known about the parameter is summarized in this posterior p.d.f. $k(\theta \mid y)$. Suppose, for illustration, they were pressed into making a point estimate of the parameter θ. They would note that they would be guessing the value of a random variable, here θ, given its p.d.f. $k(\theta \mid y)$. There are many ways that this could be done: for illustrations, the mean, the median, or the mode of that distribution would be reasonable guesses. However, in the final analysis, the best guess would clearly depend upon the penalties for various errors created by incorrect guesses. For illustration, if we were penalized by taking the square of the error between the guess, say $w(y)$, and real value of the parameter θ, clearly we would use the conditional mean

$$w(y) = \int_{-\infty}^{\infty} \theta k(\theta \mid y) \, d\theta$$

as our Bayes estimate of θ. The reason for this is that, in general, if Z is a random variable, then the function of b, $E[(Z - b)^2]$, is minimized by $b = E(Z)$. (See Example 3.2-4.) Likewise, if the penalty (loss) function is the absolute value of the error, $|\theta - w(y)|$, then we use the median of the distribution as, with any random variable Z, $E[|Z - b|]$ is minimized when b equals the median of the distribution of Z. (See Exercise 3.2-10.)

Example 9.8-2   Suppose that Y has a binomial distribution with parameters n and $p = \theta$. Thus the p.d.f. of Y, given θ, is

$$g(y \mid \theta) = \binom{n}{y} \theta^y (1 - \theta)^{n-y}, \qquad y = 0, 1, 2, \ldots, n.$$

Let us take the prior p.d.f. of the parameter to be the beta p.d.f.

$$h(\theta) = \frac{\Gamma(\alpha + \beta)}{\Gamma(\alpha)\Gamma(\beta)} \theta^{\alpha-1}(1 - \theta)^{\beta-1}, \qquad 0 < \theta < 1.$$

Such a prior p.d.f. provides a Bayesian a great deal of flexibility through the selection of the parameters α and β. Thus the joint p.d.f. is a product of a binomial p.d.f. with parameters n and θ and this beta p.d.f., namely,

$$k(y, \theta) = \binom{n}{y} \frac{\Gamma(\alpha + \beta)}{\Gamma(\alpha)\Gamma(\beta)} \theta^{y+\alpha-1}(1 - \theta)^{n-y+\beta-1},$$

on the support given by $y = 0, 1, 2, \ldots, n$ and $0 < \theta < 1$. We find

$$
\begin{aligned}
k_1(y) &= \int_0^1 k(y, \theta)\, d\theta \\
&= \binom{n}{y} \frac{\Gamma(\alpha + \beta)}{\Gamma(\alpha)\Gamma(\beta)} \frac{\Gamma(\alpha + y)\Gamma(n + \beta - y)}{\Gamma(n + \alpha + \beta)}
\end{aligned}
$$

on the support $y = 0, 1, 2, \ldots, n$ by comparing the integral to one involving a beta p.d.f. with parameters $y + \alpha$ and $n - y + \beta$. Therefore,

$$
\begin{aligned}
k(\theta \mid y) &= \frac{k(y, \theta)}{k_1(y)} \\
&= \frac{\Gamma(n + \alpha + \beta)}{\Gamma(\alpha + y)\Gamma(n + \beta - y)} \theta^{y+\alpha-1}(1 - \theta)^{n-y+\beta-1}, \qquad 0 < \theta < 1,
\end{aligned}
$$

which is a beta p.d.f. with parameters $y + \alpha$ and $n - y + \beta$. With the square error loss function we must minimize, with respect to $w(y)$, the integral

$$\int_0^1 [\theta - w(y)]^2 k(\theta \mid y)\, d\theta$$

to obtain the Bayes' solution. But, as noted earlier, if Z is a random variable with a second moment, $E[(Z - b)^2]$ is minimized by $b = E(Z)$. In the preceding display, θ is like the Z with p.d.f. $k(\theta \mid y)$, and $w(y)$ is like the b, so the minimization is accomplished by taking

$$w(y) = E(\theta \mid y) = \frac{\alpha + y}{\alpha + \beta + n},$$

which is the mean of the beta distribution with parameters $y + \alpha$ and $n - y + \beta$. (See Exercise 5.6-14). It is very instructive to note that this Bayes' solution can be written as

$$w(y) = \left(\frac{n}{\alpha + \beta + n}\right)\left(\frac{y}{n}\right) + \left(\frac{\alpha + \beta}{\alpha + \beta + n}\right)\left(\frac{\alpha}{\alpha + \beta}\right),$$

which is a weighted average of the maximum likelihood estimate y/n of θ and the mean $\alpha/(\alpha + \beta)$ of the prior p.d.f. of the parameter. Moreover the respective weights are $n/(\alpha + \beta + n)$ and $(\alpha + \beta)/(\alpha + \beta + n)$. Thus we see that α and β should be selected so that not only is $\alpha/(\alpha + \beta)$ the desired prior mean, but the sum $\alpha + \beta$ also plays a role corresponding to a sample size. That is, if we want our prior opinion to have as much weight as a sample size of 20, we would take $\alpha + \beta = 20$. So if our prior mean is 3/4, we have that α and β are selected such that $\alpha = 15$ and $\beta = 5$.

In Example 9.8-2 it is extremely convenient to note that it is not really necessary to determine $k_1(y)$ to find $k(\theta|y)$. If we divide $k(y, \theta)$ by $k_1(y)$, we get the product of a factor, which depends on y but does *not* depend on θ, say $c(y)$, and

$$\theta^{y+\alpha-1}(1 - \theta)^{n-y+\beta-1}.$$

That is,

$$k(\theta|y) = c(y)\theta^{y+\alpha-1}(1 - \theta)^{n-y+\beta-1}, \qquad 0 < \theta < 1.$$

However, $c(y)$ must be that "constant" needed to make $k(\theta|y)$ a p.d.f., namely

$$c(y) = \frac{\Gamma(n + \alpha + \beta)}{\Gamma(y + \alpha)\Gamma(n - y + \beta)}.$$

Accordingly, Bayesians frequently write that $k(\theta|y)$ is proportional to $k(y, \theta) = g(y|\theta)h(\theta)$; that is,

$$k(\theta|y) \propto g(y|\theta)h(\theta).$$

Then, to actually form the p.d.f. $k(\theta|y)$, they simply find a "constant," which is some function of y, so that the expression integrates to one.

Example 9.8-3 Suppose that $Y = \overline{X}$ is the mean of a random sample of size n that arises from the normal distribution $N(\theta, \sigma^2)$, where σ^2 is known. Then $g(y|\theta)$ is $N(\theta, \sigma^2/n)$. Further suppose that we are able to assign prior weights to θ through a prior p.d.f. $h(\theta)$, which is $N(\theta_0, \sigma_0^2)$. Then we have that

$$k(\theta|y) \propto \frac{1}{\sqrt{2\pi}(\sigma/\sqrt{n})} \frac{1}{\sqrt{2\pi}\sigma_0} \exp\left[-\frac{(y - \theta)^2}{2(\sigma^2/n)} - \frac{(\theta - \theta_0)^2}{2\sigma_0^2}\right].$$

If we eliminate all constant factors (including factors involving y only), then we have

$$k(\theta|y) \propto \exp\left[-\frac{(\sigma_0^2 + \sigma^2/n)\theta^2 - 2(y\sigma_0^2 + \theta_0\,\sigma^2/n)\theta}{2(\sigma^2/n)\sigma_0^2}\right].$$

This can be simplified, by completing the square, to read (after eliminating factors not involving θ)

$$k(\theta|y) \propto \exp\left\{ -\frac{[\theta - (y\sigma_0^2 + \theta_0\,\sigma^2/n)/(\sigma_0^2 + \sigma^2/n)]^2}{[2(\sigma^2/n)\sigma_0^2]/[\sigma_0^2 + (\sigma^2/n)]} \right\}.$$

That is, the posterior p.d.f. of the parameter is obviously normal with mean

$$\frac{y\sigma_0^2 + \theta_0\,\sigma^2/n}{\sigma_0^2 + \sigma^2/n} = \left(\frac{\sigma_0^2}{\sigma_0^2 + \sigma^2/n}\right)y + \left(\frac{\sigma^2/n}{\sigma_0^2 + \sigma^2/n}\right)\theta_0$$

and variance $(\sigma^2/n)\sigma_0^2/(\sigma_0^2 + \sigma^2/n)$. If the square error loss function is used, then this posterior mean is the Bayes' solution. Again note that it is a weighted average of the maximum likelihood estimate $y = \bar{x}$ and the prior mean θ_0. The Bayes' solution $w(y)$ will always be a value between the prior judgment and the usual estimate. Also, note here and in Example 9.8-3 that the Bayes' solution gets closer to the maximum likelihood estimate as n increases. Thus the Bayesian procedures permit the decision maker to enter his or her prior opinions into the solution in a very formal way so that the influence of those prior notions will be less and less as n increases.

In Bayesian statistics, all the information is contained in the posterior p.d.f. $k(\theta|y)$. In Examples 9.8-2 and 9.8-3 we found Bayesian point estimates using the square error loss function. It should be noted that if the loss function is $|w(y) - \theta|$, the absolute value of the error, then the Bayes' solution would be the median of the posterior distribution of the parameter, which is given by $k(\theta|y)$. Hence the Bayes' solution changes, as *it should,* with different loss functions.

Finally, if an interval estimate of θ is desired, we would find two functions of y, say $u(y)$ and $v(y)$, such that

$$\int_{u(y)}^{v(y)} k(\theta|y)\,d\theta = 1 - \alpha,$$

where α is small, say $\alpha = 0.05$. Then the observed interval $u(y)$ to $v(y)$ would serve as an interval estimate for the parameter in the sense that the posterior probability of the parameter being in that interval is $1 - \alpha$. In Example 9.8-3, where the posterior p.d.f. of the parameter was normal, the interval

$$\frac{y\sigma_0^2 + \theta_0\sigma^2/n}{\sigma_0^2 + \sigma^2/n} \pm 1.96\sqrt{\frac{(\sigma^2/n)\sigma_0^2}{\sigma_0^2 + \sigma^2/n}}$$

serves as an interval estimate for θ with posterior probability of 0.95.

In closing this short section on Bayesian estimation, it should be noted that we could begin with the sample observations, $X_1, X_2, \ldots, X_n$, rather than some statistic Y. Then, in our discussion, we would replace $g(y|\theta)$ by the likelihood function

$$L(\theta) = f(x_1|\theta)f(x_2|\theta)\cdots f(x_n|\theta)$$

which is the joint p.d.f. of $X_1, X_2, \ldots, X_n$, given θ. Thus we find that

$$k(\theta|x_1, x_2, \ldots, x_n) \propto h(\theta)f(x_1|\theta)f(x_2|\theta)\cdots f(x_n|\theta) = h(\theta)L(\theta).$$

Now $k(\theta|x_1, x_2, \ldots, x_n)$ contains all the information about θ, given the data. Depending on the loss function we would choose our Bayesian estimate of θ as some characteristic of this posterior distribution, like the mean or the median. It is interesting to observe that if the loss function is zero for some small neighborhood about the true parameter θ and is some large positive constant otherwise, then the Bayesian estimate, $w(x_1, x_2, \ldots, x_n)$, is essentially the mode of this conditional p.d.f., $k(\theta|x_1, x_2, \ldots, x_n)$. The reason for this is that we want to take the estimate, $w(x_1, x_2, \ldots, x_n)$, so that it has as much posterior probability as possible in a small neighborhood around it. Finally, note that if $h(\theta)$ if a constant (noninformative prior), then this Bayesian estimate using the mode is exactly the same as the maximum likelihood estimate. More generally, if $h(\theta)$ is not a constant, the Bayesian estimate using the mode can be thought of as a weighted maximum likelihood estimate, in which the weights reflect the Bayesian's prior opinion about θ. That is, that value of θ which maximizes $h(\theta)L(\theta)$ is the mode of the posterior distribution of the parameter given the data and can be used as the Bayesian estimate associated with the appropriate loss function.

Exercises

9.8-1 Let $X_1, X_2, \ldots, X_n$ be a random sample from a distribution $b(1, \theta)$. Let the prior p.d.f. of θ be a beta one with parameters α and β. Show that the posterior p.d.f.

$$k(\theta|x, x_2, \ldots, x_n)$$

is exactly the same as $k(\theta|y)$ given in Example 9.8-2. This demonstrates that we often get exactly the same result whether we begin with a good statistic or with the sample observations.

HINT: Note that $k(\theta|x_1, x_2, \ldots, x_n)$ is proportional to the product of the joint p.d.f. of $X_1, X_2, \ldots, X_n$ and the prior p.d.f. of θ.

9.8-2 Let Y be the sum of the observations of a random sample from a Poisson distribution with mean θ. Let the prior p.d.f. of θ be a gamma one with parameters α and β.
(a) Find the posterior p.d.f. of θ, given $Y = y$.
(b) If the loss function is $[w(y) - \theta]^2$, find the Bayesian point estimate $w(y)$.
(c) Show that this $w(y)$ is a weighted average of the maximum likelihood estimate y/n and the prior mean $\alpha\beta$, with respective weights of $n/(n + 1/\beta)$ and $(1/\beta)/(n + 1/\beta)$.

9.8-3 In Example 9.8-2 take $n = 30$, $\alpha = 15$, and $\beta = 5$.
(a) Using the square error loss, compute the expected loss (risk function) associated with the Bayes' solution $w(Y)$.

(b) The risk function associated with the usual estimator Y/n is, of course, $\theta(1 - \theta)/30$. Find those values of θ for which the risk function in part (a) is less than $\theta(1 - \theta)/30$. In particular, if the prior mean $\alpha/(\alpha + \beta) = 3/4$ is a reasonable guess, then the risk function in part (a) is the better of the two (i.e., it is smaller in a neighborhood of $\theta = 3/4$).

9.8-4   Let the conditional p.d.f. of X, given θ, be $N(0, \sigma^2 = 1/\theta)$. Assume that the unknown θ is a value of a random variable that has a gamma distribution with parameters $\alpha = r/2$ and $\theta = 2/r$ (here this θ is that associated with the gamma distribution, not the θ in $\sigma^2 = 1/\theta$). Show that the marginal p.d.f. of X is that of a Student's t with r degrees of freedom. This procedure is often called compounding and is frequently used by Bayesians in introducing the t-distribution, as well as other distributions.

10

Nonparametric Methods

10.1 Order Statistics

The **order statistics** are the items of the random sample arranged, or ordered, in magnitude from the smallest to the largest. In recent years, the importance of order statistics has increased owing to the more frequent use of nonparametric inferences and robust procedures. However, order statistics have always been prominent because, among other things, they are needed to determine rather simple statistics such as the sample median, the sample range, and the empirical distribution function. Recall that in Section 1.3 we discussed observed order statistics in connection with descriptive and exploratory statistical methods. We will consider certain interesting aspects about their distributions in this section; and in Section 10.2 we will demonstrate how they can be used to construct confidence intervals for percentiles.

In most of our discussions about order statistics, we will assume that the random sample arises from a continuous-type distribution. This means, among other things, that the probability of any two sample items being equal is zero. That is, the probability is *one* that the items can be ordered from smallest to largest without having two equal values. Of course, in practice, we do frequently observe *ties;* but if the probability of this is small, the following distribution theory will hold approximately. Thus, in the discussion here, we are assuming that the probability of ties is zero.

589

Example 10.1-1   The values $x_1 = 0.62$, $x_2 = 0.98$, $x_3 = 0.31$, $x_4 = 0.81$, and $x_5 = 0.53$ are the $n = 5$ observed values of five independent trials of an experiment with p.d.f. $f(x) = 2x$, $0 < x < 1$. The observed order statistics are

$$y_1 = 0.31 < y_2 = 0.53 < y_3 = 0.62 < y_4 = 0.81 < y_5 = 0.98.$$

Recall from Section 1.3 that the middle observation in the ordered arrangement, here $y_3 = 0.62$, is called the sample median and the difference of the largest and the smallest, here

$$y_5 - y_1 = 0.98 - 0.31 = 0.67,$$

is called the sample range.

If $X_1, X_2, \ldots, X_n$ are observations of a random sample of size n from a continuous-type distribution, we let the random variables

$$Y_1 < Y_2 < \cdots < Y_n$$

denote the order statistics of that sample. That is,

$$Y_1 = \text{smallest of } X_1, X_2, \ldots, X_n,$$
$$Y_2 = \text{second smallest of } X_1, X_2, \ldots, X_n,$$
$$\vdots$$
$$Y_n = \text{largest of } X_1, X_2, \ldots, X_n.$$

There is a very simple method for determining the distribution function of the rth order statistic, Y_r. This procedure depends on the binomial distribution and is illustrated in Example 10.1-2.

Example 10.1-2   Let $Y_1 < Y_2 < Y_3 < Y_4 < Y_5$ be the order statistics of a random sample X_1, X_2, X_3, X_4, X_5 of size $n = 5$ from the distribution with p.d.f. $f(x) = 2x, 0 < x < 1$. Consider $P(Y_4 \leq 1/2)$. For the event $Y_4 \leq 1/2$ to occur, at least four of the random variables X_1, X_2, X_3, X_4, X_5 must be less than $1/2$ because Y_4 is the fourth smallest among the five observations. Thus if the event $X_i \leq 1/2$, $i = 1, 2, \ldots, 5$, is called "success," we must have at least four successes in the five mutually independent trials, each of which has probability of success

$$P\left(X_i \leq \frac{1}{2}\right) = \int_0^{1/2} 2x \, dx = \left(\frac{1}{2}\right)^2 = \frac{1}{4}.$$

Thus

$$P\left(Y_4 \leq \frac{1}{2}\right) = \binom{5}{4}\left(\frac{1}{4}\right)^4\left(\frac{3}{4}\right) + \left(\frac{1}{4}\right)^5 = 0.0156.$$

In general, if $0 < y < 1$, then the distribution function of Y_4 is

$$G(y) = P(Y_4 \leq y) = \binom{5}{4}(y^2)^4(1 - y^2) + (y^2)^5,$$

since this represents the probability of at least four "successes" in five independent trials, each of which has probability of success

$$P(X_i \le y) = \int_0^y 2x\, dx = y^2.$$

The p.d.f. of Y_4 is therefore, for $0 < y < 1$,

$$g(y) = G'(y) = \binom{5}{4} 4(y^2)^3 (2y)(1 - y^2) + \binom{5}{4}(y^2)^4(-2y) + 5(y^2)^4(2y)$$

$$= \frac{5!}{3!1!}(y^2)^3(1 - y^2)(2y), \qquad 0 < y < 1.$$

Note that in this example, the distribution function of each X is $F(x) = x^2$ when $0 < x < 1$. Thus

$$g(y) = \frac{5!}{3!1!}[F(y)]^3[1 - F(y)]f(y), \qquad 0 < y < 1.$$

The preceding example should make the following generalization easier to read. Let $Y_1 < Y_2 < \cdots < Y_n$ be the order statistics of a random sample of size n from a distribution of the continuous type with distribution function $F(x)$ and p.d.f. $F'(x) = f(x)$, where $0 < F(x) < 1$ for $a < x < b$ and $F(a) = 0$, $F(b) = 1$. (It is possible that $a = -\infty$ and/or $b = +\infty$.) The event that the rth order statistic Y_r is at most y, $\{Y_r \le y\}$, can occur if and only if at least r of the n observations are less than or equal to y. That is, here the probability of "success" on each trial is $F(y)$ and we must have at least r successes. Thus

$$G_r(y) = P(Y_r \le y) = \sum_{k=r}^{n} \binom{n}{k}[F(y)]^k[1 - F(y)]^{n-k}.$$

That is, rewriting this slightly, we have

$$G_r(y) = \sum_{k=r}^{n-1} \binom{n}{k}[F(y)]^k[1 - F(y)]^{n-k} + [F(y)]^n.$$

Thus the p.d.f. of Y_r is

$$g_r(y) = G_r'(y) = \sum_{k=r}^{n-1} \binom{n}{k}(k)[F(y)]^{k-1}f(y)[1 - F(y)]^{n-k}$$

$$+ \sum_{k=r}^{n-1} \binom{n}{k}[F(y)]^k(n - k)[1 - F(y)]^{n-k-1}[-f(y)]$$

$$+ n[F(y)]^{n-1}f(y). \qquad (10.1\text{-}1)$$

But since

$$\binom{n}{k}(k) = \frac{n!}{(k-1)!(n-k)!} \qquad \text{and} \qquad \binom{n}{k}(n-k) = \frac{n!}{k!(n-k-1)!},$$

we have that the p.d.f. of Y_r is

$$g_r(y) = \frac{n!}{(r-1)!(n-r)!}[F(y)]^{r-1}[1-F(y)]^{n-r}f(y), \qquad a < y < b,$$

which is the first term of the first summation in $g_r(y) = G'_r(y)$. The remaining terms in $g_r(y) = G'_r(y)$ sum to zero because the second term of the first summation (when $k = r + 1$) equals the negative of the first term in the second summation (when $k = r$), and so on. Finally, the last term of the second summation equals the negative of $n[F(y)]^{n-1}f(y)$. To see this clearly, the student is urged to write out a number of terms in these summations (see Exercise 10.1-4).

It is worth noting that the p.d.f. of the smallest order statistic is

$$g_1(y) = n[1 - F(y)]^{n-1}f(y), \qquad a < y < b,$$

and the p.d.f. of the largest order statistic is

$$g_n(y) = n[F(y)]^{n-1}f(y), \qquad a < y < b.$$

REMARK There is one very satisfactory way to construct heuristically the expression for the p.d.f. of Y_r. To do this, we must recall the multinomial probability and then consider the probability element $g(y)(\Delta y)$ of Y_r. If the length Δy is *very* small, $g(y)(\Delta y)$ represents approximately the probability

$$P(y < Y_r \le y + \Delta y).$$

Thus we want the probability $g(y)(\Delta y)$ that $(r - 1)$ items fall less than y, $(n - r)$ items are greater than $y + \Delta y$, and one item falls between y and $y + \Delta y$. Recall that the probabilities on a single trial are

$$P(X \le y) = F(y)$$
$$P(X > y + \Delta y) = 1 - F(y + \Delta y) \approx 1 - F(y)$$
$$P(y < X \le y + \Delta y) \approx f(y)(\Delta y).$$

Thus the multinomial probability is approximately

$$g(y)(\Delta y) = \frac{n!}{(r-1)!1!(n-r)!}[F(y)]^{r-1}[1-F(y)]^{n-r}[f(y)(\Delta y)].$$

If we divide both sides by the length Δy, the formula for $g(y)$ results.

Example 10.1-3 Let $Y_1 < Y_2 < Y_3 < Y_4$ be the order statistics of a random sample of size $n = 4$ from a distribution with the uniform of p.d.f. $f(x) = 1$, $0 < x < 1$. Since $F(x) = x$, when $0 \le x < 1$, the p.d.f. of Y_3 is

$$g(y) = \frac{4!}{2!1!}y^2(1 - y), \qquad 0 < y < 1.$$

Thus, for illustration,

$$P\left(\frac{1}{3} < Y_3 < \frac{2}{3}\right) = \int_{1/3}^{2/3} 12y^2(1-y)\, dy$$

$$= [4y^3 - 3y^4]_{1/3}^{2/3} = \frac{13}{27}.$$

Example 10.1-4 Let $Y_1 < Y_2 < \cdots < Y_7$ be the order statistics of a random sample of size $n = 7$ from a distribution with p.d.f. $f(x) = 3(1-x)^2$, $0 < x < 1$. Compute the probability that the sample median is less than $1 - \sqrt[3]{0.6}$; that is, find $P(Y_4 < 1 - \sqrt[3]{0.6})$. We could find the p.d.f. of Y_4. However, note that the probability of a single observation being less than $1 - \sqrt[3]{0.6}$ is

$$\int_0^{1-\sqrt[3]{0.6}} 3(1-x)^2\, dx = [-(1-x)^3]_0^{1-\sqrt[3]{0.6}}$$

$$= 1 - (\sqrt[3]{0.6})^3 = 0.4.$$

Thus, from Table II in the Appendix,

$$P(Y_4 < 1 - \sqrt[3]{0.6}) = \sum_{k=4}^{7} \binom{7}{k}(0.4)^k(0.6)^{7-k} = 1.0000 - 0.7102 = 0.2898.$$

Now, in Example 10.1-4, it is easy enough to look up the resulting probability in Table II in the Appendix. Suppose that the sample size is much larger and we wish to know certain probabilities associated with a given order statistic, such as $P(Y_r < y)$. Of course, these probabilities can be expressed in terms of binomial probabilities, which may be difficult to evaluate but in turn can be approximated by normal probabilities, provided that n is large. This is illustrated in Example 10.1-5.

Example 10.1-5 Let $Y_1 < Y_2 < \cdots < Y_{100}$ be the order statistics of a random sample of size $n = 100$ from a continuous distribution having median $m = 68.1$ (i.e., the probability that a single observation X is less than 68.1 is $1/2$). What then is the probability that the 55th order statistic, Y_{55}, is less than $m = 68.1$? Since a binomial distribution with $n = 100$ and $p = 1/2$ has mean 50 and standard deviation 5, we have that

$$P(Y_{55} < 68.1) = \sum_{k=55}^{100} \binom{100}{k}\left(\frac{1}{2}\right)^k\left(\frac{1}{2}\right)^{100-k}$$

$$= 1 - \sum_{k=0}^{54} \binom{100}{k}\left(\frac{1}{2}\right)^k\left(\frac{1}{2}\right)^{100-k}$$

$$\approx 1 - \Phi\left(\frac{54.5 - 50}{5}\right) = 1 - \Phi(0.9)$$

$$= 1 - 0.8159 = 0.1841.$$

Recall that, in Theorem 4.2-2, we proved that if X has a distribution function $F(x)$ of the continuous type, then $F(X)$ has a uniform distribution on the interval zero to one. If $Y_1 < Y_2 < \cdots < Y_n$ are the order statistics of a random sample $X_1, X_2, \ldots, X_n$ of size n, then

$$F(Y_1) < F(Y_2) < \cdots < F(Y_n)$$

because F is a nondecreasing function and the probability of an equality is again zero. Note that this last display could be looked upon as an ordering of the mutually independent random variables $F(X_1), F(X_2), \ldots, F(X_n)$, each of which is $U(0, 1)$. That is,

$$W_1 = F(Y_1) < W_2 = F(Y_2) < \cdots < W_n = F(Y_n)$$

can be thought of as the order statistics of a random sample of size n from that uniform distribution. Since the distribution function of $U(0, 1)$ is $G(w) = w$, $0 < w < 1$, the p.d.f. of the rth order statistic $W_r = F(Y_r)$ is

$$h_r(w) = \frac{n!}{(r-1)!(n-r)!} w^{r-1}(1-w)^{n-r}, \qquad 0 < w < 1.$$

Of course, the mean, $E(W_r) = E[F(Y_r)]$ of $W_r = F(Y_r)$, is given by the integral

$$E(W_r) = \int_0^1 w \frac{n!}{(r-1)!(n-r)!} w^{r-1}(1-w)^{n-r}\, dw.$$

This can be evaluated by integrating by parts several times, but it is easier to obtain the answer if we rewrite it as follows:

$$E(W_r) = \left(\frac{r}{n+1}\right) \int_0^1 \frac{(n+1)!}{r!(n-r)!} w^r (1-w)^{n-r}\, dw.$$

The integrand in this last expression can be thought of as the p.d.f. of the $(r+1)$st order statistic of a random sample of size $n+1$ from a $U(0, 1)$ distribution. Hence the integral must equal one and

$$E(W_r) = \frac{r}{n+1}, \qquad r = 1, 2, \ldots, n.$$

There is an extremely interesting interpretation of $W_r = F(Y_r)$. Note that $F(Y_r)$ is the cumulated probability up to and including Y_r or, equivalently, the area under $f(x) = F'(x)$ but less than Y_r. Hence $F(Y_r)$ can be treated as a random area. Since $F(Y_{r-1})$ is also a random area, $F(Y_r) - F(Y_{r-1})$ is the random area under $f(x)$ between Y_{r-1} and Y_r. The expected value of the random area between any two adjacent order statistics is then

$$E[F(Y_r) - F(Y_{r-1})] = E[F(Y_r)] - E[F(Y_{r-1})]$$

$$= \frac{r}{n+1} - \frac{r-1}{n+1} = \frac{1}{n+1}.$$

Also, it is easy to show (see Exercise 10.1-6)

$$E[F(Y_1)] = \frac{1}{n+1} \quad \text{and} \quad E[1 - F(Y_n)] = \frac{1}{n+1}.$$

That is, the order statistics $Y_1 < Y_2 < \cdots < Y_n$ partition the support of X into $n + 1$ parts and thus create $n + 1$ areas under $f(x)$ and above the x axis. "On the average," each of the $n + 1$ areas equals $1/(n + 1)$.

If we recall that the $(100p)$th percentile π_p is such that the area under $f(x)$ to the left of π_p is p, the preceding discussion suggests that we let Y_r be an estimator of π_p, where $p = r/(n + 1)$. For this reason, we define the **(100p)th percentile of the sample** as Y_r, where $r = (n + 1)p$. In case $(n + 1)p$ is not an integer, we use a weighted average (or an average) of the two adjacent order statistics Y_r and Y_{r+1}, where r is the greatest integer $[\![(n + 1)p]\!]$ in $(n + 1)p$. In particular, the sample median is

$$\tilde{m} = \begin{cases} Y_{(n+1)/2}, & \text{when } n \text{ is odd,} \\ \dfrac{Y_{n/2} + Y_{(n/2)+1}}{2}, & \text{when } n \text{ is even.} \end{cases}$$

Example 10.1-6 Let X equal the weight of soap in a "1000-gram" bottle. A random sample of $n = 12$ observations of X yielded the following weights that have been ordered:

$$\begin{array}{cccccc} 1013 & 1019 & 1021 & 1024 & 1026 & 1028 \\ 1033 & 1035 & 1039 & 1040 & 1043 & 1047 \end{array}$$

Since $n = 12$ is even, the sample median is

$$\tilde{m} = \frac{y_6 + y_7}{2} = \frac{1028 + 1033}{2} = 1030.5.$$

The location of the 25th percentile (or first quartile) is

$$(n + 1)(0.25) = (12 + 1)(0.25) = 3.25.$$

Thus the first quartile, using a weighted average, is

$$\tilde{q}_1 = y_3 + (0.25)(y_4 - y_3) = (0.75)y_3 + (0.25)y_4$$
$$= (0.75)(1021) + (0.25)(1024) = 1021.75.$$

Similarly, the 75th percentile (or third quartile) is

$$\tilde{q}_3 = y_9 + (0.75)(y_{10} - y_9) = (0.25)y_9 + (0.75)y_{10}$$
$$= (0.25)(1039) + (0.75)(1040) = 1039.75$$

because $(12 + 1)(0.75) = 9.75$.

Since $(12 + 1)(0.60) = 7.8$, the 60th percentile is

$$\tilde{\pi}_{0.60} = (0.2)y_7 + (0.8)y_8 = (0.2)(1033) + (0.8)(1035) = 1034.6.$$

We close this section with examples, one of the type that will be used in Section 10.2 to find confidence intervals for percentiles and a second concerning a maximum likelihood estimator.

Example 10.1-7 Let $Y_1 < Y_2 < \cdots < Y_{13}$ be the order statistics of a random sample of size $n = 13$ from a continuous-type distribution with 35th percentile $\pi_{0.35} = 73.247$. To find $P(Y_3 < \pi_{0.35} < Y_7)$, note that the event $Y_3 < \pi_{0.35} < Y_7$ happens if and only if there are at least three but less than seven "successes," where the probability of success is $p = 0.35$. Thus

$$P(Y_3 < \pi_{0.35} < Y_7) = \sum_{k=3}^{6} \binom{13}{k}(0.35)^k(0.65)^{13-k}$$

$$= 0.8705 - 0.1132 = 0.7573,$$

from Table II in the Appendix.

Example 10.1-8 Let $X_1, X_2, \ldots, X_n$ denote a random sample from a distribution with p.d.f.

$$f(x; \theta) = e^{-(x-\theta)}, \qquad \theta \le x < \infty,$$

where $0 < \theta < \infty$. The likelihood function is

$$L(\theta) = e^{-n\bar{x}+n\theta}, \qquad \theta \le x_i, \ i = 1, 2, \ldots, n,$$

and zero otherwise. Thus

$$\ln L(\theta) = -n\bar{x} + n\theta, \qquad \theta \le x_i, \ i = 1, 2, \ldots, n,$$

and the derivative

$$D_\theta[\ln L(\theta)] = n$$

is positive for all $\theta \le x_i, \ i = 1, 2, \ldots, n$. That is, $\ln L(\theta)$ and hence $L(\theta)$ is an increasing function of θ. To achieve the maximum value of $L(\theta)$, we accordingly make θ as large as possible. But since $\theta \le x_i, \ i = 1, 2, \ldots, n$, the largest value of θ is the smallest x-value. That is, the maximum likelihood estimator $\widehat{\theta} = Y_1$, the first order statistic.

Exercises

10.1-1 One of the tasks performed by a computer operator is that of testing tapes in order to detect bad records (i.e., records that could not store information in the form of positive or negative charges). Let X equal the distance in feet between bad records. Ten observations of X are

$$67 \quad 6 \quad 7 \quad 35 \quad 78 \quad 28 \quad 74 \quad 5 \quad 9 \quad 37$$

(a) Find the order statistics.
(b) Find the median and 80th percentile of the sample.

(c) Determine the first and third quartiles (i.e., 25th and 75th percentiles) of the sample.

10.1-2 Let X equal the forced vital capacity (the volume of air a person can expel from their lungs) for a male freshman. Seventeen observations of X, which have been ordered, are

3.7	3.8	4.0	4.3	4.7	4.8	4.9	5.0
5.2	5.4	5.6	5.6	5.6	5.7	6.2	6.8	7.6

(a) Find the median, the first quartile, and the third quartile.
(b) Find the 35th and 65th percentiles.

10.1-3 Let $Y_1 < Y_2 < Y_3 < Y_4 < Y_5$ be the order statistics of a random sample of size $n = 5$ from an exponential distribution that has a mean of $\theta = 3$.
(a) Find the p.d.f. of the sample median Y_3.
(b) Compute the probability that Y_4 is less than 5.
(c) Determine $P(1 < Y_1)$.

10.1-4 In the expression for $g_r(y) = G_r'(y)$ in equation (10.1-1) let $n = 6$, $r = 3$, and write out the summations, showing that the "telescoping" suggested in the text is achieved.

10.1-5 Let $Y_1 < Y_2 < \cdots < Y_8$ be the order statistics of a random sample of size $n = 8$ from a continuous-type distribution with 70th percentile $\pi_{0.7} = 27.3$.
(a) Determine $P(Y_7 < 27.3)$.
(b) Find $P(Y_5 < 27.3 < Y_8)$.

10.1-6 Let $W_1 < W_2 < \cdots < W_n$ be the order statistics of a random sample of size n from a $U(0, 1)$ distribution.
(a) Find the p.d.f. of W_1 and that of W_n.
(b) Use the results of (a) to verify that $E(W_1) = 1/(n + 1)$ and $E(W_n) = n/(n + 1)$.

10.1-7 Let $Y_1 < Y_2 < \cdots < Y_{72}$ be the order statistics of a random sample of size $n = 72$ from a distribution of the continuous type having the $100(1/3)$th percentile $\pi_{1/3} = 7.2$.
(a) Approximate the probability that $Y_{20} < \pi_{1/3} = 7.2$.
(b) Find approximately the probability $P(Y_{18} < \pi_{1/3} < Y_{30})$ by first noting that the event $Y_{18} < \pi_{1/3} < Y_{30}$ means that we must have at least 18 but less than 30 "successes."

10.1-8 Let $W_1 < W_2 < \cdots < W_n$ be the order statistics of a random sample of size n from a $U(0, 1)$ distribution.
(a) Show that $E(W_r^2) = r(r + 1)/(n + 1)(n + 2)$ using a technique similar to that used in determining that $E(W_r) = r/(n + 1)$.
(b) Find the variance of W_r.

10.1-9 Let $Y_1 < Y_2 < \cdots < Y_{19}$ be the order statistics of a random sample of size $n = 19$ from the standard normal distribution $N(0, 1)$. What is the value of $E[F(Y_{19})]$?

10.1-10   Let $Y_1 < Y_2 < \cdots < Y_n$ be the order statistics of a random sample of size n from the uniform distribution $U(0, \theta)$.
   (a) Show that the maximum likelihood estimator of θ is Y_n.
   (b) Show that $E(Y_n) = n\theta/(n + 1)$ and $\mathrm{Var}(Y_n) = n\theta^2/(n + 1)^2(n + 2)$.
   (c) Find a constant c so that cY_n is an unbiased estimator of θ.
   (d) Show that a $100(1 - \alpha)\%$ confidence interval for θ is $[y_n, y_n/\alpha^{1/n}]$.

10.1-11   Let $Y_1 < Y_2 < Y_3$ be the order statistics of a random sample X_1, X_2, X_3 of size 3 from the uniform distribution $U(\theta - 1/2, \theta + 1/2)$. Let $W_1 = \overline{X} = (1/3) \sum_{i=1}^{3} X_i$, the sample mean; $W_2 = Y_2$, the sample median; and $W_3 = (Y_1 + Y_3)/2$, the sample midrange.
   (a) Give a reason why each of W_1, W_2, and W_3 is a possible estimator of θ.
   (b) Given that $E(W_1) = E(W_2) = E(W_3) = \theta$ and $\mathrm{Var}(W_1) = 1/36$, $\mathrm{Var}(W_2) = 1/20$, $\mathrm{Var}(W_3) = 1/40$, which of these three estimators is the best?
   (c) The following data give 20 sets of observations of y_1, y_2, y_3 that were generated using $\theta = 1/2$. For each replication, calculate the values of $w_1 = \overline{y} = \overline{x}$, $w_2 = y_2$, and $w_3 = (y_1 + y_3)/2$.
   (d) Find the sample mean and sample variance of the 20 observations of (i) W_1, (ii) W_2, and (iii) W_3.
   (e) From this small set of empirical evidence, which estimator seems to be the best? Why?

Replication	y_1	y_2	y_3
1	0.3098	0.3146	0.6144
2	0.0887	0.2601	0.3247
3	0.3638	0.7087	0.7960
4	0.2159	0.7038	0.8607
5	0.6016	0.7653	0.7908
6	0.4995	0.6317	0.9305
7	0.0341	0.2097	0.9026
8	0.1400	0.6444	0.9620
9	0.1485	0.5929	0.6025
10	0.1668	0.3718	0.5973
11	0.0371	0.4175	0.9272
12	0.2395	0.3198	0.3275
13	0.3199	0.3566	0.7422
14	0.7278	0.8002	0.8480
15	0.0294	0.5282	0.9613
16	0.0108	0.2354	0.4948
17	0.0981	0.4908	0.8840
18	0.1937	0.4930	0.7527
19	0.0436	0.7345	0.9981
20	0.5599	0.9046	0.9591

10.1-12 Let $Y_1 < Y_2 < \cdots < Y_m < \cdots < Y_n$ be the order statistics of a random sample of size n from a continuous distribution that has p.d.f. $f(x)$ and a distribution function $F(x)$, where $0 < F(x) < 1$ for $a < x < b$, $F(a) = 0$, $F(b) = 1$. This exercise outlines a procedure for simulating the first m order statistics in an ordered manner; in other words, first the value of y_1, then y_2, and so on. Prove the following facts.

HINT: A satisfactory way to do this is by using the heuristic method like that used to find the p.d.f. of Y_r in the Remark of this section.

(a) The joint marginal p.d.f. of Y_i and Y_{i+1} is

$$g_{i, i+1}(y_i, y_{i+1}) =$$

$$\frac{n!}{(i - 1)!(n - i - 1)!}[F(y_i)]^{i-1}[1 - F(y_{i+1})]^{n-i-1}f(y_i)f(y_{i+1}),$$

$$a < y_i < y_{i+1} < b.$$

(b) The conditional p.d.f. of Y_{i+1}, given $Y_i = y_i$, is

$$h(y_{i+1}|y_i) = \frac{(n - i)[1 - F(y_{i+1})]^{n-i-1}f(y_{i+1})}{[1 - F(y_i)]^{n-i}}, \qquad a < y_i < y_{i+1} < b.$$

(c) The conditional distribution function of Y_{i+1}, given $Y_i = y_i$, is

$$H(y_{i+1}|y_i) = 1 - \left[\frac{1 - F(y_{i+1})}{1 - F(y_i)}\right]^{n-i}, \qquad a < y_i < y_{i+1} < b.$$

(d) The distribution function of Y_1, the first order statistic, is

$$v = 1 - [1 - F(y_1)]^n, \qquad a < y_1 < b.$$

(e) When V is $U(0, 1)$, $y_1 = F^{-1}(1 - [1 - v]^{1/n})$ gives an observed value of Y_1.

(f) Let $v = H(y_{i+1}|y_i)$. If V is $U(0, 1)$, then

$$y_{i+1} = H^{-1}(v|y_i) = F^{-1}[1 - \{1 - F(y_i)\}\{1 - v\}^{1/(n-i)}]$$

is the observed value of Y_{i+1} for $i = 1, 2, \ldots, n - 1$.

10.1-13 Use the "random numbers" $v_1 = 0.2360$, $v_2 = 0.0914$, $v_3 = 0.8903$, $v_4 = 0.3195$ and the method outlined in Exercise 10.1-12 to simulate $y_1 < y_2 < y_3 < y_4$ in a random sample of size $n = 10$ from
(a) The exponential distribution with mean $\theta = 10$.
(b) The uniform distribution $U(0, 1)$.

10.1-14 Let $X_1, X_2, \ldots, X_n$ be a random sample of size n from a two-parameter exponential distribution with p.d.f.

$$f(x; \alpha, \theta) = \frac{1}{\theta}e^{-(x-\alpha)/\theta}, \qquad \alpha \leq x < \infty,$$

where $-\infty < \alpha < \infty$, $0 < \theta < \infty$.
(a) Find the mean and variance of this distribution.

(b) Find the maximum likelihood estimators for α and θ.

(c) Correct the maximum likelihood estimators to obtain unbiased estimators of α and θ.

(d) Use the following data to give unbiased point estimates of α and θ.

$$18.218 \quad 16.117 \quad 8.609 \quad 85.387 \quad 20.668 \quad 34.232 \quad 37.941 \quad 31.183$$

10.1-15 Let X equal the weight in grams of peanuts in a "14-gram" snack package. The order statistics of a random sample of size $n = 16$ are:

$$
\begin{array}{cccccccc}
13.9 & 14.4 & 14.6 & 14.7 & 14.7 & 15.2 & 15.2 & 15.2 \\
15.3 & 15.4 & 15.4 & 15.5 & 15.6 & 15.6 & 15.9 & 16.4
\end{array}
$$

(a) Find the 30th percentile of this sample.

(b) Find $P(Y_2 < \pi_{0.30} < Y_8)$, where Y_2 and Y_8 are the respective second and eighth order statistics in the sample before the sample is observed.

10.2 Confidence Intervals for Percentiles

In Section 10.1 we defined the sample percentiles in terms of the order statistics and noted that the sample percentiles can be used to estimate the corresponding distribution percentiles. In this section we use the order statistics to construct confidence intervals for the unknown distribution percentiles. Since little is assumed about the underlying distribution in the construction of these confidence intervals, they are often called **distribution-free confidence intervals**.

If $Y_1 < Y_2 < Y_3 < Y_4 < Y_5$ are the order statistics of a random sample of size $n = 5$ from a continuous-type distribution, then the sample median Y_3 could be thought of as an estimator of the distribution median $\pi_{0.5}$. We shall let $m = \pi_{0.5}$. We could simply use the sample median Y_3 as an estimator of the distribution median m. However, we are certain that all of us recognize that, with only a sample of size 5, we would be quite lucky if the observed $Y_3 = y_3$ were very close to m. Thus we now describe how a confidence interval can be constructed for m.

Instead of simply using Y_3 as an estimator of m, let us also compute the probability that the random interval (Y_1, Y_5) includes m. That is, let us determine $P(Y_1 < m < Y_5)$. This is easy if we follow the procedure given in Section 10.1. Again say that we have success if an individual item, say X, is less than m; thus the probability of success on one of the independent trials is $P(X < m) = 0.5$. In order for the first order statistic Y_1 to be less than m and the last order statistic Y_5 to be greater than m, we must have at least one success but not five successes. That is,

$$
P(Y_1 < m < Y_5) = \sum_{k=1}^{4} \binom{5}{k} \left(\frac{1}{2}\right)^k \left(\frac{1}{2}\right)^{5-k}
$$

$$
= 1 - \left(\frac{1}{2}\right)^5 - \left(\frac{1}{2}\right)^5 = \frac{15}{16}.
$$

So the probability that the random interval (Y_1, Y_5) includes m is $15/16 \approx 0.94$. Suppose that this random sample is actually taken and the order statistics are observed to equal $y_1 < y_2 < y_3 < y_4 < y_5$, respectively. Then (y_1, y_5) is a 94% confidence interval for m.

It is interesting to note what happens as the sample size increases. Let $Y_1 < Y_2 < \cdots < Y_n$ be the order statistics of a random sample of size n from a distribution of the continuous type. Thus $P(Y_1 < m < Y_n)$ is the probability that there is at least one "success" but not n successes, where the probability of success on each trial is $P(X < m) = 0.5$. Consequently,

$$P(Y_1 < m < Y_n) = \sum_{k=1}^{n-1} \binom{n}{k} \left(\frac{1}{2}\right)^k \left(\frac{1}{2}\right)^{n-k}$$

$$= 1 - \left(\frac{1}{2}\right)^n - \left(\frac{1}{2}\right)^n = 1 - \left(\frac{1}{2}\right)^{n-1}.$$

This probability increases as n increases so that the corresponding confidence interval (y_1, y_n) would have a very large confidence coefficient, $1 - (1/2)^{n-1}$. Unfortunately, the interval (y_1, y_n) tends to get wider as n increases, and thus we are not "pinning down" m very well. However, if we used the interval (y_2, y_{n-1}) or (y_3, y_{n-2}), we would obtain shorter intervals but also smaller confidence coefficients. Let us investigate this possibility further.

With the order statistics $Y_1 < Y_2 < \cdots < Y_n$ associated with a random sample of size n from a continuous-type distribution, consider $P(Y_i < m < Y_j)$, where $i < j$. For example, we might want

$$P(Y_2 < m < Y_{n-1}) \quad \text{or} \quad P(Y_3 < m < Y_{n-2}).$$

On each of the n independent trials we say that we have success if that X is less than m; thus the probability of success on each trial is $P(X < m) = 0.5$. Consequently, to have the ith order statistic Y_i less than m and the jth order statistic greater than m, we must have at least i successes but fewer than j successes (or else $Y_j < m$). That is,

$$P(Y_i < m < Y_j) = \sum_{k=i}^{j-1} \binom{n}{k} \left(\frac{1}{2}\right)^k \left(\frac{1}{2}\right)^{n-k} = 1 - \alpha.$$

For particular values of n, i, and j, this probability, say $1 - \alpha$, which is the sum of probabilities from a binomial distribution, can be calculated directly or approximated by an area under the normal p.d.f. provided n is large enough. The observed interval (y_i, y_j) could then serve as a $100(1 - \alpha)\%$ confidence interval for the unknown distribution median.

Example 10.2-1
The lengths in centimeters of $n = 9$ fish of a particular species (*nezumia*) captured off the New England coast were 32.5, 27.6, 29.3, 30.1, 15.5, 21.7, 22.8, 21.2, 19.0, Thus the observed order statistics are

$$15.5 < 19.0 < 21.2 < 21.7 < 22.8 < 27.6 < 29.3 < 30.1 < 32.5.$$

Before the sample is drawn, we know that

$$P(Y_2 < m < Y_8) = \sum_{k=2}^{7} \binom{9}{k} \left(\frac{1}{2}\right)^k \left(\frac{1}{2}\right)^{9-k} = 0.9805 - 0.0195 = 0.9610,$$

from Table II in the Appendix. Thus the confidence interval ($y_2 = 19.0$, $y_8 = 30.1$) for m, the median of the lengths of all fish of this species, has a 96.1% confidence coefficient.

So that the student need not compute many of these probabilities, we give in Table 10.2-1 the necessary information for constructing confidence intervals of the form (y_i, y_{n+1-i}) for the unknown m for sample sizes $n = 5, 6, \ldots, 20$. The subscript i is selected so that the confidence coefficient $P(Y_i < m < Y_{n+1-i})$ is greater than 90% and as close to 95% as possible.

For sample sizes larger than 20, we approximate those binomial probabilities with areas under the normal curve. To illustrate how good these approximations are, we compute the probability corresponding to $n = 16$ in Table 10.2-1. Here, using Table II, we have

$$1 - \alpha = P(Y_5 < m < Y_{12}) = \sum_{k=5}^{11} \binom{16}{k} \left(\frac{1}{2}\right)^k \left(\frac{1}{2}\right)^{16-k}$$

$$= P(W = 5, 6, \ldots, 11) = 0.9616 - 0.0384 = 0.9232,$$

where W is $b(16, 1/2)$. The normal approximation gives

$$1 - \alpha = P(4.5 < W < 11.5) = P\left(\frac{4.5 - 8}{2} < \frac{W - 8}{2} < \frac{11.5 - 8}{2}\right)$$

TABLE 10.2-1

n	$(i, n + 1 - i)$	$P(Y_i < m < Y_{n+1-i})$
5	(1, 5)	0.9376
6	(1, 6)	0.9688
7	(1, 7)	0.9844
8	(2, 7)	0.9296
9	(2, 8)	0.9610
10	(2, 9)	0.9786
11	(3, 9)	0.9346
12	(3, 10)	0.9614
13	(3, 11)	0.9776
14	(4, 11)	0.9426
15	(4, 12)	0.9648
16	(5, 12)	0.9232
17	(5, 13)	0.9510
18	(5, 14)	0.9692
19	(6, 14)	0.9364
20	(6, 15)	0.9586

because W has mean $np = 8$ and variance $np(1 - p) = 4$. The standardized variable $Z = (W - 8)/2$ has an approximate normal distribution. Thus

$$1 - \alpha \approx \Phi\left(\frac{3.5}{2}\right) - \Phi\left(\frac{-3.5}{2}\right) = \Phi(1.75) - \Phi(-1.75)$$

$$= 0.9599 - 0.0401 = 0.9198.$$

This compares very favorably with the probability 0.9232 recorded in Table 10.2-1.

The argument used to find a confidence interval for the median m of a distribution of the continuous type can be applied to any percentile π_p. In this case we say that we have success on a single trial if that X is less than π_p. Thus the probability of success on each of the independent trials is $P(X < \pi_p) = p$. Accordingly, with $i < j$, $1 - \alpha = P(Y_i < \pi_p < Y_j)$ is the probability that we have at least i successes but fewer than j successes. Thus

$$1 - \alpha = P(Y_i < \pi_p < Y_j) = \sum_{k=i}^{j-1} \binom{n}{k} p^k (1 - p)^{n-k}.$$

Once the sample is observed and the order statistics determined, the known interval (y_i, y_j) could serve as a $100(1 - \alpha)\%$ confidence interval for the unknown distribution percentile π_p.

Example 10.2-2 Let the following numbers represent the order statistics of the $n = 27$ observations obtained in a random sample from a certain population of incomes (measured in hundreds of dollars):

161	180	192	205	229	264
169	183	193	213	241	291
171	184	196	221	243	317
174	186	200	222	256	376
179	187	204			

Say we are interested in estimating the 25th percentile $\pi_{0.25}$ of the population. Since $(n + 1)p = 28(1/4) = 7$, the 7th order statistic, namely $y_7 = 183$, would be a point estimate of $\pi_{0.25}$. To find a confidence interval for $\pi_{0.25}$, let us move down and up a few order statistics from y_7, say to y_4 and y_{10}. What is the confidence coefficient associated with the interval (y_4, y_{10})? Before the sample was drawn, we had

$$1 - \alpha = P(Y_4 < \pi_{0.25} < Y_{10}) = \sum_{k=4}^{9} \binom{27}{k}(0.25)^k(0.75)^{27-k}$$

$$= P(3.5 < W < 9.5),$$

where W is $b(27, 1/4)$ with mean $27/4 = 6.75$ and variance $81/16$. Hence

$$1 - \alpha \approx \Phi\left(\frac{9.5 - 6.75}{9/4}\right) - \Phi\left(\frac{3.5 - 6.75}{9/4}\right)$$

$$= \Phi\left(\frac{11}{9}\right) - \Phi\left(-\frac{13}{9}\right) = 0.8149.$$

Thus ($y_4 = 174$, $y_{10} = 187$) serves as an 81.49% confidence interval for $\pi_{0.25}$. It should be noted that we could choose other intervals, such as ($y_3 = 171$, $y_{11} = 192$), and these would have different confidence coefficients. The persons involved in the study must select the desired confidence coefficient, and then the appropriate order statistics are taken, usually fairly symmetrically about the $(n + 1)p$th order statistic.

   When the number of observations is large, it is important to be able to determine rather easily the order statistics. A stem-and-leaf diagram, as introduced in Section 1.3, can be helpful in determining the needed order statistics. This is illustrated in the next example.

Example 10.2-3   The measurements of butterfat produced by $n = 90$ cows during a 305-day milk production period following their first calf are given in Exercise 1.2-10. The measurements are summarized in the following ordered stem-and-leaf diagram in which each leaf consists of two digits.

Stem	Leaf									Frequency
2s	74									1
2•										0
3*										0
3t	27	39								2
3f	45	50								2
3s										0
3•	80	88	92	94	95					5
4*	17	18								2
4t	21	22	27	34	37	39				6
4f	44	52	53	53	57	58				6
4s	60	64	66	70	70	72	75	78		8
4•	81	86	89	91	92	94	96	97	99	9
5*	00	00	01	02	05	09	10	13	13	16	10
5t	24	26	31	32	32	37	37	39		8
5f	40	41	44	55						4
5s	61	70	73	74						4
5•	83	83	86	93	99					5
6*	07	08	11	12	13	17	18	19		8
6t	27	28	35	37						4
6f	43	43	45							3
6s	72									1
6•	91	96								2

From this display it is quite easy to see that $y_8 = 392$. It takes a little more work to show that $y_{38} = 494$ and $y_{53} = 526$. The interval (494, 526) serves as a confidence interval for the unknown median m of all butterfat production for the given breed of cows. Its confidence coefficient is

$$P(Y_{38} < m < Y_{53}) = \sum_{k=38}^{52} \binom{90}{k} \left(\frac{1}{2}\right)^k \left(\frac{1}{2}\right)^{90-k}$$

$$\approx \Phi\left(\frac{52.5 - 45}{\sqrt{22.5}}\right) - \Phi\left(\frac{37.5 - 45}{\sqrt{22.5}}\right)$$

$$= \Phi(1.58) - \Phi(-1.58) = 0.8858.$$

Similarly, $(y_{17} = 437, y_{29} = 470)$ is a confidence interval for the first quartile, $\pi_{0.25}$ with confidence coefficient

$$P(Y_{17} < \pi_{0.25} < Y_{29}) \approx \Phi\left(\frac{28.5 - 22.5}{\sqrt{16.875}}\right) - \Phi\left(\frac{16.5 - 22.5}{\sqrt{16.875}}\right)$$

$$= \Phi(1.46) - \Phi(-1.46) = 0.8558.$$

Exercises

10.2-1 Let $Y_1 < Y_2 < Y_3 < Y_4 < Y_5 < Y_6$ be the order statistics of a random sample of size $n = 6$ from a distribution of the continuous type having $(100p)$th percentile π_p. Compute
(a) $P(Y_2 < \pi_{0.5} < Y_5)$.
(b) $P(Y_1 < \pi_{0.25} < Y_4)$.
(c) $P(Y_4 < \pi_{0.9} < Y_6)$.

10.2-2 The following are $n = 12$ test scores that were selected at random from a large collection of such scores and then ordered: 41, 52, 58, 64, 66, 69, 74, 75, 80, 83, 83, 97.
(a) Find a 96.14% confidence interval for m.
(b) The interval $(y_1 = 41, y_7 = 74)$ could serve as a confidence interval for $\pi_{0.3}$. What is its confidence coefficient?

10.2-3 The following data represent a random sample of weights of 10 female college students.

120	137	109	136	140	110	130	131	126	133

(a) Find the endpoints for an approximate 95% confidence interval for the median weight, m, of all female college students.
(b) Give the exact confidence level for your interval.

10.2-4 Let m denote the median enrollment at small liberal arts colleges. Use the following order statistics of a random sample of $n = 14$ enrollments to find an approximate 95% confidence interval for m:

651	682	716	740	822	1026	1139
1162	1254	1279	1400	1917	1926	2198

10.2-5 Use the following weights of an observed random sample of $n = 11$ pieces of candy to find an approximate 95% confidence interval for the median weight, m. Also give the exact confidence level.

$$
\begin{array}{cccccc}
2.76 & 2.96 & 2.67 & 2.97 & 2.77 & 2.72 \\
2.74 & 2.84 & 2.61 & 2.82 & 2.72 &
\end{array}
$$

10.2-6 Use the 81 weights of mints listed in Exercise 1.3-2 to find an approximate 95% confidence interval for

(a) $\pi_{0.25}$.

(b) $\pi_{0.5}$.

(c) $\pi_{0.75}$.

10.2-7 A biologist who studies spiders selected a random sample of 20 male green lynx spiders (a spider that does not weave a web but chases and leaps on its prey) and measured in millimeters the lengths of one of the front legs of the 20 spiders. Use the following measurements to construct a confidence interval for m that has a confidence coefficient about equal to 0.95:

$$
\begin{array}{ccccc}
15.10 & 13.55 & 15.75 & 20.00 & 15.45 \\
13.60 & 16.45 & 14.05 & 16.95 & 19.05 \\
16.40 & 17.05 & 15.25 & 16.65 & 16.25 \\
17.75 & 15.40 & 16.80 & 17.55 & 19.05
\end{array}
$$

10.2-8 The biologist (Exercise 10.2-7) also selected a random sample of 20 female green lynx spiders and measured in millimeters the length of one of their front legs. Use the following data to construct a confidence interval for m that has a confidence coefficient about equal to 0.95:

$$
\begin{array}{ccccc}
15.85 & 18.00 & 11.45 & 15.60 & 16.10 \\
18.80 & 12.85 & 15.15 & 13.30 & 16.65 \\
16.25 & 16.15 & 15.25 & 12.10 & 16.20 \\
14.80 & 14.60 & 17.05 & 14.15 & 15.85
\end{array}
$$

10.2-9 The following 25 observations give the time in seconds between submissions of computer programs to a printer queue:

$$
\begin{array}{ccccccccc}
79 & 315 & 445 & 350 & 136 & 723 & 198 & 75 & 161 \\
13 & 215 & 24 & 57 & 152 & 238 & 288 & 272 & \\
9 & 315 & 11 & 51 & 98 & 620 & 244 & 34 &
\end{array}
$$

(a) Construct a stem-and-leaf diagram.

(b) Give the values of y_3, y_9, y_{10}, y_{13}, and y_{17}.

(c) Give point estimates of $\pi_{0.25}$, m, and $\pi_{0.75}$.

(d) Find the following confidence intervals and give the confidence level (using Table II) for

(i) (y_3, y_{10}), a confidence interval for $\pi_{0.25}$.

(ii) (y_9, y_{17}), a confidence interval for the median m.

(iii) (y_{16}, y_{23}), a confidence interval for $\pi_{0.75}$.

10.2-10 Let X equal the weight in grams of a miniature candy bar. A random sample of candy bar weights yielded the following observations of X:

22.4	23.7	22.4	23.6	24.0	24.5	23.4	23.9
22.8	22.5	23.4	23.0	22.8	22.6	24.0	23.3

(a) Give a point estimate of the median, $m = \pi_{0.50}$.
(b) Find a confidence interval for m with confidence level between 90% and 95%. Give the value of the exact confidence.

10.2-11 Let X equal the amount of fluoride in a certain brand of toothpaste. The specifications are $0.85 - 1.10$ mg/g. The following are 25 observations of X.

0.98	0.94	1.06	0.95	1.02	0.88	0.98	1.00	1.01
0.95	0.99	0.92	0.93	0.92	0.89	0.92	0.98	
0.90	0.98	1.00	0.97	0.87	0.95	0.91	0.90	

(a) Construct an ordered stem-and-leaf diagram.
(b) Give a point estimate of the median, $m = \pi_{0.50}$.
(c) Find an approximate 95% confidence interval for the median m. Give the exact confidence level.
(d) Give a point estimate for the first quartile.
(e) Find an approximate 90% confidence interval for the first quartile and give the exact confidence level.
(f) Give a point estimate for the third quartile.
(g) Find an approximate 95% confidence interval for the third quartile and give the exact confidence level.

10.2-12 When placed in solutions of varying ionic strength, paramecia grow blisters in order to counteract the flow of water. The following 60 measurements in microns are blister lengths.

7.42	5.73	3.80	5.20	11.66	8.51	6.31	8.49
10.31	6.92	7.36	5.92	6.74	8.93	9.61	11.38
12.78	11.43	6.57	13.50	10.58	8.03	10.07	8.71
10.09	11.16	7.22	10.10	6.32	10.30	10.75	11.51
11.55	11.41	9.40	4.74	6.52	12.10	6.01	5.73
7.57	7.80	6.84	6.95	8.93	8.92	5.51	6.71
10.40	13.44	9.33	8.57	7.08	8.11	13.34	6.58
8.82	7.70	12.22	7.46				

(a) Construct an ordered stem-and-leaf diagram.
(b) Give a point estimate of the median, $m = \pi_{0.50}$.
(c) Find an approximate 95% confidence interval for m. Give the exact confidence level.
(d) Give a point estimate for the 40th percentile, $\pi_{0.40}$.
(e) Find an approximate 90% confidence interval for $\pi_{0.40}$ and give the exact confidence level.

10.3   Binomial Tests for Percentiles

In Section 10.2 we found ways of constructing interval estimates for certain characteristics of the distribution that did not depend too much on distributional assumptions. That is, we found distribution-free interval estimates for the $(100p)$th percentile π_p of a distribution of the continuous type. We now take up some tests of statistical hypotheses that are **distribution free**, provided the null hypotheses are true. In this section we find a distribution-free test of the hypothesis $H_0: \pi_p = \pi_0$, where π_0 is given. It should be noted here, however, that the powers of these tests are not distribution free when the null hypotheses are false. Moreover, the resulting nonnull distribution theory is frequently quite complicated, so little will be said here about the power functions.

We begin by considering a hypothesis about the median, $m = \pi_{0.5}$, of a distribution of the continuous type. Further, we assume that the median m is unique. Suppose it is known from past experience that the median length of sunfish in a particular polluted like was $m = 3.7$ inches. During the past two years the lake was "cleaned up," and the conjecture is made that now $m > 3.7$ inches. We consider a procedure for testing this conjecture. In particular, we describe a procedure for testing the null hypothesis $H_0: m = 3.7$ against the alternative hypothesis $H_1: m > 3.7$.

Let X denote the length of a sunfish selected at random from the lake. If the null hypothesis $H_0: m = 3.7$ is true, $P(X \leq 3.7; H_0) = 0.5$. However, if the alternative hypothesis $H_1: m > 3.7$ is true, $P(X \leq 3.7; H_1) < 0.5$. If we take a random sample of $n = 10$ fish, we would expect about half the fish to be shorter than 3.7 inches if H_0 is true. However, if H_1 is true, we would expect less than half the fish to be shorter than 3.7 inches. We shall decide on the basis of the number, say Y, of fish shorter than 3.7 inches whether to accept H_0 or H_1. Thus Y can be thought of as the number of "successes" in 10 Bernoulli trials with probability of success given by $p = P(X \leq 3.7)$. If H_0 is true, $p = 1/2$ and Y is $b(10, 1/2)$; whereas if H_1 is true, $p < 1/2$ and Y is $b(10, p)$. We reject H_0 and accept H_1 if and only if the observed value y of Y is sufficiently small, say $y \leq c$. From Table II in the Appendix we find that $P(Y \leq 2; H_0) = 0.0547$. Thus, if we let the critical region be defined by $C = \{y: y \leq 2\}$, then $\alpha = 0.0547$.

Suppose that the lengths of 10 sunfish selected at random from this like were 5.0, 3.9, 5.2, 5.5, 2.8, 6.1, 6.4, 2.6, 1.7, and 4.3 inches. Since $y = 3$ of these lengths are shorter than 3.7, we would not reject $H_0: m = 3.7$ at the $\alpha = 0.0547$ significance level by considering only these few data.

Example 10.3-1 Let X denote the length of time in seconds between two calls entering a college switchboard. Let m be the unique median of this continuous-type distribution. We test the null hypothesis $H_0: m = 6.2$ against the alternative hypothesis $H_1: m < 6.2$. If Y is the number of lengths of time in a random sample of size 20 that are less than 6.2, the critical region $C = \{y: y \geq 14\}$ has a significance level of $\alpha = 0.0577$ using Table II. A random sample of size 20 yielded the following data:

6.8	5.7	6.9	5.3	4.1	9.8	1.7	7.0
2.1	19.0	18.9	16.9	10.4	44.1	2.9	2.4
4.8	18.9	4.8	7.9				

Since $y = 9$, the null hypothesis is not rejected.

In many places in the literature the test we have just described is called the **sign test**. The reason for this terminology is that the test is based on a statistic Y that is equal to the number of negative signs among (here m_0 is the hypothesized median)

$$X_1 - m_0, X_2 - m_0, \ldots, X_n - m_0.$$

The sign test can also be used to test the hypothesis that two continuous-type random variables X and Y are such that $p = P(X > Y) = 1/2$. To test the hypothesis $H_0: p = 1/2$ against an appropriate alternative hypothesis, consider the independent pairs $(X_1, Y_1), (X_2, Y_2), \ldots, (X_n, Y_n)$. Let W denote the number of pairs for which $X_i - Y_i > 0$. When H_0 is true, W is $b(n, 1/2)$, and the test can be based upon the statistic W. It is important to recognize that X_i and Y_i do not need to be independent and are frequently not independent in practice. If they were independent, then there are many ways in which $X_1, X_2, \ldots, X_n$ and $Y_1, Y_2, \ldots, Y_n$ can be paired up. If, however, X is the length of the right foot of a person and Y the length of the corresponding left foot, there is a natural pairing. Here $H_0: p = P(X > Y) = 1/2$ suggests that either foot of a particular individual is equally likely to be longer.

Example 10.3-2 Freshmen in a health dynamics course have their percentage of body fat measured at the beginning (x) and at the end (y) of the semester. These measurements are given for 26 students in Table 10.3-1. Also recorded is a plus sign when $x_i > y_i$ and a minus sign with $x_i < y_i$. Note that, in theory, the case $x_i = y_i$ should not happen. However, because measurements are rounded off, ties do occur in applications. When a tie does occur, the subject is dropped and the test is run with a reduced sample size. We test the hypothesis $H_0: p = P(X > Y) = 1/2$ against the one-sided alternative H_1: $p > 1/2$. That is, the alternative hypothesis is that the students' percentage of body fat decreases during the semester. Let the critical region be defined by $C = \{w: w \geq 17\}$, where w is the number of "plus signs." The distribution of W is $b(25, 1/2)$ when H_0 is true, so we see that $\alpha = P(W \geq 17; p = 1/2) = 0.0539$ from Table II. Since $w = 16$, the null hypothesis is not rejected with these data; although they suggest, with more data of the same type, we might be able to reject H_0. The p-value of this test is

$$p\text{-value} = P(W \geq 16; p = 1/2) = 0.1148.$$

The sign test can also be used in comparisons when the outcomes are not numerical. For example, each of a number of coffee drinkers selected at random is asked if some freshly perked coffee is preferred over some instant coffee.

There are some objections to the sign test, among them being that the data must be paired and that the test does not take into account the magnitude of differences.

TABLE 10.3-1

x	y	Sign	x	y	Sign
35.4	33.6	+	22.4	21.0	+
28.8	31.9	−	23.5	24.5	−
10.6	10.5	+	24.1	21.9	+
16.7	15.6	+	22.5	21.7	+
14.6	14.0	+	17.5	17.9	−
8.8	13.9	−	16.9	14.9	+
17.9	8.7	+	11.7	17.5	−
17.8	17.6	+	8.3	11.7	−
9.3	8.9	+	7.9	10.2	−
23.6	23.6	0	20.7	17.7	+
15.6	13.7	+	26.8	24.1	+
24.3	24.7	−	20.6	20.4	+
23.8	25.3	−	25.1	21.9	+

How do we decide when to use the sign test? In the situation in which we need a test about a location parameter of a distribution, would we use the sign test to test $m = m_0$ or a test based on $\overline{X}$ to test $\mu = \mu_0$? In particular, if the distribution is symmetric, then $m = \mu$, and hence each procedure would be testing the hypothesis that the common value of m and μ is equal to a known constant. In more advanced theory and practice, we discover that if there are outliers (i.e., extreme x values that deviate greatly from most of the other values), the sign test is usually better than that based upon $\overline{X}$. This certainly appeals to our intuition because a few extreme x values can influence the average $\bar{x}$ a great deal; but, in the sign test, each extreme value is associated with only one sign no matter how far it is from m. Hence, in cases with highly skewed data or data with outliers, the distribution-free sign test would probably be preferred to that based on $\overline{X}$. In the next section, we consider another distribution-free test, first proposed by Wilcoxon, which does account somewhat for the magnitudes of the deviations $X_1 - m_0, X_2 - m_0, \ldots, X_n - m_0$ and is an excellent test in many situations.

Another advantage of the sign test is the fact that it can easily be generalized to percentiles other than the median. To test the hypothesis that the $(100p)$th percentile π_p of a continuous-type distribution is equal to a specified value π_0, let Y be the number of the items of the random sample of size n that are less than π_0. If H_0: $\pi_p = \pi_0$ is true, then Y is $b(n, p)$; where here, of course, p is a known probability. If, for example, the alternative hypothesis is H_1: $\pi_p > \pi_0$, then the critical region would be of the form $C = \{y: y \le c\}$.

Example 10.3-3 Suppose, from past testing, we know that the 25th percentile of ninth-grade general mathematics students taking a standardized examination is 62.4. Statisticians believe that by introducing more statistics involving real problems into such a course, the students will see the usefulness of mathematics. It is hoped that this approach will motivate the poorer students in these classes and increase some of the lower percentiles. In particular, they believe that if these courses were changed as suggested, then H_1: $\pi_{0.25} > 62.4$. To test H_0: $\pi_{0.25} = 62.4$ against H_1, 192 general mathematics students were selected at random and given this new type of course using more statistics. Let Y be the number of students scoring less than 62.4. Of course, if H_0 is true, then Y is $b(192, 1/4)$. If H_1 is true, we would expect smaller values of Y, and thus the critical region is of the form $y \leq c$. To determine c, we can use the fact that Y, under H_0, has an approximate normal distribution with mean $192(1/4) = 48$ and variance $192(1/4)(3/4) = 36$. Thus, for the significance level to be about 0.05, we want

$$c \approx 48 - (1.645)(6) = 38.13.$$

Accordingly, if we take $c = 38$, then the significance level is

$$P(Y \leq 38) = P(Y < 38.5) = P\left(\frac{Y - 48}{6} < \frac{-9.5}{6}\right)$$
$$\approx \Phi(-1.583) = 0.0567.$$

Suppose that after the course is over, these 192 students take the standardized examination and only 31 have grades less than 62.4. We would reject H_0: $\pi_{0.25} = 62.4$ and accept H_1: $\pi_{0.25} > 62.4$; that is, it seems as if the 25th percentile is greater in the new type of general mathematics course than that associated with the old one.

Exercises

10.3-1 On any given day at a candy warehouse, 13 trucks are loaded with candy. It is claimed that the median weight m of each load of candy is 40,000 pounds.
(a) Test, at the $\alpha = 0.0461$ significance level, the null hypothesis H_0: $m = 40,000$ against the one-sided alternative hypothesis H_1: $m < 40,000$ using the following observations:

41,195	39,485	41,229	36,840	38,050	40,890	38,345
34,930	39,245	31,031	40,780	38,050	30,906	

(b) What is the p-value of this test?

10.3-2 The Forced Vital Capacity (FVC) gives a measurement of the amount of air that a person can exhale from his lungs. Let m equal the median FVC in liters for male college freshmen. Use the following 17 observations to test, at the $\alpha = 0.0768$ significance level, the null hypothesis H_0: $m = 4.7$ against the two-sided alternative hypothesis H_1: $m \neq 4.7$ (deleting any observation that is equal to 4.7):

4.3	5.0	5.7	6.2	4.8	4.7	5.6	5.2	3.7
4.0	5.6	6.8	4.9	7.6	5.4	3.8	5.6	

10.3-3 A course in economics was taught to two groups of students, one in a classroom situation and the other by TV. There were 24 students in each group. These students were first paired according to cumulative grade point averages and background in economics, and then assigned to the courses by a flip of a coin (this was repeated 24 times). At the end of the course each class was given the same final examination. Use the sign test to test ($\alpha = 0.0662$) that the two methods of teaching are equally effective against a two-sided alternative. The differences in final scores for each pair of students, the TV student's score having been subtracted from the corresponding classroom student's scores, were as follows:

14	-4	-6	-2	-1	18
6	12	8	-4	13	7
2	6	21	7	-2	11
-3	-14	-2	17	-4	-5

10.3-4 The administration claims that professor A is more popular than professor B. We shall test the hypothesis H_0 of no difference in popularity against the administration's claim H_1 at an approximate 0.05 significance level. If 14 out of 20 students prefer professor A to professor B, is H_0 rejected and the administration's claim supported?

10.3-5 In a certain region, it was known that about 20% of the drivers exceeded the speed limit of 55 by more than 5 miles per hour. The highway patrol then established a new enforcement procedure and claimed that this percentage had been reduced substantially. How would you test the null hypothesis H_0: $\pi_{0.8} = 60$ against the patrol's claim H_1: $\pi_{0.8} < 60$ at $\alpha = 0.05$ if you could observe the speeds of 100 drivers selected at random? If, in fact, out of the 100 drivers, you found that the speeds of 89 were under 60 miles per hour, what decision would you make?

10.3-6 To test whether a golf ball of brand A can be hit a greater distance off the tee than a golf ball of brand B, each of 17 golfers hit a ball of each brand, eight hitting ball A before ball B and nine hitting ball B before ball A. Let m_D denote the median of the differences of the distances of ball A and ball B. Use the sign test to test the null hypothesis H_0: $m_D = 0$ against H_1: $m_D > 0$.

HINT: In case of ties, delete that pair of observations and use a reduced sample size.

Golfer	Distance for Ball A	Distance for Ball B
1	265	252
2	272	276
3	246	243
4	260	246

Golfer	Distance for Ball A	Distance for Ball B
5	274	275
6	263	246
7	255	244
8	258	245
9	276	259
10	274	260
11	274	267
12	269	267
13	244	251
14	212	222
15	235	235
16	254	255
17	224	231

10.3-7 In Exercise 10.3-6 assume that the differences for each golfer, distance for ball A minus distance for ball B, are observations of a normal random variable with mean μ_D. Test the null hypothesis H_0: $\mu_D = 0$ against the alternative hypothesis H_1: $\mu_D > 0$ using a t-test with the 17 differences.

10.3-8 For each student who takes Introductory Statistics, let X and Y denote the students' ACT Math and ACT Verbal scores, respectively. Let $D = X - Y$ and m_D denote the median of the differences. Test H_0: $m_D = 0$ against the two-sided alternative hypothesis H_1: $m_D \neq 0$ at an approximate $\alpha = 0.10$ significance level.

Student	ACT Math	ACT Verbal
1	16	19
2	18	17
3	22	18
4	20	23
5	17	20
6	25	21
7	21	24
8	23	18
9	24	18
10	31	25
11	27	29
12	28	24
13	30	24
14	27	23
15	28	24

10.3-9 In Exercise 7.3-19, a company and a vendor measured the fat content of samples of milk from the same lot. Let $D = X - Y$ and let m_D denote the median of the differences.

   (a) Test H_0: $m_D = 0$ against H_1: $m_D > 0$ using the sign test. Let $\alpha = 0.05$, approximately.

   (b) How does this conclusion compare with that of Exercise 7.3-19?

10.3-10 Let m equal the median FVC in liters for college freshmen women. (See Exercise 10.3-2.) We shall use the sign test to test the null hypothesis H_0: $m = 3.05$ against the alternative hypothesis H_1: $m > 3.05$ using 25 observations.

   (a) Define a critical region that has an approximate significance level of $\alpha = 0.05$.

   (b) What is your conclusion using the following data?

4.2	5.3	3.5	4.3	3.7	3.5	3.5	2.8	3.5
3.7	3.5	3.7	2.7	2.8	3.3	2.7	3.0	
3.1	3.0	3.7	3.3	3.4	2.3	2.6	3.2	

   (c) What is the p-value of your test?

10.3-11 Let U and V equal a student's ACT score in Social Science and in Natural Science, respectively. Let $p = P(U > V)$. We shall use the sign test to test the null hypothesis H_0: $p = 0.5$ against a two-sided alternative hypothesis using the following 15 pairs, (u, v):

(32, 28)	(30, 27)	(26, 32)	(23, 25)	(17, 23)
(16, 22)	(23, 24)	(20, 30)	(21, 28)	(23, 32)
(17, 18)	(24, 31)	(26, 31)	(18, 18)	(30, 26)

   (a) Define a critical region that has a significance level that is close to $\alpha = 0.05$. Give the exact significance level.

   (b) What is your conclusion?

10.3-12 Thirteen tons of Wisconsin cheese were packaged in "22-pound" wheels and stored underground in old gypsum mines. The manufacturer claimed that the median weight of each wheel is 22 pounds while the Michigan Department of Agriculture suspects that the median is less than 22 pounds.

   (a) Define the null (manufacturer's) and alternative (Department of Agriculture's) hypotheses.

   (b) Based on $n = 9$ observations, define a critical region that has an approximate 10% significance level.

   (c) Given the weights 21.50, 18.95, 18.55, 19.40, 19.15, 22.35, 22.90, 22.20, 23.10, what is your conclusion?

   (d) What is the p-value of this test?

10.4 The Wilcoxon Test

Let X be a continuous-type random variable. Let m denote the median of X. To test the hypothesis H_0: $m = m_0$ against an appropriate alternative hypothesis, the sign test, discussed in the previous section, can be used. That is, if $X_1, X_2, \ldots, X_n$

denote the items of a random sample from this distribution and if we let Y equal the number of negative differences among $X_1 - m_0, X_2 - m_0, \ldots, X_n - m_0$, then Y has the binomial distribution $b(n, 1/2)$ under H_0 and is the test statistic for the sign test. One major objection to this test is that it does not take into account the magnitude of the differences $X_1 - m_0, \ldots, X_n - m_0$.

In this section we shall discuss a test that does take into account the magnitude of the differences $|X_i - m_0|$, $i = 1, 2, \ldots, n$. However, in addition to the assumption that the random variable X is of the continuous type, we must also assume that the p.d.f. of X is symmetric about the median in order to find the distribution of this new statistic. Because of the continuity assumption, we assume, in the following discussion, that no two observations are equal and that no observation is equal to the median.

We are interested in testing the hypothesis H_0: $m = m_0$ where m_0 is some given constant. With our random sample $X_1, X_2, \ldots, X_n$, we rank the differences $X_1 - m_0, X_2 - m_0, \ldots, X_n - m_0$ in ascending order according to magnitude. That is, for $i = 1, 2, \ldots, n$, we let R_i denote the rank of $|X_i - m_0|$ among $|X_1 - m_0|$, $|X_2 - m_0|, \ldots, |X_n - m_0|$. Note that $R_1, R_2, \ldots, R_n$ is a permutation of the first n positive integers, $1, 2, \ldots, n$. Now with each R_i we associate the sign of the difference $X_i - m_0$; that is, if $X_i - m_0 > 0$, we use R_i, but if $X_i - m_0 < 0$, we use $-R_i$. The Wilcoxon statistic W is the sum of these n signed ranks.

> **Example 10.4-1** Consider the sunfish example in the last section. There we considered testing H_0: $m = 3.7$ against the alternative hypothesis H_1: $m > 3.7$. The observed lengths of the $n = 10$ fish were
>
> $$x_i\colon\ 5.0,\ 3.9,\ 5.2,\ 5.5,\ 2.8,\ 6.1,\ 6.4,\ 2.6,\ 1.7,\ 4.3.$$
>
> Thus we have

$x_i - m_0$:	1.3,	0.2,	1.5,	1.8,	−0.9,	2.4,	2.7,	−1.1,	−2.0,	0.6
$|x_i - m_0|$:	1.3,	0.2,	1.5,	1.8,	0.9,	2.4,	2.7,	1.1,	2.0,	0.6
Ranks:	5,	1,	6,	7,	3,	9,	10,	4,	8,	2
Signed Ranks:	5,	1,	6,	7,	−3,	9,	10,	−4,	−8,	2

> Therefore, the Wilcoxon statistic is equal to
>
> $$W = 5 + 1 + 6 + 7 - 3 + 9 + 10 - 4 - 8 + 2 = 25.$$
>
> Incidentally, the positive answer seems reasonable because the number of the 10 lengths that are less than 3.7 is 3, which is the statistic used in the sign test.

If the hypothesis H_0: $m = m_0$ is true, about one-half of the differences would be negative and thus about one-half of the signs would be negative. Thus it seems that the hypothesis H_0: $m = m_0$ is supported if the observed value of W is close to zero. If the alternative hypothesis is H_1: $m > m_0$, we would reject H_0 if the observed $W = w$ is too large, since, in this case, the larger deviations $|X_i - m_0|$ would usually be associated with observations for which $x_i - m_0 > 0$. That is, the critical region would be of the form $\{w: w \geq c_1\}$. If the alternative hypothesis is H_1: $m <$ m_0, the critical region would be of the form $\{w: w \leq c_2\}$. Of course, the critical region would be of the form $\{w: w \leq c_3 \text{ or } w \geq c_4\}$ for a two-sided alternative hypothesis H_1: $m \neq m_0$. In order to find the values of c_1, c_2, c_3, c_4 that yield desired significance levels, it is necessary to determine the distribution of W under H_0. We now consider certain characteristics of this distribution.

When H_0: $m = m_0$ is true,

$$P(X_i < m_0) = P(X_i > m_0) = \frac{1}{2}, \qquad i = 1, 2, \ldots, n.$$

Hence the probability is 1/2 that a negative sign is associated with the rank R_i of $|X_i - m_0|$. Moreover, the assignments of these n signs are independent because X_1, $X_2, \ldots, X_n$ are mutually independent. In addition, W is a sum that contains the integers $1, 2, \ldots, n$, each integer with a positive or negative sign. Since the underlying distribution is symmetric, it seems intuitively obvious that W has the same distribution as the random variable

$$V = \sum_{i=1}^{n} V_i,$$

where $V_1, V_2, \ldots, V_n$ are independent and

$$P(V_i = i) = P(V_i = -i) = \frac{1}{2}, \qquad i = 1, 2, \ldots, n.$$

That is, V is a sum that contains the integers $1, 2, \ldots, n$, and these integers receive their algebraic signs by independent assignments.

Since W and V have the same distribution, their means and variances are equal, and we can easily find those of V. Now the mean of V_i is

$$E(V_i) = -i\left(\frac{1}{2}\right) + i\left(\frac{1}{2}\right) = 0$$

and thus

$$E(W) = E(V) = \sum_{i=1}^{n} E(V_i) = 0.$$

The variance of V_i is

$$\mathrm{Var}(V_i) = E(V^2) = (-i)^2\left(\frac{1}{2}\right) + (i)^2\left(\frac{1}{2}\right) = i^2.$$

Thus

$$\mathrm{Var}(W) = \mathrm{Var}(V) = \sum_{i=1}^{n} \mathrm{Var}(V_i) = \sum_{i=1}^{n} i^2 = \frac{n(n+1)(2n+1)}{6}.$$

We shall not try to find the distribution of W in general, since that p.d.f. does not have a convenient expression. However, we demonstrate how we could find the distribution of W (or V) with enough patience and computer support. Recall that the moment-generating function of V_i is

$$M_i(t) = e^{t(-i)}\left(\frac{1}{2}\right) + e^{t(+i)}\left(\frac{1}{2}\right) = \frac{e^{-it} + e^{it}}{2}, \qquad i = 1, 2, \ldots, n.$$

Let $n = 2$; so the moment-generating function of $V_1 + V_2$ is

$$M(t) = E[e^{t(V_1 + V_2)}].$$

From the independence of V_1 and V_2, we obtain

$$\begin{aligned} M(t) &= E(e^{tV_1})E(e^{tV_2}) \\ &= \left(\frac{e^{-t} + e^{t}}{2}\right)\left(\frac{e^{-2t} + e^{2t}}{2}\right) \\ &= \frac{e^{-3t} + e^{-t} + e^{t} + e^{3t}}{4}. \end{aligned}$$

This means that each of the points -3, -1, 1, 3 in the support of $V_1 + V_2$ has probability 1/4.

Next let $n = 3$, so the moment-generating function of $V_1 + V_2 + V_3$ is

$$\begin{aligned} M(t) &= E[e^{t(V_1 + V_2 + V_3)}] \\ &= E[e^{t(V_1 + V_2)}]E(e^{tV_3}) \\ &= \left(\frac{e^{-3t} + e^{-t} + e^{t} + e^{3t}}{4}\right)\left(\frac{e^{-3t} + e^{3t}}{2}\right) \\ &= \frac{e^{-6t} + e^{-4t} + e^{-2t} + 2e^{0} + e^{2t} + e^{4t} + e^{6t}}{8}. \end{aligned}$$

Thus the points -6, -4, -2, 0, 2, 4, 6 in the support of $V_1 + V_2 + V_3$ have the respective probabilities 1/8, 1/8, 1/8, 2/8, 1/8, 1/8, 1/8. Obviously, this procedure can be continued for $n = 4, 5, 6, \ldots$, but it is rather tedious. Fortunately, however, even though $V_1, V_2, \ldots, V_n$ are not identically distributed random variables, the sum V of them still has an approximate normal distribution. To obtain this normal approximation for V (or W), a more general form of the Central Limit Theorem, due to Liapounov, can be used that allows us to say that the standardized random variable

$$Z = \frac{W - 0}{\sqrt{n(n + 1)(2n + 1)/6}}$$

is approximately $N(0, 1)$ when H_0 is true. We accept this without proof, and thus we can approximate probabilities such as $P(W \geq c; H_0) \approx P(Z \geq z_\alpha; H_0)$ when the sample size n is sufficiently large using this normal distribution.

Example 10.4-2   Let m be the median of a symmetric distribution of the continuous type. To test the hypothesis H_0: $m = 160$ against the alternative hypothesis H_1: $m > 160$, we take a random sample of size $n = 16$. For an approximate significance level of $\alpha = 0.05$, H_0 is rejected if the computed $W = w$ is such that

$$z = \frac{w}{\sqrt{16(17)(33)/6}} \geq 1.645,$$

or

$$w \geq 1.645 \sqrt{\frac{16(17)(33)}{6}} = 63.626.$$

Say the observed values of a random sample are 176.9, 158.3, 152.1, 158.8, 172.4, 169.8, 159,7, 162.7, 156.6, 174.5, 184.4, 165.2, 147.8, 177.8, 160.1, 160.5. In Table 10.4-1 the magnitudes of the differences $|x_i - 160|$ have been ordered and ranked. Those differences $x_i - 160$ that were negative have been underlined and the ranks are under the ordered values. For this set of data

$$w = 1 - 2 + 3 - 4 - 5 + 6 + \cdots + 16 = 60.$$

Since $60 < 63.626$, H_0 is not rejected at the 0.05 significance level. It is interesting to note that H_0 would have been rejected at $\alpha = 0.10$, since the approximate p-value is, making a half-unit correction for continuity,

$$p\text{-value} = P(W \geq 60)$$

$$= P\left(\frac{W - 0}{\sqrt{\dfrac{(16)(17)(33)}{6}}} \geq \frac{59.5 - 0}{\sqrt{\dfrac{(16)(17)(33)}{6}}} \right)$$

$$\approx P(Z \geq 1.538) = 0.0620.$$

This would indicate to us that the data are too few to reject H_0, but if the pattern continues, we shall most certainly reject with a larger sample size.

Although theoretically we could ignore the possibilities that $x_i = m_0$, for some i, and that $|x_i - m_0| = |x_j - m_0|$ for some $i \neq j$, these situations do occur in applica-

TABLE 10.4-1

0.1	0.3	0.5	1.2	1.7	2.7	3.4	5.2
1	2	3	4	5	6	7	8
7.9	9.8	12.2	12.4	14.5	16.9	17.8	24.4
9	10	11	12	13	14	15	16

tions. Usually, in practice, if $x_i = m_0$ for some i, that observation is deleted and the test is performed with a reduced sample size. If the absolute values of the differences from m_0 of two or more observations are equal, each observation is assigned the average of the corresponding ranks. The change this causes in the distribution of W is not very great, and thus we continue using the same normal approximation.

We now give an example that has some tied observations. This example also illustrates that the Wilcoxon rank sum test can be used to test hypotheses about the median of paired data.

Example 10.4-3 In Example 10.3-2 we gave some paired data for percentage of body fat measured at the beginning and the end of a semester. Let m equal the median of the differences, $x - y$. We shall use the Wilcoxon statistic to test the null hypothesis H_0: $m = 0$ against the alternative hypothesis H_1: $m > 0$. Since there are $n = 25$ nonzero differences, we reject H_0 if

$$z = \frac{w - 0}{\sqrt{\dfrac{(25)(26)(51)}{6}}} \geq 1.645$$

or, equivalently, if

$$w \geq 1.645 \sqrt{\frac{(25)(26)(51)}{6}} = 122.27$$

at an approximate $\alpha = 0.05$ significance level. From Table 10.3-1 we can calculate the differences, yielding:

1.8	−3.1	0.1	1.1	0.6	−5.1	9.2	0.2	0.4
0.0	1.9	−0.4	−1.5	1.4	−1.0	2.2	0.8	−0.4
2.0	−5.8	−3.4	−2.3	3.0	2.7	0.2	3.2	

In Table 10.4-2 we list the ordered absolute values, underlining those that were originally negative. The rank is under each observation. Note that in the case of ties, the average of the ranks of the tied measurements is given. The value of the Wilcoxon statistic is

$$w = 1 + 2.5 + 2.5 + 5 - 5 - 5 + \cdots + 25 = 51.$$

TABLE 10.4-2

0.1	0.2	0.2	0.4	0.4	0.4	0.6	0.8	1.0	1.1	1.4	1.5	1.8
1	2.5	2.5	5	5	5	7	8	9	10	11	12	13
1.9	2.0	2.2	2.3	2.7	3.0	3.1	3.2	3.4	5.1	5.8	9.2	
14	15	16	17	18	19	20	21	22	23	24	25	

Since $51 < 122.27$, we fail to reject the null hypothesis. The approximate p-value of this test is

$$p\text{-value} = P(W \geq 51) = P(W \geq 50.5)$$

$$\approx P\left(Z \geq \dfrac{50.5 - 0}{\sqrt{\dfrac{(25)(26)(51)}{6}}}\right)$$

$$= P(Z \geq 0.679) = 0.2486.$$

Exercises

10.4-1 In Exercise 10.3-1 it is claimed that the median weight m of each load of candy is 40,000 pounds.

    (a) Use those 13 observations and the Wilcoxon statistic to test, at an approximate significance level of $\alpha = 0.05$, the null hypothesis H_0: $m = 40{,}000$ against the one-sided alternative hypothesis H_1: $m < 40{,}000$. For convenience, we list those 13 observations again.

41,195	39,485	41,229	36,840	38,050	40,890	38,345
34,930	39,245	31,031	40,780	38,050	30,906	

    (b) What is the approximate p-value of this test? (How does it compare with the p-value in Exercise 10.3-1?)

10.4-2 In Exercise 10.3-2 we tested the null hypothesis that the median, m, of the forced vital capacity for male college freshmen is equal to 4.7 using the sign test. Now use the Wilcoxon statistic to test H_0: $m = 4.7$ against H_1: $m \neq 4.7$ using the $n = 17$ observations in Exercise 10.3-2. Use an approximate significance level of $\alpha = 0.05$.

10.4-3 With the data in Exercise 10.3-3, use the Wilcoxon statistic to test the hypothesis that the median of the differences in final scores equals zero.

10.4-4 In Table 4.5-1 we listed the outcomes on $n = 10$ simulations of a Cauchy random variable. They were -1.9415, 0.5901, -5.9848, -0.0790, -0.7757, -1.0962, 9.3820, -74.0216, -3.0678, 3.8545. For the Cauchy distribution, the mean does not exist; but, for this one, the median is believed to equal zero. Use the Wilcoxon test and these data to test H_0: $m = 0$ against the alternative hypothesis H_1: $m \neq 0$. Let $\alpha = 0.05$, approximately.

10.4-5 For the golf ball example in Exercise 10.3-6, use the Wilcoxon statistic to test the hypothesis H_0: $m_D = 0$ against H_1: $m_D > 0$. Let $\alpha = 0.05$, approximately.

10.4-6 For the milk fat content example in Exercises 7.3-19 and 10.3-9, use the Wilcoxon statistic to test the hypothesis H_0: $m_D = 0$ against H_1: $m_D > 0$. Let $\alpha = 0.05$, approximately.

10.4-7 Let x equal a student's GPA in the fall semester and y the same student's GPA in the spring semester. Let m equal the median of the differences, $x - y$.

We shall test the null hypothesis H_0: $m = 0$ against an appropriate alternative hypothesis that you select based on your past experience. Use a Wilcoxon test and the given 15 observations of paired data to test this hypothesis.

x	y	x	y
2.88	3.22	3.98	3.76
3.67	3.49	4.00	3.96
2.76	2.54	3.39	3.52
2.34	2.17	2.59	2.36
2.46	2.53	2.78	2.62
3.20	2.98	2.85	3.06
3.17	2.98	3.25	3.16
2.90	2.84		

10.4-8 Let m equal the median of the posttest grip strengths in the right arms of male freshmen in Health Dynamics. We shall use observations on $n = 15$ such students to test the null hypothesis H_0: $m = 50$ against the alternative hypothesis H_1: $m > 50$.
  (a) Using the sign test, define a critical region that has an approximate significance level of $\alpha = 0.05$.
  (b) Given the observed values,

$$58 \quad 52.5 \quad 46 \quad 57.5 \quad 52 \quad\quad 45.5 \quad 65.5 \quad 71$$
$$57 \quad 54 \quad\quad 48 \quad 58 \quad\quad 35.5 \quad 44 \quad\quad 53$$

  what is your conclusion?
  (c) What is the p-value of this test?
  (d) Perform the same test using the Wilcoxon statistic. Let $\alpha = 0.05$. What is your conclusion?
  (e) What is the approximate p-value using the Wilcoxon statistic?

10.4-9 Let X equal the weight in pounds of a "1-pound" bag of carrots. Let m equal the median weight of a population of these bags. Test the null hypothesis H_0: $m = 1.14$ against the alternative hypothesis H_1: $m > 1.14$.
  (a) With a sample of size $n = 14$, use the sign test to define a critical region. Use $\alpha = 0.10$, approximately.
  (b) What is your conclusion if the observed weights were

$$1.12 \quad 1.13 \quad 1.19 \quad 1.25 \quad 1.06 \quad 1.31 \quad 1.12$$
$$1.23 \quad 1.29 \quad 1.17 \quad 1.20 \quad 1.11 \quad 1.18 \quad 1.23$$

  (c) What is the p-value of your test?
  (d) Use the Wilcoxon statistic to perform the same test. What is your conclusion?
  (e) What is the approximate p-value using the Wilcoxon statistic?

10.4-10   In Exercise 10.3-10, we used the sign test to test H_0: $m = 3.05$ against the alternative hypothesis H_1: $m > 3.05$.

    (a) For $n = 25$ observations, define a critical region using the Wilcoxon statistic with a significance level of $\alpha = 0.05$.

    (b) What is your conclusion using the data in Exercise 10.3-10?

    (c) What is the approximate p-value using the Wilcoxon statistic?

10.5   Two-Sample, Distribution-Free Tests

In Sections 10.3 and 10.4 we considered some distribution-free tests based on a sample from one distribution or a sample of paired dependent random variables. In this section we consider corresponding tests associated with the characteristics of two independent distributions.

The first test corresponds to the sign test and is called the **median test**. Let m_X and m_Y be the respective medians of two independent distributions of the continuous type. By taking random samples $X_1, X_2, \ldots, X_{n_1}$ and $Y_1, Y_2, \ldots, Y_{n_2}$ from these two independent distributions, respectively, we wish to test the hypothesis H_0: $m_X = m_Y$. To do this, combine the two samples and count the number, say V, of X values in the lower half of this combined sample. If H_0: $m_X = m_Y$ is true, then we would expect V to equal some number around $n_1/2$. If, as an alternative, $m_X < m_Y$, we would expect V to be somewhat larger; and, of course, the alternative $m_X > m_Y$ would suggest a smaller value of V.

Let us see what this means in terms of the distribution functions $F(x)$ and $G(y)$ of the respective distributions. Of course, if $F(z) = G(z)$, then H_0: $m_X = m_Y$ is true. Since we cannot find the distribution of V knowing only that $m_X = m_Y$, we shall find its distribution assuming that $F(z) = G(z)$. If $F(z) \geq G(z)$, $m_X \leq m_Y$, as depicted in Figure 10.5-1. If the observed value of V is quite large—that is, if the number of values of X falling below the median of the combined sample is large—we would suspect that $m_X < m_Y$. Thus the critical region for testing H_0: $m_X = m_Y$ against H_1: $m_X < m_Y$ is of the form $v \geq c$, where c is to be determined to yield the desired significance level [when $F(z) = G(z)$]. Similarly, the critical region for testing H_0: $m_X = m_Y$ against H_1: $m_X > m_Y$ is of the form $v \leq c$.

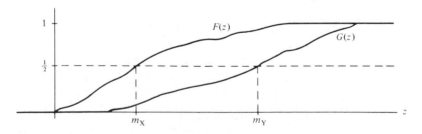

Figure 10.5-1

When $F(z) = G(z)$ is true and still assuming continuous-type distributions, we shall argue that V has a hypergeometric distribution. To simplify the discussion, say that $n_1 + n_2 = 2k$, where k is a positive integer. To compute $P(V = v)$, we need the probability that exactly v of $X_1, X_2, \ldots, X_{n_1}$ are in the lower half of the ordered combined sample. (Under our assumptions, the probability is zero that any two of the $2k$ random variables are equal.) The smallest k of the $n_1 + n_2 = 2k$ items can be selected in any one of $\binom{2k}{k}$ ways, each having the same probability, provided that $F(z) = G(z)$. Of these $\binom{2k}{k}$ ways, the number in which exactly v of the n_1 values of X and $k - v$ of the n_2 values of Y appear in the lower k items is $\binom{n_1}{v}\binom{n_2}{k - v}$.

Hence

$$h(v) = P(V = v) = \frac{\binom{n_1}{v}\binom{n_2}{k - v}}{\binom{n_1 + n_2}{k}}, \quad v = 0, 1, \ldots, n_1,$$

with the understanding that $\binom{j}{i} = 0$ if $i > j$.

Example 10.5-1 Let X and Y denote the weights of ground cinnamon in "115-gram" tins packaged by companies A and B, respectively. We shall test the hypothesis $H_0: m_X = m_Y$ against the one-sided alternative hypothesis H_1: $m_X < m_Y$. The weights of $n_1 = 8$ and $n_2 = 8$ tins of cinnamon packaged by companies A and B, respectively, selected at random, yielded the following observations of X:

117.1   121.3   127.8   121.9   117.4   124.5   119.5   115.1,

and the following observations of Y:

123.5   125.3   126.5   127.9   122.1   125.6   129.8   117.2.

The critical region is of the form $v \geq c$. To determine the value of c when $F(z) = G(z)$, we compute $P(V = v)$ for $v = 6, 7$, and 8. Using Table I in the Appendix,

$$\binom{n_1 + n_2}{k} = \binom{8 + 8}{8} = 12{,}870.$$

Thus

$$h(8) = \frac{\binom{8}{8}\binom{8}{0}}{12{,}870} = \frac{(1)(1)}{12{,}870} = \frac{1}{12{,}870};$$

$$h(7) = \frac{\binom{8}{7}\binom{8}{1}}{12{,}870} = \frac{(8)(8)}{12{,}870} = \frac{64}{12{,}870};$$

$$h(6) = \frac{\binom{8}{6}\binom{8}{2}}{12{,}870} = \frac{(28)(28)}{12{,}870} = \frac{784}{12{,}870}.$$

Since

$$h(8) + h(7) + h(6) = \frac{849}{12{,}870} = 0.066,$$

we shall reject H_0 if $v \geq 6$ at an $\alpha = 0.066$ significance level. The combined ordered sample of weights is listed below with the observations of X under-lined.

115.1	117.1	117.2	117.4	119.5	121.3	121.9	122.1
123.5	124.5	125.3	125.6	126.5	127.8	127.9	129.8

We see that $v = 6$, and thus H_0 is rejected with $\alpha = 0.066$. Note that the p-value of this test is

$$p\text{-value} = P(V \geq 6) = 0.066.$$

Of course, the median test can be generalized as easily as the sign test. Instead of letting V be the number of X values in the lower half of the combined sample of size $n_1 + n_2$, let V be the number of X values in the lower i values of the combined sample. Thus, if $i/(n_1 + n_2) = p$, we would use V to test the equality of the $(100p)$th percentiles of the two distributions. If V is much larger than $n_1 p$, we would suspect that the $(100p)$th percentile of the X distribution is smaller than that of the Y distribution. If V is much smaller than $n_1 p$, we would guess that it would be the other way around. Since, under $F(z) = G(z)$, all orderings of the n_1 values of X and n_2 values of Y have the same probability, the p.d.f. of V is

$$h(v) = P(V = v) = \frac{\binom{n_1}{v}\binom{n_2}{i-v}}{\binom{n_1 + n_2}{i}}, \qquad v = 0, 1, 2, \ldots, n_1.$$

Example 10.5-2   Using the data in Example 10.5-1 we shall test the null hypothesis that the first quartiles of X and Y are equal against the alternative hypothesis that the first quartile of X is less than the first quartile of Y. Let V equal the number of X values in the lower four values of the combined sample because $i/(n_1 + n_2) = 4/(8 + 8) = 0.25$. The critical region is of the form $v \geq c$. To determine the value of c, we use the p.d.f. of V given by

$$h(v) = \frac{\binom{8}{v}\binom{8}{4-v}}{\binom{16}{4}}, \qquad v = 0, 1, 2, 3, 4.$$

Now

$$h(4) = \frac{70}{1820} = 0.038;$$

$$h(3) = \frac{448}{1820} = 0.246.$$

Thus we take for our critical region $C = \{v: v \geq 4\}$, which gives a significance level of $\alpha = 0.038$. From Example 10.5-1 we see that three of the first four observations in the ordered arrangement are values of X and thus $v = 3$, so we fail to reject the null hypothesis. Furthermore, the p-value of this test is

$$p\text{-value} = P(V \geq 3) = 0.246 + 0.038 = 0.284.$$

There is another method (due to Wilcoxon) for testing the equality of the medians of two distributions of the continuous type that uses the magnitudes of the observations. For this test it is assumed that the populations have similar shapes. Order the combined sample of $X_1, X_2, \ldots, X_{n_1}$ and $Y_1, Y_2, \ldots, Y_{n_2}$ in increasing order of magnitude. Assign to the ordered values the ranks $1, 2, 3, \ldots, n_1 + n_2$. In the case of ties, assign the average of the ranks associated with the tied values. Let W equal the sum of the ranks of $Y_1, Y_2, \ldots, Y_{n_2}$. If the distribution of Y is shifted to the right of that of X, the values of Y would tend to be larger than the values of X and W would usually be larger than expected when $F(z) = G(z)$. Thus the critical region for testing $H_0: m_X = m_Y$ against $H_1: m_X < m_Y$ would be of the form $w \geq c$. Similarly, if the alternative hypothesis is $m_X > m_Y$, the critical region would be of the form $w \leq c$.

We shall not derive the distribution of W. However, if n_1 and n_2 are both greater than 7, a normal approximation can be used. With $F(z) = G(z)$, the mean and variance of W are

$$\mu_W = \frac{n_2(n_1 + n_2 + 1)}{2}$$

and

$$\text{Var}(W) = \frac{n_1 n_2(n_1 + n_2 + 1)}{12};$$

thus the statistic

$$Z = \frac{W - n_2(n_1 + n_2 + 1)/2}{\sqrt{n_1 n_2(n_1 + n_2 + 1)/12}}$$

is approximately $N(0, 1)$.

Example 10.5-3 We illustrate the two-sample Wilcoxon test using the data given in Example 10.5-1. The critical region for testing H_0: $m_X = m_Y$ against H_1: $m_X < m_Y$ is of the form $w \geq c$. Since $n_1 = n_2 = 8$, at an approximate $\alpha = 0.05$ significance level, H_0 is rejected if

$$z = \frac{w - 8(8 + 8 + 1)/2}{\sqrt{[(8)(8)(8 + 8 + 1)]/12}} > 1.645;$$

that is, if

$$w > 1.645 \sqrt{\frac{(8)(8)(17)}{12}} + 4(17) = 83.66.$$

From the data we see that the computed W is

$$w = 3 + 8 + 9 + 11 + 12 + 13 + 15 + 16 = 87 > 83.66.$$

Thus H_0 is rejected, an action consistent with that of the median test. The p-value of this test is, making a half-unit correction for continuity,

$$p\text{-value} = P(W \geq 87)$$
$$= P\left(\frac{W - 68}{\sqrt{90.667}} \geq \frac{86.5 - 68}{\sqrt{90.667}}\right)$$
$$\approx P(Z \geq 1.943) = 0.0260.$$

REMARKS It should be mentioned that, shortly after Wilcoxon proposed his test, Mann and Whitney suggested a test based on the estimate of the probability

$$P(X < Y).$$

Namely, they let U equal the number of times that $X_i < Y_j$, $i = 1, 2, \ldots, n_1$ and $j = 1, 2, \ldots, n_2$. Using the data in Example 10.5-1, we find that the computed U is $u = 51$ among all $n_1 n_2 = (8)(8) = 64$ pairs of (X, Y). Thus the estimate of $P(X < Y)$ is $51/64$ or, in general, $u/n_1 n_2$. At the time of the Mann–Whitney suggestion, it was noted that U was just a linear function of Wilcoxon's W and hence really provided the same test. That relationship is

$$U = W - \frac{n_2(n_2 + 1)}{2},$$

which in our special case is

$$51 = 87 - \frac{8(9)}{2} = 87 - 36.$$

Thus we often read about the test of Mann, Whitney, and Wilcoxon.

It should be noted here that the median and Wilcoxon tests are much less sensitive to extreme values than is Student's test based on $\overline{X} - \overline{Y}$. Therefore, if there is much skewness or contamination, these proposed distribution-free tests are much safer. In particular, that of Wilcoxon is quite good and does not lose too much in case the distributions are close to normal ones. It is important to note that the one-sample Wilcoxon requires symmetry of the underlying distribution, but the two-sample Wilcoxon does not and thus can be used for skewed distributions.

The next example illustrates that it is possible to accept the null hypothesis that the medians of two distributions are equal, when in fact the two distributions could be different. A quantile–quantile plot is used to help to detect this difference.

Example 10.5-4 Let X have a Cauchy distribution and let Y have a standard normal distribution $N(0, 1)$. Of course, the assumptions of the Wilcoxon test are not satisfied, but the distributions of X and Y have "similar shapes." We shall simulate on the computer random samples of sizes $n_1 = n_2 = 12$ from these distributions and use the Wilcoxon statistic to test the null hypothesis H_0: $m_x = m_y$ against a two-sided alternative hypothesis. The $n_1 = 12$ observations of X, which have been ordered, are

-9.7465	-1.9458	-1.5203	-1.3990	-0.7215	-0.1802
0.1219	0.2104	0.9365	2.3121	7.5518	13.7523

The $n_2 = 12$ observations of Y, which have been ordered, are

-1.5124	-1.3661	-1.0712	-0.6178	-0.3631	-0.2465
0.1484	0.2688	0.4528	1.0480	1.2914	1.7841

The combined ordered samples, with the observations of X underlined and the ranks of the observations listed below each observation, are

$\underline{-9.7465}$	$\underline{-1.9458}$	$\underline{-1.5203}$	-1.5124	$\underline{-1.3990}$	-1.3661
1	2	3	4	5	6
-1.0712	$\underline{-0.7215}$	-0.6178	-0.3631	-0.2465	$\underline{-0.1802}$
7	8	9	10	11	12
$\underline{0.1219}$	0.1484	$\underline{0.2104}$	0.2688	0.4528	$\underline{0.9365}$
13	14	15	16	17	18
1.0480	1.2914	1.7841	$\underline{2.3121}$	$\underline{7.5518}$	$\underline{13.7523}$
19	20	21	22	23	24

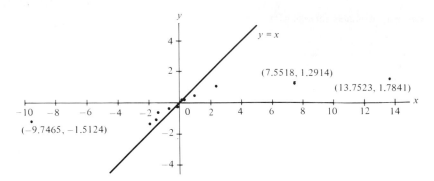

FIGURE **10.5-2**

The value of the Wilcoxon statistic, the sum of the Y ranks, is

$$w = 4 + 6 + 7 + 9 + 10 + 11 + 14 + 16 + 17 + 19 + 20 + 21$$
$$= 154.$$

Of course, the sum of the X ranks is $300 - 154 = 146$. Now.

$$z = \frac{154 - [12(25)/2]}{\sqrt{\dfrac{(12)(12)(25)}{12}}}$$

$$= \frac{154 - 150}{\sqrt{300}} = 0.23,$$

so that H_0 is clearly accepted. That is, we accept the null hypothesis that the medians of X and Y are equal, as we expected. However, if we look at the data more closely by making a q–q plot (Figure 10.5-2) it is clear that the distributions of X and Y are not equal because the distribution of X has "thicker tails" than does the distribution of Y.

Exercises

10.5-1   Let us compare the failure times of a certain type of light bulb produced by two different manufacturers, X and Y, by testing 10 bulbs selected at random from each of the outputs. The data, in hundreds of hours used before failure, are

X:	5.6	4.6	6.8	4.9	6.1	5.3	4.5	5.8	5.4	4.7
Y:	7.2	8.1	5.1	7.3	6.9	7.8	5.9	6.7	6.5	7.1

(a) Use the one-sided median test at an approximate 9% significance level. What is the p-value?

(b) Use the Wilcoxon test to test the equality of medians of the two processes at the approximate 5% significance level. What is the p-value?

(c) Modify the median test to test the equality of the 25th percentiles of the two processes.

HINT: Let V equal the number of X scores in the lower five values of the combined sample.

(d) Construct and interpret a q–q plot of these data.

10.5-2   A pharmaceutical company is interested in testing the effect of humidity on the weight of pills that are sold in aluminum packaging. Let X and Y denote the respective weights of pills and their packaging when the packaging is good and when it is defective after the pill has spent 1 week in a chamber containing 100% humidity and heated to 30°C.

(a) Use the Wilcoxon test to test H_0: $m_X = m_Y$ against H_1: $m_X - m_Y < 0$ using the following random samples of $n_1 = 12$ observations of X and $n_2 = 12$ observations of Y. What is the p-value?

X:	0.7565	0.7720	0.7776	0.7750	0.7494	0.7615
	0.7741	0.7701	0.7712	0.7719	0.7546	0.7719
Y:	0.7870	0.7750	0.7720	0.7876	0.7795	0.7972
	0.7815	0.7811	0.7731	0.7613	0.7816	0.7851

(b) Use the median test. What is the p-value?

(c) Construct and interpret a q–q plot of these data.

10.5-3   Let X and Y equal the sizes of grocery orders from a southside and a northside foodstore of the same chain, respectively. We shall test the null hypothesis H_0: $m_X = m_Y$ against a two-sided alternative using the following observations that have been ordered:

X:	5.13	8.22	11.81	13.77	15.36
	23.71	31.39	34.65	40.17	75.58
Y:	4.42	6.47	7.12	10.50	12.12
	12.57	21.29	33.14	62.84	72.05

(a) Use the median test.

(b) Use the Wilcoxon test.

(c) Construct a q–q plot and interpret it.

10.5-4   Let X and Y denote the heights of blue spruce trees, measured in centimeters, in two large fields. We shall compare these heights by measuring 12 trees selected at random from each of the fields. Use the statistic W, the sum of the ranks of the observations of Y in the combined sample, to test at $\alpha = 0.05$, approximately, the hypothesis H_0: $m_X = m_Y$, against the alternative hypothesis H_1: $m_X < m_Y$, based on $n_1 = 12$ observations of X:

90.4	77.2	75.9	83.2	84.0	90.2
87.6	67.4	77.6	69.3	83.3	72.7

and $n_2 = 12$ observations of Y:

92.7	78.9	82.5	88.6	95.0	94.4
73.1	88.3	90.4	86.5	84.7	87.5

10.5-5   A company manufactures and packages soap powder in 6-pound boxes. The quality assurance department was interested in comparing the fill weights of packages from the East and West lines. Taking random samples from the two lines they obtained the following weights:

East line (X):	6.06	6.04	6.11	6.06	6.06
	6.07	6.06	6.08	6.05	6.09
West line (Y):	6.08	6.03	6.04	6.07	6.11
	6.08	6.08	6.10	6.06	6.04

Let m_X and m_Y denote the median weights for the East and West lines, respectively. Test H_0: $m_X = m_Y$ against a two-sided alternative hypothesis using
(a) The median test with $\alpha = 0.025$, approximately.
(b) The Wilcoxon test with $\alpha = 0.05$, approximately.
(c) Construct and interpret a q–q plot of these data.

10.5-6   A charter bus line has 48-passenger and 38-passenger buses. Let m_{48} and m_{38} denote the median number of miles traveled per day by the respective buses. With $\alpha = 0.05$, use the Wilcoxon statistic to test H_0: $m_{48} = m_{38}$ against the one-sided alternative H_1: $m_{48} > m_{38}$ using the following data, which give the numbers of miles traveled per day for respective random samples of sizes 9 and 11:

48-passenger buses:	331	308	300	414	253	
	323	452	396	104		
38-passenger buses:	248	393	260	355	279	184
	386	450	432	196	197	

10.5-7   (a) Use the Wilcoxon statistic to test H_0: $m_X = m_Y$ against a two-sided alternative using the following observations of X and Y that have been ordered for your convenience:

X:	-2.3864	-2.2171	-1.9148	-1.9097	-1.4883
	-1.2007	-1.1077	-0.3601	0.4325	1.0598
	1.3035	1.5241	1.7133	1.7656	2.4912
Y:	-1.7613	-0.9391	-0.7437	-0.5530	-0.2469
	0.0647	0.2031	0.3219	0.3579	0.6431
	0.6557	0.6724	0.6762	0.9041	1.3571

(b) Construct a q–q plot and interpret it.
NOTE: These data were simulated on a computer. The distribution of X is $U(-2.5, 2.5)$ and the distribution of Y is $N(0, 1)$.

10.5-8   Let X and Y equal the CPU time in days per month for two different computers on a college campus. Let m_X and m_Y equal the medians of X and Y. Using $n_1 = n_2 = 14$ observations of each of X and Y, test the null hypothesis H_0: $m_X = m_Y$ against the alternative hypothesis H_1: $m_X < m_Y$. Use a normal approximation and the two-sample Wilcoxon test. Let $\alpha = 0.05$, approximately.

X:	7.1	8.5	9.9	13.8	2.2	2.4	2.3
	1.0	3.9	4.6	5.3	5.0	3.9	6.6
Y:	7.9	15.4	16.2	9.6	4.9	8.9	9.4
	15.2	12.1	4.3	5.0	2.4	2.3	5.4

10.5-9 Let X equal the playing time in minutes for a $33\frac{1}{3}$ rpm long-playing album and let Y equal the playing time for a compact disk. We shall test the null hypothesis H_0: $m_X = m_Y$ against the alternative hypothesis H_1: $m_X < m_Y$. Give the p-value for each of the two tests. The observed data are

X:	40.83	43.18	35.72	38.68	37.17		
	39.75	24.76	34.58	33.98			
Y:	42.82	64.42	56.92	39.92	72.38	47.26	64.58
	38.20	72.75	39.09	39.07	33.70	62.02	

What is your conclusion using
(a) The median test?
(b) The Wilcoxon test?

10.5-10 In Exercise 7.4-8 a comparison was made between a new and an old process for applying adhesive. Let X and Y equal the number of pounds of pressure needed to pull the glued parts apart using the new and the old process, respectively. Let m_X and m_Y equal the respective medians. We shall test H_0: $m_X = m_Y$ against the alternative H_1: $m_X > m_Y$ using 24 observations of each of X and Y.
(a) At an $\alpha = 0.05$ significance level, what is your conclusion using the Wilcoxon statistic?
(b) What is the approximate p-value of this test?
(c) How does this result compare with that in Exercise 7.4-8?

10.6   Run Test and Test for Randomness

Under the assumption that the random variables X and Y are of the continuous type and have distribution functions $F(x)$ and $G(y)$, respectively, we describe another test of the hypothesis H_0: $F(z) = G(z)$. This new test can also be used to test for randomness. For these tests we need the concept of runs, which we now define.

Suppose that we have n_1 observations of the random variable X and n_2 observations of the random variable Y. The combination of two sets of observations into one collection of $n_1 + n_2$ observations, placed in ascending order of magnitude, might yield an arrangement

$$\underline{y\,y\,y}\ \underline{x\,x}\ \underline{y}\ \underline{x}\ \underline{y}\ \underline{x\,x}\ \underline{y\,y},$$

where x denotes an observation of X and y an observation of Y in the ordered arrangement. We have underlined groups of successive values of X and Y. Each underlined group is called a **run**. Thus we have a run of three values of Y, followed by a run of two values of X, followed by a run of one value of Y, and so on. In this example there are seven runs.

We give two more examples to show what might be indicated by the number of runs. If the five x's and seven y's had the orderings

$$x\ x\ x\ x\ \underline{y}\ \underline{x}\ \underline{y\ y\ y\ y\ y\ y},$$

we might suspect that $F(z) \geq G(z)$. (See Figure 10.5-1.) Note that there are four runs in this ordering. The ordered arrangement

$$\underline{y\ y\ y}\ \underline{x\ x}\ \underline{y}\ \underline{x\ x\ x}\ \underline{y\ y\ y}$$

might suggest that the medians of the two distributions are equal but that the spread of the Y distribution is greater than the spread of the X distribution, for example, that $\sigma_Y > \sigma_X$. These examples suggest that the hypothesis $F(z) = G(z)$ should be rejected if the number of runs is too small, where a small number of runs could be caused by differences in the location or in the spread of the two distributions.

Let the random variable R equal the number of runs in the combined ordered sample of n_1 observations of X and n_2 observations of Y. We shall find the distribution of R when $F(z) = G(z)$ and then describe a test of the hypothesis $H_0: F(z) = G(z)$.

Under H_0, all permutations of the n_1 observations of X and n_2 observations of Y have equal probabilities. We can select the n_1 positions for the n_1 values of X in

$$\binom{n_1 + n_2}{n_1}$$

ways, the probability of each arrangement being

$$\frac{1}{\binom{n_1 + n_2}{n_1}}.$$

To find $P(R = r)$, we must determine the number of permutations that yield r runs.

First suppose that $r = 2k$, where k is a positive integer. In this case the n_1 ordered values of X and the n_2 ordered values of Y must each be separated into k runs. We can form k runs of the n_1 values of X by inserting $k - 1$ dividers into the $n_1 - 1$ spaces between the values of X, with no more than one divider per space. This can be done in

$$\binom{n_1 - 1}{k - 1}$$

ways (see Exercise 10.6-4). Similarly, k runs of the n_2 values of Y can be formed in

$$\binom{n_2 - 1}{k - 1}$$

ways. These two sets of runs can be placed together to form $r = 2k$ runs, of which

$$\binom{n_1 - 1}{k - 1}\binom{n_2 - 1}{k - 1}$$

begin with a run of x's and

$$\binom{n_2 - 1}{k - 1}\binom{n_1 - 1}{k - 1}$$

begin with a run of y's. Thus

$$P(R = 2k) = \frac{2\binom{n_1 - 1}{k - 1}\binom{n_2 - 1}{k - 1}}{\binom{n_1 + n_2}{n_1}}, \qquad (10.6\text{-}1)$$

where $2k$ is an element of the space of R.

When $r = 2k + 1$, it is possible to have $k + 1$ runs of the ordered values of X and k runs of the ordered values of Y or k runs of X's and $k + 1$ runs of Y's. We can form $k + 1$ runs of the n_1 values of X by inserting k dividers into the $n_1 - 1$ spaces between the values of X with no more than one divider per space in

$$\binom{n_1 - 1}{k}$$

ways. Similarly, k runs of n_2 values of Y can be done in

$$\binom{n_2 - 1}{k - 1}$$

ways. These two sets of runs can be placed together to form $2k + 1$ runs in

$$\binom{n_1 - 1}{k}\binom{n_2 - 1}{k - 1}$$

ways. In addition, $k + 1$ runs of the n_2 values of Y and k runs of the n_1 values of X can be placed together to form

$$\binom{n_2 - 1}{k}\binom{n_1 - 1}{k - 1}$$

sets of $2k + 1$ runs. Hence

$$P(R = 2k + 1) = \frac{\binom{n_1 - 1}{k}\binom{n_2 - 1}{k - 1} + \binom{n_1 - 1}{k - 1}\binom{n_2 - 1}{k}}{\binom{n_1 + n_2}{n_1}} \qquad (10.6\text{-}2)$$

for $2k + 1$ in the space of R.

A test based on the number of runs can be used for testing the hypothesis $H_0\colon F(z) = G(z)$. The hypothesis is rejected if the observed number of runs r is too small. That is, the critical region is of the form $r \leq c$, where the constant c is determined by using the p.d.f. of R to yield the desired significance level. The run test is sensitive to both differences in location and differences in spread of the two distributions.

> **Example 10.6-1** Let X and Y equal the percentages of body fat for freshman women and men, respectively, with distribution functions $F(x)$ and $G(y)$. We shall use the run test to test the hypothesis H_0: $F(z) = G(z)$ against the alternative hypothesis H_1: $F(z) < G(z)$. (That is, the alternative hypothesis is that the X distribution is to the right of the Y distribution.) Ten observations of both X and Y that have been ordered are:

X:	16.6	16.7	18.5	19.2	21.5
	22.4	22.6	23.2	24.2	26.3
Y:	9.4	9.7	11.3	11.8	13.3
	15.6	16.1	16.5	18.2	21.7

The critical region is of the form $r \le c$. To determine the value of c, we use formulas (10.6-1) and (10.6-2) with $n_1 = n_2 = 10$. Table I in the Appendix is very useful for evaluating these probabilities. We have

$$P(R = 2) = \frac{2}{184{,}756}; \quad P(R = 3) = \frac{18}{184{,}756};$$

$$P(R = 4) = \frac{162}{184{,}756}; \quad P(R = 5) = \frac{648}{184{,}756};$$

$$P(R = 6) = \frac{2592}{184{,}756}; \quad P(R = 7) = \frac{6048}{184{,}756}.$$

The sum of these six probabilities is $9470/184{,}756 = 0.051$, so we can take for our critical region $C = \{r: r \le 7\}$ with a significance level of $\alpha = 0.051$. To determine the number of runs, we order the combined samples and underline adjacent x and y values.

9.4	9.7	11.3	11.8	13.3	15.6	16.1	16.5	16.6	16.7
18.2	18.5	19.2	21.5	21.7	22.4	22.6	23.2	24.2	26.3

We see that the number of runs is $r = 6$ so we reject the null hypothesis. Note that the p-value of this test is

$$p\text{-value} = P(R \le 6) = \frac{3422}{184{,}756} = 0.0185.$$

When n_1 and n_2 are large, say each is at least equal to 10, R can be approximated with a normally distributed random variable. That is, it can be shown that

$$\mu_R = E(R) = \frac{2n_1 n_2}{n_1 + n_2} + 1,$$

$$\text{Var}(R) = \frac{(\mu_R - 1)(\mu_R - 2)}{n_1 + n_2 - 1} = \frac{2n_1 n_2 (2n_1 n_2 - n_1 - n_2)}{(n_1 + n_2)^2 (n_1 + n_2 - 1)},$$

and

$$Z = \frac{R - \mu_R}{\sqrt{\text{Var}(R)}}$$

is approximately $N(0, 1)$. The critical region for testing the null hypothesis H_0: $F(z) = G(z)$ is of the form $z \leq -z_\alpha$, where α is the desired significance level.

Example 10.6-2 We use the normal approximation to calculate the significance level and the p-value for Example 10.6-1. With $n_1 = n_2 = 10$,

$$\mu_R = \frac{2(10)(10)}{10 + 10} + 1 = 11; \qquad \sigma_R^2 = \frac{(11 - 1)(11 - 2)}{19} = \frac{90}{19}.$$

With the critical region $C = \{r: r \leq 7\}$, the approximate significance level, using a half unit correction for continuity, is

$$\alpha = P(R \leq 7) = P\left(\frac{R - 11}{\sqrt{90/19}} \leq \frac{7.5 - 11}{\sqrt{90/19}}\right)$$

$$\approx P(Z \leq -1.608) = 0.0539.$$

Note that this value compares very favorably with $\alpha = 0.051$ given in Example 10.6-1. Since $r = 6$, the approximate p-value, using a normal approximation, is

$$p\text{-value} = P(R \leq 6)$$

$$\approx P\left(Z \leq \frac{6.5 - 11}{\sqrt{90/19}}\right) = P(Z \leq -2.068) = 0.0193,$$

which is close to the p-value given in Example 10.6-1.

Applications of the run test include tests for randomness. Analysis of runs can also be useful in quality-control studies. To illustrate these applications, let x_1, x_2, . . . , x_k be the observed values of a random variable X, where the subscripts now designate the order in which the outcomes were observed and the observations are *not* arranged in the order of magnitude. It is possible, in a quality-control situation, that the observations are made systematically every hour, for example. Assume that k is even. The median divides the k numbers into a lower and an upper half. Replace each observation by L if it falls below the median and by U if it falls about the median. Then, for example, a sequence such as

$$U \ U \ U \ L \ U \ L \ L \ L$$

might suggest a trend toward decreasing values of X. If trend is the alternative hypothesis to randomness, the critical region would be of the form $r \leq c$. On the other hand, if we have a sequence such as

$$U \ L \ U \ L \ U \ L \ U \ L,$$

we would suspect a cyclic effect and would reject the hypothesis of randomness if r were too large. To test both for trend and cyclic effect, the critical region for testing the hypothesis of randomness is of the form $r \leq c_1$ or $r \geq c_2$.

If the sample size k is odd, the number of observations in the "upper half" and "lower half" will differ by one. In this case we will always put the extra observation in the upper group and, of course, $n_2 = n_1 + 1$. If the median is equal to a value that is tied with other values, we will again put the tied values in the upper group and then perform the test in which n_1 and n_2 are not equal to each other.

> **Example 10.6-2** We shall use a sample of size $k = 14$ to test for both trend and cyclic effect. To determine the critical region for rejecting the hypothesis of randomness, we use the p.d.f. of R with $n_1 = n_2 = 7$. Since
>
> $$P(R = 2) = P(R = 14) = \frac{2}{3432},$$
>
> $$P(R = 3) = P(R = 13) = \frac{12}{3432},$$
>
> $$P(R = 4) = P(R = 12) = \frac{72}{3432},$$

the critical region $\{r: r \leq 4 \text{ or } r \geq 12\}$ would yield a test at a significance level of $\alpha = 172/3432 = 0.05$. The 14 observations are

81.4	76.3	85.6	76.4	88.4	80.2	85.6
84.6	78.3	82.8	88.1	85.4	87.7	86.6

The median of these outcomes is $(84.6 + 85.4)/2 = 85.0$. Replacing each outcome with L if it falls below 85.0 and U if it falls above 85.0 yields the sequence.

L L U L U L U L L L U U U U.

Since $r = 8$, the hypothesis of randomness is not rejected.

Exercises

10.6-1 Let the total lengths of the male and female trident lynx spiders be denoted by X and Y, respectively, with corresponding distribution functions $F(x)$ and $G(y)$. Measurement of the lengths, in millimeters, of eight male and eight female spiders yielded the following observations of X:

5.40   5.55   6.00   5.00   5.70   5.20   5.45   4.95

and of Y:

6.20   6.25   5.75   5.85   6.55   6.05   5.50   6.65

Use these data to test the hypothesis $H_0: F(z) = G(z)$. Let $\alpha = 0.10$, approximately.

10.6-2 Let X and Y denote the times in hours per week that students in two different schools watch television. Let $F(x)$ and $G(y)$ denote the respective distribution functions. To test the hypothesis $H_0: F(z) = G(z)$, a random sample of eight students was selected from each school. Test this hypothesis using the following observations:

X:	16.75	19.25	22.00	20.50	22.50	15.50	17.25	20.75
Y:	24.75	21.50	19.75	17.50	22.75	23.50	13.00	19.00

10.6-3 A parade consists of 6 bands and 12 floats. How many different parade line-ups are possible if the parade begins and ends with a band and there is at least 1 float between each pair of bands?

10.6-4 Given six values of x, list the $\binom{5}{3} = 10$ ways in which three dividers can be inserted between the six values of x, no more than one divider per space.

10.6-5 Use the run test to test $H_0: F(z) = G(z)$ using the observations of X and Y that are given in Exercise 10.5-7.
(a) Show that if $C = \{r: r \le 11\}$, then the significance level is $\alpha = 0.0473$, approximately.
(b) Calculate the value of r and state your conclusion.
(c) What is the p-value of this test?

10.6-6 Use the run test to test $H_0: F(z) = G(z)$ using the following observations of X and Y that have been ordered for your convenience. Let $\alpha = 0.05$, approximately.

X:	-2.0482	-1.5748	-0.8797	-0.7170	-0.4907
	-0.2051	0.1651	0.2893	0.3186	0.3550
	0.4056	0.6975	0.7113	0.7377	1.7356
Y:	-1.2311	-1.0228	-0.8836	-0.6684	-0.6157
	-0.5755	-0.1019	-0.0297	0.3781	0.7400
	0.8479	1.0901	1.1397	1.1748	1.2921

NOTE: These data were simulated on a computer using the standard normal distribution for X and a $U(-1.5, 1.5)$ for Y.

10.6-7 Use the following sample of size $k = 14$ to test the hypothesis of randomness against the alternative hypothesis of a trend effect at an $\alpha = 0.025$ significance level.

12.4	14.2	11.7	14.0	12.7	15.7	12.8
14.1	17.9	18.4	17.5	20.2	20.8	20.3

10.6-8 Use the following sample of size $k = 16$ to test the hypothesis of randomness against the alternative hypothesis of a cyclic effect:

12.4	31.8	22.2	24.5	17.9	24.6	15.7	27.3
22.7	26.0	14.5	22.0	21.8	31.9	11.5	28.3

What is your conclusion if (a) $\alpha = 0.0317$? (b) $\alpha = 0.10$?

10.6-9 A powdered soap manufacturer checks periodically throughout a day the weights of soap in the company's 6-pound boxes. At each of 22 times, four boxes are selected at random, and the average of the weights of the soap in the boxes is recorded. Use the following 22 average weights to test the hypothesis of randomness against the alternative hypothesis of a trend at an approximate significance level of $\alpha = 0.025$.

6.050	6.038	6.003	6.015	6.025	6.063	6.033	6.010
5.995	6.020	6.060	6.060	6.065	6.050	6.043	6.040
6.045	6.065	6.055	6.060	6.060	6.070		

10.6-10 Each hour a manufacturer of mints selects four mints at random from the production line and finds the average weight in grams. For 1 week the following average weights were observed. Use these weights to test the hypothesis of randomness against the alternative hypothesis of a cyclic effect.

21.2	21.7	21.3	21.4	21.8	21.9	21.6	21.7	21.3	21.9
21.3	22.0	21.3	21.5	21.6	21.3	21.6	21.9	21.3	21.6
21.9	22.0	21.9	21.4	21.4	21.3	21.7	21.6	21.5	21.7
21.3	21.7	21.0	21.3	21.3	21.6	20.9	21.4		

10.6-11 It is claimed that X is $U(0, 1)$. A random sample of 14 observations of X yielded the following data:

0.15	0.67	0.05	0.47	0.29	0.23	0.10
0.01	0.96	0.92	0.51	0.73	0.91	0.82

(a) Test H_0: $m = 0.5$ against the two-sided alternative hypothesis

$$H_1: m \neq 0.5$$

using the sign test with significance level $\alpha = 0.0574$.

(b) Use a run test with these observations of X to test for both trend and cyclic effect at the approximate significance level $\alpha = 0.05$.

10.6-12 On an introductory statistics test, 27 form A and 29 form B tests were used. The order in which the tests were returned is listed below.

A	B	B	A	B	A	B	A	A	B	B	A	B	B
A	B	B	A	A	A	B	A	B	A	A	A	B	B
B	A	B	A	B	A	B	A	A	A	B	A	B	B
A	B	A	A	B	A	B	B	A	B	A	B	B	B

(a) Use a run test to test the null hypothesis that this is a random sequence against the alternative hypothesis that there is a cyclic effect. Use the normal approximation and an $\alpha = 0.01$ significance level.

(b) Based on your conclusion, what advice would you give to the professor who gave this test?

10.6-13 In Exercise 1.5-5 the United States birth rates are given for 1960–1987. In order, they are

23.7	23.3	22.4	21.7	21.1	19.4	18.4
17.8	17.6	17.9	18.4	17.2	15.6	14.8
14.8	14.6	14.6	15.1	15.0	15.6	15.9
15.8	15.9	15.5	15.5	15.8	15.6	15.7

Use the run test to test whether this is a random sequence against the alternative that there is a trend effect. Give the *p*-value of your test.

10.6-14　A window is manufactured that will be inserted into a car. For attaching the window, five studs are located in the frame in locations A, B, C, D, and E. Periodically, a window is selected randomly from the production line and a standard "stud pullout test" is performed that measures the force required to pull a stud out of the window. For each window, the tests are always performed in the order A, B, C, D, and E. Seventy observations that resulted from 14 windows were

140	159	138	102	84	126	147	126	103	92
149	155	135	120	94	149	143	109	101	86
144	154	120	105	97	149	151	140	103	99
157	140	120	96	87	146	137	120	93	89
149	154	139	100	84	148	142	130	98	81
112	135	109	84	87	112	135	109	84	87
126	118	135	90	77	125	126	131	78	75

For these data, the median is equal to 120. Use the run test to test whether this is a random sequence against the alternative hypothesis of a cyclic effect. Give the approximate *p*-value of the test. Interpret your conclusion. (See Exercise 1.5-8.)

10.7　Kolmogorov–Smirnov Goodness of Fit Test

In this section we discuss a test that considers the goodness of fit between a hypothesized distribution function and an empirical distribution function. The definition of the empirical distribution function, which was given in Section 1.2, is repeated here in terms of the order statistics. Let $y_1 < y_2 < \cdots < y_n$ be the observed values of the order statistics of a random sample $x_1, x_2, \ldots, x_n$ of size n. When no two observations are equal, the empirical distribution function is defined by

$$F_n(x) = \begin{cases} 0, & x < y_1, \\ \dfrac{k}{n}, & y_k \le x < y_{k+1}, \quad k = 1, 2, \ldots, n-1, \\ 1, & y_n \le x. \end{cases}$$

In this case the empirical distribution function has a jump of magnitude $1/n$ occurring at each observation. If n_k observations are equal to x_k, a jump of magnitude n_k/n occurs at x_k.

Suppose that a random sample of size n is taken from a distribution of the continuous type that has the distribution function $F(x)$. How can we measure the "closeness" of $F(x)$ and the empirical distribution function $F_n(x)$? How does the sample size affect this closeness? We give some theoretical results to help answer these questions and then give a test for goodness of fit.

Let $X_1, X_2, \ldots, X_n$ denote a random sample of size n from a distribution of the continuous type with the distribution function $F(x)$. Consider a fixed value of x. Then $W = F_n(x)$, the value of the empirical distribution function at x, can be thought of as a random variable that takes on the values $0, 1/n, 2/n, \ldots, 1$. Now $nW = k$ if, and only if, exactly k observations are less than or equal to x (say success) and $n - k$ observations are greater than x. The probability that an observation is less than or equal to x is given by $F(x)$. That is, the probability of success is $F(x)$. Because of the independence of the random variables $X_1, X_2, \ldots, X_n$, the probability of k successes is given by the binomial distribution, namely

$$P(nW = k) = P\left(W = \frac{k}{n} \right)$$

$$= \binom{n}{k} [F(x)]^k [1 - F(x)]^{n-k}, \qquad k = 0, 1, 2, \ldots, n.$$

Since nW has a binomial distribution with $p = F(x)$, the mean and variance of nW are given by

$$E(nW) = nF(x) \qquad \text{and} \qquad \mathrm{Var}(nW) = n[F(x)][1 - F(x)].$$

Hence the mean and the variance of $W = F_n(x)$ are

$$E[F_n(x)] = E(W) = F(x)$$

and

$$\mathrm{Var}[F_n(x)] = \mathrm{Var}(W) = \frac{F(x)[1 - F(x)]}{n}.$$

Since the variance of $F_n(x)$ gets nearer to zero as n becomes large, $F_n(x)$ and its mean $F(x)$ tend to be closer with large n. As a matter of fact, there is a theorem by Glivenko, the proof of which is beyond the level of this book, which states that with probability one, $F_n(x)$ converges to $F(x)$ uniformly in x as $n \to \infty$.

Because of the convergence of the empirical distribution function to the theoretical distribution function, it makes sense to construct a goodness of fit test based on the closeness of the empirical and a hypothesized distribution function, say $F_n(x)$ and $F_0(x)$, respectively. We shall use the Kolmogorov–Smirnov statistic defined by

$$D_n = \sup_x [|F_n(x) - F_0(x)|].$$

That is, D_n is the least upper bound of all pointwise differences $|F_n(x) - F_0(x)|$.

The exact distribution of the statistic D_n can be derived. We shall not derive this distribution but do give some values of the distribution function of D_n, namely $P(D_n \le d)$, in Table VIII in the Appendix that will be used for goodness of fit tests.

We would like to point out that the distribution of D_n does not depend on the particular function $F_0(x)$ of the continuous type. [This is essentially due to the fact that $Y = F_0(X)$ has a uniform distribution $U(0, 1)$.] Thus D_n can be thought of as a distribution-free statistic.

We are interested in using the Kolmogorov–Smirnov statistic D_n to test the hypothesis $H_0: F(x) = F_0(x)$ against all alternatives, $H_1: F(x) \neq F_0(x)$, where $F_0(x)$ is some specified distribution function. Intuitively, we accept H_0 if the empirical distribution function $F_n(x)$ is sufficiently close to $F_0(x)$, that is, if the value of D_n is sufficiently small. The hypothesis H_0 is rejected if the observed value of D_n is greater than the critical value selected from Table VIII in the Appendix, where this critical value depends upon the desired significance level and sample size.

The use of the Kolmogorov–Smirnov statistic is illustrated by two examples.

Example 10.7-1 We shall test the hypothesis $H_0: F(x) = F_0(x)$ against $H_1: F(x) \neq F_0(x)$, where

$$F_0(x) = \begin{cases} 0, & x < 0, \\ x, & 0 \leq x < 1, \\ 1, & 1 \leq x. \end{cases}$$

That is, the null hypothesis is that X is $U(0, 1)$. If the test is based on a sample of size $n = 10$ and if $\alpha = 0.10$, the critical region is

$$C = \{d_{10}: d_{10} \geq 0.37\},$$

where d_{10} is the observed value of the Kolmogorov–Smirnov statistic D_{10}. Suppose that the observed values of the random sample are 0.62, 0.36, 0.23, 0.76, 0.65, 0.09, 0.55, 0.26, 0.38, and 0.24. In Figure 10.7-1 we have plotted the empirical and hypothesized distribution functions for $0 \leq x \leq 1$. We see that $d_{10} = F_{10}(0.65) - F_0(0.65) = 0.25$ and hence H_0 is not rejected.

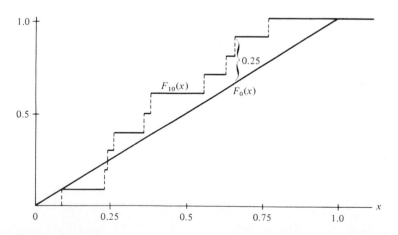

Figure 10.7-1

Note that the more carefully you plot the empirical and theoretical distribution functions, the more you can minimize the number of points at which you should calculate the possible value of the Kolmogorov–Smirnov goodness of fit statistic.

Example 10.7-2   When observing a Poisson process with a mean rate of arrivals $\lambda = 1/\theta$, the random variable W, which denotes the waiting time until the αth arrival, has a gamma distribution. The p.d.f. of W is

$$f(w) = \frac{w^{\alpha-1}e^{-w/\theta}}{\Gamma(\alpha)\theta^{\alpha}}, \qquad 0 \le w < \infty.$$

A Geiger counter was set up to record the waiting time W in seconds to observe $\alpha = 100$ alpha particle emissions of barium 133. It is claimed that the number of counts per second has a Poisson distribution with $\lambda = 14.7$ and hence $\theta = 0.068$. We shall test the hypothesis

$$H_0: F(w) = \int_{-\infty}^{w} f(t)\, dt,$$

where $f(t)$ is the gamma p.d.f. with $\theta = 0.068$ and $\alpha = 100$. Based on 25 observations, H_0 is rejected if $d_{25} \ge 0.24$ for $\alpha = 0.10$. For 25 observations, the empirical and theoretical distribution functions are depicted in Figure 10.7-2. For these data (Exercise 4.3-6), $d_{25} = 0.117$ and hence H_0 is not rejected.

You will note that we have been assuming that $F(x)$ is a continuous function. That is, we have only considered random variables of the continuous type. This proce-

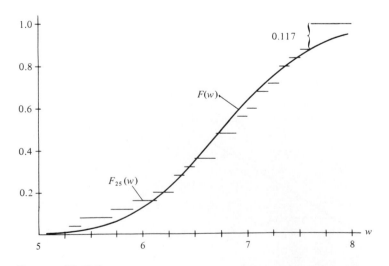

FIGURE 10.7-2

dure may also be applied in the discrete case. However, in the discrete case, the true significance level will be at most α. That is, the resulting test will be conservative.

Another application of the Kolmogorov–Smirnov statistic is in forming a confidence band for an unknown distribution function $F(x)$. To form a confidence band based on a sample of size n, select a number d such that

$$P(D_n \geq d) = \alpha.$$

Then

$$
\begin{aligned}
1 - \alpha &= P[\sup_x |F_n(x) - F(x)| \leq d] \\
&= P[|F_n(x) - F(x)| \leq d \text{ for all } x] \\
&= P[F_n(x) - d \leq F(x) \leq F_n(x) + d \text{ for all } x].
\end{aligned}
$$

Let

$$
F_L(x) = \begin{cases} 0, & F_n(x) - d \leq 0 \\ F_n(x) - d, & F_n(x) - d > 0, \end{cases}
$$

and

$$
F_U(x) = \begin{cases} F_n(x) + d, & F_n(x) + d < 1, \\ 1, & F_n(x) + d \geq 1. \end{cases}
$$

The two-step functions $F_L(x)$ and $F_U(x)$ yield a $100(1 - \alpha)\%$ confidence band for the unknown distribution function $F(x)$.

Example 10.7-3 A random sample of size $n = 15$ from an unknown distribution yielded the sample values 3.88, 3.97, 4.03, 2.49, 3.18, 3.08, 2.91, 3.43, 2.41, 1.57, 3.78, 3.25, 1.29, 2.57, and 3.40. Now

$$P(D_{15} \geq 0.30) = 0.10.$$

A 90% confidence band for the unknown distribution function $F(x)$ is depicted in Figure 10.7-3.

Exercises

10.7-1 Five observations of X, which have been ordered, are 0.40, 0.51, 0.53, 0.62, 0.74. Use the Kolmogorov–Smirnov statistic to test the hypothesis that X is $U(0, 1)$. Let $\alpha = 0.20$.

10.7-2 Select 10 sets of 10 random numbers from Table IX in the Appendix. For each set of 10 random numbers, calculate the value of the Kolmogorov–Smirnov statistic d_{10} with $F_0(x) = x$, $0 \leq x \leq 1$. Is it true that about 20% of the observations of D_{10} are greater than 0.32? (Compare your results with those of other students, provided that they selected different random numbers.)

10.7-3 A doctor of obstetrics used an ultrasound examination between the 16th and 25th weeks of pregnancy to measure in millimeters the widest diameter of the

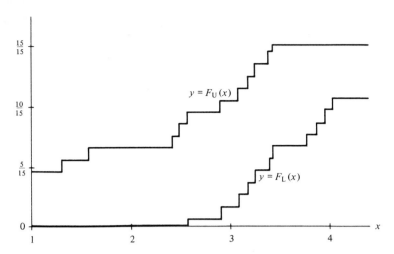

Figure 10.7-3

fetal head. Let X equal this diameter. Use the following 10 observations of X to test the hypothesis that the distribution of X is $N(60, 100)$.

$$56 \quad 65 \quad 47 \quad 57 \quad 62 \quad 48 \quad 68 \quad 75 \quad 79 \quad 49$$

10.7-4 In Table 4.5-1 and Exercise 10.4-4 are listed the outcomes of 10 simulations of a Cauchy random variable.

(a) Sketch the empirical distribution function and superimpose the Cauchy distribution function.

(b) Use the Kolmogorov–Smirnov statistic to test whether the 10 observations represent a random sample from a Cauchy distribution. Let $\alpha = 0.10$.

10.7-5 Let X equal the distance in feet between bad records on a used computer tape. Test the hypothesis that the distribution of X is exponential with a mean of $\theta = 40$ using the following observations of X.

$$18 \quad 6 \quad 1 \quad 32 \quad 116 \quad 23 \quad 12 \quad 58 \quad 101 \quad 68$$

10.7-6 Construct a 90% confidence band for the unknown distribution function $F(x)$ using the following 15 observations of X:

$$20.2 \quad 85.4 \quad 59.9 \quad 72.7 \quad 88.0 \quad 33.7 \quad 87.1 \quad 99.5 \quad 93.8$$
$$18.4 \quad 60.6 \quad 98.9 \quad 90.9 \quad 86.9 \quad 74.2$$

10.7-7 While testing a computer tape for bad records, the computer operator counted the number of flaws per 100 feet. Let X equal this number and test the hypothesis that the distribution of X is Poisson with a mean of $\lambda = 2.4$ using the following 40 observations of X. Let $\alpha = 0.10$, approximately.

x	Frequency
0	5
1	7
2	12
3	9
4	5
5	1
6	1
	40

10.7-8 Let X equal the number of chocolate chips in a chocolate-chip cookie. Use the Kolmogorov–Smirnov goodness of fit statistic and the following data to test the hypothesis that the distribution of X is Poisson with a mean of $\lambda = 5.6$. Let $\alpha = 0.10$, approximately.

x	Frequency
0	0
1	0
2	2
3	8
4	7
5	13
6	13
7	10
8	4
9	4
10	1
	62

10.7-9 In Exercise 10.6-14, 70 observations of a random variable X are listed that give the force required to pull a stud out of a window.
   (a) Test the null hypothesis that the distribution of X is $N(117.5, 25^2)$. Use the Kolmogorov–Smirnov statistic with a significance level of $\alpha = 0.20$. What is your conclusion?
   (b) Carefully analyzing the graphs of the empirical and theoretical distribution functions or by constructing a histogram, do you agree with the conclusion in part (a)? Why?
   (c) Construct and interpret a q–q plot for these data.

10.7-10   In Exercise 10.1-15 the weights in grams of "14-gram" snack packages of peanuts were given as follows:

13.9	14.4	14.6	14.7	14.7	15.2	15.2	15.2
15.3	15.4	15.4	15.5	15.6	15.6	15.9	16.4

Test the null hypothesis that these weights are observations of a normally distributed random variable with mean $\mu = 15.3$ and standard deviation $\sigma = 0.6$. Use a 10% significance level.

10.7-11   In Exercise 4.2-17 we were asked whether the times that bees spend in high-density flower patches are observations of an exponential random variable. We shall now use the Kolmogorov–Smirnov statistic to help answer this question. Use $\alpha = 0.05$.

(a) Test whether the times are observations of a one-parameter exponential distribution. Estimate θ with $\bar{x}$.

(b) Test whether the times are observations of a two-parameter exponential distribution. Estimate α and θ with their unbiased estimators. That is, letting y_1 equal the minimum, unbiased estimates of θ and α are

$$\hat{\theta} = [n/(n - 1)][\bar{x} - y_1] \quad \text{and} \quad \hat{\alpha} = y_1 - \hat{\theta}/n.$$

10.7-12   In Exercises 4.2-16 and 9.5-14 we were asked whether the times that bees spend in low-density flower patches are observations of an exponential random variable. We shall now use the Kolmogorov–Smirnov statistic to help answer this question. Use $\alpha = 0.05$.

(a) Test whether the times are observations of a one-parameter exponential distribution.

(b) Test whether the times are observations of a two-parameter exponential distribution.

10.7-13   *Parade Magazine*, March 29, 1992, reported the results of a study by Runzheimer International who gathered data from around the world on the cost in U.S. dollars per gallon for the least-expensive gasoline available. They compared each cost in December 1990 (before the Gulf war) with the cost in December 1991. Let X equal the difference, the 1991 cost minus the 1990 cost. Assume that the following $n = 18$ differences represent a random sample of observations of X. How is X distributed? Include estimates of the parameters of X.

−0.28	−0.40	0.06	0.04	−0.29	0.38
−0.08	−0.02	−0.22	−0.28	0.11	0.44
−0.17	−0.39	−0.36	−0.18	−0.39	−0.31

APPENDIX A

Review of Selected Mathematical Techniques

A.1 Algebra of Sets

The totality of objects under consideration is called the **universal set** and is denoted S. Each object in S is called an **element** of S. If a set A is a collection of elements that are also in S, then A is said to be a **subset** of S. In applications of probability, S will usually denote the **sample space**. An **event** A will be a collection of possible outcomes of the experiment and will be a subset of S. We say that event A *has occurred* if the outcome of the experiment is an element of A. The set or event A may be described by listing all of its elements or by defining the properties that its elements must satisfy.

Example A.1-1 A four-sided die, called a tetrahedron, has four faces that are equilateral triangles. These faces are numbered 1, 2, 3, 4. When the tetrahedron is rolled, the outcome of the experiment is the number of the face that is down. If this tetrahedron is rolled twice and we keep track of the first roll and the second roll, then the sample space is that displayed in Figure A.1-1.

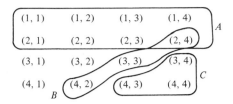

FIGURE A.1-1

Let A be the event that the first roll is a 1 or a 2. That is,

$$A = \{(x, y): x = 1 \text{ or } x = 2\}.$$

Let

$$B = \{(x, y): x + y = 6\} = \{(2, 4), (3, 3), (4, 2)\},$$

and let

$$C = \{(x, y): x + y \geq 7\} = \{(3, 4), (4, 3), (4, 4)\}.$$

Events A, B, and C are shown in Figure A.1-1.

When a is an element in A, we write $a \in A$. When a is not an element in A, we write $a \notin A$. So, in Example A.1-1, we have $(1, 3) \in A$ and $(3, 1) \notin A$. If every element of a set A is also an element in a set B, then A is a **subset** of B. We write $A \subset B$. In probability, if event B occurs whenever event A occurs, then $A \subset B$. The two sets A and B are equal (i.e., $A = B$), if $A \subset B$ and $B \subset A$. Note that it is always true that $A \subset A$ and $A \subset S$, where S is the universal set. We denote the subset that contains no elements by ϕ. This set is called the **null** or **empty** set. For all sets A, $\phi \subset A$.

The set of elements in either A and B or possibly both A and B is called the **union** of A and B and is denoted $A \cup B$. The set of elements in both A and B is called the **intersection** of A and B and is denoted $A \cap B$. The **complement** of a set A is the set of elements in the universal set S that are not in the set A and is denoted A'. In probability, if A and B are two events, the event that at least one of the two events has occurred is denoted by $A \cup B$, and the event that both events have occurred is denoted by $A \cap B$. The event that A has not occurred is denoted by A', and the event that A has not occurred but B has occurred is denoted by $A' \cap B$. If $A \cap B = \phi$, we say that A and B are **mutually exclusive**. In Example A.1-1, $B \cup C = \{(x, y): x + y \geq 6\}$, $A \cap B = \{(2, 4)\}$, and $A \cap C = \phi$. Note that A and C are mutually exclusive. Also, $C' = \{(x, y): x + y \leq 6\}$.

The operations of union and intersection may be extended to more than two sets. Let $A_1, A_2, \ldots, A_n$ be a finite collection of sets. Then the **union**

$$A_1 \cup A_2 \cup \cdots \cup A_n = \bigcup_{k=1}^{n} A_k$$

is the set of elements that belong to at least one A_k, $k = 1, 2, \ldots, n$. The **inter-section**

$$A_1 \cap A_2 \cap \cdots \cap A_n = \bigcap_{k=1}^{n} A_k$$

is the set of all elements that belong to every A_k, $k = 1, 2, \ldots, n$. Similarly, let $A_1, A_2, \ldots, A_n, \ldots$ be a countable collection of sets. Then x belongs to the **union**

$$A_1 \cup A_2 \cup A_3 \cup \cdots = \bigcup_{k=1}^{\infty} A_k$$

if x belongs to at least one A_k, $k = 1, 2, 3, \ldots$ Also x belongs to the **intersection**

$$A_1 \cap A_2 \cap A_3 \cap \cdots = \bigcap_{k=1}^{\infty} A_k$$

if x belongs to every A_k, $k = 1, 2, 3, \ldots$

Example A.1-2 Let

$$A_k = \left\{ x \colon \frac{10}{k+1} \le x \le 10 \right\}, \qquad k = 1, 2, 3, \ldots$$

Then

$$\bigcup_{k=1}^{8} A_k = \left\{ x \colon \frac{10}{9} \le x \le 10 \right\};$$

$$\bigcup_{k=1}^{\infty} A_k = \{ x \colon 0 < x \le 10 \}.$$

Note that the number zero is not in this latter union, since it is not in one of the sets $A_1, A_2, A_3, \ldots$ Also

$$\bigcap_{k=1}^{8} A_k = \{ x \colon 5 \le x \le 10 \} = A_1;$$

$$\bigcap_{k=1}^{\infty} A_k = \{ x \colon 5 \le x \le 10 \} = A_1,$$

since $A_1 \subset A_k$, $k = 1, 2, 3 \ldots$

A convenient way to illustrate operations on sets is with a **Venn** diagram. In Figure A.1-2 the universal set S is represented by the rectangle and its interior and

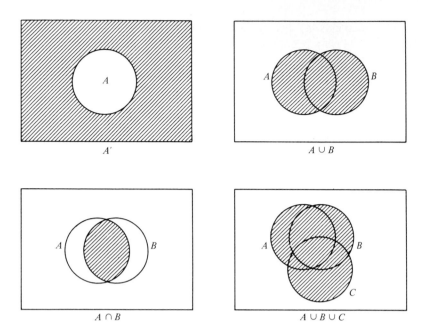

Figure A.1-2

the subsets of S by the points enclosed by the circles. The sets under consideration are the shaded regions.

Set operations satisfy several properties. For example, if A, B, and C are subsets of S, we have the following:

Commutative Laws

$$A \cup B = B \cup A$$
$$A \cap B = B \cap A.$$

Associative Laws

$$(A \cup B) \cup C = A \cup (B \cup C)$$
$$(A \cap B) \cap C = A \cap (B \cap C).$$

Distributive Laws

$$A \cap (B \cup C) = (A \cap B) \cup (A \cap C)$$
$$A \cup (B \cap C) = (A \cup B) \cap (A \cup C).$$

De Morgan's Laws

$$(A \cup B)' = A' \cap B'$$
$$(A \cap B)' = A' \cup B'.$$

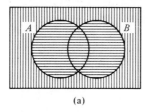 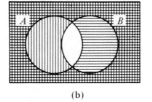

(a)   (b)

FIGURE A.1-3

A Venn diagram will be used to justify the first of De Morgan's laws. In Figure A.1-3(a), $A \cup B$ is represented by horizontal lines, and thus $(A \cup B)'$ is the region represented by vertical lines. In Figure A.1-3(b), A' is indicated with horizontal lines, and B' is indicated with vertical lines. An element belongs to $A' \cap B'$ if it belongs to both A' and B'. Thus the crosshatched region represents $A' \cap B'$. Clearly, this crosshatched region is the same as that shaded with vertical lines in Figure A.1-3(a).

A.2 Mathematical Tools for the Hypergeometric Distribution

Let X have a hypergeometric distribution. That is, the p.d.f. of X is

$$f(x) = \frac{\binom{n_1}{x}\binom{n_2}{r-x}}{\binom{n_1+n_2}{r}}, \quad x \le r,\ x \le n_1,\ r-x \le n_2.$$

To show that $\sum_{x=0}^{r} f(x) = 1$ and to find the mean and variance of X, we use the following theorem.

THEOREM A-1

$$\binom{n}{r} = \sum_{x=0}^{r} \binom{n_1}{x}\binom{n_2}{r-x},$$

where $n = n_1 + n_2$ and it is understood that $\binom{k}{j} = 0$ if $j > k$.

PROOF: Because $n = n_1 + n_2$, we have the identity

$$(1+y)^n \equiv (1+y)^{n_1}(1+y)^{n_2}.$$

We will expand each of these binomials; and, since the polynomials on each side are identically equal, the coefficients of y^r on each side of this equation must be equal.

Using the binomial expansion, we find that the expansion of the left side of the equation is

$$(1 + y)^n = \sum_{k=0}^{n} \binom{n}{k} y^k$$

$$= \binom{n}{0} + \binom{n}{1} y + \cdots + \binom{n}{r} y^r + \cdots + \binom{n}{n} y^n.$$

The right side of the equation becomes

$$(1 + y)^{n_1}(1 + y)^{n_2} = \left[\binom{n_1}{0} + \binom{n_1}{1} y + \cdots + \binom{n_1}{r} y^r + \cdots + \binom{n_1}{n_1} y^{n_1} \right]$$

$$\times \left[\binom{n_2}{0} + \binom{n_2}{1} y + \cdots + \binom{n_2}{r} y^r + \cdots + \binom{n_2}{n_2} y^{n_2} \right].$$

The coefficient of y^r in this product is

$$\binom{n_1}{0}\binom{n_2}{r} + \binom{n_1}{1}\binom{n_2}{r-1} + \cdots + \binom{n_1}{r}\binom{n_2}{0} = \sum_{x=0}^{r} \binom{n_1}{x}\binom{n_2}{r-x}$$

and this sum must equal $\binom{n}{r}$, the coefficient of y^r on the left side of the equation.

□

Using Theorem A-1, we find that it follows that if X has a hypergeometric distribution with p.d.f. $f(x)$, then

$$\sum_{x=0}^{r} f(x) = \sum_{x=0}^{r} \frac{\binom{n_1}{x}\binom{n_2}{r-x}}{\binom{n}{r}} = 1.$$

To find the mean and variance of a hypergeometric random variable, it is useful to note that, with $r > 0$,

$$\binom{n}{r} = \frac{n!}{r!(n-r)!} = \frac{n}{r} \cdot \frac{(n-1)!}{(r-1)!(n-r)!} = \frac{n}{r}\binom{n-1}{r-1}.$$

The mean of a hypergeometric random variable X is

$$\mu = \sum_{x=0}^{r} xf(x)$$

$$= \frac{\displaystyle\sum_{x=1}^{r} x \cdot \frac{n_1!}{x!(n_1-x)!} \cdot \frac{n_2!}{(r-x)!(n_2-r+x)!}}{\binom{n}{r}}$$

$$= \frac{n_1 \sum_{x=1}^{r} \frac{(n_1 - 1)!}{(x - 1!)(n_1 - x)!} \cdot \frac{n_2!}{(r - x)!(n_2 - r + x)!}}{\binom{n}{r}}.$$

If we now make the change of variables $k = x - 1$ in the summation, and replace

$$\binom{n}{r} \quad \text{with} \quad \binom{n}{r}\binom{n-1}{r-1}$$

in the denominator, this becomes

$$\mu = \frac{n_1 \sum_{k=0}^{r-1} \frac{(n_1 - 1)!}{k!(n_1 - 1 - k)!} \frac{n_2!}{(r - k - 1)!(n_2 - r + k + 1)!}}{\binom{n}{r}\binom{n-1}{r-1}}$$

$$= r\left(\frac{n_1}{n}\right) \frac{\sum_{k=0}^{r-1} \binom{n_1 - 1}{k}\binom{n_2}{r - 1 - k}}{\binom{n-1}{r-1}} = r\left(\frac{n_1}{n}\right)$$

because the summation in the expression for μ is equal to $\binom{n-1}{r-1}$ from Theorem A-1.

Note that

$$\begin{aligned}
\text{Var}(X) = \sigma^2 &= E[(X - \mu)^2] \\
&= E[X^2] - \mu^2 \\
&= E[X(X - 1)] + E(X) - \mu^2.
\end{aligned}$$

So to find the variance of X, we first find $E[X(X - 1)]$:

$$E[X(X - 1)] = \sum_{x=0}^{r} x(x - 1)f(x)$$

$$= \sum_{x=2}^{r} \frac{x(x - 1) \frac{n_1!}{x!(n_1 - x)!} \frac{n_2!}{(r - x)!(n_2 - r + x)!}}{\binom{n}{r}}$$

$$= n_1(n_1 - 1) \frac{\sum_{x=2}^{r} \frac{(n_1 - 2)!}{(x - 2)!(n_1 - x)!} \frac{n_2!}{(r - x)!(n_2 - r + x)!}}{\binom{n}{r}}.$$

In the summation, let $k = x - 2$, and in the denominator, note that

$$\binom{n}{r} = \frac{n!}{r!(n-r)!} = \frac{n(n-1)}{r(r-1)}\binom{n-2}{r-2}.$$

Thus

$$E[X(X-1)] = \frac{n_1(n_1-1)}{\dfrac{n(n-1)}{r(r-1)}} \sum_{k=0}^{r-2} \frac{\dbinom{n_1-2}{k}\dbinom{n_2}{r-2-k}}{\dbinom{n-2}{r-2}}$$

$$= \frac{n_1(n_1-1)(r)(r-1)}{n(n-1)},$$

from Theorem A-1. Thus the variance of a hypergeometric random variable is, after some algebra,

$$\sigma^2 = \frac{n_1(n_1-1)(r)(r-1)}{n(n-1)} + \frac{rn_1}{n} - \left(\frac{rn_1}{n}\right)^2$$

$$= r\left(\frac{n_1}{n}\right)\left(\frac{n_2}{n}\right)\left(\frac{n-r}{n-1}\right).$$

A.3   Limits

We refer the reader to the many fine books on calculus for the definition of a limit and the other concepts used in that subject. Here we simply remind you of some of the techniques we find most useful in probability and statistics.

Early in a calculus course the existence of the following limit is discussed and it is denoted by the letter e:

$$e = \lim_{t\to 0} (1+t)^{1/t} = \lim_{n\to\infty}\left(1 + \frac{1}{n}\right)^n.$$

Of course, e is an irrational number, which to six significant figures equals 2.71828.

Often it is rather easy to see the value of certain limits. For example, with $-1 < r < 1$, the sum of the geometric progression allows us to write

$$\lim_{n\to\infty} (1 + r + r^2 + \cdots + r^{n-1}) = \lim_{n\to\infty}\left(\frac{1-r^n}{1-r}\right) = \frac{1}{1-r}.$$

That is, the limit of the ratio $(1 - r^n)/(1 - r)$ is not difficult to find because $\lim_{n\to\infty} r^n = 0$.

However it is not that easy to determine the limit of every ratio; for example, consider

$$\lim_{b \to \infty} (be^{-b}) = \lim_{b \to \infty} \left(\frac{b}{e^b} \right).$$

Since both the numerator and the denominator of the latter ratio are unbounded, we can use **L'Hospital's rule**, taking the limit of the ratio of the derivative of the numerator and the derivative of the denominator. We have

$$\lim_{b \to \infty} \left(\frac{b}{e^b} \right) = \lim_{b \to \infty} \left(\frac{1}{e^b} \right) = 0.$$

This result can be used in the evaluation of the integral

$$\int_0^\infty xe^{-x} \, dx = \lim_{b \to \infty} \int_0^b xe^{-x} \, dx$$

$$= \lim_{b \to \infty} [-xe^{-x} - e^{-x}]_0^b$$

$$= \lim_{b \to \infty} [1 - be^{-b} - e^{-b}] = 1.$$

Note that

$$D_x[-xe^{-x} - e^{-x}] = xe^{-x} - e^{-x} + e^{-x} = xe^{-x};$$

that is, $-xe^{-x} - e^{-x}$ is the antiderivative of xe^{-x}.

Another limit of importance is

$$\lim_{n \to \infty} \left(1 + \frac{b}{n} \right)^n = \lim_{n \to \infty} e^{n\ln(1+b/n)},$$

where b is a constant.

Since the exponential function is continuous, the limit can be taken to the exponent. That is,

$$\lim_{n \to \infty} \exp[n \ln(1 + b/n)] = \exp[\lim_{n \to \infty} n \ln(1 + b/n)].$$

The limit in the exponent is equal to

$$\lim_{n \to \infty} \frac{\ln(1 + b/n)}{1/n} = \lim_{n \to \infty} \frac{\dfrac{-b/n^2}{1 + b/n}}{-1/n^2} = \lim_{n \to \infty} \frac{b}{1 + b/n} = b$$

by L'Hospital's rule. Since this limit is equal to b, the original limit is

$$\lim_{n \to \infty} \left(1 + \frac{b}{n} \right)^n = e^b.$$

Applications of this limit in probability occur with $b = -1$ yielding

$$\lim_{n \to \infty} \left(1 - \frac{1}{n} \right)^n = e^{-1}.$$

A.4   Infinite Series

A function $f(x)$ possessing derivatives of all orders at $x = b$ can be expanded in the following **Taylor's series**:

$$f(x) = f(b) + \frac{f'(b)}{1!}(x - b) + \frac{f''(b)}{2!}(x - b)^2 + \frac{f'''(b)}{3!}(x - b)^3 + \cdots.$$

If $b = 0$, we obtain the special case that is often called **Maclaurin's series**;

$$f(x) = f(0) + \frac{f'(0)}{1!}x + \frac{f''(0)}{2!}x^2 + \frac{f'''(0)}{3!}x^3 + \cdots.$$

For example, if $f(x) = e^x$ so that all derivatives of $f(x) = e^x$ are $f^{(r)}(x) = e^x$, then $f^{(r)}(0) = 1$, for $r = 1, 2, 3, \ldots$. Thus the Maclaurin's series expansion of $f(x) = e^x$ is

$$e^x = 1 + \frac{x}{1!} + \frac{x^2}{2!} + \frac{x^3}{3!} + \frac{x^4}{4!} + \cdots.$$

The **ratio test**

$$\lim_{n \to \infty} \left| \frac{x^n/n!}{x^{n-1}/(n - 1)!} \right| = \lim_{n \to \infty} \left| \frac{x}{n} \right| = 0$$

shows that this series expansion of e^x converges for all real values of x.

Note, for examples, that

$$e = 1 + \frac{1}{1!} + \frac{1}{2!} + \frac{1}{3!} + \cdots$$

and

$$e^{-1} = 1 - \frac{1}{1!} + \frac{1}{2!} - \frac{1}{3!} + \cdots + \frac{(-1)^n}{n!} + \cdots.$$

For another example, consider

$$h(w) = (1 - w)^{-r},$$

where r is a positive integer. Here

$$h'(w) = r(1 - w)^{-(r+1)}$$
$$h''(w) = (r)(r + 1)(1 - w)^{-(r+2)}$$
$$h'''(w) = (r)(r + 1)(r + 2)(1 - w)^{-(r+3)}$$
$$\vdots$$

In general, $h^{(k)}(0) = (r)(r + 1) \cdots (r + k - 1) = (r + k - 1)!/(r - 1)!$. Thus

$$(1 - w)^{-r} = 1 + \frac{(r + 1 - 1)!}{(r - 1)!1!}w + \frac{(r + 2 - 1)!}{(r - 1)!2!}w^2 + \cdots + \frac{(r + k - 1)!}{(r - 1)!k!}w^k + \cdots$$

$$= \sum_{k=0}^{\infty} \binom{r + k - 1}{r - 1} w^k.$$

This is often called the negative binomial series. Using the ratio test,

$$\lim_{n \to \infty} \left| \frac{w^n(r + n - 1)!/(r - 1)!n!}{w^{n-1}(r + n - 2)!/(r - 1)!(n - 1)!} \right| = \lim_{n \to \infty} \left| \frac{w(r + n - 1)}{n} \right| = |w|.$$

Thus the series converges when $|w| < 1$ or $-1 < w < 1$.

A negative binomial random variable receives its name from this negative binomial series. Before showing that relationship, we note that, for $-1 < w < 1$,

$$h(w) = \sum_{k=0}^{\infty} \binom{r + k - 1}{r - 1} w^k = (1 - w)^{-r}$$

$$h'(w) = \sum_{k=1}^{\infty} \binom{r + k - 1}{r - 1} k w^{k-1} = r(1 - w)^{-r-1}$$

$$h''(w) = \sum_{k=2}^{\infty} \binom{r + k - 1}{r - 1} k(k - 1) w^{k-2} = r(r + 1)(1 - w)^{-r-2}.$$

The p.d.f. of a negative binomial random variable X is

$$g(x) = \binom{x - 1}{r - 1} p^r q^{x-r}, \quad x = r, r + 1, r + 2, \cdots.$$

In the series expansion for $h(w) = (1 - w)^{-r}$, let $x = k + r$. Then

$$\sum_{x=r}^{\infty} \binom{x - 1}{r - 1} w^{x-r} = (1 - w)^{-r}.$$

Letting $w = q$ in this equation, we see that

$$\sum_{x=r}^{\infty} g(x) = \sum_{x=r}^{\infty} \binom{x - 1}{r - 1} p^r q^{x-r} = p^r(1 - q)^{-r} = 1.$$

That is, $g(x)$ does satisfy the properties of a p.d.f.

To find the mean of X, we first find

$$E(X - r) = \sum_{x=r}^{\infty} (x - r) \binom{x - 1}{r - 1} p^r q^{x-r} = \sum_{x=r+1}^{\infty} (x - r) \binom{x - 1}{r - 1} p^r q^{x-r}.$$

Letting $k = x - r$ in this latter summation and using the expansion of $h'(w)$ gives us

$$E(X - r) = \sum_{k=1}^{\infty} (k) \binom{r + k - 1}{r - 1} p^r q^k$$

$$= p^r q \sum_{k=1}^{\infty} \binom{r + k - 1}{r - 1} k q^{k-1}$$

$$= p^r q r (1 - q)^{-r-1} = r \left(\frac{q}{p} \right).$$

Thus

$$E(X) = r + r\left(\frac{q}{p}\right)$$

$$= r\left(1 + \frac{q}{p}\right) = r\left(\frac{1}{p}\right).$$

Similarly, you can show, using $h''(w)$, that

$$E[(X - r)(X - r - 1)] = \left(\frac{q^2}{p^2}\right)(r)(r + 1)$$

and thus

$$\text{Var}(X) = \text{Var}(X - r) = \left(\frac{q^2}{p^2}\right)(r)(r + 1) + r\left(\frac{q}{p}\right) - r^2\left(\frac{q^2}{p^2}\right) = r\left(\frac{q}{p^2}\right).$$

A very special case of the negative binomial series occurs when $r = 1$, obtaining the well known geometric series

$$(1 - w)^{-1} = 1 + w + w^2 + w^3 + \cdots$$

provided that $-1 < w < 1$.

The geometric series gives its name to the geometric probability distribution. Perhaps you recall the geometric series as being written, for $-1 < r < 1$,

$$g(r) = \sum_{k=0}^{\infty} ar^k = \frac{a}{1 - r}.$$

To find the mean and the variance of a geometric random variable X, simply let $r = 1$ in the formulas for the mean and variance of a negative binomial random variable. However, if you want to find the mean and variance directly, you can use

$$g'(r) = \sum_{k=1}^{\infty} akr^{k-1} = \frac{a}{(1 - r)^2}$$

and

$$g''(r) = \sum_{k=2}^{\infty} ak(k - 1)r^{k-2} = \frac{2a}{(1 - r)^3}$$

to find $E(X)$ and $E[X(X - 1)]$, respectively.

In applications associated with the geometric random variable, it is also useful to recall that the nth partial sum of a geometric series is

$$S_n = \sum_{k=0}^{n-1} ar^k = \frac{a(1 - r^n)}{1 - r}.$$

A bonus in this section is the following logarithmic series that produces a useful tool in daily life. Consider

$$f(x) = \ln(1 + x)$$
$$f'(x) = (1 + x)^{-1}$$
$$f''(x) = (-1)(1 + x)^{-2}$$
$$f'''(x) = (-1)(-2)(1 + x)^{-3}$$
$$\vdots$$

Thus $f^{(r)}(0) = (-1)^{r-1}(r - 1)!$ and

$$\ln(1 + x) = \frac{0!}{1!}x - \frac{1!}{2!}x^2 + \frac{2!}{3!}x^3 - \frac{3!}{4!}x^4 + \cdots$$

$$= x - \frac{x^2}{2} + \frac{x^3}{3} - \frac{x^4}{4} + \cdots$$

which converges for $-1 < x \le 1$.

Now consider the following question: "How long does it take money to double in value if the interest rate is i?" Assuming the compounding is on an annual basis and that you begin with \$1, after one year you have \$$(1 + i)$ and after 2 years the number of dollars that you have is

$$(1 + i) + i(1 + i) = (1 + i)^2.$$

Continuing this process, the equation that we have to solve is

$$(1 + i)^n = 2,$$

the solution of which is

$$n = \frac{\ln 2}{\ln(1 + i)}.$$

To approximate the value of n, recall that $\ln 2 \approx 0.693$ and use the series expansion of $f(x) = \ln(1 + x)$ to obtain

$$n \approx \frac{0.693}{i - \dfrac{i^2}{2} + \dfrac{i^3}{3} - \cdots}.$$

Due to the alternating series in the denominator, the denominator is a little less than i. Frequently brokers increase the numerator a little (say to 0.72) and simply divide by i, obtaining the "well known Rule of 72," namely

$$n \approx \frac{72}{100i}.$$

For example, if $i = 0.08$, then $n \approx 72/8 = 9$ provides an excellent approximation (the answer is about 9.006). Many persons find that the Rule of 72 is extremely useful when dealing with money matters.

A.5   Integration

Say $F'(t) = f(t)$, $a \le t \le b$, then

$$\int_a^b f(t)\,dt = F(b) - F(a).$$

Thus if $u(x)$ is such that $u'(x)$ exists and $a \le u(x)$, then

$$\int_a^{u(x)} f(t)\,dt = F[u(x)] - F(b).$$

Taking derivatives of this latter equation, we obtain

$$D_x\left[\int_a^{u(x)} f(t)\,dt\right] = F'[u(x)]u'(x) = f[u(x)]u'(x).$$

For example, with $0 < v$,

$$D_v\left[2\int_0^{\sqrt{v}} \frac{1}{\sqrt{2\pi}} e^{-z^2/2}\,dz\right] = \left(\frac{2}{\sqrt{2\pi}} e^{-v/2}\right)\frac{1}{2\sqrt{v}} = \frac{v^{(1/2)-1} e^{-v/2}}{\sqrt{\pi}\, 2^{1/2}}.$$

This is needed in proving that if Z is $N(0, 1)$, then Z^2 is $\chi^2(1)$.

The preceding example could be worked by first changing variables in the integral. That is, first using the fact that

$$\int_a^b f(x)\,dx = \int_{u(a)}^{u(b)} f[w(y)]w'(y)\,dy$$

where the monotone increasing (decreasing) function $x = w(y)$ has derivative $w'(y)$ and inverse function $y = u(x)$. In that example, $a = 0$, $b = \sqrt{v}$, $z = \sqrt{t}$, $z' = 1/2\sqrt{t}$, and $t = z^2$ so that

$$2\int_0^{\sqrt{v}} \frac{1}{\sqrt{2\pi}} e^{-z^2/2}\,dz = 2\int_0^{v} \frac{1}{\sqrt{2\pi}} e^{-t/2}\left(\frac{1}{2\sqrt{t}}\right)dt.$$

The derivative of the latter, by one form of the fundamental theorem of calculus, is

$$2\frac{1}{\sqrt{2\pi}} e^{-v/2}\left(\frac{1}{2\sqrt{v}}\right) = \frac{v^{(1/2)-1} e^{-v/2}}{\sqrt{\pi}\, 2^{1/2}}.$$

Integration by parts is frequently needed. It is based upon the derivative of the product of two functions of x, say $u(x)$ and $v(x)$, namely

$$D_x[u(x)v(x)] = u(x)v'(x) + v(x)u'(x).$$

Thus

$$[u(x)v(x)]_a^b = \int_a^b u(x)v'(x)\,dx + \int_a^b v(x)u'(x)\,dx$$

or, equivalently,

$$\int_a^b u(x)v'(x)dx = [u(x)v(x)]_a^b - \int_a^b v(x)u'(x)dx.$$

For example, by letting $u(x) = x$ and $v'(x) = e^{-x}$, we have that

$$\int_0^b xe^{-x}dx = [-xe^{-x}]_0^b - \int_0^b (1)(-e^{-x})dx$$
$$= -be^{-b} + [-e^{-x}]_0^b = 1 - e^{-b} - be^{-b}$$

because $u'(x) = 1$ and $v(x) = -e^{-x}$.

With some thought about the product rule of differentiation, it is not always necessary to assign $u(x)$ and $v'(x)$, however. For illustration, an integral like

$$\int_0^b x^3 e^{-x}dx$$

would require integration by parts three times, the first of which would assign $u(x) = x^3$ and $v'(x) = e^{-x}$. But note that

$$D_x(-x^3 e^{-x}) = x^3 e^{-x} - 3x^2 e^{-x}.$$

That is, $-x^3 e^{-x}$ is "almost" the antiderivative of $x^3 e^{-x}$ except for the undesirable term, $-3x^2 e^{-x}$. Clearly,

$$D_x(-x^3 e^{-x} - 3x^2 e^{-x}) = x^3 e^{-x} - 3x^2 e^{-x} + 3x^2 e^{-x} - 6xe^{-x} = x^3 e^{-x} - 6xe^{-x}.$$

So we eliminated that undesirable term $-3x^2 e^{-x}$, but got another one, namely $-6xe^{-x}$. However,

$$D_x(-x^3 e^{-x} - 3x^2 e^{-x} - 6xe^{-x}) = x^3 e^{-x} - 6e^{-x}$$

and finally

$$D_x(-x^3 e^{-x} - 3x^2 e^{-x} - 6xe^{-x} - 6e^{-x}) = x^3 e^{-x}.$$

That is,

$$-x^3 e^{-x} - 3x^2 e^{-x} - 6xe^{-x} - 6e^{-x}$$

is the antiderivative of $x^3 e^{-x}$ and can be written down without ever assigning u and v.

As practice in this technique, consider

$$\int_0^{\pi/2} x^2 \cos x \, dx = [x^2 \sin x + 2x \cos x - 2 \sin x]_0^{\pi/2}.$$

Now $x^2 \sin x$ is our first guess because we obtain $x^2 \cos x$ when we differentiate the $\sin x$ factor. But we get the undesirable term $2x \sin x$. That is why we add $2x \cos x$ as the derivative of the $\cos x$ is $-\sin x$ and $-2x \sin x$ eliminates $2x \sin x$. But the second term of the derivative of $2x \cos x$ is $2 \cos x$, which we get rid of by taking the derivative of the next term, $-2 \sin x$.

Possibly the best advice is to take the derivative of the righthand member, here

$$x^2 \sin x + 2x \cos x - 2 \sin x,$$

and note how the terms cancel leaving only $x^2 \cos x$. Then practice on integrals like

$$\int x^4 e^{-x} \, dx, \quad \int x^3 \sin x \, dx, \quad \int x^5 e^x \, dx.$$

A.6   Multivariate Calculus

We really only make some suggestions about functions of two variables, say

$$z = f(x, y).$$

But these remarks can be extended to more than two variables. The two first *partial derivatives* with respect to x and y, denoted by $\dfrac{\partial z}{\partial x}$ and $\dfrac{\partial z}{\partial y}$, can be found in the usual manner of differentiating by treating the "other" variable as a constant. For illustration,

$$\frac{\partial(x^2 y + \sin x)}{\partial x} = 2xy + \cos x$$

and

$$\frac{\partial(e^{xy^2})}{\partial y} = (e^{xy^2})(2xy).$$

The second partial derivatives are simply first partial derivatives of the first partial derivatives. If $z = e^{xy^2}$, then

$$\frac{\partial}{\partial x}\left(\frac{\partial z}{\partial y}\right) = \frac{\partial}{\partial x}(2xye^{xy^2}) = 2xye^{xy^2}(y^2) + 2ye^{xy^2}.$$

For notation we use

$$\frac{\partial}{\partial x}\left(\frac{\partial z}{\partial x}\right) = \frac{\partial^2 z}{\partial x^2}, \quad \frac{\partial}{\partial x}\left(\frac{\partial z}{\partial y}\right) = \frac{\partial^2 z}{\partial x \partial y},$$

$$\frac{\partial}{\partial y}\left(\frac{\partial z}{\partial x}\right) = \frac{\partial^2 z}{\partial y \partial x}, \quad \frac{\partial}{\partial y}\left(\frac{\partial z}{\partial y}\right) = \frac{\partial^2 z}{\partial y^2}.$$

In general,

$$\frac{\partial^2 z}{\partial x \partial y} = \frac{\partial^2 z}{\partial y \partial x},$$

provided the partial derivatives involved are continuous functions.

As you might guess, at a relative maximum or relative minimum of $z = f(x, y)$, we have

$$\frac{\partial z}{\partial x} = 0 \quad \text{and} \quad \frac{\partial z}{\partial y} = 0,$$

provided that the derivatives exist. To assure us that we have a maximum or minimum, we need

$$\left(\frac{\partial^2 z}{\partial x \partial y}\right)^2 - \left(\frac{\partial^2 z}{\partial x^2}\right)\left(\frac{\partial^2 z}{\partial y^2}\right) < 0.$$

Moreover, it is a relative minimum if $\dfrac{\partial^2 z}{\partial x^2} > 0$ and a relative maximum if $\dfrac{\partial^2 z}{\partial x^2} < 0$.

A major problem in statistics, called least squares, is to find a and b to minimize

$$K(a, b) = \sum_{i=1}^{n} (y_i - a - bx_i)^2.$$

Thus the solution of the two equations

$$\frac{\partial K}{\partial a} = \sum_{i=1}^{n} 2(y_i - a - bx_i)(-1) = 0$$

$$\frac{\partial K}{\partial b} = \sum_{i=1}^{n} 2(y_i - a - bx_i)(-x_i) = 0$$

would possibly give us a point (a, b) that minimizes $K(a, b)$. Taking second partial derivatives,

$$\frac{\partial^2 K}{\partial a^2} = \sum_{i=1}^{n} 2(-1)(-1) = 2n > 0$$

$$\frac{\partial^2 K}{\partial b^2} = \sum_{i=1}^{n} 2(-x_i)(-x_i) = 2\sum_{i=1}^{n} x_i^2 > 0$$

$$\frac{\partial^2 K}{\partial a \partial b} = \sum_{i=1}^{n} 2(-1)(-x_i) = 2\sum_{i=1}^{n} x_i$$

we note that

$$\left(2\sum_{i=1}^{n} x_i\right)^2 - (2n)\left(2\sum_{i=1}^{n} x_i^2\right) < 0$$

because $(\Sigma x_i)^2 < n \Sigma x_i^2$ provided all x_i are not equal. Noting that $\dfrac{\partial^2 z}{\partial x^2} > 0$, we see

that the solution of the two equations, $\dfrac{\partial K}{\partial a} = 0$ and $\dfrac{\partial K}{\partial b} = 0$, provides the only minimizing solution.

The value of the *double integral*

$$\int_A \int f(x, y) \, dxdy$$

can usually be evaluated by an iterated process; that is, evaluating two successive single integrals. For illustration, say $A = \{(x, y): 0 \le x \le 1, 0 \le y \le x\}$ as given in Figure A.6-1.

Then

$$\int_A \int (x + x^3y^2) \, dxdy = \int_0^1 \left[\int_0^x (x + x^3y^2)dy \right] dx$$

$$= \int_0^1 \left[xy + \frac{x^3y^3}{3} \right]_0^x dx$$

$$= \int_0^1 \left(x^2 + \frac{x^6}{3} \right) dx = \left[\frac{x^3}{3} + \frac{x^7}{3 \cdot 7} \right]_0^1$$

$$= \frac{1}{3} + \frac{1}{21} = \frac{8}{21}.$$

When placing the limits on the iterated integral, note that for each fixed x between zero and one, y is restricted to the interval zero to x. Also in the inner integral on y, x is treated as a constant.

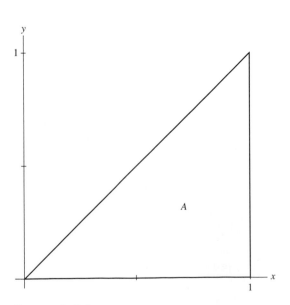

FIGURE A.6-1

In evaluating this double integral we could have restricted y to the interval zero to one, then x would be between y and one. That is, we would have evaluated the iterated integral

$$\int_0^1 \left[\int_y^1 (x + x^3 y^2)\, dx \right] dy = \int_0^1 \left[\frac{x^2}{2} + \frac{x^4 y^2}{4} \right]_y^1 dy$$

$$= \int_0^1 \left[\frac{1}{2} + \frac{y^2}{4} - \frac{y^2}{2} - \frac{y^6}{4} \right] dy$$

$$= \left[\frac{y}{2} - \frac{y^3}{3 \cdot 4} - \frac{y^7}{7 \cdot 4} \right]_0^1$$

$$= \frac{1}{2} - \frac{1}{12} - \frac{1}{28} = \frac{8}{21}.$$

Finally, we can *change variables* in a double integral

$$\int\int_A f(x, y)\, dxdy.$$

If $f(x, y)$ is a joint p.d.f. of random variables X and Y of the continuous type, then the double integral represents $P[(X, Y) \in A]$. Consider only one-to-one transformations, say $z = u_1(x, y)$ and $w = u_2(x, y)$ with inverse transformation given by $x = v_1(z, w)$ and $y = v_2(z, w)$. The determinant of order 2

$$J = \begin{vmatrix} \dfrac{\partial x}{\partial z} & \dfrac{\partial x}{\partial w} \\[2mm] \dfrac{\partial y}{\partial z} & \dfrac{\partial y}{\partial w} \end{vmatrix}$$

is called the *Jacobian* of the inverse transformation. Moreover, say the region A maps onto B in the (z, w) space. Since we are usually dealing with probabilities in this book, we fixed the *sign* of the integral so that it is positive (by using the absolute value of the Jacobian). Then it is true that

$$\int\int_A f(x, y)\, dxdy = \int\int_B f[v_1(z, w), v_2(z, w)]|J|\, dzdw.$$

For illustration, let

$$f(x, y) = \frac{1}{2\pi} e^{-(x^2 + y^2)/2}, \quad -\infty < x < \infty, \quad -\infty < y < \infty,$$

which is the joint p.d.f of two independent normal variables, each with mean zero and variance one. Say $A = \{(x, y): 0 \le x^2 + y^2 \le 1\}$ and consider

$$P(A) = \int\int_A f(x, y)\, dxdy.$$

This is impossible to deal with directly in the x, y variables. However, consider the inverse transformation to polar coordinates, namely

$$x = r \cos \theta, \; y = r \sin \theta$$

with Jacobian

$$J = \begin{vmatrix} \cos \theta & -r \sin \theta \\ \sin \theta & r \cos \theta \end{vmatrix} = r(\cos^2 \theta + \sin^2 \theta) = r.$$

Since A maps onto $B = \{(r, \theta): 0 \le r \le 1, \; 0 \le \theta < 2\pi\}$, we have that

$$
\begin{aligned}
P(A) &= \int_0^{2\pi} \left(\int_0^1 \frac{1}{2\pi} e^{-r^2/2} r \, dr \right) d\theta \\
&= \int_0^{2\pi} \left[-\frac{1}{2\pi} e^{-r^2/2} \right]_0^1 d\theta \\
&= \int_0^{2\pi} \frac{1}{2\pi} (1 - e^{-1/2}) \, d\theta \\
&= \frac{1}{2\pi} (1 - e^{-1/2}) 2\pi = (1 - e^{-1/2}).
\end{aligned}
$$

References

ASPIN, A. A., "Tables for Use in Comparisons Whose Accuracy Involves Two Variances, Separately Estimated," *Biometrika*, **36** (1949), pp. 290–296.

BOX, G. E. P., W. G. HUNTER, and J. S. HUNTER, *Statistics for Experimenters*, John Wiley & Sons, Inc., New York, 1978.

BOX, G. E. P., and M. E. MULLER, "A Note on the Generation of Random Normal Deviates," *Ann. Math. Statist.*, **29** (1958), p. 610.

CRAMÉR, H., *Mathematical Methods of Statistics*, Princeton University Press, Princeton, N.J., 1946.

FELLER, W., *An Introduction to Probability Theory and Its Applications*, Vol. 1, 3rd ed., John Wiley & Sons, Inc., New York, 1968.

FERGUSON, T. S., *Mathematical Statistics*, Academic Press, Inc., New York, 1967.

GUENTHER, WILLIAM C., "Shortest Confidence Intervals," *Amer. Statist.*, **23**, 1(1969), p. 22.

HOGG, R. V., and J. LEDOLTER, *Applied Statistics for Engineers and Physical Scientists*, 2nd ed., Macmillan Publishing Company, New York, 1992.

HOGG, R. V., and A. T. CRAIG, *Introduction to Mathematical Statistics*, 4th ed., Macmillan Publishing Company, New York, 1978.

HOGG, R. V., and A. T. CRAIG, "On the Decomposition of Certain Chi-Square Variables," *Ann. Math. Statist.*, **29** (1958), p. 608.

LINDGREN, B. W., *Statistical Theory*, 3rd ed., Macmillan Publishing Company, New York, 1976.

MONTGOMERY, D. C., *Design and Analysis of Experiments*, 2nd ed., John Wiley & Sons, Inc., New York, 1984.

MOOD, A. M., F. A. GRAYBILL, and D. C. BOES, *Introduction to the Theory of Statistics*, 3rd ed., McGraw-Hill Book Company, New York, 1974.

PEARSON, K., "On the Criterion That a Given System of Deviations from the Probable in the Case of a Correlated System of Variables Is Such That It Can Be Reasonably Supposed to Have Arisen from Random Sampling," *Phil. Mag.*, Series 5, **50** (1900), p. 157.

667

SNEE, R. D., L. B. HARE, and J. R. TROUT, *Experiments in Industry,* American Society of Quality Control, Milwaukee, Wis., 1985.

TATE, R. F., and G. W. KLETT, "Optimum Confidence Intervals for the Variance of a Normal Distribution," *J. Am. Statist. Assoc.,* **54** (1959), p. 674.

TUKEY, JOHN W., *Exploratory Data Analysis,* Addison-Wesley Publishing Company, Reading, Mass., 1977.

WILCOXON, F., "Individual Comparisons by Ranking Methods," *Biometrics Bull.,* **1** (1945), p. 80.

WILKS, S. S., *Mathematical Statistics,* John Wiley & Sons, Inc., New York, 1962.

Appendix Tables

TABLE I
Binomial Coefficients

$$\binom{n}{r} = \frac{n!}{r!(n-r)!} = \binom{n}{n-r}$$

n	$\binom{n}{0}$	$\binom{n}{1}$	$\binom{n}{2}$	$\binom{n}{3}$	$\binom{n}{4}$	$\binom{n}{5}$	$\binom{n}{6}$	$\binom{n}{7}$	$\binom{n}{8}$	$\binom{n}{9}$	$\binom{n}{10}$	$\binom{n}{11}$	$\binom{n}{12}$	$\binom{n}{13}$
0	1													
1	1	1												
2	1	2	1											
3	1	3	3	1										
4	1	4	6	4	1									
5	1	5	10	10	5	1								
6	1	6	15	20	15	6	1							
7	1	7	21	35	35	21	7	1						
8	1	8	28	56	70	56	28	8	1					
9	1	9	36	84	126	126	84	36	9	1				
10	1	10	45	120	210	252	210	120	45	10	1			
11	1	11	55	165	330	462	462	330	165	55	11	1		
12	1	12	66	220	495	792	924	792	495	220	66	12	1	
13	1	13	78	286	715	1,287	1,716	1,716	1,287	715	286	78	13	1
14	1	14	91	364	1,001	2,002	3,003	3,432	3,003	2,002	1,001	364	91	14
15	1	15	105	455	1,365	3,003	5,005	6,435	6,435	5,005	3,003	1,365	455	105
16	1	16	120	560	1,820	4,368	8,008	11,440	12,870	11,440	8,008	4,368	1,820	560
17	1	17	136	680	2,380	6,188	12,376	19,448	24,310	24,310	19,448	12,376	6,188	2,380
18	1	18	153	816	3,060	8,568	18,564	31,824	43,758	48,620	43,758	31,824	18,564	8,568
19	1	19	171	969	3,876	11,628	27,132	50,388	75,582	92,378	92,378	75,582	50,388	27,132
20	1	20	190	1,140	4,845	15,504	38,760	77,520	125,970	167,960	184,756	167,960	125,970	77,520
21	1	21	210	1,330	5,985	20,349	54,264	116,280	203,490	293,930	352,716	352,716	293,930	203,490
22	1	22	231	1,540	7,315	26,334	74,613	170,544	319,770	497,420	646,646	705,432	646,646	497,420
23	1	23	253	1,771	8,855	33,649	100,947	245,157	490,314	817,190	1,144,066	1,352,078	1,352,078	1,144,066
24	1	24	276	2,024	10,626	42,504	134,596	346,104	735,471	1,307,504	1,961,256	2,496,144	2,704,156	2,496,144
25	1	25	300	2,300	12,650	53,130	177,100	480,700	1,081,575	2,042,975	3,268,760	4,457,400	5,200,300	5,200,300
26	1	26	325	2,600	14,950	65,780	230,230	657,800	1,562,275	3,124,550	5,311,735	7,726,160	9,657,700	10,400,600

For $r > 13$ you may use the identity $\binom{n}{r} = \binom{n}{n-r}$.

TABLE II
The Binomial Distribution

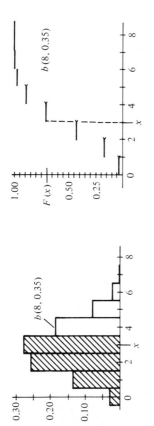

$$F(x) = P(X \le x) = \sum_{k=0}^{x} \frac{n!}{k!(n-k)!} p^k (1-p)^{n-k}$$

									p					
n	x	0.05	0.10	0.15	0.20	0.25	0.30	0.35	0.40	0.45	0.50
2	0	0.9025	0.8100	0.7225	0.6400	0.5625	0.4900	0.4225	0.3600	0.3025	0.2500
	1	0.9975	0.9900	0.9775	0.9600	0.9375	0.9100	0.8775	0.8400	0.7975	0.7500
	2	1.0000	1.0000	1.0000	1.0000	1.0000	1.0000	1.0000	1.0000	1.0000	1.0000
3	0	0.8574	0.7290	0.6141	0.5120	0.4219	0.3430	0.2746	0.2160	0.1664	0.1250
	1	0.9928	0.9720	0.9392	0.8960	0.8438	0.7840	0.7182	0.6480	0.5748	0.5000
	2	0.9999	0.9990	0.9966	0.9920	0.9844	0.9730	0.9571	0.9360	0.9089	0.8750
	3	1.0000	1.0000	1.0000	1.0000	1.0000	1.0000	1.0000	1.0000	1.0000	1.0000
4	0	0.8145	0.6561	0.5220	0.4096	0.3164	0.2401	0.1785	0.1296	0.0915	0.0625
	1	0.9860	0.9477	0.8905	0.8192	0.7383	0.6517	0.5630	0.4752	0.3910	0.3125
	2	0.9995	0.9963	0.9880	0.9728	0.9492	0.9163	0.8735	0.8208	0.7585	0.6875
	3	1.0000	0.9999	0.9995	0.9984	0.9961	0.9919	0.9850	0.9744	0.9590	0.9375
	4	1.0000	1.0000	1.0000	1.0000	1.0000	1.0000	1.0000	1.0000	1.0000	1.0000

TABLE II (continued)

$$F(x) = P(X \le x) = \sum_{k=0}^{x} \frac{n!}{k!(n-k)!} p^k (1-p)^{n-k}$$

n	x	0.05	0.10	0.15	0.20	0.25	0.30	0.35	0.40	0.45	0.50
							p				
5	0	0.7738	0.5905	0.4437	0.3277	0.2373	0.1681	0.1160	0.0778	0.0503	0.0312
	1	0.9774	0.9185	0.8352	0.7373	0.6328	0.5282	0.4284	0.3370	0.2562	0.1875
	2	0.9988	0.9914	0.9734	0.9421	0.8965	0.8369	0.7648	0.6826	0.5931	0.5000
	3	1.0000	0.9995	0.9978	0.9933	0.9844	0.9692	0.9460	0.9130	0.8688	0.8125
	4	1.0000	1.0000	0.9999	0.9997	0.9990	0.9976	0.9947	0.9898	0.9815	0.9688
	5	1.0000	1.0000	1.0000	1.0000	1.0000	1.0000	1.0000	1.0000	1.0000	1.0000
6	0	0.7351	0.5314	0.3771	0.2621	0.1780	0.1176	0.0754	0.0467	0.0277	0.0156
	1	0.9672	0.8857	0.7765	0.6553	0.5339	0.4202	0.3191	0.2333	0.1636	0.1094
	2	0.9978	0.9842	0.9527	0.9011	0.8306	0.7443	0.6471	0.5443	0.4415	0.3438
	3	0.9999	0.9987	0.9941	0.9830	0.9624	0.9295	0.8826	0.8208	0.7447	0.6562
	4	1.0000	0.9999	0.9996	0.9984	0.9954	0.9891	0.9777	0.9590	0.9308	0.8906
	5	1.0000	1.0000	1.0000	0.9999	0.9998	0.9993	0.9982	0.9959	0.9917	0.9844
	6	1.0000	1.0000	1.0000	1.0000	1.0000	1.0000	1.0000	1.0000	1.0000	1.0000
7	0	0.6983	0.4783	0.3206	0.2097	0.1335	0.0824	0.0490	0.0280	0.0152	0.0078
	1	0.9556	0.8503	0.7166	0.5767	0.4449	0.3294	0.2338	0.1586	0.1024	0.0625
	2	0.9962	0.9743	0.9262	0.8520	0.7564	0.6471	0.5323	0.4199	0.3164	0.2266
	3	0.9998	0.9973	0.9879	0.9667	0.9294	0.8740	0.8002	0.7102	0.6083	0.5000
	4	1.0000	0.9998	0.9988	0.9953	0.9871	0.9712	0.9444	0.9037	0.8471	0.7734
	5	1.0000	1.0000	0.9999	0.9996	0.9987	0.9962	0.9910	0.9812	0.9643	0.9375
	6	1.0000	1.0000	1.0000	1.0000	0.9999	0.9998	0.9994	0.9984	0.9963	0.9922
	7	1.0000	1.0000	1.0000	1.0000	1.0000	1.0000	1.0000	1.0000	1.0000	1.0000
8	0	0.6634	0.4305	0.2725	0.1678	0.1001	0.0576	0.0319	0.0168	0.0084	0.0039
	1	0.9428	0.8131	0.6572	0.5033	0.3671	0.2553	0.1691	0.1064	0.0632	0.0352
	2	0.9942	0.9619	0.8948	0.7969	0.6785	0.5518	0.4278	0.3154	0.2201	0.1445
	3	0.9996	0.9950	0.9786	0.9437	0.8862	0.8059	0.7064	0.5941	0.4770	0.3633

n	k										
	4	0.6367	0.7396	0.8263	0.8939	0.9420	0.9727	0.9896	0.9971	0.9996	1.0000
	5	0.8555	0.9115	0.9502	0.9747	0.9887	0.9958	0.9988	0.9998	1.0000	1.0000
	6	0.9648	0.9819	0.9915	0.9964	0.9987	0.9996	0.9999	1.0000	1.0000	1.0000
	7	0.9961	0.9983	0.9993	0.9998	0.9999	1.0000	1.0000	1.0000	1.0000	1.0000
	8	1.0000	1.0000	1.0000	1.0000	1.0000	1.0000	1.0000	1.0000	1.0000	1.0000
9	0	0.0020	0.0046	0.0101	0.0207	0.0404	0.0751	0.1342	0.2316	0.3874	0.6302
	1	0.0195	0.0385	0.0705	0.1211	0.1960	0.3003	0.4362	0.5995	0.7748	0.9288
	2	0.0898	0.1495	0.2318	0.3373	0.4628	0.6007	0.7382	0.8591	0.9470	0.9916
	3	0.2539	0.3614	0.4826	0.6089	0.7297	0.8343	0.9144	0.9661	0.9917	0.9994
	4	0.5000	0.6214	0.7334	0.8283	0.9012	0.9511	0.9804	0.9944	0.9991	1.0000
	5	0.7461	0.8342	0.9006	0.9464	0.9747	0.9900	0.9969	0.9994	0.9999	1.0000
	6	0.9102	0.9502	0.9750	0.9888	0.9957	0.9987	0.9997	1.0000	1.0000	1.0000
	7	0.9805	0.9909	0.9962	0.9986	0.9996	0.9999	1.0000	1.0000	1.0000	1.0000
	8	0.9980	0.9992	0.9997	0.9999	1.0000	1.0000	1.0000	1.0000	1.0000	1.0000
	9	1.0000	1.0000	1.0000	1.0000	1.0000	1.0000	1.0000	1.0000	1.0000	1.0000
10	0	0.0010	0.0025	0.0060	0.0135	0.0282	0.0563	0.1074	0.1969	0.3487	0.5987
	1	0.0107	0.0233	0.0464	0.0860	0.1493	0.2440	0.3758	0.5443	0.7361	0.9139
	2	0.0547	0.0996	0.1673	0.2616	0.3828	0.5256	0.6778	0.8202	0.9298	0.9885
	3	0.1719	0.2660	0.3823	0.5138	0.6496	0.7759	0.8791	0.9500	0.9872	0.9990
	4	0.3770	0.5044	0.6331	0.7515	0.8497	0.9219	0.9672	0.9901	0.9984	0.9999
	5	0.6230	0.7384	0.8338	0.9051	0.9527	0.9803	0.9936	0.9986	0.9999	1.0000
	6	0.8281	0.8980	0.9452	0.9740	0.9894	0.9965	0.9991	0.9999	1.0000	1.0000
	7	0.9453	0.9726	0.9877	0.9952	0.9984	0.9996	0.9999	1.0000	1.0000	1.0000
	8	0.9893	0.9955	0.9983	0.9995	0.9999	1.0000	1.0000	1.0000	1.0000	1.0000
	9	0.9990	0.9997	0.9999	1.0000	1.0000	1.0000	1.0000	1.0000	1.0000	1.0000
	10	1.0000	1.0000	1.0000	1.0000	1.0000	1.0000	1.0000	1.0000	1.0000	1.0000
11	0	0.0005	0.0014	0.0036	0.0088	0.0198	0.0422	0.0859	0.1673	0.3138	0.5688
	1	0.0059	0.0139	0.0302	0.0606	0.1130	0.1971	0.3221	0.4922	0.6974	0.8981
	2	0.0327	0.0652	0.1189	0.2001	0.3127	0.4552	0.6174	0.7788	0.9104	0.9848
	3	0.1133	0.1911	0.2963	0.4256	0.5696	0.7133	0.8389	0.9306	0.9815	0.9984
	4	0.2744	0.3971	0.5328	0.6683	0.7897	0.8854	0.9496	0.9841	0.9972	0.9999
	5	0.5000	0.6331	0.7535	0.8513	0.9218	0.9657	0.9883	0.9973	0.9997	1.0000
	6	0.7256	0.8262	0.9006	0.9499	0.9784	0.9924	0.9980	0.9997	1.0000	1.0000
	7	0.8867	0.9390	0.9707	0.9878	0.9957	0.9988	0.9998	1.0000	1.0000	1.0000

Table II (continued)

$$F(x) = P(X \le x) = \sum_{k=0}^{x} \frac{n!}{k!(n-k)!} p^k (1-p)^{n-k}$$

n	x	0.05	0.10	0.15	0.20	0.25	0.30	0.35	0.40	0.45	0.50
	8	1.0000	1.0000	1.0000	1.0000	0.9999	0.9994	0.9980	0.9941	0.9852	0.9673
	9	1.0000	1.0000	1.0000	1.0000	1.0000	1.0000	0.9998	0.9993	0.9978	0.9941
12	0	0.5404	0.2824	0.1422	0.0687	0.0317	0.0138	0.0057	0.0022	0.0008	0.0002
	1	0.8816	0.6590	0.4435	0.2749	0.1584	0.0850	0.0424	0.0196	0.0083	0.0032
	2	0.9804	0.8891	0.7358	0.5583	0.3907	0.2528	0.1513	0.0834	0.0421	0.0193
	3	0.9978	0.9744	0.9078	0.7946	0.6488	0.4925	0.3467	0.2253	0.1345	0.0730
	4	0.9998	0.9957	0.9761	0.9274	0.8424	0.7237	0.5833	0.4382	0.3044	0.1938
	5	1.0000	0.9995	0.9954	0.9806	0.9456	0.8822	0.7873	0.6652	0.5269	0.3872
	6	1.0000	0.9999	0.9993	0.9961	0.9857	0.9614	0.9154	0.8418	0.7393	0.6128
	7	1.0000	1.0000	0.9999	0.9994	0.9972	0.9905	0.9745	0.9427	0.8883	0.8062
	8	1.0000	1.0000	1.0000	0.9999	0.9996	0.9983	0.9944	0.9847	0.9644	0.9270
	9	1.0000	1.0000	1.0000	1.0000	1.0000	0.9998	0.9992	0.9972	0.9921	0.9807
	10	1.0000	1.0000	1.0000	1.0000	1.0000	1.0000	0.9999	0.9997	0.9989	0.9968
	11	1.0000	1.0000	1.0000	1.0000	1.0000	1.0000	1.0000	1.0000	0.9999	0.9998
	12	1.0000	1.0000	1.0000	1.0000	1.0000	1.0000	1.0000	1.0000	1.0000	1.0000
13	0	0.5133	0.2542	0.1209	0.0550	0.0238	0.0097	0.0037	0.0013	0.0004	0.0001
	1	0.8646	0.6213	0.3983	0.2336	0.1267	0.0637	0.0296	0.0126	0.0049	0.0017
	2	0.9755	0.8661	0.6920	0.5017	0.3326	0.2025	0.1132	0.0579	0.0269	0.0112
	3	0.9969	0.9658	0.8820	0.7473	0.5843	0.4206	0.2783	0.1686	0.0929	0.0461
	4	0.9997	0.9935	0.9658	0.9009	0.7940	0.6543	0.5005	0.3530	0.2279	0.1334
	5	1.0000	0.9991	0.9924	0.9700	0.9198	0.8346	0.7159	0.5744	0.4268	0.2905
	6	1.0000	0.9999	0.9987	0.9930	0.9757	0.9376	0.8705	0.7712	0.6437	0.5000
	7	1.0000	1.0000	0.9998	0.9988	0.9944	0.9818	0.9538	0.9023	0.8212	0.7095
	8	1.0000	1.0000	1.0000	0.9998	0.9990	0.9960	0.9874	0.9679	0.9302	0.8666

p

Binomial cumulative probability table (continued).

n	k										
	9	0.9539	0.9797	0.9922	0.9975	0.9993	0.9999	1.0000	1.0000	1.0000	1.0000
	10	0.9888	0.9959	0.9987	0.9997	0.9999	1.0000	1.0000	1.0000	1.0000	1.0000
	11	0.9983	0.9995	0.9999	1.0000	1.0000	1.0000	1.0000	1.0000	1.0000	1.0000
	12	0.9999	1.0000	1.0000	1.0000	1.0000	1.0000	1.0000	1.0000	1.0000	1.0000
	13	1.0000	1.0000	1.0000	1.0000	1.0000	1.0000	1.0000	1.0000	1.0000	1.0000
14	0	0.0001	0.0002	0.0008	0.0024	0.0068	0.0178	0.0440	0.1028	0.2288	0.4877
	1	0.0009	0.0029	0.0081	0.0205	0.0475	0.1010	0.1979	0.3567	0.5846	0.8470
	2	0.0065	0.0170	0.0398	0.0839	0.1608	0.2811	0.4481	0.6479	0.8416	0.9699
	3	0.0287	0.0632	0.1243	0.2205	0.3552	0.5213	0.6982	0.8535	0.9559	0.9958
	4	0.0898	0.1672	0.2793	0.4227	0.5842	0.7415	0.8702	0.9533	0.9908	0.9996
	5	0.2120	0.3373	0.4859	0.6405	0.7805	0.8883	0.9561	0.9885	0.9985	1.0000
	6	0.3953	0.5461	0.6925	0.8164	0.9067	0.9617	0.9884	0.9978	0.9998	1.0000
	7	0.6047	0.7414	0.8499	0.9247	0.9685	0.9897	0.9976	0.9997	1.0000	1.0000
	8	0.7880	0.8811	0.9417	0.9757	0.9917	0.9978	0.9996	1.0000	1.0000	1.0000
	9	0.9102	0.9574	0.9825	0.9940	0.9983	0.9997	1.0000	1.0000	1.0000	1.0000
	10	0.9713	0.9886	0.9961	0.9989	0.9998	1.0000	1.0000	1.0000	1.0000	1.0000
	11	0.9935	0.9978	0.9994	0.9999	1.0000	1.0000	1.0000	1.0000	1.0000	1.0000
	12	0.9991	0.9997	0.9999	1.0000	1.0000	1.0000	1.0000	1.0000	1.0000	1.0000
	13	0.9999	1.0000	1.0000	1.0000	1.0000	1.0000	1.0000	1.0000	1.0000	1.0000
	14	1.0000	1.0000	1.0000	1.0000	1.0000	1.0000	1.0000	1.0000	1.0000	1.0000
15	0	0.0000	0.0001	0.0005	0.0016	0.0047	0.0134	0.0352	0.0874	0.2059	0.4633
	1	0.0005	0.0017	0.0052	0.0142	0.0353	0.0802	0.1671	0.3186	0.5490	0.8290
	2	0.0037	0.0107	0.0271	0.0617	0.1268	0.2361	0.3980	0.6042	0.8159	0.9638
	3	0.0176	0.0424	0.0905	0.1727	0.2969	0.4613	0.6482	0.8227	0.9444	0.9945
	4	0.0592	0.1204	0.2173	0.3519	0.5155	0.6865	0.8358	0.9383	0.9873	0.9994
	5	0.1509	0.2608	0.4032	0.5643	0.7216	0.8516	0.9389	0.9832	0.9978	0.9999
	6	0.3036	0.4522	0.6098	0.7548	0.8689	0.9434	0.9819	0.9964	0.9997	1.0000
	7	0.5000	0.6535	0.7869	0.8868	0.9500	0.9827	0.9958	0.9994	1.0000	1.0000
	8	0.6964	0.8182	0.9050	0.9578	0.9848	0.9958	0.9992	0.9999	1.0000	1.0000
	9	0.8491	0.9231	0.9662	0.9876	0.9963	0.9992	0.9999	1.0000	1.0000	1.0000
	10	0.9408	0.9745	0.9907	0.9972	0.9993	0.9999	1.0000	1.0000	1.0000	1.0000
	11	0.9824	0.9937	0.9981	0.9995	0.9999	1.0000	1.0000	1.0000	1.0000	1.0000
	12	0.9963	0.9989	0.9997	0.9999	1.0000	1.0000	1.0000	1.0000	1.0000	1.0000
	13	0.9995	0.9999	1.0000	1.0000	1.0000	1.0000	1.0000	1.0000	1.0000	1.0000
	14	1.0000	1.0000	1.0000	1.0000	1.0000	1.0000	1.0000	1.0000	1.0000	1.0000
	15	1.0000	1.0000	1.0000	1.0000	1.0000	1.0000	1.0000	1.0000	1.0000	1.0000

TABLE II (continued)

$$F(x) = P(X \le x) = \sum_{k=0}^{x} \frac{n!}{k!(n-k)!} p^k(1-p)^{n-k}$$

							p				
n	x	0.05	0.10	0.15	0.20	0.25	0.30	0.35	0.40	0.45	0.50
16	0	0.4401	0.1853	0.0743	0.0281	0.0100	0.0033	0.0010	0.0003	0.0001	0.0000
	1	0.8108	0.5147	0.2839	0.1407	0.0635	0.0261	0.0098	0.0033	0.0010	0.0003
	2	0.9571	0.7892	0.5614	0.3518	0.1971	0.0994	0.0451	0.0183	0.0066	0.0021
	3	0.9930	0.9316	0.7899	0.5981	0.4050	0.2459	0.1339	0.0651	0.0281	0.0106
	4	0.9991	0.9830	0.9209	0.7982	0.6302	0.4499	0.2892	0.1666	0.0853	0.0384
	5	0.9999	0.9967	0.9765	0.9183	0.8103	0.6598	0.4900	0.3288	0.1976	0.1051
	6	1.0000	0.9995	0.9944	0.9733	0.9204	0.8247	0.6881	0.5272	0.3660	0.2272
	7	1.0000	0.9999	0.9989	0.9930	0.9729	0.9256	0.8406	0.7161	0.5629	0.4018
	8	1.0000	1.0000	0.9998	0.9985	0.9925	0.9743	0.9329	0.8577	0.7441	0.5982
	9	1.0000	1.0000	1.0000	0.9998	0.9984	0.9929	0.9771	0.9417	0.8759	0.7728
	10	1.0000	1.0000	1.0000	1.0000	0.9997	0.9984	0.9938	0.9809	0.9514	0.8949
	11	1.0000	1.0000	1.0000	1.0000	1.0000	0.9997	0.9987	0.9951	0.9851	0.9616
	12	1.0000	1.0000	1.0000	1.0000	1.0000	1.0000	0.9998	0.9991	0.9965	0.9894
	13	1.0000	1.0000	1.0000	1.0000	1.0000	1.0000	1.0000	0.9999	0.9994	0.9979
	14	1.0000	1.0000	1.0000	1.0000	1.0000	1.0000	1.0000	1.0000	0.9999	0.9997
	15	1.0000	1.0000	1.0000	1.0000	1.0000	1.0000	1.0000	1.0000	1.0000	1.0000
	16	1.0000	1.0000	1.0000	1.0000	1.0000	1.0000	1.0000	1.0000	1.0000	1.0000
20	0	0.3585	0.1216	0.0388	0.0115	0.0032	0.0008	0.0002	0.0000	0.0000	0.0000
	1	0.7358	0.3917	0.1756	0.0692	0.0243	0.0076	0.0021	0.0005	0.0001	0.0000
	2	0.9245	0.6769	0.4049	0.2061	0.0913	0.0355	0.0121	0.0036	0.0009	0.0002
	3	0.9841	0.8670	0.6477	0.4114	0.2252	0.1071	0.0444	0.0160	0.0049	0.0013
	4	0.9974	0.9568	0.8298	0.6296	0.4148	0.2375	0.1182	0.0510	0.0189	0.0059
	5	0.9997	0.9887	0.9327	0.8042	0.6172	0.4164	0.2454	0.1256	0.0553	0.0207
	6	1.0000	0.9976	0.9781	0.9133	0.7858	0.6080	0.4166	0.2500	0.1299	0.0577
	7	1.0000	0.9996	0.9941	0.9679	0.8982	0.7723	0.6010	0.4159	0.2520	0.1316
	8	1.0000	0.9999	0.9987	0.9900	0.9591	0.8867	0.7624	0.5956	0.4143	0.2517
	9	1.0000	1.0000	0.9998	0.9974	0.9861	0.9520	0.8782	0.7553	0.5914	0.4119

n	x										
	10	0.5881	0.7507	0.8725	0.9468	0.9829	0.9961	0.9994	1.0000	1.0000	1.0000
	11	0.7483	0.8692	0.9435	0.9804	0.9949	0.9991	0.9999	1.0000	1.0000	1.0000
	12	0.8684	0.9420	0.9790	0.9940	0.9987	0.9998	1.0000	1.0000	1.0000	1.0000
	13	0.9423	0.9786	0.9935	0.9985	0.9997	1.0000	1.0000	1.0000	1.0000	1.0000
	14	0.9793	0.9936	0.9984	0.9997	1.0000	1.0000	1.0000	1.0000	1.0000	1.0000
	15	0.9941	0.9985	0.9997	1.0000	1.0000	1.0000	1.0000	1.0000	1.0000	1.0000
	16	0.9987	0.9997	1.0000	1.0000	1.0000	1.0000	1.0000	1.0000	1.0000	1.0000
	17	0.9998	1.0000	1.0000	1.0000	1.0000	1.0000	1.0000	1.0000	1.0000	1.0000
	18	1.0000	1.0000	1.0000	1.0000	1.0000	1.0000	1.0000	1.0000	1.0000	1.0000
	19	1.0000	1.0000	1.0000	1.0000	1.0000	1.0000	1.0000	1.0000	1.0000	1.0000
	20	1.0000	1.0000	1.0000	1.0000	1.0000	1.0000	1.0000	1.0000	1.0000	1.0000
25	0	0.0000	0.0000	0.0000	0.0000	0.0001	0.0008	0.0038	0.0172	0.0718	0.2774
	1	0.0000	0.0000	0.0001	0.0003	0.0016	0.0070	0.0274	0.0931	0.2712	0.6424
	2	0.0000	0.0001	0.0004	0.0021	0.0090	0.0321	0.0982	0.2537	0.5371	0.8729
	3	0.0000	0.0005	0.0024	0.0097	0.0332	0.0962	0.2340	0.4711	0.7636	0.9659
	4	0.0005	0.0023	0.0095	0.0320	0.0905	0.2137	0.4207	0.6821	0.9020	0.9928
	5	0.0020	0.0086	0.0294	0.0826	0.1935	0.3783	0.6167	0.8385	0.9666	0.9988
	6	0.0073	0.0258	0.0736	0.1734	0.3407	0.5611	0.7800	0.9305	0.9905	0.9998
	7	0.0216	0.0639	0.1536	0.3061	0.5118	0.7265	0.8909	0.9745	0.9977	1.0000
	8	0.0539	0.1340	0.2735	0.4668	0.6769	0.8506	0.9532	0.9920	0.9995	1.0000
	9	0.1148	0.2424	0.4246	0.6303	0.8106	0.9287	0.9827	0.9979	0.9999	1.0000
	10	0.2122	0.3843	0.5858	0.7712	0.9022	0.9703	0.9944	0.9995	1.0000	1.0000
	11	0.3450	0.5426	0.7323	0.8746	0.9558	0.9893	0.9985	0.9999	1.0000	1.0000
	12	0.5000	0.6937	0.8462	0.9396	0.9825	0.9966	0.9996	1.0000	1.0000	1.0000
	13	0.6550	0.8173	0.9222	0.9745	0.9940	0.9991	0.9999	1.0000	1.0000	1.0000
	14	0.7878	0.9040	0.9656	0.9907	0.9982	0.9998	1.0000	1.0000	1.0000	1.0000
	15	0.8852	0.9560	0.9868	0.9971	0.9995	1.0000	1.0000	1.0000	1.0000	1.0000
	16	0.9461	0.9826	0.9957	0.9992	0.9999	1.0000	1.0000	1.0000	1.0000	1.0000
	17	0.9784	0.9942	0.9988	0.9998	1.0000	1.0000	1.0000	1.0000	1.0000	1.0000
	18	0.9927	0.9984	0.9997	1.0000	1.0000	1.0000	1.0000	1.0000	1.0000	1.0000
	19	0.9980	0.9996	0.9999	1.0000	1.0000	1.0000	1.0000	1.0000	1.0000	1.0000
	20	0.9995	0.9999	1.0000	1.0000	1.0000	1.0000	1.0000	1.0000	1.0000	1.0000
	21	0.9999	1.0000	1.0000	1.0000	1.0000	1.0000	1.0000	1.0000	1.0000	1.0000
	22	1.0000	1.0000	1.0000	1.0000	1.0000	1.0000	1.0000	1.0000	1.0000	1.0000
	23	1.0000	1.0000	1.0000	1.0000	1.0000	1.0000	1.0000	1.0000	1.0000	1.0000
	24	1.0000	1.0000	1.0000	1.0000	1.0000	1.0000	1.0000	1.0000	1.0000	1.0000
	25	1.0000	1.0000	1.0000	1.0000	1.0000	1.0000	1.0000	1.0000	1.0000	1.0000

TABLE III
The Poisson Distribution

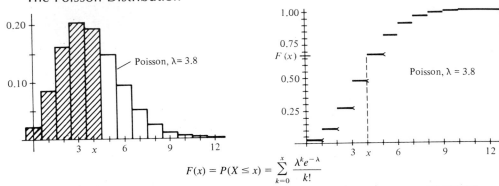

Poisson, λ= 3.8

Poisson, λ = 3.8

$$F(x) = P(X \le x) = \sum_{k=0}^{x} \frac{\lambda^k e^{-\lambda}}{k!}$$

					$\lambda = E(X)$					
x	0.1	0.2	0.3	0.4	0.5	0.6	0.7	0.8	0.9	1.0
0	0.905	0.819	0.741	0.670	0.607	0.549	0.497	0.449	0.407	0.368
1	0.995	0.982	0.963	0.938	0.910	0.878	0.844	0.809	0.772	0.736
2	1.000	0.999	0.996	0.992	0.986	0.977	0.966	0.953	0.937	0.920
3	1.000	1.000	1.000	0.999	0.998	0.997	0.994	0.991	0.987	0.981
4	1.000	1.000	1.000	1.000	1.000	1.000	0.999	0.999	0.998	0.996
5	1.000	1.000	1.000	1.000	1.000	1.000	1.000	1.000	1.000	0.999
6	1.000	1.000	1.000	1.000	1.000	1.000	1.000	1.000	1.000	1.000

x	1.1	1.2	1.3	1.4	1.5	1.6	1.7	1.8	1.9	2.0
0	0.333	0.301	0.273	0.247	0.223	0.202	0.183	0.165	0.150	0.135
1	0.699	0.663	0.627	0.592	0.558	0.525	0.493	0.463	0.434	0.406
2	0.900	0.879	0.857	0.833	0.809	0.783	0.757	0.731	0.704	0.677
3	0.974	0.966	0.957	0.946	0.934	0.921	0.907	0.891	0.875	0.857
4	0.995	0.992	0.989	0.986	0.981	0.976	0.970	0.964	0.956	0.947
5	0.999	0.998	0.998	0.997	0.996	0.994	0.992	0.990	0.987	0.983
6	1.000	1.000	1.000	0.999	0.999	0.999	0.998	0.997	0.997	0.995
7	1.000	1.000	1.000	1.000	1.000	1.000	1.000	0.999	0.999	0.999
8	1.000	1.000	1.000	1.000	1.000	1.000	1.000	1.000	1.000	1.000

x	2.2	2.4	2.6	2.8	3.0	3.2	3.4	3.6	3.8	4.0
0	0.111	0.091	0.074	0.061	0.050	0.041	0.033	0.027	0.022	0.018
1	0.355	0.308	0.267	0.231	0.199	0.171	0.147	0.126	0.107	0.092
2	0.623	0.570	0.518	0.469	0.423	0.380	0.340	0.303	0.269	0.238
3	0.819	0.779	0.736	0.692	0.647	0.603	0.558	0.515	0.473	0.433
4	0.928	0.904	0.877	0.848	0.815	0.781	0.744	0.706	0.668	0.629
5	0.975	0.964	0.951	0.935	0.916	0.895	0.871	0.844	0.816	0.785
6	0.993	0.988	0.983	0.976	0.966	0.955	0.942	0.927	0.909	0.889
7	0.998	0.997	0.995	0.992	0.988	0.983	0.977	0.969	0.960	0.949
8	1.000	0.999	0.999	0.998	0.996	0.994	0.992	0.988	0.984	0.979
9	1.000	1.000	1.000	0.999	0.999	0.998	0.997	0.996	0.994	0.992

TABLE III (continued)

10	1.000	1.000	1.000	1.000	1.000	1.000	0.999	0.999	0.998	0.997
11	1.000	1.000	1.000	1.000	1.000	1.000	1.000	1.000	0.999	0.999
12	1.000	1.000	1.000	1.000	1.000	1.000	1.000	1.000	1.000	1.000

x	4.2	4.4	4.6	4.8	5.0	5.2	5.4	5.6	5.8	6.0
0	0.015	0.012	0.010	0.008	0.007	0.006	0.005	0.004	0.003	0.002
1	0.078	0.066	0.056	0.048	0.040	0.034	0.029	0.024	0.021	0.017
2	0.210	0.185	0.163	0.143	0.125	0.109	0.095	0.082	0.072	0.062
3	0.395	0.359	0.326	0.294	0.265	0.238	0.213	0.191	0.170	0.151
4	0.590	0.551	0.513	0.476	0.440	0.406	0.373	0.342	0.313	0.285
5	0.753	0.720	0.686	0.651	0.616	0.581	0.546	0.512	0.478	0.446
6	0.867	0.844	0.818	0.791	0.762	0.732	0.702	0.670	0.638	0.606
7	0.936	0.921	0.905	0.887	0.867	0.845	0.822	0.797	0.771	0.744
8	0.972	0.964	0.955	0.944	0.932	0.918	0.903	0.886	0.867	0.847
9	0.989	0.985	0.980	0.975	0.968	0.960	0.951	0.941	0.929	0.916
10	0.996	0.994	0.992	0.990	0.986	0.982	0.977	0.972	0.965	0.957
11	0.999	0.998	0.997	0.996	0.995	0.993	0.990	0.988	0.984	0.980
12	1.000	0.999	0.999	0.999	0.998	0.997	0.996	0.995	0.993	0.991
13	1.000	1.000	1.000	1.000	0.999	0.999	0.999	0.998	0.997	0.996
14	1.000	1.000	1.000	1.000	1.000	1.000	0.999	0.999	0.999	0.999
15	1.000	1.000	1.000	1.000	1.000	1.000	1.000	1.000	1.000	0.999
16	1.000	1.000	1.000	1.000	1.000	1.000	1.000	1.000	1.000	1.000

x	6.5	7.0	7.5	8.0	8.5	9.0	9.5	10.0	10.5	11.0
0	0.002	0.001	0.001	0.000	0.000	0.000	0.000	0.000	0.000	0.000
1	0.011	0.007	0.005	0.003	0.002	0.001	0.001	0.000	0.000	0.000
2	0.043	0.030	0.020	0.014	0.009	0.006	0.004	0.003	0.002	0.001
3	0.112	0.082	0.059	0.042	0.030	0.021	0.015	0.010	0.007	0.005
4	0.224	0.173	0.132	0.100	0.074	0.055	0.040	0.029	0.021	0.015
5	0.369	0.301	0.241	0.191	0.150	0.116	0.089	0.067	0.050	0.038
6	0.527	0.450	0.378	0.313	0.256	0.207	0.165	0.130	0.102	0.079
7	0.673	0.599	0.525	0.453	0.386	0.324	0.269	0.220	0.179	0.143
8	0.792	0.729	0.662	0.593	0.523	0.456	0.392	0.333	0.279	0.232
9	0.877	0.830	0.776	0.717	0.653	0.587	0.522	0.458	0.397	0.341
10	0.933	0.901	0.862	0.816	0.763	0.706	0.645	0.583	0.521	0.460
11	0.966	0.947	0.921	0.888	0.849	0.803	0.752	0.697	0.639	0.579
12	0.984	0.973	0.957	0.936	0.909	0.876	0.836	0.792	0.742	0.689
13	0.993	0.987	0.978	0.966	0.949	0.926	0.898	0.864	0.825	0.781
14	0.997	0.994	0.990	0.983	0.973	0.959	0.940	0.917	0.888	0.854
15	0.999	0.998	0.995	0.992	0.986	0.978	0.967	0.951	0.932	0.907
16	1.000	0.999	0.998	0.996	0.993	0.989	0.982	0.973	0.960	0.944
17	1.000	1.000	0.999	0.998	0.997	0.995	0.991	0.986	0.978	0.968
18	1.000	1.000	1.000	0.999	0.999	0.998	0.996	0.993	0.988	0.982
19	1.000	1.000	1.000	1.000	0.999	0.999	0.998	0.997	0.994	0.991
20	1.000	1.000	1.000	1.000	1.000	1.000	0.999	0.998	0.997	0.995
21	1.000	1.000	1.000	1.000	1.000	1.000	1.000	0.999	0.999	0.998
22	1.000	1.000	1.000	1.000	1.000	1.000	1.000	1.000	0.999	0.999
23	1.000	1.000	1.000	1.000	1.000	1.000	1.000	1.000	1.000	1.000

TABLE III (continued)

x	11.5	12.0	12.5	13.0	13.5	14.0	14.5	15.0	15.5	16.0
0	0.000	0.000	0.000	0.000	0.000	0.000	0.000	0.000	0.000	0.000
1	0.000	0.000	0.000	0.000	0.000	0.000	0.000	0.000	0.000	0.000
2	0.001	0.001	0.000	0.000	0.000	0.000	0.000	0.000	0.000	0.000
3	0.003	0.002	0.002	0.001	0.001	0.000	0.000	0.000	0.000	0.000
4	0.011	0.008	0.005	0.004	0.003	0.002	0.001	0.001	0.001	0.000
5	0.028	0.020	0.015	0.011	0.008	0.006	0.004	0.003	0.002	0.001
6	0.060	0.046	0.035	0.026	0.019	0.014	0.010	0.008	0.006	0.004
7	0.114	0.090	0.070	0.054	0.041	0.032	0.024	0.018	0.013	0.010
8	0.191	0.155	0.125	0.100	0.079	0.062	0.048	0.037	0.029	0.022
9	0.289	0.242	0.201	0.166	0.135	0.109	0.088	0.070	0.055	0.043
10	0.402	0.347	0.297	0.252	0.211	0.176	0.145	0.118	0.096	0.077
11	0.520	0.462	0.406	0.353	0.304	0.260	0.220	0.185	0.154	0.127
12	0.633	0.576	0.519	0.463	0.409	0.358	0.311	0.268	0.228	0.193
13	0.733	0.682	0.628	0.573	0.518	0.464	0.413	0.363	0.317	0.275
14	0.815	0.772	0.725	0.675	0.623	0.570	0.518	0.466	0.415	0.368
15	0.878	0.844	0.806	0.764	0.718	0.669	0.619	0.568	0.517	0.467
16	0.924	0.899	0.869	0.835	0.798	0.756	0.711	0.664	0.615	0.566
17	0.954	0.937	0.916	0.890	0.861	0.827	0.790	0.749	0.705	0.659
18	0.974	0.963	0.948	0.930	0.908	0.883	0.853	0.819	0.782	0.742
19	0.986	0.979	0.969	0.957	0.942	0.923	0.901	0.875	0.846	0.812
20	0.992	0.988	0.983	0.975	0.965	0.952	0.936	0.917	0.894	0.868
21	0.996	0.994	0.991	0.986	0.980	0.971	0.960	0.947	0.930	0.911
22	0.998	0.997	0.995	0.992	0.989	0.983	0.976	0.967	0.956	0.942
23	0.999	0.999	0.998	0.996	0.994	0.991	0.986	0.981	0.973	0.963
24	1.000	0.999	0.999	0.998	0.997	0.995	0.992	0.989	0.984	0.978
25	1.000	1.000	0.999	0.999	0.998	0.997	0.996	0.994	0.991	0.987
26	1.000	1.000	1.000	1.000	0.999	0.999	0.998	0.997	0.995	0.993
27	1.000	1.000	1.000	1.000	1.000	0.999	0.999	0.998	0.997	0.996
28	1.000	1.000	1.000	1.000	1.000	1.000	0.999	0.999	0.999	0.998
29	1.000	1.000	1.000	1.000	1.000	1.000	1.000	1.000	0.999	0.999
30	1.000	1.000	1.000	1.000	1.000	1.000	1.000	1.000	1.000	0.999
31	1.000	1.000	1.000	1.000	1.000	1.000	1.000	1.000	1.000	1.000
32	1.000	1.000	1.000	1.000	1.000	1.000	1.000	1.000	1.000	1.000
33	1.000	1.000	1.000	1.000	1.000	1.000	1.000	1.000	1.000	1.000
34	1.000	1.000	1.000	1.000	1.000	1.000	1.000	1.000	1.000	1.000
35	1.000	1.000	1.000	1.000	1.000	1.000	1.000	1.000	1.000	1.000

Table IV
The Chi-Square Distribution

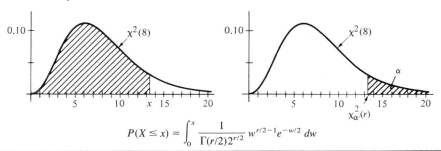

$$P(X \le x) = \int_0^x \frac{1}{\Gamma(r/2)2^{r/2}} w^{r/2-1}e^{-w/2} \, dw$$

				$P(X \le x)$				
	0.010	0.025	0.050	0.100	0.900	0.950	0.975	0.990
r	$\chi^2_{0.99}(r)$	$\chi^2_{0.975}(r)$	$\chi^2_{0.95}(r)$	$\chi^2_{0.90}(r)$	$\chi^2_{0.10}(r)$	$\chi^2_{0.05}(r)$	$\chi^2_{0.025}(r)$	$\chi^2_{0.01}(r)$
1	0.000	0.001	0.004	0.016	2.706	3.841	5.024	6.635
2	0.020	0.051	0.103	0.211	4.605	5.991	7.378	9.210
3	0.115	0.216	0.352	0.584	6.251	7.815	9.348	11.34
4	0.297	0.484	0.711	1.064	7.779	9.488	11.14	13.28
5	0.554	0.831	1.145	1.610	9.236	11.07	12.83	15.09
6	0.872	1.237	1.635	2.204	10.64	12.59	14.45	16.81
7	1.239	1.690	2.167	2.833	12.02	14.07	16.01	18.48
8	1.646	2.180	2.733	3.490	13.36	15.51	17.54	20.09
9	2.088	2.700	3.325	4.168	14.68	16.92	19.02	21.67
10	2.558	3.247	3.940	4.865	15.99	18.31	20.48	23.21
11	3.053	3.816	4.575	5.578	17.28	19.68	21.92	24.72
12	3.571	4.404	5.226	6.304	18.55	21.03	23.34	26.22
13	4.107	5.009	5.892	7.042	19.81	22.36	24.74	27.69
14	4.660	5.629	6.571	7.790	21.06	23.68	26.12	29.14
15	5.229	6.262	7.261	8.547	22.31	25.00	27.49	30.58
16	5.812	6.908	7.962	9.312	23.54	26.30	28.84	32.00
17	6.408	7.564	8.672	10.08	24.77	27.59	30.19	33.41
18	7.015	8.231	9.390	10.86	25.99	28.87	31.53	34.80
19	7.633	8.907	10.12	11.65	27.20	30.14	32.85	36.19
20	8.260	9.591	10.85	12.44	28.41	31.41	34.17	37.57
21	8.897	10.28	11.59	13.24	29.62	32.67	35.48	38.93
22	9.542	10.98	12.34	14.04	30.81	33.92	36.78	40.29
23	10.20	11.69	13.09	14.85	32.01	35.17	38.08	41.64
24	10.86	12.40	13.85	15.66	33.20	36.42	39.36	42.98
25	11.52	13.12	14.61	16.47	34.38	37.65	40.65	44.31
26	12.20	13.84	15.38	17.29	35.56	38.88	41.92	45.64
27	12.88	14.57	16.15	18.11	36.74	40.11	43.19	46.96
28	13.56	15.31	16.93	18.94	37.92	41.34	44.46	48.28
29	14.26	16.05	17.71	19.77	39.09	42.56	45.72	49.59
30	14.95	16.79	18.49	20.60	40.26	43.77	46.98	50.89
40	22.16	24.43	26.51	29.05	51.80	55.76	59.34	63.69
50	29.71	32.36	34.76	37.69	63.17	67.50	71.42	76.15
60	37.48	40.48	43.19	46.46	74.40	79.08	83.30	88.38
70	45.44	48.76	51.74	55.33	85.53	90.53	95.02	100.4
80	53.34	57.15	60.39	64.28	96.58	101.9	106.6	112.3

This table is abridged and adapted from Table III in *Biometrika Tables for Statisticians*, edited by E. S. Pearson and H. O. Hartley. It is published here with the kind permission of the *Biometrika* Trustees.

TABLE Va
The Normal Distribution

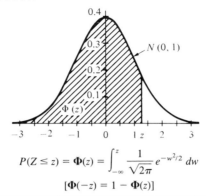

$$P(Z \le z) = \Phi(z) = \int_{-\infty}^{z} \frac{1}{\sqrt{2\pi}} e^{-w^2/2} \, dw$$

$$[\Phi(-z) = 1 - \Phi(z)]$$

z	0.00	0.01	0.02	0.03	0.04	0.05	0.06	0.07	0.08	0.09
0.0	0.5000	0.5040	0.5080	0.5120	0.5160	0.5199	0.5239	0.5279	0.5319	0.5359
0.1	0.5398	0.5438	0.5478	0.5517	0.5557	0.5596	0.5636	0.5675	0.5714	0.5753
0.2	0.5793	0.5832	0.5871	0.5910	0.5948	0.5987	0.6026	0.6064	0.6103	0.6141
0.3	0.6179	0.6217	0.6255	0.6293	0.6331	0.6368	0.6406	0.6443	0.6480	0.6517
0.4	0.6554	0.6591	0.6628	0.6664	0.6700	0.6736	0.6772	0.6808	0.6844	0.6879
0.5	0.6915	0.6950	0.6985	0.7019	0.7054	0.7088	0.7123	0.7157	0.7190	0.7224
0.6	0.7257	0.7291	0.7324	0.7357	0.7389	0.7422	0.7454	0.7486	0.7517	0.7549
0.7	0.7580	0.7611	0.7642	0.7673	0.7703	0.7734	0.7764	0.7794	0.7823	0.7852
0.8	0.7881	0.7910	0.7939	0.7967	0.7995	0.8023	0.8051	0.8078	0.8106	0.8133
0.9	0.8159	0.8186	0.8212	0.8238	0.8264	0.8289	0.8315	0.8340	0.8365	0.8389
1.0	0.8413	0.8438	0.8461	0.8485	0.8508	0.8531	0.8554	0.8577	0.8599	0.8621
1.1	0.8643	0.8665	0.8686	0.8708	0.8729	0.8749	0.8770	0.8790	0.8810	0.8830
1.2	0.8849	0.8869	0.8888	0.8907	0.8925	0.8944	0.8962	0.8980	0.8997	0.9015
1.3	0.9032	0.9049	0.9066	0.9082	0.9099	0.9115	0.9131	0.9147	0.9162	0.9177
1.4	0.9192	0.9207	0.9222	0.9236	0.9251	0.9265	0.9279	0.9292	0.9306	0.9319
1.5	0.9332	0.9345	0.9357	0.9370	0.9382	0.9394	0.9406	0.9418	0.9429	0.9441
1.6	0.9452	0.9463	0.9474	0.9484	0.9495	0.9505	0.9515	0.9525	0.9535	0.9545
1.7	0.9554	0.9564	0.9573	0.9582	0.9591	0.9599	0.9608	0.9616	0.9625	0.9633
1.8	0.9641	0.9649	0.9656	0.9664	0.9671	0.9678	0.9686	0.9693	0.9699	0.9706
1.9	0.9713	0.9719	0.9726	0.9732	0.9738	0.9744	0.9750	0.9756	0.9761	0.9767
2.0	0.9772	0.9778	0.9783	0.9788	0.9793	0.9798	0.9803	0.9808	0.9812	0.9817
2.1	0.9821	0.9826	0.9830	0.9834	0.9838	0.9842	0.9846	0.9850	0.9854	0.9857
2.2	0.9861	0.9864	0.9868	0.9871	0.9875	0.9878	0.9881	0.9884	0.9887	0.9890
2.3	0.9893	0.9896	0.9898	0.9901	0.9904	0.9906	0.9909	0.9911	0.9913	0.9916
2.4	0.9918	0.9920	0.9922	0.9925	0.9927	0.9929	0.9931	0.9932	0.9934	0.9936
2.5	0.9938	0.9940	0.9941	0.9943	0.9945	0.9946	0.9948	0.9949	0.9951	0.9952
2.6	0.9953	0.9955	0.9956	0.9957	0.9959	0.9960	0.9961	0.9962	0.9963	0.9964
2.7	0.9965	0.9966	0.9967	0.9968	0.9969	0.9970	0.9971	0.9972	0.9973	0.9974
2.8	0.9974	0.9975	0.9976	0.9977	0.9977	0.9978	0.9979	0.9979	0.9980	0.9981
2.9	0.9981	0.9982	0.9982	0.9983	0.9984	0.9984	0.9985	0.9985	0.9986	0.9986
3.0	0.9987	0.9987	0.9987	0.9988	0.9988	0.9989	0.9989	0.9989	0.9990	0.9990

α	0.400	0.300	0.200	0.100	0.050	0.025	0.010	0.005	0.001	
z_α	0.253	0.524	0.842	1.282	1.645	1.960	2.326	2.576	3.090	
$z_{\alpha/2}$	0.842	1.036	1.282	1.645	1.960	2.240	2.576	2.807	3.291	

TABLE Vb
The Normal Distribution

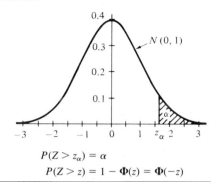

$$P(Z > z_\alpha) = \alpha$$
$$P(Z > z) = 1 - \Phi(z) = \Phi(-z)$$

z_α	0.00	0.01	0.02	0.03	0.04	0.05	0.06	0.07	0.08	0.09
0.0	0.5000	0.4960	0.4920	0.4880	0.4840	0.4801	0.4761	0.4721	0.4681	0.4641
0.1	0.4602	0.4562	0.4522	0.4483	0.4443	0.4404	0.4364	0.4325	0.4286	0.4247
0.2	0.4207	0.4168	0.4129	0.4090	0.4052	0.4013	0.3974	0.3936	0.3897	0.3859
0.3	0.3821	0.3783	0.3745	0.3707	0.3669	0.3632	0.3594	0.3557	0.3520	0.3483
0.4	0.3446	0.3409	0.3372	0.3336	0.3300	0.3264	0.3228	0.3192	0.3156	0.3121
0.5	0.3085	0.3050	0.3015	0.2981	0.2946	0.2912	0.2877	0.2843	0.2810	0.2776
0.6	0.2743	0.2709	0.2676	0.2643	0.2611	0.2578	0.2546	0.2514	0.2483	0.2451
0.7	0.2420	0.2389	0.2358	0.2327	0.2296	0.2266	0.2236	0.2206	0.2177	0.2148
0.8	0.2119	0.2090	0.2061	0.2033	0.2005	0.1977	0.1949	0.1922	0.1894	0.1867
0.9	0.1841	0.1814	0.1788	0.1762	0.1736	0.1711	0.1685	0.1660	0.1635	0.1611
1.0	0.1587	0.1562	0.1539	0.1515	0.1492	0.1469	0.1446	0.1423	0.1401	0.1379
1.1	0.1357	0.1335	0.1314	0.1292	0.1271	0.1251	0.1230	0.1210	0.1190	0.1170
1.2	0.1151	0.1131	0.1112	0.1093	0.1075	0.1056	0.1038	0.1020	0.1003	0.0985
1.3	0.0968	0.0951	0.0934	0.0918	0.0901	0.0885	0.0869	0.0853	0.0838	0.0823
1.4	0.0808	0.0793	0.0778	0.0764	0.0749	0.0735	0.0721	0.0708	0.0694	0.0681
1.5	0.0668	0.0655	0.0643	0.0630	0.0618	0.0606	0.0594	0.0582	0.0571	0.0559
1.6	0.0548	0.0537	0.0526	0.0516	0.0505	0.0495	0.0485	0.0475	0.0465	0.0455
1.7	0.0446	0.0436	0.0427	0.0418	0.0409	0.0401	0.0392	0.0384	0.0375	0.0367
1.8	0.0359	0.0351	0.0344	0.0336	0.0329	0.0322	0.0314	0.0307	0.0301	0.0294
1.9	0.0287	0.0281	0.0274	0.0268	0.0262	0.0256	0.0250	0.0244	0.0239	0.0233
2.0	0.0228	0.0222	0.0217	0.0212	0.0207	0.0202	0.0197	0.0192	0.0188	0.0183
2.1	0.0179	0.0174	0.0170	0.0166	0.0162	0.0158	0.0154	0.0150	0.0146	0.0143
2.2	0.0139	0.0136	0.0132	0.0129	0.0125	0.0122	0.0119	0.0116	0.0113	0.0110
2.3	0.0107	0.0104	0.0102	0.0099	0.0096	0.0094	0.0091	0.0089	0.0087	0.0084
2.4	0.0082	0.0080	0.0078	0.0075	0.0073	0.0071	0.0069	0.0068	0.0066	0.0064
2.5	0.0062	0.0060	0.0059	0.0057	0.0055	0.0054	0.0052	0.0051	0.0049	0.0048
2.6	0.0047	0.0045	0.0044	0.0043	0.0041	0.0040	0.0039	0.0038	0.0037	0.0036
2.7	0.0035	0.0034	0.0033	0.0032	0.0031	0.0030	0.0029	0.0028	0.0027	0.0026
2.8	0.0026	0.0025	0.0024	0.0023	0.0023	0.0022	0.0021	0.0021	0.0020	0.0019
2.9	0.0019	0.0018	0.0018	0.0017	0.0016	0.0016	0.0015	0.0015	0.0014	0.0014
3.0	0.0013	0.0013	0.0013	0.0012	0.0012	0.0011	0.0011	0.0011	0.0010	0.0010
3.1	0.0010	0.0009	0.0009	0.0009	0.0008	0.0008	0.0008	0.0008	0.0007	0.0007
3.2	0.0007	0.0007	0.0006	0.0006	0.0006	0.0006	0.0006	0.0005	0.0005	0.0005
3.3	0.0005	0.0005	0.0005	0.0004	0.0004	0.0004	0.0004	0.0004	0.0004	0.0003
3.4	0.0003	0.0003	0.0003	0.0003	0.0003	0.0003	0.0003	0.0003	0.0003	0.0002

TABLE VI

The *t* Distribution

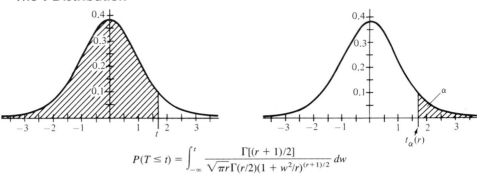

$$P(T \le t) = \int_{-\infty}^{t} \frac{\Gamma[(r+1)/2]}{\sqrt{\pi r}\,\Gamma(r/2)(1 + w^2/r)^{(r+1)/2}}\, dw$$

$$[P(T \le -t) = 1 - P(T \le t)]$$

				$P(T \le t)$			
	0.60	0.75	0.90	0.95	0.975	0.99	0.995
r	$t_{0.40}(r)$	$t_{0.25}(r)$	$t_{0.10}(r)$	$t_{0.05}(r)$	$t_{0.025}(r)$	$t_{0.01}(r)$	$t_{0.005}(r)$
1	0.325	1.000	3.078	6.314	12.706	31.821	63.657
2	0.289	0.816	1.886	2.920	4.303	6.965	9.925
3	0.277	0.765	1.638	2.353	3.182	4.541	5.841
4	0.271	0.741	1.533	2.132	2.776	3.747	4.604
5	0.267	0.727	1.476	2.015	2.571	3.365	4.032
6	0.265	0.718	1.440	1.943	2.447	3.143	3.707
7	0.263	0.711	1.415	1.895	2.365	2.998	3.499
8	0.262	0.706	1.397	1.860	2.306	2.896	3.355
9	0.261	0.703	1.383	1.833	2.262	2.821	3.250
10	0.260	0.700	1.372	1.812	2.228	2.764	3.169
11	0.260	0.697	1.363	1.796	2.201	2.718	3.106
12	0.259	0.695	1.356	1.782	2.179	2.681	3.055
13	0.259	0.694	1.350	1.771	2.160	2.650	3.012
14	0.258	0.692	1.345	1.761	2.145	2.624	2.997
15	0.258	0.691	1.341	1.753	2.131	2.602	2.947
16	0.258	0.690	1.337	1.746	2.120	2.583	2.921
17	0.257	0.689	1.333	1.740	2.110	2.567	2.898
18	0.257	0.688	1.330	1.734	2.101	2.552	2.878
19	0.257	0.688	1.328	1.729	2.093	2.539	2.861
20	0.257	0.687	1.325	1.725	2.086	2.528	2.845
21	0.257	0.686	1.323	1.721	2.080	2.518	2.831
22	0.256	0.686	1.321	1.717	2.074	2.508	2.819
23	0.256	0.685	1.319	1.714	2.069	2.500	2.807
24	0.256	0.685	1.318	1.711	2.064	2.492	2.797
25	0.256	0.684	1.316	1.708	2.060	2.485	2.787
26	0.256	0.684	1.315	1.706	2.056	2.479	2.779
27	0.256	0.684	1.314	1.703	2.052	2.473	2.771
28	0.256	0.683	1.313	1.701	2.048	2.467	2.763
29	0.256	0.683	1.311	1.699	2.045	2.462	2.756
30	0.256	0.683	1.310	1.697	2.042	2.457	2.750
∞	0.253	0.674	1.282	1.645	1.960	2.326	2.576

This table is taken from Table III of Fisher and Yates: *Statistical Tables for Biological, Agricultural, and Medical Research,* published by Longman Group Ltd., London (previously published by Oliver and Boyd, Edinburgh), by permission of the authors and publishers.

TABLE VII
The F Distribution

$$P(F \le f) = \int_0^f \frac{\Gamma[(r_1 + r_2)/2](r_1/r_2)^{r_1/2}w^{r_1/2-1}}{\Gamma(r_1/2)\Gamma(r_2/2)(1 + r_1w/r_2)^{(r_1+r_2)/2}}\, dw$$

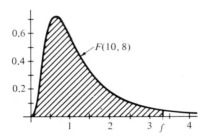

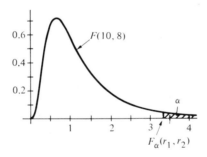

TABLE VII (continued)

$$P(F \le f) = \int_0^f \frac{\Gamma[(r_1 + r_2)/2](r_1/r_2)^{r_1/2}w^{r_1/2-1}}{\Gamma(r_1/2)\Gamma(r_2/2)(1 + r_1w/r_2)^{(r_1+r_2)/2}}\,dw$$

α	$P(F \le f)$	Den. d.f. r_2	Numerator Degrees of Freedom, r_1									
			1	2	3	4	5	6	7	8	9	10
0.05	0.95	1	161.4	199.5	215.7	224.6	230.2	234.0	236.8	238.9	240.5	241.9
0.025	0.975		647.79	799.50	864.16	899.58	921.85	937.11	948.22	956.66	963.28	968.63
0.01	0.99		4052	4999.5	5403	5625	5764	5859	5928	5981	6022	6056
0.05	0.95	2	18.51	19.00	19.16	19.25	19.30	19.33	19.35	19.37	19.38	19.40
0.025	0.975		38.51	39.00	39.17	39.25	39.30	39.33	39.36	39.37	39.39	39.40
0.01	0.99		98.50	99.00	99.17	99.25	99.30	99.33	99.36	99.37	99.39	99.40
0.05	0.95	3	10.13	9.55	9.28	9.12	9.01	8.94	8.89	8.85	8.81	8.79
0.025	0.975		17.44	16.04	15.44	15.10	14.88	14.73	14.62	14.54	14.47	14.42
0.01	0.99		34.12	30.82	29.46	28.71	28.24	27.91	27.67	27.49	27.35	27.23
0.05	0.95	4	7.71	6.94	6.59	6.39	6.26	6.16	6.09	6.04	6.00	5.96
0.025	0.975		12.22	10.65	9.98	9.60	9.36	9.20	9.07	8.98	8.90	8.84
0.01	0.99		21.20	18.00	16.69	15.98	15.52	15.21	14.98	14.80	14.66	14.55
0.05	0.95	5	6.61	5.79	5.41	5.19	5.05	4.95	4.88	4.82	4.77	4.74
0.025	0.975		10.01	8.43	7.76	7.39	7.15	6.98	6.85	6.76	6.68	6.62
0.01	0.99		16.26	13.27	12.06	11.39	10.97	10.67	10.46	10.29	10.16	10.05
0.05	0.95	6	5.99	5.14	4.76	4.53	4.39	4.28	4.21	4.15	4.10	4.06
0.025	0.975		8.81	7.26	6.60	6.23	5.99	5.82	5.70	5.60	5.52	5.46
0.01	0.99		13.75	10.92	9.78	9.15	8.75	8.47	8.26	8.10	7.98	7.87
0.05	0.95	7	5.59	4.74	4.35	4.12	3.97	3.87	3.79	3.73	3.68	3.64
0.025	0.975		8.07	6.54	5.89	5.52	5.29	5.12	4.99	4.90	4.82	4.76
0.01	0.99		12.25	9.55	8.45	7.85	7.46	7.19	6.99	6.84	6.72	6.62
0.05	0.95	8	5.32	4.46	4.07	3.84	3.69	3.58	3.50	3.44	3.39	3.35
0.025	0.975		7.57	6.06	5.42	5.05	4.82	4.65	4.53	4.43	4.36	4.30
0.01	0.99		11.26	8.65	7.59	7.01	6.63	6.37	6.18	6.03	5.91	5.81
0.05	0.95	9	5.12	4.26	3.86	3.63	3.48	3.37	3.29	3.23	3.18	3.14
0.025	0.975		7.21	5.71	5.08	4.72	4.48	4.32	4.20	4.10	4.03	3.96
0.01	0.99		10.56	8.02	6.99	6.42	6.06	5.80	5.61	5.47	5.35	5.26
0.05	0.95	10	4.96	4.10	3.71	3.48	3.33	3.22	3.14	3.07	3.02	2.98
0.025	0.975		6.94	5.46	4.83	4.47	4.24	4.07	3.95	3.85	3.78	3.72
0.01	0.99		10.04	7.56	6.55	5.99	5.64	5.39	5.20	5.06	4.94	4.85

df	q	α										
12	0.95	0.05	4.75	3.89	3.49	3.26	3.11	3.00	2.91	2.85	2.80	2.75
	0.975	0.025	6.55	5.10	4.47	4.12	3.89	3.73	3.61	3.51	3.44	3.37
	0.99	0.01	9.33	6.93	5.95	5.41	5.06	4.82	4.64	4.50	4.39	4.30
15	0.95	0.05	4.54	3.68	3.29	3.06	2.90	2.79	2.71	2.64	2.59	2.54
	0.975	0.025	6.20	4.77	4.15	3.80	3.58	3.41	3.29	3.20	3.12	3.06
	0.99	0.01	8.68	6.36	5.42	4.89	4.56	4.32	4.14	4.00	3.89	3.80
20	0.95	0.05	4.35	3.49	3.10	2.87	2.71	2.60	2.51	2.45	2.39	2.35
	0.975	0.025	5.87	4.46	3.86	3.51	3.29	3.13	3.01	2.91	2.84	2.77
	0.99	0.01	8.10	5.85	4.94	4.43	4.10	3.87	3.70	3.56	3.46	3.37
24	0.95	0.05	4.26	3.40	3.01	2.78	2.62	2.51	2.42	2.36	2.30	2.25
	0.975	0.025	5.72	4.32	3.72	3.38	3.15	2.99	2.87	2.78	2.70	2.64
	0.99	0.01	7.82	5.61	4.72	4.22	3.90	3.67	3.50	3.36	3.26	3.17
30	0.95	0.05	4.17	3.32	2.92	2.69	2.53	2.42	2.33	2.27	2.21	2.16
	0.975	0.025	5.57	4.18	3.59	3.25	3.03	2.87	2.75	2.65	2.57	2.51
	0.99	0.01	7.56	5.39	4.51	4.02	3.70	3.47	3.30	3.17	3.07	2.98
40	0.95	0.05	4.08	3.23	2.84	2.61	2.45	2.34	2.25	2.18	2.12	2.08
	0.975	0.025	5.42	4.05	3.46	3.13	2.90	2.74	2.62	2.53	2.45	2.39
	0.99	0.01	7.31	5.18	4.31	3.83	3.51	3.29	3.12	2.99	2.89	2.80
60	0.95	0.05	4.00	3.15	2.76	2.53	2.37	2.25	2.17	2.10	2.04	1.99
	0.975	0.025	5.29	3.93	3.34	3.01	2.79	2.63	2.51	2.41	2.33	2.27
	0.99	0.01	7.08	4.98	4.13	3.65	3.34	3.12	2.95	2.82	2.72	2.63
120	0.95	0.05	3.92	3.07	2.68	2.45	2.29	2.17	2.09	2.02	1.96	1.91
	0.975	0.025	5.15	3.80	3.23	2.89	2.67	2.52	2.39	2.30	2.22	2.16
	0.99	0.01	6.85	4.79	3.95	3.48	3.17	2.96	2.79	2.66	2.56	2.47
∞	0.95	0.05	3.84	3.00	2.60	2.37	2.21	2.10	2.01	1.94	1.88	1.83
	0.975	0.025	5.02	3.69	3.12	2.79	2.57	2.41	2.29	2.19	2.11	2.05
	0.99	0.01	6.63	4.61	3.78	3.32	3.02	2.80	2.64	2.51	2.41	2.32

TABLE VII (continued)

$$P(F \le f) = \int_0^f \frac{\Gamma[(r_1+r_2)/2](r_1/r_2)^{r_1/2}\, w^{r_1/2-1}}{\Gamma(r_1/2)\Gamma(r_2/2)(1 + r_1 w/r_2)^{(r_1+r_2)/2}}\, dw$$

α	$P(F \le f)$	Den. d.f. r_2	12	15	20	24	30	40	60	120	∞
							Numerator Degrees of Freedom, r_1				
0.05	0.95	1	243.9	245.9	248.0	249.1	250.1	251.1	252.2	253.3	254.3
0.025	0.975		976.71	984.87	993.10	997.25	1001.4	1005.6	1009.8	1014.0	1018.3
0.01	0.99		6106	6157	6209	6235	6261	6287	6313	6339	6366
0.05	0.95	2	19.41	19.43	19.45	19.45	19.46	19.47	19.48	19.49	19.50
0.025	0.975		39.42	39.43	39.45	39.46	39.47	39.47	39.48	39.49	39.50
0.01	0.99		99.42	99.43	99.45	99.46	99.47	99.47	99.48	99.49	99.50
0.05	0.95	3	8.74	8.70	8.66	8.64	8.62	8.59	8.57	8.55	8.53
0.025	0.975		14.34	14.25	14.17	14.12	14.08	14.04	13.99	13.95	13.90
0.01	0.99		27.05	26.87	26.69	26.60	26.50	26.41	26.32	26.22	26.13
0.05	0.95	4	5.91	5.86	5.80	5.77	5.75	5.72	5.69	5.66	5.63
0.025	0.975		8.75	8.66	8.56	8.51	8.46	8.41	8.36	8.31	8.26
0.01	0.99		14.37	14.20	14.02	13.93	13.84	13.75	13.65	13.56	13.46
0.05	0.95	5	4.68	4.62	4.56	4.53	4.50	4.46	4.43	4.40	4.36
0.025	0.975		6.52	6.43	6.33	6.28	6.23	6.18	6.12	6.07	6.02
0.01	0.99		9.89	9.72	9.55	9.47	9.38	9.29	9.20	9.11	9.02
0.05	0.95	6	4.00	3.94	3.87	3.84	3.81	3.77	3.74	3.70	3.67
0.025	0.975		5.37	5.27	5.17	5.12	5.07	5.01	4.96	4.90	4.85
0.01	0.99		7.72	7.56	7.40	7.31	7.23	7.14	7.06	6.97	6.88
0.05	0.95	7	3.57	3.51	3.44	3.41	3.38	3.34	3.30	3.27	3.23
0.025	0.975		4.67	4.57	4.47	4.42	4.36	4.31	4.25	4.20	4.14
0.01	0.99		6.47	6.31	6.16	6.07	5.99	5.91	5.82	5.74	5.65
0.05	0.95	8	3.28	3.22	3.15	3.12	3.08	3.04	3.01	2.97	2.93
0.025	0.975		4.20	4.10	4.00	3.95	3.89	3.84	3.78	3.73	3.67
0.01	0.99		5.67	5.52	5.36	5.28	5.20	5.12	5.03	4.95	4.86
0.05	0.95	9	3.07	3.01	2.94	2.90	2.86	2.83	2.79	2.75	2.71
0.025	0.975		3.87	3.77	3.67	3.61	3.56	3.51	3.45	3.39	3.33
0.01	0.99		5.11	4.96	4.81	4.73	4.65	4.57	4.48	4.40	4.31

df											
10	0.95	0.05	2.54	2.58	2.62	2.66	2.70	2.74	2.77	2.85	2.91
	0.975	0.025	3.08	3.14	3.20	3.26	3.31	3.37	3.42	3.52	3.62
	0.99	0.01	3.91	4.00	4.08	4.17	4.25	4.33	4.41	4.56	4.71
12	0.95	0.05	2.30	2.34	2.38	2.43	2.47	2.51	2.54	2.62	2.69
	0.975	0.025	2.72	2.79	2.85	2.91	2.96	3.02	3.07	3.18	3.28
	0.99	0.01	3.36	3.45	3.54	3.62	3.70	3.78	3.86	4.01	4.16
15	0.95	0.05	2.07	2.11	2.16	2.20	2.25	2.29	2.33	2.40	2.48
	0.975	0.025	2.40	2.46	2.52	2.59	2.64	2.70	2.76	2.86	2.96
	0.99	0.01	2.87	2.96	3.05	3.13	3.21	3.29	3.37	3.52	3.67
20	0.95	0.05	1.84	1.90	1.95	1.99	2.04	2.08	2.12	2.20	2.28
	0.975	0.025	2.09	2.16	2.22	2.29	2.35	2.41	2.46	2.57	2.68
	0.99	0.01	2.42	2.52	2.61	2.69	2.78	2.86	2.94	3.09	3.23
24	0.95	0.05	1.73	1.79	1.84	1.89	1.94	1.98	2.03	2.11	2.18
	0.975	0.025	1.94	2.01	2.08	2.15	2.21	2.27	2.33	2.44	2.54
	0.99	0.01	2.21	2.31	2.40	2.49	2.58	2.66	2.74	2.89	3.03
30	0.95	0.05	1.62	1.68	1.74	1.79	1.84	1.89	1.93	2.01	2.09
	0.975	0.025	1.79	1.87	1.94	2.01	2.07	2.14	2.20	2.31	2.41
	0.99	0.01	2.01	2.11	2.21	2.30	2.39	2.47	2.55	2.70	2.84
40	0.95	0.05	1.51	1.58	1.64	1.69	1.74	1.79	1.84	1.92	2.00
	0.975	0.025	1.64	1.72	1.80	1.88	1.94	2.01	2.07	2.18	2.29
	0.99	0.01	1.80	1.92	2.02	2.11	2.20	2.29	2.37	2.52	2.66
60	0.95	0.05	1.39	1.47	1.53	1.59	1.65	1.70	1.75	1.84	1.92
	0.975	0.025	1.48	1.58	1.67	1.74	1.82	1.88	1.94	2.06	2.17
	0.99	0.01	1.60	1.73	1.84	1.94	2.03	2.12	2.20	2.35	2.50
120	0.95	0.05	1.25	1.35	1.43	1.50	1.55	1.61	1.66	1.75	1.83
	0.975	0.025	1.31	1.43	1.53	1.61	1.69	1.76	1.82	1.95	2.05
	0.99	0.01	1.38	1.53	1.66	1.76	1.86	1.95	2.03	2.19	2.34
∞	0.95	0.05	1.00	1.22	1.32	1.39	1.46	1.52	1.57	1.67	1.75
	0.975	0.025	1.00	1.27	1.39	1.48	1.57	1.64	1.71	1.83	1.94
	0.99	0.01	1.00	1.32	1.47	1.59	1.70	1.79	1.88	2.04	2.18

TABLE VIII
Kolmogorov–Smirnov Acceptance Limits

$$D_n = \sup_x [|F_n(x) - F_0(x)|]$$
$$\alpha = 1 - P(D_n \leq d)$$

	α			
n	0.20	0.10	0.05	0.01
1	0.90	0.95	0.98	0.99
2	0.68	0.78	0.84	0.93
3	0.56	0.64	0.71	0.83
4	0.49	0.56	0.62	0.73
5	0.45	0.51	0.56	0.67
6	0.41	0.47	0.52	0.62
7	0.38	0.44	0.49	0.58
8	0.36	0.41	0.46	0.54
9	0.34	0.39	0.43	0.51
10	0.32	0.37	0.41	0.49
11	0.31	0.35	0.39	0.47
12	0.30	0.34	0.38	0.45
13	0.28	0.32	0.36	0.43
14	0.27	0.31	0.35	0.42
15	0.27	0.30	0.34	0.40
16	0.26	0.30	0.33	0.39
17	0.25	0.29	0.32	0.38
18	0.24	0.28	0.31	0.37
19	0.24	0.27	0.30	0.36
20	0.23	0.26	0.29	0.35
25	0.21	0.24	0.26	0.32
30	0.19	0.22	0.24	0.29
35	0.18	0.21	0.23	0.27
40	0.17	0.19	0.21	0.25
45	0.16	0.18	0.20	0.24
Large n	$\dfrac{1.07}{\sqrt{n}}$	$\dfrac{1.22}{\sqrt{n}}$	$\dfrac{1.36}{\sqrt{n}}$	$\dfrac{1.63}{\sqrt{n}}$

TABLE IX
Random Numbers on the Interval (0, 1)

3407	1440	6960	8675	5649	5793	1514
5044	9859	4658	7779	7986	0520	6697
0045	4999	4930	7408	7551	3124	0527
7536	1448	7843	4801	3147	3071	4749
7653	4231	1233	4409	0609	6448	2900
6157	1144	4779	0951	3757	9562	2354
6593	8668	4871	0946	3155	3941	9662
3187	7434	0315	4418	1569	1101	0043
4780	1071	6814	2733	7968	8541	1003
9414	6170	2581	1398	2429	4763	9192
1948	2360	7244	9682	5418	0596	4971
1843	0914	9705	7861	6861	7865	7293
4944	8903	0460	0188	0530	7790	9118
3882	3195	8287	3298	9532	9066	8225
6596	9009	2055	4081	4842	7852	5915
4793	2503	2906	6807	2028	1075	7175
2112	0232	5334	1443	7306	6418	9639
0743	1083	8071	9779	5973	1141	4393
8856	5352	3384	8891	9189	1680	3192
8027	4975	2346	5786	0693	5615	2047
3134	1688	4071	3766	0570	2142	3492
0633	9002	1305	2256	5956	9256	8979
8771	6069	1598	4275	6017	5946	8189
2672	1304	2186	8279	2430	4896	3698
3136	1916	8886	8617	9312	5070	2720
6490	7491	6562	5355	3794	3555	7510
8628	0501	4618	3364	6709	1289	0543
9270	0504	5018	7013	4423	2147	4089
5723	3807	4997	4699	2231	3193	8130
6228	8874	7271	2621	5746	6333	0345
7645	3379	8376	3030	0351	8290	3640
6842	5836	6203	6171	2698	4086	5469
6126	7792	9337	7773	7286	4236	1788
4956	0215	3468	8038	6144	9753	3131
1327	4736	6229	8965	7215	6458	3937
9188	1516	5279	5433	2254	5768	8718
0271	9627	9442	9217	4656	7603	8826
2127	1847	1331	5122	8332	8195	3322
2102	9201	2911	7318	7670	6079	2676
1706	6011	5280	5552	5180	4630	4747
7501	7635	2301	0889	6955	8113	4364
5705	1900	7144	8707	9065	8163	9846
3234	2599	3295	9160	8441	0085	9317
5641	4935	7971	8917	1978	5649	5799
2127	1868	3664	9376	1984	6315	8396

TABLE X

Divisors for the Confidence Interval for σ of Minimum Length

Let $X_1, X_2, \ldots, X_n$ be a random sample of size n from $N(\mu, \sigma^2)$; *let* $\bar{x} = (1/n) \sum_{i=1}^n x_i$ and $s^2 = (1/[n-1]) \sum_{i=1}^n (x_i - \bar{x})^2$. A $(100\gamma)\%$ confidence interval for σ which has minimum length is given by $[\sqrt{(n-1)s^2/b}, \sqrt{(n-1)s^2/a}]$. In the table, a is the upper number, b the lower number; $r = n - 1$, the number of degrees of freedom; and $\gamma = 1 - \alpha$, the confidence level.

	$\gamma = 1 - \alpha$				$\gamma = 1 - \alpha$		
r	0.90	0.95	0.99	r	0.90	0.95	0.99
2	0.206	0.101	0.020	16	8.774	7.604	5.649
	12.521	15.111	20.865		29.233	32.072	38.097
3	0.565	0.345	0.114	17	9.505	8.282	6.226
	13.153	15.589	20.973		30.480	33.362	39.469
4	1.020	0.692	0.294	18	10.242	8.969	6.814
	14.180	16.573	21.838		31.721	34.647	40.835
5	1.535	1.109	0.546	19	10.986	9.663	7.413
	15.350	17.743	22.985		32.959	35.927	42.195
6	2.093	1.578	0.837	20	11.736	10.365	8.021
	16.581	18.996	24.262		34.192	37.202	43.550
7	2.683	2.085	1.214	21	12.492	11.073	8.638
	17.839	20.286	25.602		35.420	38.472	44.899
8	3.298	2.623	1.611	22	13.253	11.788	9.264
	19.110	21.595	26.975		36.646	39.738	46.243
9	3.934	3.187	2.039	23	14.019	12.509	9.898
	20.385	22.912	28.364		37.867	41.000	47.586
10	4.588	3.773	2.496	24	14.790	13.236	10.539
	21.660	24.230	29.760		39.084	42.257	48.914
11	5.257	4.377	2.976	25	15.565	13.968	11.186
	22.933	25.548	31.158		40.299	43.510	50.243
12	5.940	4.997	3.477	26	16.344	14.704	11.841
	24.202	26.862	32.554		41.509	44.760	51.566
13	6.634	5.631	3.997	27	17.127	15.446	12.501
	25.467	28.172	33.947		42.717	46.006	52.886
14	7.338	6.278	4.533	28	17.914	16.192	13.168
	26.727	29.477	35.336		43.922	47.248	54.200
15	8.052	6.936	5.084	29	18.705	16.942	13.840
	27.982	30.777	36.719		45.123	48.487	55.511

TABLE XI

Distribution Function of the Correlation Coefficient R

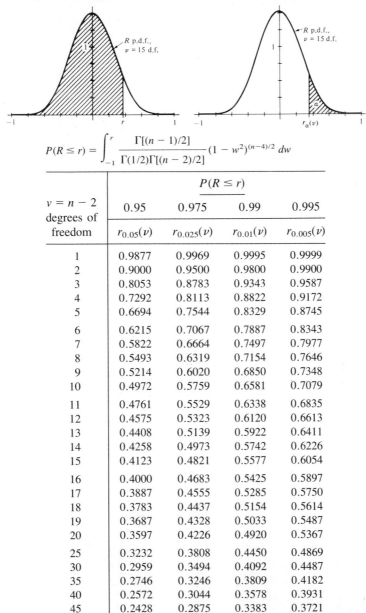

$$P(R \le r) = \int_{-1}^{r} \frac{\Gamma[(n-1)/2]}{\Gamma(1/2)\Gamma[(n-2)/2]} (1 - w^2)^{(n-4)/2} \, dw$$

$v = n - 2$ degrees of freedom	$P(R \le r)$			
	0.95	0.975	0.99	0.995
	$r_{0.05}(v)$	$r_{0.025}(v)$	$r_{0.01}(v)$	$r_{0.005}(v)$
1	0.9877	0.9969	0.9995	0.9999
2	0.9000	0.9500	0.9800	0.9900
3	0.8053	0.8783	0.9343	0.9587
4	0.7292	0.8113	0.8822	0.9172
5	0.6694	0.7544	0.8329	0.8745
6	0.6215	0.7067	0.7887	0.8343
7	0.5822	0.6664	0.7497	0.7977
8	0.5493	0.6319	0.7154	0.7646
9	0.5214	0.6020	0.6850	0.7348
10	0.4972	0.5759	0.6581	0.7079
11	0.4761	0.5529	0.6338	0.6835
12	0.4575	0.5323	0.6120	0.6613
13	0.4408	0.5139	0.5922	0.6411
14	0.4258	0.4973	0.5742	0.6226
15	0.4123	0.4821	0.5577	0.6054
16	0.4000	0.4683	0.5425	0.5897
17	0.3887	0.4555	0.5285	0.5750
18	0.3783	0.4437	0.5154	0.5614
19	0.3687	0.4328	0.5033	0.5487
20	0.3597	0.4226	0.4920	0.5367
25	0.3232	0.3808	0.4450	0.4869
30	0.2959	0.3494	0.4092	0.4487
35	0.2746	0.3246	0.3809	0.4182
40	0.2572	0.3044	0.3578	0.3931
45	0.2428	0.2875	0.3383	0.3721
50	0.2306	0.2732	0.3218	0.3541
60	0.2108	0.2500	0.2948	0.3248
70	0.1954	0.2318	0.2736	0.3017
80	0.1829	0.2172	0.2565	0.2829
90	0.1725	0.2049	0.2422	0.2673
100	0.1638	0.1946	0.2300	0.2540

TABLE XII
Discrete Distributions

Probability Distribution and Parameter Values	Probability Density Function	Moment-Generating Function	Mean $E(X)$	Variance $Var(X)$	Examples
Bernoulli $0 \le p \le 1$ $q = 1 - p$	$p^x q^{1-x}, x = 0, 1$	$q + pe^t$	p	pq	An experiment with two possible outcomes, say success and failure, p the probability of success
Binomial $n = 1, 2, 3, \ldots$ $0 \le p \le 1$ $q = 1 - p$	$\binom{n}{x} p^x q^{n-x},$ $x = 0, 1, \ldots, n$	$(q + pe^t)^n$	np	npq	The number of successes in a sequence of n Bernoulli trials with p the probability of success
Geometric $0 \le p \le 1$ $q = 1 - p$	$q^{x-1} p,$ $x = 1, 2, \ldots$	$pe^t/(1 - qe^t)$	$1/p$	q/p^2	The number of trials to obtain the first success in a sequence of Bernoulli trials
Negative binomial $r = 1, 2, 3, \ldots$ $0 \le p \le 1$ $q = 1 - p$	$\binom{x-1}{r-1} p^r q^{x-r},$ $x = r, r+1, \ldots$	$(pe^t)^r/(1 - qe^t)^r$	r/p	rq/p^2	The number of trials to obtain the rth success in a sequence of Bernoulli trials
Hypergeometric $x \le r$ $x \le n_1$ $r - x \le n_2$ $n = n_1 + n_2$	$\dfrac{\binom{n_1}{x}\binom{n_2}{r-x}}{\binom{n}{r}}$		$r\left(\dfrac{n_1}{n}\right)$	$r\left(\dfrac{n_1}{n}\right)\left(\dfrac{n_2}{n}\right)\left(\dfrac{n-r}{n-1}\right)$	Selecting r objects at random without replacement from a set composed of two types of objects
Poisson $0 \le \lambda$	$\dfrac{\lambda^x e^{-\lambda}}{x!}$ $x = 0, 1, \ldots$	$e^{\lambda(e^t - 1)}$	λ	λ	Number of events occurring in a unit interval when events are occurring randomly at a mean rate of λ per unit interval

Table XIII
Continuous Distributions

Probability Distribution and Parameter Values	Probability Density Function	Moment-Generating Function	Mean $E(X)$	Variance $\mathrm{Var}(X)$	Examples
Uniform $-\infty < a < b < \infty$	$\dfrac{1}{b-a},\ a \leq x \leq b$	$\dfrac{e^{tb}-e^{ta}}{t(b-a)},\ t \neq 0$ $1,\ t = 0$	$\dfrac{a+b}{2}$	$\dfrac{(b-a)^2}{12}$	A point is selected at random from the interval $[a, b]$
Exponential $0 < \theta$	$\dfrac{1}{\theta}e^{-x/\theta},\ 0 \leq x < \infty$	$\dfrac{1}{1-\theta t},\ t < \dfrac{1}{\theta}$	θ	θ^2	Waiting time until first arrival when observing a Poisson process with a mean rate of arrivals equal to $\lambda = 1/\theta$
Gamma $0 < \alpha$ $0 < \theta$	$\dfrac{x^{\alpha-1}e^{-x/\theta}}{\Gamma(\alpha)\theta^{\alpha}},\ 0 \leq x < \infty$	$\dfrac{1}{(1-\theta t)^{\alpha}},\ t < \dfrac{1}{\theta}$	$\alpha\theta$	$\alpha\theta^2$	Waiting time until αth arrival when observing a Poisson process with a mean rate of arrivals equal to $\lambda = 1/\theta$
Chi-square $r = 1, 2, \ldots$	$\dfrac{x^{r/2-1}e^{-x/2}}{\Gamma(r/2)2^{r/2}},\ 0 \leq x < \infty$	$\dfrac{1}{(1-2t)^{r/2}},\ t < \dfrac{1}{2}$	r	$2r$	Gamma distribution with $\theta = 2$, $\alpha = r/2$; sum of squares of r independent $N(0, 1)$ random variables
Normal $-\infty < \mu < \infty$ $0 < \sigma$	$\dfrac{e^{-(x-\mu)^2/2\sigma^2}}{\sigma\sqrt{2\pi}},\ -\infty < x < \infty$	$\exp[\mu t + \sigma^2 t^2/2]$	μ	σ^2	Error in measurements; heights and weights of children
Beta $0 < \alpha$ $0 < \beta$	$\dfrac{\Gamma(\alpha+\beta)}{\Gamma(\alpha)\Gamma(\beta)}x^{\alpha-1}(1-x)^{\beta-1},$ $0 < x < 1$		$\dfrac{\alpha}{\alpha+\beta}$	$\dfrac{\alpha\beta}{(\alpha+\beta+1)(\alpha+\beta)^2}$	$X = X_1/(X_1 + X_2)$, where X_1 and X_2 have independent gamma distributions with same θ

TABLE XIV
Confidence Intervals

Distribution	Confidence Interval For	Variable Used: W	$\gamma = 1 - \alpha$ Confidence Interval Endpoints	Comments
$N(\mu, \sigma^2)$ σ^2 known	μ	$\dfrac{\bar{X} - \mu}{\sigma/\sqrt{n}}$	$\bar{x} \pm z_{\alpha/2}\dfrac{\sigma}{\sqrt{n}}$	W is $N(0, 1)$; $P(W \geq z_{\alpha/2}) = \alpha/2$
$N(\mu, \sigma^2)$ σ^2 unknown	μ	$\dfrac{\bar{X} - \mu}{S/\sqrt{n}}$	$\bar{x} \pm t_{\alpha/2}(n-1)\dfrac{s}{\sqrt{n}}$	W has a t distribution with $r = n - 1$ degrees of freedom; $P[W \geq t_{\alpha/2}(n-1)] = \alpha/2$
Any distribution, known variance, σ^2	μ	$\dfrac{\bar{X} - \mu}{\sigma/\sqrt{n}}$	$\bar{x} \pm z_{\alpha/2}\dfrac{\sigma}{\sqrt{n}}$	W has an approximate $N(0, 1)$ distribution for n sufficiently large
$N(\mu_X, \sigma_X^2)$ $N(\mu_Y, \sigma_Y^2)$ σ_X^2, σ_Y^2 known	$\mu_X - \mu_Y$	$\dfrac{\bar{X} - \bar{Y} - (\mu_X - \mu_Y)}{\sqrt{\sigma_X^2/n + \sigma_Y^2/m}}$	$\bar{x} - \bar{y} \pm z_{\alpha/2}\sqrt{\dfrac{\sigma_X^2}{n} + \dfrac{\sigma_Y^2}{m}}$	W is $N(0, 1)$; if variances are unknown and sample sizes are large, replace σ_X^2 with s_X^2 and σ_Y^2 with s_Y^2 for an approximate $(100\gamma)\%$ confidence interval
$N(\mu_X, \sigma_X^2)$ $N(\mu_Y, \sigma_Y^2)$ $\sigma_X^2 = \sigma_Y^2$, unknown	$\mu_X - \mu_Y$	$\dfrac{\bar{X} - \bar{Y} - (\mu_X - \mu_Y)}{\sqrt{\dfrac{(n-1)S_X^2 + (m-1)S_Y^2}{n+m-2}\left(\dfrac{1}{n} + \dfrac{1}{m}\right)}}$	$\bar{x} - \bar{y}$ $\pm t_{\alpha/2}(n+m-2)\sqrt{\dfrac{(n-1)s_X^2 + (m-1)s_Y^2}{n+m-2}\left(\dfrac{1}{n} + \dfrac{1}{m}\right)}$	W has a t distribution with $r = n + m - 2$ degrees of freedom

$N(\mu, \sigma^2)$ μ unknown	σ^2	$\dfrac{(n-1)S^2}{\sigma^2} = \dfrac{\sum_1^n (X_i - \bar{X})^2}{\sigma^2}$	$\dfrac{(n-1)s^2}{\chi^2_{\alpha/2}(n-1)}, \dfrac{(n-1)s^2}{\chi^2_{1-\alpha/2}(n-1)}$	W is $\chi^2(n-1)$, $P[W \le \chi^2_{1-\alpha/2}(n-1)] = \alpha/2$, $P[W \ge \chi^2_{\alpha/2}(n-1)] = \alpha/2$

$N(\mu, \sigma^2)$
μ unknown
σ^2
$$\frac{(n-1)S^2}{\sigma^2} = \frac{\sum_1^n (X_i - \bar{X})^2}{\sigma^2}$$
$$\frac{(n-1)s^2}{\chi^2_{\alpha/2}(n-1)}, \frac{(n-1)s^2}{\chi^2_{1-\alpha/2}(n-1)}$$
W is $\chi^2(n-1)$,
$P[W \le \chi^2_{1-\alpha/2}(n-1)] = \alpha/2$
$P[W \ge \chi^2_{\alpha/2}(n-1)] = \alpha/2$

$N(\mu, \sigma^2)$
μ unknown
σ
$$\frac{(n-1)S^2}{\sigma^2}$$
$$\sqrt{\frac{(n-1)s^2}{b}}, \sqrt{\frac{(n-1)s^2}{a}}$$
W is $\chi^2(n-1)$. Select a and b from Table X in the Appendix for interval of minimum length

$N(\mu_X, \sigma_X^2)$
$N(\mu_Y, \sigma_Y^2)$
μ_X, μ_Y unknown
$$\dfrac{\sigma_X^2}{\sigma_Y^2}$$
$$\frac{S_Y^2/\sigma_Y^2}{S_X^2/\sigma_X^2}$$
$$\frac{s_x^2/s_y^2}{F_{\alpha/2}(n-1, m-1)}, \; F_{\alpha/2}(m-1, n-1)\frac{s_x^2}{s_y^2}$$
W has an F distribution with $m-1$ and $n-1$ degrees of freedom

$b(n, p)$
p
$$\frac{Y/n - p}{\sqrt{\left(\dfrac{Y}{n}\right)\left(1 - \dfrac{Y}{n}\right)\Big/ n}}$$
$$y/n \pm z_{\alpha/2}\sqrt{\left(\frac{y}{n}\right)\left(1 - \frac{y}{n}\right)\Big/ n}$$
W is approximately $N(0, 1)$ for n sufficiently large

$b(n_1, p_1)$
$b(n_2, p_2)$
$p_1 - p_2$
$$\frac{(Y_1/n_1) - (Y_2/n_2) - (p_1 - p_2)}{\sqrt{\dfrac{(Y_1/n_1)(1 - Y_1/n_1)}{n_1} + \dfrac{(Y_2/n_2)(1 - Y_2/n_2)}{n_2}}}$$
$$\frac{y_1}{n_1} - \frac{y_2}{n_2} \pm z_{\alpha/2}\sqrt{\frac{(y_1/n_1)(1 - y_1/n_1)}{n_1} + \frac{(y_2/n_2)(1 - y_2/n_2)}{n_2}}$$
W is approximately $N(0, 1)$ when n_1 and n_2 are sufficiently large

Answers to Odd-Numbered Exercises

CHAPTER 1

1.1-1 (a) $S = \{1, 2, 3, \ldots, 36\}$;

(b) $S = \{w: 0 < w \le 5\}$, where w is the weight in ounces;

(c) $S = \{\text{HHH, HHT, HTH, THH, HTT, THT, TTH, TTT}\}$;

(d) $S = \{(s, w): s = M \text{ or } F, 10 \le w \le 17\}$, where w is length in centimeters.

1.1-3 (a) $S = \{\text{H, T}\}$, $p = \frac{1}{2}$;

(b) $S = \{1, 2, 3, 4, 5, 6\}$, $p = \frac{4}{6}$;

(c) $S = \{\text{listing of 52 cards}\}$, $p = \frac{13}{52} = \frac{1}{4}$;

(d) $S = \{(x, y): 0 \le x \le 1, 0 \le y \le 1\}$, $p = \frac{9}{32}$.

1.1-5 (a) $S = \{(1, 2), (1, 3), (1, 4), (1, 5), (2, 3), (2, 4), (2, 5), (3, 4), (3, 5), (4, 5)\}$;

(b) (i) $\frac{1}{10}$; (ii) $\frac{5}{10}$.

1.1-7 (a) $S = \{0, 1, 2, \ldots, 9\}$;

(b) $P(A) = \frac{3}{10}$.

1.1-9 $\frac{2}{3}$.

1.1-11 $P(A_x) = \dfrac{1}{2^x}$, $x = 1, 2, 3, \ldots$.

1.2-1 (a) 1.1;
(b) 0.035;
(c) 0.1871.

1.2-3 (a) 3, 2, 7, 7, 5, 4, 2;
(b) 2.9667;
(c) 2.7920;
(d) 1.6709.

1.2-5 2.6, 5.2, 6.6, 9.0, 3.6.

1.2-7 (a) 4, 1, 2, 5, 5, 3;
(c) 36.5675, 13.321;
(d) 35.995, 13.947.

1.2-9 (a) 2, 8, 15, 13, 5, 6, 1;
(d) 8.785, 0.352;
(e) 15.5 to 17.

1.3-1 (a)

Stem	Leaf	Frequency
127	8	1
128	8	1
129	5 8 9	3
130	8	1
131	2 3 4 4 5 5 7	7
132	2 7 7 8	4
133	7 9	2
134	8	1

(b) (i) 131.3; (ii) 131.43; (iii) 7.0; (iv) 2.575; (v) 131.45; (vi) 131.47; (vii) 3.034.

1.3-3 (a)

Stem	Leaf	Frequency
11t	2	1
11f		0
11s	7	1
11•		0
12*	0 1	2
12t	2 2 3 3	4
12f	4 4 4 4 5 5 5	7
12s	6 6 7 7 7 7	6
12•	8 8 8 8 8 8 9	7
13*	0	1
13t		0
13f	5	1
13s	6	1
13•	8	1

1.3-5 (a)

Stem	Leaf	Frequency
0	5 5 5 7 7 7 7 7 9 9 9 9 9 9	14
1	1 1 1 1 3 3 3 3 5 5 7 9 9 9 9 9	16
2	1 1 3 3 5 5 5 7 7	9
3	3 3 5 5 7 9	6
4	1 3 3 5 7 9 9	7
5	1 3 7 7	4
6	1 1 3 3 3 3 5 5 5 5 7	11
7	1 1 5 5 5	5
8	3 3 5 7 9	5
9	1 3 5 5	4
10	1 9	2
11		0
12	7	1
13	1 1 5	3

Possible outliers: 177, 213, 247, 307, 413, 443, 471, 507, 515, 615, 703, 877, and 1815.

(b) (i) $y_1 = 5$; (ii) $y_{100} = 1815$; (iii) $\tilde{q}_1 = 17.2$, $\tilde{q}_3 = 86.5$; (iv) $\tilde{m} = 48$;

(d) inner fence: 190; outer fence: 293.5; suspected outliers: 213, 247; outliers: 307, 413, 443, 471, 507, 515, 615, 703, 877, 1815;

(e) 113.12.

1.3-7 (a)

Stem	Leaf	Frequency
101	7	1
102	0 0 0	3
103		0
104		0
105	8 9	2
106	1 3 3 6 6 7 7 8 8	9
107	3 7 9	3
108	8	1
109	1 3 9	3
110	0 2 2	3

(b) 101.7, 106, 106.7, 108.95, 110.2;

(c) No.

1.3-9 (a)

Stem	Leaf	Frequency
1∗	295 495	2
1•	595 795	2
2∗		0
2•	595 900	2
3∗	395	1
3•	700 735 895 995	4
4∗	100 200 265 330 400	5
4•	995	1
5∗	200	1
5•	550 700	2

(b) 12.95, 26.7125, 39.45, 43.825, 57.00;
(c) No.

1.4-1 (a)

Frequency	≤ 5 Visit Leaves	Stem	≥ 6 Visit Leaves	Frequency
1	9	4		0
1	2	5		0
0		6		0
0		7		0
1	2	8	7	1
2	6 3	9	3 7 8	3
2	8 1	10	6 8	2
5	6 4 4 4 0	11	0 6 9 9	4
1	0	12	9	1
1	4	13	1 3	2
0		14		0
0		15	3	1

(b) ≤ 5 visits: 49, 84.75, 109, 114.5, 134;
≥ 6 visits: 87, 97.75, 113, 129.5, 153.

1.4-3 (a)

Frequency	Female Leaves	Stem	Male Leaves	Frequency
2	9 8	0•		0
6	4 4 2 2 1 0	1*		0
11	9 9 9 8 8 7 6 6 5 5 5	1•		0
5	3 1 1 0 0	2*		0
1	7	2•	7 8	2
0		3*	1 2 4	3
0		3•	6 7 7 7 8 9	6
0		4*	0 1 2 2 2 2 3	7
0		4•	5 6 7 9	4
0		5*	2 2 2	3

(b) Female: 0.8, 1.3, 1.6, 1.95, 2.7; male: 2.7, 3.65, 4.1, 4.55, 5.2;
(e) The male weights are significantly higher than the female weights.

1.4-5 (a)

Frequency	$(0.5)10^{-4}$ Leaves	Stem	10^{-4} Leaves	Frequency
1	1	0*		0
4	9 9 8 5	0•	8	1
4	4 2 0 0	1*	0 1 2 3 4	5
2	8 7	1•	6 8 9	3
0		2*	0 4	2
0		2•	5 6	2

(b) 0.1, 0.8, 1.0, 1.4, 1.8; 0.8, 1.15, 1.6, 2.2, 2.6;
(e) Growth is greater using the 10^{-4} concentration of IAA.

1.4-7 (a)

Frequency	Female Leaves	Stem	Male Leaves	Frequency
1	2	11t		0
0		11f		0
1	7	11s		0
0		11•		0
2	1 0	12*		0
4	3 3 2 2	12t		0
7	5 5 5 4 4 4 4	12f		0
6	7 7 7 7 6 6	12s		0
7	9 8 8 8 8 8 8	12•		0
1	0	13*	0	1
0		13t		0
1	5	13f	4 5 5 5	4
1	6	13s	6 6 7	3
1	8	13•	8 8 8 9	4
0		14*	0 0 1 1 1	5
0		14t	2 3 3	3
0		14f	4 4 4 5	4
0		14s		0
0		14•	8	1

(b) 11.2, 12.325, 12.6, 12.8, 13.8; 13.0, 13.6, 14.0, 14.3, 14.8;

(e) Most male grackles have longer wing chords than female grackles.

1.4-9 (a) 42.36, 42.72, 42.87, 43.055, 43.32; 41.42, 42.115, 42.39, 42.47, 42.81; 41.97, 42.22, 42.57, 42.895, 43.16.

1.5-1 Data are cyclic and becoming less variable.

1.5-3 Bags weighed by scale #5 have much more variation than those weighed by scale #6.

1.5-5 Birth rates decreased until 1975, with a slight increase in 1968–1970, and now show random fluctuations.

1.5-7 Plot shows random fluctuations.

1.5-9 Plot shows random fluctuations.

1.6-1 (a) $\hat{y} = 1.401x + 0.257$;
(b) 0.686.

1.6-3 (a) $\hat{y} = -1.518x + 54.186$;
(b) -0.951;
(d) Yes.

1.6-5 (b) $\hat{y} = 0.89x + 0.02$.

1.6-7 (a) $\hat{y} = 52.809x + 30.285$;
(b) 0.899;
(c) Yes.

1.6-9 (b) $\hat{y} = 0.694x + 73.698$.

1.6-13 (a) $\hat{y} = 88.900 + 1.761x + 0.161x^2$;
(c) 109.1.

1.6-15 (b) $\hat{y} = 0.825x + 2.208$.

CHAPTER 2

2.1-1 (a) 12/52; (b) 2/52; (c) 16/52; (d) 1; (e) 0.

2.1-3 27/64.

2.1-5 (a) (i) 4/8, (ii) 2/8, (iii) 3/8, (iv) 4/8;
(b) B and C, A and D;
(c) (i) 1/8, (ii) 0, (iii) 2/8;
(d) (i) 5/8, (ii) 5/8, (iii) 5/8.

2.1-7 (a) 3/4, 1/4;
(b) 3/8, 5/8.

2.1-9 0.1.

2.1-11 (a) $3(1/3) - 3(1/3)^2 + (1/3)^3$;
(b) $P(A_1 \cup A_2 \cup A_3) = 1 - [1 - 3(1/3) + 3(1/3)^2 - (1/3)^3] = 1 - (1 - 1/3)^3$.

2.1-13 (a) $1 - (\frac{1}{2})^{10}$; (b) $1 - (\frac{1}{2})^{20}$; (c) $1 - (\frac{1}{2})^{20}$; (d) $1 - (\frac{1}{2})^{10}$; (e) $(\frac{1}{2})^{10} - (\frac{1}{2})^{20}$; (f) $(\frac{1}{2})^{20}$.

2.2-1 4096.

2.2-3 (a) 6,760,000; (b) 17,567,000.

2.2-5 (a) 24; (b) 256.

2.2-7 (a) 0.0024; (b) 0.0012; (c) 0.0006; (d) 0.0004.

2.2-9 (a) 2; (b) 8; (c) 20; (d) 40.

2.2-11 (a) 96; (b) 393,216; (e) 2,973,696.

2.2-13 (a) 362,880; (b) 84; (c) 512.

2.2-15 (a) 0.00024; (b) 0.00144; (c) 0.02113; (d) 0.04754; (e) 0.42257.

2.2-17 (a) 0.2917; (b) 0.0062.

2.2-19 (a) 80; (b) 256; (c) 20.

2.3-1 (a) 5000/1,000,000; (b) 78,515/1,000,000; (c) 73,630/995,000; (d) 4885/78,515.

2.3-3 (a) 26/58; (b) 48/58; (c) 26/38; (d) 10/22.

2.3-5 (a) 3/10; (b) 3/5; (c) 3/7.

2.3-7 (a) $S = \{(R, R), (R, W), (W, R), (W, W)\}$; (b) 1/3.

2.3-9 (a) 1/33; (b) 5/66.

2.3-11 (a) 1/220; (b) 3/110; (c) 9/55.

2.3-13 (b) 8/36; (c) 5/11; (e) $\frac{8}{36} + 2[(\frac{5}{36})(\frac{5}{11}) + (\frac{4}{36})(\frac{4}{10}) + (\frac{3}{36})(\frac{3}{9})] = 0.49293$.

2.3-15 (a) 365^r; (b) $P(365, r)$; (c) $1 - P(365, r)/365^r$; (d) 23.

2.3-17 (a) 49/153; (b) 28/49.

2.3-19 23/40.

2.4-1 (a) 0.14; (b) 0.76; (c) 0.86.

2.4-3 (a) 1/6; (b) 1/12; (c) 1/4; (d) 1/4; (e) 1/2.

2.4-5 Yes.

2.4-7 0.398.

2.4-9 (a) 9/16; (b) 9/16; (c) 10/16.

2.4-11 (a) $(\frac{1}{2})^3(\frac{1}{2})^2$; (b) $(\frac{1}{2})^3(\frac{1}{2})^2$; (c) $(\frac{1}{2})^3(\frac{1}{2})^2$; (d) $\dfrac{5!}{3!2!}(\frac{1}{2})^3(\frac{1}{2})^2$.

2.4-13 (a) 0.936; (b) 0.9993

2.4-15 (a) 5/9; (b) 3/5.

2.4-17 (b) $1 - 1/e$.

2.5-1 (a) 21/32; (b) 16/21.

2.5-3 (a) 11/24; (b) 2/11.

2.5-5 (a) 1/2; (b) 9/10.

2.5-7 (a) 3/10; (b) 7/10; (c) 1/2.

CHAPTER 3

3.1-1 (b)

$$F(x) = \begin{cases} 0, & x < 2, \\ 2/9, & 2 \le x < 3, \\ 5/9, & 3 \le x < 4, \\ 1, & 4 \le x. \end{cases}$$

3.1-3 (a) 10; (b) 1/55; (c) 3; (d) 1/30; (e) $n(n + 1)/2$.

3.1-5 (b)

$$F(x) = \begin{cases} 0, & x < 1, \\ 0.4, & 1 \le x < 2, \\ 0.7, & 2 \le x < 3, \\ 0.9, & 3 \le x < 4, \\ 1.0, & 4 \le x. \end{cases}$$

3.1-7 (a) $f(x) = \dfrac{2x - 1}{36}$, $x = 1, 2, 3, 4, 5, 6$;

(b)

$$F(x) = \begin{cases} 0, & x < 1, \\ \dfrac{[x]^2}{36}, & 1 \le x < 6, \\ 1, & 6 \le x. \end{cases}$$

3.1-9 (b)

$$F(x) = \begin{cases} 0, & x < 1, \\ 3/11, & 1 \le x < 2, \\ 5/11, & 2 \le x < 3, \\ 6/11, & 3 \le x < 4, \\ 8/11, & 4 \le x < 5, \\ 1, & 5 \le x. \end{cases}$$

3.1-11 (a)

$$F(x) = \begin{cases} 0, & x < 3, \\ 1, & 3 \le x. \end{cases}$$

(b)

$$F(x) = \begin{cases} 0, & x < 1, \\ 1/3, & 1 \le x < 2, \\ 2/3, & 2 \le x < 3, \\ 1, & 3 \le x. \end{cases}$$

(c)

$$F(x) = \begin{cases} 0, & x < 1, \\ 1/15, & 1 \le x < 2, \\ 3/15, & 2 \le x < 3, \\ 6/15, & 3 \le x < 4, \\ 10/15, & 4 \le x < 5, \\ 1, & 5 \le x. \end{cases}$$

(d)

$$F(x) = \begin{cases} 0, & x < 0, \\ 1 - (3/4)^{k+1}, & k \le x < k + 1, \end{cases}$$

where k is a nonnegative integer.

3.2-1 $5.

3.2-3 (a) 3; (b) 2; (c) $\frac{11}{3}$; (d) 3.

3.2-5 -30.33 cents.

3.2-7 $-$0.0787.

3.2-9 $E(X) = \sum_{x=1}^{n} \dfrac{1}{x}\left(\dfrac{1}{n}\right)$; 0.4567; $\dfrac{1}{100}\left[\dfrac{\ln(101) + 1 + \ln(100)}{2}\right]$.

3.2-11 (a) $0.46; (b) lose $0.54.

3.2-13 $-$0.01414.

3.3-1 1.4; 1.44; 1.2.

3.3-3 (a) 15; 50; (b) 5; 0; (c) 5/3; 5/9.

3.3-5 (a) 3; 28/11; (b) 1; 2/3.

3.3-7 (a) 16; (b) 6; (c) 16.

3.3-9 (a) (ii) 1.9844; (iii) 1.7966; (iv) 1.6635;
(b) (ii) 6.0156; (iii) 1.7966; (iv) -1.6635.

3.3-11 (a) 25/8, 55/64, $\sqrt{55}/8$;
(b) 3.16, 0.8024, 0.8958.

3.3-13 (b) 75.

3.4-1 (i) $f(x) = 0.001$, $x = 000,001, \ldots, 999$; (ii) 499.5; (iii) 83,333.25;
(iv) 288.675.

3.4-3 (a) 6.567; (b) 12.114; (c) 3.481; (e) They have approximately the same
distribution.

3.4-5 (a) $g(y) = \dfrac{1}{b - a + 1}$, $a \le y \le b$, y an integer;
(b) $\dfrac{a + b}{2}$; (c) $\sigma_Y^2 = \dfrac{(b - a + 1)^2 - 1}{12}$; $\sigma_Y = \sqrt{\sigma_Y^2}$.

3.4-7 (a) 0.400; (b) 0.902; (c) 0.6, 0.140.

3.4-9 0.416.

3.4-11 (a) $f(x) = \dfrac{\dbinom{6}{x}\dbinom{43}{6 - x}}{\dbinom{49}{6}}$, $x = 0, 1, 2, 3, 4, 5, 6$;

(b) $36/49 = 0.7347$; 0.5776, 0.7600;
(c) 2.98, 4.30, 5.61, 6.93;

(d) 0;

(e) Expected values are 1,853,292; 101,705; 1937; 7.5.

3.4-13 (a) 0.0128; (b) 0.1247; (c) 0.7427.

3.4-15 1.000, 0.967, 0.909, 0.834.

3.5-1 $f(x) = (\frac{7}{18})^x(\frac{11}{18})^{1-x}$, $x = 0, 1$; $\mu = \frac{7}{18}$; $\sigma^2 = \frac{77}{324}$.

3.5-3 (a) $(\frac{1}{5})^2(\frac{4}{5})^3 = 0.0205$; (b) $\dfrac{5!}{2!3!}(\frac{1}{5})^2(\frac{4}{5})^3 = 0.2048$.

3.5-5 (a) 0.2459; (b) 0.5501; (c) 0.2040; (d) 11.2, 3.36, 1.833.

3.5-7 (a) $b(2000, \pi/4)$; (b) 1570.796, 337.096, 18.360; (c) π.

3.5-9 (a) $b(9, 0.10)$; (b) 0.9, 0.81, 0.9; (c) (i) 0.1722, (ii) 0.2252.

3.5-11 (a) 0.8725; (b) 0.5841; (c) 0.1597.

3.5-13 0.1268.

3.5-15 (a) $\mu = 7.2$, $\sigma^2 = 1.44$, $\sigma = 1.2$;

(c) $\bar{x} = 7.32$, $s^2 = 1.4117$, $s = 1.188$.

3.5-17 (a) 0.6513; (b) 0.7941.

3.5-19

x	$b(8, 0.2)$	$n_1 = 8, n_2 = 32$	$n_1 = 16, n_2 = 64$
0	0.1678	0.1368	0.1527
1	0.3355	0.3501	0.3429
2	0.2936	0.3299	0.3104
3	0.1468	0.1466	0.1473
4	0.0459	0.0327	0.0399
5	0.0092	0.0036	0.0063
6	0.0011	0.0002	0.0006
7	0.0001	0.0000	0.0000
8	0.0000	0.0000	0.0000

3.6-1 (a) 1/8; (b) 7/8.

3.6-3 $(\frac{4}{5})^4(\frac{1}{5})$.

3.6-5 (a) 0.4604; (b) 0.5580; (c) 0.0184.

3.6-7 (a) 3000; (b) 0.3679; (c) 0.4866.

3.6-9 1/4.

3.6-11 (a) $f(x) = (1/2)^{x-1}$, $x = 2, 3, 4, \ldots$;

(b) 3, 2, 1.414;

(c) (i) 3/4, (ii) 1/8, (iii) 1/4.

3.6-13 25/3.

3.6-15 33.2187.

3.6-17 (a) 0.0670; (b) 0.4823; (c) 0.1319.

3.6-21 (a) (i) 16.667, (ii) 11.111, (iii) 3.333

(b) 0.124.

3.7-1 (a) 0.693; (b) 0.762; (c) 0.433.

3.7-3 0.540.

3.7-5 (a) 0.543; (b) 0.410.

3.7-7 0.607.

3.7-9 (a) 0.040; (b) 0.497.

3.7-11 0.938.

3.7-13 (a) 17, 47, 63, 63, 49, 28, 21, 11, 1;
(b) 3.03, 3.193, yes;
(c) These look like observations of a Poisson random variable.

3.7-15 (b) Yes.

3.7-17 (b) Yes.

3.7-19 If $n = 80$, $Ac = 3$, $Oc(0.02) \approx 0.92$, $Oc(0.08) \approx 0.12$.

3.8-1 (a) $M(t) = \frac{1}{3}e^{t} + \frac{1}{3}e^{2t} + \frac{1}{3}e^{3t}$; (b) $M(t) = e^{5t}$;
(c) $M(t) = 0.4e^{t} + 0.3e^{2t} + 0.2e^{3t} + 0.1e^{4t}$;
(d) $M(t) = 0.7 + 0.3e^{t}$.

3.8-5 $M(t) = e^{t}/(3 - 2e^{t})$, $t < \ln(3/2)$; $\mu = 3$; $\sigma^2 = 6$.

3.8-7 (a) (i) $b(5, 0.7)$; (ii) $\mu = 3.5$, $\sigma^2 = 1.05$; (iii) 0.1607;
(b) (i) Poisson, $\lambda = 4$; (ii) $\mu = 4$, $\sigma^2 = 4$; (iii) 0.220;
(c) (i) geometric, $p = 0.3$; (ii) $\mu = 10/3$, $\sigma^2 = 70/9$; (iii) 0.51;
(d) (i) Bernoulli, $p = 0.55$; (ii) $\mu = 0.55$, $\sigma^2 = 0.2475$; (iii) 0.55;
(e) (ii) $\mu = 2.1$, $\sigma^2 = 0.89$; (iii) 0.7;
(f) (i) negative binomial, $p = 0.6$, $r = 2$; (ii) $\mu = 10/3$, $\sigma^2 = 20/9$; (iii) 0.36;
(g) (i) discrete uniform on 1, 2, . . . , 10; (ii) 5.5, 8.25; (iii) 0.2.

3.8-9 $M(t) = e^{5t}$; $f(5) = 1$.

CHAPTER 4

4.1-1 (b) $F(x) = \begin{cases} 0, & x < 0, \\ x(2 - x), & 0 \le x < 1, \\ 1, & 1 \le x. \end{cases}$
(c) (i) 0.75, (ii) 0.50, (iii) 0, (iv) 0.0625.

4.1-3 (a) (i) 3; (ii) $F(x) = x^4$, $0 \le x \le 1$;
(b) (i) 3/16; (ii) $F(x) = (1/8)x^{3/2}$, $0 \le x \le 4$;
(c) (i) 1/4; (ii) $F(x) = x^{1/4}$, $0 \le x \le 1$.

4.1-5 (a) (i) 4/5; (ii) 0.0168; (ii) 0.1295;
(b) (i) 12/5; (ii) 1.0971; (iii) 1.0474;
(c) (i) 1/5; (ii) 64/225; (iii) 8/15.

4.1-7 (a) $\mu = \pi/2$; (b) $\sigma^2 = (\pi^2 - 8)/4$;

4.1-9 (a) $d = 2$; (b) $E(Y) = 2$; (c) $E(Y^2)$ is unbounded.

4.1-11 (a) $\mu = 0$, $\sigma^2 = 3/5$;
(b) $\mu = 0$, $\sigma^2 = 1/3$;
(c) $\mu = 0$, $\sigma^2 = 1/6$.

4.1-13 (a) $\mu = 2$; (b) $\sigma^2 = 2$.

4.1-15 $f(-x) = \dfrac{e^{x}}{(1 + e^{x})^2} \dfrac{e^{-2x}}{e^{-2x}} = \dfrac{e^{-x}}{(e^{-x} + 1)^2} = f(x)$.

4.1-17 (a) $\pi_{0.5} = 0$; (b) $\pi_{0.25} = -0.50$; (c) $\pi_{0.90} = 0.80$.

4.1-19 (a) $F_n(x) = \begin{cases} 0, & x < 0, \\ nx, & 0 \le x < 1/n, \\ 1, & 1/n \le x. \end{cases}$

4.2-3 (a) $f(x) = 1/10$, $0 < x < 10$; (b) 0.2; (c) 0.6;
(d) $\mu = 5$; (e) $\sigma^2 = 25/3$.

4.2-5 (a) $G(w) = (w - a)/(b - a)$, $a \le w \le b$;
(b) $U(a, b)$.

4.2-7   (a) 2; 4; $M(t) = (1 - 2t)^{-1}$, $t < 1/2$;
     (b) $e^{-3/2}$; (c) $e^{-3/2}$.

4.2-9   (a) $f(x) = (1/3)e^{-x/3}$, $0 < x < \infty$; 3; 9;
     (b) $f(x) = 3e^{-3x}$, $0 < x < \infty$; 1/3; 1/9.

4.2-11 $P(X > x + y \mid X > x) = \dfrac{P(X > x + y)}{P(X > x)} = \dfrac{e^{-(x+y)/\theta}}{e^{-x/\theta}} = P(X > y)$.

4.2-13   (a) $\lambda = 0.025$; (b) exponential distribution, $\theta = 40$;
     (c) $\mu = 40$, $\sigma^2 = 1600$; (d) (i) $1 - e^{-1/2}$; (ii) e^{-1}; (iii) e^{-1}.

4.2-15   (a) 10.524; 9.320; yes.

4.2-17   (a) $F(x) = \begin{cases} 0, & -\infty < x < \alpha \\ 1 - e^{-(x-\alpha)/\theta}, & \alpha \le x < \infty; \end{cases}$
     (b) $\theta + \alpha$; θ^2.

4.2-19   (a) If Y is $U(0, 1)$, then $X = -3 \ln (1 - Y)$ has an exponential distribution with
     $\theta = 3$.

4.2-21 $q_3 = \theta \ln 4$; 0.8965.

4.2-23   (c) e.

4.3-1   (a) $f(x) = \dfrac{1}{\Gamma(10)(3/2)^{10}} x^9 e^{-2x/3}$, $0 \le x < \infty$;
     (b) $M(t) = (1 - 3t/2)^{-10}$, $t < 2/3$; $\mu - 15$; $\sigma^2 = 45/2$.

4.3-5 $f(x) = \dfrac{1}{\Gamma(20)7^{20}} x^{19} e^{-x/7}$, $0 \le x < \infty$; $\mu = 140$; $\sigma^2 = 980$.

4.3-7   (a) Gamma distribution with $\alpha = 3$, $\theta = 40$;
     (b) $\mu = 120$, $\sigma^2 = 4800$, (c) 0.4562.

4.3-9   (a) 0.025; (b) 0.05; (c) 0.94; (d) 8.672; (e) 30.19.

4.3-11   (a) 0.80; (b) a $= 11.69$, b $= 38.08$;
     (c) $\mu = 23$, $\sigma^2 = 46$; (d) 35.17, 13.09.

4.3-13 $r - 2$.

4.3-15   0.9444.

4.3-17   (c) $\bar{x} = 9.362$, $s^2 = 19.6410$;
     (d) Compare 0.06 with 0.05; 0.12 with 0.10.

4.4-1   (a) 0.2784; (b) 0.7209; (c) 0.3007; (d) 0.9616; (e) 0.0019; (f) 0.9500; (g) 0.6826;
     (h) 0.9544; (i) 0.9974.

4.4-3   (a) 2.326; (b) -2.576; (c) 1.67; (d) -2.17.

4.4-5   (a) 1.96; (b) 1.96; (c) 1.645; (d) 1.645.

4.4-7   (a) 0.3849; (b) 0.5403; (c) 0.0603; (d) 0.0013; (e) 0.6826; (f) 0.9544; (g) 0.9974.

4.4-9   (a) 0.6326; (b) 50.

4.4-15   (a) 0.8673; (b) 0.6026; (c) 0.3103; (d) 0.6821.

4.4-17   Set $f''(x) = 0$ and solve for x.

4.4-19   0.075.

4.5-1   (b) β; (c) (i) $0.8578/\beta$; (ii) $1.5228/\beta$; (iii) $2.2400/\beta$.

4.5-3   (a) $e^{-(5/\beta)^2} = e^{-1/4}$ implies that $\beta = 10$;
     (b) let $w = 10\sqrt{-\ln (1 - u)}$.

4.5-5   0.00006744; 0.0543.

4.5-7 $g(w) = (ae^{bw} + c)e^{-(a/b)e^{bw}-cw+(a/b)}, \; 0 < w < \infty;$
      $G(w) = 1 - e^{-(a/b)e^{bw}-cw+(a/b)}, \; 0 \le w < \infty.$

4.5-9 $m = e^{\mu}$, mode $= e^{\mu - \sigma^2}$;
      (a) $m = e^2 = 7.389$ mode $= 6.9414$;   max $= 0.2228$;
      (b) $m = e^2$;   mode $= 5.7546$;   max $= 0.1224$;
      (c) $m = e^2$;   mode $= 4.2102$;   max $= 0.0954$;
      (d) $m = e^2$;   mode $= 2.7183$;   max $= 0.0890$;
      (e) $m = e = 2.718$;   mode $= 1.0000$;   max $= 0.2420$;
      (f) $m = e^3 = 20.086$;   mode $= 7.3891$;   max $= 0.0327$.

4.5-11   (a) 0.25; (b) 0.063; (c) 0.032.

4.6-1   (a) 0.5; (b) 0; (c) 0.25; (d) 0.75; (e) 0.625.

4.6-3   (b) $\mu = 31/24$, $\sigma^2 = 167/567$;
       (c) 15/64; 1/4; 0; 11/16.

4.6-5   (b) \$2.25.

4.6-7   (b) 1/4; (c) 3/8; (d) 19/24; (e) 191/576.

CHAPTER 5

5.1-1   (a) $f_1(x) = (2x + 5)/16$, $x = 1, 2$;
       (b) $f_2(y) = (2y + 3)/32$, $y = 1, 2, 3, 4$;
       (c) 3/32; (d) 9/32; (e) 3/16; (f) 1/4; (g) dependent.

5.1-3

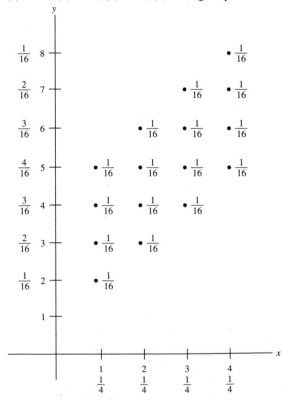

(e) dependent; the space is not rectangular.

5.1-5 (a) 0.065; (b) 0.021.

5.1-7 (a) $f(x_1, x_2, x_3) = \dfrac{208!}{x_1!x_2!x_3!x_4!} \left(\dfrac{9}{16}\right)^{x_1} \left(\dfrac{3}{16}\right)^{x_2} \left(\dfrac{3}{16}\right)^{x_3} \left(\dfrac{1}{16}\right)^{x_4}$,

$0 \le x_1 + x_2 + x_3 \le 208$ and $x_4 = 208 - x_1 - x_2 - x_3$;

(b) $b(208, 9/16)$;

5.1-9 $f_1(x) = 2e^{-2x}$, $0 < x < \infty$; $f_2(y) = 2e^{-y}(1 - e^{-y})$, $0 < y < \infty$; no.

5.1-11 $f_1(x) = 1/2$, $0 \le x \le 2$; $f_2(y) = 1/2$, $0 \le y \le 2$; yes.

5.1-13 (a) 1/2; (b) e^{-2}; (c) 0; (d) $1 - e^{-2}$; (e) 1/2; (f) 1/4.

5.2-1

y	2	3	4	5	6
$g(y)$	1/36	4/36	10/36	12/36	9/36

; $\mu = 14/3$; $\sigma^2 = 10/9$.

5.2-3 (a)

y	3	4	5	6	7
$g(y)$	1/64	3/64	6/64	10/64	12/64

y	8	9	10	11	12
$g(y)$	12/64	10/64	6/64	3/64	1/64

5.2-5 $\mu_Y = 1$; $\sigma_Y^2 = 61$.

5.2-7 (a) $M(t) = e^{7(e^t - 1)}$; (b) Poisson, $\lambda = 7$; (c) 0.800.

5.2-9 (a) $M(t) = (pe^t)^5/(1 - qe^t)^5$, $t < -\ln q$;

(b) negative binomial with parameters $r = 5$ and p.

5.2-11 (a) $M(t) = 1/(1 - 5t)^{21}$, $t < 1/5$;

(b) gamma distribution, $\alpha = 21$, $\theta = 5$.

5.2-13 (a) $(1/24)(e^{2t} + 2e^{3t} + 3e^{4t} + 4e^{5t} + 4e^{6t} + 4e^{7t} + 3e^{8t} + 2e^{9t} + e^{10t})$;

(b)

w	2	3	4	5	6	7	8	9	10
$g(w)$	1/24	2/24	3/24	4/24	4/24	4/24	3/24	2/24	1/24

5.2-15 (a)

w_1	0	1	2	3	4
$h(w_1)$	1/36	4/36	10/36	12/36	9/36

(b) $h(w_2) = h(w_1)$;

(c)

w	0	1	2	3	4	5
$h(w)$	1/1296	1/162	1/36	13/162	107/648	13/54

w	6	7	8
$h(w)$	1/4	1/6	1/16

5.3-1 (a) 0.4772; (b) 0.8561.

5.3-3 (a) 46.58; (b) 2.56; (c) 0.8447.

5.3-5 0.925.

5.3-7 (b) 0.05466; 0.3102.

5.3-9 (a) 0.8962; (b) 0.8962; (c) 87.31; 229.58.

5.3-11 0.8997.

5.3-13 (a) 0.3085; (b) 0.2267.

5.4-1 0.4772.

5.4-3 0.8185.

5.4-5 (a) $\chi^2(18)$; (b) 0.0756, 0.9974.

5.4-7 0.6247.

5.4-9 0.95.

5.5-1 (a) 0.2878, 0.2881; (b) 0.4428, 0.4441; (c) 0.1550, 0.1554.

5.5-3 0.9258.

5.5-5 0.6915.

5.5-7 0.3085.

5.5-9 0.6247.

5.5-11 (a) 0.5548; (b) 0.3823; (c) 0.6026.

5.5-13 (a) 0.5640; (b) 0.0111.

5.5-15 (a) 0.2417; (b) 0.220; (c) 0.2244.

5.5-17 (a) 0.4290; (b) 0.5710; (c) 0.5642.

5.6-1 (a) 0.9984; (b) 0.998.

5.7-1 (a) 0.025; (b) 0.975; (c) 0.05; (d) 0.94; (e) 0.09.

5.7-3 (a) 1.771; (b) 2.602; (c) -1.740; (d) -2.571.

5.7-7 (a) 0.279, 4.15; (b) 0.157, 8.10.

5.7-11 (a) 0.8997; (b) 2.042.

5.7-15 (f) $E(U) = 0$, $E(W) = 0$, $\text{Var}(W) = 1$.

5.8-1 (a) 158.97, 12.1525, 30.55; (f) yes.

5.8-3 (a) 335.176, 0.5214, 1.29; (f) no.

5.8-5 Yes.

5.8-7 (a) 0, 5.82; (b) yes.

5.9-1 $g(y) = y/8$, $0 < y < 4$.

5.9-3 $g(y) = 1$, $0 < y < 1$.

5.9-5 $g(y_1, y_2) = (1/4)e^{-y_2/2}$, $0 < y_1 < y_2 < \infty$;
$g_1(y_1) = (1/2)e^{-y_1/2}$, $0 < y_1 < \infty$;
$g_2(y_2) = (y_2/4)e^{-y_2/2}$, $0 < y_2 < \infty$; no.

5.9-7 (a) $g(y_1, y_2) = \dfrac{1}{\sqrt{2\pi}\,\Gamma\left(\dfrac{r}{2}\right)2^{r/2}} y_2^{r/2-1} e^{(-y_2/2)(1+y_1^2/r)}\left(\dfrac{\sqrt{y_2}}{\sqrt{r}}\right),$

$-\infty < y_1 < \infty$, $0 < y_2 < \infty$.

(b) $g_1(y_1) = \dfrac{\Gamma[(r+1)/2]}{\sqrt{\pi r}\,\Gamma(r/2)(1+y_1^2/r)^{(r+1)/2}}$, $\quad -\infty < y_1 < \infty$.

5.9-15 840.

5.9-17 (a) 0.1792; (b) 0.1792.

CHAPTER 6

6.1-1 (a) $E(X) = \theta$ so $E(\overline{X}) = \theta$;
(b) $\mathrm{Var}(X) = \theta^2$ so $\mathrm{Var}(\overline{X}) = \theta^2/n$;
(c) 3.48.

6.1-3 (a) $E(\overline{X}) = E(Y/n) = np/n = p$;
(b) $p(1-p)/n$;
(c) $\mathrm{Var}(\overline{X}) = \mathrm{Var}(Y/n) = np(1-p)/n^2 = p(1-p)/n$;
(d) $(1/n)[np/n - \{np(1-p) + n^2p^2\}/n^2] = (n-1)p(1-p)/n^2$;
(e) $c = 1/(n-1)$.

6.1-5 (a) $8/3\sqrt{2\pi}$; (b) $3\sqrt{5\pi}/8\sqrt{2}$.

6.1-7 $\widetilde{\alpha} = \overline{X}^2/V$; $\widetilde{\theta} = V/\overline{X}$.

6.1-9 (a) (i) $\widetilde{\theta} = 0.5975$, (ii) $\widetilde{\theta} = 2.4004$, (iii) $\widetilde{\theta} = 0.8646$.

6.1-11 7.2.

6.1-13 [71.35, 76.25].

6.1-15 [2.03, 2.15].

6.1-17 [48.467, 72.267].

6.2-1 (a) 7.465, 0.063; (b) [7.420, 7.510].

6.2-3 (a) 46.42; (b) [40.26, 52.58].

6.2-5 [1.739, 1.911].

6.2-7 (a) 14.72, 1.381; (b) [14.292, 15.148].

6.2-9 (a) 1.753σ; (b) 2.334σ.

6.2-11 [6.049, 6.051].

6.2-13 [−74.527, 63.857].

6.2-15 (a) 393.314; (b) [179.149, 607,479].

6.2-17 (d) $[\overline{X} - \overline{Y} - (\mu_X - \mu_Y)] \Big/ \sqrt{\dfrac{(n-1)S_Y^2/d + (m-1)S_Y^2}{n+m-2}\left(\dfrac{d}{n} + \dfrac{1}{m}\right)}$ has a t distribution with $r = n + m - 2$ degrees of freedom.

6.2-19 (a) 0.079; (b) [−0.029, 0.187]; (c) not necessarily.

6.3-1 (a) 6.144; (b) [4.405, 10.142] or [4.107, 9.521].

6.3-3 (a) 273.04, 3155.54; (b) [2079.43, 5468.08]; (c) [45.60, 73.95]; (d) [44.02, 71.56];
(e) yes.

6.3-5 $\left[\dfrac{\sum\limits_{i=1}^{n}(X_i - \mu)^2}{\chi^2_{\alpha/2}(n)}, \dfrac{\sum\limits_{i=1}^{n}(X_i - \mu)^2}{\chi^2_{1-\alpha/2}(n)} \right].$

6.3-7 (a) 0.5263; (b) [0.121, 2.158].

6.3-9 [0.383, 0.976].

6.3-11 (a) [0.560, 6.020]; (b) [0.748, 2.454].

6.4-1   (a) 0.0374; (b) [0.0227, 0.0521]; (c) [0.0252, 0.0550].

6.4-3   (a) 0.69; (b) [0.6613, 0.7187].

6.4-5   (a) 0.506; (b) [0.461, 0.551]; (c) not necessarily.

6.4-7   [0.074, 0.286].

6.4-9   (a) 0.680; (b) [0.646, 0.714].

6.4-11 (a) 0.144; (b) [0.095, 0.193]; (c) 0.0764; (d) [0.014, 0.139].

6.4-13 [0.011, 0.089].

6.4-15 (a) $[-0.013, 0.111]$; (b) yes, the interval includes zero.

6.5-1   117.

6.5-3   (a) 1083; (b) [6.047, 6.049]; (c) $58,800; (d) 0.0145.

6.5-5   (a) 257; (b) yes.

6.5-7   (a) 1068; (b) 2401; (c) 752.

6.5-9   2035.

6.5-11 451.

6.5-13 (a) 38; (b) [0.621, 0.845].

6.5-15 601.

6.5-17 235.

6.6-3   (b) 2.225.

6.6-5 $\bar{x} = 33.43$, $v = 5.098$.

6.6-7 $1/9.5 = 0.1053$.

6.7-3   (a) $p(1 - p)/n$; (b) 100%.

6.7-5   (a) $\theta^2/2n$; (b) $\theta^2/3n$; (c) θ^2/n.

6.8-1   (a) 0.84; (b) 0.082.

6.8-3 $k = 1.464$; 8/15.

6.8-5   (a) 0.25; (b) 0.85; (c) 0.925.

6.8-7   0.75.

CHAPTER 7

7.1-1 $\alpha = 7/27$; $\beta = 7/27$.

7.1-3   (a) 0.3032 using $b(100, 0.08)$, 0.313 using Poisson approximation, 0.2902 using normal approximation;

   (b) 0.1064 using $b(100, 0.04)$, 0.111 using Poisson approximation, 0.1010 using normal approximation.

7.1-5   (a) $\alpha = 0.1056$; (b) $\beta = 0.3524$.

7.1-7   (a) $z = 2.266 > 1.645$, reject H_0;

   (b) $z = 2.266 < 2.326$, do not reject H_0.

7.1-9   (a) $z = 1.752 > 1.645$, reject H_0;

   (b) $z = 1.752 < 1.96$, do not reject H_0.

7.1-11 (a) $z = \dfrac{y/n - 0.40}{\sqrt{(0.40)(0.60)/n}} \geq 1.645$;

   (b) $z = 2.215 > 1.645$, reject H_0.

7.1-13 (a) H_0: $p = 0.40$, H_1: $p > 0.40$, $z \geq 2.326$;

   (b) $2.236 < 2.326$, do not reject H_0.

7.1-15 (a) $|z| \geq 1.96$; (b) 1/20; (c) 1/20; (d) 19/20; (e) no.

7.1-17 (a) $z \leq -1.645$; (b) $-3.058 < -1.645$, reject H_0.

7.2-1 (a) $K(\mu) = \Phi\left(\dfrac{22.5 - \mu}{3/2}\right)$; $\alpha = 0.0478$;

(b) $\bar{x} = 24.1225 > 22.5$, do not reject H_0;

(c) 0.2793.

7.2-3 (a) $K(\mu) = \Phi\left(\dfrac{510.77 - \mu}{15}\right)$;

(b) $\alpha = 0.10$;

(c) 0.5000;

(e) (i) 0.0655, (ii) 0.0150.

7.2-5 $n = 25$, $c = 1.6$.

7.2-7 $n = 40$, $c = 678.38$.

7.2-9 (a) $K(p) = \displaystyle\sum_{y=14}^{25} \binom{25}{y} p^y(1 - p)^{25-y}$, $0.40 \leq p \leq 1.0$;

(b) $\alpha = 0.0778$;

(c) 0.1827, 0.3450, 0.7323, 0.9558, 0.9985, 1.0000;

(e) yes;

(f) 0.0344.

7.3-1 (a) $1.4 < 1.645$, do not reject; (b) $1.4 > 1.282$, reject H_0;

(c) p-value $= 0.0808$.

7.3-3 (a) $z = (\bar{x} - 170)/2$, $z \geq 1.645$;

(b) $1.260 < 1.645$, do not reject H_0;

(c) 0.1038.

7.3-5 (a) $t = (\bar{x} - 3.315)/(s/\sqrt{30}) \leq -1.699$;

(b) $-1.414 > -1.699$, do not reject H_0;

(c) $0.05 < p$-value < 0.10 or p-value ≈ 0.08.

7.3-7 (a) $t = (\bar{x} - 47)/(s/\sqrt{20}) \leq -1.729$;

(b) $-1.789 < -1.729$, reject H_0;

(c) $0.025 < p$-value < 0.05, p-value ≈ 0.037.

7.3-9 (a) $t \geq 2.764$; (b) $4.028 > 2.764$, reject H_0; (c) p-value < 0.005; (d) $\chi^2 \leq 3.940$;

(e) $4.223 > 3.940$, do not reject H_0; (f) $0.05 < p$-value < 0.10.

7.3-11 (a) $\chi^2 = 24s^2/140^2$, $\chi^2 \geq 36.42$;

(b) $29.18 < 36.42$, do not reject H_0.

7.3-13 (a) $59.53 < 64.28$, reject H_0;

(b) $0.025 < p$-value < 0.05.

7.3-15 (a) $0.015 < 0.058 < 0.078$, do not reject H_0;

(b) $[0.030, 0.158]$, yes.

7.3-17 $1.607 < 1.645$, accept H_0.

7.3-19 $1.477 < 1.833$, do not reject H_0.

7.4-1 (a) $t \geq -1.734$; (b) $-2.221 < -1.734$, reject H_0.

7.4-3 (a) $t \leq -1.703$; (b) $-0.869 > -1.703$, do not reject H_0;

(c) $0.10 < p$-value 0.25;

(d) $0.818 < 2.96$ and $1.222 < 3.18$, do not reject H_0.

7.4-5   (a) $|t| \geq 2.101$; (b) $|-2.151| > 2.101$, reject H_0;
      (c) $0.01 < p$-value < 0.025;
      (e) $1.318 < 4.03$ and $0.759 < 4.03$, do not reject H_0 at $\alpha = 0.05$ significance level.

7.4-7   (a) $3.247 < 4.32$ and $0.308 < 5.52$, do not reject H_0;
      (b) $|4.683| > 2.131$, reject H_0; (c) no, yes, yes.

7.4-9   (a) $0.84 < 2.53$, $1.19 < 2.91$, do not reject H_0;
      (b) $3.403 > 2.326$, reject H_0.

7.4-11 (a) $z \leq -1.96$; (b) $-8.98 < -1.96$, reject H_0.

7.4-13 $2.42 < 3.28$, do not reject H_0.

7.5-1   (a) $L(80)/L(76) = \exp\left[\dfrac{6}{128}\sum_{1}^{n} x_i - \dfrac{624n}{128}\right] \leq k$ or $\bar{x} \leq c$;

      (b) 43, 78.

7.5-3   (a) $L(3)/L(5) \leq k$ if and only if $\sum_{1}^{n} x_i \geq (-15/2)[\ln(k) - \ln(5/3)^n] = c$;

      (b) $\bar{x} \geq 4.15$;
      (c) $\bar{x} \geq 4.15$;
      (d) yes.

7.5-5   (a) $\dfrac{L(50)}{L(\mu_1)} \leq k$ if and only if $\bar{x} \leq \dfrac{(-72)\ln(k)}{2n(\mu_1 - 50)} + \dfrac{50 + \mu_1}{2} = c$;

7.5-7   (a) $\dfrac{L(0.5)}{L(\mu)} \leq k$ if and only if $\sum_{i=1}^{n} x_i \geq \dfrac{\ln(k) + n(0.05 - \mu)}{\ln(0.5/\mu)} = c$;

      (b) $y \geq 9$.

7.6-1   (a) $|-1.80| > 1.645$, reject H_0;
      (b) $|-1.80| < 1.96$, do not reject H_0;
      (c) 0.0718.

7.6-3   (a) $\bar{x} \geq 230 + 10c/\sqrt{n}$ or $\dfrac{\bar{x} - 230}{10/\sqrt{n}} \geq z_\alpha$;

      (b) yes; (c) $1.04 < 1.282$, accept H_0; (d) 0.1492.

7.6-5 $|2.10| < 2.306$, do not reject H_0; $0.05 < p$-value < 0.10.

7.6-7 $2.20 > 1.282$, reject H_0; 0.0139.

CHAPTER 8

8.1-1 $7.875 > 4.26$, reject H_0.

8.1-3 $13.773 > 4.07$, reject H_0.

8.1-5 $5.171 > 3.25$, reject H_0.

8.1-7 $14.757 > 4.43$, reject H_0.

8.1-9   (a) $F \geq 4.07$;
      (b) $4.106 > 4.07$, reject H_0;
      (c) $4.106 < 5.42$, do not reject H_0;
      (d) $0.025 < p$-value < 0.05, p-value ≈ 0.05.

8.1-11 $10.224 > 4.26$, reject H_0.

8.1-13 (a) $F \geq 5.85$;
      (b) $6.337 > 5.85$, reject H_0.

8.2-1 $18.00 > 5.14$, reject H_A.

8.2-3 (a) $7.624 > 4.46$, reject H_A;
(b) $15.538 > 3.84$, reject H_B.

8.2-5 (a) $1.723 < 2.90$, accept H_{AB};
(b) $5.533 > 4.15$, reject H_A;
(c) $28.645 > 2.90$, reject H_B.

8.2-7 $1.686 < 2.37$, accept H_{AB};
$2.179 < 3.27$, do not reject H_A;
$2.057 < 2.87$, do not reject H_B.

8.2-9 $0.111 < 3.89$, accept H_{AB};
$9.692 > 4.75$, reject H_A;
$2.606 < 3.89$, do not reject H_B.

8.3-1

2^2 Design

Run	A	B	AB	Observations
1	−	−	+	X_1
2	+	−	−	X_2
3	−	+	−	X_3
4	+	+	+	X_4

(a) $[A] = (-X_1 + X_2 - X_3 + X_4)/4$,
$[B] = (-X_1 - X_2 + X_3 + X_4)/4$,
$[AB] = (X_1 - X_2 - X_3 + X_4)/4$.
(b) It is sufficient to compare the coefficients on both sides of the equations of X_1^2, X_1X_2, X_1X_3, and X_1X_4 which are $3/4$, $-1/2$, $-1/2$, $-1/2$, respectively.
(c) Each is $\chi^2(1)$.

8.3-3 $[A]$ is $N(0, \sigma^2/2)$ so $E([A]^2) = E[(X_2 - X_1)^2/4] = \sigma^2/2$ or $E[(X_2 - X_1)^2/2] = \sigma^2$.

8.3-5 (a) $[A] = -4$, $[B] = 12$, $[C] = -1.125$, $[D] = -2.75$, $[AB] = 0.5$, $[AC] = 0.375$, $[AD] = 0$, $[BC] = -0.625$, $[BD] = 2.25$, $[CD] = -0.125$, $[ABC] = -0.375$, $[ABD] = 0.25$, $[ACD] = -0.125$, $[BCD] = -0.375$, $[ABCD] = -0.125$.
(b) There is clearly a temperature (B) effect. There is also a catalyst charge (A) effect and probably a concentration (D) and a temperature-concentration (BD) effect.

8.4-1 (a) $\widehat{y} = 86.8 + (421/414.5)(x - 74.5)$;
(c) $\widehat{\sigma^2} = 17.9998$.

8.4-3 (a) $\widehat{y} = 4.483x + 6.483$.

8.4-5 Solve for α: $P\left[-t_{\alpha/2}(n-2) \le \dfrac{\widehat{\alpha} - \alpha}{\sqrt{\widehat{\sigma^2}/(n-2)}} \le t_{\alpha/2}(n-2) \right]$.

8.4-7 $[83.341, 90.259]$, $[0.478, 1.553]$, $[10.262, 82.568]$.

8.4-9 (a) $\widehat{y} = 15.695 + 0.620x$;
(c) $\widehat{\sigma^2} = 0.260$;
(d) $[0.326, 0.914]$.

8.4-11 (a) $\widehat{y} = 1.896 + 0.538x$;
(b) $\widehat{\alpha} = 6.967$, $\widehat{\beta} = 0.538$, $\widehat{\sigma^2} = 0.0549$;
(d) $[6.826, 7.107]$, $[0.208, 0.868]$, $[0.033, 0.164]$.

8.5-1 (a) $4.359 > 2.306$, reject H_0;

(b) $19.005 > 5.32$, reject H_0;

(c) $[73.867, 86.530]$, $[83.838, 90.777]$, $[89.107, 99.728]$;

(d) $[68.206, 92.190]$, $[75.833, 98.783]$, $[82.258, 106.577]$.

8.5-3 (a) $2.148 < 2.160$, fail to reject H_0;

(b) $4.615 < 4.68$, fail to reject H_0;

(c) $[62.793, 129.495]$, $[101.854, 135.266]$, $[119.483, 162.470]$;

(d) $[28.926, 163.361]$, $[57.856, 179.265]$, $[78.784, 203.169]$.

8.5-5 (a) $[4.897, 8.444]$, $[9.464, 12.068]$, $[12.718, 17.004]$;

(b) $[1.899, 11.442]$, $[6.149, 15.383]$, $[9.940, 19.782]$.

8.5-7 (a) $2.369 > 1.771$, reject H_0;

(b) $[19.669, 26.856]$, $[22.122, 27.441]$, $[24.048, 28.551]$, $[25.191, 30.445]$, $[25.791, 32.882]$;

(c) $[15.530, 30.996]$, $[17.306, 32.256]$, $[18.915, 33.684]$, $[18.711, 36.926]$, $[19.923, 38.750]$.

8.5-9 $\widehat{y} = 0.8082 + 0.2630x - 0.3178x^2 + 1.1673x^3$.

8.5-11 Since $\mu(x)$ is not linear in β_1 and β_2, it is not a linear model. However, $\ln \mu(x) = \ln \beta_1 + \beta_2 x = \beta_3 + \beta_2 x$ is linear in β_2 and β_3, so it is a linear model.

CHAPTER 9

9.1-1 (a) $\dfrac{\dbinom{30}{1}\dbinom{23}{1}\dbinom{22}{1}\dbinom{9}{1}}{\dbinom{84}{4}} = 0.0708 = 136{,}620/1{,}929{,}501$;

(b) $\dfrac{\dbinom{23}{4}}{\dbinom{84}{4}} = 0.0046 = 8855/1{,}929{,}501$.

9.1-3 (a) $f(x_1, x_2, x_3) = \dbinom{13}{x_1}\dbinom{13}{x_2}\dbinom{13}{x_3}\dbinom{13}{13 - x_1 - x_2 - x_3} \bigg/ \dbinom{52}{13}$, $0 \leq x_1 + x_2 + x_3 \leq 13$;

(b) $f_{12}(x_1, x_2) = \dbinom{13}{x_1}\dbinom{13}{x_2}\dbinom{26}{13 - x_1 - x_2} \bigg/ \dbinom{52}{13}$, $0 \leq x_1 + x_2 \leq 13$;

(c) $f_3(x_3) = \dbinom{13}{x_3}\dbinom{39}{13 - x_3} \bigg/ \dbinom{52}{13}$, $0 \leq x_3 \leq 13$.

9.1-5 (a) $f(x, y) = \dfrac{15!}{x!y!(15 - x - y)!}\left(\dfrac{6}{10}\right)^x \left(\dfrac{3}{10}\right)^y \left(\dfrac{1}{10}\right)^{15-x-y}$, $0 \leq x + y \leq 15$;

(b) no because it is not rectangular;

(c) 0.0735;

(d) X is $b(15, 0.6)$;

(e) 0.9095.

9.1-7 $\mu_X = 25/16$, $\mu_Y = 45/16$, $\sigma_X^2 = 63/256$, $\sigma_Y^2 = 295/256$, $\rho = -0.037$, dependent.

9.1-9 (a)

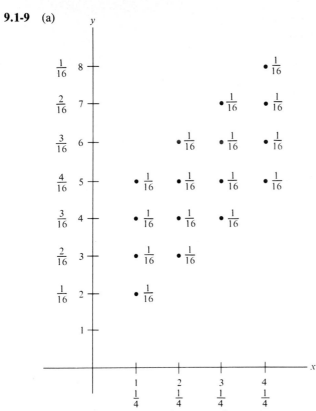

(b) $\mu_X = 2.5$, $\mu_Y = 5$, $\sigma_X^2 = 5/4$, $\sigma_Y^2 = 5/2$, $\text{Cov}(X, Y) = 5/4$, $\rho = \sqrt{2}/2$;
(c) $y = x + 2.5$.

9.1-11 $b = \text{Cov}(X, Y)/\sigma_X^2$, $a = \mu_Y - \mu_X b$.

9.1-13 (a) No; (b) $\text{Cov}(X, Y) = 0$, $\rho = 0$.

9.1-15

w	0	1	2	3	4
$f_2(w)$	16/81	32/81	24/81	8/81	1/81

; $\rho = \sqrt{2}/2$.

9.2-1 (a)

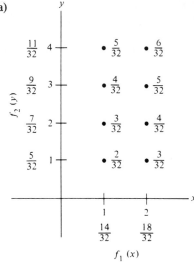

(b)

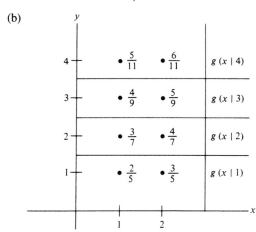

(c)

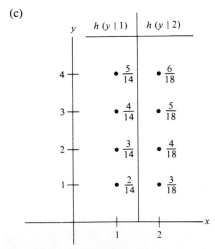

(d) (i) 7/14; (ii) 7/18; (iii) 5/9;

(e) 20/7, 55/49.

9.2-3 (a) $f(x, y) = \dfrac{50!}{x!y!(50 - x - y)!}(0.02)^x(0.90)^y(0.08)^{50-x-y}$, $0 \le x + y \le 50$;

(b) X and Y have a trinomial distribution with parameters $n = 50$, $p_1 = 0.02$, $p_2 = 0.90$;

(c) $b(47, 0.09/0.98)$; (d) 43.163; (e) −0.429.

9.2-5 (a) $E(Y|x) = 2(2/3) - (2/3)x$; (b) Yes.

9.2-7 (a) $E(Y|x) = x + 2.5$, $x = 1, 2, 3, 4$; yes; yes.

9.2-9 (a) $f_1(x) = 1/8$, $x = 0, 1, \ldots, 7$;

(b) $h(y|x) = 1/3$, $y = x, x + 1, x + 2$, for $x = 0, 1, \ldots, 7$;

(c) $E(Y|x) = x + 1$;

(d) $\sigma^2_{Y|x} = 2/3$;

(e) $f_2(y) = \begin{cases} 1/24, & y = 0, 9, \\ 2/24, & y = 1, 8, \\ 3/24, & y = 2, 3, 4, 5, 6, 7. \end{cases}$

9.2-11 (b) $f_1(x) = (3 - x)/6$, $x = 0, 1, 2$; $f_2(y) = (3 - y)/6$, $y = 0, 1, 2$;

(c) $E(Y|x) = 1 - (1/2)x$;

(d) $E(X|y) = 1 - (1/2)y$;

(e) $\rho = -1/2$.

9.2-13 (b) $f_1(x) = 1/10$, $0 \le x \le 10$;

(c) $h(y|x) = 1/4$, $10 - x \le y \le 14 - x$ for $0 \le x \le 10$;

(d) $E(Y|x) = 12 - x$.

9.2-15 (a) $f(x, y) = 1/(2x^2)$, $0 < x < 2$, $0 < y < x^2$;

(b) $f_2(y) = (2 - \sqrt{y})/(4\sqrt{y})$, $0 < y < 4$;

(c) $E(X|y) = [2\sqrt{y} \ln (2/\sqrt{y})]/[2 - \sqrt{y}]$;

(d) $E(Y|x) = x^2/2$.

9.2-17 (a) $h(y|x) = 1/e^x = e^{-x}$, $0 < y < e^x$ for $0 < x < 1$;

(b) $E(Y|x) = e^x/2$, $0 < x < 1$;

(c) $f(x, y) = e^{-x}$, $0 < x < 1$, $0 < y < e^x$;

(d) $f_2(y) = \begin{cases} 1 - 1/e, & 0 < y < 1; \\ 1/y - 1/e, & 1 \le y < e. \end{cases}$

9.3-1 (a) 0.6006; (b) 0.7888; (c) 0.8185; (d) 0.9371.

9.3-3 (a) 0.5746; (b) 0.7357.

9.3-5 (a) $N(86.4, 40.96)$; (b) 0.4192.

9.3-7 (a) 0.8248;

(b) $E(Y|x) = 457.1735 - 0.2655x$;

(c) $\text{Var}(Y|x) = 645.9375$;

(d) 0.8079.

9.3-9 $a(x) = x - 11$, $b(x) = x + 5$.

9.3-11 (a) 0.3785; (b) $\mu_{Y|x} = -0.2x + 4.7$;

(c) $\sigma^2_{Y|x} = 8.0784$; (d) 0.4230.

9.4-1 $-0.45 < -0.3808$, reject H_0.

9.4-3 [0.419, 0.802].

9.4-5 $0.283 > 0.2500$, reject H_0.

9.4-9 $u(R) = c \ln [(1 + R)/(1 - R)]$.

9.4-11 (a) $C = \{r\colon |r| \geq 0.4258\}$, $r = -0.4906$, reject H_0;
 (b) $C = \{r\colon |r| \geq 0.4973\}$, $r = -0.4906$, do not reject H_0.

9.4-13 $r = 0.5491$, $0.01 < p$-value < 0.025, reject H_0 if $\alpha = 0.025$.

9.5-1 $6.25 < 7.815$, do not reject if $\alpha = 0.05$; p-value ≈ 0.10.

9.5-3 $7.60 < 16.92$, do not reject H_0.

9.5-5 $3.65 < 7.815$, do not reject H_0.

9.5-7 (a) $q_3 \geq 7.815$;
 (b) $q_3 = 1.744 < 7.815$; do not reject H_0.

9.5-9 $6.072 < 16.92$, do not reject H_0.

9.5-11 $3.841 < 14.07$, do not reject H_0.

9.5-13 $2.89 < 16.92$, do not reject H_0.

9.5-15 Using 10 classes with equal probabilities, the frequencies are 6, 7, 3, 6, 5, 7, 8, 7, 6, 5; $q = 3.00 < 14.07$, do not reject H_0; using 10 classes with boundaries 123.5, 129.5, 135.5, and so on, the frequencies are 3, 2, 5, 6, 11, 13, 10, 8, 1, 1, $q = 5.59 < 14.07$, do not reject H_0.

9.5-17 (a) Frequencies are 2, 4, 21, 33, 20, 14, 4, 1, 1;
 (b) $q = 4.854 < 9.488$, do not reject H_0.

9.5-19 (a) $658.93 < 28.87$, reject;
 (b) $32.44 > 28.87$, reject;
 (c) $12.55 < 28.87$, do not reject;
 (d) $n \geq 10$.

9.6-1 $3.23 < 11.07$, do not reject H_0.

9.6-3 $2.40 < 5.991$, do not reject H_0.

9.6-7 $8.410 < 9.488$, do not reject H_0, $0.05 < p$-value < 0.10.

9.6-9 $4.268 > 3.841$, reject H_0, $0.025 < p$-value < 0.05.

9.6-11 $7.683 < 9.210$, reject H_0, $0.01 < p$-value < 0.025.

9.6-13 $9.488 < 10.076 < 11.14$, $0.025 < p$-value < 0.05.

9.6-15 $8.79 > 7.378$, reject H_0.

9.6-17 $4.242 < 4.605$, do not reject H_0.

9.7-1 (a) $f(x; p) = e^{x \ln (1-p) + \ln [p/(1-p)]}$; $K(x) = x$; $\displaystyle\sum_{i=1}^{n} X_i$ is sufficient;
 (b) $\overline{X}$.

9.7-3 (a) $\displaystyle\sum_{i=1}^{n} X_i^2$; (b) $\widehat{\sigma^2} = (1/n) \displaystyle\sum_{i=1}^{n} X_i^2$; (c) yes.

9.8-1 $k(\theta | x_1, x_2, \ldots , x_n) \propto \theta^{\Sigma x_i}(1 - \theta)^{n - \Sigma x_i} \left[\dfrac{\Gamma(\alpha + \beta)}{\Gamma(\alpha)\Gamma(\beta)} \right] \theta^{\alpha - 1}(1 - \theta)^{\beta - 1}$
 $\propto \theta^{\Sigma x_i + \alpha - 1}(1 - \theta)^{n - \Sigma x_i + \beta - 1}$.

9.8-3 (a) $E[\{w(Y) - \theta\}^2] = \{E[w(Y) - \theta]\}^2 + \text{Var}[w(Y)]$
 $= (74\theta^2 - 114\theta + 45)/500$;
 (b) $\theta = 0.569$ to $\theta = 0.872$.

CHAPTER 10

10.1-1 (a) $y_1 = 5$, $y_2 = 6$, $y_3 = 7$, $y_4 = 9$, ..., $y_{10} = 78$;
(b) $\tilde{m} = 31.5$, $\tilde{\pi}_{0.80} = 72.6$;
(c) $\tilde{q}_1 = 6.75$, $\tilde{q}_3 = 68.75$.

10.1-3 (a) $g_3(y) = 10(1 - e^{-y/3})^2 e^{-y}$, $0 < y < \infty$;
(b) $5(1 - e^{-5/3})^4 e^{-5/3} + (1 - e^{-5/3})^5 = 0.7599$;
(c) $e^{-5/3} = 0.1889$.

10.1-5 (a) 0.2553; (b) 0.7483.

10.1-7 (a) 0.8697; (b) 0.8634.

10.1-9 0.95.

10.1-11 (a) $E(W_1) = E(W_2) = E(W_3) = \theta$;
(b) W_3;
(d) (i) 0.5097, 0.0275; (ii) 0.5242, 0.0433; (iii) 0.5026, 0.0249;
(e) W_3.

10.1-13 (a) 0.2692, 0.3757, 3.1382, 3.6881;
(b) 0.0266, 0.0369, 0.2694, 0.3085.

10.1-15 (a) 14.75;
(b) 0.8995.

10.2-1 (a) 0.7812; (b) 0.7844; (c) 0.4528.

10.2-3 (a) ($y_2 = 110$, $y_9 = 137$); (b) 97.86%.

10.2-5 (2.72, 2.84), 93.46%.

10.2-7 (15.40, 17.05).

10.2-9 (a)

Stem	Leaf	Frequency
0	09 11 13 24 34 51 57 75 79 98	10
1	36 52 61 98	4
2	15 38 44 72 88	5
3	15 15 50	3
4	45	1
5		0
6	20	1
7	23	1

(b) $y_3 = 13$, $y_9 = 79$, $y_{10} = 98$, $y_{13} = 161$, $y_{17} = 244$;
(c) $\tilde{\pi}_{0.25} = 54$, $\tilde{m} = 161$, $\tilde{\pi}_{0.75} = 301.5$;
(d) (i) (13, 98), 89.66%; (ii) (79, 244), 89.22%; (iii) (238, 455), 89.66%.

10.2-11 (a)

Stem	Leaf	Frequency
0.8•	7 8 9	3
0.9*	0 0 1 2 2 2 3 4	8
0.9•	5 5 5 7 8 8 8 8 9	9
1.0*	0 0 1 2	4
1.0•	6	1

(b) $\tilde{m} = 0.95$;
(c) (0.92, 0.98), 95.68%

(d) $\bar{q}_1 = 0.915$;

(e) $(0.89, 0.93)$, 89.66%;

(f) $\bar{q}_3 = 19.5$;

(g) $(0.97, 1.02)$; 96.33%.

10.3-1 (a) $9 < 10$, do not reject H_0; (b) p-value $= 0.1334$.

10.3-3 $C = \{w: w \le 7 \text{ or } w \ge 17\}$, $\alpha \approx 0.0662$, $w = 13$, do not reject.

10.3-5 Let Y equal number of drivers going 60 mph or less. Then $C = \{y: y \ge 87\}$, $\alpha \approx 0.0521$, reject H_0.

10.3-7 $2.16 > 2.12$, reject H_0.

10.3-9 (a) $7 < 8$, do not reject H_0 at $\alpha = 0.0547$;

(b) $1.477 < 1.833$, do not reject H_0.

10.3-11 (a) $C = \{w: w \le 3 \text{ or } w \ge 11\}$, $\alpha = 0.0574$;

(b) $w = 3$, reject H_0.

10.4-1 (a) $-55 < -47.08$, reject H_0; (b) 0.0285.

10.4-3 $w = 132$, p-value $= 0.0302$ for a one-sided alternative, p-value $= 0.0604$ for a two-sided alternative.

10.4-5 $74 > 63.6$, reject H_0; p-value ≈ 0.0287.

10.4-7 $w = 54$, $z = 1.533$, do not reject H_0.

10.4-9 (a) Let Y equal the number of weights less than 1.14 pounds. Then $C = \{y: y \le 4\}$, $\alpha = 0.0898$;

(b) $y = 5$, fail to reject H_0;

(c) p-value $= 0.2120$;

(d) $z = 2.072 > 1.282$, reject H_0;

(e) p-value ≈ 0.0191.

10.5-1 (a) $v = 8 > 7$, reject H_0; p-value $= 0.0115$;

(b) $w = 145 > 126$ or $z = 3.02 > 1.645$, reject H_0; p-value ≈ 0.0014;

(c) $v = 1$, p-value $= 0.1517$, do not reject.

10.5-3 (a) $C = \{v: v \le 2 \text{ or } v \ge 8\}$, $\alpha = 0.023$, $v = 4$, do not reject H_0;

(b) $C = \{w: w \le 79 \text{ or } w \ge 131\}$, $\alpha = 0.054$, $w = 95$, do not reject H_0.

10.5-5 (a) $C = \{v: v \le 2 \text{ or } v \ge 8\}$, $\alpha = 0.023$, $v = 6$, do not reject H_0;

(b) $C = \{w: w \le 79 \text{ or } w \ge 131\}$, $\alpha \approx 0.054$, $w = 107.5$, do not reject H_0.

10.5-7 (a) $C = \{w: w \le 184 \text{ or } w \ge 280\}$, $\alpha \approx 0.05$, $w = 241$, do not reject H_0.

19.5-9 (a) $C = \{v: v \ge 7\}$, $\alpha = 0.04$, $v = 7$, p-value $= 0.04$, reject H_0;

(b) $C\{w: w \ge 174\}$, $\alpha \approx 0.05$, $w = 188$, p-value ≈ 0.0055, reject H_0.

10.6-1 $C = \{r: r \le 6\}$, $\alpha = 0.10$, $r = 6$, reject H_0.

10.6-3 $(6!)(12!)\binom{11}{4} = 113{,}810{,}780{,}160{,}000$.

10.6-5 (b) $r = 11$, reject H_0; (c) p-value ≈ 0.0473.

10.6-7 $r = 4$, p-value $= 86/3432 = 0.025$, reject the hypothesis of randomness.

$r = 7$, p-value $= 15{,}972/705{,}432 = 0.0226$, reject the hypothesis of randomness.

10.6-9 (a) $C = \{y: y \le 3 \text{ or } y \ge 11\}$, $y = 7$, do not reject H_0;

10.6-11 (b) $C = \{r: r \le 4 \text{ or } r \ge 12\}$, $r = 4$, reject the hypothesis of randomness. There appears to be a trend effect.

10.6-13 $C = \{r: r \le 10\}$, $\alpha \approx 0.04$, $r = 6$, reject H_0, p-value ≈ 0.0005.

10.7-1 $d_5 = 0.40 < 0.45$, do not reject H_0.

10.7-3 $d_{10} = 0.1643$ at $x = 49$, do not reject.

10.7-5 $d_{10} = 0.1654$ at $x = 58$, do not reject.

10.7-7 $d_{40} = 0.046$, do not reject.

10.7-9 (a) $d_{70} = 0.1152 < 1.07/\sqrt{70} = 0.1279$, do not reject;

(b) The histogram is bimodal;

(c) The data appear to be more uniform than normal.

10.7-11 (a) $\hat{\theta} = 208.436$, $d_{39} = 0.1353$ at $x = 95$;

(b) $\hat{\alpha} < 0$, which is not compatible with the experiment.

10.7-13 Two-parameter exponential, -0.416, 0.286.

Index